THE SEEDS OF DICOTYLEDONS
VOLUME 1

THE SEEDS OF DICOTYLEDONS

E. J. H. CORNER, F.R.S.

EMERITUS PROFESSOR OF TROPICAL BOTANY, UNIVERSITY OF CAMBRIDGE

VOLUME 1

CAMBRIDGE UNIVERSITY PRESS

CAMBRIDGE

LONDON · NEW YORK · MELBOURNE

Published by the Syndics of the Cambridge University Press
The Pitt Building, Trumpington Street, Cambridge CB2 IRP
Bentley House, 200 Euston Road, London NW1 2DB
32 East 57th Street, New York, NY 10022, USA
296 Beaconsfield Parade, Middle Park, Melbourne 3206, Australia

First published 1976

Printed in Great Britain
at the
University Printing House, Cambridge
(Euan Phillips, University Printer)

Library of Congress Cataloguing in Publication Data

Corner, Edred John Henry.
The seeds of dicotyledons.

Includes bibliographical references and index.
1. Dicotyledons. 2. Seeds. I. Title.
QK495.A12C67   583'.04' 16   74-14434
ISBN: 0 521 20688 x (vol. 1)

CONTENTS

VOLUME 1

CONTENTS

The Figures are in Volume 2

PREFACE

Open a book on flowering plants and find how little there is about their seeds! The subject is abstruse with little to commend it unless for the identification of the seeds of commerce but, pursued botanically, it is absorbing, penetrating and illuminating. From seeds came the plants which made fruits (angiosperms) and those which added flowers (anthophytes). Modern theories of the origin of these plants dwell on flowers. The fruit was the subject of the durian-theory. I turn now to their source in seeds.

When the late Professor Kwan Koriba came from Kyoto to Singapore in 1942 to direct the Botanical Gardens during the war, we turned in our confinement to the botany of trees. In 1945 when lies, starvation and chaos were around, we discovered this interest. We did not know of Netolitzky's book, which I have since studied and re-studied. I have pondered why such genius has passed unrecognized. His book on the seed-structure of angiosperms summarizes knowledge up till the end of 1923; no detail has escaped; the erudition is profound; and he came to the brink of the discovery that seed-structure should be the basis of the natural classification of flowering plants. Perhaps the terse descriptions and the sketchy illustrations failed to convey the message. Probably the weight of authority overwhelmed him. Engler, Wettstein, Warming, Lotsy, Hallier and other great exponents of classification failed to perceive the importance of the researches into seed-structure which French, German and Italian schools had begun to explore last century.

How to follow, for there is still an enormous amount of research to be undertaken, I have contemplated. A modern encyclopedia might be planned but parts exist, knowledge is inadequate, and patience would be unrewarded. It is possible to continue, however, where Netolitzky left off. I have built, then, on his text and borne in mind

five considerations. (1) There is great pleasure in discovering how a seed is made; it is the most elaborate part of the plant. (2) Families have characteristic seeds by which they may be related in orders. (3) Compared with monocotyledons, the orders of dicotyledons are unsatisfactory, but they have the greater variety of seed-structure. (4) Conviction calls for ample illustration. (5) A prototype must have existed for this variety if, as the intricacy of their seeds seems to prove, angiosperms were monophyletic.

The outcome is an account of the seed in those families of dicotyledons for which there is something known about the microscopic structure; the pocket-lens description is totally inadequate. It is illustrated mainly from my researches on tropical seeds, these being the less known. There is an attempt to prove the importance of microscopic structure in the ordinal classification of families, a consideration of the prototype, and a general introduction. The presentation has largely been reversed for the very reason that, in all systems of dicotyledonous classification, on which the botany of flowering plants depends, the grouping of families into orders is uncertain, even arbitrary and artificial, and for the most part unsatisfactory; most orders do not fit the precision in seed-structure which Netolitzky exposed. The family becomes of necessity the unit of description though some families are at fault. Descriptions have been assembled, therefore, in the last but major section of the work where the arrangement is alphabetical. To have assembled them by similarity in seed-structure, though ideal, would have been bewildering because there would be constant distraction through reference to an index; Vitaceae would come near Magnoliaceae with Winteraceae as far away as alphabetically; Convolvulaceae would follow Paeoniaceae; Proteaceae would accompany Papaveraceae; Cruciferae would not go with Capparidaceae, nor

[vii]

Geraniaceae with Linaceae; thus one can appreciate Netolitzky's desistence. Moreover the time is not ripe. The seeds of many families are too little known or too problematic, e.g. Droseraceae. Before this main section there comes, therefore, the criticism of the customary orders. Here, with so much detail under review, it is necessary to pass continually to and fro between the orders and the families which they are supposed to embrace. I tried various methods to facilitate this perpetual motion and found none more satisfactory than the alphabetical; it does not burden the mind.

Lack of knowledge and of space have forced me to forgo an attempt to define generic differences. Even in such well-studied families as Rosaceae and Scrophulariaceae too few species have been examined, and where, as with Cruciferae, there is more certainty, the modern text can be cited.

Seed-studies begin with the ovule and end with germination. Flower and fruit pass in review. Ovary and pericarp must be sectioned. Vascular bundles must be traced. Lignification, suberization, cutinization and mucilaginization must be investigated, as well as cell-contents; all contribute to the hardening of the seed-coat which becomes its character. It is easy to stray into a multitude of problems from floral to physiological and biochemical evolution. I have left these in other hands and concentrated on the gross form of the seed, its cellular construction, vascular supply, and general lignification. The chemical characters may be as important but it soon becomes clear that, for the moribund tissues of the seed-coat, no ordinary stains or reagents are discriminatory enough in the common round of anatomy. A palisade so characteristic as the outer epidermis of the Leguminous seed or the hypodermis of the Convolvulaceous seed may or may not be lignified, and yet it is the family character. Probably the seed-coat of *Eucalyptus* has been examined in greatest detail but it affords no chemical satisfaction.

The first reaction will be to suppose that I have exaggerated the importance of seed-structure. With increasing familiarity it will be realized that seed-structure is a prime inheritance. Its histological detail expresses genetic character. The many variations cannot be ascribed to environmental selection. The profoundly different seeds of crucifer, mallow, speedwell and chenopodium arrive on the same waste ground. The profoundly different seeds of *Magnolia, Sterculia, Garcinia* and *Tabernaemontana*

sprout together on the floor of the tropical forest. Drupes and nuts protect seeds with thick outer lignification, yet the intrinsic details of the seed-structure may be retained, e.g. *Canarium, Elaeocarpus, Grewia, Scaevola* and *Terminalia*. Magnoliaceous seeds are complicated, those of Sympetalae are simplified. The main trend in seed-evolution has been simplification by reduction in complexity and size, e.g. *Myristica* compared with *Begonia* or *Bellis*, as palm to orchid. Embryologists have established the advance from the bitegmic to the unitegmic ovule and from the crassinucellate to the tenuinucellate. Simplification becomes the theme, and from the whittling away of the Magnolialean complexity there emerges the idea of a complicated prototype. Palaeobotany would not be able to recognize a primitive dicotyledonous seed. In a recent paper (Boumann 1971) phyletic emphasis has been laid not so much on the final structure of the seed-coat as on the manner in which the integuments may arise in the ovule. Both matters are important but, since the final structure of the differentiated integument is so much more explicit, it cannot be discarded.

This work is also a vindication of that part of the durian-theory which discovered a primitive factor in the arillate seed. The conclusion has been derided through prejudice and ignorance, but what is laughed at commonly comes to be taken seriously. All along the evidence accumulates in favour of the primitive envelope; it is woven into these pages in the continual effort to show that classification without seed-structure is unsound and, consequently, our knowledge of the evolution of flowering plants.

It is a pleasure to thank the many botanists who have supplied me with material, in particular H.K. Airy Shaw, J. A. R. Anderson, P. S. Ashton and F. Hallé. On two short visits to Ceylon in 1968 and 1972, I was enabled to collect a large number of critical seeds through the kindness of Professor B. A. Abeywickrama and his colleagues Dr Dassanayake and Dr Balasubramaniam of the University of Ceylon; indeed this work, begun in Singapore, has matured in Ceylon. Many slides have been prepared for me in the last three years by P. Mohana Rao, of the University of Delhi. With E. C. Bate-Smith I have had numerous discussions on the biochemical classification of flowering plants; the subject is beyond most provincial schools of botany but of increasing concern to the chemist (Bate-Smith 1972).

Dr Gh Dihoru, of the Institut de Biologie 'Tr. Savulesiu', Bucharest, has kindly supplied me with the following notes on the life of Netolitzky; they are taken from the obituary by Popovici (1947). Netolitzky was born at Zwickau, Bohemia, on 1 October 1875 and died in Vienna on 5 January 1945. He was a student in the universities of Prague, Strassburg and Vienna where, as a junior assistant from 1 January 1896 to 31 August 1899, he took his doctorate in medicine in 1899. He moved to the University of Innsbruck, 1 November 1899 to 30 April 1904, then to the University of Graz, 1 May 1904 to 31 January 1910. From 14 February 1910 to 30 September 1912 he was an assistant in the chemistry of foodstuffs at the University of Cernauţi. In 1912 he became an associate professor of the University and in 1919 was made Professor of Pharmacognosy and Plant Anatomy, and Director of the Institute of Plant Anatomy and Physiology. As pensioner in 1940, he entered a professorship at the University of Iati before he retired to Vienna in the capacity of Professor of Pharmaceutical Medicine. A profound knowledge of botany illuminates his writings.

E.J.H.C.

TO THE MEMORY OF

DR FRITZ NETOLITZKY

(1 OCTOBER 1875–5 JANUARY 1945)

Professor of Pharmacognosy and Plant Anatomy in the University of Cernauţi, Rumania
Professor of Pharmaceutical Medicine in the University of Vienna

PART ONE

1. Material and method

It is impossible to over-emphasize the desirability of the simplest approach, which is to study living ovules and seeds by means of free-hand sections mounted in water, cleared in lactophenol, and stained for lignin. Practice improves until sections can be cut with precise orientation under the binocular microscope. The lengthy procedure of embedding for the microtome is avoided with much saving in time, cost and result. Ovules and seeds are often oblique to the axis of ovary and fruit. The microtome supplies a large number of exasperatingly oblique sections, to which one correctly orientated free-hand section is preferable. Then, as the seed matures, its tissues become too hard for the microtome; they fracture under the blade, and the final and most characteristic features must be studied with free-hand shavings. This was the method of the early investigators, though unsupplied with modern binocular dissecting microscopes, by means of which they drew up and illustrated some of the best descriptions of seeds, e.g. Meunier on Papaveraceae. In modern research the structure of the seed-coat is often an adjunct to embryological details which require the microtome, and the results have seldom been outstanding. The microtome may seem more suitable for immature seeds because it allows microphotography, but extremely few photographs have been published which are so clear and convincing as drawings; background opacity, out of focus, blurs essentials. Moreover until structures have been followed cell by cell with the pencil, they are not appreciated. Then, a great advantage of the free-hand method is the thick, unstained but cleared, section which enables one to observe in depth and to follow oblique surfaces or strands. The advantages of studying living tissues are many; chlorenchyma, aerenchyma and mucilage-spaces are as clear as vascular bundles and lignified layers; integuments are separable when fixation causes them to adhere; and critical stages are quickly obtained. Colour, translucency and texture reveal at once important points and remind one that the seed is a growing photosynthetic structure. Few botanists realize that ligneous tissues pass through a highly aqueous phase and that water is the medium for lignification.

The best place to work is in a botanical garden in the tropics where so many seeds require investigation. For most genera and species, however, there will be only preserved material gathered on outings. It rarely supplies a full series from ovule to ripe seed, unless it has been gathered for this purpose. Nevertheless, botanists who visit the tropics and subtropics must be urged to preserve flowers and fruits in alcohol or other fixative, in addition to the usual and, now, often unnecessary dried material. My own researches have been helped in this way by students, though their gatherings may represent only one stage in seed-development. Dried material can be used when one is sufficiently acquainted with the general structure in a family or genus but it often does not enable one to distinguish the precise layering of the tissues into their derivation from one or other integument, which is important. The distinction between the products of the inner epidermis of the outer integument and those of the outer epidermis of the inner integument is usually decisive.

For the preliminary examination of a seed, the first command is to hunt the micropyle. It may be seen externally or it may be found internally from the direction of the radicle, though this is not infallible as the seed of *Sterculia* will show. The seed must be sectioned longitudinally and transversely so as to include micropyle and chalaza, because at these places, even in mature seeds, the separation of the integuments may still be discerned. Then tangential or paradermal sections must be made to determine the shapes of the cells; a layer of fibres cut transversely may give the appearance of a palisade of columnar cells; alternatively, a palisade of cells may have elongate facets or the facets may have the stellate lobing of the epidermal cells of many leaves. The need for such sections is often overlooked, but becomes a first consideration with the free-hand method. If the seed is embedded in thick woody endocarp, this may be cut cleanly as with a sharp blow on a knife or it may be sawn with a fine band-saw; in either case it is usually possible to extract parts of the seed and to study them in the normal way. Whatever method is employed, the later stages of hardening seeds must not be overlooked because definitive characters are frequently the last to appear.

For illustration I have preferred line-drawing made with the *camera lucida*. In cases of low magnification without cell-details, I have indicated endosperm with stippling, sclerenchyma with coarse stippling or speckling, palisade-layers with striation, and vascular bundles with broken lines. In high-power drawings I have often exaggerated slightly by means of heavy black lines the separation of the integuments and nucellus. Air-spaces have been shown in black.

2. Seed-form

The form of the seed, though neither its size nor its differentiation, is set usually by that of the ovule. As this organ is described in most books on plant-anatomy, it is necessary only to emphasize one point: the ovule is the embryonic seed. The cells of the ovule are small, thin-walled and isodiametric; they have large nuclei and few small vacuoles; vascular bundles are mainly procambial; air-spaces, if any, are slight; stomata, if present on the seed, are rudimentary or unformed. Then on fertilization the cells renew their growth; they divide, enlarge and differentiate; vascular bundles function; aeren-chyma is formed; most seeds are photosynthetic. At length the tissues around the endosperm and embryo die; mechanical layers have been formed in certain parts of the seed-coats with characteristic position and construction; an elaborate vascular supply may have perfused the testa; micropyle, chalaza and hilum may be stoppered; many complicated chemical changes may be the result of this senescence. In like manner the ovary is the embryonic fruit but, with exposed surface and functional style, it is partly adult. Nevertheless most of the inner tissue, particularly towards the base of the syncarpous ovary, is embryonic; on fertilization it proliferates and differentiates, while the precocious style withers or is discarded.

There is an interesting paragraph on this matter in an article by Croizat (1947a), which begins 'It is curious that it never seems to have occurred to orthodox morphologists that the *flower itself is an embryonal structure*, and that in most cases fertilization reaches the flower in *its embryonal stage*.' The author compares the adult fruit with the embryonal ovary, as if it were branch to twig, and concludes that to understand the flower the fruit must be known. The paragraph is worth study because the germ of truth is hidden in a confusion between the adult and functional parts of the flower, which are external to the ovary, and its embryonic parts which are no twig but the curtailed core of the reproductive bud. In his *Principia Botanica* (Croizat 1960), the message is forgotten and the author proceeds to estrange ovules from their nest, as if they were adult, and to doodle with them in outline on paper just as the orthodox morphologists of his complaint in their theories of ovular evolution without seed or fruit. The evolutionary process seems to have been the neotenic functioning in part of the reproductive bud whereby the divergent outer scales mature into the relatively small and caducous parts of the open flower around its embryonal centre; then, with sepals, petals, and stamens over, the centre grows into the massive fruit (Corner 1964). The primitive reproductive character of the flowering plant lies in the delayed expression of fruit and seed. The delay is successful because it avoids the expensive outlay in massive construction which would be forced upon the reproductive bud if pollination did not occur until fruit and seed were fully formed; little neotenic parts effect pollination in anticipation of the outlay and, if ineffective, the loss is minimal. The simplicity of the ovule is not primitive but primordial. Prime characters for the classification of flowering plants should lie therefore in the construction of fruit and seed, which are the parts so universally omitted by theorists. In the long run seeds also become neotenic, small and simple, and after fertilization merely enlarge the cells of the integuments to become, as it were, just adult ovules, e.g. *Begonia*, *Digitalis*. Alternatively, and as successfully, the number of seeds in a fruit reduces until the one-seeded fruit comes to function as a seed; the integuments may not differentiate in the seed-coat and the endocarp performs this duty. Thus, small or simple seeds are no more primitive than ovules. Primitive families, as those of Magnoliales, have fruits and seeds of great complexity; advanced sympetalous families simplify both. 'The difference cannot be capital between *Gnetum* and *Mezzettia* when the ovular structure of both genera is virtually identical tegument by tegument' (Croizat 1960, p. 397). The fallacy is clear. The seed of *Gnetum* has no fruit, and the fruit and seed of *Mezzettia* (Annonaceae) are exceedingly different; their primordia are similar.

The bitegmic anatropous seed

Ordinarily this seed is merely the regular enlargement and differentiation of the ovule. The integuments cover the seed except at the small part of the chalaza round which they are attached; raphe and antiraphe are almost equally long and the sides of the seed are identical. In certain cases, however, the enlargement is unequal; parts become displaced; the shape of the seed differs from that of the ovule, and the manner of differentiation may be varied (Fig. 1). Then in other cases, even in families, these alterations appear neotenically in the ovule which, again, prepares the shape of the seed. It is commonly assumed that these distinctions, particularly of shape, arise in the ovule and are conveyed into the seed, but the study of seeds suggests that the reverse has been the evolutionary procedure, e.g. Bixaceae–Cistaceae. As these changes happen independently in different families, there are many differences in detail and, as they are often intricate, it is necessary to consult the description and illustrations of individual cases. The present account is but a general outline.

Campylotropous seeds

In some species, if not genera, the anatropous ovule develops the antiraphe more extensively than the raphe, and the seed becomes curved or campylotropous, e.g. *Psidium* (Fig. 428). In Capparidaceae the seed is campylotropous and the transition to the neotenically campylotropous ovule can be seen in *Capparis* and *Crataeva* (Figs. 65, 69) with suborthotropous ovule. The alteration in shape is so gradual that it becomes impossible to employ accurately the various terms that have been suggested for the intermediate stages. Other examples will be found in Papaveraceae, Cactaceae, and Leguminosae.

Obcampylotropous seeds

This form (Fig. 1a) is the converse in which the raphe of the anatropous ovule enlarges more than the antiraphe. It occurs in *Bauhinia* (Fig. 322) and *Barklya* (Fig. 320), the seeds of which appear campylotropous in external view like those of

Papilionaceae. Their ally *Cercis* (Fig. 334) retains the anatropous seed. Such seeds occur also in Vitaceae.

Hilar seeds

In these (Fig. 1c) the greater part of the circumference of the seed, which is usually flattened, is made up of the extended hilum, e.g. *Mucuna* (Papilionaceae; Corner 1951). The ovule is campylotropous and undergoes this curious deformation as it enlarges into the seed; raphe and antiraphe remain short. Again various degrees of hilar development are found in other Papilionaceous genera, such as *Canavalia* and *Erythrina*, and a peculiarity of several lies in the development of most vascular bundles for the testa from the recurrent bundles of the hilum. In Meliaceae and other families with sessile arillate seeds there is often a short expansion of the hilum, e.g. *Aesculus*.

Pre-raphe seeds

In these the very short distance which usually separates the beginning of the raphe from the micropyle, and which causes the micropyle of most seeds to be adjacent to the hilum, is here lengthened (Fig. 1d). In consequence the micropyle is far removed from the hilum and what appears to be the raphe is actually the pre-raphe or the part between the micropyle and the hilum. This construction is characteristic of Connaraceae (Figs. 136–155); it is more or less pre-formed in the ovule, and the pre-raphe has a longitudinal vascular bundle similar to that of the raphe which is variously shortened. Many Connaraceous seeds seem to resemble Papilionaceous seeds until, as the first requirement of seed-study, the micropyle is found. The Meliaceous *Dysoxylon cauliflorum* has a kind of adnate pre-raphe which, if free of the placenta, would make the ovule and seed Connaraceous.

This kind of seed is clearly on the way to becoming orthotropous. Thus it figures in various Urticaceae which, as Conocephaloideae, are intermediate between the anatropous Moraceae and the typically orthotropous Urticaceae. Possibly it is the construction, also, in some hemi-anatropous Proteaceae. In Euphorbiaceae–Crotonoideae there may be a short pre-raphe with the hilum central on the adaxial side of the seed, e.g. *Croton laevifolium* (Fig. 227).

Orthotropous seeds

These are developed from orthotropous ovules and occur in several and diverse families such as Urticaceae, Proteaceae, Flacourtiaceae, Piperaceae, and Polygonaceae (Fig. 1e). They have been assumed to be primitive through analogy with gymnosperms, but the evidence of angiosperms points to derivation from the anatropous, either directly (as in *Chisocheton*, Meliaceae) or through the pre-raphe seeds as suborthotropous seeds for which Urticaceae, in no way primitive, is a fair example; it is true also of Proteaceae. It seems that the simple orthotropous shape is determined by the position of the ovule and the direction in which the ovarian loculus is extended by intercalary growth; the ovule-primordium ascends or descends directly in accordance, or fails to curve. The result is a radially symmetrical seed with the micropyle at the opposite end from the hilum. Post-chalazal vascular bundles may then permeate the testa, as in *Myrica*, but this genus is unitegmic and may have a pachychalazal seed. Orthotropous ovules do not occur in families which, according to the structure of the fruit or flower, are regarded as primitive, e.g. Magnoliales, Dilleniaceae, Mimosaceae, Theales, or Clusiaceae.

The dorsal raphe

A deceptive form of the suspended and anatropous ovule is that with a dorsal raphe that curves over the adaxial micropyle. It distinguishes certain families as Lauraceae, Monimiaceae–Monimioideae, Buxaceae, Ebenaceae, and some genera of other families as Anacardiaceae, Celastraceae, Theaceae, and Proteaceae. The seed is suspended in the same manner. The relation of this ovule to the ordinary anatropous form with abaxial micropyle is uncertain.

Perichalazal seeds

The ovule in this case appears to be anatropous but, internally, the inner integument is attached to the outer along the whole course of the vascular bundle which, in ovule and seed, extends round the periphery from funicle to micropyle (Fig. 1g). In place of the punctiform chalaza opposed to the micropyle, a perichalaza surrounds the nucellus as a hoop or band. Instead of an extended hilum for the periphery of the seed, as in *Macuna*, there is an extended chalaza; and the complexity of the seed is rendered apparent by the manner in which intercalary growth of the ovule is prompted. Perichalazal construction distinguishes the ovule and seed in Annonaceae (Corner 1949b); in a few genera it is connected with the development of a middle integument. How frequently the construction may occur in other families is uncertain, e.g. *Hortonia* in Monimiaceae (Fig. 394) and Ebenaceae, but there is a partial perichalaza at the chalazal end of the seed of *Cryptocarya* (Lauraceae, Fig. 304), along the raphe in the seed of *Swietenia* (Meliaceae, Fig. 389), in more or less complete form in some species of *Aglaia* and *Lansium* in Meliaceae and also in some seeds of Vitaceae (Figs. 616, 622, 630); yet the ovule in these cases is not perichalazal. In the Vitaceous genera the ovule first becomes obcampylotropous and then more or less perichalazal in the developing seed. It is possible that the intrusive raphe in Convolvulaceous seeds and the intrusive hilum or placenta of Apocynaceous seeds are cognate.

Pachychalazal seeds

The chalaza of the perichalazal seed is extended in the median plane. In the pachychalazal seed it develops in all directions and builds by intercalary growth a new container for the endosperm and embryo (Fig. 1h). The wall of the container is single and is constructed by the multiplication of the cells where the two integuments adjoin the nucellus and chalaza; generally it becomes highly vascular from extensions of the chalazal vascular supply. The two integuments persist at the micropylar end of the seed in a more or less vestigial state. The ovule is anatropous and the resulting seed appears normal until its structure has been followed in development.

The expression 'pachychalazal' was introduced by Periasamy (1962b). The construction has been found in a variety of families and, as it has certainly been overlooked, it may occur in many others. Periasamy considered that it accompanied the ruminations of the endosperm, but the instance of Annonaceae and Myristicaceae with ruminate endosperm, yet neither pachychalazal, forbids such a generalization; there are also families with pachychalazal seeds without rumination, e.g. Meliaceae, Sapindaceae. The construction is, in fact, another instance of that intercalary growth which is so disconcerting for morphologists because it supplies in place of the growth of free parts an intercalated sheet or tube that simulates the original, cf. the syncarpous or intercalary ovary, the leaf-sheath, the pitchers of *Nepenthes*, and indeed the lamina of entire leaves. As a basipetal growth it is the antithesis of the primitive acropetal growth of dicotyledonous organs, and the pachychalaza appears as a polyphyletic advance in seed-construction.

Pachychalazal seeds have been found in Balsaminaceae, Flacourtiaceae, Lauraceae, Meliaceae, Ochnaceae, Rosaceae, Sapindaceae, Simarubaceae, Ebenaceae and Euphorbiaceae. They may occur in Anacardiaceae, Combretaceae, Icacinaceae, Polygonaceae, Proteaceae (*Macadamia*), Rhamnaceae, Apocynaceae and, indeed, in *Nelumbo*. In some of these the characters of the testa differentiate in the single coat of the pachychalaza, e.g. *Taraktogenos* (Flacourtiaceae, Figs. 277–281). In the arillate seeds of Meliaceae and Sapindaceae, the outer part of the pachychalaza may be fleshy like the arillar tissue while the inner part may have a sclerotic layer (Figs. 374, 501). These are the sarcotestal seeds which van der Pijl has confused with the truly sarcotestal seeds of Magnoliaceae, and of course they are not primitive.

In these pachychalazal seeds of Meliaceae the fibrous exotegmen, characteristic of the family, can be found in the free tegmen round the micropyle, but is absent from the wall of the pachychalaza, e.g. *Aphanamixis* (Figs. 379, 380). In some of these, moreover, the pachychalaza is partial and affects only the dilated hilar side of the seed, e.g. *Dysoxylon*; this is the case also in the Anacardiaceous *Campnosperma* (Fig. 13). By contrast, in *Taraktogenos* there is no trace of the fibrous exotegmen, which is distinctive of Flacourtiaceae, and the affinity of this genus and its allies with the rest of the family is not certain. The condition in Euphorbiaceae is also different because the pachychalaza affects only the tegmen, e.g. *Cleidion* (Fig. 222) and *Ricinus* (Fig. 248). Such Euphorbiaceous seeds have a typical testa, and the tegmen is covered both in its free and pachychalazal part with the exotegmic palisade distinctive of the family. A further complication in Euphorbiaceae is the need to distinguish the vascular tegmen from the vascular pachychalaza.

In the preceding examples the ovule is normally bitegmic; the pachychalaza develops after fertilization. In *Ochna* (Fig. 437) the ovule is already pachychalazal and the short vestigial integuments take no part in the formation of the seed-coat. This condition was described in detail for many Rosaceae by Péchoutre (1902). It occurs in Tropaeolaceae, some Balsaminaceae, and in *Phytocrene* (Icacinaceae). It is the intermediate state to the unitegmic ovule and seed in which, it is said or assumed, the integuments have fused. There is no fusion but a substitution of the free growth of the integuments by a basal intercalary region with the thickness of both integuments. If, as in some Rosaceae and perhaps Icacinaceae, the combination is congenital with the inception of the integuments, there results the unitegmic seed. This seems to explain the unitegmic ovule of Limnanthaceae in its relation to the pachychalazal ovules of Balsaminaceae and Tropaeolaceae. The knowledge of such seeds is slender. The single seed-coat, though it is not truly integumentary, may be described as testal, as will be explained in the next chapter.

Now it is doubtful if the vascular pachychalaza of *Ochna* or *Tropaeolum* represents the original construction of the free testa in the primitive bitegmic seeds of their families. The point is important because pachychalazal ovules may signify the origin of the massively unitegmic ovule of most sympetalous families (p. 50); their seeds generally lack the complications of the polypetalous.

Alate seeds

The wing of the seed is a local outgrowth of the testa or, in the unitegmic seed, of the seed-coat. It displays the local morphogenetic potentialities of the ovule for it may arise from different parts. But wings are also connected with the manner in which the ovary enlarges into the fruit and the consequent change in shape of the loculus. Fruit-factors must be even more varied than those which control the development of the seed, and they remain to be analysed. Their interest for the study of seeds lies in the relation between the primitively arillate seed and the alate as an intermediate derivative (Corner 1954; Forman 1965). The generalized and extensive growth of the arillate follicle or capsule seems firstly to become constrained in a way that flattens the seed, perhaps with additional crowding through increase in number of the ovules, and then it is further narrowed to make slits into which the wings may extend. They may be completely peripheral as in Bignoniaceae, or restricted to the raphe, chalaza, antiraphe, hilum (Vochysiaceae), funicle, and even along the three angles of a plump seed (*Moringa*); the aril itself seems not to be involved but to disappear. *Catha* (Celastraceae), however, has an alate aril (Fig. 83). Various genera in one family have these differences. In Theaceae *Stewartia* (Fig. 592) is peripherally alate; *Gordonia* (Fig. 587) has a raphe-wing; and *Schima* (Fig. 589) is peripherally alate except for the extended hilum and pre-raphe. These facts are seldom co-ordinated into the generic character which, at the customary level of the pocket-lens, is unable to distinguish them and to relate them into evolutionary sequence.

Some alate seeds with the wing extended from the chalazal end of the raphe carry the vascular bundle of the raphe round the periphery of the wing, e.g. *Cratoxylon* (Fig. 292), and this loop may be retained as a vestige in diminutive seeds, e.g. *Tetracentron* and *Trochodendron* (Fig. 611). Yet, it is absent from such seeds as *Ixonanthes* (Fig. 299) and *Lagerstroemia* (Fig. 352). All these details render alate seeds among the more interesting.

Pleurogrammatic seeds

Most modifications of the ovule in the course of its development into the seed affect the hilum, the chalaza or the periphery. Few affect the sides, and these I call pleurogrammatic because they leave a lateral mark of some kind. The best known is that which I called the pleurogram in the Mimosoid seed, though in this case the expression *linea fissura* or *linea sutura* has precedence (see p. 163). The mark is an alteration in the surface of the seed where, possibly, raphe–antiraphe factors impinge on others arising on the sides of the growing seed. The Mimosoid *linea fissura* is a break in the construction of the exotestal palisade. The pleurogram of *Cassia* (Figs. 329–333) is caused by a difference in the palisade-cells themselves. The complicated pleurograms of many Cucurbitaceous seeds result from differences in the layering of the complex outer part of the testa (p. 112). The feature is generally rare and, thus, helpful in seed-identification. Its physiological basis is unknown.

3. Seed-coats

Testa and tegmen

To avoid the ambiguity that arises when the integuments of the ovule and seed are called by the same names, I refer to those of the ovule as the outer integument (o.i.) and the inner (i.i.); the product of o.i. then becomes the *testa*, that of i.i. the *tegmen*. Seeds with characteristic testa can be called *testal*, those with characteristic tegmen *tegmic*. The outer and inner epidermal layers of each integument become:

    o.e. (o.i.), i.e. (o.i.), o.e. (i.i.) and i.e. (i.i.)

    o.e. (testa), i.e. (testa), o.e. (tegmen), i.e. (tegmen).

It is easy to confuse the first line of symbols. They cannot be applied in the same way in other languages, and in the older reports on seeds o.i. would be i.e. (*integumentum externum*).

The middle layers are referred to as mesophyll (o.i., i.i., testa, or tegmen). The outer and inner hypodermal layers, which sometimes need to be distinguished, become o.h. and i.h.

The only possible confusion may be with the words bitegmic and unitegmic which refer to ovules or seeds with two integuments or one (p. 49).

Multiplicative integuments and overgrown seeds

In many ovules, especially those which become large seeds, the cells of both integuments divide after fertilization and form both more cells in a layer by anticlinal division and more cell-layers by periclinal division. Both methods of growth may occur or one or other may predominate, but finally the cells enlarge, become adult, and differentiate. The second method by periclinal division I call the *multiplicative*; it results in seeds with massive and, usually, complicated seed-coats, e.g. Magnoliaceae, Myristicaceae, Annonaceae, Clusiaceae, Leguminosae, Bombacaceae. The contrast is the non-multiplicative integument which may extend by anticlinal cell-divisions, but does not develop more cell-layers. It produces the simpler seed and, eventually, the simplified seed in which the cells of the integuments merely enlarge and differentiate without further cell-division. Such small seeds have often a few large external cells which with thickened walls give the reticulate, punctate, or verrucose surface of the testa, e.g. Caryophyllaceae, Papaveraceae, Ranunculaceae, Gesneriaceae, Scrophulariaceae. In some ovules both integuments are multiplicative; in others either the outer or the inner is more or less non-multiplicative. These distinctions are important both for the accurate description of seeds and for their theoretical consideration. Multiplicative seeds occur in families acknowledged to be primitive and show from what complexity the small seeds of advanced families have been simplified neotenically. Seed-evolution has caused in the main loss of multiplication but also, as a diversion, an excess, and such seeds with excessive multiplication are what I have called overgrown seeds (Corner 1951).

The cell-layers of the seed-coats of the overgrown seeds multiply without tissue-differentiation. The seeds come to fill the loculus of the fruit as if they were tumours. They fail to differentiate the family-character of the seed-coats which, if they are not finally crushed by the enlarging endosperm or embryo, as in *Anisophyllea, Barringtonia, Mangifera, Persea* or *Arachis*, remain as a more or less thin-walled jacket round the seed. Overgrown seeds occur in a variety of families, especially those with drupes or nuts. In a way they are a nuisance because they often occur in critical genera of uncertain classification for which a definite structure of the seed-coats would be decisive, e.g. various Swartzieae (Leguminosae) and *Irvingia*. Fagaceae and most Anacardiaceae seem to have overgrown seeds. It may be the case with Ochnaceae and Simarubaceae but in these families there is also the problem of the pachychalazal seed with its overgrowth. Research is needed into all these seeds by comparison with related genera, e.g. *Aesculus*, *Aldina*, *Avicennia*, *Arachis*, *Persea* and *Rhizophora*. Not only the smallest seeds are the most advanced but also the largest, as shown by *Orchis* and *Cocos*.

Factors in the formation of seeds

The chief factors in the growth of the ovule-wall into the seed-coat may be summarized as follows.

(1) Enlargement:
    (*a*) By cell-division:
        (i) periclinal, increasing the number of cell-layers,
        (ii) anticlinal, increasing the number of cells per layer,
        (iii) by a special meristematic layer of cells, usually epidermal or hypodermal.
    (*b*) By cell-enlargement:
        (i) uniform,
        (ii) radial elongation to give a palisade-layer of prismatic cells with hexagonal facets,
        (iii) tangential elongation to give tubular cells with stellate outline, or fibres parallel or transverse to the longitudinal axis of the seed.
(2) Differentiation into tissue-layers:
    (*a*) epidermal,
    (*b*) chlorenchyma,
    (*c*) aerenchyma,
    (*d*) sclerenchyma or collenchyma, as the mechanical tissue,
    (*e*) vascular bundles,
    (*f*) chalazal elaboration.
(3) Outgrowths:
    (*a*) aril (in the general sense),
    (*b*) wings,
    (*c*) hairs.
(4) Funicle:
    (*a*) elongation,
    (*b*) special developments (such as jaculators).

Multiplication of the cell-layers is often diffuse throughout the testal mesophyll, in some cases also in the tegmic, but a few families or genera are specialized in this respect and show that the process needs more careful investigation than has hitherto been given. Periclinal divisions of o.e. (o.i.) produce the firm outer layer of the sarcotesta of Magnoliaceae; those of o.e. (i.i.) make the layers of fibres in the tegmen of *Capparis*, but the fibres of Annonaceae result from diffuse divisions in the testal mesophyll. Periclinal divisions of o.h. (o.i.) make inwardly the ridges of firm tissue in the testa of Caricaceae. Similarly divisions of o.h. and i.h. (o.i.), perhaps also of i.e. (o.i.), make the massive mesophyll of *Calophyllum* (to 100 cells thick), in *Moringa* (40–50 cells thick), and in *Taraktogenos* (40–70 cells

thick). Localized periclinal divisions of i.h. (o.i.) make small ridges in the inner testal layer of *Averrhoa* (Oxalidaceae). Periclinal divisions of i.e. (o.i.) make the woody endotesta or Magnoliaceae, Rutaceae–Rutoideae, and Vitaceae. The most striking case is that of Cucurbitaceae in which, as is now well-established, the durable parts of the testa are made from fairly precise periclinal divisions of o.e. (o.i.); they result in three cell-layers which may be further multiplicative, each in its own manner (p. 113).

Description of the seed-coats

Seeds may vary much in form, size, colour and arillar investment within a single large and manifold family, e.g. Clusiaceae, Myrtaceae, Rosaceae, Rutaceae, Sapindaceae, or Theaceae. It is the microscopic structure of the seed-coats which supplies the critical detail, and even such manifold families have a basic microscopic character. There is no *a priori* reason for this. It happens that the genera of a natural family have the same microscopic construction. Thus, the Magnoliaceous is not found in the Rosaceous or Leguminous, the Annonaceous is not Myristicaceous and the Tiliaceous is not Celastraceous. But Bombacaceae, Euphorbiaceae, Malvaceae, Sterculiaceae and Thymelaeaceae agree with Tiliaceae; Connaraceae, Flacourtiaceae, Meliaceae and Sapindaceae agree with Celastraceae; Lecythidaceae, Myrtaceae, and Rosaceae agree with Theaceae. By exploring the microscopic detail in a sample of genera from a family, one discovers this great and unexplained peculiarity of the angiosperm seed. If the structure were the same in all, the subject would not arise. Netolitzky demonstrated it but, probably for lack of conveniently brief and unambiguous notation, it has been neglected. As it is, the Magnolialean families have distinctive and complex seed-coats, of which none is so intricate as that of Myristicaceae. Advanced families with inferior ovary or sympetalous corolla have uniform and simplified seeds. Intermediate and, as it were, exploratory families display the variety that may reveal how the first turned into the advanced.

The distinctive character of the seed-coat lies in the position and structure of the main mechanical, thick-walled but not necessarily lignified, layer. It may be a palisade of radially elongate cells, a layer of fibres, or a layer of cuboid sclerotic cells. It may be one or more cells thick, but whether the seed-coats are multiplicative or not appears to be less significant. Therefore, with the distinction between testal and tegmic seeds, I have invented the following, simple and self-explanatory method of description. It is based firstly on the position of the mechanical layer and then on its nature, because this appears to be their order of importance so far as may be judged from family-alliances.

(A) Testal seeds

(1) *Exotestal* seeds, with the mechanical layer in o.e.

(*a*) As a palisade of radially elongate cells, e.g. Ranunculaceae, Rhamnaceae, Leguminosae, Winteraceae.

(*b*) As a layer of fibres (in various genera of different families).

(2) *Mesotestal* seeds, with the mechanical layer in the mesophyll.

(*a*) As an outer hypodermal palisade, e.g. Paeoniaceae.

(*b*) As cuboid sclerotic cells in the mesophyll generally, e.g. Myrtaceae, Rosaceae, Theaceae.

(*c*) As fibres in the mesophyll generally (Annonaceae).

(3) *Endotestal* seeds, with the mechanical layer in i.e.

(*a*) As a palisade, e.g. Myristicaceae, Dilleniaceae, Cruciferae.

(*b*) Tracheidal, the tangentially elongate cells with spiral or annular thickening, e.g. Lauraceae, Monimiaceae, Calycanthaceae.

(*c*) As a multiple layer of sclerotic cells, e.g. Magnoliaceae, Vitaceae.

(*d*) As a single layer of cuboid sclerotic cells (Grossulariaceae).

(B) Tegmic seeds

(1) *Exotegmic* with the mechanical layer in o.e.

(*a*) As a palisade, e.g. Malvales.

(*b*) As a layer of fibres, e.g. Celastraceae.

(*c*) With stellately lobed cells, e.g. Clusiaceae, Hypericaceae, Geraniaceae.

(2) *Mesotegmic* seeds, with the mechanical layer in the mesophyll (rare, in conjunction with exotegmic or endotegmic seeds).

(3) *Endotegmic* seeds, with the mechanical layer in i.e., as a short palisade or as cuboid sclerotic cells, e.g. Piperaceae, Saururaceae, Nandinaceae.

The majority of families may be placed in one or other category because they have one predominant feature, but some have a double character or even a treble, e.g. Myristicaceae, Cucurbitaceae, ? Droseraceae. In monocotyledons there is much less diversity; there are exotestal and endotestal seeds but few, if any, that are tegmic (? Bromeliaceae, ? Dioscoreaceae, ? Taccaceae).

Exotestal seeds

In this kind the outer epidermis of the testa forms, typically, a rigid palisade. The cells are thick-walled, lignified or not, and usually columnar, but they may be shortened into cuboid cells or elongate longitudinally almost into fibres. Typically the tegmen and the rest of the testa have little or no mechanical specialization, and these inner tissues are generally crushed by endosperm or embryo. As represented by Winteraceae, Ranunculaceae, Nymphaeaceae, Leguminosae and Rhamnaceae, the exotestal seed suggests the primitive construction in which the exposed surface is the protective layer. But, as the feature is pursued, it is found in various families combined with other specializations. It occurs with sclerotic mesophyll (Buxaceae, Clusiaceae *pr. p.*, Rosaceae, Staphyleaceae), with exotegmic fibres (Celastraceae, Connaraceae, Meliaceae, Sapindaceae), with tracheidal endotesta (Calycanthaceae), with tracheidal tegmen (Rutaceae), and as a reduced palisade in endotestal Myristicaceae. Indeed the hard exotesta occurs in some degree in most arillate seeds, the seed-coats of which may have almost any internal organization. It becomes the character, also, of most unitegmic sympetalous families with their simplified seeds (p. 49). The exotestal character, therefore, seems to be but part of the primitive construction of the seed-coat which commonly persists because of its serviceability. That it can result from simplification is proved by its occurrence in certain genera or species of Connaraceae, Meliaceae and Sapindaceae in which the characteristic fibrous exotegmen fails to develop.

A full and unqualified list of families with exotestal seeds would be long and confusing. I have made two short lists for those with bitegmic seeds in which the character predominates. The first, with albuminous seed and minute embryo, contrasts with other Magnoliales, but even here Illiciaceae–Schisandraceae indicate a deviation from the mesotestal construction. The second, with large embryo, is the mixture which will be sorted out when the knowledge of seed-evolution is improved. In this list Podophyllaceae–Lardizabalaceae–Berberidaceae, perhaps with Melianthaceae, make an alliance in which the epidermal cells tend to have elongate facets, even to become fibriform. Centrospermae appear to have been primitively exotestal and arillate (cf. Nymphaeaceae, Podophyllaceae). Leguminosae, also arillate in origin, provide no evidence of a more complicated seed-ancestry.

(1) Exotestal seed, bitegmic, albuminous; embryo minute.

Nymphaeaceae, Ranunculaceae, Sarraceniaceae (? unitegmic), Winteraceae (but lignified endotegmen in *Belliolum*);
Podophyllaceae (exotesta fibriform in some genera);
Illiciaceae–Schisandraceae (1–3 layers of sclerotic cells in the mesotesta);
? Droseraceae (cf. complicated seed in *Drosophyllum*);
? Polygonaceae.

(2) Exotestal seed, bitegmic, albuminous or not; embryo well-developed.

Begoniaceae, Berberidaceae, Cactaceae, Centrospermae, Elaeagnaceae, Fouquieraceae, Frankeniaceae, Lardizabalaceae, Leguminosae, Melianthaceae, Melastomataceae, Myrsinaceae, Plumbaginaceae, Podostemaceae, Primulaceae, Rhamnaceae, Salicaceae, Salvadoraceae, Saxifragaceae, Tamaricaceae; ? Caryocaraceae (cf. *Anthodiscus*), ? Zygophyllaceae (also fibrous endotegmen);
also families with more or less sclerotic mesotesta: Buxaceae (cf. *Sarcococca*), Ebenaceae, Lecythidaceae, Myrtaceae, Rosaceae, Rutaceae, Theaceae, Sapindales (Aceraceae, Akaniaceae, Hippocastanaceae, Sapindaceae, Staphyleaceae, ? Melianthaceae);
also families with fibrous exotegmen: Capparidaceae, Celastraceae, Connaraceae, Meliaceae, Proteaceae, Sapindaceae, Scyphostegiaceae.

Mesotestal seeds

In this case the mesotesta becomes more or less sclerotic with cuboid or shortly tangentially elongate cells, and the rest of the seed-coat is unspecialized. It seems to be the character of the seeds of such large or important families as Hamamelidaceae, Lecythidaceae, Myrtaceae, Rosaceae and Theaceae. It distinguishes the relict families Stachyuraceae with arillate seeds and Moringaceae, and it occurs in the Clusiaceae–Calophylleae. Therefore it must be emphasized. It draws *Moringa* out of any close alliance with Capparidaceae (tegmic fibres) and Cruciferae (endotestal). Nevertheless the character evidently passes by simplification into the exotestal. *Buxus* is plainly exotestal but *Sarcococca* is mesotestal. Some species of *Diospyros* are exotestal, others also mesotestal. Similarly there is a transition in Lecythidaceae, Melastomataceae and Myrtaceae. It also appears in the pachychalazal construction of Flacourtiaceae (*Hydnocarpus* etc.) and Sapindaceae; and this mesotestal and pachychalazal combination may be

the explanation of the hard seed of *Macademia* (Proteaceae).

Then, to avoid proliferation of categories at this stage of insufficient knowledge, I have included as mesotestal (1) Annonaceae with fibrous mesotesta, (2) Paeoniaceae with an outer hypodermal palisade, and (3) Cucurbitaceae with a secondary outer testa developed from o.e. (o.i.) into sclerotic layers. Thus the mesotestal seed is undoubtedly a grade of seeds derived from various phyletic lines, but its occurrence in Illiciaceae, Schisandraceae, Rosaceae, Stachyuraceae, Theaceae, Hamamelidaceae, Buxaceae, Lecythidaceae, Myrtaceae and Melastomataceae may well indicate one phyletic line to which Moringaceae and Ebenaceae may belong. A fibrous mesotesta has been described for *Magonia* (Sapindaceae) and *Casimiroa* (Rutaceae), the last with crossed fibres as in Annonaceae.

Mesotestal bitegmic seed: Annonaceae, Buxaceae, Clusiaceae–Calophylleae, Ebenaceae, Hamamelidaceae, Lecythidaceae, Melastomataceae, Moringaceae, Myrtaceae, Rosaceae, Rutaceae, Stachyuraceae, Theaceae;
of pachycaul construction, Flacourtiaceae *pr. p.* (*Hydnocarpus* etc.), Proteaceae (? *Macadamia*), Sapindaceae: Cucurbitaceae, Paeoniaceae.

Endotestal seeds; crystal-cells and raphid-cells

The inner epidermis of the testa becomes the mechanical layer in these seeds. The absence of lignification in the outer part of the testa leads to the development of the sarcotesta in such families as Magnoliaceae, Degeneriaceae, Grossulariaceae, Vitaceae and some Dilleniaceae. It might be assumed that sarcotesta and endotesta, as in Magnoliaceae, were indeed primitive marks of a primitive seed but, if this argument is extended to Grossulariaceae (arillate) and Vitaceae, difficulties are encountered which suggest, rather, that the sarcotestal construction is an outcome of endotestal specialization. Thus, there are considerable differences in the detailed structure of both sarcotesta and endotesta in these families; it is impossible to draw a line between the pulpy sarcotesta and the dry, but thin-walled, outer testa in Dilleniaceae and Myristicaceae; and both these families with arillate seeds have other features of internal construction. The implication is that the endotestal seed is the polyphyletic simplification from more complicated states, and that it contrasts the internal mechanical layer with the external of the polyphyletic exotestal seeds. A fibrous exotegmen occurs also, for instance, in Onagraceae, Papaveraceae and some Chloranthaceae, Myristicaceae and Proteaceae. In Vitaceae the tegmen is tracheidal, as the exotegmen is in Dilleniaceae. In some Myrtaceae (*Arillastrum, Eucalyptus spp., Rhodomyrtus*) and, perhaps, Polygalaceae the derivation is from mesotestal seeds.

The few families which fall strictly into this category can be separated into groups according to the shape of the endotestal cells, their manner of lignification or thickening, and their arrangement whether in a single layer or, through periclinal division of i.e. (o.i.), in several layers. These points are set out in the following classification which, in bringing together Degeneriaceae–Magnoliaceae and Calycanthaceae–Lauraceae–Monimiaceae, both being accepted alliances, proves the value of the endotestal character. It is most complicated in Myristicaceae, the seed of which may also be exotestal and exotegmic with fibres or tracheidal cells; yet the remarkably thick endotestal palisade of prismatic cells (0.2–1 mm high) must place the family here.

(A) Endotestal cells thin-walled but strengthened by an internal fibrillar reticulum pervading the cell-cavity and lignified or not.

(1) Endotesta multiple: Degeneriaceae, Magnoliaceae.

(2) Endotesta as a single layer of cells: Chloranthaceae, Papaveraceae, Proteaceae (all these families with fibrous tegmen in some genera at least).

(B) Endotestal cells thick-walled, generally lignified, pitted or with spiral-annular thickening, without the fibrous endoreticulum.

(1) Endotestal cells tabular, with stellate-undulate facets: Onagraceae (*Jussieua*, with fibrous exotegmen).

(2) Endotestal cells rhomboid or longitudinally elongate, not radially elongate, tracheidal with spiral, annular or reticulate thickening, in a single cell-layer or multiple: Calycanthaceae, Lauraceae, Monimiaceae, Burseraceae, Combretaceae (also with fibrous or tracheidal exotegmen).

(3) Endotestal cells cuboid or radially elongate and prismatic, not tracheidal.

(*a*) Endotesta multiple: Vitaceae (also with tracheidal tegmen).

(*b*) Endotesta as a single cell-layer.

(i) Cells cuboid or shortly radially elongate, with thickened inner and radial walls (U-shaped thickening): Cruciferae, Grossulariaceae, Myrtaceae *pr.p.*, Onagraceae, Polygalaceae *pr.p.*

(ii) Cells thickened all round.

Cells more or less cuboid-substellate: Dilleniaceae (with tracheidal exotegmen).

Cells prismatic in a palisade: Myristicaceae (fibrous or tracheidal exotegmen), Polygalaceae.

The endotestal character was not emphasized by Netolitzky who drew attention, instead, to the fact that the inner epidermis of the testa is distinguished in many families (though not necessarily in all their genera) by the presence of one or more angular crystals of calcium oxalate in the cells. Such crystal-cells may occur in the mesophyll of the testa in other families, e.g. Theaceae, but seem rarely to occur in the tegmen, even in its lignified parts. Nevertheless, it seems that the presence of crystals in the endotesta is but a step in the process of lignification in the testa and, in the case of the endotestal families, it leads to the inclusion of the crystals in their thickened endotestal cells; exceptions are Dilleniaceae and the five families with the tracheidal endotesta (Burseraceae, Calycanthaceae, Combretaceae, Lauraceae, Monimiaceae). The presence of crystals in the unthickened cells of the endotesta in other families seems to indicate a loss of the primitive lignification. Thus, of the families cited by Netolitzky as possessing crystal-cells in the endotesta, the following have thin-walled endotestal cells and are essentially tegmic in seed-construction:

Aristolochiaceae, Caricaceae, Geraniaceae, Lythraceae, Malvaceae, Meliaceae, Oxalidaceae, Resedaceae, Tiliaceae, Violaceae; to these may be added Connaraceae and Elaeocarpaceae. (Annonaceae have crystal-cells in the outer hypodermis of the testa.)

That lignification in the testa is not necessarily the same procedure as in the tegmen is proved by the peculiar thickening of the endotestal cells with fibrillar endoreticulum in the five families listed in group A of the preceding classification. It is a peculiar process in need of ultramicroscopic investigation; it does not occur in the tegmen or, indeed, elsewhere in the testa unless in some outer epidermal cells of *Chloranthus*. If this peculiarity be taken as a sign of affinity, then it means that the truly Magnolialean order of families is extremely limited and that the Papaveraceous seed shows better than the Magnoliaceous the derivation from a more complicated arillate seed with fibrous tegmen. The failure of the tegmen to become an organized structure in the seed of *Magnolia* testifies against the primitive nature of its seed (p. 52).

Raphid-cells, as distinct from crystal-cells, occur sporadically in a few families, namely Actinidiaceae (in aril and seed-coat), Balsaminaceae (in testa), Dilleniaceae (in aril), Marcgraviaceae (in testa), Onagraceae (*Boisduvalia*, in the chalazal region) and Vitaceae (in testa and ? tegmen).

Endotestal cells must be distinguished carefully from the exotegmic. Both epidermal layers become pressed tightly together as the seed grows and, without a series of developmental stages, it may be difficult or impossible to distinguish them. This is the trouble, generally, with herbarium-material, but in the living seeds or those preserved in alcohol it is usually possible to distinguish the two epidermal layers at the chalaza and micropyle. As an instance of the mistakes that can arise, the thick testa of *Calophyllum* (p. 99) has been mistaken for endocarp; and it is very easy to mistake the endocarp of Lauraceae for the exotesta. For some time I made the mistake of supposing Clusiaceae–Clusieae to be endotestal instead of exotegmic which, indeed, they are in affinity with Hypericaceae.

Exotegmic seeds with a palisade

In contrast with testal seeds, the mechanical tissue develops in the tegmen of tegmic seeds. It usually forms in the outer epidermis to produce exotegmic seeds, rarely in the inner epidermis as endotegmic seeds. Mesotegmic seeds in which neither epidermis partakes in the mechanical construction seem not to exist, but the mesophyll may be extensively fibrous or tracheidal in certain families with these kinds of exotegmen.

Exotegmic seeds fall into two kinds which block out two large groups of families, though a few uncertain intermediates may occur. The outer epidermal cells develop either into a palisade of radially elongate prismatic cells, the simplest state of which is the cuboid cell without radial elongation, or into tangentially elongate fibres which are usually longitudinal on the seed. How these two forms may be related, if at all, is not clear but the problem is discussed after the account of the fibrous exotegmic seeds (p. 16).

The families with an exotegmic palisade can be separated into two groups according to the shape of the palisade-cells, but both may be composite as indicated in the following outline:

(A) Palisade cells of the exotegmen more or less prismatic with angular, isodiametric or shortly oblong facets.
  (1) Bombacaceae, Dipterocarpaceae, Euphorbiaceae–Crotonoideae (Phyllanthoideae *pr. p.*), Gonystylaceae, Malvaceae, Sterculiaceae, Tiliaceae, Thymelaeaceae.
  (2) Bixaceae, Cistaceae; with complicated chalaza.
  (3) Passifloraceae, *Perrotetia* (Celastraceae), ? Turneraceae.

  (4) Piperaceae.
(B) Palisade-cells tubular or radially elongate but with stellate–undulate or lobate facets.
  (1) Clusiaceae–Clusieae, Elatinaceae, Geraniaceae, Hypericaceae.
  (2) *Actephila* (Euphorbiaceae–Phyllanthoideae).

The importance of this construction is shown by the first group (A1) which is one of the major contingents of dicotyledons. With two exceptions, so far as known, the character distinguishes this acknowledged alliance. It may be lost, however, in some Bombacaceae as it is in the majority of Dipterocarpaceae. The default emphasizes the need for caution where knowledge is insufficient. The exceptions are *Leptonychia* which is customarily assigned to Sterculiaceae (p. 34), and that difficult group Euphorbiaceae–Phyllanthoideae (p. 138); both have exotegmic fibres. Conversely there is the case of *Perrotetia* with exotegmic palisade, but customarily assigned to Celastraceae with fibrous exotegmen. Bixaceae and Cistaceae are another acknowledged alliance which, because of parietal placentation, has been confused with Flacourtiaceae and other families placed in Violales with exotegmic fibres (p. 30). Piperaceae are, as yet, scarcely understood (p. 43).

In the second main group (B) the shape of the exotegmic cells is so characteristic and unusual that it is not surprising that it should confirm the acknowledged alliance of Clusiaceae with Hypericaceae. It can be extended to Elatinaceae but doubt may arise with Geraniaceae, as discussed later (p. 35). *Actephila* seems outside this compass and related in this peculiarity with the problem of *Drypetes* (p. 138); it is not sure that the exotegmic cells of *Actephila* agree exactly with the more or less tabular cells of Clusiaceae–Hypericaceae. However, this shape of cell, which has more or less isodiametric facets in Clusiaceae–Hypericaceae, becomes shortly oblong in the longitudinal direction of the seed in Elatinaceae, and this seems to lead in turn to the longer, almost fibriform, cells of Bonnetiaceae, Ixonanthaceae and Erythroxylaceae. Future investigations may test this possibility.

The seeds of most families with exotegmic construction bear evidence of arillate ancestry. Hence it is not surprising that many should have a firm exotesta, e.g. various Bombacaceae, Euphorbiaceae, Malvaceae, Sterculiaceae, Thymelaeaceae, Clusiaceae and Hypericaceae. In a few the outer part of the testa is thin-walled and slightly pulpy, but it usually dries up as the seed ripens; such a pulpy layer has been mistaken for an aril in some Euphorbiaceae–Crotonoideae as *Aleurites*; a rather dry kind of sarcotesta occurs in some species of *Sterculia*. Ecologically this simple testa may be comparable with the sarcotesta of Magnoliaceae, as van der Pijl has insisted, but not morphologically; the first is the whole testa and the Magnoliaceous is a special product of the outer epidermis.

Exotegmic seeds with fibres

In this construction the outer epidermal cells of the tegmen elongate tangentially into longitudinal woody fibres. They are typically contiguous and, in transverse section, give the appearance of a short palisade, for which they have been mistaken by some authors. The deception is greatest when the fibres are laterally compressed, ribbon-like, and with the longer diameter radial, e.g. Celastraceae, Euphorbiaceae–Phyllanthoideae *pr. p.*, Flacourtiaceae, Linaceae, Meliaceae, Staphyleaceae–Tapisceae, and Violaceae. Broadly interpreted, this construction occurs in the following 34 or 35 families; in some (marked with an asterisk) there are additional layers of fibres in the outer hypodermis and mesophyll of the tegmen.

Such a varied list suggests that the exotegmic fibrous seed has been polyphyletic. The conclusion appears to be supported by differences in the shape, size, manner of thickening, lignification, and layering of the fibres. On closer inspection it becomes impossible, however, to draw any sharp distinction between such differences. They may occur with gradations in several natural families, e.g. Capparidaceae, Lythraceae, Meliaceae and Myristicaceae. But the general character of the fibrous exotegmen is remarkable and separates all these families from those with other kinds of seeds. I incline, therefore, to regard them as monophyletic. They have advanced and diversified floral characters. The affinity of several has been uncertain, e.g. Connaraceae, Euphorbiaceae–Phyllanthoideae, and Proteaceae. The character does not occur among the families of Magnoliales and Ranales which, on other grounds, are considered to be primitive, except the Myristicaceae, the source of which would mark the beginning of the alliance.

The most conspicuous difference lies in the form and size of the fibres. The large ribbon-like fibres, which are laterally compressed, reach 0.3–1 mm long, 25–90 μ in radial width, and 12–35 μ in tangential. At the other extreme there are narrow cylindric fibres merely 6–12 μ wide, e.g. Aristolochiaceae, Erythroxylaceae, Lythraceae, Oxalidaceae, Papaveraceae, Proteaceae and Sauvagesiaceae. Intermediate states with wider subcylindric fibres occur in Caricaceae, Chloranthaceae, Elaeocarpaceae and Ixonanthaceae. All shapes and sizes can be found in Capparidaceae, Connaraceae, Malpighiaceae, Meliaceae and Onagraceae.

In some families the fibres are short, 50–300 μ long, and suggest that the transition to radially elongate palisade-cells may not have been impossible, e.g. Elaeocarpaceae (compared with exotegmic palisade-cells of Gonystylaceae and Tiliaceae), some Euphorbiaceae–Phyllanthoideae, and Violaceae *pr. p.* In one species of *Trichilia* (Meliaceae, Fig. 391) short rows of cuboid sclerotic cells are scattered among the fibres; yet an exotegmic palisade is not known in Meliaceae. Nevertheless fibres shorten generally at the chalaza and micropyle, where the integuments are not stretched, to become more or less radially elongate palisade-cells. This may indicate their origin without necessarily allying the families one by one with those which have the exotegmic palisade. The point is taken up on p. 16.

Another variation is the conversion of one or more outer cell-layers of the mesotegmen into fibres or, if the tegmen is merely 2–3 cells thick, both exotegmen and endotegmen may be fibrous. Here, again, there are differences. The inner fibres may lie across the outer longitudinal fibres, e.g. Aristolochiaceae, *Carica*, Oxalidaceae, and Proteaceae. In contrast, all the fibres are longitudinal in *Ascarina* (Chloranthaceae), Capparidaceae, *Cylicomorpha* (Caricaceae), Elaeocarpaceae, Euphorbiaceae–Phyllanthoideae *pr. p.*, Lythraceae, Meliaceae, *Panda*, and *Rinorea* (Violaceae). *Aristolochia* and, perhaps, *Hakea* have an entirely fibrous tegmen. Then from some genera or species of most, if not all, of the families the fibres are missing, or they may be so widely spaced among undifferentiated cells as to imply that the character degenerates. Exotegmic fibres are so generally lacking from Sapindaceae that the seeds of most genera would be considered exotestal or mesotestal with sclerotic cells, but the deficiency is associated also with their pachychalazal construction.

Yet another diversification which appears significant is the manner of thickening of the fibre-wall. Generally the thickening has fine or coarse pores, as if it were heavily reticulate, but in certain families

TABLE I. *Comparison of seed-coats in families with exotegmic fibres*

	Exotestal	Endotestal	Endotegmic	Exotegmic fibres
Capparidaceae	+	+	+	w, n, *
Celastraceae	+	−	+	w
Connaraceae	+	+c	−	w, n
Elaeocarpaceae	+	+c	−	w, *
Erythroxylaceae	−	−	−	n
Euphorb. Phyllanth.	−	−	−	w, *
Flacourtiaceae	−	?	?	w
Ixonanthaceae	+	−	−	w
Legnotidaceae	+	+c	−	w
Linaceae	+	+	−	w
Malpighiaceae	−	−	+	n
Meliaceaeae	+	+c	−	w, n
Myristicaceae	+	+	−	w
Pandaceae	−	−	−	w, *
Resedaceae	+	+c	+	w
Sapindaceae	+	−	−	w, or none
Sauvagesiaceae	−	+c	−	n
Scyphostegiaceae	+	−	−	w
Violaceae	+	+c	+	w, *
Staphyleaceae Tapisceae	+	−	−	w
Leptonychia	+	−	−	w
Aristolochiaceae	+	+c	?	n, *
Caricaceae	+	+c	+	n, *
Oxalidaceae	+	+c	−	n, *, t
Droseraceae ?	+	+c	+	n
Combretaceae	−	+	−	n, t
Lythraceae	+	+c	−	n, *,t
Onagraceae	+	+c	−	n, *, t
Punicaceae	−	−	−	n, t
Sonneratiaceae	−	+c	−	n, t
Trapaceae	+	−	−	n, t
Chloranthaceae	+	+c	+	n, *
Fumariaceae	+	+c	−	n
Papaveraceae	+	+c	+	n, *
Proteaceae	+	+c	−	n, *

NOTE. The families arranged as in the groups on pp. 46–47.
* signifies with multiple exotegmic fibres; c, with endotestal crystals; n, with narrow fibres; t, with tracheidal fibres; w, with wide fibres.

with narrow fibres the thickening is spiral or annular, as if the fibres were elongate tracheids. Such tracheidal fibres are distinctive of Combretaceae, Punicaceae and Trapaceae. They occur in some genera of Lythraceae, Onagraceae and Oxalidaceae (*Averrhoa*), but other genera have narrow pitted fibres. Then this character seems to merge into that which I have called the tracheidal exotegmen (p. 17). This has short, rhomboid not fibriform, tracheidal cells with spiral-annular thickening in the exotegmen. Such are the exotegmic cells of *Sonneratia*, though *Duabanga* of the same family has coarsely pitted fibres. Thus the question arises whether such families with fibrous exotegmen have not arisen from the stock of those with tracheidal exotegmen, e.g. Dilleniaceae, Rutaceae, Vitaceae. One is dealing with problematic affinities, for which the construction of the seed-coat seems more precise and critical than polyphyletic floral states. If *Piper* can be related to *Magnolia*, then *Dillenia* may be related to *Sonneratia*. The intricacies of the seed-coats have yet to be explored. The tracheidal exotegmen lights up the dubious assemblage of Myrtales (p. 37). Combretaceae with short-celled tracheidal endotesta (cf. Lauraceae) and fibriform tracheidal exotegmen are lifted from the rut.

There are many other details of the seed-coats among these families with fibrous exotegmen which need investigation. Many have arillate genera or species with a firm, even palisade-like, exotesta, e.g. Celastraceae, Connaraceae, Meliaceae, Papaveraceae, Sapindaceae, Scyphostegiaceae, and the problematic *Leptonychia* (Sterculiaceae). A conspicuous sarcotesta occurs in many Euphorbiaceae–Phyllanthoideae, where there seems to be no arillate representative. Similarly there is the case of *Punica*, which seems to fit closely with Combretaceae. But these kinds of sarcotesta are differently constructed and bear no evidence of a primitive nature. In various families as Lythraceae, Punicaceae, Meliaceae, Sapindaceae and Sonneratiaceae the mesotesta is sclerotic. The endotesta is crystalliferous in Aristolochiaceae, Caricaceae, Connaraceae, Elaeocarpaceae, some Lythraceae, Meliaceae, Onagraceae, Oxalidaceae, Resedaceae, Sauvagesiaceae, and Violaceae. Here, perhaps, is a genetic vestige of a lignified endotesta. In Papaveraceae, Chloranthaceae (*Ascarina, Chloranthus*) and some Proteaceae these crystal-cells are pervaded with fibrous and lignified endoreticulum which, as already mentioned, characterizes the Magnoliaceous endotesta. The three families are both endotestal and fibrous exotegmic, as if half-way between Magnoliaceae and Myristicaceae. In some Capparidaceae, Caricaceae, Linaceae (*Hugonia*, Fig. 345), Lythraceae, Onagraceae and Oxalidaceae (*Averrhoa*) the endotestal cells have thick walls. All these details, which are summarized in the table on p. 15, seem to indicate that the exotegmic fibrous seed is a refinement of a more complicated construction. Caricaceae and *Averrhoa* reveal this in their curiously multiplicative testas, and Droseraceae in their minutely varied seeds. The Magnolialean answer is Myristicaceous.

It is convenient here, to review the problem whether the two kinds of exotegmic seed, namely that with a palisade and that with fibres, have had independent origins or one has been derived from the other. Four facts seem to support the derivation of the fibrous from the palisade:

(1) In most families with fibrous exotegmen the fibres shorten into cuboid or radially elongate sclerotic cells at the chalaza and at the endostome, e.g. Figs. 68, 100, 101, 219, 275, 276, 379, 380, 534. The transformation suggests that the fibres are palisade-cells elongated longitudinally through lack of anticlinal cell-divisions over the extending body of the young seed.

(2) A slight tangential and longitudinal broadening of the palisade-cells distinguishes Tiliaceae in the Malvalean alliance. It is more pronounced in *Gonystylus* which, in this respect, seems intermediate with the fibriform exotegmen of Elaeocarpaceae. Similar transitional states occur in Euphorbiaceae–Phyllanthoideae (e.g. *Glochidion–Breynia–Cicca–Baccaurea*). There may be a comparable process in the exotesta because longitudinally elongate, almost fibriform, cells occur in certain genera of several families where exotestal cells with isodiametric facets preponderate, e.g. Campanulaceae, Ebenaceae, Lardizabalaceae (*Akebia*), Loasaceae (*Blumenbachia*), Podophyllaceae, Rubiaceae (*Coffea*), Rutaceae (*Citrus, Murraya*) and Tiliaceae (*Grewia*).

(3) A similar transition may occur from the stellate-undulate exotegmic cells of Clusiaceae–Clusieae, which are tabular with isodiametric facets on the sides of the seed but radially elongate at the endostome, into the more elongate, oblong or fibriform exotegmic cells of Bonnetiaceae and Ixonanthaceae with undulate radial walls (p. 29).

(4) Several pairs of families or subfamilies, which are considered on other grounds to be closely related, display this difference; thus:

With palisade	*With fibres*
Euphorb. Crotonoid.	Euphorb. Phyllanthoid.
Passifloraceae	Caricaceae
Tiliaceae	Malpighiaceae, Elaeocarpaceae
Hypericaceae	Sauvagesiaceae

I have failed, however, to find convincing proof of this possibility. Indeed, as noted under Euphorbiaceae (p. 130) and Violaceae (p. 276), the opposite may have occurred and the palisade may have come from the fibrous if, indeed, the two constructions are consecutive. The double or multiple state of the fibrous tegmen, which has no parallel in the palisade, supports this opposite view and implies not that the single fibrous layer is derived from the single palisade but that it is the simplification of the multiple or wholly fibrous tegmen (cf. *Sloanea*, Elaeocarpaceae, p. 124). The tracheidal tegmen offers a similar problem.

Two explanations can be given. A wholly fibrous tegmen is similar to a wholly fibrous testa, e.g. Annonaceae. If, by transference of function, the fibrous character became tegmic and, similarly, the palisade-endotesta (e.g. Myristicaceae) became tegmic, there would be independent origins for both kinds of tegmic seed. Both Annonaceae and Myristicaceae, when placed in Magnoliales, take on a prime character in the evolution of dicotyledons and this must embrace the seed-coats as well as the flower and fruit, xylem, pollen, endosperm, embryo and other features which have evolved. These thoughts lead to the second explanation which stems from Myristicaceae. The endotesta in this family is a palisade of strong character which continues shortly into the exotegmen at the chalaza where it becomes rather abruptly the fibrous exotegmen. Continuation of this palisade over the whole tegmen and its loss from the endotesta would give the palisade exotegmen; loss of palisade would leave the fibrous exotegmen. Both kinds of exotegmic seed may, therefore, have diverged into the two orders of families.

Though unanswered, the question is important because Magnoliales, Dilleniales, Bixales, Capparidales, Celastrales, Ranales, Rhoeadales, Theales, Tiliales, and Violales figure constantly as stock-sources for other orders without attention to the construction of the seed which, as discussed in Chapter 4, is not constant in any of the orders. The problem may not be so intractable because the small sampling of genera to date already reveals a few which, on the structure of the seed-coats, have clearly been misplaced in their supposed families; they are the clues which should be followed, e.g. *Huertea* and *Tapiscia* in Staphyleaceae, Legnotidaceae, *Leptonychia*, *Perrotetia*, *Turnera* and Sauvagesiaceae.

I incline to the view that there was an early dichotomy in the evolution of the arillate seed into two phyletic lines, one with exotegmic palisade and the other with exotegmic fibres; and in this light the classification of families into orders needs to be revised. Aristolochiaceae (fibrous tegmen) have been considered derivatives from Magnoliales or Berberidales, and Proteaceae (fibrous tegmen) from Thymelaeales or Elaeagnales, none of which supposed sources has a fibrous tegmen. I see no more difficulty in associating them with Caricaceae, Myristicaceae or Papaveraceae when the ancestral states, not the modern and specialized derivatives, of their families are the objects of consideration.

Exotegmic tracheidal seeds

In three families in which the tegmen is not the main mechanical layer it develops, nevertheless, a characteristic tracheidal structure. The cells are short, rhomboid, and closely packed, with lignified spiral or annular thickening. They lie tangentially as an inner casket around the endosperm and are neither radially elongate nor fibriform. The families are Dilleniaceae, Rutaceae and Vitaceae. In Dilleniaceae only the exotegmen is tracheidal, possibly as the most constant feature of the family. In Vitaceae the outer hypodermis is also tracheidal. In Rutaceae the whole tegmen may be tracheidal, e.g. *Toddalia*, *Xanthoxylum*. The Vitaceous has a multiple endotesta of sclerotic crystal-cells, reminiscent of Magnoliaceae but without the fibrous endoreticulum. The Rutaceous seed is mesotestal with sclerotic cells, reminiscent of the Myrtaceous. The tracheidal exotegmen appears to be a relic of some more complicated construction. The example, again, comes from Myristicaceae where *Knema* has a tegmen of short sclerotic cells, not fibres, and in one species, at least, the cells are tracheidal (Fig. 414). Possibly the tracheidal exotegmen is cognate with the slight scalariform thickenings of the otherwise unmodified endotegmic cells, which occurs in many families as a final trace of the primitive elaboration (p. 18).

Then there arises the question whether the fibrous exotegmen with tracheidal fibres, as in Combretaceae, Lythraceae and Punicaceae (p. 37), may not be derived from the lengthening of the tracheidal cells. As the seed enlarges, there may be fewer anticlinal divisions in the exotegmic cells which become lengthened into the tracheidal fibres. There is the case of Sonneratiaceae in this alliance where *Sonneratia* has a typically short-celled tracheidal exotegmen and the other genus of the family *Duabanga* has a fibrous exotegmen with reticulate pittings on the walls, yet in *D. grandiflora* the endostome is said to be composed of tracheidal cells (Fig. 546). Another point is introduced by Combretaceae with fibrous-tracheidal exotegmen but shortly tracheidal endotesta as in Lauraceae and Monimiaceae. One thinks of a transference of function from endotesta to exotegmen in this intricate dissociation of genetic factors.

At present the phyletic significance of the tracheidal exotegmen is indeed obscure. Further investigation may bring to light other examples which may assist. But this kind of tegmen has to be taken into account when using Dilleniaceae, as many have, as a focus for the origin of other families (p. 29), when assigning Rutaceae to Sapindales, or the sarcotestal Vitaceae to the exotestal Rhamnales.

Endotegmic seeds

Apart from Piperaceae, Saururaceae, Rafflesiaceae, Hydnoraceae, and Podostemaceae, in which both exotegmen and endotegmen participate in the mechanical function of the seed-coat, there are three instances of truly endotegmic seeds.

Firstly, Nandinaceae are exceptional in the exotestal Berberidaceae, where they have usually been placed, for the endotegmic seed (Fig. 433). It differs from that family also in having endotestal crystal-cells.

Secondly, what little is known of Zygophyllaceae reveals the endotegmen as a lignified layer of longitudinally elongate, rectangular cells, not fibriform with pointed ends. The construction is variously joined in different genera with an exotestal palisade and an endotesta of crystal-cells (lignified in *Tribulus*). The family is placed usually near Malpighiaceae which have primarily exotegmic seeds with fibres and lack the crystal-cells of the endotesta. But the fibres are evidently lost in some Malpighiaceous seeds developed in indehiscent fruits, and one of these (*Heteropterys*) has lignified, but not elongate, endotegmic cells; another (*Tristellateia*, with tardily dehiscent samaras) has such lignified cells shortly elongate. Hence it is likely that the Zygophyllaceous seed has lost the exotegmic fibres, even from the dehiscent fruits. If so, the family shows the endotegmic seed as a final product of seed-evolution similar to the exotestal.

The third example is the curious case of *Symphonia* (Clusiaceae–Garcinieae) in which the testa and most of the tegmen are thin-walled, but some thick-walled fibres occur in the inner part of the tegmen and the whole of the inner epidermis is converted into hair-like fibres. An explanation

may be forthcoming when related genera can be studied.

The endotegmen should not, however, be dismissed. Netolitzky emphasized the slightly thickened bars or streaks on the radial walls of the endotegmic cells of several families, e.g. Dilleniaceae, Elaeagnaceae, Ranunculaceae, Rhamnaceae, and Resedaceae; they occur also in Anacardiaceae, Bixaceae, Cruciferae, Caparidaceae, Centrospermae, Elaeocarpaceae, Euphorbiaceae, Geraniaceae, Leitneriaceae, Linaceae, Malvales, Podostemaceae, Rutaceae, Simarubaceae and Sterculiaceae. These are more or less advanced families with specialized seed-coats in which vestigial features, such as these thickenings, may be expected. Similar thickenings occur on otherwise unspecialized endotestal cells of various exotestal and exotegmic seeds, e.g. Berberidaceae, Podophyllaceae. Then in many families the endostome may be lignified, whether the tegmen persists or not, e.g. Annonaceae, Podophyllaceae, and various genera as *Helleborus*, *Hugonia*, and *Jollydora*. Such details are easily overlooked; they may not appear until the last stages of the hardening of the seed-coats.

Undifferentiated seed-coats

In a number of families, unfortunately, for which one would like to know peculiarities of the seed in order to relate them, the seed-coats fail to differentiate any character. Seeds of this kind generally occur in indehiscent fruits, especially the drupaceous, but it must not be supposed, as I once mistakenly thought, that all such fruits have undifferentiated seed-coats. Aquifoliaceae, Burseraceae, Combretaceae, Elaeagnaceae, Elaeocarpaceae, Monimiaceae and Tiliaceae have corrected this impression. The seed-coats may or may not be multiplicative; if multiplicative, then obviously they grade into the overgrown seeds (p. 10). Some families with well-formed seed-coats show this undifferentiated character sporadically, e.g. Celastraceae, Connaraceae, Meliaceae, Monimiaceae, Proteaceae, Rosaceae, Rutaceae, Thymelaeaceae and, perhaps, Sapindaceae. Those in which no differentiation, other than a slight exotestal tendency, has been found are listed below; it is to be hoped that some of their less advanced genera may yet reveal a distinction:

Aceraceae, Anacardiaceae, Araliaceae, Betulaceae, Cannabiaceae, Caryocaraceae, Casuarinaceae, Clusiaceae–Garcinieae, Cornaceae and allies, Daphniphyllaceae,

Eucommiaceae, Fagaceae, Juglandaceae, Julianaceae, Menispermaceae, Moraceae, Myricaceae, Ochnaceae, Olacales, Santalales, Simarubaceae, Ulmaceae, Umbelliferae, and Urticaceae.

Cell-form

Generally it is not difficult to recognize a palisade-layer in the seed-coat, yet there are degrees of perfection and, indeed, there are intermediates with other forms of cell. The palisade-cell in its most typical form, as the Malpighian cell of the Leguminous exotesta, of the Myristicaceous endotesta, and of the Euphorbiaceous exotegmen, may have been part of the primitive seed of the flowering plant or it may have evolved in subsequent phyletic lines. The first possibility is discussed on p. 55, and the second is discussed in connection with the tegmen of *Durio* on p. 33. Several genera in such palisade-families have apparently simple cell-forms but their systematic occurrence in advanced tribes or genera indicate that they are not primitive antecedents but neotenic, or imperfectly developed, states of typical palisade-cells, e.g. some Euphorbiaceae, Leguminosae, Malvaceae, Sterculiaceae and Thymelaeaceae.

A perplexing case is the stellate cell which suddenly appears, as it were, in an alliance distinguished by regular polygonal facets to the epidermal cells. Thus, stellate cells occur in *Cardiospermum*, *Paullinia* and *Urvillea* (Sapindaceae) and in *Elaeagnus* in contrast with *Hippophae*; stellate endotestal cells occur in *Jussieua* (Onagraceae); and stellate exotegmic cells occur in *Actephila* (Euphorbiaceae–Phyllanthoideae). Stellate, undulate, and polygonal cells occur in the characteristic cell-layers of various genera of Convolvulaceae, Cucurbitaceae, Lecythidaceae, Loganiaceae, Nymphaeaceae, Solanaceae and Staphyleaceae. These variations may be regarded as examples of the alternation, however little understood, between polygonal and stellate or undulate epidermal cells that occur in the leaf of many large genera of dicotyledons, particularly in the lower epidermis. In leaves the polygonal facet is associated with the development of a hypodermis, but it may occur also without this construction in the same manner as the stellate facet, and this is how it occurs in seeds. Then, in certain families, the stellate cell becomes the characteristic form of the exotegmic cells, e.g. Clusiaceae–Clusieae, Geraniaceae, Hypericaceae and Elatinaceae (p. 29), or of the exotestal cells in *Illicium*. Some prismatic palisade-cells have a regular cylindric lumen; others have a stellate lumen, as seen in t.s., in contrast with the hexagonal outline of the facet. Whether this internal feature is connected with the external stellate form is unknown, but in Cistaceae the Malpighian cells have stellate outer ends.

The relation between palisade-cells and the fibrous has been discussed. Every large family presents such detailed problems wherein it is necessary to distinguish neotenic developmental forms of cell from the primitive and the more specialized.

Aerenchyma and stomata

The aeration of the growing seed is a problem that becomes apparent on the sectioning of living fruits. Thick sections, mounted in water, show the air-spaces which are lost in fixed and infiltrated material; in such it is often impossible to know if gaps between cells or cell-layers have been filled with air, water or mucilage. In the illustrations in this work air-spaces have been shown in black. Commonly there is a narrow air-gap between the two integuments and, even, between the inner integument and the nucellus; these gaps, which usually close in fixed material, have been over-emphasized in many of my illustrations in order to delimit clearly the structural layers. There appear to be no air-gaps within the nucellus or in the endosperm, which are not photosynthetic (except for the endosperm of *Viscum*); but air-gaps appear in the photosynthetic embryo. The mesophyll of the tegmen often has no air-gaps, but they occur in the thick tegmen of many tegmic seeds. The mesophyll of the testa is generally, even conspicuously, aerenchymatous. Presumably aeration takes place by diffusion from the photosynthetic mesophyll of the integuments, unless respiration in the inner parts of the seeds is anaerobic.

Stomata occur on the outer epidermis of the testa in some species of some genera (as listed below). No genus has yet been checked thoroughly, and the presence of stomata is not known to be a constant character of any family. Only one instance of stomata on the inner epidermis of the testa is known, in the Rosaceous *Purshia* (cited by Netolitzky). They do not occur on the tegmen or the nucellus. A few seeds have an aerenchymatous outer epidermis to the testa in which the epidermal cells have short arms surrounding air-gaps into the mesophyll, e.g. *Caryocar* (Fig. 76), some Euphorbiaceae as *Jatropha*, and perhaps many overgrown seeds.

The pericarp is photosynthetic and commonly supplied with stomata, but by no means always, e.g. *Buxus*, Chloranthaceae, and *Illicium*. Presumably it helps to aerate the loculi of the fruit which in many cases enlarge to full-size while the growing seeds are still small, e.g. Connaraceae, Papilionaceae; but in other cases the loculi are occluded by endocarp, e.g. *Cassia*.

Stomata have been recorded in the following 19 families, to which Netolitzky added Ranunculaceae. The only sympetalous family is Myrsinaceae. Otherwise they occur in the 'lower orders' of dicotyledons with endotestal (Magnoliaceae, Myristicaceae) or exotegmic seeds. These facts suggest that the presence of stomata may indicate a primitive or unspecialized state of the outer epidermis of the testa which, according to the classical theory of the carpel, represents the underside of a pinna. Stomata do not occur when the exotesta is ordered into a close palisade, as in Leguminosae.

Families (genera) with stomata in the testa: Bixaceae (*Cochlospermum*), Bombacaceae (*Bombax*, *Ceiba*), Capparidaceae (*Capparis*, *Cleome*, *Isomeris*), Dipterocarpaceae (*Dipterocarpus*), Euphorbiaceae (*Ricinus*), Flacourtiaceae (*Flacourtia*), Geraniaceae (*Geranium*), Krameriaceae (*Krameria*), Juglandaceae (*Juglans*), Magnoliaceae (*Magnolia*, *Michelia*), Malpighiaceae (*Tristellateia*), Malvaceae (*Gossypium*, *Hibiscus*), Meliaceae (*Melia* ?, *Swietenia*), Myristicaceae (*Knema spp.*, *Myristica*), Myrsinaceae (*Embelia*), Papaveraceae (*Argemone*, *Eschscholtzia*), Polygalaceae (*Polygala*), Rosaceae (*Cowania*, *Purshia*), Sterculiaceae (*Scaphium*), Violaceae (*Rinorea*, *Viola*).

The vascular supply of the seed

The small seeds of which botany is chiefly cognisant have a single vascular bundle which joins funicle to chalaza in the raphe. This leads to the assumption that so minute a strand was the original state of the vascular supply in the primitive seed. With bigger seed and embryo, then more complicated systems of vasculature would have been evolved. The study of larger seeds, particularly within the limits of a family of tropical trees, prompts the contrary view that small and simple seeds are neotenic products of the larger, scarcely developed beyond the embryonic state of the ovule, though the largest seeds are certainly specializations at the opposite extreme, e.g. Papilionaceae, Euphorbiaceae, or Rubiaceae. Many seeds of medium-size, 5–10 mm long, have a more profuse vascular supply the procambial strands of which are initiated in the

ovule which no longer has such simplicity. Hence it is not established that the single bundle of the raphe was the primitive condition. From my own studies I find the evidence growing to prove the theory of reduction, but there is not enough knowledge to decide in any family from what elaboration the simple may have been derived. A few botanists have studied the subject (Le Monnier 1872, v. Tieghem 1872, Kühn 1927). Netolitzky preferred the standard view that what is simple is primitive, and gave little attention to the vascular supply. It is scarcely studied nowadays. The best material is the living seed before maturity, when vascular bundles appear as pale or dark streaks from which the surrounding tissue can be scraped; then transverse sections are readily interpreted; but anatomists seem to avoid the living. In all the seeds which I have studied, I have endeavoured to reconstruct the vascular supply.

First proof that the standard theory is mistaken comes from Magnoliaceae. The supposedly primitive seed with much endosperm and microscopic embryo is moderately large and the raphe-bundle has two or three post-chalazal extensions, visible in the ovule. This is a common condition, developed in other seeds into many post-chalazal bundles which reach even to the micropylar region and, frequently, branch or anastomose. The mesophyll of the testa has the construction of the mesophyll of the dicotyledonous leaf. In overgrown seeds an intricate vascular network may develop as in the areoles of the leaf. Some of these overgrown seeds are pachychalazal and the network of vascular bundles can be traced to an expansion of the chalazal supply. In certain families there is but one post-chalazal bundle and the seed is surrounded in the median plane with an unbranched vascular bundle from hilum to chalaza and micropyle, e.g. Annonaceae, Convolvulaceae, Cucurbitaceae, Caesalpiniaceae, Mimosaceae, Vitaceae. This is the construction in many thousands of species for which there is no evidence that it was ever more restricted; indeed, the vascular supply of some may be more complicated. The restriction comes in Papilionaceae when, as the more advanced Leguminosae, the vascular bundle may terminate at the chalaza and, in small seeds, this is the simple supply to the seed-coat, apart from the hilum. Here the Papilionaceous seed with extended hilum develops two characteristic, if small, recurrent bundles from the beginning of the raphe; in some large seeds they are exploited to give the main supply, even a reticulum, to the seed-coat. There is no knowledge of the primitive complexity of the vascular supply in Papilionaceous seeds (about 500 genera, 12000 species). Another peculiarity is the pre-raphe bundle in Connaraceae, which extends from the funicle to the micropyle in the opposite direction from the short raphe. Various Connaraceous seeds have branches from the chalaza, raphe or pre-raphe, and again there is no knowledge of the primitive organization in this family (24 genera, about 400 species). There is a similar problem in *Dysoxylon* (Meliaceae).

The single raphe-bundle of small seeds has a very simple construction. In larger seeds with post-chalazal branches the raphe-bundle is larger and centric. In seeds which are yet larger, but not overgrown, there may be a massive and compound raphe-bundle from which, at the junction of the funicle, vascular bundles pass into the aril and, even, from the chalaza into the tegmen, e.g. *Durio*, *Gonystylus*, Myristicaceae, and various Clusiaceae and Lecythidaceae. This compound raphe-bundle may be primitive and correspond with the central strand of a fertile petiolule modified into an ovule. Limitation in size of the strand would limit also the size of the seed; alternatively, the small size of the seed may fail to induce a massive procambium, cf. the complicated vascular supply in the seeds of *Gustavia* (Fig. 314) and *Pterygota* (Fig. 565).

Tegmic vascular bundles are known in Myristicaceae, *Dipterocarpus*, *Chisocheton* (Meliaceae), *Elaeocarpus*, *Gonystylus*, and some Euphorbiaceae-Crotonoideae, in which subfamily they appear related with a special chalazal expansion of the vascular supply in pachychalazal seeds (p. 130). Nucellar tracheids have been found in the following families, but not organized into vascular bundles (except in Thymelaeaceae): Amentiferae (*Carpinus*, *Castanea*, *Casuarina*), Asclepiadaceae, Capparidaceae, Olacaceae, Resedaceae, and Thymelaeaceae (Figs. 600, 601). Though regarded by some as relics of an extensive nucellar vascular supply, they have not been reported in any more primitive families. Others regard them as abnormalities in differentiation. They prove, at least, that the proliferating mesophyll in all parts of the seed has the innate propensity to develop vascular bundles.

Remarkable properties are attributed to vascular bundles in modern theories of the flower. Being conspicuous, their differentiation is thought to initiate surface-outgrowths but, as they are intrusive,

they are at best incursive. If there are several longitudinal bundles in the carpel, ovary, stamen, petal or sepal, they are taken to indicate a primitively branched state of those organs supposed, originally, to have consisted of segments with a single longitudinal bundle; affinities with pteridosperms can then be imposed. Lateral connections between longitudinal bundles are overlooked because lateral chorisis is unpopular. Other theories postulate 3, 5, 7 or more longitudinal bundles for the primitive state of those floral parts, either as a phyllodic base to a reduced leaf or as vestiges of the lost branch-system which has been planated. Yet there are outgrowths without vascular bundles, e.g. the sepals of *Mimosa* and the ovary-wall of *Chenopodium*; there are single outgrowths which take a double vascular supply, e.g. the tepals of Petaloideae and of *Saraca* (Caesalpiniaceae); and there are longitudinal vascular bundles without corresponding outgrowths, e.g. the vascular supply of the missing petals and stamens in *Saraca* (Corner 1958).

All these problems are found in the seed. The time will come when post-chalazal branches are taken to imply as many free segments to a foliaceous testa; the seed will become a unisporangiate sorus integumented with reduced pinnae having each a single vascular bundle, perhaps on the lines of the Marsileaceous sporocarp. The aril of *Myristica* will supply that primitive chorisis, as in the pteridosperm cupule, for it is webbed or consolidated in various genera of the family with loss of vascular branching. Similarly both testa and tegmen in that family will become consolidated laciniate cupules around the undivided nucellus. The picture is already seen in the non-vascular lobings of the micropyle (v. Heel 1970a, 1971a). This explanation cannot account, however, for the recurrent vascular bundles in the Papilionaceous hilum, e.g. *Mucuna* (Corner 1951, figs. 20, 21), and, if this is allowed, it cannot hold for others, e.g. the pre-raphe bundle of Connaraceae (Figs. 136, 146). What is overlooked is the capacity of the dicotyledonous mesophyll to differentiate a reticulum of vascular bundles. The same process occurs in aril, testa and tegmen if they are sufficiently multiplicative; the vascular supply lessens in complexity as the meristematic activity produces fewer procambial strands. We still lack full specific analysis of the vascular supply of seeds in a large genus with diversity in this respect, but it will surely supply the explanation. Nevertheless, the fact that the Myristicaceous seed has a massive vascular supply in its three integuments implies that it is more primitive than the Magnoliaceous, inasmuch as it is less reduced. To suppose that the Myristicaceous seed is an elaboration of the Magnoliaceous is to deny the significance of the vascular supply whether it be taken to represent relics of branching or relics of intercalary mesophyll.

Forest-seeds have an immediate supply of water in their tissues for germination. Desiccated seeds, dried in the ripening fruit, have the problem of water-absorption for germination. The micropyle is one means, the mucilage of the testa another, but it is possible that the broken end of the funicle or hilum may offer passage for water along the vascular tissue into the interior of the seed (Berggren 1963).

Hairs

The development of hairs from the outer epidermis of the testa in certain genera or species of many families is a phyletic problem. The best known example is *Gossypium*. Are the hairs innovations or inherited? On the vegetative parts of dicotyledons lack of hairs is the derived state as shown by the occurrence of glabrous varieties in many species; it is not the primitive state because it cannot be maintained that such species are now inventing the hairs that are the character of the family. With seeds, however, the glabrous state appears to be primitive. Hairs do not occur on arillate seeds or the sarcotestal, whichever is held to be primitive. Very complicated hairs occur in some genera, e.g. *Fouquiera*, *Limonia*, *Strychnos*; but there is no comparative anatomy to explain their evolution. Peltate scales occur on the testa of *Scaphium* (Sterculiaceae), stellate hairs in *Elaeocarpus*. It seems that seeds acquire the epidermal hairs of the leaf and modify them as mucilage-hairs.

Chalaza

Strictly the chalaza may be the termination of the vascular bundle of the raphe at the nucellus, but a name is needed for the whole region where the integuments originate, and I use it in this sense. It is the most intricate part of the seed where many factors work in closest proximity. In small seeds the chalaza appears simple, but that of *Bixa* opens the mind (Figs. 37, 38); it is one of the more detailed examples of tissue-differentiation in the plant-kingdom. The embryonic simplicity in the ovule of

Bixa (Fig. 36) resembles that of small and neotenic seeds. The adult structure shows the differential paths from the vascular termination to the layering of the tegmen, in a lesser degree of the testa, for the Bixaceous seed is essentially tegmic, and to the nucellus. In the allied Cistaceae (Fig. 104), the construction is simplified in the sense that, being less developed, it is neotenic. The relation between the two families implies that, as with the complexity of the testa and tegmen, so that of the chalaza has in the course of evolution been simplified. In *Magnolia* the woody endotesta forms a tube round the chalaza and at maturity the soft inner tissue dries up to form the heteropyle or false micropyle. In *Myristica* and *Virola* (Figs. 425, 426) the endotestal palisade turns out into the beginning of the exotegmen and the obliquity of the vascular bundle obscures the massive heteropyle. In other families sclerotic tissue may be massed round the vascular bundle, as in Celastraceae (Fig. 96), Chloranthaceae (Fig. 102), Clusiaceae (Fig. 116), Euphorbiaceae (Fig. 258), Flacourtiaceae (Fig. 276), and *Huertea* (Fig. 554). Details of the elaborate chalaza are seldom reported in modern studies; the tissue generally becomes very hard and almost impossible to section.

What physiological processes are involved in the chalaza on the differentiation of the integuments have not been explored. In many seeds it is clear from the direction of development that the factors may start from the chalaza and proceed towards the micropyle, e.g. the differentiation of the endotesta or the exotegmen. In many, if not most, seeds with more or less complicated chalaza the tissue of the hypostase cuts off the chalaza and, presumably, the operation of these factors from the nucellus, though the hypostase has also been regarded as the barrier to excessive growth of the endosperm and embryo. The hypostase consists, typically, of a few layers of small, angular, contiguous cells without cell-spaces, which become brown and suberized or lignified but lack differentiation in form. I use the name in a general sense for this plate or disc at the base of the nucellus (Singh and Gupta 1967). A seed which will help in the investigation of the morphogenetic processes arising in the chalaza is that of *Sterculia* (Fig. 567). Here a false chalaza is established at the base of the anatropous seed, next to the micropyle, while the true chalaza is scarcely differentiated; in consequence the embryo is inverted. Other useful material may come from pachychalazal seeds in which the chalaza may form the greater part of the seed-coat, e.g. Flacourtiaceae, Meliaceae, and Sapindaceae.

Aril

In common with most botanists, and botanical dictionaries, I have used this name for the pulpy structure which grows from some part of the ovule or funicle after fertilization and invests part or the whole of the seed. Recently this use has been criticized magisterially (v.d. Pijl 1952, 1955, 1957, 1966). Thereby he insists profound differences are obscured because it is phyletically significant whether the structure grows from the funicle, exostome, raphe or some other part. These criticisms were aimed, of course, at the durian-theory (Corner 1949a) which brought to light, among other matters, the primitive nature of arillate seeds, whereas on v.d. Pijl's theorizing the sarcotestal seed is primitive. This issue is discussed on p. 51, where the sarcotesta is shown to be equally diverse in construction and in most cases a derivative state.

A good review of the use of the name *aril* was given by de Lanesson (1876). Linnaeus used the name for the seed-coat. Gaertner introduced the current meaning, which I have followed. Planchon was impressed by differences in the precise situation from which the aril grew but he finally admitted that there were intermediates which gave intermediate products. Baillon realized this more fully and, preferring the general term aril, qualified it as funicular, hilar, exostomal or micropylar, chalazal and of raphe-origin. Planchon had distinguished the *true aril* as an outgrowth of the funicle and the *arillode* as an outgrowth of the exostome. Now v.d. Pijl has to coin a new name *arilloid* for outgrowths from other parts of the seed and as a general term for arils of, as yet, uncertain origin. Presumably one should write, therefore, of an exarilloid seed for the naked seed in general, instead of the customary exarillate. But confusion is inevitable when such similar words are employed for distinctions so slight. A displacement of a few cells where the limits of funicle and testa are indistinct is supposed to mark another phyletic category. Thus, v.d. Pijl (1966) writes 'the green durian has no true aril but a combined aril-arillode', when previously (1955) he had classified *Durio* among the seeds with an arilloid arillode or, possibly, a true aril. Lanesson had written 'Il était donc inutile d'inventer pour cette production le nom nouveau d'*arillode*, et il est plus simplement de la désigner, avec M. Baillon, sous celui d'arille micropylaire qui indique à la fois sa nature et son siège'.

The point is that among the genera of certain families, perhaps even among the species of a large genus such as *Clusia*, the aril develops in different ways, e.g. Celastraceae, Clusiaceae, Meliaceae, Myristicaceae, Sapindaceae, Violaceae. As an outgrowth it is a phyletic character of the family; its precise manner of development seems to be generic or subgeneric. In Myristicaceae the aril develops in two parts, as an exostomal flap (or arillode) and as a crescentic flange round the base of the ovule (? true aril, but there is no exact funicle in this case); both outgrowths combine and build the large entire or laciniate aril of the mature seed. Clusiaceae present the funicular aril, the exostomal and intermediates which are impossible to distinguish, as Planchon realized, because they variously cover both extremes. Sapindaceae have exostomal, funicular and raphe-arils. Raphe-chalazal arils occur in Connaraceae, Elaeocarpaceae, *Gonystylus*, Meliaceae and Violaceae. In *Durio* the aril is a complete ring developed simultaneously from the exostome, its junction with the funicle and all round the funicle; it cannot be subdivided morphologically. *Sarawakodendron* (Celastraceae, Fig. 94) has separate exostomal aril (arillode) and funicular aril (true aril) on the one seed; there can be no phyletic distinction here between the two kinds of aril. If *Leptonychia* belongs in Sterculiaceae, which is doubtful, the family has exostomal, funicular and antiraphe-arils. In all cases the result is an edible structure of general similarity, external to the testa and not to be confused with it in the manner implied by v.d. Pijl when he attempts to derive arils, arillodes and arilloids from the sarcotesta. What is confusing is the confluence of the aril and the pulpy outside of pachychalazal seeds which fail to develop a true sarcotesta, e.g. Connaraceae, Meliaceae and Sapindaceae; herein lies v.d. Pijl's error.

In some families, as Clusiaceae, Dilleniaceae, Nymphaeaceae, Stachyuraceae and Turneraceae, the aril is already present in embryonic state in the unpollinated flower; yet the aril is not normally a feature of the ovule. It is a special product that requires a general name in the same way as the alate seed, though the wings have different origins, and indeed the ovule which may be derived from various parts of the ovary, or the fruit which may be superior or inferior, or the tree which may have grown from a root-sucker. Thus, Cucurbitaceae and, perhaps, Apocynaceae have placental arils, and *Siparuna* (Monimiaceae) provides the instance of a carpellary aril or arillate drupelet that functions as a seed in a receptacular fruit (p. 195).

Various diminutive and vestigial arils, acting as elaiosomes for ants, have been studied in structural and physiological detail by Bresinsky (1963); he reviewed the genera *Asarum, Chelidonium, Corydalis, Euphorbia, Helleborus, Mercurialis, Moehringia, Polygala, Primula, Reseda, Sarothamnus, Stylophora,* and *Viola*. But he is unacquainted with tropical floras, eschews durianology, and prefers without reason the sarcotestal theory of v.d. Pijl. There are, of course, many elaiosomes for ants in the tropics where ants abound, and the bigger the ant the bigger the elaiosome. Big ants and little ones eat the arils of *Durio* and *Myristica*.

Sarcotesta

This is a convenient descriptive term for the pulpy and edible testa or outer part of it, which simulates the aril. Its constructure varies. In *Magnolia* and its allies, often regarded as typical, though a pomegranate is more familiar, the sarcotesta is the pulpy mesotesta bounded internally by the woody endotesta and externally by a pellicle of several layers of small cells with tough walls, derived from o.e. (o.i.). *Nandina* is similar. Ribesiaceae and Vitaceae are also similar but lack the pellicle; in both, o e. (o.i.) becomes pulpy. In *Garrya*, o.e. (o.i.) is multiplicative and the testal epidermis is pulpy. In Caricaceae and Cucurbitaceae the pulpy layer is but part of o.e. (o.i.); periclinal divisions give a seed-hypodermis as the sarcotesta in *Carica* and a seed-epidermis as such in Cucurbitaceae. In tegmic seeds the whole of the testa may be pulpy, e.g. *Aporosa* (Euphorbiaceae), or only the inner hypodermis, as in *Baccaurea* of the same family. In other cases there are transitions to a dry pellicular testa (various Euphorbiaceae–Crotonoideae), or to a scarcely pulpy testa which is starchy and sugary but hardly seems edible, e.g. *Sterculia* with the aril variously vestigial. Such undifferentiated states of the testa are not morphologically equivalent to the preceding though, since something edible may be scraped from them, they can be classed ecologically with sarcotestal and arillate seeds.

In certain families such as Connaraceae, Meliaceae and Sapindaceae, the expanded pachychalaza has a pulpy outer layer similar to the aril and continuous with it. Though called sarcotestal seeds, or bacciform (baccate) if most of the seed-surface is pulpy, there is no testa in the sense of a free integument in this part of the seed. Baillon regarded this sarcotesta as the generalised aril but he was unaware of the pachychalazal construction. Radlkofer in his studies of Sapindaceae called it the adnate aril through similar misunderstanding. Likewise v.d. Pijl has built this pachychalazal sarcotesta into his theory of the derivation of the aril from the Magnoliaceous sarcotesta without realizing the complete absence of homology. These points are taken up in further detail on p. 53, and under each family.

4. Criticism of the arrangement of dicotyledonous families into orders

The claim is that every natural family of dicotyledons has a characteristic seed. It was the view of Netolitzky. It is substantiated at once by the families to which a primitive position is generally assigned, namely Magnoliaceae, Monimiaceae, Lauraceae, Myristicaceae, Annonaceae, Dilleniaceae, Winteraceae, Ranunculaceae and Nymphaeaceae. It is true also of more advanced families such as Mimosaceae, Papilionaceae, Malvaceae, Sterculiaceae, Bixaceae, Violaceae, Capparidaceae, Cruciferae, Papaveraceae, Hypericaceae, Geraniaceae, Oxalidaceae, Rhamnaceae, Lythraceae, Cucurbitaceae, Cactaceae, Connaraceae and Convolvulaceae. In other large and advanced families the characteristics may deteriorate and, as Netolitzky observed, they may disappear from the seeds of indehiscent fruits; examples are Clusiaceae, Rosaceae, Rutaceae, Meliaceae, Sapindaceae, Proteaceae and Myrtaceae. There are other large families, however, in which such distinct kinds of seeds are found that there is reason to doubt their unity. I separate Bonnetiaceae from Theaceae, Legnotidaceae from Rhizophoraceae, and Sauvagesiaceae from Ochnaceae. Flacourtiaceae fall into two groups of seeds. In Euphorbiaceae Crotonoideae are uniform in general character but Phyllanthoideae are diverse. Unfortunately there are large families with indehiscent fruits in which the pericarp functions as the seed-coat, and the seed-coat differentiates no character, e.g. Anacardiaceae, Fagaceae, Simarubaceae, and Urticales.

The same diversity of the seeds that occurs in the primitive families extends also to the advanced. As the characteristics of the seeds were inherited in the original diversity, so they must have been inherited in the advanced. The classification of families into orders should, therefore be consistent with their seeds. It is from this standpoint that I examine them. The evidence is detailed in the second part of this work where the characters of the families and their generic diversification are described. Here the main points are assembled for a test that is undoubtedly disturbing. But the unsatisfactory nature of ordinal classification is proved by the number of systems that have been proposed. I shall consider only the more recent, for my purpose is not historical, and confine myself mainly to the systems of Cronquist, Hutchinson, Takhtajan and Thorne. Beginning with Magnoliales–Ranales, the test reveals these orders as a grade of floral evolution cut already by the seed-characters into separate phyletic lines. Proceeding, it will pick up the clues in other grades which pass as orders. In justification, the better known families will be re-assembled according to their agreement in seed-structure. New orders will not be proposed because there are enough names. Exceptional genera will be mentioned; and their presence, even among the few genera which have been more or less satisfactorily studied, proves the importance of the research.

Magnoliales–Ranales

The apocarpous and polypetalous families, classified as Ranales, Magnoliales and Ranales, or Annoniflorae (Thorne 1968), have anatropous, bitegmic and crassinucellate ovules. They become fairly large anatropous seeds with oily endosperm in which a small or microscopic embryo develops slowly. There are exceptions. Unitegmic tenuinucellate ovules occur in some, evidently advanced, Monimiaceae and Ranunculaceae (cf. Rosaceae). Exalbuminous seeds with large embryo distinguish Lauraceae, Calycanthaceae and Hernandiaceae; fairly large embryos occur in some Monimiaceae. Small seeds distinguish Eupteleaceae and Trochodendraceae. The seeds can be classified according to the structure of the seed-coat into three main groups with nine subdivisions; that is, there are nine kinds of seed-coat; as follows.

(1) Exotestal seeds
  (a) Simply exotestal: Berberidaceae, Lardizabalaceae, Nymphaeaceae, Podophyllaceae, Ranunculaceae, Win-

teraceae; (cf. Centrospermae, Leguminosae, and many other families).

(*b*) Also with more or less sclerotic mesotesta: Illiciaceae, Schisandraceae, (? Lardizabalaceae); (cf. Hamamelidaceae, Rosaceae, Theaceae, Myrtaceae, Ebenaceae).

(2) Mesotestal seeds without hard exotesta

(*a*) Mesotesta fibrous: Annonaceae, Eupomatiaceae; (? no derivatives, but ? Lardizabalaceae).

(*b*) Outer hypodermis of the testa as a palisade: Paeoniaceae; (cf. Convolvulaceae, Cucurbitaceae).

(3) Endotestal seeds

(*a*) Endotesta multiple, the cells filled with a fibrous lignified endoreticulum: Degeneriaceae, Magnoliaceae; (cf. Papaveraceae, Chloranthaceae, Proteaceae).

(*b*) Endotesta as a single layer of tracheidal cells: Calycanthaceae, Lauraceae, Monimiaceae, ? Hernandiaceae; (cf. Burseraceae; Combretaceae).

(*c*) Endotesta as a palisade of prismatic cells with thick walls, in a single cell-layer; exotegmen fibrous, sclerotic or tracheidal; tegmen vascular: Myristicaceae; (cf. many families with fibrous or tracheidal exotegmen, others with endotestal palisade or exotegmic palisade).

(*d*) Endotesta as a single layer of substellate sclerotic cells; exotegmen as a layer of tracheidal cells; tegmen not vascular: Dilleniaceae.

(*e*) Endotesta as a single layer of small sclerotic cells in the small seeds; ? with sclerotic exotegmen: Eupteleaceae, Trochodendraceae.

The following conclusions can be drawn.

(1) The construction of the seed-coat is essentially testal. A purely tegmic mechanical construction does not occur; this is the peculiarity of advanced orders, e.g. Bixales, Capparidales, Celastrales, Euphorbiales, Malvales, Malpighiales.

(2) Almost as a corollary, the inner integument is thin (mostly 2–4 cells thick) and is crushed sooner or later in the growth of the seed. It differentiates little or no structure. The exception is Myristicaceae with stout inner integument (7–10 cells thick), multiplicative into the massive, ruminate and vascular tegmen with lignified exotegmen. Dilleniaceae suggests a simplification with non-multiplicative integuments.

(3) The character of the seed-coat is decisive. Even if consideration is limited to the families with small embryo, the seed-coat of Annonaceae, Dilleniaceae, Magnoliaceae, Monimiaceae, Myristicaceae and Winteraceae is in all cases profoundly different. These families represent the polyphyletic diversification of the primitive dicotyledon at a grade of apocarpy.

(4) This classification by the seed is supported by the acknowledged alliances of Magnoliaceae–Degeneriaceae, Monimiaceae–Lauraceae–Hernandiaceae, Ranunculaceae–Podophyllaceae, and Winteraceae–Illiciaceae–Schisandraceae.

(5) The embryo may enlarge and modernize with depleted endosperm without alteration in the character of the seed-coat, unless it is to degenerate. Examples among exotestal seeds are Berberidaceae, Lardizabalaceae and Menispermaceae; among endotestal seeds there are Monimiaceae, Lauraceae, Hernandiaceae and Calycanthaceae.

(6) Most of these fundamental kinds of seed occur in advanced families, as indicated in brackets under the several headings. The result may bring together unusual bed-fellows but, stemming from a generalized apocarpous and follicular stock, the course of the advanced families, which have diversified in habit, flower and fruit, must be codified in some basic inheritance. The character of the seed, established in the apocarpous stock, appears to be the sign.

(7) Annonaceae, so often used as a fertile stock for the derivation of other families, seems in contrast to be a blind alley. The fibrous testa does not occur in advanced families. However, Illiciaceae have a sclerotic outer hypodermis; these cells may be elongate tangentially in Schisandraceae. Then, *Akebia* (Lardizabalaceae) has beneath its fibriform exotestal cells several layers of fibrous mesotestal cells and, of all seeds that are known in sufficient detail, that of *Akebia* is nearest to the Annonaceous.

(8) Paeoniaceae appear as another blind alley. Because of the centrifugal androecium I regarded the family as allied with Dilleniaceae (Corner 1946), and in this I have been followed by others (Cronquist 1968). But their seed-coats do not agree unless one supposes that the testa in Dilleniaceae has been reduced to such a degree that the endotesta becomes equivalent to the outer hypodermis of *Paeonia*; and this seems very unlikely, for there is no analogy among other seeds. Possibly, however, as argued later (p. 50), there is an agreement in seed-coat between *Paeonia*, Cucurbitaceae and Convolvulaceae. Originally the outer hypodermal palisade may have been derived from a pre-Illiciaceous state with multiplicative testa and lignified mesophyll. Thorne (1968) placed Dilleniaceae and Paeoniaceae in Theales, but their seed-coats do not correspond. Glaucidiaceae may be an ally of *Paeonia*.

(9) The families placed in Magnoliales–Ranales have a primitive stability in seed-construction, as

if they represented the first standardization of the variations of a complicated seed. The families do not show the diversification which, by modification of seed-form and loss of tissue-differentiation, render so perplexing the evolutionary relationships of the seeds in such advanced families as Clusiaceae, Meliaceae, Rosaceae, Rutaceae, Sapindaceae and Theaceae. This appears to have been the second main step in the evolution of the seed liberated from the primitive and elaborate construction. Some of these advanced families are almost as apocarpous in their beginnings as Magnoliales and they may clarify, when better known, the concept of the archetype. At present this seems to be fulfilled by Myristicaceae (p. 55).

(10) The few families of Magnoliales–Ranales with small seeds, e.g. Eupteleaceae, Trochodendraceae and Ranunculaceae, appear to have neotenic seeds as enlarged ovules with little or no multiplication of the cell-layers of the integuments. They have a minimum of tissue-differentiation that renders almost nugatory the interpretation of what remains. It may be that Eupteleaceae and Trochodendraceae belong with Hamamelidaceae, as placed by Croizat (1947b) and Thorne (1968); if so their seeds have been much simplified.

(11) Stomata occur in the exotesta of Magnoliaceae and some Myristicaceae. They occur, also, in advanced families (p. 20), and their presence may be facultative as well as primitive. It would be perplexing if they did not occur in any Magnolialean seeds.

(12) Magnoliales–Ranales appear as a primitive grade of apocarpous floral evolution. Seed-structure analyses this artificial grade into phyletic lines most of which lead on to advanced families which, in their turn, have been artificially classified into orders as grades of further floral evolution. The history of dicotyledonous classification has been the slow dissolution of such grades. The latest review of Magnoliales (Bhandari 1971) misses the point.

Ranunculales

If this order, as formulated by Takhtajan (1969), is taken to include Nymphaeaceae (Nymphaeales of Takhtajan), without worrying over the distinction of Berberidales (Hutchinson 1959), then there is a uniformity in the exotestal structure of the seed with its retrograde tegmen which at once distinguishes it from the endotestal and exotegmic seeds

of Rhoeadales (discussed on p. 32). The slight bar-thickenings of the endotegmen in Ranunculaceae may be a vestige of an earlier state with tracheidal endotegmen, but I can see only a parallel with Rhoeadales and no direct alliance between their families. Both stem from pachycaulous stocks with arillate fruit, as shown clearly in Papaveraceae (p. 213), but this origin is obscured in Ranales by their rhizomatous or aquatic tendency and through the evolution of leptocaul shrubs (Berberidaceae) and climbers (Clematis, Lardizabalaceae, Menispermaceae). Ranalean ancestry may have largely disappeared before the capsular Rhoeadales had evolved. The relict status of the order is shown by the recent tendency to raise aberrant genera to the rank of family. Thus, every one of the genera Barclaya, Cabomba, Ceratophyllum, Circaeaster, Euryale, Glaucidium, Hydrastis, Nandina, Nelumbo, Nuphar and Sargentodoxa may be equated systematically with Berberidaceae, Lardizabalaceae, Menispermaceae, Nymphaeaceae, Podophyllaceae and Ranunculaceae without reason for there is no theory. Is Myosurus more primitive than Victoria, Akebia than Sargentodoxa, or Nelumbo than Nuphar? Primitive characters or, in this wonderful field for experimental genetics, primitive genes are variously displayed in all these families, and it is impossible to decide which may have the most primitive genome. Those which are advanced, simplified and in their spheres of activity successful do not become a major element of vegetation, e.g. Berberidaceae, Ranunculaceae and, particularly, Menispermaceae. Of the two which appear the more primitive, namely Nymphaeaceae and Podophyllaceae, the first have become a major element of fresh-water shallows, but the key to the evolution of Ranales may be Lardizabalaceae.

According to the durian-theory, the Ranalean ancestor was a pinnate-leafed, apocarpous, spiny pachycaul of swampy forest, reproduced by arillate fleshy follicles containing exotestal seeds with oily endosperm and minute embryo, such as a cross between Nandina and Victoria might restore, or a spiny and arillate Decaisnea. Funicular arils occur in Nymphaeaceae, Podophyllaceae, and in vestigial form in at least one species of Berberidaceae, and in Akebia. Fleshy follicles occur in Akebia and Decaisnea. Pinnate leaves occur in many groups among which Ranunculus shows the modification of the primitive set of genes into simple and phyllodic leaves. Massive pachycaul construction occurs in Nymphaeaceae which add the lack of vessels and the presence of thick starchy perisperm, more primitive than in the seeds of *Magnolia* and Winteraceae. The aquatic habitat is close to the swampy forest where angiosperms may have evolved (Corner 1964). The Rhoeadalean pachycaul, however, if one can judge from present occurrence, was established on *terra firma* in the pioneer dicotyledonous forest; that is, the ancestor of Rhoeadales had progressed structurally and physiologically and had struck out on a different ecological line from the Ranalean.

Into Ranales, distinguished by flower and seed, Illiciaceae, Schisandraceae and Winteraceae must be gathered as three other relict families advanced in leptocauly. Their apparent alliance with Magnoliaceae is merely an instance of the polyphyletic evolution of the apocarpous flower and fruit. Seed-structure introduces clear distinctions. Takhtajan's super-order Ranunculanae indicates this, except for the inclusion of Papaverales. The main difference in Illiciaceae and Schisandraceae is the tendency of the mesotesta to become sclerotic; it may foreshadow the evolution of mesotestal seeds, or it may be connected with the ancestry of Lardizabalaceae.

If, now, one takes Berberidales as a separate order, then there is a detail of seed-structure to mark it, as far as knowledge goes; and this feature introduces another and brighter light into the tangled ideas of affinity. The exotestal cells in Berberidaceae, Lardizabalaceae and Podophyllaceae are more or less elongate longitudinally, either shortly oblong or lengthened into fibriform cells. In Ranales they are prismatic with isodiametric facets (but with undulate outline in Illiciaceae). The fibriform structure of the testa in *Akebia* with its very primitive follicles and subarillate seed is the only instance that I have met which is comparable with the Annonaceous testa. One is forced to consider whether a primitive fibriform testa of Annonaceous origin may have become palisade-like with oblong cells and thus lead to that of Podophyllaceae, Berberidaceae, Lardizabalaceae and, indeed, Melianthaceae (p. 31), without Ranunculalean intermediary. In favour of this view, the subpachycaul habit with pinnate leaves of *Decaisnea* and *Mahonia* and the large fleshy follicles of *Akebia* and *Decaisnea* are much more primitive than any habit or fruit in Ranunculales (as opposed to Berberidales); the point is noted by Payne and Seago (1968) in their study of *Akebia*, though unable to pursue their observations

durianologically. The survival of plants with such primitive genes may indicate a later origin than in the case of Ranunculales. Thus, three lines of pachycaul evolution appear, namely Ranunculalean and Berberidalean as well as Rhoeadalean. In the Berberidalean, Lardizabalaceae and Podophyllaceae figure clearly as the more primitive. Both have the elongate exotestal cells; both have more or less pulpy arillate seeds; both have the tendency to embed the seeds in pulpy placenta (*Akebia*, *Podophyllum*); and the Berberidaceous placentation may well have come through simplification from that of Lardizabalaceae. Hence, contrary to a disposition in Ranunculales, Podophyllaceae belong in Berberidales. I incline to add Melianthaceae, which would provide the well-developed funicular aril. The link with Magnoliales may be the relict complex of Illiciaceae–Schisandraceae; there is a close likeness between the seeds of *Akebia* and *Illicium*.

An exception in Berberidales is *Nandina*, now usually removed to its own family Nandinaceae. The seed is endotegmic. The rather pulpy testa (9–12 cells thick) remains thin-walled and collapses at maturity; the one or two outer cell-layers of the tegmen are crushed. The result is a seed-coat reminiscent of the Piperalean. The conclusion seems inevitable that *Nandina* is yet another relic of early dicotyledonous pachycaul ancestry connected with the line of Piperales. On floral grounds it lies in the Ranalean grade. Comparison with the Papaveraceous seed is helpful. If *Nandina* lost the Magnoliaceous endotesta and the Papaveraceous fibrous exotegmen, it could relate to the Papaveraceous stock. If so, the loss of aril in *Nandina* and the persistence of a sarcotesta (though apparently functionless as such) suggests that the Magnoliaceous seed itself might have lost the funicular aril.

Theales–Guttiferae–Dilleniales

These orders assume a central position in most classifications of dicotyledons. They are considered to be a focus from which many families and other orders have radiated. Seed-structure uncovers profound discrepancies. Just as it reveals Magnoliales as the early grade of polypetalous, multistaminate and apocarpous evolution at which level separate lines of dicotyledonous descent have been arrested, so it exposes the three orders as a grade in the onward passage of the flower to the syncarpous, paucistaminate and sympetalous construction. None of the three orders is uniform. When analysed, Dilleniales have endotestal seeds, Guttiferae exotegmic, and Theales mesotestal.

Theaceae, as the basis of Theales, imply a seed with sclerotic mesotesta. Stachyuraceae, in agreement with Theaceae, add the primitive character of the aril. Actinidiaceae support the conclusion though ovule and seed are unitegmic through concrescence, apparently, of the integuments. The tribe Adinandreae shows the evolution of small seeds in which the outer epidermis becomes the mechanical layer as well as the reduced mesophyll. This kind of simplified seed may have led to the exotestal construction in bitegmic Marcgraviaceae, comparable with unitegmic Actinidiaceae. Flower-structure then introduces Ericales with small or minute unitegmic and exotestal seeds. Ebenales may be the sympetalous contrast with large seeds, derived from ovules with the dorsal raphe as in *Ternstroemia*. Bonnetiaceae with exotegmic seed must be excluded. Then, if the centrifugal development of the stamens is a firm indication of common descent (Corner 1946; Cronquist 1968; Takhtajan 1969), a common origin with other mesotestal families with centripetal stamens must be ruled out, e.g. Hamamelidaceae, Rosaceae, Myrtaceae, and Rutaceae, but there is the mesotestal family Lecythidaceae with this centrifugal androecium. I retain Caryocaraceae in Theales because the undifferentiated seed-coat gives no indication of affinity.

Theales are supposed to be allied with Bixales, Capparidales, Malvales and Violales but, as all these orders are exotegmic in their particular ways defined by their type-families, it is a spurious affinity based on parallel trends in floral evolution. Dipterocarpaceae, now assigned to Theales or Ochnales, are a case in point with their firm, if vestigial, evidence of exotegmic construction.

Ochnales are classed near to or under Theales. The Ochnaceous seed appears to be an undifferentiated pachychalazal seed which resembles that of Simarubaceae and, as far as known, is not Theaceous. However, the idea of Ochnales is confused because Sauvagesiaceae with fibrous exotegmen must be removed.

Guttiferae, based on Clusiaceae and Hypericaceae, are submerged in Theales by Cronquist, Takhtajan and Thorne. The seed, however, has a unique structure with large, lignified, and tabular, stellate cells in the exotegmen, discernible in such diverse plants as *Clusia*, *Allanblackia*, *Cratoxylon* and *Hypericum*. If this endotestal character marks

the close affinity of these two families, it adds also Elatinaceae, Geraniaceae and, perhaps, Bonnetiaceae as the Thealean error. It excludes any immediate phyletic connection with Theales. Both orders are primitively arillate (Clusiaceae–Clusieae and Stachyuraceae) and in many ways such as the simple sclerophyllous leaf run a parallel course, noticeable in the evolution of the hypocotylar embryo, but no Thealean fruit is as primitive as that of *Clusia*. There is, however, the problem of Clusiaceae-Calophylloideae which have mesotestal seeds very similar to those of Theaceae. They may be a Clusiaceous evolution in parallel, which future investigation of Clusiaceous seeds may disclose, but the possibility that Calophylloideae and Theaceae are related must be entertained.

The seed of Bonnetiaceae, if *Ploiarium* is representative (p. 82), introduces an elongation of the exotegmic cell which becomes impossible to distinguish from the undulate exotegmic fibre of Ixonanthaceae and, even, Erythroxylaceae. It is possible that these families are a line of descent from Clusiaceae or Hypericaceae with their small seeds. The winged seeds of *Cratoxylon* and *Ploiarium* indicate arillate ancestry as shown in Ixonanthaceae. Otherwise Guttiferales are a restricted line of evolution.

Derivation of Theales from apocarpous or slightly syncarpous Dilleniales is suggested in classification but the very uniform endotestal seed of Dilleniaceae differs in most details from the Theaceous. The Dilleniaceous seed has a thin testa with sclerotic, often substellate, endotestal cells and a characteristically tracheidal exotegmen. The campylotropous form may suggest affinity with Theaceae–Adinandreae but the seed of this tribe proves the deception in its simplified mesotestal construction. Expanded, the relatively small Dilleniaceous seed may have resembled that of *Paeonia* with outer hypodermal palisade, but of this I am doubtful (p. 211), or it may have been Myristicaceous; neither of these possibilities applies to the Theaceous. The alliance, formerly supposed between Dilleniaceae and Actinidiaceae, is now discredited. The occurrence of raphid-cells in both is hardly sufficient evidence. Thus Dilleniales are a short blind alley of evolution. Theales may lead to Ericales and Ebenales, Guttiferae to minor families of Malpighialean confusion, but there seems to be no consequence of Dilleniaceous existence unless it is in connection with the endotestal families

Cruciferae, Grossulariaceae and Polygalaceae; they lack the tracheidal exotegmen of Dilleniaceae.

The question remains from what Magnolialean lines these orders may have descended. Dilleniaceae suggest simplification from Myristicaceous ancestry with modification of the endotesta and retention of the exotegmic tracheids, minute embryo, and aril. The massive Dilleniaceous flower is more primitive than the Myristicaceous and would relate to the extinct pachycaul ancestry of an order Myristicales. Guttiferae, as an exotegmic order, have no immediate representation in Magnoliales but, if the Myristicaceous seed were divested of the endotesta and its sclerotic nature transferred to the exotegmen, the arillate Clusiaceous seed might have been produced. The highly multiple endotesta of *Allanblackia* (p. 98) recalls the multiple endotesta of Magnoliaceae but the characteristic lignified intracellular fibrils are absent from the Clusiaceous exotegmen. Nevertheless the ancestry of Guttiferae may have been with the pachycaul and arillate progenitors of Magnoliaceae–Myristicaceae.

The sclerotic mesotesta of Theales relates to the mesotesta, though fibrous, of Annonaceae and the mesotestal possibilities of Illiciaceae and Schisandraceae. In vegetative construction, ecological evolution and geographical distribution the second possibility seems more likely.

The massive seeds of Guttiferae, Theales and Ochnales grow from tenuinucellate ovules, the Dilleniaceous from crassinucellate. As the tenuinucellate seems to have been derived from the crassinucellate in parallel in several lines of dicotyledonous evolution, it is not necessarily a mark of affinity but it points to the earlier status of Dilleniaceae along with the other features of seed and flower.

Bixales–Violales

Bixales have been submerged by Cronquist and Takhtajan in Violales most of which have been submerged by Thorne in Cistales. With Hutchinson they agree in suggesting derivation from Dilleniales, close affinity with Theales, and a central position for Flacourtiaceae. None of these conclusions is supported by seed-structure because they concern grades and not phyla. So long as Magnoliales, Dilleniales, Theales, Cistales or Violales are treated as artificial grades in floral evolution through which the families have passed on their own lines of descent, there is no need to disagree; the misapprehension enters in the subtle supposition that families so

graded have a common evolutionary descent among themselves. Bixales and Violales introduce exotegmic seeds which differ from the testal seed of Magnoliales, Dilleniales and Theales. They fall into two categories according to the nature of the exotegmen and this upsets the attempt to unite the two orders. Thus:

(1) Exotegmen as a layer of lignified palisade-cells: Bixaceae (incl. Cochlospermaceae), Cistaceae; as Bixales.
(2) Exotegmen as a layer of longitudinal fibres: Flacourtiaceae, Violaceae; as Violales.

The alliance of Bixales is with Malvales and Tiliales (p. 33); that of Violales is with Capparidales, Celastrales, Euphorbiaceae–Phyllanthoideae, Meliaceae and many other families with the fibrous exotegmen (p. 14). Theales have neither feature and their mesotestal seeds are in strong contrast. The fact seems to be that Dilleniales, Bixales and Violales, in spite of some agreement in the centrifugal stamens, have progressed in parallel similarity to a Parietalean floral grade which the structure of the seed-coat analyses into phyletic lines. Cronquist adds numerous families to Violales and Thorne to Cistales, some of which again divide into the exotegmic palisade (Passifloraceae, Turneraceae) and the fibrous (Caricaceae). Even Flacourtiaceae is suspect of phyletic uniformity (p. 143).

The Bixalean seed appears to be a derivative of Myristicaceous ancestry specialized, as in Malvales, into the palisade-exotegmen and parallel with Guttiferae.

Rutales–Sapindales

About twenty families have been assigned to these orders. With such large families as Anacardiaceae, Burseraceae, Meliaceae, Rutaceae, Simarubaceae and Sapindaceae, sharp distinctions of ordinal rank are not easily made. Rutales are not recognized by Cronquist who assigns all to Sapindales. Hutchinson and Takhtajan retain Rutales for Burseraceae, Rutaceae and Simarubaceae with a few additions over which there is some disagreement. Meliaceae become an order of their own in Hutchinson's classification; Takhtajan and Thorne refer the family to Rutales and add Anacardiaceae. An alliance between all the families seems, nevertheless, accepted. Seed-structure is by no means in accord.

The Rutaceous seed-coat is distinguished by a tracheidal tegmen, or exotegmen, as in Dilleniaceae. The Burseraceous seed, though much simplified in the woody endocarp, retains a more or less tracheidal endotesta which seems to ally the family with Lauraceae and Monimiaceae. The Meliaceous seed has a fibrous exotegmen as in Celastraceae, Connaraceae and Flacourtiaceae. Sapindaceous seeds, though predominantly meso-testal in pachychalazal construction, have evidence of a similar fibrous exotegmen. Staphyleaceae are problematic; either primarily with fibrous exotegmen (Tapisceae) or mesotestal in Staphyleae without an exotegmic derivation. Anacardiaceae and Simarub-aceae have practically unspecialized, thin-walled seed-coats which supply no criteria for their classification; Simarubaceae seem, rather, to agree with Ochnaceae, but there is the problematic, if common, genus *Irvingia* (p. 155). All the families with positive features of the seed-coat have some genera in which they deteriorate or are absent; that the families can be distinguished by these features, however, is proof of their importance.

The fibrous exotegmen of Meliaceae, characteristic even in its reduced state round the micropyle of pachychalazal seeds, relates the family with Celastraceae, Connaraceae, Flacourtiaceae, Oxalid-aceae, Violaceae and others with the same feature. None of these has the Rutaceous tegmen. Aver-rhoaceae, put by Hutchinson in Rutales, has the fibrous (not shortly tracheidal) exotegmen and must be returned to Oxalidaceae.

Another problem arises with Melianthaceae. Though placed usually near Sapindaceae, a Rosalean affinity has been suggested by Cronquist. The erect ovule is Sapindaceous but the albuminous seed, fibriform exotesta, and thin unspecialized tegmen, as well as the stipulate leaves, are not Sapindaceous. The tegmen is so reduced in the arillate seed of *Bersama*, that it is difficult to believe that the family ever possessed the fibrous exotegmen. The chief microscopic feature of the seed is the fibriform exotestal cells which appear as a palisade in t.s. If a search is made for this kind of seed among other families, it is found in Lardizabalaceae. The exarillate seed of *Melianthus*, with the fibres deeply elongate radially, resembles that of *Decaisnea*; the seed of *Bersama* resembles that of *Akebia*, which has a vestigial aril, and both have the raphe-bundle divided into chalazal branches without post-chalazal extension, though *Bersama* lacks the fibrous outer mesophyll of *Akebia*. So many details in common indicate the same seed-ancestry for both families. It bears comparison, also, with the Leguminous, so much confused with Rosaceae but not mesotestal. If there has been this Leguminous connection, the stipulate leaves are explicable; yet there is also *Mahonia*.

Most Sapindaceae and Staphyleaceae have exotestal or mesotestal seeds with unspecialized tegmen. Therefore their affinity would be sought in the first place with Hamamelidales–Rosales–Theales. But three surprising facts have come to light. *Alectryon* (Sapindaceae) has a large vascular aril which, on the durian-theory, would give the seed a primitive status. The exotegmen of *Alectryon* is fibrous. *Huertea* and *Tapiscia* (Staphyleaceae) have also this fibrous exotegmen, coupled in *Huertea* with sclerotic cells in the mesotesta, and these genera may supply the primitive character of the staphyleaceous seed, as *Alectryon* in Sapindaceae. Moreover, the complicated chalaza of *Huertea* (Fig. 554) suggests the primitive state which has simplified and expanded into a pachychalaza in Sapindaceae. But, if Staphyleaceae divide on seed-structure into two separate phyletic groups, Tapiscieae must go with Sapindales and Staphy-leaceae, as Thorne has indicated, with Rosales. Sapindaceae introduce a large concatenation to overgrown seeds (Hippocastanaceae) and unspecial-ized seeds (Aceraceae), but these obvious allies must be removed from Rutalean affinity and joined with Sapindaceae and Meliaceae.

Cronquist has added Connaraceae to Sapindales. This accords with the seed-construction, for Connaraceae have a fibrous exotegmen, and emphasizes the feature as a primary character of Sapindales. The classification of Connaraceae has been a problem. Hutchinson placed the family in Dilleniales, but this may be ruled out because the endotestal seed of Dilleniaceae has no affinity with the exotegmic pre-raphe seed of Connaraceae. Similarly a connection with exo-mesotestal Rosaceae and exotestal Leguminosae or Saxifragales can be ruled out. With exstipulate pinnate leaves, apocar-pous flowers and arillate follicles Connaraceae fit well a side-branch of Meliaceous–Sapindaceous ancestry. There is a tendency to the Connaraceous seed with pre-raphe in *Dysoxylon* (Meliaceae).

Anacardiaceae are most problematic. The un-specialized testa may degenerate to give unitegmic seeds. In some the inner epidermis of the tegmen becomes thick-walled. Neither feature accords with Rutales or Sapindales and, in the absence of positive evidence, the family seems to lie with Simarubaceae.

The addition of Zygophyllaceae to Sapindales by Cronquist (Malpighiales in Hutchinson's classification, Geraniales in Takhtajan's) seems also out of place; the seed is exotestal with somewhat fibrous endotegmen, and may belong with Malpighiales (p. 18).

Thus, for the families in which the seed-structure is known, the disposition appears to be as follows:

Rutales: Rutaceae, ? Vitaceae.
Sapindales: Aceraceae, Connaraceae, Hippocastanaceae, Meliaceae, Sapindaceae, Staphyleaceae: Oxalidaceae.
Ochnales: ? Anacardiaceae, ? Simarubaceae.
Malpighiales: ? Zygophyllaceae.
Berberidales: ? Melianthaceae.

Celastrales

Of the many families attributed to this order the only alliance to be supported by seed-structure is that between Celastraceae and Scyphostegiaceae; their arillate seeds have the fibrous exotegmen as in Sapindales and Violales. The unitegmic families are merely exotestal, namely Aquifoliaceae, Icacinaceae, and the two families Cyrillaceae and Empetraceae which have been returned to Ericales (Takhtajan 1969). So little is known of the seeds of the other bitegmic families that useful criticism cannot be made unless to emphasize that the seeds of Salvadoraceae and Stackhousiaceae lack entirely the Celastraceous character. The inner integument is evanescent in Stackhousiaceae. The two integuments of Aquifoliaceae may be combined, as in Rosaceae, into a unit; the family is placed by Thorne in Theales where there is this unitegmic tendency. In Icacinaceae there are two free integuments at the micropyle of *Phytocrene* and the seeds seem in general to be pachychalazal. The disappearance of the tegmen does not accord with its predominance in Celastraceae.

If an order so much reduced can be retained, Celastrales relate to other families with the fibrous exotegmen. Affinity with Santalales (Thorne 1968) is not supported by seed-structure.

Three genera have particular interest. *Catha* may show how the arillate seed can become alate (Fig. 83). *Turraea* (Meliaceae) has the Meliaceous flower but the Celastraceous fruit and seed (Fig. 392). *Perrotetia* with exotegmic palisade is anomalous; it bears comparison with Legnotidaceae, and both with Turneraceae (p. 38).

Capparidales–Cruciales–Rhoeadales

An alliance between these orders is often postulated. Seed-structure supplies important distinctions. Cruciferae with endotestal seeds are sharply delimited from Capparidaceae, Fumariaceae, Papaveraceae and Resedaceae with fibrous tegmen. Though many botanists have insisted on a close affinity between Cruciferae and Capparidaceae, they were regarded by Hutchinson (1959) as convergent from a Papaveraceous and a Bixaceous or Flacourtiaceous ancestry respectively. Later, Hutchinson (1967) separated the herbaceous Cleomoideae from the woody Capparidaceae and associated them with Cruciferae. But Cleomoideae have the fibrous tegmen of Capparidaceae, which is entirely lacking from Cruciferae. This family, through the researches of Vaughan and Whitehouse (1971), is now the best known of all dicotyledons in regard to seed-structure. These authors uphold Hutchinson's first contention, but the question of the origin of Cruciferae is unsolved. The conclusion is inevitable that Cleomoideae are herbaceous Capparidaceae convergent with herbaceous Cruciferae; the seeds of Cleomoideae are reduced and simplified compared with *Capparis*, but those of *Crataeva* bridge the difference. A parallel is supplied by Resedaceae which, with fibrous tegmen, are herbaceous derivatives of Violaceous ancestry, where *Rinorea* has the more complicated seed as in *Capparis*.

Accordingly, I remove Cruciferae from the large group of families with fibrous tegmen and place them, problematically, with the few advanced endotestal families, where they may have some affinity with Polygalaceae. Certainly the fibrous tegmen is lacking in some genera of families that can be defined on this character, but loss of the feature does not lead to the endotestal construction. The convergence of herbaceous Cruciferae, Capparidaceae, Papaveraceae and Resedaceae is an example of parallel evolution from pachycaul ancestors distinguishable by seed-construction, and for which Papaveraceae supply the arillate *Bocconia*. The mistaken alliance should shatter faith in floral classification.

Fumariaceae and Papaveraceae share with Chloranthaceae and Proteaceae a double construction. The fibrous tegmen has on its outside an endotesta of crystal-cells permeated with internal cellulose fibrils which, at least in Chloranthaceae, become lignified. This is the character of the multiplicative endotesta of Magnoliaceae. It has not been found in other families with fibrous tegmen and it suggests that these four families are survivors of the ancestors that connected Myristicaceae (endotestal palisade of thick-walled cells, exotegmic fibres) with Magnoliaceae (woody endotesta, no specialized tegmen). In any case the association of Papaveraceae with exotestal Ranunculaceae must be ruled out. Thorne, in placing both Chloranthaceae and Papaveraceae in his Annoniflorae, as the equivalent of Magnoliales, has realized their primitive nature.

Moringaceae, put habitually in Capparidales, are another surprise. The mesotestal seeds with highly multiplicative testa and unspecialized tegmen of short duration remove the family from Capparidales, Cruciferae and Papaverales or Rhoeadales to the contingent of Myrtales, Rosales and Theales. But the seeds of the Madagascan pachycaul species must be studied.

Euphorbiales–Malvales–Thymelaeales–Tiliales

This alliance, generally admitted for the main families, is distinguished by the exotegmic palisade. The testa may be thinly pulpy or, in varying measure, it is pellicular and dries on to the hard tegmen. Even in the arillate seeds the exotesta is not usually much indurated, and this detail emphasizes the tegmic construction. The cells of the exotegmen become radially elongate, lignified, Malpighian cells with hexagonal facets; the lumen is more or less obliterated except towards the outer end where the nucleus resides. The height of the cells varies greatly, from 30μ to 1 mm or more, which is as good a measure of any microscopic element of the dicotyledon, and this height may vary in parts of the same seed. Clusters of short cells and others of long cells give the reticulate or pitted surface of the tegmen. The cells are strictly radial or, especially in Euphorbiaceae–Crotonoideae, they are curved with the narrower inner end displaced slightly towards the micropyle; here and at the chalaza the cells are rectified.

Typically this construction contrasts strongly with that of the fibrous exotegmen but there are variations which lead to intermediate forms of cell, and two problems emerge. Either these variations are minor consequences of the typical palisade or they are phyletic intermediates. The evidence has been reviewed on p. 16.

In Tiliaceae the palisade-cells are rather short and, often, slightly widened tangentially so as to have rhombic facets. The tangential lengthening is more pronounced in *Gonystylus* where the cells appear intermediate between the prismatic form and the shortly fibrous (as in Elaeocarpaceae). The large arillate seed of *Gonystylus* appears primitive in Thymelaeales and its exact distinction in tegmen from such Elaeocarpaceae as *Sloanea* needs attention. By contrast the typical prismatic cell may shorten and become subcuboid or, as I have called it, merely sclerotic. In some genera as *Pachira* (Bombacaceae), *Theobroma* (Sterculiaceae) and *Grewia* (Tiliaceae), the subcuboid cell appears to be neotenic. The palisade-layer is the last cell-layer to differentiate in the seed-coat; it undergoes much anticlinal cell-division before its cells lengthen and lignify, and it is the layer most likely to be arrested, or stunted, on precocious maturation. Yet, it is not clear that all short or subcuboid palisade-cells are neotenic.

The outer ends of the palisade-cells are slightly wider than the inner and usually they are rather coarsely pitted from the minutely stellate lumen. The cell-body may be finely pitted throughout, as in Euphorbiaceae–Crotonoideae, or the pits of the outer end extend as longitudinal grooves of the lumen during the lengthening of the cell; they then become occluded as the wall thickens from the inner end of the cell outwards. The outer ends then appear fenestrate in side-view and the cell-body has no distinct pits. This seems to be the general state in Bombabaceae, Malvaceae, Sterculiaceae, Thymelaeaceae and Tiliaceae. However, in the Bombacaceous *Durio* and *Cullenia* (Fig. 52), the short palisade-cells are coarsely pitted throughout and appear, not so much as neotenic cells but as primitive sclerotic cells as yet unspecialized into the Malpighian form. This kind of sclerotic cell occurs in those few genera of Dipterocarpaceae which retain a vestige of the tegmic differentiation, namely *Dipterocarpus* and *Vatica*. It is to this kind of cell, rather than to the perfect Malpighian cell, that the oblong sclerotic cells of *Gonystylus* may make the transition from the fibrous.

On the whole, I regard Bombacaceae, Dipterocarpaceae, Euphorbiaceae–Crotonoideae, Gonystylaceae, Huaceae (Baas 1972), Malvaceae, Sterculiaceae, Thymelaeaceae and Tiliaceae (excluding Elaeocarpaceae) as an order, or super-order, distinguished by the exotegmic palisade from that with the fibrous exotegmen which comprises Celastraceae, Connaraceae, Elaeocarpaceae, Euphorbiaceae–Phyllanthoideae *pr. p.*, Flacourtiaceae, Meliaceae, Sapindaceae and Violaceae. The first, or Malvalean, complex is distinguished by the basically palmate form of the leaf; the second, or Sapindalean complex, is distinguished by the pinnate form. In both the compound leaf becomes webbed into a simple leaf or reduced to a terminal leaflet. Thus Bixales and Passiflorales, with basically palmate leaf and exotegmic palisade, fit the Malvalean complex; the pinnately phyllomorphic ramuli of various Phyllanthoideae and Flacourtiaceae fit the Sapindalean.

The conclusion dismisses such suggestions as the derivation of the Malvalean complex from Dilleniales (endotestal) or Violales. It disposes of attempts to place Dipterocarpaceae in Ochnales (undifferentiated testa, evanescent tegmen) or in Theales (mesotestal with evanescent tegmen). It excludes several other families from the Malvalean complex where they have been tentatively referred, namely Buxaceae (exo-mesotestal), Daphniphyllaceae (undifferentiated), Elaeocarpaceae (fibrous exotegmen), and Nyctaginaceae (exotestal). The problem of Phyllanthoideae and of Pandaceae is so intricate and uncertain through lack of knowledge that I have deferred it to the detailed account of Euphorbiaceae (p. 138). A problematic genus, comparable with *Perrotetia* (exotegmic palisade) in Celastraceae (fibrous exotegmen), is *Leptonychia*; with fibrous exotegmen and peculiar aril it is clearly misplaced in Sterculiaceae (exotegmic palisade). Then Turneraceae must be mentioned because the seeds of *Turnera* have such oblique palisade-cells that they are in a way intermediate between the palisade-cells of Passifloraceae and the fibres of Celastraceae or, indeed, of Caricaceae (p. 38).

For the origin of the Malvalean seed one must look to an arillate seed with multiplicative tegmen. Among Magnoliales–Ranales this distinction belongs to Myristicaceae. The tegmic vascular bundles of *Dipterocarpus*, various Euphorbiaceae–Crotonoideae, and *Gonystylus* may be homologous with those of Myristicaceae. The Myristicaceous chalaza, penetrated by the vascular bundle of the raphe, is the Malavalean heteropyle which persists, if modified, even in the seed of *Sterculia* with its inverted embryo. The difficulty is to find the derivation of the exotegmic palisade, but the Myristicaceous seed provides also a unique solution. At the Myristicaceous chalaza there is an inflexure of the endotestal palisade which starts as an exotegmic palisade though this quickly gives place to the fibrous, sclerotic or tracheidal exotegmen; if prolonged, however, to the micropyle it would give the Malvalean construction. Then, presumably, the Myristicaceous endotestal palisade disappeared in Malvalean ancestry. It is remarkable that the endotesta persists as a thin-walled palisade of short cells, similar to the inner hypodermal palisade of Myristicaceae, in many Euphorbiaceae–Crotonoideae. A shift inwards, therefore, of these two Myristicaceous cell-layers *via* the chalaza would give the Malvalean seed. There are two alternatives. Either the Malvalean seed has been derived from some unknown Magnolialean source with exotegmic palisade, or this palisade has been transformed from the fibrous exotegmen which is, really, a paraphrase of the first suggestion. Thus Malvalean and Sapindalean roots seem to be with Myristicaceous ancestry.

For convenience of ready reference, I give below the main distinctions of the seeds of the Malvalean alliance:

Bombacaceae. Seeds medium-size to large; seed-coats equally developed, multiplicative; ovules transverse or erect; possibly the most generalized construction.

Malvaceae. Seeds small to medium-size, the smaller usually campylotropous; testa with few cell-layers, not or slightly multiplicative; tegmen thicker and usually multiplicative; exotegmic palisade-cells long (Hibisceae) or short; endotegmic cells with pitted radial walls; endotestal cells with crystals (Hibisceae); ovules transverse, erect or suspended.

Sterculiaceae. Seeds small to large; testa generally with fewer cell-layers than the tegmen, both usually multiplicative; endotesta and endotegmen with or without crystals; ovules more or less erect or transverse.

Tiliaceae. Seeds small to medium-size; testa with fewer cell-layers than the tegmen, both multiplicative; exotegmic palisade-cells rather short, generally slightly elongate longitudinally; exotesta generally with the cells longitudinally elongate; endotesta with crystals; ovules various.

Euphorbiaceae–Crotonoideae. Seeds small to large; testa generally with fewer cell-layers than the tegmen, both more or less multiplicative; exotegmic palisade-cells finely pitted, often curved; endotesta often as a thin-walled palisade; raphe v.b. entering the chalazal heteropyle, with tegmic v.b. or pachychalazal complex in some genera; ovule 1 per loculus, suspended.

Gonystylaceae. Seeds large, exalbuminous; seed-coats multiplicative, the testa with fewer cell-layers than the tegmen; exotegmic palisade-cells shortly longitudinally elongate, even subfibriform; v.b. numerous in the raphe, tegmen and aril; ovule suspended.

Thymelaeaceae. Seeds small to medium-size, with the endosperm generally much reduced; seed-coats usually not multiplicative, the testa with fewer cell-layers than the tegmen; endotestal and endotegmic cells lignified and pitted; ovule suspended.

Dipterocarpaceae. Seeds large, exalbuminous; testa multiplicative or not, with fewer cell-layers than the multiplicative tegmen; exotegmic palisade-cells short (*Dipterocarpus*, *Vatica*); tegmic v.b. in *Dipterocarpus*; ovules suspended.

Geraniales–Malpighiales–Polygalales

The modern tendency associates the three orders with Sapindales or Rutales, if they are distinguished. Hutchinson (1959, 1967) regards the three as polyphyletic; Geraniales are referred to an origin from Caryophyllales or Ranales, Malpighiales from Tiliales, and Polygalales from Capparidales. Seed-structure reinforces the polyphyletic conclusion but does not accord with the derivations.

Firstly, Polygalales are endotestal with a lignified palisade and their arillate seeds, though much simplified, bear upon the Myristicaceous and Cruciferous. The position of Vochysiaceae has yet to be determined. The seed-structure of the order has no characteristic of the Capparidaceous (fibrous exotegmen) or the Sapindaceous (fibrous exotegmen) or the Rutaceous (tracheidal tegmen).

Secondly, the seed of Malpighiales according to Erythroxylaceae, Ixonanthaceae, Linaceae and Malpighiaceae is exotegmic with fibres. It agrees, therefore, with the Sapindalean and not with the Tilialean (exotegmic palisade). Exceptions, however, are Zygophyllaceae, *Irvingia* and *Balanites*, discussed on p. 75. Unfortunately the seed of Humiriaceae is not known.

Thirdly, Geraniales offer a mixture with different seed-coats in Oxalidaceae, Geraniaceae and Balsaminaceae–Limnanthaceae–Tropaeolaceae. They appear to be convergent in floral character. The Oxalidaceous seed has a fibrous exotegmen as in Malpighiales. The Geraniaceous seed is also exotegmic but not fibrous; the outer epidermal cells become stellately lobed and sclerotic in the same way as in Hypericaceae. They are not a regular palisade of prismatic cells, though in *Biebersteinia* the arms of the cells become irregularly radially elongate to suggest that the more tabular stellate cell may have been derived from or have evolved into a regular

palisade-cell. The problem of the stellate cell is discussed on p. 19. It is noteworthy that in Cistaceae the prismatic cells of the exotegmic palisade dilate at their outer ends into substellate facets. If this lobing predominated, the stellate sclerotic exitegmen of Geraniaceae might result. Neither a Bixaceous nor a Hypericaceous affinity explains the endotestal crystal-cells of Geraniaceae.

Balsaminaceae, Limnanthaceae and Tropaeolaceae lack entirely the exotegmic character, and they have more or less completely unitegmic and pachychalazal seeds. In some species this pachychalazal seed-coat may be more or less sclerotic, but there is no evidence of differentiation in the tegmen in the cases where it persists. Possibly these families have been derived from the beginning of the main line of unitegmic Sympetalous evolution, or they are Rosalean.

Seed-structure distinguishes four polyphyletic lines in these orders as they are generally understood:

(1) Polygalales, endotestal.

(2) Malpighiales with fibrous exotegmen: Erythroxylaceae, Ixonanthaceae, Linaceae, Malpighiaceae, Oxalidaceae, ? Zygophyllaceae.

(3) Geraniales, exotegmic with stellate cells, (Geraniaceae).

(4) Balsaminaceae, Limnanthaceae and Tropaeolaceae, pachychalazal, more or less unitegmic.

No Rutalean affinity appears.

Hamamelidales–Rosales–Leguminosae

When fully differentiated, the seeds of Hamamelidaceae and Rosaceae are mesotestal. The character allies them with Myrtales and Theales. Yet the modern trend is to derive Rosales from a Dillenialean stock (Hutchinson, Takhtajan) and also Hamamelidaceae (Hutchinson) or to refer Hamamelidales to a more primitive Magnolialean stock (Takhtajan). The seed-structure is precise because it denies the Dillenialean derivation and refers both families to the exo-mesotestal line of Illiciaceae and Schisandraceae. Calycanthaceae (Rosales of Hutchinson) are placed by Cronquist and Takhtajan among the families assigned to Laurales, near to Lauraceae and Monimiaceae; this disposition is confirmed by their tracheidal endotesta (p. 12), and it emphasizes the precision of seed-structure. Thus, if the seeds of Dichapetalaceae differentiate a character, though they are said to be unitegmic, it will be decisive for their classification; the family hovers between Rosales, Celastrales and Euphorbiales.

Among other families placed in Hamamelidales, the exo-mesotestal construction occurs in Stachyuraceae and Buxaceae, which forbids the disposition of the first in Violales (Takhtajan) and the second its confusion with Euphorbiales (Cronquist, Takhtajan). Daphniphyllaceae bring uncertainty because the seed-coats are unspecialized. But Hamamelidales are considered to be the stock from which such Amentiferous orders as Salicales, Fagales, Casuarinales and Urticales have evolved and, as these also lack specialization of the seed-coats, Daphniphyllaceae may belong with them; like Buxaceae, however, they are exstipulate and the detail may indicate Myrtalean or Thealean phylogeny.

Connaraceae and Leguminosae are also placed in or near to Rosales. The first, however, as already explained, has the fibrous exotegmen and belongs with Meliaceae and Sapindaceae (p. 31). Leguminosae, in contrast, are one of the most typical exotestal orders. They supply their own evidence for derivation from a multistaminate and multicarpellary pachycaul with arillate follicles of far more primitive construction than is known in Rosales. Leguminosae reveal the exotestal seed as a primary feature of dicotyledonous origins in the same way as Winteraceae, Ranunculaceae and Nymphaeaceae. Hence I see no reason to subordinate Leguminous ancestry to Rosalean. I purposely prefer the name Leguminosae because Fabales suggests a taxonomic equivalent with Casuarinales or Hippuridales which is ridiculous. Leguminosae make one of the great phyletic lines of land-plants and have radiated into nearly every habitat.

Several families, assigned variously to Cunoniales and Saxifragales, are associated with Rosales. The seed-structure of most is not known, but the more typical Saxifragalean families (Crassulaceae and Saxifragaceae) have small exotestal seeds. Whether this is a primary feature relating the order with Winteraceae or secondary and derived from a mesotestal seed (as in Melastomataceae and Myrtaceae) or some other construction is not known. In any case, the endotestal Grossulariaceae must be removed from these alliances. *Ribes* has an arillate sarcotestal seed with woody and uniseriate endotesta of short crystal-cells; in some species the aril is variously degenerate. The nearest fit in seed-structure is supplied by Vitaceae with thicker sarcotesta, multiplicative endotesta, tracheidal tegmen and exarillate seeds which become ruminate and perichalazal. *Ribes* may have lost the tracheidal

tegmen, as happens in Rutaceae, but, if originally devoid of it, its affinity lies with Cruciferae or Polygalaceae. The seeds of both Grossulariaceae and Vitaceae are remarkably similar to those of *Magnolia* but their endotestal cells lack the internal fibrils of *Magnolia*. *Ribes* proves that aril and sarcotesta can co-exist, and that loss of the aril leaves a sarcotestal seed, perhaps in explanation of *Magnolia* and *Nandina*.

Rhamnales–Proteales

It is usual to associate in Rhamnales both Rhamnaceae and Vitaceae. Elaeagnaceae may be assigned to this order or to Proteales or elevated to Elaeagnales in conjunction with Proteales. A common origin with Celastrales is generally postulated for all. A connection with Thymelaeaceae has been suggested for Proteales. Seed-structure is definite. It unites Rhamnaceae and Elaeagnaceae as exotestal; it separates both Vitaceae (thick-walled multiplicative endotesta, tracheidal tegmen) and Proteaceae (internally fibrillar endotestal cells, fibrous tegmen). Celastraceous affinity lies, if at all, with Proteaceae; there is no Thymelaeaceous (exotegmic palisade).

The seeds of Rhamnaceae and Elaeagnaceae are fairly well known. Degenerate trends can be detected, e.g. *Ventilago* (Rhamnaceae), but the seeds of both capsular and indehiscent fruits differentiate in minute detail. The tegmen of both families has at most some pitting on the anticlinal walls of the endotegmen. Therefore the families must be assigned to the unsatisfactory category of the exotestal seed. They may be singled out through the tendency of the palisade-cells to have a stellate lumen, if not a more or less stellate facet.

Vitaceae introduce several problems, foremost among which is the construction of the seed. With such advanced habit, flowers and fruits the family appears in no way primitive; yet, its seed is scarcely improved on that of *Magnolia* and may, even, be more primitive. Certainly there is no affinity in this regard with Rhamnales. Points of agreement with Magnoliaceae and Degeneriaceae are these:

(1) a testa composed of a multiple sarcotesta and a multiple woody endotesta: (2) a thin tegmen: (3) sclerotic exostome: (4) massive nucellus: (5) copious oily endosperm, ruminate in Degeneriaceae and Vitaceae: (6) a small embryo late in developing: (7) no aril.

If all these features are considered primitive

inasmuch as they are Magnoliaceous, then the seed of Vitaceae is equally primitive. There are certainly differences, as set out below; most seem of minor consequence or to be such advances as are found in other phyletic lines, but that of the lignification of the endotestal cells is a critical distinction. Thus:

(1) The sarcotesta of Magnoliaceae is stratified into a pink horny pellicle (2–6 cells thick with stomata), derived from o.e. (o.i.) by periclinal division as a multiple hypodermis, and an inner layer with oil-cells derived from the mesophyll (o.i.). In Vitaceae there is no such dual structure; there are raphid-cells but neither oil-cells nor stomata.

(2) The endotestal cells of Magnoliaceae have internal fibrillar lignification. The Vitaceous have thick, finely pitted, lignified walls and, usually, large crystals in the lumen.

(3) The vascular bundle has two post-chalazal branches in Magnoliaceae. In Vitaceae it is undivided without post-chalazal extension, but the chalaza itself may elongate into a perichalaza. It is possible that Degeneriaceae may be perichalazal.

(4) The tegmen is unspecialized in Magnoliaceae. In Vitaceae the outer epidermis and, sometimes, the outer hypodermis become a layer of short tracheids; the inner epidermis has somewhat thickened or slightly pitted walls which may be somewhat lignified.

The last point introduces the tracheidal tegmen which distinguishes Rutaceae. The elaborate seed of *Xanthoxylum* (Fig. 499) bears comparison with that of *Vitis* (Fig. 632). The main differences are the sclerotic testa of *Xanthoxylum*, its larger embryo and absence of ruminations. If the two families are allied, as seed-structure suggests, then Vitaceae have diverged with Magnoliaceous testa and perichalazal growth, while Rutaceae have enlarged the embryo and simplified the testa. The question arises whether the tracheidal tegmen is primitive and has disappeared from Magnoliaceae or it represents the tracheidal endotesta (as in Lauraceae and Monimiaceae) transferred to the tegmen. If the first supposition is held, because there is no evidence for the second, then Vitaceae are descendants of a phyletic line from which Magnoliaceae and Rutaceae have diverged. To this stock Rutaceae will add the aril. Reference of Vitaceae to Cornales (Thorne 1968) inverts the problem.

Lythrales–Myrtales

These orders, if parted in classification, are usually neighbours. Hutchinson had endeavoured to estrange Lythraceae as herbaceous, but he seems to forget such tropical trees as *Lafoensia*, *Lagerstroemia* and, indeed, both *Sonneratia* and *Duabanga* which may be removed with considerable doubt as Sonneratiaceae. *Lawsonia* and *Pemphis* are the shrubs or small trees leading to the suffrutescent herbs of more familiar Lythraceae. Nevertheless, the seed-coat intervenes and distinguishes three states in this assemblage:

(1) Seeds with a fibrous tegmen composed of narrow pitted fibres or elongate tracheidal fibres; often with sclerotic mesotesta: Combretaceae, Lythraceae, Onagraceae, Punicaceae, Sonneratiaceae, Trapaceae (? Legnotidaceae).

(2) Seeds with sclerotic mesotesta and without a special exotegmen: Lecythidaceae, Melastomataceae, Myrtaceae, Penaeaceae, Rhizophoraceae (excluding Legnotidaceae).

(3) Seeds lacking both sclerotic mesotesta and fibrous exotegmen: Callitrichaceae, Halorrhagaceae.

The first group is also tenuinucellate, exalbuminous, exarillate and provided with straight or slightly curved embryos with plane or contort cotyledons. *Punica* with epidermal sarcotesta seems to provide the least specialized and reduced seed. Yet, as discussed on p. 17, Combretaceae possess the tracheidal endotegmen of Lauraceae and Monimiaceae. The alliance of these families is generally conceded. Recently Mayr (1969) has shown that in floral construction they must be separated from Myrtaceae. Their origin may be from Myristicaceous or Rutaceous stock. Lythrales are certainly not allied with the exotestal Caryophyllales (Hutchinson 1959) or the Thymelaeaceae with exotegmic palisade (Cronquist 1968). For *Lagerstroemia* and *Punica*, Mayr reports the centrifugal androecium.

The second group, which is Myrtales in the strict sense, seems to lack any particularity of the tegmen. Conceivably such may have been lost, as in most Sapindaceae, but in the absence of evidence the seed-structure of Myrtales resembles that of Hamamelidaceae, Rosaceae and Theaceae. Seven families, commonly allied in systems of classification, with this sclerotic mesotesta can be subdivided into two groups:

(1) Crassinucellate; androecium centripetal: Hamamelidaceae, Melastomataceae, Myrtaceae, Rosaceae, Stachyuraceae.

(2) Tenuinucellate; androecium centrifugal: Lecythidaceae, Theaceae.

The endosperm in six of these families is nuclear; cellular endosperm has also been reported for Hamamelidaceae; it has not been investigated in Stachyuraceae. The arillate example is *Stachyurus*, but vestiges of the aril may occur in Lecythidaceae.

The third group of, mainly, marsh- or water-plants has no character of the seed-coats to assist in classification. Callitrichaceae (unitegmic) have been removed to Lamiales (Takhtajan 1969, Thorne 1968).

Cucurbitales–Passiflorales

Whatever may be the rank employed to segregate the families of these orders, it is a general conclusion that Passifloraceae, Caricaceae and Cucurbitaceae are products of one short branch of dicotyledonous evolution. Seed-structure exposes disturbing differences and the alliance, if truly natural, bears on the value of seed-structure, whether it is a detail as subject to parallel evolution as any other or the most fundamental in classification.

Firstly the testa in Caricaceae and Cucurbitaceae is highly multiplicative and becomes in each family peculiarly complicated; that of Passifloraceae is non-multiplicative and simple. Secondly, though the tegmen is not multiplicative in any, it becomes exotegmic with a palisade of cells in Passifloraceae, exotegmic with a layer of longitudinal fibres in Caricaceae (with mesophyll fibres also in some cases), and unspecialized in Cucurbitaceae where it soon degenerates in the growth of the seed. Either the three families are fundamentally distinct on these grounds, which correspond with Bixales, Violales and Theales, or their seed-coats have undergone a divergence which is not encountered in other natural alliances. Hutchinson derives the three families from Bixales or Flacourtiales, which are united by Thorne as Cistales (cf. p. 30), and this at once introduces the antithesis between the palisade and the fibrous exotegmen. Takhtajan derives them from Violales (fibrous exotegmen) where Cronquist places them. Yet only Caricaceae have the tegmen of Violales. For convenience the differences in seed-structure are given in Table 2.

Among the other families placed in the two orders, for which the seed-structure is known, Begoniaceae and Datiscaceae do not assist because their very small seeds are exotestal and seem to indicate a very different line of evolution. Turneraceae, however, are important. They are placed in Passiflorales by Takhtajan, but in Loasales by Hutchinson, though without any agreement in seed with the unitegmic and exostestal Loasaceae. The testa is simple, 2 cells thick, and unspecialized as in Passifloraceae, and the exotegmen becomes a layer of more or less radial, sclerotic cells as in Passifloraceae, but these cells are so oblique as to appear transitional to fibres. They have the obliquity seen in many Euphorbiaceae-Crotonoideae but more exaggerated; nevertheless, they rectify into radial cells at the chalaza and micropyle (Fig. 612). The coarsely pitted character of the exotegmic cells resembles that of *Durio* and *Dipterocarpus*. Therefore, it could be supposed that the Turneraceous exotegmen was intermediate between the Passifloraceous and the Caricaceous with long fibres. Alternatively the obliquity of the Turneraceous cells may be merely a peculiarity of their small seeds, unconnected phyletically with the fibrous exotegmen. This view accords better with the advanced status of the family in other respects. I incline to consider the arillate Passifloraceae and Turneraceae as belonging to an order distinct from Caricaceae and Cucurbitaceae, which refers back to Bixalean or Malvalean origin.

Caricaceae with fibrous exotegmen seem to be the Flacourtiaceous connection which others have discerned. The Caricaceous testa, however, multiplies into a pulpy sarcotesta (from the outer epidermis), a somewhat thick-walled but unlignified outer mesotesta (from the outer hypodermis), and a thin-walled inner mesotesta (from the inner layers of the mesophyll), while the inner epidermis remains as a single layer of crystal-cells. The outer mesotesta forms the spongy, firm seed-coat after the slippery sarcotesta has been removed, and it is reticulate from the localized differences in thickening of the outer hypodermis. The result is a testa from 16 cells thick (*Jacaratia*) to more than 60 (*Carica*), thus entirely different from Begoniaceae and Passifloraceae. As yet, no evidence has come to explain the evolution of this testa, unless from *Averrhoa* (Oxalidaceae) which has also the fibrous exotegmen. The testa of *Averrhoa* multiplies to a thickness of 10–13 cells but in places to 25 cells because there are, as in *Carica*, reticulate wedges of cell-multiplication derived in *Averrhoa*, however, from the inner hypodermis. In most Oxalidaceae the small seed

TABLE 2. *Seed-coats of Passifloraceae, Caricaceae and Cucurbitaceae*

	Passifloraceae		Caricaceae		Cucurbitaceae	
	Ovule	Seed	Ovule	Seed	Ovule	Seed
o.i.	3 cells	Non-multiplicative, unspecialized	4–6 cells	Non-multiplicative, not lignified	6–10 cells	Multiplicative
			o.e.	Single or multiple sarcotesta	o.e.	Seed-epidermis (*e*-layer) Seed-hypodermis (*e''*-layer) Sclerotic layer (*e'*-layer)
			o.h.	Outer mesotesta		
			Meso-phyll	Inner mesotesta	Meso-phyll	Aerenchyma collapsing
			i.e.	Crystal-cells	i.e.	Unspecialized
i.i.	3 cells	Non-multiplicative	4–5 cells	Non-multiplicative	1–3 cells	Non-multiplicative, collapsing
	o.e.	Lignified palisade	o.e.	Lignified fibres		
	i.e.	Inner walls lignified	i.e.	Unspecialized		
v.b.	Raphe only		Raphe only		Raphe–antiraphe	
Aril	Funicular		None (? vestige)		Placental or none	
Endosperm	Persistent		Persistent		Slight or none	

is simplified, and the bearing of the ancestral seed, indicated by *Averrhoa*, on that of *Carica* has yet to be investigated.

The testa of Cucurbitaceae is multiplicative from the outer epidermis of the ovule which, practically, makes the mature seed-coat, and the short-lived tegmen has no particular structure. Seeds of several species in most common genera have been studied but there are still many uninvestigated, especially among those with alate seeds. It is possible, therefore, that a fibrous tegmen may be found to show, as in Sapindaceae, this construction has been lost in most genera. If so, the Cucurbitaceous seed could be regarded as the Caricaceous in which the multiplication of the testa has become peripheral and the character of the tegmen has been transferred to this peripheral differentiation. The inner sclerotic layer could be regarded as the transferred exotegmen, as the placental aril of some Cucurbitaceae may be the transferred funicular aril which is constant in Passifloraceae but, at most,

vestigial in Caricaceae. Then this sclerotic layer introduces the antithesis, just as the exotegmen, for some genera have radially directed cells and others tangentially as short fibres, and which may be primitive is not clear. Fibriform cells occur in the highly multiplicative and pseudo-arillate seeds of *Momordica* with its dehiscent fruit to suggest that they were the primitive form in agreement with Caricaceae. An analogy for this outward displacement of factors is supplied by *Siparuna* (Monimiaceae) and *Artocarpus* (Moraceae; Corner 1962).

If, however, there is no evidence forthcoming to prove a primitively tegmic construction for Cucurbitaceae, its testal derivation must be considered. If the outer epidermal cells of the outer integument divided periclinally once, instead of twice, the sclerotic layer would be hypodermal as in Paeoniaceae which is a strangely isolated family; and this interpretation would lend support to the contrary view that the palisade-cell was antecedent to the fibriform. The Passifloraceous seed could, then, be regarded as Cucurbitaceous with the sclerotic factor transferred to the exotegmen, as in Bixales, and with the testa non-multiplicative. The Paeoniaceous seed would supply the post-chalazal extension of the raphe vascular bundle which is characteristic of Cucurbitaceae.

In conclusion there are three possible alliances for these families, thus:

(1) Passifloraceae and Turneraceae relate with Bixales, Caricaceae with Violales, Cucurbitaceae with Paeoniales.

(2) Cucurbitaceae as Caricaceae deprived of the fibrous tegmen, and both related with Violales.

(3) Cucurbitaceae with exotegmic heritage leading to Bixales and Passiflorales, but modified in the multiplicative testa.

The first possibility for Cucurbitaceae is pursued further in connection with Convolvulaceae (p. 50). It implies that the floral characters now employed to align the main families of these orders are the result of convergent evolution as shown by the seed-structure; in any case Caricaceae stand apart from Passiflorales. Thus the study of seeds probes into the conclusions of floral classification. For Cucurbitaceae, Hutchinson (1967) wrote 'their nearest relatives are the comparatively more primitive Passifloraceae'. For both, therefore seed and fruit would have commenced with the arillate capsule. This fruit is connected in the durian-theory with pachycauly and, in the whole complex of these orders, *Carica* stands out as the pachycaul tree with traces of arillate capsule and, probably, a truly primitive seed-structure. The question is what was the structure of the seed of the Passifloraceous and Cucurbitaceous pachycauls.

Aristolochiales–Nepenthales–Rafflesiales–Sarraceniales

Six families figure in this complex. That *Thottea, Nepenthes, Rafflesia, Sarracenia* and *Drosera* may be related appears quixotic; yet, divested of their peculiarities of growth, they offer enough technical resemblance in floral construction to perplex those systematists who have never seen *Rafflesia*. The four recent modes of classification are set out in Table 3, together with the facts of seed-structure which reveal such differences that the technical resemblances suggest parallel floral evolution rather than consanguinity.

Aristolochiaceae have the fibrous exotegmen that distinguishes the Sapindalean complex and, parti-

cularly, that of Oxalidaceae, Linaceae and Caricaceae. The construction forbids alliance with Magnoliales, unless with Myristicaceae. Nepenthaceae and Sarraceniaceae fit the exotestal category of Ranunculaceae and Saxifragaceae, where unitegmic seeds, as in Sarraceniaceae, are known to occur. The Droseraceous seed, though minute, is complicated but, if it has a fibrous exotegmen (as reported in *Drosophyllum*), then it must go with Aristolochiaceae. The seeds of Rafflesiaceae agree with none of the others, except the unitegmic Hydnoraceae, but suggest the Piperaceous where *Peperomia* is also unitegmic (p. 43).

I conclude that there are three separate alliances, namely Aristolochiales (? with Droseraceae), Rafflesiales, and Sarraceniales (including Nepenthaceae).

Araliales–Cornales–Caprifoliaceae

Under one or other of these titles, the following six families are usually placed: Araliaceae, Alangiaceae, Cornaceae, Garryaceae, Nyssaceae, and Umbelliferae. All are unitegmic, exotestal, albuminous and with small embryo. There are exceptions for the embryo may be large in Alangiaceae and Cornaceae; the endosperm is usually oily but may be starchy in Cornaceae; and the ovule may be bitegmic in Nyssaceae, the seed-structure of which is little known but appears to be very simple.

Apart, therefore, from Nyssaceae, the families have a uniform and simplified seed-structure. The nature of the single integument is not known but, as it is massive in some cases as Cornaceae and Garryaceae, it may be the result of connation of the two integuments rather than the loss of one or other; in fact the seeds may be pachychalazal. On the whole the orders seem to be unitegmic derivatives of Hamamelidalean and Rosalean ancestry. Takhtajan favours a Saxifragalean ancestry, Cronquist a Myrtalean, and Hutchinson a Rosalean or Cunonialean. These suggestions refer to exo-mesotestal seeds and confirm the importance of the construction.

Caprifoliaceae are added to Araliales by Hutchinson. Cronquist and Takhtajan refer the family to Dipsacales which Takhtajan relates to Cornales. The unitegmic Caprifoliaceous seed presents, however, a less simplified structure; it is more strongly exotestal, even with a palisade of lignified cells much as in Rhamnaceae, and *Symphoricarpus* introduces hypodermal fibres. There must, clearly,

TABLE 3. *Classification of the Aristolochialean complex, with suggested alliances (in brackets) and indications of seed-structure*

Families	Orders			
	Hutchinson	Takhtajan	Cronquist	Thorne
	1959	1969	1968	1968
Aristoloch. (tegmic fibres)	Aristoloch. (Berberid. exotestal)	Aristoloch. (Magnoliales endotestal)	Aristoloch. (Magnolial. Piperales, tegmic)	Aristoloch. (Annonales mesotestal)
Raffles. (sclerotic tegmen)	Aristoloch. (Berberid. exotestal)	Raffles. (Aristoloch.)	Raffles. (Santal. atestal)	Raffles.
Hydnorac. (unitegmic sclerotic)	Aristoloch. (Berberid. exotestal)	Raffles. (Aristoloch.)	Raffles. (Santal. atestal)	Raffles.
Nepenth. (exotestal)	Aristoloch. (Berberid. exotestal)	Nepenth. (Saxifrag. exotestal)	Sarracen. (Theales mesotestal)	Nepenth.
Droserac. (exotestal, ? fibrous exotegmic)	Sarracen. (Saxifrag. exotestal)	Nepenth. (Saxifrag. exotestal)	Sarracen. (Theales mesotestal)	Rosales (Mesotestal)
Sarracen. (unitegmic exotestal)	Sarracen. (Saxifrag. exotestal)	Sarracen. (Ranuncul. exotestal)	Sarracen. (Theales mesotestal)	Sarracen.

be a fuller investigation into the seeds before the position of Caprifoliaceae can be gauged.

Centrospermae

The main families of this alliance, ordered sometimes into Caryophyllales (Takhtajan 1969), agree in the shape and structure of the ovule and seed. Derivation from exotestal Ranales is generally accepted. The small seeds of Centrospermae, however, with a relatively large embryo are reduced and simplified. The integuments consist generally of the minimal two cell-layers, as the attenuation suitable for herbaceous growth. Unitegmic seeds, apparently without an inner integument, occur in *Thelygonum* (Cynocrambaceae) and in two genera of Nyctaginaceae (*Abronia, Boerhaavia*). The seed-coat is commonly reduced to the exotesta and the endotegmen, but this inner layer may also be lacking. In two respects the Centrospermous seed is more primitive than the Ranunculaceous. In several Phytolaccaceae with dehiscent fruits and thick testa (4–7 cells thick) there is a well-developed aril; it occurs also in several genera of Aizoaceae, but is usually absent from indehiscent fruits. Then the seeds contain starchy perisperm; the nuclear endosperm is reduced to a thin layer round the radicle. If perisperm is a primitive feature of the angiosperm seed, the Centrospermous implies a derivation from a Ranalean source such as ancestral Nymphaeaceae, both arillate and perispermous, rather than Ranunculaceous. The exceptionally large seed of *Pisonia longirostris* and its allies is certainly not Ranunculaceous (Fig 434).

Here may enter the Madagascan Didiereaceae. They were classified with Sapindales because of the trilocular ovary, the absence of perisperm and the practically exalbuminous seed. The seed-coat structure is Centrospermous in its simplicity. There is no vestige of the fibrous exotegmen but, of course, it may have disappeared as it seems to have done from most Sapindaceae. I can find no criterion in seed-structure to decide on the position of

Didieraceae. The cactoid habit, as a xerophytic modification of the primitive pachycaul, may be phyletic with Cactaceae or evolved in parallel in the deserts of Pangaea.

The idea of Ranales springs from the apocarpous flower. It must have been the antecedent of all phyletic lines of angiosperms. Thus it represents not only the beginning of floral evolution but the first grade through which all lines of descent passed in their special evolution, not the least being that indicated by the structure of the seed. Such a line appears to be the Centrospermous. The Cactaceous flower, except for its ovary, is more primitive than many Ranalean, especially the Ranunculaceous. Cactaceae with primitive pachycauly indicate the xerophilous tendency which has caused Centrospermae to deviate ecologically from Ranunculaceae though, ultimately, phyllodic *Ranunculus* and *Stellaria* grow together. Buxbaum (1961) considered Phytolaccaceae as the most primitive Centrospermous family, related perhaps with Illiciaceae; it certainly introduces both pachycauly and the arillate seed.

Tamaricales

It has been suggested that Fouquieraceae, Frankeniaceae and Tamaricaceae, which comprise this order, are related to Bixales or Violales. Both these orders have exotegmic seeds in contrast with the exotestal Tamaricales. Seed-structure does not, therefore, support the suggestion but prefers that which allies Tamaricales with Salicales and, indeed, Centrospermae. Thorne (1968) places Fouquieraceae in Solanales but the bitegmic seed requires explanation.

Amentiferae

This general term for the catkin-bearing trees covers the eight orders Balanopsidales, Casuarinales, Fagales, Juglandales, Leitneriales, Myricales, Salicales and Urticales. About seventeen families are attributed to them but, as there is no sharp distinction between Urticaceae and Moraceae, being the two larger, or between Betulaceae and Corylaceae, as two of the smaller, the number is arbitrary. They have been regarded as primitive dicotyledons of gymnosperm-like simplicity, e.g. *Casuarina*, though not *Ficus* or *Quercus*. Modern opinion prefers their reduction in flower, fruit, leaf and seed from Rosalean or Hamamelidalean ancestry; yet the primitive status has its advocates (Meeuse and Houthuesen

1964) who see in the flower of *Engelhardtia* a decorated *Gnetum*. Now, many well-known families such as Anacardiaceae, Euphorbiaceae, Leguminosae and Sapindaceae provide ample evidence for the simplification of the flower in condensed inflorescences, and this evidence exists in Urticales to relate *Ulmus* with the naked uni-ovulate ovary in some dioecious *Ficus*. The evidence has never been refuted, chiefly because it seems never to have been contemplated. One must observe that there is an incredibly large amount of living evidence to show how flowering plants have evolved, but they become important to phylogenists, unaware of ecological evolution, only when devitalized, fossil and derelict. In contrast there is no evidence to explain how the tardily dehiscent drupaceous capsules of Juglandaceae can be related to the seed of *Gnetum* or how the primitive pachycaul tree with pinnate leaf (*Artocarpus, Juglans*), pinnatifid leaf (*Quercus*), or palmate leaf (*Ficus*) is to be related to the equisetiform *Casuarina*, except as supplied by the durian-theory.

Amentiferous seeds offer no primitive mark whether in the structure of the seed-coats, the endosperm or the embryo; and they are exarillate. Commonly the seeds are solitary in drupes or nuts containing 2–6 loculi with as many as 3–11 abortive ovules (*Castanea, Quercus*). These fruits are obviously reduced from many-seeded capsules and, in the very long run, from arillate capsules, but not from *Gnetum*-spikes. Then, unitegmic ovules occur in Betulaceae, Juglandaceae, *Balanops, Eucommia, Myrica, Nothofagus* and *Salix*, and there is evidence of this reduction from a bitegmic state in Betulaceae, Salicaceae and *Nothofagus*. Many of the seeds are exalbuminous or have merely a thin layer of endosperm round the well-developed embryo; the microscopic embryo of Magnolialean antiquity and of *Gnetum* does not occur.

The seed-coats appear to be non-multiplicative and, in their endocarp-enclosure, to have little or no specialization. The tegmen, when present, soon disappears. The testa is thin-walled and becomes more or less crushed, though the outer epidermal cells may be somewhat enlarged and thick-walled; in some of the families the testa is highly vascularized as if pachychalazal. At most the seeds may be called exotestal, which harmonizes with the supposed derivation from Rosalean ancestry and implies, of course, a primitively pinnate leaf. But it must have been a long way from that arillate ancestry to modern Rosaceae, Hamamelidaceae and Amentiferae.

There are two exceptions. *Leitneria* has a some-what thick-walled mesotesta and persistent endo-tegmen (p. 173). The second is Salicaceae, which have been referred to Flacourtiaceous ancestry Violalean and Tamaricalean. The complete absence of tegmic differentiation and, apparently, of the tegmen in *Salix* and some species of *Populus* disposes of the first two possibilities. The third seems undeniable for the Salicaceous seed strongly resembles in its smallness, hairiness and exotestal construction the Tamaricalean; and among the Amentiferous families, the Salicaceous is exceptional in being truly capsular. Thus, compared with the small seeds of other capsular families with well-developed seed-structure (Papaveraceae, Hyperic-aceae, or Saxifragaceae) there is no reason to suppose that the seed of Salicaceae was other than exotestal and hairy.

Piperales

Seed-structure in this order exposes strange consequences of dicotyledonous evolution. Custom refers to the order Chloranthaceae, Piperaceae and Saururaceae because of their minute flowers in spicate inflorescences, orthotropous ovules, minute embryos, and few- or 1-seeded fruits. The recent trend is to separate Chloranthaceae; yet all are regarded as simplifications of Magnoliales, Laurales or Ranales. None of these orders, however, display the primitive vegetative evolution from pachycaul to leptocaul which, with a primitively diffuse vascular system, is found in *Piper*. The neglect of this prolific tropical genus is responsible for much under-estimation of the order, especially when theory is based on academic details. As I have shown with *Ficus* (Corner 1967, 1970a, b), so it will be discovered that the species of *Piper* present the stages in which the primitively pachycaul dicotyle-don has entered modern vegetation; in this case it has been mainly through climbers, shrubs and herbs of the forest.

Seed-structure certainly distinguishes Chloranth-aceae. The fibrous tegmen and the firm endotestal palisade of crystal-cells filled with a fibrillar lignification unites the family, as befits the syncar-pous ovary, with Papaveraceae and Proteaceae (p. 32). This dual peculiarity does not relate to the seed-structure of any one Magnolialean family, but combines the endotestal character of Magnoli-aceae with the fibrous tegmen of Myristicaceae. Thus, in the primitively pachycaul and arillate

state, ancestral Chloranthaceae may merge with ancestral Magnoliaceae and Myristicaceae. Affinity has been suggested with Aristolochiaceae, also with fibrous tegmen, but the endotestal crystal-cells of Aristolochiaceae lack the lignified fibrils. It is impossible to decide, as yet, whether this feature has been lost or was never present; hence I leave Aristolochiaceae in the category with fibrous tegmen.

The seeds of Piperaceae and Saururaceae are much alike and differ in most details from those of Chloranthaceae. They are tegmic seeds in which (1) the thin-walled outer integument is crushed at such an early stage of seed-development that no testa is formed; (2) the tegmen forms the seed-coat by means of a thick-walled outer epidermis of wide cells and, to some extent, a more or less thick-walled and, even, palisade-like inner epidermis, but neither layer is fibrous; (3) the nucellus persists as a massive storage perisperm and the thick nucellar cuticle becomes curiously plicate against the endo-tegmen (? in Saururaceae); (4) the endosperm is a small block of cells at the micropylar end. Thus in such detail of the seed, as well as of the flower, the two families are close. Saururaceae supply the transition from an apocarpous ovary to the syncar-pous, and this is reduced to the uni-ovulate state in Piperaceae (Murty 1959a, b).

If Saururaceae can be classed in Magnoliales, it is impossible to exclude Piperaceae, yet its inclusion opens the gates of the order to many other apocar-pous families, e.g. Rosaceae, Mimosaceae, Simarub-aceae, Sterculiaceae and Apocynaceae. The artificial nature of Magnoliales, as a grade in floral evolution through which all dicotyledonous lines have passed, is obvious. How it is to be dissected into natural series is shown by the structure of the seed. Piper-ales emerge as an order distinguished by the reten-tion of perisperm and the loss of the testa. Saururaceae and Piperaceae are the specialized relics, with *Piper* yet emerging from the pachycaul state; and even this evidence will disappear as the lowland forests of the tropics are cleared in biological ignorance.

The question arises whether there are other advanced families with Piperalean seeds. One which seems to fit, though its seeds are very small and numerous, is Rafflesiaceae (p. 40). If the large flower of *Rafflesia*, which is by no means the consequence of its parasitism, had an apocarpous fruit with large arillate seeds, which does not seem impossible when *Papaver* is compared with *Bocconia*, and it were borne on a green leafy stem, there might be the pachycaulous Piperalean ancestor. The wide distribution of Rafflesiaceae bespeaks antiquity. The parasitic tendency may have begun in the primitive pachycaul forest of flowering plants and, with leptocaul evolution, have resulted in the progressively atestal and ategmic Santalales. If these plants can be conserved, experimental biology will be able to test the hypothesis. Taught in schools of temperate biology, this wonderful subject seems unaware of the jeopardy in which its great future hangs.

Another family which fits the seed-structure of Piperales is aquatic Podostemaceae. They are supposed to have had a Crassulaceous or Saxifragaceous origin but the simple exotestal seed-structure of these families is so different from that of Podostemaceae as to discredit the idea. Podostemaceae lack endosperm but have a false embryo-sac of nucellar origin which may be traced in Piperaceous diversity.

Summary of positive contributions

The introduction of this bewildering mass of detail about the seed-coat into the classification of dicotyledons will seem destructive. Gradually its purpose will be seen for it is the bunch of keys to most family-sets along the corridors of angiosperms. That there are many uncertainties in our genealogical reconstruction of their history is revealed through the many different ways in which many families may be classified. The injection of so much intrinsic detail is illuminating. It will lead, one hopes, to renewed interest and, especially, the endeavour to study the seeds of tropical genera. I list, therefore, some positive contributions.

(1) Magnoliales–Ranales are a grade of floral evolution through which the advanced families have passed. The grade is composed of eight or more distinct constructions of the seed-coat. Some of these lead on to the advanced families; others appear to have been unsuccessful diversifications of the primitive complexity.

(2) The new picture of dicotyledonous affinities is shown in the accompanying tableau, which is limited necessarily to families (*a*) in which the structure of the seed-coat is known, (*b*) in which the structure is sufficiently differentiated. Families and orders with unspecialized seed-coats have mostly been omitted.

(3) The conclusion is that the orders or super-orders should be based on the structure of the seed-coat, as one of the peculiar and intrinsic characters of angiosperms. Thus, Annonales, Laurales, Magnoliales, Myristicales, Paeoniales, and Ranales are distinct orders comparable with Centrospermae, Leguminosae and Piperales. To classify the Magnolialean–Ranalean families into one order is as artificial as the old classification into hypogynous, perigynous and epigynous groups; floral neoteny spells convergence, and the history of botanical classification has been the continual substitution of artificial classes or grades by means of genealogical lines.

(4) Most large orders, as explained in the preceding pages, are artificial. Hence comes the diversity of opinion among the systematists.

(5) New disposition of some families:

(*a*) Calycanthaceae belong in Laurales, not Rosales; so do Burseraceae and possibly Combretaceae.

(*b*) Cruciferae must be removed from Capparidales and Rhoeadales, and placed with endotestal families.

(*c*) Moringaceae must be removed from Capparidales to the category of the mesotestal seed, e.g. Theales, unless it can be shown that the family had, primitively, a fibrous exotegmen.

(*d*) Podophyllaceae belong with Berberidales, not Ranales (or Ranunculaceae).

(*e*) Ochnaceae–Luxemburgieae must be separated from Ochnaceae as Sauvagesiaceae (fibrous exotegmen) in the affinity of Flacourtiaceae and Violaceae. This clarifies Ochnaceae.

(*f*) Connaraceae (fibrous exotegmen) go likewise with Celastraceae, Flacourtiaceae, Meliaceae etc., not with Dilleniales, Leguminosae, Rosales or Saxifragales.

(*g*) Cactaceae and Nyctaginaceae go with Centrospermae.

(*h*) Punicaceae (fibrous exotegmen) go with Lythraceae and Combretaceae, not Myrtales.

(*i*) *Gynotroches* and *Pellacalyx*, as Legnotidaceae, must be removed from Rhizophoraceae (mesotestal) to the category of seeds with fibrous exotegmen.

(*j*) Buxaceae (exo-mesotestal) go with Hamamelidales–Rosales, not with Euphorbiaceae.

(*k*) Rutales (tracheidal tegmen, mesotestal) must be removed from Sapindalean affinity.

(*l*) Vitaceae (endotestal, tracheidal tegmen) must

be removed from Rhamnales (exotegmic) to the affinity of Rutales.

(*m*) Geraniaceae agree with Clusiaceae and Hypericaceae (endotesta with stellate cells). Elatinaceae belong with Hypericaceae, not Caryophyllales.

(*n*) Bonnetiaceae must be removed from Theaceae and allied either with Guttiferae or Ixonanthaceae.

(*o*) Cucurbitaceae are problematic, but there is evidence that they are part of a phyletic line that includes the problematic Paeoniaceae and Convolvulaceae.

(*p*) Most other Sympetalae seem to belong to a phyletic line of apocarpous arillate origin, now lost, which had converted the seed (exotestal as Ranales or mesotestal as Illiciaceae) into the unitegmic. In contrast, Ebenales and Ericales fit with Theales.

(*q*) Leguminosae afford a parallel with Sympetalae, as an exotestal phyletic line with apocarpous and arillate origin, but persistently bitegmic. They make one of the great phyletic lines of dicotyledons.

(*r*) Scyphostegiaceae agree closely with Celastraceae.

(*s*) Zygophyllaceae (endotegmic fibres) are problematic, comparable with Nandinaceae, but probably reduced Malpighiales.

(*t*) Chloranthaceae must be removed from Piperales and aligned with Papaveraceae and Proteaceae, for they have the Magnoliaceous endotesta and the Myristicaceous fibrous exotegmen.

(*u*) Piperales become another phyletic line, the origin of which is lost, but which seems to have led primitively to the parasitism of Rafflesiaceae and Santalaceae. Comparison with Nandinaceae suggests derivation from the Papaveraceous stock.

(*v*) Amentiferae, excluding Salicaceae, have unspecialized seed-coats which give no indication of a primitive character. Salicaceae fit Tamaricales.

(*w*) The great alliance of Bombacaceae, Malvaceae, Sterculiaceae, Tiliaceae and Thymelaeaceae is confirmed by uniformity in the exotegmic palisade. Euphorbiaceae–Crotonoideae and Dipterocarpaceae belong, possibly also Gonystylaceae, but they may connect with the category of seeds with fibrous exotegmen to which some Euphorbiaceae–Phyllanthoideae belong.

(*x*) Elaeocarpaceae (fibrous exotegmen) must be removed from the affinity of Tiliaceae.

(*y*) Relict evidence of a fibrous exotegmen affirms the affinity between Meliaceae and Sapindaceae.

(*z*) The most primitive seed-construction is found in Myristicaceae.

(6) Disposition of some genera:

(*a*) *Barklya* must be transferred from Papilionaceae–Cadieae to Caesalpiniaceae–Bauhinieae (p. 165).

(*b*) *Perrotetia* (exotegmic palisade) is anomalous in Celastraceae (exotegmic fibres).

(*c*) *Actephila* with exotegmic palisade of stellate cells (cf. Hypericaceae, Geraniaceae) is anomalous in Euphorbiaceae–Phyllanthoideae (exotegmic fibres).

(*d*) *Biebersteinia* is Geraniaceous.

(*e*) *Averrhoa* is Oxalidaceous.

(*f*) *Hydnocarpus*, *Taraktogenos*, *Scaphocalyx* and *Pangium* are anomalous in Flacourtiaceae; their pachychalazal seeds have no fibrous exotegmen.

(*g*) *Huertea* and *Tapiscia* (both with fibrous exotegmen) appear anomalous in Staphyleaceae (mesotestal) but may indicate the primitive construction in the family, similar to *Alectryon* in Sapindaceae.

(*h*) *Leptonychia* (fibrous exotegmen) is anomalous in Sterculiaceae (exotegmic palisade).

(7) Families with much diversification of the seed-coat:

Celastraceae, Clusiaceae, Flacourtiaceae (*sensu lato*), Meliaceae, Rutaceae, Sapindaceae and Theaceae.

The investigation of these families will throw light on the whole problem of the construction and evolution of the seed-coat.

(8) Families have been found to indicate kinds of seed-structure which have disappeared in the course of evolution. The ancestry of Chloranthaceae, Papaveraceae and Proteaceae points to a link, now lost, between Magnoliaceae and Myristicaceae. Oxalidaceae may link Caricaceous and Lythraceous ancestry. Trochodendraceae may be survivors of Hamamelidalean ancestry. Knowledge of the seed-structure of Glaucidiaceae will be awaited with great interest.

Classification of bitegmic dicotyledonous seeds

(A) Testal seeds
    (1) Exotestal   Group 1
    (2) Endotestal   Group 2
    (3) Mesotestal   Group 3
    (4) Hypodermal   Group 4
(B) Tegmic seeds
    (1) Exotegmic with a palisade of prismatic, cuboid or stellate cells   Group 5
    (2) Exotegmic with tangentially elongate cells or fibres   Group 6
    (3) Tegmen tracheidal   Group 7
    (4) Endotegmic seeds   Group 8

Group 1 (exotestal seeds)

(A) Exotestal cells cuboid or radially elongate, with more or less isodiametric polygonal facets, commonly with the outer wall thickened.

(1) Exotesta of Malpighian cells with linea lucida; outer hypodermis (sometimes also the inner hypodermis) composed of hourglass-cells: Leguminosae.

(2) Not as Malpighian cells.

    (*a*) Nymphaeaceae, Ranunculaceae, Winteraceae (? Schisandraceae).

    (*b*) Begoniaceae, Buxaceae (mesotestal), Capparidaceae (fibrous exotegmen), Celastraceae (fibrous exotegmen), Connaraceae (fibrous exotegmen), Coriariaceae, Crassulaceae, Datiscaceae, Didiereaceae, Droseraceae (?), Ebenaceae, Elaeagnaceae, Lecythidaceae (mesotestal), Melastomataceae (mesotestal), Meliaceae (fibrous exotegmen), Myrsinaceae, Myrtaceae (mesotestal), Nepenthaceae, Plumbaginaceae, ? Podostemaceae (tegmic), Proteaceae (fibrous exotegmen, endotestal), Rhamnaceae, Salicaceae, Salvadoraceae, Sapindaceae (fibrous exotegmen), Saxifragaceae, Tamaricales, Theaceae (mesotestal).

    (*c*) Centrospermae (including Cactaceae).

(B) Exotestal cells tangentially elongate or fibriform: Berberidaceae, Lardizabalaceae, Melianthaceae, Podophyllaceae. *Citrus* (Rutaceae), *Grewia* (Tiliaceae); (see also *Gymnacranthera* and *Knema*, Myristicaceae).

(C) Exotestal cells stellate-undulate: Illiciaceae (? Schisandraceae); Nymphaeaceae (*Brasenia*, *Nymphaea*, *Victoria*); *Elaeagnus*; Lecythidaceae (*Gustavia*); Melastomataceae (*Pternandra*); Sapindaceae (*Cardiospermum*, *Paullinia*, *Urvillea*).

Group 2 (endotestal seeds)

(A) Endotestal cells columnar, thick-walled.

(1) Endotesta as a single layer of cells: Myristicaceae (exotegmic), Cruciferae, Polygalaceae.

(2) Endotesta multiple: Vitaceae (tracheidal tegmen).

(B) Endotestal cells cuboid, thick-walled: (Cruciferae), Dilleniaceae (substellate; tracheidal exotegmen), Grossulariaceae; Eupteleaceae, Trochodendraceae.

(C) Endotestal cells stellate, thick-walled: *Jussieua* (Onagraceae, fibrous exotegmen).

(D) Endotestal cells tracheidal with spiral-annular thickening, more or less tangentially elongate: Burseraceae, Calycanthaceae, Combretaceae (fibrous exotegmen), Hernandiaceae, Lauraceae, Monimiaceae.

(E) Endotestal cells cuboid or columnar with fibrillar and, generally, lignified endoreticulum.

(1) Endotesta as a single layer of cells, also with fibrous exotegmen: Chloranthaceae, Fumariaceae, Papaveraceae, Proteaceae.

(2) Endotesta multiple; tegmen unspecialised: Degeneriaceae, Magnoliaceae.

Group 3 (mesotestal seeds)

(A) Mesotesta with fibres in longitudinal, transverse or oblique bundles: Annonaceae, Eupomatiaceae (*Akebia*, Lardizabalaceae).

(B) Mesotesta with sclerotic cells (not elongate into fibres).

(1) Illiciaceae, Schisandraceae.

(2) Balanitaceae, Buxaceae, Clusiaceae-Calophylleae, Ebenaceae, Hamamelidaceae, Lardizabalaceae, Lecythidaceae, Marcgraviaceae, Melastomataceae, Moringaceae, Myrsinaceae, Myrtaceae, Penaeaceae, Rhizophoraceae, Rosaceae, Sapindaceae *pr. p.*, Stachyuraceae, Staphyleaceae *pr. p.*

(3) Coupled with tracheidal or fibrous exotegmen: Combretaceae, Lythraceae, Onagraceae, Punicaceae, Sonneratiaceae, Trapaceae.

Group 4 (hypodermal seeds)

(A) Outer hypodermis as a single layer of thick-walled prismatic cells: Paeoniaceae.

(B) Seed-hypodermis with two or more layers of sclerotic cells, the inner layer as palisade-cells or stellate-sclerotic cells: Cucurbitaceae.

    (Unitegmic; hypodermal layer of 2 or more rows of thick-walled, radially elongate cells; Convolvulaceae.)

Group 5 (exotegmic seeds with palisade)

(A) Exotegmic cells stellate-undulate: Clusiaceae-Clusieae, Elatinaceae, Hypericaceae, Geraniaceae, ? Bonnetiaceae (cells elongate). – *Actephila* (Euphorbiaceae–Phyllanthoideae).

(B) Exotegmic cells prismatic or cuboid, the facets isodiametric and polygonal or slightly tangentially elongate.

(1) Bixaceae, Cistaceae, Passifloraceae, Turneraceae. – *Perrotetia* (Celastraceae).

(2) Bombacaceae, Dipterocarpaceae, Euphorbiaceae–Crotonoideae, Euphorb.–Phyllanthoideae *pr. p.*, Malvaceae, Sterculiaceae, Thymelaeaceae, Tiliaceae.

(3) Exotestal cells with slightly elongate facets: Gonystylaceae, Euphorbiaceae–Phyllanthoideae *pr. p.*

(4) Exotegmen and endotegmen with more or less cuboid cells: Piperaceae, Podostemaceae, Rafflesiaceae, Saururaceae.

Group 6 (exotegmic seeds with fibres)

(Fibres in a single cell-layer, s, or multiple, m)

(A) Capparidaceae (s, m), Celastraceae (s), Connaraceae (s), Elaeocarpaceae (s, m), Erythroxylaceae (s), Euphorbiaceae-Phyllanthoideae *pr. p.* (s, m), Flacourtiaceae (s), Ixonanthaceae (s), ? Legnotidaceae (s), Linaceae (s), Malpighiaceae (s), Meliaceae (s), Myristicaceae (s; endotestal), ? Pandaceae (m), Resedaceae (s), Sapindaceae (s; mostly exo-mesotestal, pachychalazal), Sauvagesiaceae (s), Scyphostegiaceae (s), Violaceae (s, m). –

Huertea, *Tapiscia* (Staphyleaceae, s), *Leptonychia* (Sterculiaceae, s), ? Turneraceae (s).

(B) Aristolochiaceae (m), Caricaceae (m), Oxalidaceae (s, m). – ? Droseraceae.

(C) Combretaceae (s), Lythraceae (s), Onagraceae (s), Punicaceae (s), Sonneratiaceae (s); (see also Group 7B).

(D) Seeds also endotestal: Chloranthaceae, Fumariaceae, Papaveraceae, Proteaceae. – Myristicaceae.

Group 7 (exotegmic seeds with tracheidal cells)
(A) Tracheidal cells not narrowly fibriform.
　(1) Dilleniaceae (tracheidal cells in a single layer; endotestal).
　(2) Rutaceae (tracheidal cells in a single or multiple layer; exo-mesotestal).

(3) Vitaceae (tracheidal cells multiple; endotesta multiple).

(B) Tracheidal cells narrowly fibriform: Combretaceae, Lythraceae *pr. p.* Onagraceae *pr. p.*, Punicaceae, Sonneratiaceae *pr. p.*, Trapaceae.

Group 8 (endotegmic seeds)
(A) Exotegmic and endotegmic with thick-walled cuboid cells: Piperaceae, Podostemaceae, Rafflesiaceae, Saururaceae.

(B) Exotestal–endotestal; endotegmic cells elongate tangentially: Zygophyllaceae. – (*Belliolum*, Winteraceae).

(C) Tegmen with the inner part and the endotegmen fibrous: *Symphonia* (Clusiaceae–Garcinieae).

(D) With only the endotegmen sclerotic: Nandinaceae.

5. Seed-evolution

The primitive dicotyledonous seed

In the preceding chapter it has been shown that the families of dicotyledons can be arranged into phyletic orders distinguished by the structure of the seed-coats. Within these orders families are, then, distinguished in the customary way by trends in vegetative, floral, and fruiting evolution; the seed, too, may be modified by overgrowth, pachychalazal construction, or simplification. The problem, now, is to consider what may have been the primitive nature of the seed. The existing variety offers much opportunity for speculation; most kinds refer back to the apocarpous fruit. I have found that it is more helpful to begin with the general conclusions, arrived at in the preceding pages, because they provide the trends of specialization, reduction and simplification that explain the modern variety. The conclusions can be listed under the following ten heads.

(1) It is unlikely that any living seed is entirely primitive.

(2) The primitive seed was a factor in the incipient dicotyledonous forest.

(3) The primitive seed was of moderate size, arillate, bitegmic, with multiplicative and mechanically differentiated testa and tegmen.

(4) Because aril, testa and tegmen were fairly massive, they were vascular.

(5) A complicated structure prevails in the seeds of Magnoliales and in the less primitive, but polypetalous, families that are generally considered to be the points of departure for the advanced, e.g. Clusiaceae, Theaceae, Flacourtiaceae, Bombacaceae, Rosaceae, Rutaceae.

(6) In contrast, simplification through loss of aril, evolution of the unitegmic character, restriction to an exotestal seed-coat, and reduction in size to ovular dimensions, distinguish many epigynous families and most Sympetalae.

(7) Proof of this simplification is given by various genera in a family, or species in a genus, which depart from the normal through the loss or modification of some factor of construction, e.g. the loss of the palisade or fibrous structure of the exotegmen, or the loss of independent integuments.

(8) The character of the seed-coats generally resides in the restriction of the mechanical tissue to certain layers, mainly the exotesta, endotesta and exotegmen, rarely the endotegmen.

(9) In many such cases there are also vestiges of mechanical construction in other layers, such as the slight thickening on the walls of the endotestal and endotegmic cells, the presence of crystals in the endotestal cells, and the presence of a palisade-like, but unlignified, endotesta in exotegmic seeds, e.g. Euphorbiaceae.

(10) The chief mechanical layers are the epidermal layers which represent the epidermis of the pinna or involucre which primitively surrounded the nucellus. If one surface had thickened cell-walls or elongate cells, then so may the others.

From these premises I argue that the primitive seed was of medium-size (5–10 mm long, see p. 54), arillate, anatropous, and supplied with multiplicative integuments that differentiated mechanical tissue in every epidermis except that of the nucellus; that it was at once exotestal befitting the aril, endotestal, exotegmic and endotegmic, perhaps even mesotestal; in consequence the chalaza was elaborate. Then to these outer properties there must be added the internal as crassinucellate, albuminous, probably perispermous, and with minute embryo. The perisperm was starchy and, perhaps, so was the endosperm but, to judge from its nature in modern families, it soon became oily on the loss of the perisperm.

The evolution of the dicotyledonous seed

Many fossil seeds have been discovered but what light they may throw on the origin of the dicotyledonous seed is uncertain for the simple reason that what to expect is undecided. The time is past when it could be assumed that the dicotyledonous evolved from a contemporary gymnosperm because no existing or fossil order of Gymnospermae provides the character of the pre-angiosperm. Parallels there are, as calamite must have been to bamboo, but not the construction of stem, leaf, flower, fruit and seed requisite for the flowering plant. Herbaceous *Ranunculus* was once regarded as primitive but there seems never to have been a herbaceous gymnosperm. Thus it is that the origin of angiosperms is unsolved. Yet I have always contended that it should be possible to discover in the living host of species that which is advanced and so to uncover that which is primitive; it should be possible in all the larger and more varied families, and the conclusions should expose the primitive core of the flowering plant unless, of course, there has been no evolution. I shall follow this plan for the seed, and the prototype will be the best fit to the extraordinary range from *Annona* to *Viscum*; the herbaceous seed is always a specialization.

Unitegmic seeds: the origin of Sympetalae

Of some 350 families of dicotyledons, about 105 appear to have only unitegmic ovules. Among the sympetalous families, which are held to be more advanced than the polypetalous or the apetalous, 56 out of 67 are unitegmic. Among the polypetalous and apetalous the character accompanies mainly indehiscent fruits or diminutive seeds. It preponderates in the saprophytic and parasitic families where it leads to the ategmic condition in Santalales. Among the Magnolialean or Ranalean families it is found only in the more advanced and extreme, namely Ceratophyllaceae, Circaeasteraceae, and a few genera of Monimiaceae and Ranunculaceae. The systematic evidence leads to the conclusion that the unitegmic ovule is not primitive in dicotyledons, but has been derived from the bitegmic.

The following 22 families have both bitegmic and unitegmic ovules: Anacardiaceae, Burseraceae, Fagaceae, Halorrhagaceae, Icacinaceae, Menispermaceae, Monimiaceae, Myrsinaceae, Myrtaceae, Nyctaginaceae, Olacaceae, Piperaceae, Polygonaceae, Primulaceae, Rafflesiaceae, Ranunculaceae, Rosaceae, Salicaceae, Sarraceniaceae, Saxifragaceae, Simarubaceae, and Styracaceae.

In Anacardiaceae, Piperaceae and Rafflesiaceae it seems to be the outer integument which disappears. In Burseraceae, Fagaceae, Monimiaceae and Salicaceae it seems to be the inner integument, e.g. *Canarium*, *Nothofagus*, *Siparuna*, *Salix*. In Rosaceae, according to the critical study of Péchoutre, either integument may disappear or both may become connate to make a pachychalazal seed. The presence of two vestigial integuments in Balsaminaceae, Tropaeolaceae, Halorrhagaceae, Icacinaceae and Ochnaceae shows that the unitegmic ovule with fairly thick integument is pachychalazal. In Rafflesiaceae the outer integument may be reduced to a thickness of a single cell or absent, and its loss seems to explain the unitegmic ovule of Hydnoraceae. Then there are such pairs of bitegmic-unitegmic allies as Myrsinaceae–Aegicerataceae, Theaceae–Actinidiaceae, Theales–Ericales, and Ebenaceae–Sapotaceae, to suggest that the inner integument, so often short-lived in the developing seed, especially the tenuinucellate, is that which has succumbed. While, therefore, there may have been a long line of unitegmic evolution which has led to the majority of Sympetalae, there have been parallel lines in other dicotyledonous series where one or other integument has been lost or they have been compounded. The unitegmic ovule is clearly secondary and polyphyletic; its seed requires analysis.

Most families with unitegmic ovules have seeds which can be called exotestal because the integument differentiates in the exotestal manner of the bitegmic seed. They can also be called unitegmic so long as it is remembered that (1) unitegmic means 'uni-integumented' and not 'provided only with a tegmen', (2) the expression does not refer to testal seeds of bitegmic ovules which lose the tegmen in course of development. It appears, therefore, that the majority of unitegmic seeds came from an exotestal (perhaps, also, mesotestal) bitegmic ancestry. One must question, accordingly, the association of the following exotestal unitegmic families with exotegmic bitegmic families:

Aquifoliaceae, Icacinaceae, placed in Celastrales; Cyrillaceae, Empetraceae, placed in Celastrales; Callitrichaceae, placed in Lythrales; Halorrhagaceae, placed in Lythrales; Limnanthaceae, placed in Geraniales; Loasaceae, associated with Turneraceae or Violales; Nyctaginaceae, associated with Thymelaeaceae; Polemoniaceae, associated with Geraniales; Scytopetalaceae, placed in Malvales or Tiliales.

There must be caution, however, in this generalization because many seeds remain to be studied and a relic of tegmic construction may yet be found at the micropyle, as in the pachychalazal seeds of Meliaceae. The unitegmic seed of *Peperomia* has the tegmic construction of the bitegmic *Piper*. The unitegmic Burseraceae (*Canarium*, *Santiria*) seem to retain the endotestal construction characteristic of their family. Among Sympetalae, the Convolvulaceae have a remarkable hypodermal construction, discussed on p. 51, and there are traces of sclerotic mesophyll in Bignoniaceae, Caprifoliaceae, Cobaeaceae, Scrophulariaceae, Solanaceae and Verbenaceae.

Most orders of Sympetalae are assembled by Takhtajan under the subclass Asteridae. They are considered to have evolved from a Magnolialean stock such as Dilleniales (endotestal) via Saxifragales (exotestal) or Theales (mesotestal) and Cornales (unitegmic with little differentiation). Such possibilities do not conform readily with seed-construction. Affinity between Ericales and Theales is generally admitted but satisfactory proof from the seed is lacking. The derivation of Ebenales from Theales is also suggested and, if the unitegmic Sapotaceae are to be derived from bitegmic ancestry of Ebenaceae, then Theales have provided two such extremely different seeds as the Ericaceous and the Sapotaceous; some would add even the bitegmic Primulaceae. An ultimate derivation of Sympetalae, or Asteridae, from Dilleniales must be ruled out on grounds of seed-structure. Thealean and Saxifragalean association is possible, but ultimately the source of Sympetalae would seem to reside in the exotestal Magnoliales or Ranales. Here enters the Apocynaceous problem for the arillate follicles in this family suggest a stock much more primitive than what is known for Magnoliales, Ranales, Saxifragales and Theales. The source of unitegmic Sympetalae may, therefore, extend as far back into the origin of dicotyledons as that of any modern sub-class or order.

Ochnaceae are placed with or near to Theales. The black drupes on red receptacles convey a modification of the arillate follicle. The seeds of *Ochna*, at least, are pachychalazal with vestigial integuments (Figs. 436–442). If ancestral Ochnaceae with arillate follicles developed the pachychalazal seed into the neotenic unitegmic ovule and exotestal unitegmic seed, there would have been a suitable ancestor for Apocynaceae and most Sympetalae. Theales, by contrast, suggest the alternative derivation of unitegmic Sympetalous families by loss of the inner integument. In seed-structure, Ochnaceae and Theaceae relate more closely with Illiciaceae than with other families of Magnoliales and Ranales, but Illiciaceae have advanced beyond the arillate follicle and are a modern and specialized relic. Perhaps, then, this lost source of Illiciaceae, Ochnaceae and Theales was that of the unitegmic Sympetalae. The thesis is that the massively constructed unitegmic ovule of Sympetalae is a neotenic pachychalazal seed; their one integument is a basal, intercalary and chalazal substitution of the primitive two. The occurrence of sclerotic cells in the mesophyll of many of their seed coats may be inherited from a sclerotic pachychalaza.

The subject will require closer analysis when the significance of the development of the endosperm is understood. There are the two categories:

Endosperm cellular: Acanthaceae, Adoxaceae, Bignoniaceae, Boraginaceae, Campanulaceae, Caprifoliaceae, Cyrillaceae, Empetraceae, Ericaceae (*Rhododendron* nuclear), Goodeniaceae, Labiatae, Lentibulariaceae, Lobeliaceae, Loganiaceae, Martyniaceae, Oleaceae, Orobanchaceae, Pedaliaceae, Phrymaceae, Plantaginaceae, Scrophulariaceae, Selaginaceae, Solanaceae, Sphenocleaceae, Stylidiaceae, Verbenaceae.

Endosperm nuclear: Apocynaceae, Asclepiadaceae, Compositae (but some cellular), Convolvulaceae, Ebenaceae, Gentianaceae, Gesneriaceae (? also cellular), Myrsinaceae, Polemoniaceae, Primulaceae, Rubiaceae, Sapotaceae.

Convolvulaceae

Whether as Solanales or Polemoniales, this family is associated in classification with Solanaceae, but there are critical differences in seed-structure. The Solanaceous seed is exotestal in common with most Sympetalae, though there are some traces of mesotestal lignification. The Convolvulaceous seed differs from all other unitegmic seeds of Sympetalae by the special differentiation of the three outer cell-layers of the integument. The innermost of these layers forms a palisade of narrow prismatic cells with thickened and, sometimes, lignified walls. It is remarkably similar to the hypodermal palisade of the testa in Paeoniaceae; if the character is used to isolate the bitegmic Paeoniaceae, it must be so employed with the unitegmic Convolvulaceae. Then on searching for similar examples, one finds a distinct resemblance with the curious bitegmic seed of Cucurbitaceae; and this family has been in and out of Sympetalae in many classifications.

In both the Convolvulaceous seed and the Cucurbitaceous, it is the outer epidermis of the ovule which multiplies into the distinctive and persistent layers of the seed-coat and, of these, the innermost becomes the mechanical palisade. The peculiarity is now well established for Cucurbitaceae but it has escaped students of Convolvulaceae because, probably, there are differences of a kind that may be expected as variations in a phyletic series, thus:

(1) On fertilization of the Cucurbitaceous ovule, the outer epidermis of the outer integument (called the e-layer) divides periclinally to form an inner e'-layer, which becomes the innermost palisade-layer of the seed-coat and it is also called the sclerotic layer. Then a second periclinal division in the e-layer forms the e''-layer as the intervening hypodermal layer of the seed-coat; it is called the seed-hypodermis because the outer hypodermis of the outer integument is left internal to these three layers and takes no part in the mechanical construction of the testa. In the Convolvulaceous seed the sequence is different; the e-layer first forms the ovule-hypodermis (e''-layer), which then divides to form the e'-layer.

(2) Whereas these divisions are post-fertilization in Cucurbitaceae, they are pre-fertilization in Convolvulaceae, though the second division to separate the e''- and the e'-layers in Convolvulaceae may also be post-fertilization, as is any subsequent multiplication of the e'-layer.

(3) The seed-hypodermis (e''-layer) of Cucurbitaceae is often multiplicative into a tissue of small sclerotic cells, to which the e-layer may also contribute in some species, but there is no further multiplication of these layers in the Convolvulaceous seed.

(4) The inner palisade-layer (e'-layer) commonly multiplies in the Convolvulaceous seed but not in the Cucurbitaceous seed, or only to double itself.

(5) The Convolvulaceous seed is neither so complex in this exotestal construction as the Cucurbitaceous nor so heavily lignified, and the cells of the Convolvulaceous seed are generally smaller.

(6) The Cucurbitaceous ovule is bitegmic and crassinucellate; the Convolvulaceous is unitegmic and tenuinucellate. In both, however, the vascular bundle of the raphe-antiraphe extends from the funicle to the micropyle around the seed.

These differences can be set out in the following manner, together with the procedure in *Paeonia*:

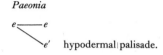

Curcurbitaceae

e———e———e (seed-epidermis).

e'' (seed-hypodermis), single layer or multiplicative.

e'———e' (seed-palisade or sclerotic layer), 1–2 cell-layers.

Convolvulaceae

e———e———e

e'———e''

e' palisade layer, generally multiplicative.

Paeonia

e———e

e' hypodermal palisade.

The Convolvulaceous seed appears as the Cucurbitaceous speeded up with neotenic epidermal divisions in the ovule and improved by riddance of the inner integument and nucellar mass. Both kinds of seed suggest derivation from a simpler construction in which one periclinal division of o.e. (o.i.) gives the truly hypodermal palisade, as in *Paeonia*.

The position of *Paeonia* in classification is uncertain. The structure of the seed throws doubt on that of Cucurbitaceae (p. 39); it disturbs the tranquillity of Convolvulaceae in Solanales. It may seem ridiculous that agreement in seed-structure should be employed to align such diverse plants in a phyletic series, as modern fragments, but it can stand enquiry. The modern theory of dicotyledons requires an apocarpous ancestry with minute embryo (*Paeonia*), then a syncarpous, bitegmic and crassinucellate state (Cucurbitaceae), and finally a syncarpous, unitegmic and tenuinucellate derivative (Convolvulaceae), even to a parasite (*Cuscuta*). To these states the durian-theory adds a pachycaul ancestor with pinnate leaf and arillate seed (*Paeonia*) with simple-leafed exarillate climbers as derivatives. The suggestion is no more absurd than that which derives *Ranunculus* from the stock of *Magnolia*, *Actinidia* from that of Theaceae, or a climbing *Piper* from that of Annonaceae. The question is why such a phyletic series did not evolve rain-forest trees, cf. *Dendrosicyos* and *Ipomaea*.

Arillate and sarcotestal seeds

The first conclusion of the durian-theory was that the exarillate seeds of angiosperms had been derived from the arillate (Corner 1949a, 1953, 1954). It is disliked chiefly, it seems, because critics are unfamiliar with Myristicaceae. This arillate family of Magnolialean trees is so universal in the lowland tropical forest that it sites the ecological origin of angiosperms; the forest could be called more exactly nutmeg-forest. No critic has met the evidence of the theory, far less refuted it, and it has begun to dawn on others (Camp and Hubbard 1963a). The evidence lies in the sytematic occurrence of the aril, especially in large genera with both arillate and exarillate species, in the occurrence of vestigial arils, particularly in indehiscent fruits, in the association of the aril with the armoured follicle or capsule, in the large size of fruit and arillate seed, and the connection of these with pachycauly and all its foliar and anatomical effects. More evidence along these lines has come to light; it is mentioned under the special accounts of families where the systematic consideration is appropriate, e.g. Clusiaceae, Connaraceae, Elaeocarpaceae, Grossulariaceae, Meliaceae, Rutaceae and Sapindaceae. The common assumption that the aril is a fortuitous ecological device, initiated *de novo* as a special creation without genetic precedent, is purely supposition and begs the question which the durian-theory sought to answer; that is, what genetic steps led to the inception of the aril. If its vestiges in *Bixa*, *Chelidonium*, *Desmodium* or *Turnera* are too slight to provoke evolutionary thought, the massive vascular arils of *Alectryon*, *Durio*, *Gonystylus* and *Myristica* invite contemplation.

An alternative theory has been proposed by van der Pijl (1952, 1955, 1957, 1958). It takes the sarcotesta as the primitive covering of the angiosperm seed, borrows the ecological consequences of the durian-theory, and assumes that from the successful sarcotesta the aril was extruded as a third integument. Naturally I had considered this possibility before arriving at the durian-theory, but I discarded it because the many difficulties which it raised proved the idea unworkable.

The sarcotestal theory is based on four assumptions. Firstly, angiosperms evolved from known gymnosperms with sarcotestal seeds. Secondly, animals became accustomed to eat the sarcotesta of pteridosperms, for it was the 'seed-food' for them

in the trees, such as they were. Thirdly, *Magnolia*, as the primitive dicotyledon, has maintained the sarcotestal predilection. Fourthly, it was so alluring that it became selected into the aril. Then it has to be admitted that trees got the better of animals and dispensed with sarcotesta and aril, in confirmation of the durian-theory. As the last point suggests there are many fallacies. With fuller knowledge of the dicotyledonous seed, they can now be taken up.

It is not known when, where or how angiosperms evolved and displaced gymnospermous forest. The pachycaul aspect of the durian-theory was an attempt, based on the evidence of living angiosperms, to supply the answer to how the gymnosperm was supplanted by the dicotyledon. The sarcotestal theory ignores this essential point. It selects one feature only, and that by partiality. If *Liriodendron*, instead of *Magnolia*, had been considered or *Casuarina* or *Trochodendron*, the samara or alate seed would have been evidence for derivation of angiosperms from anemochorous gymnosperms. Unconsidered by the sarcotestal theory, this problem was met in the durian-theory where it was shown that the systematic occurrence of the alate seed or samara in angiosperm families pointed to their derivation from arillate seeds or arillate follicles. If *Liriodendron* had an arillate ancestor, what of *Magnolia*? Then, if *Gnetum* had been considered, this pseudodicotyledonous tree would have supplied a bitegmic seed with an aril (or arilloid) but no sarcotesta; it is omitted by van der Pijl.

It is so difficult to understand how the unitegmic gymnosperm seed could have become the bitegmic dicotyledonous that the assumption that angiosperms arose from unitegmic gymnosperms appears false; indeed, bitegmic angiosperms have produced unitegmic genera and families as part of their advance, not of their heritage. The issue is avoided by van der Pijl who takes the sarcotesta of *Magnolia*, in which it is the outer part of the testa, as homologous with that of *Cycas* or of a pteridosperm without morphological explanation. Which of the two angiosperm integuments does that of *Cycas* represent? If the testa, then the invention was the tegmen but it is more or less vestigial in all Magnoliales except the arillate and non-sarcotestal Myristicaceae; *Magnolia* does not explain a tegmic innovation. If the tegmen, however, were the homologue of the gymnosperm integument, the innovation was the testa. Its successful persistence in angiosperms corroborates the point but there are consequences which belie the idea. Thus, (1) no sarcotestal tegmen occurs in angiosperms. (2) There is no reason why a testa should have been evolved, cf. unitegmic dicotyledons. (3) The sarcotestal layer of the gymnosperm must have been transferred to the new testa of the angiosperm among which, as an unlignified testa, it occurs only in such advanced tegmic families as Bombacaceae, Connaraceae, Euphorbiaceae, Meliaceae, Nandinaceae, Passifloraceae, Sapindaceae and Sterculiaceae, not in Magnoliales. (4) Eventually, to give the testal structure of Annonaceae, Magnoliaceae, Monimiaceae, and Myristicaceae, the whole integumental structure of the sarcotestal gymnosperm must have been transferred to the testal innovation, only to be returned to the tegmen in the advanced tegmic families, unless they represent a separate line of evolution apart from Magnoliales, and they are families which supply evidence of arillate ancestry. (5) There is no explanation of the exotestal seeds of Illiciaceae, Ranunculaceae and Winteraceae, except to suppose that they have lost the sarcotesta and left the zoochorous syndrome, or transferred it to an aril which had been lost.

Several of these points imply transference of function. Van der Pijl is at pains to discredit this operation; yet it seems a necessary adjunct to the sarcotestal theory, either to make the new testa sarcotestal or to make the new tegmen mechanical. But aril and sarcotesta are merely part of the problems of the angiosperm seed. There are six main problems, which are the origin of testa, tegmen and aril and the modifications of all three. To consider only the aril or the sarcotesta as the prime and over-ruling ecological adaptation to zoochory is to avoid the main problems of the seed-coats. To find that fish and alligators eat the outer parts of seeds and to regard this as the primitive past of the sarcotestal seed with low-down vertebrates is irrelevant for both eat men who eat them all. Zoochory is a fallacious guide to botanical construction, especially when there is no record of the primitive angiosperm vegetation and of the animals that lived in it.

On the morphological side there may be evidence that the two integuments differ in the manner of their inception by cell-division. In Rosaceae, according to the extensive researches of Péchoutre (1902), the outer integument starts with hypodermal cell-divisions which bulge the epidermis into the initial annulus and are then accompanied by anti-clinal divisions of the epidermis; the inner integument starts with epidermal divisions to be followed by hypodermal. Then, as is well established in many families, periclinal divisions of both epidermal and hypodermal layers make the regular cell-layers of testa and tegmen. These observations have been confirmed for Magnoliaceae (de Boer and Bouman 1972). Both Rosaceae and Magnoliaceae have the outer integument and testa more massive than the inner integument and tegmen. Whether the difference in inception holds in families in which the inner integument is also massive, and the tegmen becomes thicker than the testa, has not been ascertained; nor is it clear what is the start of the integument in families in which both are merely two cells thick and non-multiplicative. In other words, it is not clear whether the difference is a morphological peculiarity of each integument or merely the determinant of subsequent behaviour of either testa or tegmen.

The conviction that all the properties of Magnoliales and of *Magnolia* in particular are primitive is erroneous. As leptocaul trees with simple leaves, often pre-eminent in the forest, they are not primitive. If *Magnolia* retains the primitive elongate axis of the flower this cannot be said of other Magnoliales, unless Ranunculaceae are included. The fruits are variously reduced and specialized. If lack of vessels renders Winteraceae and Chloranthaceae primitive, it excludes *Magnolia* from this status. In fact, as one would expect, all the families of Magnoliales and Ranales are variously advanced with different manner of specialization. That their seeds differ so much is also proof of this ancestral specialization.

While he contends that the aril is polyphyletic, van der Pijl seems not to realize that this is true also of the sarcotesta (p. 24). Sapindaceae, upheld by van der Pijl as a primitively sarcotestal family that has evolved the aril (only to lose it), develops the sarcotesta over the pachychalazal region of the seed while the true and free part of the testa has the firm black exotestal character of the family and, especially, of its arillate seeds (p. 238, cf. *Alectryon* and *Harpullia* compared with *Guioa*, Fig. 501). This is the case in Meliaceae and some genera, at least, of Connaraceae. In these families the true testa is not sarcotestal, and the sarcotesta develops on the expanded chalaza as a modification of the bitegmic seed not homologous with the state in *Magnolia*. An analogy would make the phyllode of

Acacia homologous with that of a grass. A table, entitled the 'shift of the attractive layer in zoochorous diaspores' (v.d. Pijl 1955), which seems a paraphrase of transference of function, contains numerous errors. Thus, *Aporosa* has a dehiscent fruit; Annonaceae may have a micropylar aril, vestigial in the indehiscent *Annona* and *Canangium* (omitted from the table) and similar to the caruncle of Euphorbiaceae which are placed in a different class; the seed of *Nephelium* is not fleshy only in the basal part; the true aril is not chalazal; *Degeneria* is by no means unique in having a fruit dehiscent only after being shed, e.g. *Prunus*, *Millettia*, *Juglans*, *Nephelium*, and many others. These are not the only facts adduced in favour of the sarcotestal theory which are erroneous.

It is easy to find in the diversity of angiosperms superficial examples to support a hypothesis. The fact which militates most strongly against the sarcotestal theory is, as I have noticed previously (Corner 1953), the systematic occurrence of the feature. In exotestal and mesotestal seeds the outer epidermis is a mechanical layer, often as a rigid palisade, which is a systematic character of the family or order. But in these families there are sporadic genera or species which have a sarcotesta of some kind; it owes its existence to the absence of the mechanical layer, e.g. *Inga*, *Pithecellobium pr. p.*, *Euchresta*, *Jollydora*, *Manotes*, *Guarea* and *Nephelium*. This must mean on the sarcotestal theory that these genera possessed primitively a sarcotesta and had, then, evolved intragenerically the family character of the hard exotesta. But the sarcotestal examples in these families, as shown in the subsequent descriptions e.g. Leguminosae, Connaraceae, Meliaceae, and Sapindaceae, are not closely allied. Therefore several genera of such families must, according to the theory, be in the process of evolving independently the family or ordinal character of the exotesta. An analogy would be the evolution of follicles or capsules as family characters by genera with indehiscent, yet several-seeded, fruits. Primes of classification are not incidental but fundamental.

Some genera have both sarcotesta and aril. The thinly pulpy seeds of Passifloraceae are covered by a thin aril; the seed is exotegmic with unspecialized testa and not homologous with the construction of the seed-coat of Magnoliaceae. In *Casearia* (Flacourtiaceae) with well-developed aril, the testa may be thickly sarcotestal. In several cases such as *Bixa*, *Carica*, *Sterculia*, various Euphorbiaceae with caruncle, and some Annonaceae (if they can be considered sarcotestal), the aril is becoming vestigial. *Ribes* with sarcotesta and hard endotesta, reminiscent of *Magnolia*, but not multiplicative, has species in which the aril covers the seed, partly covers it, and is more or less completely absent; it is not impossible, therefore, the primitive Magnoliaceae may have been arillate. In Connaraceae, Meliaceae and Sapindaceae the diminishing aril gives over to the pachychalazal sarcotesta; the two are continuous and cannot be distinguished except in the course of development. Hence it has seemed to me, as previously explained (Corner 1953), that the pulpy character of the aril, which results from lack of firm thickening of the cell-walls, has spread from the attachment of the aril over the expanding chalaza. *Aphanamixis* (Meliaceae) provides a clear example with the family characters of testa and tegmen only in the dry micropylar part of the seed. In *Jollydora*, as I have since learnt, the seed has become so extensively sarcotestal that the Connaraceous exotesta occupies only a strip along the preraphe and round the micropyle. *Nephelium* is even more advanced.

I conclude that the aril and firm exotesta, as a protection for the seed, were primitive features in flowering plants; that some families combined the aril with the endotestal or tegmic construction and that, on decadence of the aril, the testa became partly or wholly sarcotestal in different ways according to the seed-construction of the family; that the sarcotesta, therefore, is polyphyletic in angiosperms and not homologous with the sarcotesta of gymnosperms. It is not clear that any truly sarcotestal dicotyledonous seed has progressed beyond this state which appears as an end-product of the particular seed-evolution. Aril has been lost; testa and tegmen have combined in the unitegmic seed of Sympetalae; but the sarcotesta is a peculiarity.

The question remains whether Magnoliaceae were primitively arillate or not. In favour of its arillate past there is the analogy of *Ribes* and there is the past history of *Nandina* to be discovered. Its seed-coat may be a derivative of the Papaveraceous (p. 28) which is intermediate between the Myristicaceous and the Magnoliaceous. The Papaveraceous seed appears to have been primitively arillate, and certainly the Myristicaceous. Then if *Nandina* was derived from the Papaveraceous stock, it too has lost the aril and become isolatedly sarcotestal.

Seed-size

The distinction between megaspermy and microspermy was discussed by me in connection with the durian-theory (Corner 1949a, 1954). Palms and orchids offer the extremes which range from 0.01 mm or less in orchids to 300 mm in length in *Lodoicea*. In dicotyledons the range may be 0.01–100 mm. Length is only a rough measure because the seeds may be flattened or narrow; weight would refer mainly to the endosperm or embryo. Then there is the structural difference. Large seeds have multiplicative integuments which form massive seed-coats. Small seeds merely enlarge the cells of the integuments with little or no subdivision. Such small seeds distinguish the advanced genera or families and the herbaceous. The large, massively constructed seed is the forest-factor which must have operated since the time when dicotyledonous trees began to take over the gymnosperm forests (Corner 1964). It is the seed of Magnolialean trees and of most tropical trees except the more highly advanced. Many diversified families from Theaceae and Mimosaceae to Myrtaceae and Apocynaceae show a range in seed-size with the smaller seeds suited to the more herbaceous or leptocaul members.

The question arises what was the size of the primitive dicotyledonous seed. The largest of seeds are generally overgrown and, without testal or tegmic specialization, are clearly exceptions in their families. The smallest are neotenic advances. Some intermediate size is indicated. Possibly the range was 5–10 mm, even 15 mm, for the length of the primitive anatropous seed with multiplicative and specialized seed-coat. It is the general range for the arillate seeds of trees and climbers that are produced in many-seeded follicles and capsules. Decrease in the number of seeds in the fruit probably led to an increase in size, as in Myristicaceae, Lauraceae, Bombacaceae and Euphorbiaceae, and this in turn may have led to the overgrown seeds in so many indehiscent fruits, as those of Anacardiaceae, Moraceae or Lecythidaceae; so *Calophyllum* may stand to *Clusia*. But many exceptions will be encountered and each family will need its own evaluation. *Sterculia* has consistently large seeds in its follicles. *Theobroma* has larger seeds in its berries. *Cassia* may have a hundred seeds of moderate size in its pods which do not open and a few small seeds in some which do. Yet it can be concluded that the moderately large seed with full complexity of seed-coat structure is the primitive form among modern families.

A Myristicalean start

No living seed fulfils all the requirements of a primitive nature deduced in the previous pages. Closest is the Myristicaceous. It lacks perisperm and a mechanical endotegmen. The exotesta is not remarkable but it is a tough layer with slightly thickened and lignified walls; in some cases it is pierced with stomata. The seed is large; the embryo, though small, is not microscopic; the endotesta is a strong palisade of lignified cells with crystals; the exotegmen is a rigid layer of woody longitudinal fibres, sclerotic cells or tracheidal cells; the endosperm is ruminate; the tegmen dries up in the ripe seed but it is thicker than the testa in the unripe; aril, testa and tegmen are vascular. The seed has the most elaborate construction among angiosperms, and it needs explanation in the front row of Magnoliales. It is not a hypothetical invention.

The seed may have enlarged as it became solitary in the fruit. Analogy with other 1-seeded follicles indicates that the Myristicaceous was primitively many-seeded, in which case the seeds may have been *c*. 10 mm long and have contained a microscopic embryo. This reconstruction seems obligatory. Then if the flower were also multicarpellary, the pre-Myristicalean fruit would have resembled a whorl of *Sterculia*-follicles or a cone of *Michelia*-follicles. The seed, be it noted, has never become overgrown; presumably the genetic code of its differentiation was too rigid for dissolution, cf. the overgrown seeds of the one-seeded fruits of such advanced families as Anacardiaceae with the mango-seed, or the less advanced Lauraceae with the avocado-seed.

There is no evidence that the Myristicaceous seed ever lacked the aril or ever was sarcotestal. The ancestors of pigeons, which are the chief nutmeg-eaters nowadays, may have selected the larger seeds in the primitive follicles which had the tendency to develop fewer seeds. The aril is both exostomal and funicular in outgrowth and disposes, at once, of the academic insistence on the distinction between true and false arils, but this double origin foreshadows the specialization into exostomal and funicular arils in other families.

Concerning the internal features of the seed, loss of endotestal and exotegmic differentiation would have produced the exotestal seed. The endotesta would supply the layer of crystal-cells emphasized by Netolitzky, and the traces of wall-thickening and lignification found in the endotesta of many families. Multiplication of the endotesta could have produced the multiple, or storied, endotesta of *Magnolia* and *Vitis*; it may be foreshadowed in the palisade-like, but thin-walled and unlignified, inner hypodermis of the Myristicaceous testa which, indeed, may have had primitively a two-layered endotesta. Loss of endotesta would give the fibrous or tracheidal exotegmic seed, which is also arillate and exotestal in many families, e.g. Celastraceae, Connaraceae, Flacourtiaceae, Meliaceae and Sapindaceae. The extension or transfer of the endotestal palisade to the exotegmen would give the exotegmic palisade of the great Malvalean alliance. It is already indicated in the Myristicaceous chalaza where this palisade curls round and passes abruptly into the exotegmic fibres. Many Euphorbiaceous seeds have the endotesta in the form of a thin-walled palisade pressed against the lignified palisade of the exotegmen so as to suggest a relic of the Myristicaceous endotesta, exactly as the inner hypodermis of the Myristicaceous testa, as if the two cell-layers had taken one step inwards. Then most exotegmic families give evidence of derivation from a seed with vascular aril. Limitation of the chalaza, as testa and tegmen simplify, would provide the simple construction of most seeds. The Bixaceous chalaza is as complicated as the Myristicaceous and it is simplified in the small seeds of Cistaceae. The endosperm of Myristicaceae has both starch and oil; it may afford the physiological explanation of the dichotomy into starchy and oily seeds in advanced families. Finally, to revert to the aril, that of the nutmeg *Virola* is constructed in the same way as the sarcotesta of *Magnolia*; if transferred to the testa in pre-Myristicaceae with many-seeded follicles it would supply the Magnoliaceous origin except, of course, for *Liriodendron* which may have lost the aril entirely.

The fibrous testa of Annonaceae is difficult to derive. It seems hardly possible that fibres could have been transferred from the Myristacaceous tegmen. Possibly fibres were part of the primitive mesotesta and have been lost in the evolution of Myristicaceae. Stout fibres are abundant in the vascular integuments of *Gnetum* and have been found in various fossil seeds (*Conostoma*, *Lagenostoma*, *Stephanospermum*, *Caytonia*, *Gristhorpia*). The fibres do not appear to be modifications of palisade-cells (p. 14).

Concerning the endosperm, it is generally assumed that rumination is an advanced feature, even connected with pachychalazal construction (Periasamy 1962b). Its presence in Annonaceae, Degeneriaceae and Myristicaceae suggests that it may be a primitive complexity, and their seeds are not pachychalazal. The evidence from living Myristicaceae is equivocal. Thus *Mauloutchia* (Madagascar) has the most primitive male flower in the family with 30–40 almost free stamens. It has been reduced to the Madagascan *Brochoneura* (Sinclair 1958), which is monoecious and has non-ruminate endosperm; but it is not clear if mature seeds of *Mauloutchia* have been found. Then *Compsoneura* and *Iryanthera* with the most advanced androecium and pollen also lack rumination. In Myristicaceae the ruminations are tegmic from the massive tegmen, but they are both testal and tegmic in Annonaceae, Canellaceae, Degeneriaceae and Eupomatiaceae which have a slight tegmen, while in Menispermaceae they are testal or even endocarpic. Thus they appear as facultative incursions into massive endosperm but, since the origin and primitive form of the endosperm is as unknown as the need for a tegmen, the origin of ruminations must remain doubtful.

The anatomist is impressed by the complexity underlying the outward simplicity of Myristicaceae (Wilson and Maculans 1967). The tissue-construction is massive. Thus, nine perianth vascular bundles are incorporated in the trimerous perianth, perhaps as a vestige of the primitive formula P $3+3+3$. The single carpel has many vascular bundles and, as with the Magnolialean stamen, it has been argued that the primitive carpel was multivascular (Sastri 1959b); the multivascular state is, of course, primitive in the follicle and will be neotenic in the ovule. The massive vascular supply distinguishes the aril, testa and tegmen of Myristicaceae. If refinement to few or one vascular bundle is progress in the simplification of sepals, petals, stamens and carpels, the same principle renders the massive Myristicaceous seed primitive and the small seed with only the raphe-bundle advanced.

There are many indications from their distribution, habitat, leptocaul habit, aromatic tissues, simple flowers, dioecism, and fruit that the wild nutmeg trees are survivors of a primitive order. They inhabit the lowland tropical forest, with a few submontane species, and they abound in the freshwater swampy forest which appears to have been the original home of angiosperms (Corner 1964). They are monopodial, shade-tolerant trees, springing up in the depths of the forest, the canopy of which a few may reach. With aerial loop-roots, as pneumatophores, and stilt-roots they endure waterlogging of the soil and inundation. The leptocaul habit with entire, exstipulate, shortly petiolate, elliptic leaves, distichous on the side-shoots, tells of a long and extinct evolution from pachycaul ancestry with compound ascending foliage. The flowers, treated as Magnolialean because of the single carpel, are reduced in size, apetalous, with sessile and connate anthers, and simplified gynoecium. The small trimerous perianth suggests the Annonaceous flower deprived of petals. The monocolpate pollen is Annonaceous (Canright 1963); so is the ruminate endosperm with small embryo, and the oil-cells. The wood is said to resemble the Magnoliaceous, though peculiar in the red or pink tanniniferous sap from tubules in the xylem-rays, whence flows the blood of the Myristicaceous blaze. The family has been summed up as transitional between 'the more primitive Degeneriaceae, on the one hand, and the more highly specialized Himantandraceae and Annonaceae on the other' (v.d. Wyk and Canright 1956). However, one needs to think not of the survivors but of their pachycaul ancestry in order to trace resemblances. Myristicaceous trees have much the same habit as the Annonaceous but there is a detail, in addition to the red sap, which suggests that they have run in parallel the common course of tree-evolution in the forest, where both have been left in the shade. The branching of the main stem is normally axillary in Myristicaceae. In Annonaceous trees and some, if not all, climbers the main branch stands at right angles to the axillary leaf (Corner 1964, p. 114, fig. 48). About 16 genera of Myristicaceae have been distinguished (Hutchinson 1964; Sinclair 1958). In habit they are so alike that they can be regarded as one genus, less varied than *Ficus*, but diversified in the male flower; and this one genus must have been evolved and distributed throughout the tropics when Pangaea was breaking up.

These thoughts reflect Myristicaceae not as modified Magnoliaceae, Degeneriaceae, Annonaceae, Himantandraceae, Lauraceae, Monimiaceae or, even, Winteraceae, but as survivors of an order Myristicales which maintained in fruit and seed the ancestral characters of dicotyledons. That, as I have remarked elsewhere, was the obstacle to their further progress into microspermous trees, climbers, shrubs and herbs. To the trunk of this order the branches of many others may eventually be traced. The one misgiving is that the story of Myristicales seems too good to be true. Yet, by a totally independent statistical correlation of characters other than those of the seed, Sporne (1956) arrived at the conclusion that 'the most primitive family appears to be Myristicaceae'. More recently he has modified this and placed Magnoliaceae in the prime position, but concluded that 'the primitive dicotyledon ovule is crassinucellate, with three envelopes (two integuments and an aril), some or all of which are provided with a vascular supply' (Sporne 1969). My only observation is to substitute 'seed' for 'ovule'.

Seed-progress

The hypothesis is that from a complicated beginning the seed-coats have become simplified, specialized and diversified in ways that help to define the phyletic series or orders through which dicotyledons have taken their modern station. In general, seed-progress has accompanied floral evolution. The most advanced families, as Compositae and Orchidaceae, have small simple seeds, but there are many instances to show that the seed has progressed without much floral advance, e.g. in Ranunculaceae or Rosaceae. There has been loss of integuments and of tissue-differentiation; small and derived seeds become spuriously similar. Though every family will be found to have its own way of seed-progress, the following survey is a general outline.

Specialization of the seed-coat: by limitation of the mechanical tissue to one cell-layer, exotestal, endotestal, exotegmic and endotegmic seeds have been produced.

Simplification of the seed-coat

(1) Loss of aril, testa and/or tegmen.

(2) Lack of tissue-differentiation, especially in indehiscent fruits.

(3) Loss of vascular ramifications, especially the post-chalazal, the tegmic and the arillar.

(4) Simplification of the chalaza.

(5) Decrease in size by loss of multiplicative cell-layers, and by neoteny with diminishing growth after fertilization.

(6) Change of the anatropous form through the campylotropous to the orthotropous.

Elaboration of the seed-coat

(1) Alate seeds follow the loss of the aril.

(2) Sarcotestal seeds result from transfer of aril factors to the more or less undifferentiated testa.

(3) Perichalazal and pachychalazal seeds result from intercalary growth of the chalaza rather than of the integuments.

(4) Overgrown seeds result from the highly multiplicative testa without differentiation except in vascular supply.

(5) Seeds have evolved with special multiplicative layers, as in Caricaceae, Convolvulaceae and Cucurbitaceae.

(6) Vascular elaboration has occurred in seeds of large size.

(7) The exotesta has been elaborated with hairs and mucilage-production; the enlarged exotestal cells of neotenic seeds make the sculpture of ridges, warts or reticulations on the dried seed-coat.

These progressions are variously accompanied by the change from the crassinucellate to the tenuinucellate state, as superfluous tissue is eliminated, and by the change from the albuminous to the exalbuminous seed with relatively large embryo. Both of these progressions have left many intermediate states: they will call for investigation when seed-physiology has advanced. Primitively the crassinucellate state fills the fully grown seed with a massive nucellus, or perisperm, which is then slowly displaced by the endosperm. In other seeds the endosperm is precocious and the nucellus neither reaches such dimensions nor functions as a storage-tissue. In the tenuinucellate ovule the embryo-sac has neotenically substituted the nucellus. In the case of the exalbuminous seed it is the embryo which gradually supplants the endosperm and becomes the storage-tissue. In both cases epidermal enzymatic layers may persist.

One feature that is outstanding and emphasizes the peculiarity of Annonaceae is the appearance of a third or middle integument in some genera. It seems an innovation; if not it is a relic of some construction similar to the Gnetalean.

Transference of function

Theoretically this principle could be employed to any desired extent in the endeavour to explain the evolution of the dicotyledonous seed. By supposing that a particular manner of cell-differentiation might have been transferred from one cell-layer to another, the four main kinds of seed from exotestal to endotegmic could be derived. There can be no doubt that transference has occurred in some measure, and the chalaza is a suitable site because it is the small active centre from which the integu-

ments arise and their cell-differentiation proceeds. Used unwarily the principle merely destroys morphology. Used cautiously, with appreciation of the conservatism of tissue-differentiation so widely established by morphology, it helps to understand the manner in which unusual constructions may have occurred. Morphology established that the functions of the seed-coats have not been transferred to or from the nucellus; it is a separate organ from the seed-coats, cut off by the hypostase. Morphology establishes that the majority of seeds in a family conforms with a certain construction which is also that of allied families. Hence the transference of function is uncommon in the case of seeds and needs to be invoked only for exceptional cases, e.g. the flower of *Saraca* (Corner 1958). The following appear to be examples among seeds.

(1) The aril may become sarcotestal, or be developed along the raphe (*Gonystylus*) or antiraphe (*Leptonychia*), even placental (Cucurbitaceae, Apocynaceae, Rubiaceae); or the colour-effect of the aril may have been transferred to the testa, as in various Leguminous seeds. Such instances have been tabulated, not altogether successfully (p. 53), as the 'shift of the attractive layer' in seeds (v.d. Pijl 1955).

(2) The Myristicaceous endotesta gives evidence that it may invade the exotegmen (p. 56).

(3) The fibrous exotegmen may have become the fibrous mesotesta in Annonaceae (p. 56).

(4) The middle integument of Annonaceae seems to be a new intercalation to which the testal characters have been transferred (Corner 1949b).

(5) *Siparuna*, discussed under Monimiaceae (p. 195), seems to be a remarkable case of the transfer of the aril to the pericarp, similar to what may have happened in *Artocarpus* (Corner 1962).

(6) The seed-coat of Cucurbitaceae is complex and difficult to assign to a standard pattern other than its own. On p. 39, I have explored the ways in which it might be brought into line through transference of function; they are devious and hypothetical but necessary to comprehend possibilities. In conclusion I do not favour them.

The origin of the angiosperm seed

The classical idea of the seed as an integumented megasporangium is the source of modern theories. They become involved with the possibilities of the cupule which invests the seed or cluster of seeds in various pteridosperms. As a truss of branchlets or pinnae, for want of precise description, planated and involucrated by intercalary growth, this laciniate, lobate or dentate cupule can be turned diagrammatically on paper into a second or outer integument. A third can then be added to explain the aril or the seed of *Gnetum*. Misconceptions of the intercalary growth in the chalazal regions of cycads has led to the interpretation of their seeds as bitegmic (Stopes 1905). It has been taken up to re-interpret the lyginopterid pteridosperm seed as the bitegmic forebear of the angiosperm (Camp and Hubbard 1963b). The issue is disputed by Long (1966) who cannot find the cleft which should separate two fossilized integuments, and he considers that the second integument arose as a dorsal outgrowth of the chalaza or first integument as the pteridosperm seed became anatropous or campylotropous. It is generally agreed, however, that the inner integument came first, then the outer (Smith 1964), and lastly, as some would have, the aril. But of all these integuments in the angiosperm the most cupular and primitively laciniate, as a truss of vascular branchlets, is the aril of *Myristica* which arises from the funicle and micropyle. The fossils which palaeobotanists have been able to study are not angiosperms. They do not meet the requirements of angiosperms between which and gymnosperms a chasm still yawns. The primitive angiosperms require many seeds, every one with three integuments, inside a follicle, as many or more bitegmic ovules inside a carpel, and numerous carpels on a conical stem-apex surrounded by floral parts as the neotenic fruit pre-disposed to a flower for pollination. The ancestor of angiosperms requires a reproductive bud that was at once flower and fruit (Corner 1964). The whole construction appears totally different from the gymnospermous succession.

Analogies may also be drawn in microscopic structure. The single integument of the pteridosperm and gymnosperm, if they are to be distinguished, contains pulpy cells, secretory cells or canals, lignified palisade-cells, sclerotic cells, lignified fibres, and vascular bundles (Schnarf 1937; Stidd and Hall 1970). In *Stephanospermum* the sclerotic seed-coat had an epidermal palisade-layer of prismatic cells about half as thick as the rest of the sclerotesta; it suggests the Leguminous testa. Most pteridosperms appear to have had an outer sarcotesta separated by a palisade-layer from the sclerotesta or sclerotic endotesta; the arrangement recalls the

testa of *Magnolia* or the testa and tegmen of Bombacaceae. The sarcotestal seeds of *Conostoma* and *Lagenostoma* and the seeds of Caytoniales (? with a thin epidermal sarcotesta, Harris 1951) had an inner layer of fibres, as in the Annonaceous testa or that of some Rutaceae, or as the testa and tegmen of Flacourtiaceae. There were in pteridosperms, as there are in gymnosperms, the microscopic ingredients of the angiospermous seed-coats but, confined to one integument, they do not fit. With living seeds it is hard enough, if not impossible, without a series of developmental stages to decide where testa ends and tegmen begins. How much more difficult it must be with the fossil!

The problem of the angiosperm does not cease, however, with the pteridosperm. Among living pteridophytes the unique sporocarp of Marsiliaceae bears a considerable microscopic resemblance to the seed of the angiosperm; as a sorus of megasporangia enclosed in a cupule of modified pinnae or branchlets, it fits the palaeobotanical requirement. Beneath an epidermis of radial cells there are lignified palisade-cells, stellate mesophyll-cells and vascular bundles. In general organization the sporocarp is more elaborate than the seed and comparable rather with the carpel; in microscopic construction it is simpler. By imagining outgrowths, ingrowths, complications and reductions, the one can be transformed into the other; there is less difficulty than in the attempt to explain how the sporophyte came from the gametophyte. What must be borne in mind is the ancestry of Marsileaceae. Fossils are known from the Upper Cretaceous or Lower Tertiary, but their form is modern (Chitaley and Paradkar 1972). As reduced aquatic herbs, an origin in swamp-forest is implied, where they may have coincided with the robust beginnings of the angiosperm.

Neoteny again

The problem of the angiosperm, as the plant with seed, fruit and flower, is to explain the two stages in its reproduction. There is the preparatory stage of the flower with ovary and ovules and the mature stage with fruit and seeds. When they are considered in the outward aspect of the flower and fruit, there is little perception of the intrinsic causes resident in the ovary which is in apical control of that bit of stem. Here the ovule is the embryonic compound of sporophytic and residual gametophytic tissues, and its development is arrested until fertilization induces the threefold complexity of sporophytic seed-coat, triploid endosperm, and new embryo-sporophyte. If there were no arrest, and fertilization did not occur until there was a full-sized seed, there would be the Cycadalean analogy of the pre-angiosperm. It would have necessitated pollination of a full-grown and dehisced fruit, because the character of the angiosperm is the enclosed seed. Thus it seems clear to me that, by gradually assuming pollination and fertilization at more and more immature states of development, the fruit became partly neotenic as the ovary, which has to function with stigma and exposed surfaces (until epigynous), and the seed becomes wholly neotenic as the ovule. The first step, according to the classical theory of the carpel, would have been the growth of the pollen on the residual, but yet living, carpel-tip of the fully developed carpel or megasporophyll. No angiosperm is known without the dual manner of reproduction. Presumably all such ancestry with its wasteful complexity has long since been extinct.

Yet there are degrees of neotenic progress still at work in the refinement of the ovule; perhaps, even, apomixis falls in this category as the state to which so many genera converge when the physiological necessity for pollination and fertilization have been eliminated. It is well established that in many orchids at the time of pollination the ovules are rudimentary, if indeed they have been initiated, and that the development of the ovule is as much post-pollination as that of the seed is post-fertilization. An example among dicotyledons is *Muntingia* (Elaeocarpaceae) the tiny seeds of which contrast with the arillate lumps of *Sloanea*. Buxaceae are another instance. The ovule of *Buxus* is well-formed at anthesis; its outer integument is 6–10 cells thick. In *Sarcococca* the ovule at this stage is rudimentary; its outer integument, at a thickness of two cells, expresses the limit of neoteny. On fertilization, the outer integument of *Buxus* merely differentiates the cell-layers without further multiplication, but that of *Sarcococca* increases to a thickness of 12–13 cells before differentiation. In *Sarcococca* the ovule is more neotenic and simplified than in *Buxus*; the seeds of both are comparable. Then, of course, there are very small ovules in many families and all degrees of development of massive

ovules 1–2 mm in length, such as are found in Clusiaceae and Dipterocarpaceae. The majority of small ovules has the integuments two cells thick, expressing the embryonic minimum as in *Sarcococca*. The small seeds which develop from them may have non-multiplicative integuments, expressive of the neotenic seed-coat, just as the minute seedling with simple cotyledons expresses the embryonic neoteny. Eventually, it seems, the entire heritage of ovular construction is eliminated except for the embryo-sac, and there is the naked seed of Loranthaceae and Santalaceae.

The course of neoteny will have changed as its direction has been polyphyletic. The evidence will be supplied by families with large and small seeds, such as Elaeocarpaceae, Leguminosae, Rutaceae, Theaceae, and such pairs of families as Bombacaceae–Malvaceae or Bignoniaceae–Scrophulariaceae. Every developmental process from ovular inception to seed-ripening is a succession of steps some of which may be shortened, deferred until after fertilization, or finally eliminated. Variations must be expected on all the processes outlined on p. 9, for the development of the complete seed, and they may not be in step. Tegmen and testa may diminish together or independently while nucellus, endosperm and embryo follow their predilections. In one family or tribe there will be one course and in another there will be different changes as the seed eliminates its ancestral complexity. Thus in so many ways the seed is the epitome of the flowering plant replete with vestigial characters. The larger seeds will supply the start and there has resulted the precise diversity of the little seeds so well instanced by *Begonia, Digitalis, Eurya, Gynotroches, Hypericum, Lawsonia, Melastoma, Muntingia, Papaver, Peperomia, Ploiarium, Thlaspi, Utricularia* and *Viola*.

TABLEAU (*opposite*)

Tableau of the relationships of dicotyledons based on the structure of the seed; the families or orders centred around Myristicaceae, as the relic of the complicated protodicotyledonous seed, and graded horizontally into (1) Magnoliales with endotestal, mesotestal and exotestal specialization, (2) advanced testal seeds, (3) advanced tegmic seeds, and (4) advanced unitegmic seeds derived from exotestal, mesotestal and endotegmic sources.

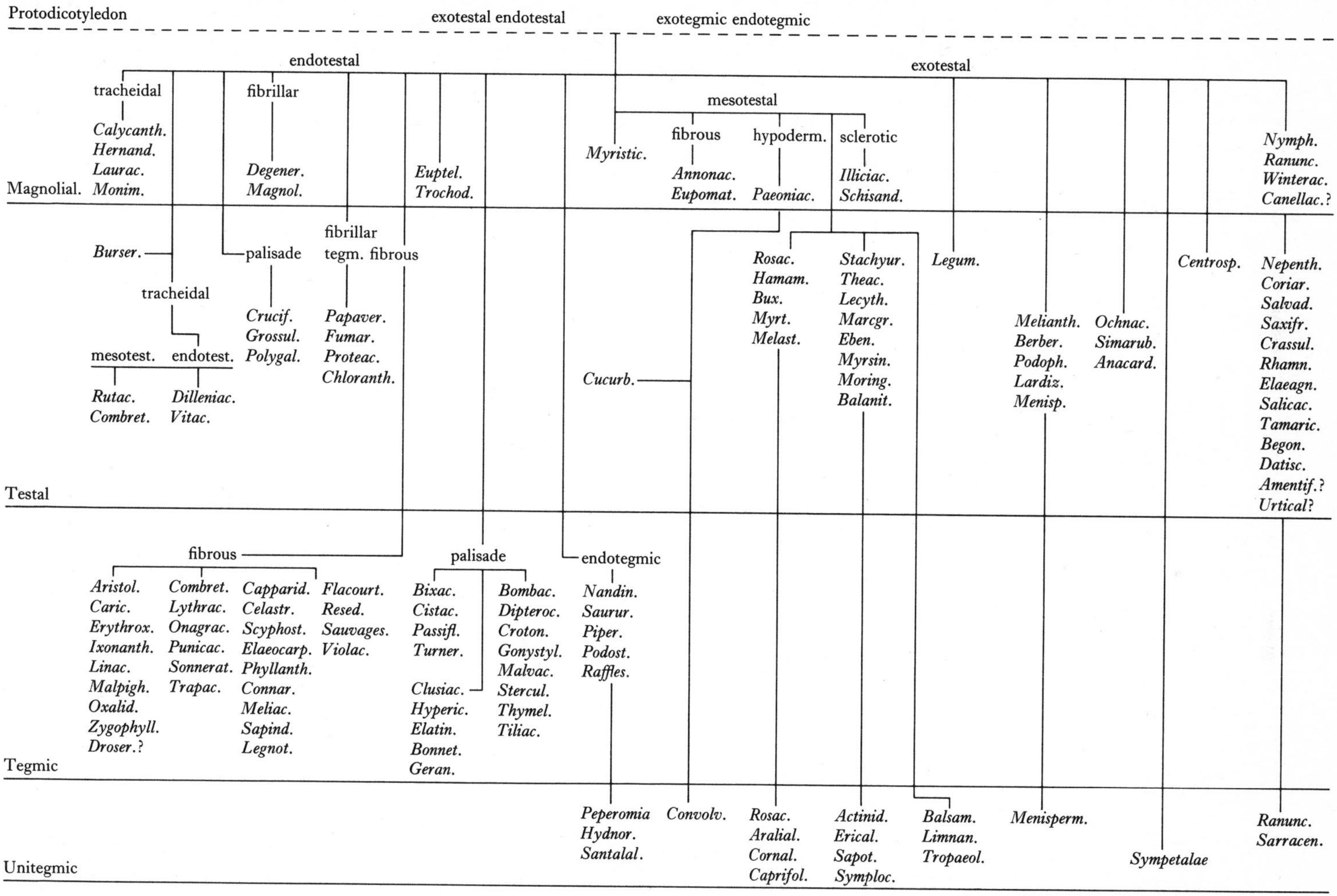

Protodicotyledon
exotestal endotestal
exotegmic endotegmic
endotestal
exotestal
tracheidal
Calycanth.
Hernand.
Laurac.
Monim.
fibrillar
Degener.
Magnol.
Euptel.
Trochod.
mesotestal
Myristic.
fibrous
Annonac.
Eupomat.
hypoderm.
Paeoniac.
sclerotic
Illiciac.
Schisand.
Nymph.
Ranunc.
Winterac.
Canellac.?
Magnolial.
Burser.
tracheidal
mesotest.
endotest.
Rutac.
Combret.
Dilleniac.
Vitac.
palisade
Crucif.
Grossul.
Polygal.
fibrillar
tegm. fibrous
Papaver.
Fumar.
Proteac.
Chloranth.
Cucurb.
Rosac.
Hamam.
Bux.
Myrt.
Melast.
Stachyur.
Theac.
Lecyth.
Marcgr.
Eben.
Myrsin.
Moring.
Balanit.
Legum.
Melianth.
Berber.
Podoph.
Lardiz.
Menisp.
Ochnac.
Simarub.
Anacard.
Centrosp.
Nepenth.
Coriar.
Salvad.
Saxifr.
Crassul.
Rhamn.
Elaeagn.
Salicac.
Tamaric.
Begon.
Datisc.
Amentif.?
Urtical?
Testal
fibrous
Aristol.
Caric.
Erythrox.
Ixonanth.
Linac.
Malpigh.
Oxalid.
Zygophyll.
Droser.?
Combret.
Lythrac.
Onagrac.
Punicac.
Sonnerat.
Trapac.
Capparid.
Celastr.
Scyphost.
Elaeocarp.
Phyllanth.
Connar.
Meliac.
Sapind.
Legnot.
Flacourt.
Resed.
Sauvages.
Violac.
palisade
Bixac.
Cistac.
Passifl.
Turner.
Clusiac.
Hyperic.
Elatin.
Bonnet.
Geran.
Bombac.
Dipteroc.
Croton.
Gonystyl.
Malvac.
Stercul.
Thymel.
Tiliac.
endotegmic
Nandin.
Saurur.
Piper.
Podost.
Raffles.
Tegmic
Peperomia
Hydnor.
Santalal.
Convolv.
Rosac.
Aralial.
Cornal.
Caprifol.
Actinid.
Erical.
Sapot.
Symploc.
Balsam.
Limnan.
Tropaeol.
Menisperm.
Sympetalae
Ranunc.
Sarracen.
Unitegmic

PART TWO

Descriptions of seeds by families

ACANTHACEAE

Ovules anatropous, campylotropous or amphitropous, unitegmic, tenuinucellate; integument thick.

Seeds small to medium-size, albuminous, rarely exalbuminous (*Blepharis*), often with the funicle developed into the jaculator (? with rudimentary aril in *Barleria* and *Dipteracanthus*), exotestal. Testa mostly crushed except for o.e., but with 2–3 inner layers of thick-walled and pitted cells in *Acanthus*; *o.e.* more or less palisade-like with thick walls or with fibrous, spiral or reticulate thickenings of the walls (? lignified), the cells often with short or long, rigid or mucilaginous hairs, in flattened seeds with a fringe of hairs round the edge, with raphids at the base of the hairs in *Barleria*; *mesophyll* with outer sclerotic layers in some cases (*Dipteracanthus*). Hypostase woody (? in all cases). Endosperm cellular, oily, ruminate in *Adhatoda*, *Andrographis* and *Elytraria*.

(Netolitzky 1926; Mullan 1936; Maheshwari and Negi 1955; Johri and Singh 1959; Phatak and Ambegaoker 1961; Mohan Ram 1960, 1962; Mohan Ram and Masand 1963; Singh, B. 1964; Jaitly 1969a; Sell 1969; Bhatnagar and Puri 1970.)

A problem in this family is the nature of the stiffened funicle or jaculator which ejects the seed. In the account given by Maheshwari and Negi for *Dipteracanthus* the jaculator resembles a modified funicular aril.

ACERACEAE

Ovules 2 per loculus, collateral or superposed, anatropous or orthotropous, erect or transverse, bitegmic, crassinucellate; o.i. and i.i. 3–5 cells thick.

Seeds becoming campylotropous with partial radicular septum, exalbuminous, exarillate. Testa thin, not multiplicative or –7 cells thick (*Acer campestre*); *o.e.* composed of shortly columnar cells with slightly thickened outer wall, strongly columnar in *A. macrophyllum*, *A. pennsylvanicum* and *A. platanoides*; *mesophyll* thin-walled; *i.e.* usually as a layer of small crystal-cells, but in some species without crystals (*A. pennsylvanicum*). Tegmen not multiplicative, eventually crushed. Embryo curved, with flat, thick, curved, or thin and plicate cotyledons. Figs. 3–5.

(Guérin 1901; Netolitzky 1926.)

As observed by Netolitzky the pericarp has taken over the protective function of the seed-coat, which has little more than the enlarged structure of the ovule but the campylotropous shape and incipient exotesta as in Sapindaceae. The family shows how the simplified seed-coat has been derived by loss of differentiation from the complicated.

Dipteronia sinensis Oliv.
(Cambridge University Botanic Garden; living material.)
Figs. 4, 5.

Ovule; o.i. and i.i. 3–5 cells thick.

Seed soon campylotropous through growth of the antiraphe. Testa unlignified; *o.e.* with very large, thin-walled cells, scarcely cuticulate; *mesophyll* with scattered crystal-cells, more abundant in i.e. Tegmen; *o.e.* and *i.e.* with more or less transversely elongate cells.

Pericarp with white latex-ducts.

ACTINIDIACEAE

Ovules numerous, anatropous, transverse, unitegmic, tenuinucellate; integument 2–9 cells thick, o.e. with enlarged cells, i.e. with the cells slightly enlarged and more or less endothelial; hypostase small.

Seeds small, 0.7–2.5 mm long, brown, with large-celled crustaceous seed-coat, arillate or not, albuminous. Seed-coat more or less multiplicative; *o.e.* composed of large cuboid cells with more or less hexagonal facets, the inner wall strongly thickened, lignified and perforate; *mesophyll* thin-walled, not lignified, often with raphid cells, more or less crushed. Vascular bundle in the raphe only. Chalaza simple. Endosperm cellular, thin-walled, oily proteinaceous, without starch. Embryo small to relatively large, straight; cotyledons short. Figs. 6–12.

(Svedelius 1911; Schnarf 1924; Netolitzky 1926; Crété 1944; Johri 1963; Vijayaraghavan 1965.)

The seeds of *Actinidia* and *Saurauia* have the same construction. Both, according to Svedelius, may have funicular arils. To separate them in two families seems as exaggerated as to separate *Pyrus* and *Rubus*, *Ceratonia* and *Cassia*, or *Ficus* and *Dorstenia*.

The family has hovered between Dilleniales and Theales, with both of which it agrees in the centrifugal stamens. Evidence favours Theales (Johri 1963;

Vijayaraghavan 1965). The exotestal unitegmic seed does not agree with Dilleniaceae, but it is similar to that of Adinandreae in Theaceae (p. 266). Nevertheless, the Actinidiaceous seed is more primitive in being arillate and anatropous. It seems that the family is a relic of the stock of this branch of Theaceous ancestry and, in the reduced unitegmic state, is comparable with the unitegmic Rosaceae in which the one integument has as many cell-layers as those of the two integuments added together. Thus the small seeds of Actinidiaceae, similar to those of various sympetalous families as Loganiaceae, are extremely neotenic. The seed-structure is not known in *Clematoclethra* or *Sladenia* which have been referred to Theaceae (Metcalfe and Chalk 1950).

Actinidia Lindl.

Ovules many, in a single row in each loculus; integument 6–9 cells thick; mesophyll with some raphid-cells near the chalaza (*A. polygama*).

Seeds 1.5–2.5 × 1–1.5 mm, oblong, dark brown, in some species enclosed in a long white pulpy aril with scattered raphid-cells. Testa evidently multiplicative, the inner cells becoming crushed, exotestal; *o.e.* composed of large cuboid cells with subhexagonal facets, the inner wall strongly thickened, lignified and finely perforate, the radial walls thin, the smooth outer wall thickened but not lignified, forming the hard crustaceous seed-coat; *mesophyll* not lignified. Vascular bundle only in the raphe.

A. chinensis Planch.

(Cambridge University Botanic Garden; material in alcohol.) Figs. 6–9.

Ovule; integument 7–9 cells thick, i.e. with slightly lignified cells; funicle well-developed.

Seeds 2–2.5 × 1.3–1.5 × 1 mm, enclosed in a thin pulpy jacket with scattered raphid-cells in the tissue. Seed-coat; *o.e.* with the cells 60 × 100–140 μ (Crété 1944.)

The jacket round the seed appears to be the aril as described by Svedelius.

A. polygama Franch. et Sav.

Ovule; integument 6–7 cells thick; mesophyll with some raphid-cells round the chalaza.

Seed-coat –12 cells thick, most of the tissue collapsing. Embryo small. Seed exarillate.

(Vijayaraghavan 1965.)

Saurauia Willd.

Saurauia sp.

(Borneo, Kinabalu, RSNB 75; flowers and fruits in alcohol.) Figs. 10–12.

Ovules very numerous on bilobed axile placentas; integument 2–3 cells thick, o.e. with large cells; funicle slender. Ovary 5-locular.

Seeds 0.7–0.8 × 0.6 mm, brown, very variable in shape, nugget-like, mucilaginous, drying pitted-reticulate, exarillate, with minute funicle; seed-coat 5–7 cells thick; *o.e.* composed of large, roughly hexagonal cells, merely enlarged from the ovule without further division, with thin mucilaginous outer wall, thick lignified and very finely pitted inner wall and with struts of varying length and similar construction at the junctions with the radial walls; *mesophyll* composed of small thin-walled cells, several cells in the third layer (from the outside) enlarged with bundles of raphids, eventually more or less collapsing; i.e. with slightly thickened and lignified walls, eventually crushed.

There were as many outer epidermal cells in the seed as in the ovule which simply enlarges into the seed with some cell-division in the mesophyll of the integument. Schnarf (1924) gives the integument in *Saurauia* as c. 7 cells thick.

ADOXACEAE

Ovules solitary, anatropous, suspended, unitegmic, tenuinucellate.

Seeds with the integument crushed except the thin-walled o.e. Endosperm cellular, oily. Embryo minute.

(Netolitzky 1926; Maheshwari, P. 1963.)

Maheshwari and Takhtajan favour alliance with Caprifoliaceae, as considered by Bentham and Hooker. Hutchinson keeps the family in Saxifragales, where unitegmic seeds also occur. The minute seed appears to offer no help.

AEXTOXICACEAE

Aextoxicum punctatum

Ovules anatropous, suspended, bitegmic, crassinucellate; o.i. 2–3 cells thick; i.i. 5–7 cells thick; endostome projecting beyond the exostome. Seed ?

(Mauritzon 1936d.)

AIZOACEAE (including Molluginaceae)

Ovules anatropous to campylotropous, bitegmic, crassinucellate; o.i. generally 2 cells thick, 3 cells (*Trianthema*); i.i. 2 cells thick; micropyle formed by the endostome; funicle often long.

Seeds small, reniform, arillate or not, with perisperm and scant endosperm; seed-coats not multiplicative, crushed except for the exotesta and endotegmen. Testa; *o.e.* composed of enlarged cells with more or less thickened and yellow-brown walls, varying shortly elongate radially or tangentially elongate, in some cases papilliform; i.e. unspecialized. Tegmen; *o.e.* unspecialized, crushed, or with persistent thick outer wall (*Mesembryanthemum*); i.e. usually persistent, with bar-like thickenings on the radial walls or uniformly thickened all round, not lignified. Perisperm on the hilar side, starchy. Endosperm nuclear. Embryo curved, well-developed.

Aril funicular, more or less developed before fertilization, white (? in all cases), 2–9 cells thick in different genera, completely covering the seed (*Sesuvium*, *Trianthema* and *Mollugo oppositifolia*), covering part of the seed (*Glinus*) or vestigial as a small outgrowth (*Gisekia*, *Orygia*, *Mollugo nudicaulis*), the cells longitudinally elongate, thin-walled, with dense contents.

(Netolitzky 1926; Woodcock 1931; Ihlenfeldt 1959; Dnyansagar and Malkhede 1962; Narayana, H. S. 1962c; Prakash 1967a, b.)

Though the seed-structure is generally uniform, there is much variety in the size and shape of the exotestal cells. The primitively arillate state of the family is shown by the degeneration and loss of the aril in various genera. It seems best developed in the capsular *Trianthema* which has the largest ovule with thickest outer integument completely covered by the aril at anthesis. At its insertion the aril is 3–5 cells thick in *Trianthema*, 5–7 cells in *Glinus*, and 7–9 cells in *Orygia*, but in all cases mostly 2 cells thick distally.

According to Narayana and Jain (1962) the embryology of *Limeum* suggests Phytolaccaceae to which the family is allied. Ihlenfeldt found the centrifugal stamens in *Caryotophora*.

AKANIACEAE

Ovules 2 per loculus, superposed, anatropous, suspended, bitegmic, crassinucellate; o.i. 5 cells thick; i.i. 2–3 cells thick; micropyle formed by the exostome.

Seeds albuminous, exarillate. Testa; o.e. with large thin-walled cells. Vascular bundle of the raphe with more or less reticulate postchalazal branches.

(Netolitzky 1926; Mauritzon 1936a.)

ALANGIACEAE

Ovules solitary, anatropous, suspended, unitegmic, crassinucellate; i.e. forming an endothelium; hypostase developed after fertilization.

Seed enclosed in the drupe. Seed-coat ? not multiplicative, thin-walled. Vascular bundle of the raphe extending almost to the micropyle. Endosperm cellular or nuclear, copious. Embryo well-developed, straight.

(Netolitzky 1926; Kapil and Mohana Rao 1966a; Eyde 1968.)

Possibly the seeds are pachychalazal.

AMARANTHACEAE

Essentially as Chenopodiaceae. Seeds with white, saccate, funicular aril (*Chamissoa*) or minute and pulvinate aril (*Allmania*, *Ptilotus*).

(Netolitzky 1926; Bhargava 1932; Woodcock 1932; Puri and Singh 1935; Bakshi 1952; Padhye 1962; Singh, B. 1964.)

ANACARDIACEAE

Ovule solitary, anatropous, apical, lateral or basal, with dorsal or ventral raphe, bitegmic or more or less unitegmic, crassinucellate; funicle often long.

Seeds medium-size to large, the integuments with little or no specialization, eventually more or less crushed. Tegmen; *i.e.* often persistent with bar-like thickenings or pittings on the radial walls. Vascular bundle often with postchalazal branches free and subdividing or anastomosing. Chalaza extended in *Campnosperma*. Endosperm nuclear, persistent as a thin layer or wholly absorbed, oily or with some starch. Embryo large.

(Netolitzky 1926; Copeland 1955, 1962; Kelkar 1958a, b; Grundwag and Fahn 1969; Vaughan 1970; van Heel 1971c.)

In this family, as Netolitzky observed, the endocarp takes over the protective function of the testa or tegmen, and the seed-coat is so little differentiated that, as far as known, there is no guide to its original construction. What remains suggests that the seed was endotegmic rather than exotestal and exotegmic as in Sapindaceae with which the family is generally allied, but there may be traces of an exotestal state in *Campnosperma*. In *Pistacia* the outer integument is vestigial (Copeland 1955; Grundwag and Fahn 1969) while the seed-coat (? tegmen or pachychalaza) has vascular bundles. This may be the case in *Lannea*, *Rhus* and *Schinus* (Kelkar 1958a, b; Copeland 1959). The strong development of the funicle in many of these unliberated seeds is peculiar and suggests that vestiges of an aril may be found to indicate where seed-structure should be studied for less degenerate states. In fruit and seed, Anacardiaceae are among the more advanced forest families.

Campnosperma Thw.

C. minor Corner

(Singapore; flowers and fruits in alcohol.)　Fig. 13.

Ovule suspended from the side of the loculus with the raphe arching over the micropyle.

Seed becoming oblong-pyriform, then curved into the loculus, at first with two slight projections at the lower end and minutely glandular hairy in this region, developing basipetally with elongate chalaza and brown hypostase extending along most of the raphe-side, the free integuments limited to the antiraphe and the sides of the seed; mature seed-coats almost entirely crushed. Testa 4–5 cells thick, 3–4 along the antiraphe, thicker along the chalazal band, thin-walled, not lignified; o.e. as a short palisade over the sides of the seed, with isodiametric facets, becoming 2–4 cells thick over the elongate chalaza through 1–3 periclinal divisions, simplified to small rounded cells over the antiraphe. Tegmen 3–4 cells thick, –5 at the junction with the elongate chalaza, unspecialized. Chalaza elongating along one side of the seed with the v.b. dividing up into a longitudinal plexus of tracheids. Hypostase several cells thick along the whole of the elongate chalaza, composed of small contiguous cells with thin brown walls. Embryo curved; radicle short; cotyledons oblong entire.

ANCISTROCLADACEAE

Ovules hemianatropous; o.i. thick; i.i. 2–3 cells thick. Endosperm ruminate.

(Netolitzky 1926.)

Some dried seeds that I examined showed only a single layer of shortly elongate, narrow cells with thick brown outer wall as the remains of the seed-coat, possibly the exotesta. Schmid (1964) elaborates the possible connection of the family with Dioncophyllaceae.

ANNONACEAE

Ovules many to one per carpel, anatropous, perichalazal, transverse or sub-basal and erect, bitegmic or tritegmic with a middle integument between o.i. and i.i., crassinucellate; o.i. 4–6 cells thick; i.i. 2–4 cells thick; middle integument as o.i.; funicle short or none; micropile formed by the endostome.

Seeds 5–30 mm long, massive, sessile, ellipsoid, mostly exarillate, a few with thin sarcotesta, albuminous and ruminate. Testa generally multiplicative except in tritegmic seeds, becoming 6–18 cells thick or more; *o.e.* composed of thin-walled, generally cuboid cells, radially elongate (*Xylopia* sp.), or shortly longitudinally elongate (*Polyalthia, Trivalvaria*), never as a thick-walled palisade, without stomata; *o.h.* as a layer of small crystal-cells or absent; *mesophyll* generally divided into several outer layers of longitudinal fibres and several inner layers of transverse or oblique fibres, pitted, lignified, as the mechanical protection of the seed; *i.e.* with small thin-walled cells, as longitudinal fibres (*Alphonsea*) or as several layers of aerenchyma (*Anaxagorea*), not as a palisade. Middle integument (*Canangium, Cyathocalyx, Mezzettia*) multiplicative and forming the main seed-coat in such seeds; *o.e.* composed of thin-walled, cuboid cells with crystals; *mesophyll* as in the testa. Tegmen not multiplicative, with or without oil-cells, eventually crushed except at the lignified endostome. Chalaza extending round the seed, with the tegmen attached to its margins, without special structure. Vascular supply as an unbranched perichalazal bundle, in some cases branched at the end of the antiraphe to form a ring round the micropyle and joining or not with the main v.b. of the perichalaza at its entry. Endosperm cellular, with thin or thick walls, oily with or without starch; ruminations as transverse folds of the tegmen or, also,

of the testa or middle integument, not vascular. Embryo small.

Aril micropylar, as two outgrowths in the perichalazal plane, more or less covering the seed (*Xylopia spp.*) or variously reduced in indehiscent fruits (*Annona, Canangium*), without v.b. Figs. 14–18.

(Netolitzky 1926; Corner 1949b; Periasamy and Swamy 1961.)

'The seeds are not a great deal of help in systematic diagnosis and it is only when they exhibit peculiarities or are reduced to one in number that they are useful as a generic character' (Sinclair 1955). Thus my contribution is overlooked, and I take this opportunity to affirm that the alliance of genera will not be established satisfactorily until the microscopic structure of the seed has been thoroughly examined. *Mezzettia* is placed by Hutchinson (1964) next to *Anaxagorea, Canangium* next to *Alphonsea*; yet the seeds of *Mezzettia* and *Canangium* have the remarkable middle integument, and *Cyathocalyx* with the same feature is far removed from both. Certain names have changed from my previous account. *Xylopia curtisii* becomes *Cyathocalyx sumatranus* Scheff., *Popowia nervosa* becomes *Trivalvaria nervosa* (Hook.f. et Th.) Sinclair, and *Polyalthia purpurea* becomes *P. sclerophylla* Hook.f. et Th. (Sinclair 1955). The first two changes accord with the differences in seed-structure but it remains to be proved whether all species of *Cyathocalyx* and *Trivalvaria* agree with them.

Annonaceae, placed in or next to Magnoliales, differ from all other families of the order, except Eupomatiaceae, in the fibrous mesophyll of the testa. The testa has no thick-walled palisade layers. The tegmen is unspecialized except for the varied occurrence of oil-cells. Van der Pijl considers the seed to have been primitively sarcotestal, then to have become arillate, then exarillate with or without sarcotesta, and finally without either feature. Thus he seeks to force the Annonaceous seed into the sarcotestal ambit of Magnoliaceae and to derive Annonaceae therefrom. The argument is specious to suit his theory and does not explain why the aril should have been evolved only to disappear. The alternative proposition, which I made in accordance with the durian-theory, is that the fleshy character of the aril has been transferred in the disuse of this structure to the testa; thus, in *Xylopia* with dehiscent fruit, the sarcotesta is better developed in the species (so far as known) which lack the aril. It should be noted that in dicotyledons structures are transferred inwards, as from testa to tegmen, or, as in Annonaceae, from testa to middle integument, and not outwards from testa to aril; eventually only endospermous seeds may result (cf. Santalaceae, Loranthaceae). Therefore, according to the durian-theory, the primitively arillate seed of Annonaceae is more primitive than the exarillate seed of Magnoliaceae. In perichalazal growth, fibrous testa, larger embryo, and, possibly, ruminate endosperm, the Annonaceous seed may be more advanced. The middle integument appears to be unique in angiosperms and connected, perhaps, with the perichalazal structure: its outer epidermis has the crystal-cells of o.h. (testa).

The fibrous character of the testa may be related with the strongly basipetal growth of the ovule and seed. The free annular parts of the integuments remain only near the micropyle, and the rest of the ovule and seed is developed by the basal perichalazal growth. Instead of radial growth, there is tangential growth such as may lead to the elongation of the cells into fibres. The original character of the Annonaceous seed may, therefore, be found in the sclerotic endostome. Nevertheless, there is a tendency to fibrous mesophyll in *Akebia* (Lardizabalaceae, Fig. 300) and, if this indicates a true alliance which the structure of the aril supports, then Annonaceae may relate with Illiciaceae, rather than with Magnoliaceae, Myristicaceae, or Lauraceae as has been suggested (Canright 1960). A separate order Annonales (Hutchinson 1964) is indicated for the evolution of this family, along with Eupomatiaceae, from the stock of the primitive angiosperm. It must be noted, however, that a fibrous testal mesophyll turns up unexpectedly in Rutaceae (*Casimiroa*, *Triphasia*).

Annonaceae resemble Myristicaceae in their ubiquitous occurrence in the lower stories of the tropical lowland rain-forest. Though apparently uniform, enquiry is beginning to show the peculiarities of Annonaceae in other ways, as pollen (Walker 1971) and pollination (Gottsberger 1970).

In this account I have not repeated the descriptions which I gave formerly (Corner 1949b) of *Alphonsea*, *Anaxagorea*, *Annona*, *Artabotrys*, *Canangium*, *Cyathocalyx*, *Desmos*, *Mezzettia*, *Polyalthia* and *Xylopia*.

Cyathocalyx Champion

Seeds tritegmic.

C. sumatranus Scheff.

(Corner 1949b, as *Xylopia curtisii* King.)

C. carinatus (Ridley) Sinclair

(Borneo, RSNB 230; flowers and fruits in alcohol.) Fig. 14.

Ovule; o.i. 3–4 cells thick; i.i. 2 cells thick; middle integument as a slight bulge of tissue.

Seeds 18–20 × 13–14 × 5–6 mm, compressed ovoid, smooth, dark brown, exarillate, with small ovate hilum 1 mm long. Testa *c*. 50 μ thick, scarcely multiplicative, 4–5 cells thick; *o.e.* composed of short tabular cells; *mesophyll* with 1–2 rows of longitudinal fibres and 1–2 rows of inner oblique fibres; *i.e.* small-celled, unspecialized. Middle integument 150–200 μ thick; *o.e.* as a more or less continuous layer of small tabular cells with an oblong crystal; *mesophyll* composed of several outer layers of longitudinal fibres and several inner

layers of oblique fibres; *i.e.*? Tegmen evidently multiplicative, –12 cells thick in the free part round the apex of the nucellus, with many oil-cells; *endostome* thick, woody, with small rounded sclerotic cells. Ruminations transverse from the sides of the seeds, somewhat anastomosing at the inner ends, composed of tegmen with a single row of oil-cells and of fibres from the middle integument. Endosperm with thick walls staining deep blue in iodine, contents oily.

The mature seed is very hard and it was not possible to make out the finer details of tegmen and nucellus.

Goniothalamus Hook. f. et Th.

Goniothalamus sp.

(Solomon Islands, RSS 2302; ripe fruits in alcohol.) Fig. 15.

Seeds 25 × 20 × 10–11 mm, ovoid-compressed, conico-attenuate to the small hilum, reddish brown, subrugose with irregular outline, the surface (in alcohol) mucilaginous, exarillate but the extreme base with a zone of mucilage-hairs. Testa *c*. 500 μ thick on the sides of the seed; *o.e.* composed of cuboid cells with thin red-brown walls, not lignified, thinly and microscopically hairy; *hairs* 100–250 × 10–15 μ, lanceolate-subcylindric to subventricose, acute or obtuse, with thick unlignified walls, smooth, aseptate; *o.h.* as small flattened cells with a crystal; *mesophyll* with *c*. 15 layers of longitudinal fibres and *c*. 20 inner layers of transverse and oblique fibres, pitted, lignified, the inner layers projecting with short flanges into the outer layers. Tegmen with oil-cells, apparently not multiplicative except at the endostome. Vascular bundle as a hoop round the seed, apparently with a narrow ring round the inner end (or base) of the endostome, with two small v.b. descending along the midline of each side of the seed nearly to the hilum. Ruminations formed by the inner layer of the testa, with oil-cells from the tegmen. Endosperm with rather thin walls turning blue-black in iodine, cell-contents oily.

Follicle 6–11 × 2.5–3 cm, subcylindric, shortly stipitate, 2–6 seeded; wall *c*. 8 mm thick, with small clusters of sclerotic cells below the outer epidermis and *c*. 2 layers of laxly arranged v.b., largely thin-walled.

This seed is peculiar in its irregular and mucilaginous surface, the minute hairs, the additional v.b. on the sides of the seed, and the prolonged, flattened, micropylar end composed of the transversely dilated tegmen (with unlignified endostome) and internal plates of fibres. Unfortunately it was difficult to make out details in these mature seeds with very hard testa. It appeared that in the subhilar part the tegmen dilates and fills the inner region between the internal plates of fibres with thin-walled tissue. The endostome is a rather wide tube on one side of which the tissue is slightly lignified. This dilated subhilar end of the tegmen is limited by thin plates of fibres, which are the inner part of the testa; the thin-walled middle part of the testa carries the small lateral v.b. descending to the hilum; the outer part is thickly fibrous and continues the fibrous mesophyll of the testa to the hilum. Thus the fibrous mesophyll of the testa becomes divided into inner and outer plates in the subhilar region, and oil-cells occur in the thin-walled tissue on both sides of the inner plates. This extensive endostomal region suggests a primitive feature in the perichalazal seed.

The mucilage on the surface of the seed may come from the epidermal cells of the testa and their hairs, but it may come from the mucilage hairs which form a narrow zone at the very base of the seed, as if on the very short funicle. These mucilage-hairs have a body similar to that of the testal hairs but filled with starch-grains and terminating in a mucilaginous extremity.

Miliusa Lesch.

M. wightiana Hook.f. et Th.
(Periasamy and Swamy 1961.)

The seed resembles that of *Polyalthia* (Corner 1949b), but oil-cells were not noted in the ruminations.

Xylopia L.

X. caudata Hook.f. et Th., *X. fusca* Maingay, *X. malayana* Hook.f. et Th. (Corner 1949b.)

X. peekelii Diels
(Solomon Islands, RSS 3; flowers and ripe fruits in alcohol.) Fig. 16.

Seeds 15–18 × 9–11 × 5–7 mm, compressed with obtuse edges and apex, slightly dilated at the base, black with a very thin waxy-grey sarcotesta, several per follicle, hanging on pseudofunicles (strands of ruptured endocarp). Testa; *o.e.* composed of shortly columnar cells with polygonal isodiametric facets, thin-walled, slightly pulpy, with oily contents, slightly but distinctly enlarged over the base of the seed; *o.h.* composed of small tabular cells with a crystal; *mesophyll* thick, with outer longitudinal and inner transverse or oblique fibres, each in several layers. Endostome columnar, woody, composed of small, rounded, sclerotic cells in several layers. Hilum with a broad white annular mass of thin-walled cells packed with starch-grains. Vascular bundle with a small annulus round the micropyle. Ruminations formed from the testa and tegmen (collapsed), without oil-cells. Endosperm with thick-walled cells, oily, without starch.

Follicles dull olive tinged bluish, usually 1(–2) per flower, splitting open very irregularly, the mauve-carmine or pink endocarp peeling away from the exocarp and splitting up but compacted at the base of the seeds into pseudofunicles.

This species resembles *X. malayana* but the larger seeds have a slightly thickened or dilated base over which the exotesta seems to have the last traces of arillar development.

Xylopia sp.
(Brazil, Mato Grosso, Chavantina, leg. E. J. H. Corner 1968; flowers and fruits in alcohol.) Figs. 17, 18.

Ovule; o.i. 5–6 cells thick; i.i. 2 cells thick; without middle integument.

Seeds 6–7 mm long, black, 1–5 per follicle, sub-compressed, with two short pulpy yellowish aril-lobes at the base. Testa hard, mainly fibrous, blackish brown; *o.e.* composed of radially elongate columnar cells with thickened outer wall, pulpy, with many starch-grains, eventually drying up; *o.h.* as small crystal-cells; *mesophyll* with *c.* 3 outer layers of longitudinal fibres and 3 inner layers of transverse or oblique fibres, much thickened in the ruminations; *i.e.* composed of small cuboid cells with slightly thickened walls. Tegmen disappearing. Ruminations without oil-cells. Nucellus becoming very large-celled, without oil-cells. Vascular bundle giving off at its entry into the hilum a plexus of small v.b. as a ring internal to the aril. Subhilar tissue becoming very woody. Endosperm with thick-walled cells, oily without starch, the walls red-brown in iodine.

Aril with a small-celled epidermis containing much starch, and an inner row of large palisade-like cells with oily contents.

This has essentially the structure of *Annona* (Corner 1949b) with the aril of *Canangium*, but the columnar cells of the exotesta form a thin pulpy sarcotesta as in *X. africana*, *X. dinklagei* and *X. humilis* (Netolitzky 1926). Except that it is pulpy, this sarcotesta bears little resemblance to that of *Magnolia*.

APOCYNACEAE

Ovules more or less anatropous, unitegmic, tenuinucellate.

Seeds small to moderately large, ellipsoid, oblong, plano-convex, or flattened and alate, arillate, mostly exarillate, some with silky hairs, often with a funicular or raphe-groove, mostly albuminous. Seed-coat multiplicative, but the mesophyll more or less crushed and only o.e. persisting; *o.e.* as a palisade of cuboid to clavate cells with various thickenings on the anticlinal and outer walls, in some cases with groups of projecting cells (*Catharanthus*) or the cells elongate into hairs (*Strophanthus*), or excrescent as folds (*Funtumia*), with crystals (*Beaumontia*), with thick pitted walls (*Thevetia*); *mesophyll* thin-walled, crushed, or with 2–3 outer hypodermal layers of thick-walled, pitted or reticulately thickened, cells (*Aspidosperma*, *Thevetia*); *i.e.* not as an endothelium. Raphe short or variously elongate in a deep groove, more or less adnate with the placenta (*Lepinia*). Vascular bundle limited to the raphe (? with postchalazal branches in *Thevetia*), the funicle with a single v.b. or several. Endosperm nuclear, horny, oily, with thin or thickened walls. Embryo straight; cotyledons flat, ligulate or broad, or much folded (*Chilocarpus pr. p.*).

Aril funicular or from the base of the seed, entire or lobulate, or transferred to a generalized placental outgrowth (*Voacanga*), vascular or not. Figs. 19–23.

(Netolitzky 1926; Andersson 1931; Periasamy 1963; Berger 1964, pp. 411–444; v. Heel 1970b; Khan 1970; Maheshwari Devi 1971.)

Because of the sympetalous corolla and obvious affinity with the elaborate flowers of Asclepiadaceae, there is a tendency to depreciate the primitive characters of the more or less free carpels, the free follicles, and the arillate seeds. In fact, this richly varied family displays much of the general evolution of the pachycaul into trees, lianes, and herbs and, as one that has these advanced and primitive features, it has an impelling interest. No one has explained why *Dyera* as an enormous tree, *Willoughbeia* as a lofty liane, and *Catharanthus* as a herb, may grow together. The family is interposed between Loganiaceae and Asclepiadaceae but in ovary, fruit and arillate seed it is more primitive. The aril is relatively rare and has been reported from *Chilocarpus* and *Tabernaemontana*. Comparison with related genera or of species within a genus will certainly show the loss of the aril and the development of the winged seed. It seems, also, that the aril-factors may be transferred to the placenta to form, as in some Rubiaceae, a generalized outgrowth among the seeds. These problems need careful investigation together with that of the feature which, for lack of better understanding, I call the raphe-groove. It occurs in *Tabernaemontana* in which the aril is attached to the top of the funicle which is extended along the raphe-groove. The funicle may become rather massive (*Chilocarpus*, Figs. 19, 20) and extend along the midline, or raphe-line, of most of the seed but the aril in this case is entirely free of the funicle. Then *Lepinia* (Fig. 21) appears to be an extension of this process with the loss of the aril (? indehiscent follicle), whereby the raphe extends nearly the whole length of the long seed (4–5 cm) and throughout its course receives additional vascular bundles from the placenta which becomes compressed in the raphe-groove. *Lepinia* has 3–4 carpels which separate widely in the fruit but remain attached by the concrescent style-base, and in this respect it is one of the more primitive genera.

There are clearly many critical generic differences and alliances in the form of the outer epidermal cells of the seed-coat and in the thickening of their walls.

Chilocarpus Bl.

Seeds dark brown to black, plano-convex, smooth or with plicate surface; aril as a fleshy cushion round the flat (or concave) side of the seed, pale orange to reddish pink. Seed-coat several cells thick, without v.b., crushed except for o.e. composed of cuboid cells with more or less thickened outer wall and hexagonal facets, slightly lignified, the walls brown. Funicle thin, flattened, extended along the flat side of the seed, traversed by several (5–10) irregular vascular strands (with central phloem and external xylem), forming a reticulum of tracheids along the flat (or concave) underside of the seed. Endosperm horny, with slightly thickened walls, without starch, in some species ruminate (*Neokeithia*). Embryo straight or slightly curved; hypocotyl long; cotyledons flat or folded in the ruminate endosperm.

Aril arising from the top of the funicle, annular or in two halves, smooth or lobulate, composed of large cells with thin lignified walls and hexagonal facets, without v.b.

(Markgraf 1971.)

C. torulosus (Boerl.) Markgr.

(Sarawak, Gunong Matang, leg. Corner 1972; seeds in alcohol.) Fig. 19.

Seeds 7–8 × 6–6.5 × 4 mm, black, smooth, with a slight micropylar prominence at one end. Seed-coat; o.e. with distinctly thickened, brown, outer wall, the inner and radial walls thin.

Aril in two halves, smooth, not lobulate, reddish pink.

Follicles c. 37 × 1.4 cm, with 33 seeds, subcylindric, scarcely moniliform, apex acute, firmly fleshy, orange.

Chilocarpus sp.

(Borneo, Kinabalu, RSNB 8326; fruits in alcohol.) Fig. 20.

Seeds 12–15 × 6–7 mm, dark brown, smooth. Seed-coat; o.e. composed of thin-walled cells, covered on the outside by a thin layer of granules.

Aril annular (? enclosing the micropyle), lobulate with crenulate–tesselate surface, pale orange.

Follicles 8–11 × 4–5 cm, fusiform, the acuminate tip 15–20 mm long, fleshy, shiny, orange, dehiscing longitudinally into two parts each filled with numerous seeds. Pericarp 5 mm thick, with a median ring of large v.b. (concentric xylem surrounding the phloem), and with one outer (subepidermal) plexus of small v.b. and two plexuses of small v.b. internal to the main ring; sclerotic cells in abundant clusters throughout, more or less continuous in the hypodermis. Peduncle with a single continuous ring of vascular tissue; sclerotic cells scattered or clustered in the cortex and pith.

Lepinia Decne

L. solomonensis Hemsl.

(Solomon Islands, RSS 234; flowers and fruits in alcohol.) Fig. 21.

Ovules hemianatropous, c. 125 μ long, several per carpel in 2 marginal rows; integument 6–9 cells thick; i.e. not as an endothelium.

Seeds 3.8–5 × 0.7 cm, 1(–2) per follicle, oblong with subacute ends, dark brown, longitudinally sulcate, with an elongate hilar slit (or raphe-groove) pervaded by the placenta along the adaxial side, exarillate, albuminous. Seed-coat multiplicative, 15–20 cells thick, thin-walled, becoming entirely crushed into a bright brown membrane except for o.e., intrusive on the endocarp between the v.b. and so developing the superficial furrows; *o.e.* composed of cuboid cells with hexagonal facets, with thin brown walls, not lignified, the outer walls convex except where compressed against the endocarp. Hilum extending as a raphe-groove for most of the length of the seed, with the placental strand attached and compressed between the bulging sides of the seed, traversed by 2–3 longitudinal anastomosing v.b. in the elongate raphe, with numerous branches from the placental v.b. along its course in contact with the seed; no v.b. to the sides of the seed. Endosperm horny, with thick pitted walls, without starch. Embryo straight, nearly as long as the seed, narrow; cotyledons ligulate, shorter than the hypocotyl, flat; radicle-tip pointing to the base of the follicle.

Ovary composed of 3–4 carpels, concrescent at base and apex, free along the ovuliferous part, the carpellary sutures concrescent in the narrow style 150 μ wide.

Pericarp consisting mainly of thin-walled tissue, rupturing readily and longitudinally between the larger v.b., with many small latex-tubes; v.b. in a single ring close to the endocarp, the larger v.b. joined by sclerotic tissue to the endocarp, the xylem more or less concentric; endocarp as a single layer of longitudinal, thick-walled, pitted and lignified fibres; o.e. with small thin-walled cells, without stomata.

(Hooker, *Ic. Pl.* ser. 4, *8* (1905) t. 2703.)

The remarkable fruit of this genus has been illustrated but the details of the seed have been obscure. The long and deep hilar or raphe-groove is traversed by the vascular bundle of the raphe which is joined by numerous branches from the main longitudinal bundle of the placenta, a flange of which extends the length of the groove. Thus this large seed employs the placenta in place of vascular branches to the integument. The groove, from which the placental strand emerges at each end, appears to be mainly an extended hilum, but I had not sufficient young seeds to decide the point. The endosperm appears shortly ruminate but the appearance is false; the seed-coat is extruded into the soft intervals between the v.b. of the pericarp and the endosperm presses into these ridges; there are no inward folds of the integument. How the seed is distributed and germinates seems not to be known.

Leuconotis Jack

L. anceps Jack

(Sarawak, S. 28471; fruits in alcohol.)

Seed-coat reduced to a thin colourless, unlignified, but toughly mucilaginous, pellicle a few cells thick, unspecialized, surrounding the large seed.

Tabernaemontana Plum.

Ovules more or less anatropous; integument *c.* 8 cells thick.

Seeds many per follicle, oblong, dark brown, longitu-dinally sulcate, with a deep raphe-groove, each surrounded by a red or orange aril, pendent from the edges of the dehisced follicle, albuminous. Seed-coat multiplicative, reduced eventually to the thick-walled o.e., the thin-walled mesophyll crushed into a brown stratum; *o.e.* as a palisade of clavate cells with brown contents, the distal end of the cells with internal, pale yellow, reticulate thickenings, the inner part of the radial walls with pale yellow laminate thickening bulging into the cell-lumen, the whole layer becoming plicate through excessive anticlinal cell-divisions, the cells in the folds more or less compressed; *i.e.* not endothelial. Vascular supply limited to the shortly extended raphe-chalaza. Endosperm thick-walled, horny, oily, peripherally ruminate with folds of the integument. Embryo straight; cotyledons subcordate, equal to or exceeding the length of the hypocotyl.

Aril waxy then mucilaginous, developed from the funicle and raphe round the base of the seed, the cells with pink or orange oil-globules; vascular in some species (? in all)   Figs. 22, 23.

(Periasamy 1963; Corner 1964, pl. 25.)

According to van der Pijl (1966), an endocarp-pulp covers the groups of seeds in some Indonesian species of *Tabernaemontana* and it is to be regarded as an outgrowth of the placenta. Therefore, splitting hairs, he denies that there is an aril in the genus. Periasamy had already described the funicular aril of *Ervatamia heyneana*, which is a species of *Tabernaemontana* in the wide sense. In all species that I have seen, an aril neatly covers every seed but, on becoming mucilaginous, there is an appearance of a generalized investment, which explains van der Pijl's error. In an unidentified species of *Tabernaemontana* Periasamy found no aril or outgrowth of placental tissue, but 'the ovary-wall becomes spongy and surrounds the seeds'.

Similar outer epidermal cells occur in the seed-coat of *Holarrhena* according to Netolitzky.

Tabernaemontana sp.

(Java, Bogor, Kebun Raya; flowers and fruits in alcohol.) Fig. 23.

Ovules anatropous, *c.* 140 μ long, sessile, crowded on the lobulate placenta.

Seeds 6–7 × 3–4 mm, oblong-ellipsoid, compressed, the decurved sides concealing the chalazal region, dark brown with minutely tesselate surface, covered by the red, thinly pulpy aril. Chalaza shortly longitudinally elongate. Vascular bundle from the endocarp traversing the thick aril and giving off numerous, apparently recurrent, branches to the aril; at the chalaza entering the testa and forming a continuous plate of tracheids along the recurved sides of the testa round the chalaza.

Aril soon developing after fertilization around the attachment of the ovule, thick, enveloping the seed, with a narrow slit-like opening over the distal side of the seed, with a loose network of v.b., ripening mucilaginous and dark red.

Follicles 35 × 15 mm, orange-red.

The seeds have an irregular form as they become variously compressed in the developing follicle; hence it is difficult to cut a truly median longitudinal section.

The ovule has no funicle and none appears after fertilization. The aril develops from the lobule of the placental ridge on which the ovule is borne. Hence the aril can be described as placental, but these lobules may be vestiges of funicles, and if the funicle is lost it is difficult to see how else the aril could develop.

Voacanga Thou.

V. grandifolia Rolfe
Similar to *Tabernaemontana* but the outgrowths round the developing seeds coalescing into a white pulpy mass embedding the seeds.

(Periasamy 1963.)

AQUIFOLIACEAE

Ovules 1–2 per loculus, anatropous, suspended with adaxial micropyle, unitegmic, tenuinucellate; funicle often with a small papillate protuberance as an obturator.

Seeds in pyrenes, with the non-multiplicative seed-coat reduced to o.e., albuminous; *o.e.* as a layer of cuboid or shortly longitudinally elongate cells with lignified and pitted radial and inner walls; *mesophyll* and *i.e.* unspecialized, crushed. Endosperm cellular, thin-walled oily. Embryo small, straight.

(Netolitzky 1926; Herr 1959, 1961; Copeland 1964.)

The customary position of this family in Celastrales finds no immediate support in the seed-structure. If the single integument represents the testa, the lignified epidermis differs from that of Celastraceae. If it represents the tegmen, the epidermis is scarcely reconcilable with the fibrous exotegmen of Celastraceae. There seems to be no indication whether the single integument is individual or, as its massive form may suggest, a combination of both integuments as in Rosaceae.

The placentation of *Ilex* appears to be free-central from the underside of a mamillon, such as occurs in Santalaceae, and the ovary is septate in the lower part.

Ilex L.

I. aquifolium L.
(England; living material.) Figs. 24–26.

Ovule; integument 12–15 cells thick, more in the raphe; vascular bundle of the raphe shortly ramified at the chalaza; obturator with papilliform cells.

Seed-coat not multiplicative, the mesophyll-cells enlarging greatly, aerenchymatous, then collapsing; *o.e.* as a layer of shortly longitudinally elongate cells with strongly lignified, thickened and pitted radial and inner walls, as the persistent layer of the seed-coat.

Ilex sp.
(Malaya, Pahang, Ulu Kali, 1800 m; fruits in alcohol.)

Seeds 1 or 2 per pyrene, oblong, more or less triangular in t.s., the single seeds 7 × 2.5–3 mm, the paired seeds 4 × 2.5–3 mm. Seed-coat 8–9 cells thick; *o.e.* composed of shortly longitudinally elongate cells with shortly fusiform facets, with a strongly thickened, pitted and lignified bar on the radial walls, the inner walls slightly thickened and lignified; mesophyll thin-walled, crushed; *i.e.* composed of rather large cells; *exostome* very woody with sclerotic cells in the mesophyll and i.e.

Fruit with 6 pyrenes; o.e. of the pericarp with scattered large stomata.

This species appears to be *I. venulosa*.

ARALIACEAE

Ovules anatropous, suspended with ventral raphe, 1 per loculus or a second vestigial, unitegmic, tenuinucellate; i.e. becoming an endothelium.

Seeds enclosed in endocarp. Seed-coat ? not multiplicative, usually crushed, with or without crystals in the cells; *o.e.* composed of cuboid or tabular cells, thin-walled, often large, or with the outer wall thickened. Vascular bundle only in the raphe. Endosperm nuclear, oily, ruminate from irregular folds of the integument or not, the cells thick-walled. Embryo small.

(Netolitzky 1926; Baumann 1946; Periasamy 1962b.)

Except for a vestige of exotestal construction, the seed-coat in this family seems to have deteriorated.

ARISTOLOCHIACEAE

Ovules anatropous, transverse, bitegmic, crassinucellate; o.i. 2 cells thick; i.i. 2–3 cells thick, or thicker near the endostome; funicle often thickened and flattened against the ovule; micropyle formed by both integuments or by the endostome.

Seeds rather small to medium-size, 3-angled, flattened or winged, albuminous, exarillate; seed-coat generally consisting of the five cell-layers of the ovule, the innermost layer sometimes crushed, the testa thickening –13 cells in some species of *Aristolochia*, the two integuments joined to the raphe for most of their length. Testa: *o.e.* with cuboid to radially elongate cells, thin-walled or with variously thickened, pitted and lignified walls, even as a compact palisade, but the cells tabular in *Asarum*; *mesophyll*, when present, also lignified; *i.e.* with cuboid cells, with thin or thick walls, not lignified, every cell with a small crystal (except *Thottea*). Tegmen with the two outer layers of crossed, lignified fibres with or without pitting; *o.e.* with longitudinal fibres; *middle layer* of more or less transversely elongate fibres, often less lignified; *i.e.* with thin-walled cells with brownish contents, or in some cases with slight reticulate or spiral thickenings on the walls. Vascular bundle terminating at the small chalaza. Endosperm cellular, thick-walled, oily and, in some cases, also starchy. Embryo small or minute and undifferentiated. Figs. 27–29.

(Netolitzky 1926; Nair and Narayanan 1961.)

The chief characters of the Aristolochiaceous seed are the attachment of the integuments along the course of the raphe and the development of two layers of crossed fibres in the tegmen, which makes the mechanical layer of the seed-coat. In some cases, especially *Aristolochia*, the exotesta may also form a thick-walled mechanical layer and, if this becomes many-celled as in *A. gigas*, it would seem to be the more important protection.

Kratzer drew attention to the multiplication of the exotesta by periclinal division in the manner of Cucurbitaceae, but there is little further resemblance in details of seed-coat structure. The tegmen of Caricaceae, however, seems strikingly similar, having the same set of crossed fibres in the tegmen. It is not impossible that the Aristolochiaceous seed is phylogenetically a simplified version of the Caricaceous.

The peculiar attachment of the integuments to the raphe seems to arise through the enlargement of the ovule mainly in the small segment of the chalaza which joins the integuments on the raphe-side; if the antiraphe were included, the seed would become perichalazal.

Apama Lamk.

A. siliquosa Lamk.
Seed with uneven seed-coat appearing in section corrugate externally and internally. Testa; *o.e.* thin-walled, pulpy.

(Periasamy 1962b.)

A. tomentosa (Bl.) Engl.
(Java, leg. Valeton 1912, Herb. Bogor.; flowers and fruits in alcohol.)

Seed oblong, pulpy with the hard tegmen angular in t.s. Testa 4–8 cells thick, thin-walled, pulpy; *o.e.* with scattered pitted sclerotic cells; *i.e.* with many pitted sclerotic cells. Tegmen ? 4 cells thick; *o.e.* with coarsely pitted, lignified, longitudinal fibres, somewhat laterally compressed; *o.h.* with narrow transverse fibres with slightly lignified walls; *i.h.* ? thin-walled and collapsed; *i.e.* composed of short cuboid cells with rectangular facets and brown contents, not lignified.

Aristolochia L.

Seeds flattened, in some species winged. Funicle thick, dilated laterally, flattened against the seed and generally larger than it, becoming rather corky, the cells of the mesophyll more or less reticulately thickened on the walls and lignified, in some cases also with oil-cells. Testa smooth or uneven; *o.e.* composed of uniformly thick-walled cells or with few thick-walled cells scattered

among the thin-walled; *mesophyll* in some species consisting of 1 (*A. fimbriata*) or 6–11 cells (*A. gigas*), formed by periclinal division of o.e., becoming thick-walled. Figs. 27–29.

A. bracteata Retz., *A. clematitis* L., *A. elegans* Mast., *A. fimbriata* Cham., *A. gigas* Lindl., *A. indica* L.

There is clearly much detail in the exotesta to be investigated.

Asarum L.

Seeds with convex antiraphe. Funicle dilated at the apex into an elaiosome, not flattened as in *Aristolochia*. Testa; *o.e.* with tangentially elongate cells, thin-walled. Tegmen with 1–3 rows of sclerotic cells (? fibres) in *A. canadense*. Elaiosome with 1–5 layers of large oil-cells beneath the small-celled epidermis.

(Kratzer 1918; Leemann 1927; Wyatt 1955; Bresinsky 1963.)

A. arifolium Michx., *A. canadense* L., *A. europaeum* L., *A. lewisii* Fern., *A. virginicum* L.

Bragantia Lour.

B. wallichii R.Br.
Ovule; o.i. 2 cells thick, 4 cells at the exostome; i.i. 3 cells thick, developing before o.i.; hypostase present.

Testa; *o.e.* composed of cuboid cells with reticulate thickening; *i.e.* with fibrillar thickening on the radial walls. Tegmen; *o.e.* as longitudinal fibres; mesophyll crushed; *i.e.* as tannin-cells.

(Nair and Narayanan 1961.)

Thottea Roxb.

Testa; *o.e.* with secretory cells between the thin-walled cells; *i.e.* without crystals, but with a strong conical thickening projecting into the cell-lumen from the inner wall.

ASCLEPIADACEAE

Ovules anatropous, often small (100 × 150 μ, *Asclepias*), unitegmic, more or less tenuinucellate; integument rather thick; i.e. not forming an endothelium.

Seeds small to large, various, mostly flattened or alate, generally with a coma of hairs (1–50 mm long) developed from the micropylar region, albuminous. Seed-coat in some cases multiplicative, reduced generally to o.e. or with a few outer layers of mesophyll as flattened crystal-cells (*Calotropis*, *Morrenia*); *o.e.* with thickened walls or thin walls (*Periploca*). Endosperm nuclear, more or less thin-walled, oily.

(Netolitzky 1926; Sabet 1931; Moore 1946; Pearson 1948; Venkata Rao 1954; Biswas 1957.)

There seems to have been little interest in recent years in the seeds of this family. Moore and Pearson have described in detail the development of the seed and its hairs in *Asclepias syriaca*. It is noteworthy that the coma, of unknown origin, is in the position of a funicular or micropylar aril.

BALANITACEAE

Ovules campylotropous, suspended, 1 per loculus, bitegmic, crassinucellate; o.i. 4–8 cells thick; i.i. 3–6 cells thick; micropyle formed by both integuments.

Seeds (*Balanites aegyptiaca*) 15 mm long, rugulose, yellow, exalbuminous. Testa 10–13 cells thick, some with single or clustered crystals, others sclerotic and pitted, pervaded by v.b.; *i.e.* as small rectangular cells. Endosperm nuclear. Embryo with thick cotyledons.

(Nair and Jain 1956; Johri 1963; Vaughan 1970.)

This family has been referred to the neighbourhood of Malpighiaceae, Simaroubaceae and Zygophyllaceae. The seed does not agree but, as it may be pachychalazal or overgrown, it may fit Simaroubaceae. Compare, however, *Irvingia* (p. 155).

BALSAMINACEAE

Ovules anatropous, several per loculus, bitegmic, tenuinucellate; integuments normal (*Impatiens pallida*, *I. parviflora*) or very short and only in the micropylar region and the rest of the ovule formed by the expanded chalaza (*Hydrocera, Impatiens spp.*); o.e. 6–9 cells thick; i.i. 4–6 cells thick, forming the micropyle; i.e. as an endothelium in the pachychalazal ovules.

Seeds small to medium-size, exarillate, exalbuminous, in most cases pachychalazal, the integument not or scarcely multiplicative. Seed-coat formed by the expanded chalaza; *o.e.* unspecialized or the cells with a papilla or a hair; mesophyll unspecialized, more or less crushed, in *Hydrocera* with o.e. and 4–5 outer cell-layers of the mesophyll thick-walled and lignified; *i.e.* endothelial, then crushed. Tegmen when present becoming crushed. Vascular bundle ending at the chalaza. Endosperm cellular, but absorbed. Embryo straight.

(Netolitzky 1926; Venkateswarlu and Lakshminarayana 1958; Narayana and Sayeeduddin 1959; Narayana 1963; Takao 1968.)

Hydrocera triflora W. et A., *Impatiens arguta* Hook.f. et Th., *I. balsamina* L., *I. inconspicua* Benth., *I. leschenaultii* Wall., *I. levengi* Gamble, *I. pallida* Nutt., *I. parviflora* L., *I. radiata* Hook.f., *I. roylei* Walp., *I. sultani* Hook.f., *I. tenella* Heyne, *I. textori* Miq., *I. tripetala* Roxb.

It appears that in *Impatiens* there will be found a complete series of species to show the origin of the unitegmic pachychalazal seed from the bitegmic. In *I. parviflora*, as described by Guignard (1893), there is no evidence of the exotegmen distinctive of Geraniaceae (p. 35).

BASELLACEAE

As Chenopodiaceae. Tegmen; *i.e.* with distinct streak-like thickenings on the radial walls. Endosperm nuclear. (Netolitzky 1926.)

BEGONIACEAE

Ovules anatropous, bitegmic, tenuinucellate; o.i. and i.i. 2 cells thick.

Seeds minute, exarillate; seed-coats not or slightly multiplicative, 3–4 cells thick, crushed except the exotesta. Testa; *o.e.* composed of much enlarged, cuboid cells with lignified and pitted inner wall, with polygonal facets. Endosperm nuclear, reduced to a single layer of cells in the seed.

(Netolitzky 1926; Mauritzon 1936b; Swamy and Parameswaran 1960.)

Begonia, Hillebrandia.

The seed is simplified and gives no evidence of its derivation.

BERBERIDACEAE

Ovules anatropous, erect, bitegmic, crassinucellate; o.i. 5–8 cells thick; i.i. 3–5 cells thick; micropyle formed by the exostome; funicle short.

Seeds 1.5–7 mm long, pale or dark, exarillate (but a vestigial aril in *Berberis umbellata*), albuminous. Testa not or slightly multiplicative; *o.e.* forming a layer of thick-walled, pitted, more or less lignified cells shortly elongate longitudinally, with oblong facets, but appearing as a palisade in t.s., without stomata; *mesophyll* aerenchymatous, drying up, often with the inner layers crushed, the walls slightly thickened and pitted (*Berberis*), not lignified; *i.e.* as a layer of small, thick-walled, cuboid cells, not or slightly lignified, continuous or the cells more or less separated by tangential expansion of the seed. Tegmen unspecialized, crushed. Vascular bundle only in the raphe or with a slight postchalazal extension. Endosperm nuclear, oily. Embryo well-developed; radicle rather long. Figs. 30–33.

(Netolitzky 1926; Johri 1935; Mauritzon 1936d; Sastri 1969.)

Berberis

Testa; *o.e.* in some species with some of the cells red from anthocyanin. Aril minute and vestigial in *B. umbellata* (Sastri 1969).

Mahonia

Seeds dark. Testa; *o.e.* with small crystals scattered along the anticlinal walls below the cuticle.

This small family, in the limits set by Hutchinson (1959), is distinguished by the shortly oblong and thick-walled cells of the exotesta and by the small, thick-walled and cuboid cells of the endotesta. As noted by Netolitzky, neither Berberidaceae nor the allied Podophyllaceae have the bar-like thickenings on the radial walls of the endotegmen.

The ovary is often treated as unicarpellary but it has no ventral suture. It has many longitudinal vascular bundles (7 in *Berberis*, up to 22 in *Mahonia*) which anastomose in a ring just below the entire stigma. The ovary develops basipetally below the primordial stigmatic ring just as the syncarpous ovary. The unilateral placentation suggests the abortion of laminar or parietal placentation to one side, thus simulating a single carpel. Sastri (1969), however, considers that the ovary may be a single multivascular carpel. The thick outer integument seems primitive. Affinity with Nandinaceae is doubtful (p. 28).

BETULACEAE

Essentially as Fagaceae but the ovules unitegmic and the seeds small. Endosperm nuclear, oily, persistent as 2–4 layers of cells.

(Netolitzky 1926; Hjelmquist 1948, 1957; Vaughan 1970.)

It has been suggested that the thick integument in Betulaceae is the result of connation of two integuments; hence the vascular seed-coat may be pachychalazal.

BIGNONIACEAE

Ovules anatropous or hemianatropous, unitegmic, tenuinucellate; i.e. as an endothelium; hypostase present.

Seeds small to large, often alate with membranous or corky wing, exalbuminous, exarillate. Seed-coat exotestal; *wing* 1–2 cells thick, the cells elongate tangentially with thickened walls; *mesophyll* crushed or with longitudinally elongate sclerotic cells in the inner tissue round the embryo (*Oroxylon*). Endosperm cellular.

(Netolitzky 1926; Govinda 1951; Ghatak 1956.)

BIXACEAE (including Cochlospermaceae)

Ovules bitegmic, anatropous, more or less transverse, crassinucellate; o.i. 3–5 cells thick; i.i. 3–5 cells thick; micropyle formed by the exostome; without chalazal complication before fertilization.

Seeds anatropous and more or less straight (*Bixa*) or reniform campylo–anatropous (*Cochlospermum*), arillate (*Bixa*) or not, with rather long funicles; chalaza internally bullate and complicated, externally umbonate or becoming slightly depressed (*Bixa*). Testa not multiplicative, thin-walled, thinly pulpy, drying up; *o.e.* with hairs and stomata (*Cochlospermum*) or not, with sclerotic cells round the chalaza (*Bixa*); *mesophyll* with oil-cells (*Bixa*). Tegmen multiplicative, 7–10 cells thick; *o.e.*

as a palisade of lignified Malpighian cells, pitted at the outer ends; *mesophyll* with enlarged cells in o.h. (*Cochlospermum*) or hour-glass cells in i.h. (*Bixa*); i.e. with pitted anticlinal walls. Chalaza with re-entrant annulus formed by the exotegmen, a thin-walled counterpalisade from the endotesta, and a core of brown cells as a hypostase–operculum (complicated in *Bixa*) occluding the heteropyle, not penetrated by the v.b. Vascular supply only in the raphe. Endosperm nuclear, starchy (*Bixa*). Embryo well-developed. Figs. 34–42.

Though it is becoming customary to separate Bixaceae from Cochlospermaceae, the differences are less than may be found in most large families and the resemblances in seed-structure are great. I treat them as one. *Amoreuxia* (Cochlospermaceae) may be intermediate with trilocular ovary. The seeds would pass as Sterculiaceous except for the complicated chalaza which allies Bixaceae with Cistaceae. Differentiation of this structure is post-fertilization in the larger seeds of Bixaceae but it is foreshadowed in the ovule of the small neotenic seeds of Cistaceae.

The mature chalaza is involved in the closure of the wide chalazal foramen in the ripened seeds. The edge of the exotegmic palisade is inflexed and the core of the ground-tissue, which connects the end of the vascular bundle with the nucellus as a massive hypostase, becomes a hard brown operculum fitting the exotegmic annulus. In this mature and desiccated state it is practically impossible to make out the microscopic details. The most elaborate structure of testa, tegmen and chalaza occurs in *Bixa* which has the vestigial aril and softly spinous fruit as durian-relics. The evidence suggests that Bixaceae–Cistaceae, now relictual, diverged from the pachycaul ancestry of Sterculiaceae. The chalaza appears as an ancestral feature from the Myristicalean source.

The tegmen of *Bixa*, with palisade and hourglass cells (cf. Malvaceae), recalls the testa of Leguminosae. It is tempting to suppose that a transference of function has brought the tegmic construction, but the palmate leaf agrees with the Sterculiaceous stock. There is no resemblance in detail with the Dilleniaceous seed and this, again, would involve an ancestral leaf of pinnate construction.

Because of the parietal placentation, Bixaceae are classified usually with Capparidaceae, Flacourtiaceae and Violaceae. The fibrous exotegmen in these families makes a sharp distinction and their floral agreement seems to be an example of parallel evolution.

Bixa L.

B. orellana L.

(Singapore; living material.) Figs. 34–38.

Ovule; o.i. 4–5 cells thick; i.i. 5 cells thick, developing before o.i.

Seed 5–7 mm long, pyriform, subangular, often slightly grooved along the raphe, light grey with a red tinge (under a lens, red-spotted from the bixin-cells), drying dull red, the chalazal end as a slight pinkish or bluish umbo turning blackish and collapsing at maturity, the seeds remaining attached to the long white funicles (4 × 0.75 mm) for many weeks after dehiscence of the fruit; *aril* as a narrow white, minutely undulate, horny rim *c.* 0.3 mm wide, round the top of the funicle, slightly incised at the micropyle. Testa 4–6 cells thick, pulpy, eventually drying up and minutely papillate from the bixin-cells; *o.e.* with pink or bluish anthocyanin in some cells (especially over the chalaza), thin-walled except for the broad ring (*c.* 5 cells wide) of sclerotic cells around the chalaza; *mesophyll* developing the large red, oily, multinucleate, bixin-cells; *i.h.* with occasional pitted cells near the chalaza; *i.e.* unspecialized except at the chalaza as a palisade of thin-walled cells. Tegmen 7–9 cells thick, the additional layers contributed from the mesophyll; *o.e.* as a palisade, 60–80 μ high, with pale yellow lignified walls and brown lumina; *o.h.* as a layer of pyriform cells, slightly thick-walled, with brown gummy contents, the outer pointed ends projecting slightly between the palisade-cells, occasionally extending half-way into the palisade as intermediate cells, not as hourglass-cells; *mesophyll* 4–5 cells thick, without bixin-cells, composed of slightly thick-walled and more or less contiguous cells with pale brown contents, eventually more or less crushed; *i.h.* as a row of well-formed, thick-walled, hourglass-cells (stellate ends and columnar body), surrounding the endotegmen except along the raphe; *i.e.* with ridge-like thickenings on the anticlinal walls (? lignified). Chalaza (see below). Nucellus enlarging greatly, then crushed by the endosperm. Endosperm with slightly thickened walls and large starch-grains (absent from the outer 2–3 layers of cells).

Aril developing long after fertilization and slowly, from the top of the funicle (not micropylar), composed of cells with slightly thickened and pitted walls; *o.e.* with thicker walls; *i.e.* with strongly thickened and pitted anticlinal walls, as in the sclerotic cells of the chalazal epidermis. Funicle composed of similar cells with horny walls.

Chalaza developing into a small knob at the obtuse end of the seed, then depressed in the pulpy testa through the collapse of the thin-walled mesophyll, consisting internally of a core (in 3 sections) surrounded by 3 annuli of special cells, eventually drying into a woody, dark brown funnel with an operculum closing the wide heteropyle; *outer core* (Fig. 37, 3) composed of thin-walled cells in rows leading from the end of the v.b. to the middle core, developing dark brown gummy contents in the cells and forming the operculum; *middle core* (Fig. 37, 2) as a cushion-like mass filling the heteropyle, the cells shortly elongate, watery, colourless, with fine spaced annular thickenings on the walls, tracheid-like; *inner core* (Fig. 37, 1) as a column of short thin-walled cells connecting the middle core with the nucellus; *outer annulus* (Fig. 37, 7) composed of thick-walled cells, isodiametric, with dark brown gummy contents at maturity; *middle annulus* (Fig. 37, 6) composed of thin-walled cells radiating from the middle core to the exotegmic palisade; *inner annulus* (Fig. 37, 5) composed of slightly elongate cells connecting with the inner core by a sheath of slightly thick-walled cells (with dark brown contents at maturity) and with the endotegmen; *mesophyll* (tegmen) thickened and stellate-celled round the chalaza; *palisade* (exotegmen) inflexed and appearing in section as a sigmoid flexure, the cells adjoining the outer annulus with widened ends and wide lumina with brown gummy contents, and connected with the middle annulus by a narrow band (zone) of short thin-walled cells; *o.h.* (tegmen) forming strongly pyriform cells round the chalaza; *i.h.* (tegmen) ending abruptly round the chalaza with 2–5 contiguous thick-walled cells; *i.e.* (testa) as a short palisade of thin-walled cells round the chalaza.

Ovary initiated as a tumid ring, then shortly cylindrical with a bilabiate slit at the apex as the vestiges of the two carpel-primordia; the shortly cylindric ovary enlarging in the middle and forming the two parietal placentas, then with two regions of basipetal intercalary growth, the upper inserting the style, the lower the sterile ovary-base; when one-quarter of the final size, then with the external bristles and peltate scales beginning to form; *bristles* initiated from a single emergent cell, dividing into a cluster of cells, extending by basipetal growth, with a slender v.b. along the centre of the larger (earlier formed) bristles, not developing under peltate scales but the tips of the larger bristles becoming subclavate and somewhat glandular, the bristles lengthening to 5–15 mm after fertilization, losing the initial bixin-colour and reddening from anthocyanin in the epidermis, soft, flexible, withering hard and brown on dehiscence of the fruit; *ovary-wall* initially with bixin-cells and mucilage-canals, the inner epidermis with large stomata of early development, 2–3 layers of inner hypodermal cells with spherical aggregates of crystals persistent in the fruit.

Pericarp without bixin-coloration but with anthocyanin in the epidermis; *mesocarp* with a peripheral plexus of v.b. supplying the bristles, also with scattered idioblasts, but without sclerotic masses; *endocarp* consisting of 3–4 layers of small, slightly thick-walled cells detaching from the mesocarp and drying into a papery membrane, neither fibrous nor lignified.

(Mauritzon 1936b; Chopra and Kaur 1965; Kaur 1969.)

This American tree, now widely cultivated throughout the tropics, is a fund of botanical curiosity. It has peltate scales, palmate stipulate leaves and mucilage-canals of the Bombacaceous–Sterculiaceous alliance, centrifugal stamens as Dilleniaceae, horseshoe-shaped anthers which have been compared with those of *Gonystylus* (Hutchinson 1967), a practically acarpous ovary with parietal placentation which puts it with Flacourtiaceae and Violaceae, a durian-type of fruit, a seed with exotegmic palisade, sarcotesta and aril that agrees with that of Passifloraceae, and a chalaza which is the most intricate cellular contraption in the plant kingdom. It is one of those successful relics which, like *Gnetum*, *Magnolia*, *Carica*, *Myristica*, *Durio* and *Nipa*, challenge classification.

If the minute aril is to be regarded as a successful ecological extrusion of the sarcotesta according to the sarcotestal theory of van der Pijl, it has either led to nothing in the exarillate Cistaceous seed or to the Passifloraceous improvement. On the durian-theory it is a relic, as so many other features of this plant. Actually the aril has the structure of the funicle from which it is an outgrowth, not of the sarcotesta.

Cochlospermum Kunth

C. religiosum (L.) Alston
(Ceylon; flowers and fruits in alcohol.)   Figs. 39–42.
   Ovule; o.i. 3 cells thick; i.i. 3–4 cells thick.
   Seeds 6–7 mm long, dark brown, reniform, densely pilose with white hairs –12 mm long along the curved dorsal side of the seed, exarillate. Testa 3–4(–5) cells thick, 7–10 at the exostome, 15–25 at the chalaza; *o.e.* on the sides and concavity of the seed composed of subcuboid cells with polygonal facets, on the convex back of the seed with transversely elongate cells; *stomata* frequent on the dorsal and dorso-lateral parts of the seed; *hairs* aseptate, thin-walled, unbranched; *mesophyll* aerenchymatous near the chalaza and micropyle, but consisting of 1–2 layers of subcuboid cells on the sides of the seed; *i.e.* composed of thin-walled subtabular cells. Tegmen 7–10 cells thick, –16 cells at the chalaza, eventually shrivelling except for the palisade and chalazal plug; *o.e.* as a palisade 180–220 μ high; *o.h.* consisting of large cells, shortly radially elongate, with brown and rather oily contents but hyaline round the chalaza; *mesophyll* aerenchymatous; *i.e.* composed of subtabular brown cells with thickened walls (? not lignified), the radial walls slightly pitted at the junction with the inner walls. Chalaza apparently less elaborate than in *Bixa*, represented as a ring of thin-walled cells with brown contents (? ultimately lignified). Endosperm without starch. Embryo curved; radicle rather short; cotyledons ligulate, flat.
   Ovary 5-celled at base and apex, the middle part with 5 parietal placentas bearing the many ovules; stylar canal hollow, compressed at the apex of the ovary and 5-rayed; stigma minutely 5-lobed; v.b. with fine branches to the cortex; hairs simple, unbranched.
   Pericarp with a peripheral zone of small sclerotic blocks, with minute v.b. at the sides of the blocks.
   This seed differs from that of *Bixa* in its curved form

with curved embryo, the hairy testa with stomata but without bixin-cells or sclerotic cells round the chalaza, the longer exotegmic palisade and less specialized tegmic mesophyll, the less complicated chalaza, the absence of starch from the endosperm, and the absence of an aril. The curved form results from the feeble enlargement of the raphe which is foreshadowed in the ovule.

BOMBACACEAE

Ovules anatropous, bitegmic, crassinucellate, erect or transverse; o.i. 5–8 cells thick; i.i. 6–9 cells thick (? less); micropyle formed by the exostome, out of line with the endostome; hypostase more or less developed.

   Seeds medium-size to large, with massive and mostly multiplicative seed-coats, arillate in *Coelostegia*, *Cullenia*, *Durio* and *Neesia* albuminous or not. Testa 6–22 cells thick; *o.e.* composed of more or less cuboid cells, with hairs in many genera, with stomata in *Bombax* and *Ceiba*, more or less unspecialized in *Pachira*; *mesophyll* aerenchymatous; *i.e.* unspecialized. Tegmen 6–33 cells thick; *o.e.* as a palisade of lignified Malpighian cells or as a palisade of shortly radially elongate pitted cells more or less lignified (*Durio*) or not (*Cullenia*); *mesophyll* aerenchymatous, the cells more or less stellate; *i.e.* unspecialized or with irregular thickenings on the anti-clinal walls (*Pachira*). Chalaza massive, unspecialized. Vascular supply in many genera elaborate with post-chalazal branches, forming a plexus in some genera (*Durio*, *Pachira*). Endosperm nuclear, oily. Embryo straight; cotyledons thin, flat, folded or thick. Aril funicular and micropylar.   Figs. 43–53.
   (Netolitzky 1926; Venkata Rao 1954.)
   This rather small family (30 genera, 225 species) is comparable in pantropical distribution with Myristic-aceae (16 genera, 380 species). The nutmeg fruit is stabilized as a one-seeded follicle in which a large aril covers the seed. It presupposes a many-seeded follicle and, though the single modern seed may be an enlargement of the original, its follicle must have been as massive as in *Sterculia*. With this limitation the family has succeeded. It is divided into genera on characters seldom more different, commonly less, then distinguish subdivisions of *Ficus* which, vegetatively, is more varied; thus, in the fruiting state, Myristicaceae present a single genus of many species, such being the uncertainty of ordinals in taxonomy. In contrast, Bombacaceae illustrate the modification of the capsule from a many-seeded, arillate and armoured beginning into diverse forms which are exarillate, few or one-seeded, unarmoured and indehiscent; the seeds may be winged, as a frequent succession from the arillate, comose, converted into almost overgrown lumps (*Pachira*), or small and simple as in Malvaceae. In tree-form, leaf and flower, Bombac-aceae convey more evidence of their ancestry than

Myristicaceae and retain fragments of extensive evolution from a palmate-leafed pachycaul. The early stages of tree-making have disappeared from Myristicaceae, as indeed from Magnoliaceae, because such beginnings probably vanished before the rise of Bombacaceae.

The Indo-Malaysian tribe Durioneae has exploited the primitive capsule and in tree-form is comparable with Myristicaceae. *Durio* (27 species, though several are dubious) and *Cullenia* (2 species), if generically distinct, possess the large, vascular and complicated aril. In *Neesia* (8 species) and *Coelostegia* (5 species), the aril is merely a thickened collar without vascular supply at the base of the seed or it may be absent, and the capsule is massively sclerotic. *Kostermansia* (1 species) drops exarillate seeds from its dry durian-capsules and successfully becomes one of the few great and gregarious trees of the forest. There is a tendency for the aril to shorten in *Durio*. In the most slender and leptocaul species (*D. griffithii*) it is a red collar at the base of the black seed in the red and shortly spinous capsule 5–7 cm long. It is more or less reduced in four other species and is absent from *D. singapurensis*; sixteen species have the fully developed aril, but in six others fruit or seed are unknown (Kostermans 1956, 1958). The other tribes have advanced fruits though a trace of the aril may persist in *Ceiba* and *Hampea*; there is no evidence of sarcotestal evolution. These tribes are the African and American elements of advanced Bombacaceae in which, like *Durio*, merely three genera (*Eriotheca*, *Matisia*, *Pseudobombax*) have attained a comparable success with 20–25 species.

Durio and *Cullenia* have a feature of the tegmen which may be primitive. The palisade-layer is composed not of long Malpighian cells with almost solid prismatic base, but of short, strongly pitted cells which seem merely to be radially elongate sclerotic cells. They may be simplifications of the Malpighian cells or their antecedents unlengthened into the characteristic base. In *Cullenia* they appear unlignified, and small patches of unlignified cells may occur in the tegmen of some species of *Durio*. Where, however, the Malpighian cells are reduced and short, as in *Pachira* (or *Sterculia*), they retain their Malpighian character. The less advanced state in *Durio* and *Cullenia* may relapse more readily into undifferentiated parenchyma in the enlarged seeds, but this cannot be assumed as the primitive condition for the family. Thus the most successful genus *Durio*, as judged from specific numbers, and its ally *Cullenia* may be profoundly distinct both in the elaboration of the arillate capsule and in the tegmic construction from other Bombacaceae, for which *Coelostegia*, *Kostermansia* and *Neesia* are the relics. The tegmic structure of *Durio*, the vascular aril and the complex funicular vascular supply bear strongly on the origin of the arillate capsule of the Malaysian *Gonystylus* and its phyletic connection with the exarillate, drupaceous Thymelaeaceae.

Kostermans (1958) and Hutchinson (1967) comment adversely on the durian-theory, but offer no explanation of the origin of the durian fruit which is one of the most complicated among angiosperms. I am led to suppose that they rely on some mystic creation without genetic antecedent by the ecological summons which van der Pijl invokes (Corner 1952). Kostermans is mystified because he has twice published the anachronism that arils and, presumably, the very complicated capsules of arillate seeds may be both innovations and relics, and concludes that there is as much evidence for, as against, the durian-theory. He is worried that squirrels should gnaw unripe durians, without considering the anguish of a famished creature; rodents, of course, do much more damage to the unarmed coconut. Hutchinson imputes to the durian-theory the absurdity that '*Durio* and its fruit, and indeed all Bombacaceae' are primitive among flowering plants. Neither author contemplates the status of *Durio* in the great alliance of Bombacaceae, Tiliaceae, Sterculiaceae, Euphorbiaceae and Thymelaeaceae, where durian-relics are frequent, or the origin of Bombacaceae, or the primitive nature of the arillate seed of Myristicaceae. The evolution of the neotenic flower-bud should not be confused with the heritage of the ancestral fruit that announced the angiosperm.

Adansonia L.

A. digitata L.
Seed reniform. Testa 6–8 cells thick, thin-walled, the epidermal hairs 2–5-celled. Tegmen 6–10 cells thick, the palisade of Malpighian cells –1 mm high.
(Guignard 1893, p. 151; Venkata Rao 1954.)

Bombax L.

B. malabaricum DC
Seed-structure as in *Ceiba*, the testa also with stomata.
(Vaughan 1970.)

Ceiba Mill.

C. pentandra Gaertn.
Seed-structure very similar to that of *Gossypium*, the testa with stomata.
(Vaughan 1970.)

Coelostegia Benth.

C. griffithii Benth.
(Singapore; living material.)   Figs. 43–45.
Seed *c.* 3.5 cm long, narrowly clavate, attached to the basal axile part of the loculus, 1–3(–4) per loculus, dark brown, shiny, with an orange-yellow waxy aril as a basal cushion. Testa many cells thick; *o.e.* composed of shortly radially elongate cells with colourless lumen and strongly thickened brown outer wall, without stomata; *mesophyll* thick composed of more or less stellate cells with slightly thickened brownish arms, thicker along the raphe; *i.e.* unspecialized. Tegmen many cells

thick; *o.e.* as a palisade of colourless Malpighian cells –500 μ long, forming a cylindrical shell round the seed except for a wide opening at the chalaza; *mesophyll* composed of substellate cells with distinctly thickened and pitted walls, the inner tissue with brownish walls and crushed; *i.e.* unspecialized. Vascular supply in the funicle as a compound bundle surrounded by numerous small v.b., continued in the raphe with fine side-branches ramifying over the seed in the testal mesophyll, breaking up at the chalaza into a fan of fine bundles, several ending blindly in the thick chalazal tissue, others passing to the antiraphe side of the seed, without a main postchalazal bundle. Endosperm soft, very oily, thin-walled, the cells with fine oil-globules and with abundant starch (as in the embryo).

Aril derived from the exostome and from the junction of the seed with the funicle, breaking from the funicle by the disruption of its cortex and remaining attached to the seed, turning orange-brown and horny with soft interior at maturity on exposure; *o.e.* with thick brownish orange walls, without oil-drops in the cells, the outer walls finely pitted; *mesophyll* thin-walled, the cells full of colourless, yellow and orange, large and small oil-drops, without v.b. but the funicular v.b. with a bulge or lobe towards the micropyle before entering the raphe.

Capsule 11–15 cm wide, globose, ripening yellow-green to dull black on some trees and green to reddish brown then amber or purple to dull black on other trees (? same species), studded closely with conical angular spines 7–10 mm high, 10–20 mm wide at the base (the trees with reddening fruits with longer, narrower spines, finely grooved, –15 × 5–10 mm); loculicidally dehiscent about halfway from the apex to the base, the broad acute woody valves gaping slightly to show the lurid yellow and orange interior and the tops of the brown seeds. Pericarp very woody, with soft pockets in the septa drying up and rotting away in fallen fruits to form the slots in the septa; *spines* traversed by numerous fine v.b., bundles of fibres, and mucilage-canals; *exocarp* consisting of a single layer of epidermal cells and a narrow photosynthetic cortex, with mucilage-canals and short groups of woody fibres, the cells of both layers drying up at maturity with dark brown gummy contents to give the colour of the fruit; *mesocarp* thick, very woody, composed mainly of bundles of long fibres variously directed and spaced by groups of pitted tracheid-like cells; *endocarp* developing into a thick soft oily, pale orange-yellow tissue filling the loculi, pale pinkish orange near the septa, composed of thin-walled cells filled with oil-globules, turning lurid orange and yellow on exposure to the air on dehiscence; *septa* consisting of stout, uneven, somewhat flattened, woody bars passing obliquely from the centre of the fruit to the outer wall, separated by thin areas of soft tissue as in the septal pockets; *septal pockets* in pairs at the expansion of the

septa into the fruit-wall, consisting of soft, thin-walled tissue, unlignified but permeated by mucilage-canals, irregular radial rods of woody tissue, and the main v.b. of the pericarp; *axis* of fruit composed in the proximal third or quarter of the stout woody cylinder continuous from the fruit-stalk, tapering distally to a thin woody strand, the woody septal bars attached to the stout axial core at the base of the fruit and preventing its full dehiscence (cf. *Kostermansia* without columella or septal bars, and widely dehiscent).

This great fruit of detailed complexity falls to the ground where ants, chiefly, consume the soft oily tissue, carry off the seeds or bite away the aril, and leave the skeleton for the botanist. Is this an ecological device?

Cullenia Wight

Tegmen without lignified palisade. Vascular bundles in the testa apparently without anastomosis.

Aril microscopically longitudinally plicate, coarsely fimbriate at the distal end, but absent from *C. exarillata*.

(Kostermans 1956; Robyns 1970.)

C. zeylanica (Gardn.) K.Schum.

(Ceylon, Peradeniya; mature fruits in alcohol.) Figs. 46, 47, 52.

Seed 25–30 × 13–16 mm, oblong, with obtuse chalazal end, smooth, reddish brown, covered by the white to brownish aril, 1(–2) seeds per loculus, usually 2 sterile loculi in the fruit. Testa 20–30 cells thick; *o.e.* composed of cuboid cells, not lignified, with brown walls; *mesophyll* aerenchymatous, orange in the living tissue; *i.e.* unspecialized. Tegmen 30–34 cells thick, the inner layers becoming crushed; *o.e.* as a short palisade *c.* 35 μ high of thin-walled cells, slightly radially elongate, not or very slightly lignified, the radial walls eventually crumpled; *mesophyll* aerenchymatous. Vascular supply as a stout centric bundle in the raphe, dividing into 10–12 branches at the stout chalaza, the branches descending the sides of the seed in the testal mesophyll, ramifying in places but with little or no anastomosis, the micropylar ends of the branches joining the v.b. plexus in the funicle.

Aril 35–40 × 17–20 mm, 1.5–3 mm thick, developing from the exostome and the very short funicle, covering the seed, fimbriate-lobate at the distal end, white then brownish, the outer and inner surfaces finely longitudinally striate-plicate; o.e. composed of longitudinally elongate cells; v.b. very numerous, small, in a single layer in the central tissue, longitudinal with little (if any) anastomosis, arising from the funicular v.b.

Fruit 8–10 cm wide, subglobose, often irregular from sterile loculi, stoutly and rigidly spiny; *exocarp* traversed radially by many fibrovascular bundles entering the spines; *mesocarp* pulpy with many large mucilage-canals and a peripheral network of v.b., cream-white; *endocarp* membranous, unlignified; *spines* with many stout fibrovascular bundles.

The recent discovery of the exarillate species, *C. exarillata* of south India (Robyns 1970), confirms the postulate of the durian-theory that the aril is a primitive feature in Bombacaceae because it renders negligible the likelihood that *Coelostegia, Cullenia, Durio* and *Neesia* have independently evolved the highly complicated arillate capsule of the family.

Durio Adans.

Ovules sessile, anatropous, erect.

Seeds large, arillate or not, pale brown to black. Testa 15–22 cells thick, with a close plexus of v.b. in the mesophyll; *i.e.* unspecialized. Tegmen 14–30 cells thick; *o.e.* as a short palisade of more or less lignified, strongly pitted cells, not as radially elongate Malpighian cells, in some species incomplete with small undifferentiated patches of thin-walled cells showing as pale wavy lines on the surface of the testa; *mesophyll* aerenchymatous, the subglobose cells with short stellate arms, those of the outer part of the mesophyll with brown contents, eventually more or less crushed; *i.e.* unspecialized.

Aril covering the seed in most species, partly in others (*D. acutifolius, D. griffithii. D. malaccensis, D. oblongus*), absent from *D. singapurensis*, unknown in 6 species, red, yellow or white, apparently neither plicate nor fimbriate, pulpy or waxy, with a single median ring of small longitudinal v.b., not anastomosing; *o.e.* with the cells more or less longitudinally elongate; v.b. derived from the funicular ring of v.b.

(Kostermans 1958; Kostermans and Soegeng Reksodihardjo 1958.)

D. acutifolius (Mast.) Kosterm.
(Sarawak, S 28268; ripe seeds in alcohol.) Fig. 52.

Seed 23–27 × 11–12 mm, glossy black, about three-quarters covered by the dark red aril. Testa 15–18 cells thick; *o.e.* with slightly thickened external walls; *i.h.* and *i.e.* colourless, the rest of the mesophyll with brownish cell-contents. Tegmen 14–17 cells thick; *o.e.* as a palisade *c.* 70 μ high, with pitted cells slightly lignified in the inner halves. Vascular supply as in *D. zibethinus*.

Aril *c.* 1 mm thick at the base, thinner to the subcrenulate edge, not plicate or fimbriate.

D. zibethinus Murr.
(Singapore; living material.) Figs. 48–51.

Seed 3–5 × 2–3 cm, light brown, smooth, shiny, sessile, ellipsoid-subcylindric, with broad white mealy hilum, covered by the white, pale yellow or rich yellow aril. Testa reaching 1 mm thick in developing seeds with 16 cell-layers, shrinking to 0.5 mm in the ripe seed; *o.e.* short-celled, the somewhat thickened outer wall pale brown, not as a palisade, without stomata; *mesophyll* aerenchymatous, the cells substallate with slightly thickened and brownish walls, without mucilage-canals but with many small v.b. anastomosing over the whole seed; *i.e.* unspecialized. Tegmen white, reaching 1.5–1.7 mm thick, shrinking to 0.2 mm in the ripe seed, developing abundant mucilage-canals in the mesophyll (–16 cells thick), without v.b.; *o.e.* as a short palisade *c.* 35 μ high, colourless, the cells with strongly thickened and pitted anticlinal and inner walls, lignified, in places lacking. Chalaza unspecialized, with numerous divergent v.b. Micropyle occluded by the aril, the testa very thin at the exostome on joining the aril, but the tegmen here thickened with more cell-layers and almost entirely sclerotic with thick-walled pitted cells. Nucellus persisting as a thin layer at the micropyle. Endosperm completely absorbed. Placental tissue becoming mealy-friable and oily, permeated with fine v.b. ramifying into the aril.

Aril 2–15 mm thick in different parts of the seed, waxy then creamy-oily, arising after fertilization from the exostome and from the broad base of the enlarging ovule, permeated by fine v.b. in a single circle (in t.s.); *o.e.* composed of shortly longitudinally elongate cells with firm walls; *mesophyll* composed of elongate cells with mucilaginous walls; both sides of the aril, but chiefly the outer epidermis, secreting a thin yellowish waxy-resinous crust lining the loculus.

Vascular supply to the seed consisting of a single stout bundle in the raphe, forming a plexus over the seed and round the micropyle, the placental v.b. joining with those of the aril and testal mesophyll.

Pericarp with many radiating v.b. entering the spines, firm but not woody (except in the spines); endocarp not becoming pulpy or waxy.

(Corner 1949a.)

Durio sp.
(Sarawak, S 29470; ripe seeds in alcohol.) Fig. 52.

Seeds 30–32 × 16–18 mm, light brown, covered by the longer, thick aril. Testa 18–22 cells thick; *o.e.* with strongly thickened outer walls to the cells and reduced lumen, not lignified. Tegmen *c.* 30 cells thick, mostly crushed; *o.e.* as a palisade 45–60 μ high, composed of rather strongly lignified, pitted cells, absent in subcircular patches 0.1–1.4 mm wide, these patches appearing on dissection as minute holes on the sides of the seed. Vascular supply as in *D. zibethinus*.

Aril *c.* 40 × 20–23 mm, 2–3 mm thick, neither plicate nor fimbriate.

Neesia Bl.

Seeds as in *Coelostegia* but the testa with only a raphe v.b. (? in all species); aril absent from some species, practically vestigial.

Neesia sp.
(Sarawak, S 28431; full-grown but immature seeds in alcohol.)

Seeds 22 × 12 mm, cylindric with obtuse chalazal
end, dark brown, smooth, with a slight white pulpy
annulus round the micropyle, slightly keeled along the
raphe. Testa 11–14 cells thick; *o.e.* composed of small
cuboid cells with dark brown congealed contents, not as
a palisade, with isodiametric facets; *mesophyll* aerenchymatous, the thin-walled cells with brown contents; *i.e.*
composed of rather small, colourless cells, many with
a crystal. Tegmen 28–33 cells thick; *o.e.* as a palisade of
colourless lignified cells 350 μ high, as in *Coelostegia*;
o.h. colourless; *mesophyll* aerenchymatous, the cells in
the outer half with brown contents, the colourless inner
cells collapsing. Vascular supply only as the raphe-
bundle, ending in a broad network at the chalaza with
3–4 short branches not descending the sides of the seed.

This seed resembles that of *Coelostegia* and both are
in marked contrast with *Durio* in respect of the exotegmic
palisade. The aril is vestigial and represented merely
by the rather pulpy, colourless thickening round the
micropyle. The large woody capsule of *Neesia*, beset
on the inside with stiff and irritant hairs, is not an ecological device for dispersal but a sclerotic relic that pro-
tects the developing seeds. No doubt rummaging ants
make use of it as they will of any corpse.

Pachira Aubl.

P. insignis Savign.
(Brazil, Goiania; seeds in alcohol.)   Fig. 53.

Seeds very massive, rounded compressed or subangular with thick seed-coat 2.3–2.5 mm; both integuments with abundant starch. Testa 16–20 cells thick,
aerenchymatous with more or less stellate cells; *o.e.*
uneven the cells more or less separated, evidently thickening by periclinal division forming loose rows of 2–4
cells terminating in a rounded or clavate unspecialized
cell, without cuticular membrane; *i.e.* as a row of small
thin-walled cells somewhat conical or conico-truncated
in t.s. of the seed; *v.b.* forming a network in the mesophyll, bordered by large round brownish cells. Tegmen
25–30 cells thick, strongly aerenchymatous, the cells in
the middle region inflated with rather long, narrow arms,
becoming thick-walled but not lignified; *o.e.* as a short
palisade 70–90 μ high of prismatic cells, thickening
from the inner end to the outer testal side, lignifying
from the middle lamella; *i.e.* composed tabular, polygonal cells with lignified, often irregular, thickenings on
the anticlinal walls; *mesophyll* without v.b., more or less
collapsing.

BONNETIACEAE

If the seed-structure of *Ploiarium* is typical of this
family, then its alliance is not with Theaceae but with
Hypericaceae and Clusiaceae, as Takhtajan (1969) has
suggested. With the ripe seeds of *Ploiarium* that were
available to me, it was not possible to decide whether the

lignified, undulate-oblong cells that make the second
layer of the seed-coat were endotestal or exotegmic. The
seed agrees generally, however, with the Hypericaceous,
and the small gaps between the bases of the exotestal
cells seem to indicate an inner thin-walled endotesta
that had collapsed, as in that family. On the other hand,
the elongate form of these exotegmic cells links the
Hypericaceous or Elatinaceous seed with that of Ixonanthaceae (p. 29). If the layer is endotestal, then the alliance
must be with Dilleniaceae, which seems improbable.

Ploiarium Korthals

P. alternifolium Melchior
(Malaya; ripe capsules in alcohol.)   Fig. 54.

Seeds 2.5–3 × 0.25–0.3 mm, many per loculus,
anatropous, erect, subcylindric with a short extension
or flange at each end, whitish. Seed-coat composed of
two layers of longitudinally elongate, oblong cells;
outer layer (exotesta) thin-walled, not lignified, with
small gaps between the inner ends of the cells (as seen in
t.s.), the cells more or less cuboid in the terminal flanges;
inner layer (? exotegmen) composed of longitudinally
elongate cells with thickened, pitted and lignified radial
and inner walls, the radial walls undulate, giving
elongate undulate facets. Vascular bundle only in the
raphe, recurved at the chalazal end, slender. Endosperm
none or reduced to a single layer of cells at the ends of
the seed. Embryo straight; radicle long; cotyledons
short.

Capsule with lignified endocarp as a valve to each
loculus, consisting of an outer palisade of sclerotic cells
and an inner layer, *c*. 2 cells thick, of transverse sclerotic
cells.

The extension at the micropylar end of the seed may
be a vestigial aril.

BORAGINACEAE

Ovules anatropous to hemianatropous, suspended,
unitegmic, tenuinucellate; integument 5–14 cells thick,
massive; i.e. as an endothelium (*Heliotropium*) or not;
hypostase present.

Seeds exotestal with the mesophyll more or less
crushed, albuminous; *o.e.* with thickened, lignified and
pitted outer wall, the cells not as a palisade, or with the
inner and radial walls thickened (*Ehretia, Heliotropium,
Mertensia*); *mesophyll* thin-walled, a few cell-layers
persisting or not (Heliotropeae). Vascular supply as a
short raphe-bundle, simple (*Heliotropium*), or dividing
on entry into the seed into 4–7 prechalazal branches to the
sides of the seed (*Cynoglossum, Echium, Lithospermum,
Pulmonaria*), or with 3–7 or more postchalazal branches
(*Anchusa, Borago, Nonnea, Symphytum*), one such branch
almost reaching the micropyle. Endosperm cellular,
helobial, or nuclear, as several cell-layers in the seed
(Heliotropeae) or reduced to a single cell-layer.

(Guignard 1893; Netolitzky 1926; Kühn 1927; Millsaps 1940; Johri and Vasil 1956; Pal 1963; Khanna 1964a.)

The details of the exotesta, of the vascular supply, and of the endosperm will undoubtedly assist classification.

BRETSCHNEIDERACEAE

The red seed of this monotypic family may decide whether its affinity is Sapindaceous, Capparidaceous, Moringaceous, or even Leguminous, as has been suggested.

BRUNIACEAE

Ovules anatropous, suspended, 1–2 per loculus, unitegmic, ? crassinucellate or with a small evanescent nucellus; integument arising by fusion of two primordia.

(Netolitzky 1926.)

BURSERACEAE

Ovules anatropous to campylotropous, suspended, bitegmic, (? unitegmic in *Canarium* and *Santiria*), crassinucellate; micropyle formed by the endostome.

Seeds enclosed in the woody endocarp, exarillate, albuminous or practically exalbuminous. Testa evidently multiplicative, unlignified except i.e.; *o.e.* as a continuous layer of cells shortly radially elongate, cuboid or tabular; *mesophyll* aerenchymatous, becoming crushed; *i.e.* as a pellicle of small, cuboid or tabular, lignified cells, sometimes shortly tangentially elongate but not fibriform, with fine spiral, gyrose, annular or subscalariform thickenings of the wall. Tegmen soon crushed, unspecialized. Chalaza ? more or less pachychalazal. Perisperm persistent as a thin, more or less disorganized, cuticulate layer. Endosperm nuclear, thin-walled, oily, reducing to one or a few layers of cells. Embryo large; radicle short; cotyledons often folded or palmately lobed.

(Netolitzky 1926; Narayana 1960; Vaughan 1970.)

The development of the seed appears not to have been followed in this family. The study of the adult seed suggests a tracheidal endotesta as in Lauraceae and Monimiaceae. The unitegmic condition, said to occur in *Canarium* and *Santiria*, may result from loss of the tegmen, because I find the tracheidal endotesta in *Canarium*, or through pachychalazal construction.

Canarium Stickm.

C. indicum L.
(Ceylon, Peradeniya; ripe fruits in alcohol.) Fig. 55.

Testa 11–15 cells thick, not lignified except i.e.; *o.e.* as a layer of cells shortly radially elongate, with slightly thickened outer wall and slight thickenings on the radial walls; *mesophyll* aerenchymatous with substellate cells, ? eventually crushed; *i.e.* as a pellicle of rhomboid cells, distinctly lignified, with very fine, irregular,

subspiral or subgyrose, thickenings on the radial and outer walls. Chalaza elongate as a broad band along the adaxial side of the seed. Hypostase composed of small, thin-walled, brown cells. Vascular bundle of the raphe dividing in the distal part of the chalaza and giving off a few branches to the side of the seed (? not anastomosing). Perisperm with distinct cuticle, more or less crushed. Endosperm thin, thin-walled, oily.

Commiphora Jacq.

C. caudata (Wight et Arn.) Engler
(Ceylon, Dambulla; ripe fruits in alcohol.) Fig. 55.

Testa 7–9 cells thick; *o.e.* as a layer of cuboid cells, thin-walled; *mesophyll* more or less crushed; *i.e.* as a pellicle of rhomboid lignified cells with rather irregular scalariform thickening on the outer and radial walls. Chalaza broad, ovoid, on the adaxial side of the seed, with the very short raphe v.b. branched in the chalaza. Hypostase composed of thin-walled brown cells. Perisperm more or less crushed. Endosperm thin, oily.

Protium Burm. f.

P. unifoliolatum Engl.
(Brazil, Mato Grosso, Chavantina, leg. Corner 1968; fruits in alcohol.) Fig. 56.

Seeds 5 mm wide, enclosed in the woody endocarp covered by the white arilloid mesocarp. Testa 8–10 cells thick, thicker in the raphe; *o.e.* as a pavement of tabular cells with sinuous facets and irregular radial walls, some with short flange-like extensions into the cell-lumen, the walls strongly thickened, not lignified; *mesophyll* aerenchymatous with substellate cells, thin-walled, collapsing; *i.e.* as a pavement of short, lignified tracheids with annular or subspiral thickenings, but absent along the lines of the v.b. Tegmen 4–5 cells thick, unspecialized, collapsing. Vascular bundles, one in the raphe and two on the sides of the seed (? post-chalazal). Endosperm reduced to a single layer of cells. Embryo with small radicle concealed between the basal lobes of the cotyledons.

Capsule 12–14 mm wide, subglobose, typically 3-shouldered, 1–3 seeded, red, splitting from the base into 1–3 convex valves showing the white arilloid mesocarp, the tissue with a strong resinous smell. Exocarp composed of small cells, with abundant large resin-canals, the larger subtended on the inner side by a small v.b., mostly longitudinal, some anastomosing. Mesocarp splitting from the exocarp on dehiscence of the capsule, composed of rows of radially elongate, pulpy cells, mostly 3–4 in a row, terminating at the boundaries with exocarp and endocarp in a row of small cells, the cell-walls slightly thickened, neither pitted nor lignified. Endocarp composed of longitudinal and oblique fibres, very thick-walled, lignified.

Unfortunately the seeds were too few and too mature to make out exactly the vascular supply in the testa.

BUXACEAE

Ovules anatropous, suspended with the micropyle towards the axis (*Buxus*) or away from it (*Sarcococca*), bitegmic, crassinucellate; o.i. of variable thickness; i.i. 2(–3) cells thick.

Seeds small to medium-size, oblong, brown or black, exarillate or with a small caruncle. Testa multiplicative (*Sarcococca*) or not (*Buxus*); o.e. as a lignified palisade; *mesophyll* thin-walled, crushed, or with 1–2 hypodermal layers of sclerotic cells (*Sarcococca*); i.e. unspecialized or shortly palisade-like and lignified at the micropylar end (*Buxus*). Tegmen not multiplicative, crushed; i.e. lignified (*Sarcococca*). Vascular bundle in the raphe only or with slight postchalazal ramification (*Buxus*), or extensive (*Simmondsia*). Endosperm oily, thin-walled, or absent (*Simmondsia*). Embryo straight or curved; radicle long (*Buxus*) or short (*Sarcococca*); cotyledons thin, flat, or thick and fleshy (*Simmondsia*).

(Netolitzky 1926.)

The exo-mesotestal structure agrees with the disposition of the family in Hamamelidales, away from the exotegmic Euphorbiaceae. The relict nature of the family, emphasized by Hutchinson (1967), is borne out by the diversity in detail of the ovule and seed. *Buxus* is peculiar in the false endostome, and it has a vestige of a funicular aril. *Sarcococca* has one or two hypodermal layers of sclerotic cells in the testa. If a fully developed aril and a fully sclerotic mesophyll were combined, then such a seed would closely resemble that of Stachyuraceae, placed near Buxaceae by Hutchinson; and *Stachyurus* fits as a keystone in the distribution of Buxaceae. The position of *Simmondsia* is considered uncertain (Wunderlich 1968), but its seed-structure is Buxaceous.

Buxus L.

B. sempervirens L.
(Cambridge; living material.)   Figs. 57, 58.

Ovule with adaxial micropyle; o.i. 6–10 cells thick, the thicker part along the raphe, the cells of o.e. shortly palisade-like; i.i. 2(–3) cells thick; v.b. already with 5 short postchalazal branches; funicle short, slightly swollen at the junction with the ovule on either side of the micropyle.

Seeds 6 × 2.5 × 2 mm, brown then black, shiny, with two small waxy-horny white lobes on either side of the lateral-subapical micropyle (as a rudimentary funicular aril or caruncle). Testa non-multiplicative; o.e. as a hard palisade of prismatic cells with thick yellow-brown walls; *mesophyll* aerenchymatous, thin-walled; i.e. unspecialized at the chalazal ends, as a short thin-walled palisade on the sides of the seed, but at the micropylar end forming a hard palisade as that of the exotesta and making a false endostome. Tegmen non-multiplicative,

unspecialized. Vascular supply as in the ovule, the post-chalazal branches extended over the base of the seed. Embryo curved, with long radicle.

Ovary constructed basipetally as a tube beneath the stigmata (as the carpel-primordia); stomata abundant on the styles, absent from the syncarpous part of the ovary; v.b. differentiating basipetally.

Pericarp; outer hypodermis sclerotic; endocarp composed of several layers of fibres (transverse in the septa, longitudinal in the valves), and with the inner epidermis as a palisade of thick-walled prismatic cells.

Sarcococca Lindl.

S. zeylanica Baill.
(Ceylon, Wiriawan n. 773; flowers and fruits in alcohol.) Figs. 59, 60.

Ovule with abaxial micropyle, at anthesis minute and rudimentary with small nucellus; o.i. 2(–3) cells thick; i.i. 2 cells thick; obturator none.

Seeds 5 × 2 × 2.5 mm, pale to dark brown, exarillate, without trace of caruncle. Testa multiplicative, 12–13 cells thick; o.e. as a palisade of thick-walled, sparsely pitted, prismatic cells with brown contents; *mesophyll* with 1–2 hypodermal layers of small thick-walled lignified cells; i.e. unspecialized. Tegmen not multiplicative; i.e. lignified with bar-like thickenings on the anticlinal walls. Embryo straight with short radicle.

Pericarp with many large irregular stone-cells. Fruits 1–2(–3) seeded. Inflorescence spikes axillary; 1–3 female flowers at the base, 5–9 male flowers distally.

Simmondsia Nutt.

S. californica Nutt.

Seeds 16 mm long, oblong, subtriquetrous, red-brown, exarillate. Testa *c.* 10 cells thick; o.e. as a palisade of thick-walled cells –120 × 25 μ, with brown contents; *mesophyll* aerenchymatous, thin-walled; i.e. with rectangular pigmented cells. Vascular bundle with postchalazal branches reaching to the micropyle. Endosperm none. Cotyledons thick.

(Netolitzky 1926; Vaughan 1970.)

BYBLIDACEAE

Ovules anatropous, bitegmic (? unitegmic in *Byblis*), tenuinucellate; i.i. with i.e. as an endothelium; micropyle formed by the endostome.

Seeds small. Testa; o.e. with tangentially and transversely elongate cells with strongly thickened and pitted walls. Endosperm ? cellular, persistent.

(Netolitzky 1926, under Droseraceae.)

The modern tendency is to place this family near to Pittosporaceae. The seed-structure, so far as known, seems to prefer the old position near Droseraceae.

CACTACEAE

Ovules anatropous to more or less campylotropous, bitegmic, crassinucellate, i.i. developing before o.i. and often forming a prominent endostome; o.i. 2 cells thick, 3–4 in *Cereus*; i.i. 2(–3) cells thick, –4 cells at the endostome; funicle often curved and more or less surrounding the ovule, often with funicular hairs; often with a conspicuous air-space between the integuments on the antiraphe-side.

Seeds more or less campylotropous, mostly (?) with non-multiplicative integuments, perispermous, albuminous, exarillate. Testa with o.e. as the mechanical layer, the cells with more or less strongly thickened, often convex, brown, outer wall or becoming entirely thick-walled; *i.e.* persistent or not, thin-walled. Tegmen crushed or with persistent i.e. as a layer of thin-walled cells or with bar-like thickenings on the radial walls. Funicle with somewhat dilated apex, the inner tissue of the apex next the seed in some cases becoming rather fibrous and forming a hilar cup on desiccation of the soft tissue, in some cases the funicle more or less covering the seed, in Opuntioideae often with papillate epidermal outgrowths becoming pulpy. Vascular bundle of the raphe ending at the chalaza. Perisperm persistent on the incurved side of the seed. Endosperm nuclear, more or less persistent, generally much reduced.

(Netolitzky 1926; Mauritzon 1934c; Neumann 1935; Archibald 1939; Maheshwari and Chopra 1955; Tiagi 1956b, 1957; Buxbaum 1958, 1961, 1968; Engelman 1960; Kapil and Prakash 1969.)

It appears from the researches of Maheshwari and Chopra, Tiagi, Kapil and Prakash, and Engelman on *Opuntia*, *Mammilaria*, *Cereus*, *Ferocactus*, *Astrophytum*, *Thelocactus* and *Toumeya* that the seed-coats are merely enlarged and differentiated integuments with little or no cell-division. As usual in Centrospermae, the seed is simply exotestal, but there are evidently many variations in the thickening of the outer wall of the exotestal cells which may correspond with those in the epidermis of the stem. The funicle seems, also, to have undergone many diversifications and introduces its own chapter of funicular evolution without, apparently, arillate accompaniment.

Recently the Cactaceous seed has been compared with those of Phytolaccaceae and Illiciaceae (Buxbaum 1961), which introduce the arillate seed. The larger Illiciaceous seed is more primitive in the multiplicative testa and tendency to a thick-walled mesotesta.

CALLITRICHACEAE

Ovules anatropous, suspended, unitegmic, tenuinucellate; integument 2 cells thick.

Seeds minute, with membranous seed-coat (? persistent o.e.). Endosperm oily, cellular.

(Netolitzky 1926; Maheshwari, P. 1963.)

This small family has been allied with Halorrhagaceae and Labiatae, as if an aquatic and apetalous *Mentha*. The seed appears to offer no help in the decision.

CALYCANTHACEAE

Ovules 2 per carpel, marginal, the upper abortive, the basal fertile, anatropous, erect, bitegmic, subcrassinucellate; o.i. 6–9 cells thick; i.i. 4–5 cells thick; nucellus narrowly cylindric, the wall 3–5 cells thick; hypostase present; micropyle closed by both endostome and exostome, both with a fringe of microscopic hairs; funicle short, slender.

Seeds 10–14 mm long, subcylindric, pale brownish, rather soft, exalbuminous, exarillate, enclosed in the hard pericarp. Testa 13–16 cells thick, –22 in *Chimonanthus*, becoming crushed except o.e. and i.i.; *o.e.* consisting of cuboid cells, often more or less projecting, thin-walled, with brownish contents, slightly lignified in *Chimonanthus*; *mesophyll* thin-walled; *i.e.* consisting of shorter, often somewhat tangentially elongate, lignified cells with reticulate thickening, forming a persistent hyaline pellicle. Tegmen with the mesophyll cells enlarging, not multiplicative, soon crushed by the endosperm. Vascular bundle of the raphe expanding in the shallowly cupular chalaza, the ground-tissue thin-walled without lignification. Endosperm cellular, becoming large-celled, then mostly crushed by the embryo. Cotyledons convolute, thin. Peduncle with inverted cortical v.b. Figs. 61, 62.

(Netolitzky 1926; Brofferio 1930.)

Both genera, *Calycanthus* and *Chimonanthus*, have almost identically the same structure of the ovule and seed. The family has been related with Monimiaceae and Rosaceae. The tracheidal form of the pellicular endotesta confirms the alliance with Monimiaceae and Lauraceae. The large embryo is Lauraceous.

CAMPANULACEAE

Ovules anatropous, unitegmic, tenuinucellate; integument 3 cells thick (*Pentaphragma*), 6 cells thick (*Downingia*).

Seeds small or minute, albuminous, exarillate. Testa with i.e. at first endothelial, then crushed with the mesophyll; *o.e.* composed of cuboid cells with strongly thickened, pitted and lignified, radial and inner walls (*Isotoma*, *Pentaphragma*) or of fibriform cells with thickened, pitted and lignified, radial walls and narrow lumen varying hourglass-shape in t.s. (*Campanula*, *Downingia*, *Wahlenbergia*). Endosperm cellular, oily, thin-walled, starchy in *Cephalostigma*.

(Netolitzky 1926; Kausik and Subramanyam 1947; Subramanyam 1948b, 1951c; Kapil and Vijayaraghavan 1965; Beltran 1970; Kaplan 1970.)

There appears to be a distinction in this family between the cuboid exotestal cells with isodiametric facets and the fibriform.

CANELLACEAE

Ovules numerous, anatropous, bitegmic, crassinucellate; o.i. 4–8 cells thick; i.i. 3 cells thick.

Seeds small, with shiny crustaceous testa, rugulose-cerebriform in *Cinnamosma*, exarillate or with a vestigial hilar aril in *Canella*, albuminous. Testa ? multiplicative; *o.e.* as a layer of sclerosed cells (except *Cinnamosma*); *mesophyll* unspecialized or with oil-cells. Tegmen apparently crushed. Endosperm oily, ruminate in *Cinnamosma*. Embryo small to moderately well developed.

(Baillon 1871; Netolitzky 1926; Melchior and Schultze-Motel 1959; Parameswaran 1961.)

The seeds appear exotestal and to indicate alliance with Winteraceae. The larger seed of *Cinnamosma* with ruminate endosperm and unspecialized testa appears to be pachychalazal according to the description given by Parameswaran. Most systematists now place the family in Magnoliales near Annonaceae and Myristicaceae, but there is no evidence of the peculiar seed-coats of those families in Canellaceae. The syncarpous ovary with parietal placentation is, however, not Magnolialean, though similar to the syncarpous ovary of the exotestal Nymphaeaceae. Baillon described a vestigial aril round the hilum of *Canella*, but later authors have not reported it.

CANNABIACEAE

Ovule anatropous, suspended, solitary, bitegmic, crassinucellate; o.i. 3–4(–8) cells thick; i.i. 3 cells thick.

Seed small, enclosed in the small nut, albuminous. Testa not multiplicative, more or less crushed; *o.e.* as a closely perforate reticulum of tangentially elongate cells with many short arms, thin-walled; *mesophyll* with substellate cells. Tegmen obliterated. Hypostase corky. Endosperm nuclear, oily. Embryo curved or involute.

(Netolitzky 1926; Mohan Ram and Nath 1964; Vaughan 1970.)

This is the Urticalean seed without specialization of the seed-coats.

CAPPARIDACEAE

Ovules anatropous or, more generally, campylotropous, bitegmic, crassinucellate; o.i. 2 cells thick; i.i. 2–6 cells thick; micropyle formed by the exostome, out of line with the endostome or formed by the endostome (*Crataeva*).

Seeds campylotropous, fibrous-exotegmic, with testal stomata in some species, albuminous, arillate in some genera. Testa 2–6 cells thick, usually non-multiplicative in small seeds; *o.e.* as an uneven palisade of columnar cells with more or less thickened walls, or with cuboid cells, or compressed, with stomata in *Capparis*, *Cleome spinosa* and *Isomeris*; *mesophyll*, when present, composed of shortly tangentially elongate, thin-walled cells at right angles to the fibres of the exotegmen; *i.e.* unspecialized or with thickened walls (*Capparis*, *Cleome*) or much enlarged (*Cleome viscosa*). Tegmen multiplicative through periclinal division of all three cell-layers (? i.e.); *o.e.* as a single layer of longitudinal, oblique, or more or less transverse fibres, thick-walled, pitted, lignified, and appearing as a palisade of columnar cells in t.s., or multiplied into 2–10 layers of subcylindric fibres (*Capparis*); *mesophyll* composed of 3–9 layers of thin-walled cells, shortly tangentially elongate at right angles to the fibres, often collapsing; *i.e.* more or less lignified with pitted or fibrous walls, at least the inner tangential wall, or unspecialized (*Cadaba*, *Crataeva*), eventually forming a white inner pellicle; the tissue of the tegmen differentiating from the micropyle towards the chalaza. Vascular bundle of the raphe dividing into two short lateral branches (? always). Perisperm often persisting as a single layer of cells. Endosperm nuclear, persisting as 1–6 layers of cells. Embryo curved. Figs. 63–70.

(Netolitzky 1926; Mauritzon 1935b; Pax and Hoffmann 1936.)

The seed-coats are complicated and require in explanation longitudinal, transverse and tangential sections; as these have not always been made known in publications, some doubts remain concerning the direction of the fibres of the tegmen and their shape. Many more genera need investigation before the family character can be truly assessed, particularly such apparently arillate genera as *Emblingia* (West Australia), *Physena* (Madagascar) and *Calyptotheca* (Africa). I give merely such distinctions as appear to be affirmed. It is not clear that the seed-structure coincides with generic limits in genera with small seeds, e.g. *Cleome*, *Gynandropsis*. The so-called third integument, described by Orr (1921a, b, c), has been shown by Raghavan (1937) to be the meso-tegmen and endotegmen which may detach in dried seeds.

There are two kinds of seed-coat. That of *Capparis* seems to distinguish the more massive seeds 5–12 mm wide. The other kind can be subdivided into the structure of the larger seeds 4–6 mm wide (*Crataeva*) and that of the smaller seeds 1–2 mm wide (*Cleome*, *Gynandropsis*), but there may be intermediates as *Cadaba* and *Isomeris*.

(1) *Capparis*: testa with unspecialized o.e., but usually with stomata; tegmen with 4–10 layers of subcylindric to subfusiform fibres, longitudinal or oblique, in t.s. subcircular and not palisade-like; endotegmen lignified.

(2) *Crataeva–Cleome*: tegmen with a single layer of flattened fibres appearing as a palisade in t.s. (? 1-2-3 layers of such fibres in *Cadaba* and *Isomeris*, or merely one layer with overlapping fibre-ends).

(*a*) *Crataeva*: exotesta as a palisade of cells with thickened walls (at least the outer wall) without stomata; exotegmic fibres transverse to the long axis of the seed; endotegmen not lignified.

(*b*) *Cleome*: exotesta unspecialized or as a palisade, with or without stomata; exotegmic fibres more or less longitudinal; endotegmen lignified: *Cleome*, *Gynandropsis*, *Polanisia*, ? *Cadaba*, ? *Isomeris*.

Cadaba Forssk.

C. indica Lamk.
Ovule campylotropous; o.i. 2 cells thick; i.i. 3 cells thick.

Seed 2.5 × 2 mm. Testa; *o.e.* composed of subcuboid cells with thick outer walls; *i.e.* with thick inner tangential walls. Tegmen 3(-4) cells thick; *o.e.* as in *Crataeva* (?), or in places 2 cells thick from periclinal division; middle layer of cells and i.e. thin-walled, crushed.

(Narayana, H. S. 1965.)

Capparis L.

C. decidua (Forssk.) Pax, *C. divaricata* Lamk., *C. frondosa* Jacq., *C. grandis* Linn.f., *C. sepiaria* L., *C. zeylanica* L.   Figs. 63–67.

Ovule campylotropous; o.i. 2 cells thick; i.i. 3–4 cells thick.

Seeds 6-12 × 5-7 mm, irregularly reniform, subcompressed, sessile, brown to blackish, smooth, embedded in the firm, pulpy or creamy (*C. zeylanica*) endocarp. Testa 0.2–0.3 mm thick, the cells at first filled with starch, eventually all the cell-walls more or less lignified, with or without distinct thickening, practically non-multiplicative, 3–4 cells thick near the chalaza and at the micropyle; *o.e.* with unspecialized cells often compressed, with air-spaces at the angles (*C. sepiaria*), with stomata (? in *C. decidua* and *C. frondosa*), some with anthocyanin in the walls; *i.e.* unspecialized, or with much thickened inner tangential walls in *C. decidua* (? lignified). Tegmen consisting of an outer fibrous and lignified layer, a thin-walled and unlignified mesophyll, and a lignified endotegmen; *fibrous layer* 4–10 cells thick (3–4 in *C. frondosa*), derived mainly from o.e., the fibres more or less longitudinal or oblique, not flattened, often irregular, pitted, sometimes with short lobes, short and isodiametric at the micropyle; *mesophyll* 5–7 cells thick, 3–5 in *C. sepiaria*, shortly elongate tangentially at right angles to the fibres; *i.e.* composed of cuboid cells firmly contiguous, with strongly thickened radial walls, the outer wall thickened (*C. grandis*) or not, the inner wall thin, with reticulate thickening or fibrous (*C. decidua*). Embryo with long stout hypocotyl curved by two bends into three parts;

radicle apicular, short; cotyledons flat, thin, set on one side of the hypocotyl.

Pericarp in *C. divaricata*, *C. grandis* and *C. zeylanica*, with the clusters of stone-cells in a double layer in the outer part of the mesocarp, one layer being external to the ring of v.b. and the other internal, but joining into a single thick sclerotic layer at maturity, the fruit red and dehiscing irregularly in *C. zeylanica*; in *C. sepiaria* with a single ring of sclerotic cells external to the v.b. in the small fruit; *endocarp* with the cells full of starch, in *C. zeylanica* developing shallow reticulate and scalariform lignified thickenings on the cell-walls.

(Mauritzon 1935b; Narayana, H. S. 1962b.)

Cleome L.

Ovules anatropous or campylotropous; o.i. 2 cells thick; i.i. 2–3–5 cells thick, but thicker at the micropyle.

Seeds small, campylotropous, compressed reniform, exarillate. Testa 2 cells thick, with stomata in *C. spinosa*. Tegmen 4–8 cells thick, the exotegmic fibres with crystals in the lumen.

There appears to be considerable variation in detail among the seeds of this genus, some species of which seem to resemble rather *Gynandropsis*, e.g. *C. chelidonii*.

C. chelidonii Linn.f.
Ovule; i.i. 5 cells thick. Testa; *o.e.* composed of thick-walled cells shortly elongate radially as a rather loose palisade, the longer cells in small groups to form the spinous processes on the seed, without stomata; *i.e.* with small thick-walled cells (? lignified). Tegmen 7–8 cells thick; *o.e.* as a palisade in t.s. (but ? fibres), the cell-walls thick, lignified with scalariform thickening, the cells containing crystals; *mesophyll* 5–6 cells thick, thin-walled; *i.e.* with rather large cuboid cells with lignified and pitted walls.

(Raghavan 1937.)

C. monophylla L.
Ovule; o.i. and i.i. each 2 cells thick. Testa; *o.e.* with thick walls. Tegmen 6–7 cells thick; *o.e.* ? (mature seeds not studied); *i.e.* with lignified walls.

(Rao, A. V. N. 1967.)

C. spinosa Jacq.
Testa with stomata in o.e. Tegmen 4 cells thick; *o.e.* as a layer of long pitted fibres, lignified.

(Orr 1921a.)

C. viscosa L.
(Ceylon; flowers and fruits in alcohol.)   Fig. 68.
Ovule anatropous to subcampylotropous; o.i. 2 cells thick; i.i. 3 cells thick.

Seeds 1.3 mm wide, compressed, dark brown to blackish, transversely rugulose. Testa 2 cells thick; not lignified; *o.e.* consisting of small cuboid cells, without stomata; *i.e.* consisting of much enlarged hyaline cells extending between the ridges of the tegmen. Tegmen 6–7 cells thick; *o.e.* with lignified, closely pitted fibres containing small tetrahedral crystals, shortened and prismatic round the micropyle, in places with the ends raised on the outside to form the stout ridges or points of the rugulose surface; *mesophyll* consisting of 4–5 layers of small cells with slightly thickened and lignified walls, elongate shortly and transversely to the exotegmic fibres, not pitted; *i.e.* as a membrane of tabular cells with thickened, slightly lignified and very finely pitted inner walls. Endosperm reduced to a single layer of cells.

Crataeva L.

Ovules subanatropous; o.i. 2 cells thick; i.i. 3 cells thick.

Seeds 4.5–6 mm wide, reniform, cochleate, rough, dark brown. Testa 3–6 cells thick, not lignified, drying pellicular, without stomata; *o.e.* as an uneven palisade of shortly columnar cells with thickened brown outer walls, the longer cells set in groups to give the rough surface; *mesophyll* 1–4 cells thick, the cells elongate tangentially and more or less at right angles to the exotegmic fibres, becoming slightly thick-walled; *i.e.* unspecialized. Tegmen 5–9 cells thick; *o.e.* as a single layer of very thick-walled fibres elongate transversely to the long axis of the seed, closely and finely pitted, lignified; *mesophyll* thin-walled, the cells shortly elongate longitudinally, not lignified; *i.e.* as a layer of thin-walled cells with hexagonal facets, not lignified. Embryo with short radicle; cotyledons thick, folded at the tip.

C. nurvala Buch. Ham.
Testa 4–6 cells thick. Tegmen 7–9 cells thick.
(Narayana, H. S. 1965.)

C. religiosa Forst. f.
(Ceylon; flowers and fruits in alcohol.) Figs. 69, 70.
Testa 3–4 cells thick, –12 cells in the parts adjoining the raphe, at first all the cells filled with starch. Tegmen 5–8 cells thick. Vascular bundle shortly bifid as in *Capparis*.

Pericarp with a single hypodermal layer of large clusters of stone-cells, with a few small clusters inside the main layer but not as a definite second layer; v.b. as an interconnecting double ring; endocarp consisting of thin-walled, radially elongate cells filled with starch-grains, the innermost cells pressed closely on the seeds.

Gynandropsis DC

Gynandropsis sp.
(Cambridge, University Botanic Garden; living material.)
Ovule; o.i. 2 cells thick; i.i. 3 cells thick.
Testa 2 cells thick; *o.e.* composed of thin-walled cuboid cells, raised here and there into papillae by local radial elongation of small groups of cells and leaving a gap in the papilla between the two cell-layers, not

lignified, without stomata; *i.e.* unspecialized, with small cells. Tegmen 6 cells thick, as in *Cleome viscosa*. Vascular bundle shortly bifid at the chalaza.

The seed much resembles that of *Cleome viscosa* but the cells of the endotesta do not enlarge and the papillae are formed only by the exotesta. In *C. chelidonii* the elongate exotestal cells do not leave a gap under the papilla.

Isomeris Nutt.

I. arborea Nutt.
Ovule subanatropous; o.i. 2–3 cells thick; i.i. 6–10 cells thick.

Seed 6 mm wide. Testa 3 cells thick, thin-walled, unspecialized; *o.e.* consisting of small cuboid cells with scattered stomata, especially near the raphe. Tegmen *c.* 15 cells thick, in two layers; *outer layer* composed of 1–4 rows of large, thick-walled, finely pitted, lignified fibres (?); *inner layer* composed of 10–12 rows of small thin-walled cells; *i.e.* unspecialized. Embryo with the radicle about equal in length to the thick, not folded, cotyledons.

(Sachar 1956.)

It is not clear whether the outer layer of the tegmen consists of several layers of fibres as in *Capparis* or merely one layer with overlapping ends as in *Crataeva*; they are figured only in t.s. as sclerotic cells.

Polanisia Rafin.

P. graveolens DC
Ovule campylotropous; o.i. 2 cells thick; i.i. 7–10 cells thick.

Seed developing mainly by cell-enlargement without multiplication. Testa thin-walled, without stomata; *i.e.* with the cells much larger than those of o.e. Tegmen; *o.e.* as a single layer of thick-walled, lignified, laterally compressed fibres (?), appearing as a palisade in t.s.; *mesophyll* thin-walled; *i.e.* unspecialized.

(Guignard 1893.)

CAPRIFOLIACEAE

Ovules anatropous, suspended, 1-several per loculus, some often sterile, unitegmic, tenuinucellate; i.e. as an endothelium.

Seeds small to medium-size. Seed-coat ? not multiplicative, more or less crushed except 1–3 outer cell-layers; *o.e.* various, the cells flat with thick outer wall (*Sambucus*), more or less radially elongate (*Viburnum*) and with thick, pitted, lignified, radial and inner walls (*Lonicera*), or very thick-walled and with 2 crossed layers of hypodermal fibres (*Symphoricarpus*). Vascular bundle ending at the chalaza (*Sambucus*) or generally extending partly or almost to the micropyle. Endosperm cellular, copious, subruminate (*Viburnum*) from folds of the endocarp. Embryo rather small, straight.

(Netolitzky 1926; Kühn 1927; Periasamy 1962b.)

CARICACEAE

Ovules anatropous, bitegmic, crassinucellate; o.i. and i.i. 4–6 cells thick; micropyle formed by the exostome; funicle well developed.

Seeds medium-size, with a thick gelatinous pellicle round the tough spongy, hygroscopic, often verrucose or subreticulate mesotesta; exotegmen fibrous, lignified; albuminous, exarillate (occasionally a funicular aril in *Carica*). Testa strongly multiplicative, not lignified except i.e.; *o.e.* consisting of enlarged cuboid cells with thick outer wall (*Jacaratia*) or proliferated by periclinal division in the intervals between the ridges of the mesotesta into 2–4 rows of radially elongate, thin-walled, pulpy then gelatinous cells (*Carica, Cylicomorpha*); *mesotesta* differentiated into an outer and inner layer; *outer mesotesta* formed by periclinal division of o.h. (o.i.) on its inner side, regularly round the seed (*Jacaratia*) or in longitudinal bands or cones (*Carica, Cylicomorpha*), the cells enlarged, subaerenchymatous, thin-walled or becoming somewhat thick-walled (*Cylicomorpha*), colourless; *inner mesotesta* derived from the inner 1–3 layers of the mesophyll (o.i.), composed of small cells in few layers (*Jacaratia*) to many layers (*Carica, Cylicomorpha*), with firm brown walls not or scarcely thickened, without air-spaces (in *Carica* without cellulose-reaction at maturity), this layer carrying the raphe v.b.; *i.e.* as a single layer of crystal-cells, thin or slightly thick-walled (lignified in *Cylicomorpha*). Tegmen not multiplicative (*Carica*) or slightly; *o.e.* as a single layer of thick-walled, lignified, rather finely pitted, longitudinal fibres, not lobulate, shortened into radial sclerotic cells at the micropyle and chalaza; *mesophyll* composed of a few rows of transversely elongate sclerotic cells (*Carica*) or only the outer 1–2 layers modified into thinly sclerotic cells shortly longitudinally elongate (*Cylicomorpha, Jacaratia*); *i.e.* composed of subcuboid cells with thin or slightly thickened walls and brownish tannin-contents, with an inner cuticle. Vascular bundle only in the raphe, entering the chalazal heteropyle and spreading over the base of the suberized or brown hypostase. Endosperm nuclear, thin-walled, oily. Embryos straight; cotyledons broad, flat. Figs. 71–73.

The position of Caricaceae in relation to Cucurbitaceae, Passifloraceae and Oxalidaceae is discussed on p. 38. The seeds of *Averrhoa* and *Carica* have points in common.

Carica L.

C. cundinamarcensis Hook.f.
Similar to *C. papaya*.
(Dathan and Singh 1969b.)

C. papaya L.
Seed 5–6 × 4–5 mm, blackish beneath the pellicle. Testa; *o.e.* with inward proliferations and separating from the mesotesta in the intervals between the ridges; *outer mesotesta* –60 cells thick, very massive; *inner mesotesta* many cells thick. Tegmen 4–5 cells thick (? more; Foster 1943); *mesophyll* transformed into shortly and transversely elongate cells with thickened lignified walls. Funicle stout, with trabeculate cortex, the head of the funicle occasionally enlarged and fleshy as a short funicular aril. Endosperm becoming cellular from the micropyle to the chalaza 11–12 weeks after pollination.
(Stephens 1910; Kratzer 1918; Foster 1943; Singh, D. 1960.)

Cylicomorpha Urb.

C. parvifolia Urb.
(Kenya, leg. D. Mabberley 1970; dried seeds.) Fig. 73.
Seeds 5–6 × 3.5 mm (dried), as in *Carica*. Testa; *outer mesotesta* thinner than in *Carica*, the cells developing firm thickened walls; *inner mesotesta* thinner than in *Carica*; *i.e.* as small crystal-cells with thickened and lignified radial and inner walls. Tegmen 7–9 cells thick; *mesophyll* with o.h. and, in places, the second layer of cells shortly elongate longitudinally with rather thin, lignified walls, scarcely fibrous, the inner cell-layers collapsed. Chalaza with thick sclerotic ring round the heteropyle, derived from the thickened mesophyll of the tegmen, the cells isodiametric.

Jacaratia Endl.

J. conica Kerber
Fig. 72.
Seed apparently smooth. Testa; *o.e.* not or slightly multiplicative, the outer cells shortly radially elongate with thin radial walls; *outer mesotesta* c. 12 cells thick, not forming ridges; *inner mesotesta* 3–4 cells thick. Tegmen c. 7 cells thick; *o.e.* composed, apparently, of laterally compressed fibres appearing as a palisade in t.s.; *mesophyll* ? not sclerotic.
(Kratzer 1918.)

CARYOCARACEAE

This small and remarkable family is placed in Theales but it is clearly a relic of an ancestral state far removed from modern Theales. The two genera, *Anthodiscus* and *Caryocar*, show in their divergence how many ancestral forms must have become extinct. *Caryocar* has a massive drupe with 1–4 complex pyrenes enclosing a massive hypocotyl-embryo; the drupe is derived from a 4–6-locular ovary. *Anthodiscus* has a small 8–20-locular fruit with thin sclerenchymatous pyrenes, very large-celled mesocarp, and a small thin seed; the slender embryo is evidently hypocotylous but to a much less advanced degree than in *Caryocar*. The complicated structure of the pyrene of *Caryocar* was described briefly by Wittmack (Netolitzky 1926) who concluded that the woody internal processes in the pyrene of *C. nuciferum*

(Fig. 75) were equivalent to the needles on the other species, e.g. *C. butyrosum* (Fig. 74). These processes and the dark stony endocarp to which they are attached are composed of closely cemented long fibres with dark brown walls; the fibres are arranged transversely in the endocarp and deviate outwards to make the processes. Such rudimentary deviations occur in *C. butyrosum* and appear to correspond with the processes of *C. nuciferum* rather than the needles, which are embedded in the colourless sclerotic tissue of the inner mesocarp and are free from the endocarp though composed, like it, of brown, thick-walled, cemented fibres; the needles seem to correspond with the slight fibres of the mesocarp between the endocarp processes of *C. nuciferum*. In *Anthodiscus* the thin sclerotic endocarp is also composed of similar transverse brown fibres and the large cells of the mesocarp are thinly lignified and pitted. Thus far these fruits agree but there seems to be no account of their development to explain how they are formed or how that of *Caryocar* may be derived as one of the more detailed structures in flowering plants. There appears to be no morphological explanation for the hard needles which certainly must deter animals from chewing the fallen or germinating nuts of delicious flavour. The needles may be the equivalent of the radiating fibro-vascular bundles in the mesocarp of so many durian-kinds of fruit, e.g. *Durio, Sloanea, Taraktogenos*.

The thick testa of *C. butyrosum* is strongly aerenchymatous, even the epidermal cells having short stellate arms. There is no distinctive mechanical structure to assist in classification. That of *Anthodiscus* is also simple and much thinner, but it appears to have a firm compact outer epidermis.

(Netolitzky 1926.)

Anthodiscus C. F. W. Mey.

A. obovatus Benth.
(Brazil, Spruce 8146; dried specimen, CGE.) Fig. 77.
Fruit 8-celled, with one seed (immature) in each loculus; *exocarp* with small thin-walled cells, with a v.b. at each angle; *mesocarp* very large-celled, with slightly thickened, lignified and pitted walls; *endocarp* woody, thin, consisting of transverse, thick-walled, brown fibres lining each loculus; columella? (the tissue wholly agglutinated).

Testa with thin-walled brown epidermal cells, not aerenchymatous.

Caryocar L.

C. butyrosum Willd.
(Ceylon, Peradeniya; nearly ripe fruit in alcohol.) Figs. 74, 76.
Ovules anatropous, bitegmic.

Seed-coat formed by the thin-walled testa *c.* 0.4 mm thick, aerenchymatous, even the cells of o.e. with short arms and wide air-spaces, the thin walls brownish; *i.e.*

composed of small thin-walled brownish cells with rectangular facets; *mesotesta* eventually collapsing. Tegmen hyaline, collapsed. Endosperm as a trace near the chalaza, thin-walled, oily. Embryo as a massive hypocotyl; epicotyl with 3–4 pairs of scale-leaves, on the chalazal side of the hypocotyl.

Fruit developing one seed with three abortive seeds, every one in its own pyrene; *exocarp* massive, thin-walled, oily, composed of a soft outer layer and a firmer inner layer adherent to the mesocarp; *mesocarp* 2-layered, the outer layer bulging slightly into the exocarp with small convex facets and composed of rather loosely adherent, isodiametric, colourless, sclerotic cells with thick walls, and the inner layer composed of radially elongate, thick-walled, colourless sclerotic cells separating the embedded needles; *needles* 6–7 × 0.13–0.25mm, dark brown, hard, fusiform, acute, radiating, composed of closely cemented, parallel fibres with dark brown, thick walls without pits, the acute ends of the needles minutely jagged with the projecting bases of short fibres (? cells of the ground-tissue); *endocarp* very woody, blackish brown, composed of very compact, tangentially and transversely orientated, long fibres with thick, dark brown walls (not pitted), with short irregular and equally dense fibrous processes into the mesocarp; v.b. ?

C. nuciferum L.
(Singapore, Botanical Gardens; living material.) Figs. 75, 77.
(No microscopic observation.)
Germination at first slow but rapid on emergence of the radicle; radicle developing from one end of the swollen hypocotyl, the other end lengthening into a rather slender stem carrying the cotyledons and plumule into the air; cotyledons small, shrivelling; second pair of leaves as foliage leaves, simple or with 2 leaflets; adventitious roots arising from the base of the emergent hypocotyl.

CARYOPHYLLACEAE

Ovules anatropous or campylotropous, bitegmic, crassinucellate; o.i. 2 cells thick, or more (*Agrostemma, Alsine, Cucubalus, Viscaria*); i.i. 2 cells thick.

Seeds small, reniform, often verruculose or reticulate from exotestal markings, brown or black, perispermous, exarillate (except *Moehringia*), the integuments not multiplicative. Testa reduced to o.e., or with persistent i.e.; *o.e.* as a firm layer of cuboid cells with thickened outer wall and undulate facets, the cells in some cases papilliform; *mesophyll* when present unspecialized; *i.e.* composed of cuboid cells with thickened inner wall or thin-walled and more or less crushed. Tegmen unspecialized, crushed or i.e. persistent with thin walls or thickened (*Agrostemma, Dianthus, Tunica, Vaccaria*). Perisperm persistent, starchy. Endosperm nuclear, slight. Embryo generally curved round the perisperm.

Aril as a funicular elaiosome in *Moehringia*.

(Netolitzky 1926; Woodcock 1927, 1928; Buell 1952; Bresinsky 1963; Singh, B. 1964; Rohweder 1971.)

This family has the more simplified seed of Centrospermae.

CASUARINACEAE

Ovules more or less anatropous, suspended, paired, bitegmic, crassinucellate; o.i. 3 cells thick; i.i. 2–3 cells thick, very soon crushed except at the micropyle; nucellus projecting in *Casuarina montana*, with nucellar tracheids in some species.

Seeds small, enclosed in the samara, exalbuminous. Testa not multiplicative, unspecialized, crushed or shrivelled, or o.e. with thickened outer walls. Endosperm nuclear, absorbed. Embryo straight.

(Netolitzky 1926; Swamy 1948a.)

This family may belong with Hamamelidales, but its peculiarities have never been explained. The cone is certainly comparable with the cones of Hamamelidaceae and the seed could be a simplification.

CELASTRACEAE

Ovules anatropous, bitegmic, crassinucellate or tenuinucellate, 1–18 per loculus, erect or suspended with dorsal or ventral raphe; o.i. 3–8 cells thick; i.i. 2–4 cells thick.

Seeds small to fairly large, arillate or not, in some genera alate; aril exostomal and/or funicular, or as a double structure in *Sarawakodendron*. Testa 5–12 cells thick, multiplicative or not; o.e. often more or less as a palisade with thick cuticle and outer wall, not lignified; mesophyll aerenchymatous, sometimes with scattered sclerotic cells; i.e. unspecialized. Tegmen 3–10 cells thick, multiplicative or not; o.e. typically as a lignified layer of laterally compressed, longitudinal, pitted, ribbon-like fibres, cylindrical in *Catha*, absent in some species or genera (*Kokoona, Lophopetalum*); mesophyll crushed; i.e. as a layer of small cells, often shortly transversely elongate with somewhat thickened, unlignified walls, in some cases as an endothelium in the young seed. Vascular supply with or without postchalazal branches. Chalaza simple or with a woody plug. Endosperm nuclear, oily, thin-walled, in some cases absent. Embryo with flat, often green, cotyledons.

Aril non-vascular, mostly post-fertilization but visible as a swelling of the funicle before pollination in *Euonymus latifolius*, *Celastrus paniculatus* and *Moya*, often double with filamentous or pulpy mesophyll. Figs. 78–99.

(Netolitzky 1926; Andersson 1931; Mauritzon 1936a, c; Ding Hou 1962.)

In this family there is much need to investigate the seed-structure in tropical genera. It has three characters but all may disappear; the result is merely an exotestal seed or an overgrown seed without specialization of the seed-coat. First the exotegmen has ribbon-like fibres as in some Capparidaceae, Euphorbiaceae–Phyllanthoideae, Flacourtiaceae and Meliaceae. Then the funicular or exostomal aril is often double as in some Clusiaceae, and lastly there is the palisade of the exotesta which usually accompanies the arillate seed. The fibrous exotegmen may fail to differentiate even in the species of a genus, e.g. *Euonymus*. The aril may be a micropylar arillode or a true funicular aril or an arilloid (v.d. Pijl 1955, 1957); such variation shows that the distinction is largely academic, e.g. the double aril of *Sarawakodendron*. The aril is absent from genera with alate seeds but the state of *Catha* suggests how the arillar extension may be converted into a winged seed. The ovules are variously placed and oriented in a diversity that requires more attention (Andersson 1931). The testa is rather thick and little, if at all, multiplicative, while the thin tegmen seems generally to be multiplicative. Durianrelics appear in the 'ornamented' capsules of various species of *Euonymus* (Blakelock 1951; Corner 1954), as well as in the winged seeds. *Sarawakodendron* seems to be a remnant of the ancestry of the *Lophopetalum*-alliance with thickly valved capsules replete with large seeds, comparable with *Pterygota* in Sterculiaceae. The aril enters as the key to this alliance, as well as that of the family, while the sarcotesta, supposed to be primitive by van der Pijl, has no place. Compare *Scyphostegia* as an arillate ally of *Sarawakodendron* and, perhaps, Legnotidaceae.

Perrotetia has been omitted from the seed-description of the family because its seed is not Celastraceous; it could be mistaken for that of Euphorbiaceae–Crotonoideae. The thinly pulpy testa is little developed; there is no aril; the exotegmen becomes a palisade of Malpighian cells projecting to different levels and giving the rough surface to the inner layer of the seed; the cells of the endotesta have peculiar small rod-like structures on the inner tangential walls which project into the cell-cavity. The two basal and erect ovules per loculus are Celastraceous, not Euphorbiaceous, but the exotegmic palisade indicates that the genus has evolved in parallel with these families. The closest resemblance is the seed of Turneraceae.

The position of Hippocrateaceae is uncertain. The seed appears to lack any definitive character. According to Mauritzon (1936c), the tegmen disappears as the endosperm forms.

Bhesa Buch. Ham.

Ovules 2 per loculus, erect.

Seeds albuminous, arillate. Tegmen with fibrous outer layer, appearing as a palisade in t.s. Vascular supply with 5–8 postchalazal branches. Aril arising after fertilization from the exostome and the apex of the funicle.

B. paniculata Arn.
(Singapore; living material.)   Figs. 78–80.

Ovule; o.i. 6–8 cells thick; i.i. 4 cells thick; micropyle formed by the exostome (3 cells thick).

Seed 9–11 × 6–8 mm, pale brown. Testa 6–8 cells thick; *o.e.* composed of large, colourless cells, occasionally with red anthocyanin, the outer wall thickened; *mesophyll* aerenchymatous, collapsing. Tegmen 7–8 cells thick, thin-walled and collapsing except for the dark brown fibrous exotegmen. Chalaza with a small heteropyle surrounded by the short sclerotic cells of the exotegmen thickened into 3–4 cell-layers. Vascular bundle of the raphe with 7–9 postchalazal branches, simple or branched, and with a ring-like extension from the raphe round the micropyle supplying short fine simple branches to the testa, often also with a few short branches from the raphe to the testa.

Aril crimson from anthocyanin in the epidermal cells; mesophyll thick, white, pulpy, slimy from the swelling of the cell-walls, with abrupt transition to the tissue of the testa.

Pericarp yellow suffused reddish from the base, rather pithy-succulent, thin-walled except for 4–5 layers of longitudinal lignified fibres in the inner hypodermis; splitting irregularly loculicidally and detaching from the base of the fruit to leave the red arillate seed on the axis.

B. robusta (Roxb.) Ding Hou
(Singapore; living material.)   Figs. 81, 82.

Seeds 20 × 9 mm, dark shiny brown, blackish at the chalazal end, hard, with darker stripes indicating the v.b. Testa 5–7 cells thick; *o.e.* composed of shortly radially elongate cells with strongly thickened brown walls and thick yellow cuticle; mesophyll aerenchymatous, drying up. Tegmen 5 cells thick, as in *B. paniculata*, but with pale yellow fibres. Chalaza as in *B. paniculata*. Vascular bundle of the raphe with 5–8 postchalazal branches, the main branch almost reaching the micropyle. Micropyle embedded in the thickened core of the fruit-pedicel.

Aril orange, fleshy; epidermal cells cuboid, colourless; mesophyll thin-walled, with orange-yellow oil-drops, with gradual transition to the tissue of the testa.

Fruit 1-seeded, dehiscing by one suture, then rupturing round the base to leave the thickly arillate seed on the fruit-pedicel.

The embryo of this species is rather small when the seed is shed but develops rapidly in the fallen seed and quickly germinates without dormancy.

Catha Forssk.
C. edulis Forssk.
(Java, Tjibodas, cult.; flowers and fruits in alcohol.) Fig. 83.

Ovules 2 per loculus, anatropous, ascending; o.i. 2 cells thick; i.i. 5 cells thick.

Seeds 3–3.5 × 1.3 mm, ovoid, brown, minutely papillate, with a dry white aril 5–7 mm long, 1–2 per loculus and per fruit. Testa 3 cells thick; *o.e.* as a continuous layer of conical projecting cells with striate cuticle, scarcely thick-walled, not lignified; *mesophyll* and *i.e.* thin-walled, unspecialized, eventually crushed. Tegmen 5 cells thick, not multiplicative; *o.e.* as a continuous layer of longitudinal, strongly pitted, narrow, lignified fibres, not elongate radially, shortened at the micropyle and chalaza; *mesophyll* and *i.e.* unspecialized, thin-walled. Funicle rather long, appearing lateral. Endosperm oily, thin-walled. Embryo green; radicle short; cotyledons flat.

Aril arising from the funicle and the side of the exostome next to the funicle, shortly embracing the micropylar end of the seed, elongated into a membranous wing filling the basal part of the loculus, 2–3 cells thick along its attachment, thinning to one cell in the wing-like part, the middle tissue more or less disrupted, thin-walled, not lignified.

The aril seems not to be edible but to serve as a wing.

Celastrus L.

C. paniculatus Willd.
(Cambridge, University Botanic Garden; living material.)

Testa; *o.e.* with the cells shortly longitudinally elongate, some slightly projecting. Tegmen with fibrous exotegmen, the fibres 160–600 × 30 × 10 μ; *i.e.* as a regular layer of small cells with tannin.

Aril as a small hump on the raphe-side at the junction with the funicle before fertilization, then becoming funicular–exotestal after fertilization, folded, wrinkled, scarlet.

(Adatia and Gavde 1962; Vaughan 1970.)

Euonymus L.

Ovary 4–5 locular, mostly with 2 ovules per loculus, 3–12 in some species; ovules basal and erect or apical and suspended.

E. europeus L.
Figs. 84–88.

Ovule; o.i. 5–6 cells thick; i.i. 2–3 cells thick, the cells longitudinally elongate; micropyle formed by the endostome.

Seeds 4–5 mm wide, shortly oblong, pale brown; aril orange-red, crenulate, waxy, developed after fertilization from the exostome, the short funicle and the raphe, investing the seed, developing as 1–2–3 folds, much plicate through intercalary growth, the apical (chalazal) pore occluded. Testa 7–12 cells thick; *o.e.* composed of cuboid cells with slightly thickened outer walls and in places rugose cuticle, not lignified, the facets polygonal; *mesophyll* unspecialized, the cells shortly transversely elongate, becoming slightly thick-walled; *i.e.* unspecialized. Tegmen 3–4 cells thick but ultimately represented only by the fibres; *o.e.* composed of ribbon-like, lignified fibres with scalariform–reticulate thickening, short at the endostome as a cone 10–12 cells thick with short tracheidal cells similarly pitted; *i.e.* with slightly thickened inner walls, then crushed. Chalaza with small heteropyle closed by a small cone of short-celled tracheid-like sclerotic cells (as at the endostome), not quite reaching the end of the v.b.

Aril composed of thin-walled cells; epidermis shortly palisade-like with orange oil-globules; mesophyll soon lacunose–aerenchymatous with the cells pulled apart into rows by the intercalary extension of the epidermis, eventually the aril-folds consisting of the opposed epidermal layers separated by one or a few strands of compressed mesophyll cells.

E. glandulosus (Merr.) Ding Hou
(Borneo, Kinabalu, leg. E. J. H. Corner 1964; fruits in alcohol.) Fig. 89.

Testa 7–9 cells thick; *o.e.* as a tough palisade, the cells with strongly thickened outer walls and cuticle, not lignified; *mesophyll* subaerenchymatous, thin-walled but with scattered sclerotic cells with brownish pitted and lignified walls, mostly transversely elongate; *i.e.* composed of thin-walled cells, longitudinally elongate. Tegmen more or less collapsed into a brownish layer, without fibrous exotegmen.

Aril derived entirely from the stout funicle; epidermal cells cuboid; mesophyll filamentous with extended cells, thin-walled.

This species was originally ascribed to *Glyptopetalum* the seed-structure of which is not known. The lack of tegmic fibres and, perhaps, the funicular aril are exceptional in *Euonymus*.

E. japonicus Linn.f.
Ovules pendulous from the apex of the ovary with dorsal raphe and adaxial micropyle.

Tegmen evidently fibrous as in *E. europeus*.
(Copeland 1967.)

Glyptopetalum Thwaites
Raphe with two long postchalazal branches of the v.b. extending almost to the micropyle and up to 5 short branches.
(Ding Hou 1962.)

Gymnosporia Benth. et Hook.f.
Ovules tenuinucellate.

G. rothiana Wight et Arn., *G. spinosa* Fiori
Testa 7 cells thick; *o.e.* with large cells. Tegmen 7–8 cells thick, crushed except i.e., without a fibrous exotegmen.
(Adatia and Gavde 1962.)

Hippocratea L.
Ovules tenuinucellate. Seeds exarillate.

H. grahamii Wight
Ovule; o.i. 3–4 cells thick; i.i. 3 cells thick; micropyle formed by the endostome. Seeds exalbuminous.
(Adatia and Gavde 1962.)

Kokoona Thwaites
K. ceylanica Thw.
(Ceylon, CP 2584; dried material, CGE.)
The seeds, which were not mature, had no sign of a fibrous exotegmen.

Lophopetalum Wight
Lophopetalum sp.
(Penang, leg. E. J. H. Corner 1972; ripe seeds in alcohol.)
Fig. 90.

Seeds 7–10 × 2.5 cm, flat, elliptic, very thin, pale buff-ochraceous, with lateral hilum. Testa several cells thick, not lignified; *o.e.* composed of large cells, cuboid in t.s. but with rhombic facets (shortly elongate towards the periphery of the wing), without stomata, the cuticle minutely rough with subgyroso–reticulate or asperate projections; *mesophyll* aerenchymatous, the globose cell-body with rather long arms, this tissue prolonged into the wings, mostly crushed over the seed-body; *i.e.* unspecialized. Tegmen crushed into a thin brown layer, without fibres. Vascular bundle surrounding the seed-body as a centric v.b., unbranched. Endosperm none. Cotyledons flat; radicle rather long.

As with *Kokoona* there is no lignified tissue in this seed except the xylem of the v.b. which surrounds the seed-body as if it were perichalazal.

Microtropis Wall.
M. platyphylla Merr.
(Borneo, Kinabalu, leg. E. J. H. Corner 1964; seeds in alcohol.) Fig. 91.

Seeds 15–17 × 8–10 mm, oblong, subacute at each end, orange, exarillate. Testa 7–10 cells thick, pulpy; *o.e.* composed of rather irregular cells (as seen in surface-view) with orange brown contents; *mesophyll* subaerenchymatous, permeated with the network of v.b.; *i.e.* unspecialized. Tegmen crushed except for the fibrous exotegmen composed of small, rather coarsely pitted, oblique or longitudinal fibres, very thick-walled but slightly lignified, not laterally compressed, straight or subsinuous (not lobate). Vascular supply as a copious network of rather large v.b. without distinct raphe.

Endosperm with rather thick walls, oily, the cells in radial rows.

Fruit 20–23 × 11–13 mm, dull orange, 1-seeded; pericarp with the outer 8–12 cell-layers (including the epidermis) composed of thin-walled cells with orange-brown contents; mesocarp with longitudinal v.b. and many groups of sclerotic cells; endocarp composed of 8–10 layers of lignified cells with somewhat thickened walls, forming two valves lining the fruit-cavity.

It appears that the thick testa has been mistaken for an aril (Ding Hou 1962). The sarcotestal seed comes in with reduction of the fruit-capacity, the loss of the aril and, possibly, some peculiar state of the ovules (Ding Hou 1962).

Perrotetia HBK

Ovules 2 per loculus, basal, erect.

Fruit baccate, 2–4 seeded, with membranous endocarp around the pairs of exarillate seeds.

P. alpestris (Bl.) Loes. ssp. philippinensis (Vidal) Ding Hou

(Borneo, Kinabalu, RSNB 5950; fruits in alcohol.) Figs. 92, 93.

Seeds 2 × 1.8 × 1.4 mm, pale brown. Testa 3–5 cells thick, pulpy, transparent thin-walled; o.e. not as a palisade; i.e. more or less adherent to the tegmen, with abundant rod-like small crystals (?) attached to the inner wall. Tegmen 4–5 cells thick, dark brown or blackish, reticulately asperate; o.e. as a compact palisade of prismatic Malpighian cells with thick, brown, lignified walls, sparsely and finely pitted, curved with the outer ends displaced towards the chalaza, projecting in groups of longer cells to form the superficial asperities; mesophyll thin-walled, collapsing; i.e. composed of thin-walled tannin-cells with rectangular facets. Vascular supply only in the raphe, decurved to the chalaza, bordered round the heteropyle with coarsely reticulately pitted, radially elongate, sclerotic cells in contact with the exotegmic palisade. Hypostase none. Endosperm thin-walled, oily. Embryo straight with short cotyledons.

If this seed were solitary in the loculus and suspended, instead of basal and erect, it would pass for one the Euphorbiaceae–Crotonoideae. It has been described as arillate (Ding Hou 1962) but the thin sarcotesta must have been mistaken for an aril and the sclerotic tegmen for the testa. It is entirely anomalous in Celastraceae. Compare Turneraceae and Legnotidaceae.

Salacia L.

Ovules slightly crassinucellate; o.i. 4–5 cells thick; i.i. 2–3 cells thick. Tegmen crushed and absorbed by the endosperm.

(Mauritzon 1936a.)

Salacia sp.

(Sarawak, S 28855; ripe seeds in alcohol.)

The large seeds appeared to be typically overgrown with unspecialized, though highly vascular, testa round the large embryo without endosperm. The thick testa (? 15–20 cells thick, though with most of the inner tissue crushed) consisted of cells with thin, brown, unlignified walls. There was no trace of tegmen, as Mauritzon reported. The embryo was composed entirely of swollen hypocotyl, as in Garcinia or Bertholletia, without trace of cotyledons. Possibly Salacia has a pachychalazal seed. It has, of course, a non-ecological sarcotesta.

Sarawakodendron Ding Hou

Ovary 3-locular; ovules 8 per loculus, biseriate, horizontal.

Capsule trigonous, 3-locular, septicidal with thick valves. Seeds 6–8 per loculus, descending, fusiform, doubly arillate with a caruncle-like body surrounded by simple filaments often connate and appearing branched, albuminous.

S. filamentosum Ding Hou

(Sarawak, leg. P. S. Ashton; fruits in alcohol.) Figs. 94–99.

Seeds 21–24 × 4–5 mm, blackish brown, irregularly rugulose. Testa 5–8 cells thick; o.e. as a palisade of large columnar cells with strongly thickened brown outer wall and slightly thickened radial and inner wall, not lignified, the cuticle thick and more or less reticulately marked on the outside; mesophyll aerenchymatous, the cells transversely elongate, thin-walled; i.e. unspecialized. Tegmen 7–10 cells thick; o.e. as a compact layer of longitudinal, flattened, ribbon-like, lignified fibres, sparsely pitted on the sides, strongly reticulately pitted on the inner wall; mesophyll crushed round the micropyle, in other parts more or less dissolved into a gummy residue; i.e. as a compact layer of small thin-walled cells, closely adherent to the endosperm, unlignified. Chalaza with a long plug of reticulately thickened, lignified tracheids between the end of the raphe v.b. and the nucellar remains, the tracheidal tissue ending at different levels in the chalaza, filling the heteropyle; hypostase none. Vascular bundle only in the raphe. Endosperm thin-walled, oily, longitudinally rugulose on the outside.

Aril double, as a caruncle and as filaments, ? colour. Caruncle 6–7 mm wide, developed from the exostome and funicle round the base of the seed; cortex composed of thick-walled, unlignified, tangentially elongate cells in 3–4 layers, firm; mesophyll more or less lacunose, with elongate, shortly armed cells, without v.b. Filaments 15–20 mm long, tapering, flexuous, rather uneven, arising from a single or double ring (or band) at the base of the funicle, solid, without v.b., with a thin-walled epidermis of longitudinal cells and rather compact, not lacunose, rows of similar internal cells.

Fruit-valves with heavily lignified endocarp and thinly lignified mesocarp, the main longitudinal v.b. at the angles of the valves, with a network of v.b. in the intervening mesophyll but not in one row.

(Ding Hou 1967.)

This singular plant should be compared with the equally curious and Bornean *Scyphostegia* (p. 23).

CERATOPHYLLACEAE

Ovule anatropous, unitegmic, crassinucellate; integument 4 cells thick at the base, thinning to a single cell thick distally.

Seed minute, enclosed in the more or less spiny achene, exalbuminous. Integument obliterated. Endosperm cellular, absorbed. Embryo well-developed, green; plumule conspicuous.

(Netolitzky 1926.)

Ceratophyllum is regarded as a unitegmic derivative of Nymphaeaceae. The spinous fruit is a durian-relic.

CERCIDIPHYLLACEAE

Ovules numerous, anatropous, bitegmic, crassinucellate; o.i. 4–5 cells thick; i.i. 2–3 cells thick.

Seeds small, alate at the chalazal end, albuminous. Testa not multiplicative, apparently unspecialized. Tegmen crushed, not multiplicative. Chalazal wing with curved or hooked vascular bundle. Endosperm cellular, oily, reduced in the ripe seed. Embryo relatively large with well-developed hypocotyl.

(Netolitzky 1926; Swamy and Bailey 1949.)

The small seeds resemble those of Trochodendraceae but lack the sclerotic endotesta and exotegmen, and the embryo is better developed. There is nothing primitive in these simplified seeds and nothing, so far as known, to indicate their original construction. The modern tendency places the family in Hamamelidales, though Hutchinson retains it in Magnoliales.

CHENOPODIACEAE

Ovule campylotropous, solitary, basal, funiculate, bitegmic, crassinucellate; o.i. and i.i. 2 cells thick, or 3 cells round the micropyle; endostome forming the micropyle.

Seeds small, reniform or lenticular, with perisperm exarillate; the integuments not multiplicative, reduced generally to the exotesta and the endotegmen, but with three cuticular layers. Testa; *o.e.* composed of cuboid or shortly radially elongate cells with more or less strongly thickened outer wall, often with dark inwardly directed striae ('stalactites') in the outer wall, in some genera the cells with a papilla or hair, generally with tannin;

i.e. unspecialized, in some cases with crystal-cells, persistent in *Bassia*. Tegmen; *o.e.* unspecialized, crushed; *i.e.* generally persistent, with fine streaks or thickenings on the radial and inner walls, lignified or not, often with dark contents. Perisperm starchy, abundant on the raphe-side of the seed. Endosperm nuclear reduced to a thin layer, even a single cell thick, round the radicle. Embryo curved, generally spirally coiled.

(Netolitzky 1926; Artschwager 1927; Bhargava 1936; Singh, B. 1964; Hindmarsh 1966; Eckardt 1967; Wunderlich 1968.)

There is much minor generic variation in the construction of the seeds. The reduced nature of the small ovary, fruit and seed is shown by Barghava's observation that the ovary-wall of *Chenopodium album* has no vascular supply; the remains of the vascular supply of the flower, after the separation of the stamen-traces, goes into the funicle. The ovule is practically terminal, as in Gramineae, and the ovary grows round it.

CHLORANTHACEAE

Ovule solitary, suspended from the apex of the loculus, orthotropous, bitegmic, crassinucellate; o.i. 2 cells thick (*Ascarina*, 3–4 cells round the chalaza), 3 cells (*Hedyosmum*), 4–6 or 6–8 cells (*Chloranthus*); i.i. 3 cells thick (*Hedyosmum*), 5–6 cells (*Chloranthus*) or 6–8 cells (*Ascarina, Chloranthus*); micropyle at first formed by the endostome, or becoming overgrown by the exostome.

Seeds 2–6 mm long, subglobose or ovoid, albuminous, exarillate, enclosed in the drupe or berry; seed-coats multiplicative or not. Testa; *o.e.* thin-walled, the cells shortly longitudinally elongate, with scattered sclerotic cells (*Chloranthus spp.*); *mesophyll* thin-walled, aerenchymatous, crushed, or with scattered sclerotic cells (*Chloranthus spp.*); *i.e.* as a palisade of almost cuboid to radially elongate cells with hexagonal facets, containing 1–several crystals, eventually permeated by an internal cellulose reticulum, becoming lignified strongly (*Chloranthus*) or not (*Ascarina*), but unspecialized in *Hedyosmum*. Tegmen thin-walled, more or less crushed, or with several layers of longitudinal fibres (*Ascarina*) or incompletely fibrous (*Chloranthus spp.*); *i.e.* as a compact layer of cells shortly longitudinally elongate with firm brown walls, not lignified, appearing almost as a palisade in t.s., often eventually crushed. Chalaza well-developed with a thick sclerotic sheath round the v.b. and continuous with the exotegmic fibres (when present). Vascular bundle short, without branches to the testa. Perisperm none as a storage-tissue, but as a trace of more or less crushed nucellar tissue next the chalaza and thin brown hypostase. Endosperm cellular, copious, oily, with thin or slightly thickened walls. Embryo microscopic, the cotyledons scarcely formed. Figs. 100–103.

(Armour 1906; Netolitzky 1926; Swamy 1953a; Yoshida 1957, 1959, 1960a; Vijayaraghavan 1964.)

Three genera have usually been distinguished, namely *Ascarina*, *Chloranthus* and *Hedyosmum*. Recently *Sarcandra* has been separated again from *Chloranthus*, chiefly through the absence of vessels from *Sarcandra*, its unistaminate male flowers and its acolpate pollen. Seed-structure indicates three states: (1) *Hedyosmum* with unspecialized seed-coat; (2) *Chloranthus* and *Sarcandra* with lignified endotestal palisade; (3) *Ascarina* with unlignified endotestal palisade but with lignified fibrous exotegmen. The distinction between the last two, however, is blurred by the incompletely fibrous tegmen of *C. elatior*. There appears to have been a simplification from the most complicated state in *Ascarina* to the undifferentiated seed-coat of *Hedyosmum* enclosed in a rigid endocarp. Loss of fibrous tegmen appears to have occurred on the sides of the seed, while the endostome and chalaza remain densely sclerotic (*Chloranthus*) or only the chalazal region (*C. glaber*).

Primitively, therefore, the Chloranthaceous seed appears to have had a lignified endotestal palisade of crystal-cells with internal reticulate fibrils (as in Magnoliaceae), a tegmen with several outer layers of lignified fibres and, possibly, a lignified exotesta. This combination occurs in Papaveraceae and Proteaceae. Their ancestors may well have led in simplification of stem, leaf, flower and fruit to Chloranthaceae, for this seems to be much less improbable than the Lauralean, Magnolialean and Ranalean ancestry suggested by others in disregard of seed-structure. A trace of a third integument (? vestigial aril) has been reported for *Sarcandra*. The arillate, 1-seeded fruit of *Bocconia*, though with anatropous ovule, might have led to the exarillate, orthotropous, drupaceous Chloranthaceae with simple, but dentate, leaves. Lack of vessels in *Sarcandra* would be persistent juvenilism or neoteny in Carlquist's sense.

Ascarina Forst.

A. maheshwarii Swamy
(Solomon Islands, RSS 1208; flowers and fruits in alcohol.) Fig. 100.

Ovule; o.i. mostly 2 cells thick, 3–4 round the chalaza; i.i. 6–8 cells thick.

Seeds 1.8–2 mm long, the seed-coats not multiplicative. Testa; *o.e.* composed of thin-walled cells, often shortly longitudinally elongate; *i.e.* as a short palisade of yellowish, cuboid cells with several crystals, filled with internal cellulose processes as a fibrillar network, not lignified. Tegmen; *o.e.* and 3–4 outer cell-layers of the mesophyll transformed into longitudinal, thick-walled, coarsely pitted, lignified fibres, the epidermal fibres 120–270 × 14–21 μ, the inner fibres shorter with straighter ends 65–100 μ long; *inner hypodermis* 1–2 cells thick, thin-walled, more or less crushed; *i.e.* as a layer of shortly longitudinally elongate cells with firm brown walls, continuous from the hypostase, at first

almost palisade-like in t.s., then crushed, not lignified. Nucellus enlarging greatly, then crushed by the endosperm.

Berry 2.5–3 × 2–2.5 mm, purplish grey then black, juicy, without special endocarp; epidermis without stomata.

Chloranthus Sw.

C. elatior R.Br.
(Java, Pelabuhanratu; ripe fruits in alcohol.) Fig. 101.

Seeds 3.2 × 3 mm, subglobose, minutely apiculate at each end, grey, invested by the thin fibrous endocarp. Testa 5–6 cells thick, 4–5 at the exostome, thin-walled, not lignified except i.e. as a palisade of stout, radially elongate cells with internal fibrillar-lamellar lignified thickenings (? crystals obscured by the fibres). Tegmen 5–6 cells thick, wholly densely sclerotic at the minutely projecting endostome; *o.e.* as a layer of longitudinal, pitted, lignified fibriform cells, compact round the endostome and chalaza but the fibres spaced with short thin-walled cells over the sides of the seed; *mesophyll* thin-walled and more or less crushed, but with scattered sclerotic cells in the o.h.; *i.e.* crushed.

Berry white, pulpy; endocarp with several layers of lignified fibres closely adherent to the seed.

The following collection with white fruit, from Penang Hill where the plant is wild, seemed to be *C. elatior* but the fruit and seed agree better with the following species *C. glaber*; it is not clear that the species are yet satisfactorily distinguished:

Seeds 4.5–5.5 × 3.7–4 mm, ovoid, brown, becoming free in the berry. Testa 4–6 cells thick; *o.e.* composed of longitudinally oblong cells with brown walls and with scattered sclerotic cells as in *C. glaber*; *mesophyll* also with scattered sclerotic cells; *i.e.* as a lignified palisade *c.* 140 μ high, as in *C. glaber*. Tegmen 7–9 cells thick, not lignified except at the chalaza, the cells longitudinally oblong, thin-walled, not fibriform except immediately round the chalaza with o.e. as lignified fibrous cells; *i.e.* with brown walls. Berry without endocarp-fibres.

C. glaber (Thunb.) Makino
(Ceylon; flowers and fruits in alcohol.) Figs. 102, 103.

Ovule; o.i. 6–8 cells thick (4–5 in Japanese material; Yoshida 1960a), with i.e. as an incipient palisade; i.i. 10–12 cells thick (6 cells in Japanese material; Yoshida 1960a), i.e. with somewhat enlarged cells.

Seeds 5–5.5 × 4 mm, ovoid, brownish, the seed-coats multiplicative. Testa 8–11 cells thick; *o.e.* composed of shortly longitudinally elongate cells, thin-walled, but many with thickened, lignified walls and irregularly internally lignified, these sclerotic cells scattered or aggregated into irregular longitudinal strips; *mesophyll* aerenchymatous, the cells longitudinally oblong, not lignified; *i.e.* as a woody palisade of radially elongate cells containing 1–several crystals, the cell-cavity becoming filled with an internal reticulum of lignin, the layer lignifying rather unevenly from the chalaza to the micropyle. Tegmen 16–20 cells thick, unspecialized, thin-walled, not lignified, the cells somewhat transversely elongate, eventually crushed, but with some sclerotic cells around the chalaza; *i.e.* developing brownish walls, then crushed.

Berry with 9–10 v.b. in the pericarp, without endocarp fibres; *o.e.* without stomata.

It is not certain that the Ceylon and Japanese material really belong to the same species.

CIRCAEASTERACEAE

Ovules orthotropous, paired, pendent from the apex of the loculus, intramarginal, the uppermost ovule aborting, unitegmic, ? tenuinucellate; integument 2 cells thick.

Achene 1-seeded. Seed-coat crushed, the endosperm in contact with the pericarp. Endosperm cellular, oily. Embryo relatively large.

Circaeaster agrestis Maxim.
(Junell 1931.)
This is a Ranunculaceous ovule and seed.

CISTACEAE

Ovules subanatropous or suborthotropous, bitegmic, crassinucellate; o.i. 2 cells thick, entirely free from the funicle and without raphe; i.i. 2–4 cells thick; chalaza with pre-formed adult structure; hypostase initiated between the shouldered base of i.i. and the nucellus; micropyle formed by the slightly projecting exostome or by both integuments.

Seeds small, albuminous, exarillate; seed-coats not multiplicative. Testa unspecialized; *o.e.* usually consisting of more or less tabular cells with rectangular to subundulate facets, or shortly prismatic cells in *Fumana*, in some cases mucilaginous, without stomata; *i.e.* collapsing, but with thickened walls in *Helianthemum*. Tegmen with the inner layers unspecialized and crushed; *o.e.* as a palisade of short, more or less lignified, Malpighian cells, the radial walls sometimes undulate or substellate in t.s. Chalaza with an inwardly directed hard bulla continuous with the exotegmen and lined by a counter-palisade of thin-walled cells, the core of cells turning brown at maturity and the whole drying into a conical plug. Endosperm nuclear, thin-walled, starchy. Embryo more or less curved.

(Netolitzky 1926; Boursnell 1950; Kapil and Maheshwari 1965.)

This family, which is little more than variations on a single generic theme, is allied with Bixaceae and has a similar, but simplified, seed derived from the ovule without multiplication of the cell-layers of the integu-

ments. The complex chalaza of *Bixa* is simplified and pre-formed in the ovule before fertilization. Thus the small seeds are another example of the neotenic ovule enlarged by cell-inflation and with the exotegmen differentiated.

Cistus L.

C. corbariensis Pourr.
(England, cultivated; living material.) Fig. 104.
Ovule; o.i. 2 cells thick; i.i. 4 cells thick.
Seed 1.5 × 1.1 mm, blackish brown, with long white funicle. Tegmen; *o.e.* with only the middle layer of the wall lignified; *i.e.* at first as a layer of large thin-walled cells.

CLUSIACEAE

Ovules anatropous, bitegmic, crassinucellate or tenuinucellate, usually erect, varying suspended or transverse, axile, basal or parietal (*Clusia sp., Pentadesma*), 1–many per loculus; o.i. 2–30 cells thick; i.i. 2–15 cells thick; micropyle formed by the exostome; aril as a rudiment on the ovule (*Clusia*).

Seeds small to large, arillate or not, exalbuminous. Testa multiplicative or not, (1) exotestal and the cuboid or oblong cells with thick outer wall (Clusieae), (2) mesotestal (Calophylleae), or (3) unspecialized (Garcinieae). Tegmen multiplicative or not; *o.e.* with stellate-undulate, thick-walled, pitted and lignified cells, tabular on the sides of the seeds and radial at the endostome, often multiple round the chalaza, closely appressed or intergrown with the endotesta (Clusieae), or unspecialized (Calophylleae, Garcinieae); *mesophyll* often with much enlarged cells, then crushed, thin-walled; *i.e.* with much enlarged, radially elongate cells or unspecialized, or fibrous (*Symphonia*). Vascular bundle of the raphe with or without postchalazal branches. Endosperm nuclear, evanescent. Embryo large, hypocotylar with minute cotyledons (Clusieae, Garcinieae, Moronobeae), or with large thick cotyledons and short radicle (Calophylleae), or with radicle of various lengths (Keilmeyeroideae). Figs. 105–133.

(Planchon et Triana 1860–62; Brandza 1908; Sprecher 1919; Netolitzky 1926.)

The family offers great variety in seed-structure. What little is known goes to show that all genera and most species are in need of investigation. Unfortunately a full series of developmental stages is hard to come by; most species flower seasonally and the full series to mature seeds may require several months to obtain. Collectors must be urged, nevertheless, to preserve material in alcohol, or suitable fixative, in addition to dried specimens. The presence of gum or latex necessitates the use of alcohol-material.

There are three basic constructions, as given in the definition of the family. For direction I turn to the arillate and capsular Clusieae. They suggest that the primitive seed of the family was arillate with moderately thick testa with a firm exotesta and lignified exotegmen of the characteristic stellate-undulate cells. These cells become so much lobed that they interdigitate with the endotestal cells to make almost a single layer which appears, accordingly, to be an endotestal construction. The manner of origin of the cells can usually be made out at the endostome where they become radially elongate and distinguishable from the endotesta. In immature seeds the exotegmic cells may remain small with dense contents for a long time and thus be mistaken for the immature endotesta. In some of these arillate seeds, however, the exotegmen fails to differentiate, e.g. *Tovomitopsis*, and this state seems to lead to the undifferentiated seed of Garcinieae. In this tribe with mostly indehiscent fruits, the large seeds come to be invested with the fleshy endocarp to make a 'sarcopyrene' in place of the numerous and individually arillate seeds. *Septogarcinia* with this modification shows that it must have evolved before the fruit became indehiscent; indeed, the genus may not be separable from *Garcinia*, in which *G. atroviridis* suggests an indehiscent *Septogarcinia*. Clusieae and Garcinieae are related also by the hypocotylar embryo which becomes more supine in the advanced Garcinieae. In Clusieae the hypocotyl elongates on germination, the small but green cotyledons expand, and the radicle becomes the main root (Brandza 1908); in Garcinieae the hypocotyl does not elongate, the cotyledons remain minute and apparently functionless, and an adventitious root from the epicotyl takes over from the short-lived radicle. Thus the *Clusia*-seed may turn into that of *Garcinia*, which resembles the Ochnaceous but is not pachychalazal. *Symphonia* with strangely fibrous endotegmen sets a problem.

Calophylleae introduce an exarillate and thickly mesotestal seed, in which the testa is highly multiplicative and sclerotic. The number of seeds per loculus of the fruit is reduced, even to a single seed per fruit as in *Calophyllum*. The seed fills the loculus and gives the appearance of an overgrown seed. The large seeds of this tribe may, therefore, be more or less overgrown and in the absence of a sclerotic endocarp to the fruit have acquired the sclerotic mesotesta. The tribe may, however, be related to Theaceae.

The aril of Clusieae may be (1) a true funicular aril (*Chrysochlamys*, *Tovomitopsis*, Fig. 129); (2) an arillode or exostomal and funicular aril (*Clusia*, *Tovomita*), in which case it may be single, double or treble (*Clusia*, *Havetia*, *Havetiopsis*); (3) a mixture of true aril and arillode (*Havetia laurifolia*); or (4) an exostomal aril combined with funicular lobes but without participation of the raphe-side of the funicle (*Clusia sp.*, p. 100). Planchon was troubled with this diversity which overstepped the morphological and academic distinction between the true and the false aril which he had drawn; he desisted.

Others, including myself, have followed Baillon's advice (p. 23). Yet, van der Pijl continues to labour this long discarded morphological abstraction, unaware of the real diversity of the aril in the allied genera of such large families as Clusiaceae and Sapindaceae. A slight displacement or inhibition of aril-outgrowth at the intricate site of origin round the micropyle and the apex of the funicle, where ovule-factors are beginning to materialize, easily explains the differences (Corner 1949a, fig. 20); it is what happens, despite the incredulity of van der Pijl, and dispels the notion that these extremes of aril-formation are phyletically distinct. However, what kind of aril was primitive in Clusiaceae cannot be decided with present meagre knowledge. Generally the aril, whether single or multiple, has thin lobes, merely two cells thick, but in *Tovomitopsis* it is 5–8 cells thick and vascular. If the vascular aril is primitive, here is another agreement with Myristicaceae. A vascular sarcotesta is said to occur in *Tovomita* but its exact nature needs investigation; the seed may be pachychalazal or, indeed, as there is but one seed per loculus, *Tovomita* may be a relic of the transition to Garcinieae. There is no evidence that this sarcotesta is primitive and inherited from pteridosperms or cycads as van der Pijl claims.

If Clusieae present the more primitive tribe of the family, the hypocotylar embryo seems advanced. Actually, as one would expect, all the tribes display primitive and advanced features. *Garcinia*, for instance, has advanced to the single ovule per loculus, the unspecialized testa, and the extreme hypocotylar embryo, but it has traces of the more primitive multicarpellary and apocarpous gynoecium in the very numerous stigmata (up to 40 or more), every one of which has the crescentic shape and adaxial cleft or suture of a carpel–primordium. And though placed relatively low in the evolution of families, the flowers of Clusiaceae have a massive and minutely elaborate structure which shows the complexity of their ancestors and the advanced simplicity of many genera. *Pentadesma* is the counter to an apetalous *Calophyllum*. Hypericaceae seem the main advance from Clusiaceous ancestry with normal embryo. Derivation from Myristicaceae, not Theales or Dilleniales, is discussed on p. 29.

Allanblackia Oliv.

There is said to be an aril in this genus. It is not described for either of the following species and it may be the thick vascular testa or endocarp-pulp. The exotegmic cells are much enlarged and become radially elongate over the sides of the seed where they combine with the multiple endotesta to unite the two integuments. There is a trace of this endotestal proliferation in *Tovomitopsis*.

A. floribunda Oliv.
Ovules with massive integuments; exostome forming the micropyle.

Seeds large, both integuments highly multiplicative, especially through periclinal division of i.e. in both to form radial rows of cells, eventually the outer part of the tegmen lignified, ? also the testa. Testa with v.b., but without special cell-layers; *i.e.* composed of small cells proliferating radially, then infiltrating between the enlarged exotegmic cells and uniting the two integuments, ? becoming lignified. Tegmen; *o.e.* developing into very large columnar cells, more or less radiating, becoming thick-walled and lignified from the micropyle to the chalaza in the order of their enlargement; *mesophyll* and *i.e.* more or less crushed. Endosperm nuclear, evanescent. Embryo hypocotylar.

(Delay et Mangenot 1960.)

A. stuhlmannii Engl.

Seed 5 cm long, ovoid, smooth, red-brown. Seed-coat composed of three layers: (1) a thin-walled outer region with large intercellular spaces, vascular bundles and canals; (2) a middle region of thick-walled cells, some pigmented; (3) an inner region, 600 μ wide, composed of stone-cells, some reaching 400 μ wide.

(Vaughan 1970.)

It seems that the third layer corresponds with the joint endotesta and exotegmen of *A. floribunda*. The result is a seed-coat which superficially resembles that of *Calophyllum*, but is made up of the two integuments.

Calophyllum L.

C. inophyllum L.

(Ceylon; flowers and fruits in alcohol.) Figs. 105–109.

Ovule anatropous, erect, large, 0.8 mm long and wide; o.i. 20–30 or more cells thick, most of the epidermal cells and some of the mesophyll cells with dark fuscous brown, glairy contents; i.i. 2–3 cells thick, very inconspicuous; nucellus occupied by the embryo-sac; micropyle formed by the stout exostome; v.b. ascending to the chalaza and dividing into 2–3 branches, themselves branching over the side of the ovule, also with slender branches from the raphe.

Seed 3 cm wide, globose, pale brownish ochraceous, somewhat uneven from the impressions of the pericarp v.b., hard. Testa 4–8 mm thick, thickest at the chalaza (distal or stylar end of the seed), becoming 100 cells thick or more; *o.e.* composed of a single layer of small, thin-walled, unlignified cells with dark contents, eventually somewhat crushed; *woody mesophyll* as an outer layer 8–12 cells thick (0.7–1 mm), composed of rows of cells elongated perpendicularly to the surface, but variously curved inwards, with much thickened, pitted and lignified walls, derived from the periclinal division of 2–3 layers of hypodermal cells in the ovule, but developing late in young seeds 4–5 mm long; *inner mesophyll* aerenchymatous, but not stellate-celled, thin-walled, with scattered bands of cells with dark glairy contents forming a network through the tissue, not

lignified but traversed throughout by a network of v.b. from the raphe and post-chalazal v.b., without gum-canals, the whole tissue drying up at maturity into a brownish felted papery stuffing; *i.e.* unspecialized, small-celled, not lignified, eventually crushed. Endosperm watery. Embryo developing late but becoming massive; radicle scarcely indicated; cotyledons thick, somewhat unequal, pressed very closely together, with secretory canals and starch; plumule minute.

Ovary syncarpous but unilocular with one ovule, without trace of other loculi; style hollow; stigma expanded and slightly 5-lobed with an apical 5-lobed perforation leading down the style to the ovary-cavity.

Fruit 4 cm wide, as a 1-seeded berry; pericarp green then often yellowish, pulpy, with many scattered v.b. not in distinct circles, the endocarp crushed by the seed, without lignified ground-tissue but with many gum-canals.

Germination slow, the hypocotyl breaking through the base of the seed as a small disc, developing the radicle at the tip of this short thick hypocotyl; cotyledons remaining in the seed; epicotyl long, the first pair of leaves or the first two alternate leaves as scale-leaves; taking *c.* 3 months to the opening of the first foliage leaves.

The apparently simple ovary, fruit and seed are indeed complicated. The woody layer in the fruit has been interpreted as the endocarp but it is the outer part of the mesophyll of the testa. This extremely massive testa, traversed by a network of v.b., is evidently typical of the genus.

Caraipa Aubl.

C. calophylla Spruce

(Brazil, Rio Negro, leg. Spruce Dec.–March 1850–51; dried material, CGE.)

Seed large. Testa undifferentiated, several cells thick, unlignified, permeated by large anastomosing v.b. Tegmen not seen.

Clusia L.

Ovules numerous, more or less transverse, axile or parietal; o.i. 2–3 cells thick or 4–5; i.i. 3 cells thick or 6–7: hypostase conspicuous at the base of the nucellus within i.i., enlarging in the seed.

Seeds rather small to medium-size, arillate. Testa not or slightly multiplicative; *o.e.* as a layer of short cells, as seen in t.s., with thick outer wall, but with more or less longitudinally elongate facets, not lignified; *mesophyll* aerenchymatous, thin-walled, more or less crushed; *i.e.* unspecialized, becoming closely adherent to the exotegmen. Tegmen not or scarcely multiplicative but the cells of the mesophyll and i.e. much enlarged in the developing seed, eventually more or less crushed; *o.e.* remaining for a long while during development of the seed as small cells with dense contents, then enlarging greatly into a firm layer of stellate–undulate, thickwalled, pitted and lignified cells, tabular over the sides of the seed, radially elongate at the endostome, often in several rows round the chalaza, becoming intimately connected with the unlignified small cells of the endotesta; *mesophyll* with large transversely elongate cells, thin-walled, not lignified; *i.e.* composed of very large, radially elongate cells with hexagonal facets, thin-walled, not lignified. Vascular bundle of the raphe entering the heteropyle, ending below the shortly columnar hypostase, without postchalazal branches. Embryo hypocotylar.

Aril 2 cells thick, simple or with basal folds, without v.b.

The seeds of this genus will certainly repay careful study. The material that I have had has been more or less fragmentary. For a long time I thought that the lignified stellate cells belonged to the endotesta but the obvious alliance of the family with Hypericaceae, and the careful account of *Allanblackia* by Delay and Mangenot (1960), caused me to re-examine the matter and correct the error. The third species which I have studied, as *Clusia sp.*, may not belong to the genus; it differs in the single seed in the loculus, the thick chalaza, and the simple aril.

C. rosea Jacq.

(Ceylon, Galaha, naturalized, leg. E. J. H. Corner 1968; immature fruits in alcohol.) Fig. 110.

Ovules 5–9 per loculus, obliquely ascending, with rudimentary aril-lobes at the base; o.i. 4–5 cells thick; i.i. 6–7 cells thick.

Seeds 1–3 per loculus, axile, arillate with many aril-folds as in *C. sellowiana*. Testa 5–11 cells thick. Tegmen 6–7 cells thick.

Fruit 3 × 2 cm, oblong–ellipsoid, 8–locular; stigmata 8. Pericarp as in *C. sellowiana*.

C. sellowiana Schlecht.

(Brazil, Mato Grosso, Chavantina, leg. E. J. H. Corner 1968; fruits in alcohol.) Figs. 111, 112.

Seeds (immature) numerous in each loculus, parietal, more or less transverse. Testa 6–8 cells thick. Tegmen 5–7 cells thick.

Aril arising as 4–9 lobes, 2 cells thick, from the exostome and the funicle (including the raphe-side), covering the young seeds with spathulate–foliaceous lobes, some of the cells with sphaerocrystals.

Fruits 20–25 × 15 mm (? half-grown), 5-locular. Pericarp with narrow and small-celled exocarp permeated by many minute latex-tubes, mostly longitudinal, and with scattered sclerotic cells; mesocarp with large cells, fewer larger latex-tubes, and v.b. in 1(–2) circles; endocarp rather thick, consisting of tangentially elongate, narrow, slightly thick-walled cells, without latex-tubes and v.b. (? forming cocci in the ripe fruit).

Clusia sp.

(Brazil, Manaus, leg. E. J. H. Corner 1948; flowers and fruits in alcohol.) Figs. 113–116.

Ovules numerous in each loculus, axile, sessile without funicle but with an aril-rim developed from the exostome before opening of the flower; o.i. 2–3 cells thick, o.e. already with distinctly thickened outer wall; i.i. 3 cells thick; nucellus small, soon absorbed by the embryo-sac, but with massive hypostase; micropyle formed by the exostome.

Seeds (immature) 1 per loculus, invested by the aril, the exotegmen developing a thick woody layer round the chalaza and containing the shortly columnar, unlignified hypostase. Testa not multiplicative; o.e. with longitudinally elongate cells, especially along the raphe, with strongly thickened outer wall and finely striate cuticle (striae transverse to the long axis of the cell). Tegmen 4 cells thick, of typical construction.

Aril arising from the exostome (pre-fertilization) and from the sides of the very short funicle (post-fertilization), not from the raphe-side, developing foliaceous lobes 2 cells thick, the sides of the funicle developing 2–3 superposed rows of folds.

Fruit 6–7 mm wide, immature, 4–locular. Pericarp with a single ring of v.b. accompanying the latex-tubes; endocarp thin (evidently not forming cocci).

Garcinia L.

Ovary 2–16 locular, with one erect ovule per loculus.

Seed large, surrounded by endocarp pulp, the exarillate testa unspecialized except for secretory canals and v.b. in a network. Embryo hypocotylar, generally with secretory canals.

The seeds in this large genus seem remarkably uniform, but there are many differences in the structure of the pericarp, particularly the endocarp, which need systematic study. Unfortunately progress with this fascinating genus has been stopped for many years through retention of the collections at Bogor.

The seeds of *G. mangostana* develop apomictically and may have 2–3 embryos.

G. mangostana L.

Ovule; o.i. 5–7 cells thick; i.i. 2–3 cells thick.

Testa 9–12 cells thick, connate with the endocarp-pulp. Tegmen not multiplicative, soon crushed.

Pericarp with (1) a suberised outer epidermis; (2) 8–10 layers of collenchymatous cells; (3) 12–15 layers of sclerotic cells; (4) thick inner mesocarp with somewhat thickened cellulose walls, slightly lignified in places; and (5) endocarp round the seeds in each loculus; v.b. throughout the tissues; secretory canals in tissues (2) and (4), 50–100 μ wide in the outer tissue, 100–200 μ wide in the inner tissue; v.b. at the base of the ovary arranged in as many circles (4–8) of *c.* 10 v.b. as loculi; each circle giving at the base of the loculus a v.b. to form in the axis of the ovary an inner circle of v.b. with the xylem orientated outwards, and one of these inner v.b. passing to an ovule; upper part of the ovary unilocular.

Endocarp of each loculus c. 10 cells thick, widening in the fruit by cell-enlargement, without air-spaces or v.b. but with secretory canals, the outer 3 layers of cells swelling enormously to make the pulp round the seed; inner epidermis eventually dividing to form 3–4 layers of small thin-walled cells pressed tightly against the testa and adhering to it.

(Sprecher 1919.)

G. xanthochymus Hook. f.
(Ceylon, Peradeniya; fruits in alcohol.)

Testa 1–2 mm thick, many cells thick, undifferentiated, permeated by a close reticulum of v.b. Embryo with secretory canals.

Fruit 5-locular; stigmata 5, flat, flap-like. Pericarp wholly thin-walled, pervaded by a network of cortical v.b. and a closer network of secretory canals; endocarp composed of tangentially elongate, collenchymatous cells, without secretory canals, the epidermal cells in the region of the micropyle projecting (? secreting the gum round this end of the seed).

Garcinia sp.
(Borneo, Kinabalu, RSNB 4085; fruits in alcohol.)
Figs. 117–120.

Seeds 10 mm long. Testa c. 8 cells thick, the cells of o.e. somewhat transversely elongate. Tegmen crushed. Embryo apparently without secretory canals.

Fruit 25–28 mm long, subglobose to broadly ellipsoid, 6-locular, surmounted by a disc of 30–38 stigmatic papillae; sepals 4. Pericarp with four layers: (1) a small-celled exocarp; (2) small-celled outer mesocarp with abundant anastomosing clusters of stone-cells forming a sclerotic zone; (3) inner mesocarp large-celled, soft, thin-walled, with few v.b.; (4) endocarp round each loculus consisting of a single layer of large contiguous secretory canals passing obliquely round the loculus from the axis of the fruit downwards and outwards to the base of the loculus, not connected with the secretory canals of the mesocarp; secretory canals and v.b. forming a network in the exocarp and mesocarp; stigmatic lobes each with a stylar canal, these canals combining into 6 main canals, each leading to a loculus.

Havetiopsis Planch. et Triana

H. flexilis Planch. et Triana
(Brazil, Manaus, Corner n. 59 and 146; fruits in alcohol.)
Figs. 121–123.

Seeds 5 × 2.2 mm, (1–)2 per loculus, suspended. Testa 5–7 cells thick; o.e. composed of cuboid cells with strongly thickened outer wall; mesophyll unspecialized, but scattered cells with sphaerocysts; *i.e.* un-

specialized. Tegmen 5–7 cells thick, as in *Clusia* with the same exotegmic layer of stellate-undulate lignified cells. Chalaza unspecialized. Raphe v.b. ending at the chalaza. Endosperm large-celled, collapsing. Embryo hypocotylar; cotyledons minute.

Aril in numerous plicate-undulate folds, derived from the exostome (not from the funicle or raphe), covering the seed except at the apex, 2 cells thick, without v.b.

Fruit 15 mm long, ellipsoid, 5-locular, surmounted by 5 flat subtriangular stigmata, septicidal, each loculus with a bony coccus more or less enclosing and holding the (1–)2 seeds detached from the strongly fluted columella. Pericarp with very thick, heavily lignified endocarp forming the cocci and evidently containing the v.b. of the pericarp. Septa with very large secretory canals.

The seeds resemble those of *Pilosperma*. The bony cocci of the fruit seem exceptional for most Clusiaceae have thin or unlignified endocarp.

Mesua L.

M. ferrea L.
Seed 26 mm long, pyriform, brown, smooth, hard, exarillate. Testa thick, evidently multiplicative, lignified; *o.e.* composed of flattened cells; *mesophyll* mostly sclerotic, the middle layer with many radial and tangential fibres (–1 mm long), the inner tissue thinner-walled and crushed.

(Vaughan 1970.)

The structure may conform with that of *Calophyllum*.

Pentadesma Sabine

P. butyracea Sabine
(Ceylon, Peradeniya; flowers and fruits in alcohol.)
Figs. 124–127.

Ovules numerous in each loculus, anatropous, variously orientated but mostly with the micropyle directed to the base of the ovary, massive, 1.4 × 1 mm, parietal (only the extreme base of the ovary completely septate), bitegmic with massive integuments, tenuinucellate; o.i. 18–22 cells thick (excluding the procambial strands), the cells of o.e. with mucilaginous outer walls, with a network of procambial strands derived from the raphe and chalaza; i.i. 10–15 cells thick; micropyle formed by the massive exostome.

Seeds 3.5–4.5 × 2.5 cm, ellipsoid or subglobose, brown with pale streaks, slightly arillate at the base. Testa c. 1 mm thick, much thicker at the hilum, leathery, composed of 40–50 layers of cells (excluding v.b.); *o.e.* composed of rather tabular cells with brown walls but with small colourless patches of cells, the outer walls slightly thickened, not lignified; *mesophyll* aerenchymatous, composed of vesicular cells stellate with short arms, many with pale brown contents, not lignified; *i.e.* unspecialized, eventually crushed. Tegmen unspecialized, crushed. Vascular supply as a stout reticulum throughout the testa. Embryo hypocotylar; cotyledons minute; without secretory canals; v.b. procambial.

Aril as a slight pulpy rim round the top of the stout funicle, developed after fertilization; outer 3–4 layers of cells colourless with pulpy mucilaginous walls, passing gradually into the tissue of the endocarp on one side of the aril-rim and, by reduction of the cell-layers, into the firm exotesta.

Ovary elongate-conic, 5–locular with 5 styles, the loculi filled with transparent pulp; ovary-wall with the complicated network of v.b. and secretory canals as in the pericarp, but many v.b. procambial.

Pericarp 8–10 mm thick, with sclerotic exocarp, thin collenchymatous endocarp, and massive thin-walled mesocarp; *exocarp* with a thin brown corky periderm, more or less cracked, and an inner diffuse layer of sclerotic cells; *mesocarp* pervaded by the network of v.b. and secretory canals, forming a fine close network just outside the endocarp; *endocarp* composed of transverse, tangentially arranged fibres, not lignified, without secretory canals. Peduncle of fruit with a network of many cortical v.b.; secretory canals very sparse in the medulla.

(Vaughan 1970.)

The microscopic structure of the massive flower is extremely complicated. The cortical network of vascular bundles in the peduncle dilates in the ovary-wall into a close and intricate network; the ring of primary v.b., just external to the loculi, is ill-defined; the placentas have many v.b. which enter as diversely into the stout funicles of the seed, the testa of which is similarly pervaded by a vascular network. Then through the meshes of the vascular system there is an even more complicated one of the secretory canals, but they do not enter the seed. The complexity extends to the stamens.

Septogarcinia Kosterm.

As *Garcinia* but the fruit dehiscent.

(Kostermans 1962.)

Septogarcinia sp.

(Solomon Islands, RSS 2454; fruits in alcohol.) Fig. 128.

Seeds 7 mm long, somewhat flattened, covered with thin yellow endocarp-pulp, hanging on strips of septal (axial) tissue. Testa 6–8 cells thick, thin-walled, unspecialized, more or less crushed. Tegmen crushed. Vascular bundles 5–6, divergent from the hilum and forking into the testa. Chalaza ?. Embryo hypocotylar without evident radicle or cotyledons, slightly curved, consisting of a solid mass of cells with a central column of thin-walled cells (? procambial or secretory canal), without v.b., with many crystals and raphids in the cells.

Fruit 15 mm wide, subglobose, the stigmata concealed in an apical slit, green, 7–9 locular, splitting widely into 7–9 acute lobes as a star 4 cm wide, hanging and the

seeds dangling. Pericarp thin-walled, succulent, without sclerotic cells; v.b. in 3–4 circles, the outermost smallest; secretory canals abundant, rather close, the outermost smallest; septa composed of plates of small radially elongate cells. Endocarp round each loculus and detaching with the seed, composed of four layers; (1) a thick layer of thin-walled cells with brown contents (in alcohol-material); (2) a thin layer of compact collenchymatous fibres not lignified, variously directed; (3) a layer of large subglobose gum-cavities (not elongate as canals); (4) a narrow inner layer, 5–6 cells thick, of thin-walled cells pressed against the testa.

This genus, perhaps no more distinct from *Garcinia* than *Wormia* from *Dillenia*, has a fruit as *Clusia* and seeds covered with endocarp, simulating the aril, as in *Garcinia*. The flowers of this species (RSS 231) were assigned to *Garcinia*.

Symphonia L.

S. clusioides Baker

Seeds 15 mm wide with brown veins (latex-ducts) radiating from the hilum. Testa *c.* 600 μ thick, unspecialized, with a network of latex-tubes filled with brown resin. Tegmen *c.* 14 cells thick, unspecialized except for lignified fibres singly or in groups in the inner part; *i.e.* with the cells elongated into thick-walled, lignified fibres (not pitted but finely and obliquely striate), making a mass of hairs round the embryo.

Fruit 20 mm, globose, 1-seeded, with 4 abortive loculi. Pericarp 2–4 mm thick, fragile, with a network of v.b. and of brown latex-tubes in *c.* 6 layers; o.e. composed of short thick-walled cells; endocarp 2–3 cells thick.

(de Cordemoy 1911.)

Remarkable for the fibrous hairy endotegmen.

Tovomitopsis Planch. et Triana

Tovomitopsis sp.

(Brazil, Manaus, Corner 255; alcohol-material.) Figs. 129–133.

Ovules evidently 1–2 per loculus, axile, anatropous, erect.

Seed 10–15 × 5–6 mm, ellipsoid or subclavate, blackish brown with pale mottling, 1–2 per loculus, detaching from the placenta but fixed in the dehiscent fruit by the aril-process at the base. Testa 7–12 cells thick; *o.e.* composed of cuboid cells with rather strongly thickened outer wall; *mesophyll* unspecialized; *i.e.* composed of small irregular cells with slightly thickened walls, not lignified. Tegmen 3–4 cells thick, the cells rather large but soon crushed. Vascular bundles arising as 6–7 bundles from the hilum, without distinct raphe, forming a loose network in the testa. Chalaza not distinct. Secretory tubes abundant in the placenta-funicle, not in the testa. Embryo hypocotylar, with abundant secretory canals; cotyledons minute.

Aril arising soon after fertilization from the very short funicle (not from the exostome), entire, covering the seed as a single cupule, often sublobate at the apex, varying more or less incised on the raphe-side and exposing the seed, scarlet, pulpy, 5–8 cells thick but 10–14 cells in some parts, with a loose irregular network of v.b. from the hilum, the epidermal cells much larger than the internal cells, with faintly striate cuticle.

Fruit 25 mm long, ellipsoid, 4-locular but often uneven with abortive loculi, green, surmounted by 4 stigmata, septicidal, the recurving lobes exposing the reddish pink inner surface. Pericarp not woody, with 2 rows of numerous small v.b. not accompanying the secretory canals; endocarp thin, not lignified; mesocarp with sphaerocrystals in many cells.

The seed-coats of the young seed are shown in Fig. 132; those of the ripe seed are similar but with cells about twice as large and with the tegmen crushed. Though young fruits were collected with the ovules just beginning to form seeds, some of these were abnormal and consisted of a lump of homogeneous tissue without integuments or aril; possibly they were the aborted ovules in the state which has practically one ovule/seed per loculus.

This small tree was not uncommon round Manaus but I cannot assign it satisfactorily to a genus. It has the four sepals and petals, the many stamens (*c.* 20) with long terete anthers and short filaments of *Tovomitopsis*; it has the funicular aril of *Chrysochlamys*, but this aril is peculiar in being thick and vascular; and, as exceptional points to these genera, the ovules are erect and superposed in pairs in the loculus. The seed appears on the way to Garcinieae.

Tsimatima Jum. et Perr.

T. pervillei Jum. et Perr.
Seed surrounded with endocarp–mesocarp pulp. Testa 0.9 mm thick, unspecialized, with v.b. (? only one integument).

(de Cordemoy 1911.)

COMBRETACEAE

Ovules anatropous, bitegmic, crassinucellate, 2–6 in the ovary, suspended from the apex of the loculus, usually with long funicles.

Seeds small to moderately large, usually retained in the indehiscent fruit, exalbuminous, exarillate: seed-coats thin, multiplicative or not (? pachychalazal in *Poivrea*). Testa; *o.e.* composed of cuboid cells with isodiametric facets, often tabular through compression, not lignified; *mesophyll* thin-walled, with scattered sclerotic or tracheidal cells with more or less reticulate thickening, in some cases spiral-annular; *i.e.* composed of sclerotic or tracheidal cells or (?) unspecialized. Tegmen; *o.e.* composed of narrow longitudinally elongate, lignified fibres with fine close spiral thickenings, as a continuous layer or interrupted with thin-walled cells; *mesophyll* and *i.e.* unspecialized, crushed. Vascular supply limited to the raphe v.b. or with several post-chalazal branches. Endosperm nuclear.

(Netolitzky 1926; Fagerlind 1941.)

Suitable material for the study of the seed-coat has been difficult to obtain. The indehiscent fruit would not appear to favour much differentiation in the thin and, so far as concerns seed-dispersal, functionless integuments. Yet, I find in *Terminalia* two remarkable and diagnostic features. The endotesta is a layer of short, tangentially elongate, tracheidal cells with spiral thickening, as in Lauraceae and Monimiaceae, and the exotegmen is a layer of tangentially elongate fibriform tracheids with close spiral thickening as in Lythraceae. The mesotesta has sclerotic cells more abundant at the chalaza and exostome, also as in Lythraceae, but Combretaceae seem to lack the crystal-cells of the Lythraceous endotesta. If Combretaceae thus combine the distinctive characters of Lauraceous and Lythraceous seeds, yet they agree with Punicaceae in the thin-walled exotesta, the sclerotic mesophyll, the exotegmic fibres, and the straight embryo with contort cotyledons. The family appears to be a pauci-ovulate derivative of Punicaceae without close affinity with Lecythidaceae, Melastomataceae and Myrtaceae, as usually maintained, unless these families have lost the tracheidal exotegmen. According to Netolitzky, *Combretum* has merely a thin-walled testa and *Lumnitzera* a tracheidal mesotesta; they, too, may have lost the tracheidal exotegmen.

Poivrea Comm.

P. coccinea DC
Ovules 3 in the ovary, suspended with long funicles; o.i. 5 cells thick; i.i. 2–3 cells thick; both integuments forming the zigzag micropyle.

Seed apparently pachychalazal, the extended chalaza forming a cup reaching a half to two-thirds the length of the seed. Testa 7–8 cells thick. Vascular bundle with 5 branches from the chalaza ascending the chalazal cup to the attachments of the integuments.

(Venkateswarlu 1952b.)

Quisqualis L.

Q. indica L.
Testa; *o.e.* with strongly thickened outer walls; *mesophyll* with an outer layer of lignified cells with bar-like thickenings on the walls, and an inner layer of thin-walled cells more or less crushed but the outer 1–2 cell-layers with dark contents.

(Netolitzky 1926.)

Terminalia L.

Seeds elongate-cylindric or fusiform, suspended by a long funicle in the small cavity of the thick woody endocarp, solitary or occasionally paired. Testa thin, pale brown, multiplicative or not; *o.e.* composed of subcuboid cells with slightly thickened brownish walls and isodiametric facets, not lignified; *mesophyll* (when multiplicative) thin-walled with scattered sclerotic cells (with subreticulate or subspiral thickening), more abundant at the chalaza and exostome; *i.e.* composed of short, tangentially placed, lignified tracheidal cells with spiral-annular thickening. Tegmen multiplicative or not; *o.e.* as a layer of narrow, longitudinally elongate, lignified tracheids with fine, close, spiral thickening, in some cases as a discontinuous layer; *mesophyll* and *i.e.* unspecialized, crushed. Embryo with contort cotyledons.

T. arjuna W. et Arn.
(Ceylon; mature fruits in alcohol.) Fig. 134.

Testa ? not multiplicative, 2(–3) cells thick; *i.e.* composed of large ellipsoid cells with strong lax spiral thickenings, often in distorted pattern, lignified. Tegmen ? 4–5 cells thick, the tracheidal cells of o.e. with 1–3 spiral bands.

T. catappa L.
(Ceylon; flowers and fruits in alcohol.)

Ovules 2; o.i. 3–4 cells thick, with rather large cells in o.e.; i.i. 2 cells thick.

Testa 7–10 cells thick; *mesophyll* with very scattered sclerotic cells, abundant at the woody exostome; *i.e.* ? unspecialized. Tegmen not multiplicative; *o.e.* as a discontinuous layer of long tracheids with spiral thickening. Vascular bundle with 3 postchalazal branches anastomosing into a fine reticulum throughout the mesotesta. Endosperm oily, eventually absorbed.
(Netolitzky 1926; Vaughan 1970.)

The seeds develop late and are still small when the fruit has reached two-thirds of its final size. The inner two layers of the testa are figured by Vaughan as composed of longitudinal tracheids with spiral-annular thickening; the inner layer is the exotegmen and the outer layer may be the endotesta, as in *T. parviflora*, unless this layer remains thin-walled and is crushed.

T. glabrata Forst. f.
Testa with the outer thin-walled mesophyll carrying the vascular network, a middle region of very thick-walled and lignified cells, and an inner compressed layer (6–7 cells thick) of thin-walled cells; *i.e.* unspecialized.
(Netolitzky 1926.)

T. parviflora Thw.
(Ceylon; ripe fruits in alcohol.) Fig. 135.

Fruit and seed much as in *T. catappa*; funicles 5–6 mm long. Testa 5–6 cells thick; *mesophyll* with scattered sclerotic cells with reticulate thickening or, in the cells next to i.e., annular-subreticulate, abundant at the chalaza and exostome; *i.e.* as a continuous layer of short,

lignified tracheidal cells with spiral to subannular thickening. Tegmen 5–6 cells thick, unspecialized and crushed except the exotegmen as a more or less continuous layer of longitudinal tracheids 150–600 × 10–14 μ, with fine close spiral thickening, shortening at the micropyle to form a woody tracheidal margin to the endostome.

The integuments of the ripe seed are so closely applied to each other over most of their length that it is impossible to distinguish their apposed surfaces, but at the micropyle both are distinct and can be separated by dissection as pellicles. The endotesta is a layer of coarse short tracheidal cells and the exotegmen is a layer of narrow long tracheids.

COMPOSITAE

Ovule anatropous, suspended, solitary, unitegmic, tenuinucellate; integument 4–18 cells thick, often rather massive; i.e. as an endothelium, in some cases dividing to form 2–3 cell-layers (*Ainsliaea*).

Seeds small, enclosed in the achene, with thin or papery seed-coat, generally albuminous. Seed-coat multiplicative or not; *o.e.* as a palisade of radially elongate prismatic cells with the walls thickened all round (*Arctium, Carduus, Carthamus, Centaurea, Cirsium*), even with a *linea lucida*, in other genera more or less flattened, with thick pitted wall (*Helianthus*), with spiral thickening (*Taraxacum*), with the inner and radial walls thickened, or thin-walled, the facets generally polygonal and isodiametric, undulate in *Guizotia*; *mesophyll* thin-walled and crushed, or 1–4 layers persistent; all the cells of the integument with a small crystal (*Podolepis*). Vascular supply various (see below). Endosperm cellular or nuclear, oily, persistent or not.
(Guignard 1893; Netolitzky 1926; Kühn 1927, p. 344; Davis 1961; Deshpande 1962a, b, 1964a, b; Kapil and Bela Sethi 1963a, b; Singh, B. 1964; Sehgal 1966; Dittrich 1968; Vaughan 1970.)

While most botanists are satisfied with the external features of the achene and a few have enquired into the development of the embryo and ovule, little attention has been given to the seed-coat since the time of Kühn. It is clear, nevertheless, that the seed-coat has not completely deteriorated in the achene but has retained an exotestal palisade (Cynareae), a variety in the thickening of the exotestal cells and in their facets, in the retention of the mesophyll, and in the vascular supply, which may be surprisingly complicated. The following three states of vascular supply have been reported:

1. Vascular supply as a single bundle extending round the seed from funicle more or less to the micropyle (*Achillea, Anacyclus, Carlina, Carthamus, Cichorium, Cnicus, Hypochaeris, Jurinea, Lactuca, Lappa, Leontodon, Madia, Microseris, Onopordon, Pamphalea, Picridium, Saussurea, Scolymus, Scorzonera, Silybum,*

Taraxacum, Trichocline, Tridax, Urospermum, Warionia, Xeranthemum).

2. Vascular supply with postchalazal branches; 2 branches extending to the micropyle (*Grindelia*); 2–3 branches reaching to the micropyle (*Calendula, Helianthus*); 9 divergent branches (*Echinops*) and up to 15 (*Crupina*).

3. Funicular v.b. giving off 2 branches on entry into the seed, these dividing once and making with the main raphe v.b. 5 v.b. descending to the chalaza, some with small recurrent branches (*Xanthium*).

In the case of Centaureae, Dittrich reports three kinds of hilum which may affect the vascular supply of the seed. Even in this advanced family there is much to be discovered of morphological and systematic importance, not the least being the derived nature of the small seed with only the raphe bundle.

CONNARACEAE

Ovules 2(–3) per carpel, collateral, hemianatropous with pronounced preraphe (with free integuments between the hilum and the micropyle on the raphe-side), bitegmic, crassinucellate; o.i. 5–7 cells thick; i.i. 3–5 cells thick; micropyle formed by the exostome; funicle none.

Seeds 1–2(–3) per follicle, large, black, arillate or sarcotestal at the chalazal base or more or less wholly sarcotestal (baccate seeds), the seed-body developed mostly from the preraphe or micropylar end of the seed distal from the hilum; hilum lateral, generally sub-basal; micropyle more or less apical, far removed from the hilum; exotestal with fibrous exotegmen more or less developed; exalbuminous or more or less albuminous (*Burttia, Cnestis, Ellipanthus, Pseudellipanthus, Pseudoconnarus*); funicle none or very short.

Testa multiplicative, 13–25 cells thick; *o.e.* as a palisade of short cells with thick brown outer walls, not lignified, ill-defined in *Cnestis*, becoming sarcotestal in the raphe–chalaza or more or less over the seed (baccate seeds); *mesophyll* strongly aerenchymatous, collapsing; *i.e.* often as a short palisade of cells with thick (*Cnestis*) or thin walls, not lignified, but with many sclerotic cells in *Jollydora*, as a layer of small crystal-cells in *Connarus spp*. Tegmen multiplicative, 8–14 cells thick, generally soon disintegrating or crushed except o.e. as a layer of longitudinal lignified fibres (*Cnestis, Connarus spp., Jollydora, Rourea*); *mesophyll* developing much enlarged aqueous cells before collapsing; *i.e.* developing more or less as an endothelium, then collapsing.

Vascular supply consisting typically (*Connarus*) of a short raphe v.b., a chalazal plexus with short branches in the sarcotestal part round the chalaza, and a long unbranched preraphe v.b. to the micropyle, without a postchalazal branch in the antiraphe; in *C. grandis* and *C. villosus* with branches of the chalazal plexus extending almost to the antiraphe; in *Jollydora* with 4–5 postchalazal v.b. extending to the micropyle and the preraphe v.b. often 1–2 times furcate; in *Cnestis* with many postchalazal v.b. more or less anastomosing and forming with branches of the preraphe v.b. a network throughout the testa.

Embryo with thick straight cotyledons, but thin in albuminous seeds; hypocotyl short; radicle minute.

Aril present to some extent in all cases, primarily as an outgrowth from the short raphe and chalaza, yellow to red, pulpy, typically (*Connarus*) with a thick, often crenulate, free border or short limb and a sarcotestal area along the raphe and chalaza; with extensive free limb (*Rourea spp.*); more or less wholly developed as a sarcotesta in genera with baccate seeds, but ? the preraphe not sarcotestal; composed of large pulpy radiating cells (? always colourless), but in *Connarus* with vermiform internal masses of deeply coloured small cells with carotenoid contents; oily and starchy; without v.b.; epidermal cells small. Figs. 136–155.

(Netolitzky 1926; Leenhouts 1958; Hutchinson 1964; Dickison 1971.)

The typical Connaraceous seed is one of the more distinct among dicotyledons. Its main features are the preraphe form which separates the micropyle so widely from the hilum, the black testa with firm but not lignified exotestal palisade, the fibrous exotegmen, the preraphe v.b., the chalazal aril, the large embryo, and often the bean-like smell of the crushed tissue. The form is set in the ovule and the great lengthening of the preraphe may render the seed almost obcampylotropous (Figs. 139, 155).

Of these characters the fibrous exotegmen must be selected as the intrinsic feature aligning the family with Celastraceae, Flacourtiaceae, Meliaceae, Violaceae and such Sapindaceous genera as *Alectryon*. Nevertheless, this feature may be lacking in some species, even among those of a genus such as *Connarus*, just as it may happen in these other families. The point needs further investigation, and so does the occurrence of the lignified endotesta (*Cnestis, Jollydora, Rourea*). Thus the Connaraceous seed may have three distinctive epidermal layers, but lacks a mechanical endotesta.

The nature of the aril has been disputed, particularly the nature of the fleshy modification of the testa. According to the durian-theory, the Connaraceous seeds were primitively arillate and the testa became gradually transformed into a sarcotesta with arillar surface, similar to the adnate aril of Sapindaceae. According to van der Pijl, the seed was primitively sarcotestal, as an inheritance from cycadalean ancestry (v.d. Pijl 1955; Leenhouts 1958). Actually both aril and sarcotesta are present but, of course, without such far-fetched homology. In *Cnestis, Connarus* and *Rourea*, perhaps in most genera, fertilization induces the short raphe and chalaza to thicken; their epidermal and hypodermal cells

divide periclinally, and around the margin of this thickening the proliferating tissue extends as a free edge to
make the short free part of the aril which overlies the
incipient testa. The free part rarely covers more than a
quarter of the seed and commonly it becomes crenulate
at the edge, but in some species of *Rourea* it invests the
whole seed though cleft along the preraphe side. It
grows into the spacious cavity of the enlarging follicle
(Fig. 141) and in no way can it be regarded as the effect
of squashing the proliferating tissue over the surface
of the testa. Neither the aril nor its thickened base (the
sarcotesta) receives vascular bundles, but both form the
same kind of pulpy tissue. If the base were less thickened
and the attachment of the aril were narrowed, the
sarcotesta would be negligible, as in most arillate seeds;
and this may be the condition in both *Rourea* and
Santaloides.

The resulting complex of aril and sarcotesta, as in
Cnestis, or *Connarus*, is comparable with the construction of the seed in such Sapindaceae as *Harpullia* and
Alectryon, even with that of *Gonystylus*. It is not a
modification of the testa which, as a free integument,
develops its own firm exotesta and aerenchymatous
mesophyll. It is not a pachychalaza, as occurs in Sapindaceae, though undoubtedly there is a certain, not
disproportionate, enlargement of the chalaza.

This is the state of the seed in most Connaraceae with
dehiscent fruits and in all which have persistent endosperm; it accompanies, that is, the more primitive
states of fruit and seed. But in a few exalbuminous
genera almost the whole of the testa becomes pulpy,
not merely the raphe and chalaza, so that the seeds,
styled baccate seeds, are sarcotestal without a free aril.
Three of these genera have dehiscent fruits (*Byrsocarpus*,
Jaundea, *Manotes*); two have indehiscent fruits (*Heman-
dradenia*, *Jollydora*). The baccate seeds accompany the
more advanced states of fruit and seed. Hence I have
always regarded the Connaraceous seed as primitively
arillate with firm black testa set in the red fleshy follicle
with mechanical protuberances (*Agelaea*, *Castanola*).
The baccate seed I regard as the derivation through
encroachment of the chalazal testa over most of the seed
from the side opposite to the preraphe. I emphasize
the detail of most of the seed because in *Byrsocarpus*,
Jollydora and *Manotes* there is left along the preraphe
and round the micropyle a strip of true firm black testal
tissue, as the vestige of the primitive testa of the family
(Fig. 146). It is probably no freak that the large aril of
Rourea minor is cleft along the preraphe side. Similar
strips of unmodified testa occur in the pachychalazal
sarcotestal Sapindaceae and there is an analogy, at least,
with the vestigial testa and tegmen at the micropyle of
the pachychalazal and sarcotestal seed of *Aphanamixis*
(Meliaceae, p. 186). Yet there is no clear evidence of
true pachychalazal construction in Connaraceae.

These points have been omitted by van der Pijl in
his contention that the sarcotesta is primitive for the
family. It is not clear that he has understood even the
free aril or aril-lobe of many genera, or ever examined
the Connaraceous seed in detail. He confuses the issue by
supposing, without proving, that the chalazal sarcotesta
is homologous with the pantestal sarcotesta of *Magnolia*
and *Cycas*. In these genera there is no strip of firm
testal tissue along a preraphe or round the micropyle.
He has failed to explain, also, the origin of the hard
testa and fibrous exotegmen which, as family marks,
link Connaraceae with Celastraceae, Meliaceae, Violaceae
and the other families with these peculiarities. His
hypothesis credits a primitive state to an indehiscent
fruit and exalbuminous seed; it fits none of the
facts.

There is, nevertheless, much to be learnt about the
seeds of the family. Many genera have not been investigated. It is not even clear that the sarcotesta of *Jollydora*
with its simple arrangement of radiating cells is equivalent to the chalazal sarcotesta of *Connarus* with its
vermiform internal structure. That of *Jollydora* suggests
the pulpy modification of the overgrown testa of *Cnestis*,
which is similar to that of its aril and, unlike that of
Connarus, similar to that of *Rourea*. Such problems must
be resolved through evaluation of the whole seed-
structure, inclusive of the vascular supply. Primitively,
to judge from *Connarus*, there would have been a short
raphe v.b. and a longer preraphe v.b. without branches
to the sides of the seed, e.g. *Rourea minor*. Then, as the
chalazal sarcotesta extended, so vascular bundles may
have proceeded from the chalaza towards the antiraphe
and, finally, from the preraphe itself, e.g. *Connarus
semidecandrus* (Fig. 140). Alternatively the last state
may have been primitive and connected with a large
lobulate and vascular aril e.g. *Alectryon* of Sapindaceae
(Fig. 503). Finally this network may have extended
through the whole testa in the manner of overgrown
seeds, for which *Cnestis* is an example. But *Jollydora*,
with several postchalazal v.b. extending without anastomosis almost to the micropyle, is different and shows that
there can not be just one all-embracing solution to the
evolution of the Connaraceous seed. If the sarcotesta of
Byrsocarpus and *Manotes* is not homologous with that
of *Jollydora*, then their seeds will have the vasculature
of *Connarus* and *Cnestis*.

A curious resemblance exists between the Connaraceous
Schellenbergia of Burma and the Sapindaceous *Guioa* in
the prolongation of the aril into a false funicle.

The resemblance in vascular supply of the carpel
between Connaraceae and Rosaceae is an example
of convergence through neotenic simplification
(Dickison 1971). Such, also, is the Papilionaceous
resemblance.

With pinnate exstipulate leaves and the same essential
seed-structure, Connaraceae come near to Meliaceae as
an offshoot of the apocarpous ancestry.

Agelaea Sol.

Agelaea sp.

(Sarawak, S 25648; alcohol-material.)

Seed 20 × 8 mm, cylindric, black; aril adnate, 5–6 mm long and wide, with a slight free margin 0.5 mm wide, more or less surrounding the raphe. Testa 17–24 cells thick; *o.e.* as a palisade 45–50 μ high, of columnar cells with thick, dark brown, outer wall and somewhat thickened radial walls; mesophyll crushed; *i.e.* not recognizable. Tegmen crushed, without trace of fibres. Vascular bundles *c.* 8 reaching from the chalaza more or less to the micropyle, not anastomosed.

Cnestis Juss.

C. palala (Lour.) Merr.

(Singapore; living material.)   Figs. 136, 137.

Seed 2.3 × 1.5 × 2.1 cm (excluding the aril), black, shiny, oblong, compressed, obtuse, without funicle; hilum 2–2.3 × 1.5 mm, elongate, pallid, dilated and deepened at the arillar end, the seed separating from the placenta and caught in the gaping follicle by the swollen aril. Testa multiplicative; *o.e.* 2–3 cells thick, the cells subcuboid to subcylindric in radial rows, filled with starch-grains, the radial walls undulate, thickened and dark yellowish brown, without stomata, not as a regular palisade; *mesophyll* as a thick layer of more or less stellate cells, somewhat in radial rows, eventually separating at the septa in the arms and the arms becoming friable; *i.e.* as a very short palisade of cells with strongly thickened, dark brown inner wall. Tegmen multiplicative but eventually reduced to a single layer of rather thick-walled, colourless, pitted, lignified fibres 9–13 μ wide, elongate longitudinally towards the micropyle, with irregular lumen. Nucellus crushed into a brown lamella. Vascular bundles in the mesotesta, forming a reticulum with a recurrent branch along the preraphe, a short raphe v.b., a short postchalazal v.b., and 4–5 main branches from the chalaza to each side of the seed. Endosperm thick, horny, the radially elongate cells with slightly thickened walls, oily, without starch.

Aril 7–8 mm thick at the base, pale bright yellow, pulpy-fleshy, cupular at the base of the seed, arising from the wide chalazal patch, with a short free, entire or crenulate, margin; composed of thin-walled pulpy cells slightly elongate radially, with many small and large pale yellow oil-globules, without v.b.; epidermal cells short, cuboid, with thickened outer walls.

Follicle red, fleshy, the wall 12–13 mm thick, composed mainly of sappy thin-walled tissue with abundant v.b.; endocarp 0.7 mm thick, as a thin valve of tracheidal cells, shortly elongate longitudinally with thickened and much pitted walls and wide lumen; on the inside of the endocarp, lining the follicle-cavity, with a small-celled, thin-walled tissue 0.1–0.2 mm thick, bearing the very abundant chaffy pale brown and bristly hairs 1–3 mm

long, with thickened walls, smooth, acute, with wide sparsely and finely septate lumen.

This is an overgrown seed, filling the follicle-cavity, and the testa undergoes repeated periclinal division of the cells to form the thick mesophyll and multiple epidermis. The species *C. palala* was said to have a red aril. *C. flammea* Griff., reduced by Hooker and Schellenberg to *C. palala*, had a yellow aril according to Griffith, and this is the Malayan plant.

Connarus L.

C. grandis Jack

(Malaya; living material.)   Fig. 138.

Seed 4–6 × 2–3.5 × 1.5–2 cm, large, flattened and bean-like but thick, black, shiny, with thin leathery testa and rich yellow aril covering the basal half or two-thirds of the seed. Testa multiplicative; *o.e.* as a palisade of shortly radially elongate cells, at first with orange-yellow oily contents (making the immature seed rich yellow as the aril), then the walls turning brown to black and the lumen colourless; *mesophyll* aerenchymatous, collapsing. Tegmen soon disappearing; fibres ? Vascular bundles as in *C. semidecandrus* but the laterals reaching the antiraphe without anastomosis. Endosperm watery, soon absorbed. Embryo rapidly growing to fill the whole seed; cotyledons greenish, internally pinkish brown flecked with red dots, with thin-walled oily cells and small clusters of red resin-cells; hypocotyl short; radicle minute.

Aril without v.b., developed from the raphe and chalaza, with short free crenulate margin, at first greenish, composed of large colourless pulpy cells with interspersed rows or sinuous masses of smaller cells with yellow, oily-granular contents.

Follicle 7–9 × 4–5 × 3–4 cm, soon enlarging, pink then scarlet, finally after several weeks turning deep blood red and ripening, dehiscing along the placental suture only, the valves catching the detached seed. Pericarp composed of six layers: (1) *o.e.* with red anthocyanin; (2) outer cortex of thin-walled cells, pale yellow, with fairly numerous stone-cells, scattered or in groups, and with scattered clusters of red resin-cells; (3) a thin fibrous-woody layer of very numerous v.b. encased in thick-walled pitted fibres and stone-cells; (4) a thin-walled white inner cortex of stellate-celled aerenchyma; (5) a thin endocarp-valve, 1–2 cells thick, of oblique or transverse fibres; (6) *i.e.* as a single layer of cells with pale pink sap and colourless thick-walled aseptate hairs, without glandular hairs.

Viviparity; almost 20% of the seeds with the embryo rupturing the testa along the thin antiraphe in the immature seed *c.* 3 cm long, the cotyledons then developing into the fruit-cavity and turning dark green, often one cotyledon freed from the testa and remaining rather small (normally with intimate contact by means of fine papillae); normally with the testal palisade preventing rupture, but the palisade developing from the hilum outwards towards the antiraphe as the last stage with thinner tissue.

C. monocarpus L.
(Malaya; living material.) Fig. 139.

Seed 1.8–2.2 × 1–1.2 cm, oblong, slightly compressed, shining blackish brown, with thin testa and rather small hilum. Testa as in *C. semidecandrus* but the palisade of o.e. scarcely differentiated, the cells of unequal size, the antiraphe very thin. Vascular supply consisting of the short raphe v.b. and chalazal plexus, and with shortly recurrent branches along the raphe, a long preraphe v.b. to the distant micropyle opposite to the hilum.

Aril pulpy, deep orange, attached only round the chalaza, without v.b., with the same structure as *C. semidecandrus*.

Follicle 2.5–3.5 × 1.4–1.8 cm, bright yellow, rather fusiform; composed of six layers as in *C. grandis* but the outer cortex rather thin, the inner cortex relatively thick, the endocarp-valve not lignified into fibres, i.e. also without glandular hairs.

The almost undifferentiated testa approaches the condition in *Cnestis*, but the vascular supply to the testa is limited.

C. salomonensis Schellenb.
(Solomon Islands, San Cristobal; fallen seeds in alcohol.) Fig. 144.

Seed 4 × 2.1–2.3 × 1.4 cm, compressed bean-shape, shiny black, smooth. Testa; *o.e.* as a short palisade of cells with strongly thickened brown outer walls, with brownish globules in the lumen; *mesophyll* aerenchymatous, more or less collapsed. Tegmen; *o.e.* as a more or less disrupted layer of longitudinal, thick-walled, coarsely pitted, lignified fibres 23–35 μ wide, often of unequal width, often sinuous, with wide lumen; *mesophyll* and *i.e.* collapsed.

Aril 10–15 mm long, 4–5 mm thick, the free limb 4–9 mm long, orange, crenulate, with obtuse edge, internally pulpy with orange lines of small cells filled with starch.

This species has a distinct layer of exotegmic fibres, compacted at the chalaza but spread out as a disrupted layer over the sides of the seed and there consisting of microscopic bundles of 2–5 fibres irregularly interlacing. The vascular supply could not be made out in the dark shrivelled tissue of the testa.

C. semidecandrus Jack
(Singapore; living material.)   Figs. 140–142.

Ovules two, collateral near the base of the loculus, only one becoming a seed, sessile, hemianatropous with the micropyle directed apically; o.i. 6 cells thick; i.i. 3–4 cells thick; micropyle formed by the exostome.

Seed c. 15 × 9 × 4 mm, bean-shaped, shiny black with whitish hilum, the micropyle at the opposite end from the chalaza. Testa thin, pellicular; *o.e.* as a short-celled palisade, the walls thickening and becoming dark brown; *mesophyll* aerenchymatous with stellate cells, drying up at maturity; *i.e.* unspecialized and forming a layer of subclavate cells abutting on the disintegrated tegmen and nucellus. Tegmen soon crushed by the enlarging cotyledons. Hilum-tissue unspecialized but the cells more compact and slightly thick-walled round the v.b. Vascular supply as a short raphe v.b. breaking up at the chalaza into 5–7 branches just internal to the pulpy tissue of the aril, the branches ramifying shortly on to the sides of the seed, not forming a network, without a postchalazal branch, with a strong preraphe v.b. Endosperm soon absorbed. Embryo quickly developing; cotyledons pale olive green, internally pinkish from groups of reddish resin-cells; hypocotyl short; radicle minute.

Aril developing after fertilization round the chalaza, orange yellow, with a short tumid crenulate lobate margin, watery pulpy; composed of large thin-walled cells with colourless contents, and pervaded by vermiform groups of smaller cells packed with fine granular-oleaginous orange-yellow cytoplasm; epidermal cells small, thin-walled; without v.b.; junction of aril and testa abrupt.

Follicle c. 2 cm long with a stalk c. 6 mm long, scarlet, inner surface yellow and finely hairy, dry, dehiscing only along the placental suture, large and inflated, constructed as in *C. grandis*; endocarp as a thin layer of stone-cells and fibres, the i.e. developing capitate glandular hairs after fertilization and a few long aseptate, thick-walled, acute hairs.

C. villosus Jack
(Sarawak, S 26829; alcohol material.)   Fig. 143.

Seed 20 × 11 × 9 mm, bean-shaped, black; aril 6–7 mm long, 10 mm wide, with free margin 1–2 mm thick, yellow. Testa many cells thick; *o.e.* as a palisade 55–70 μ high, the cells with thick brown outer walls and thinner radial walls, the brown lumen attenuate to the outside; *mesophyll* crushed; *i.e.* composed of small cells, most with a single crystal. Tegmen; *o.e.* as a compact layer of lignified, rather sparsely pitted, colourless fibres 55–350 × 35–50 × 16–17 μ, somewhat ribbon-like, forming a hard membrane.

Connarus sp.
(Brazil, Manaus, Corner 256; immature fruits in alcohol.)   Figs. 144, 145.

Seed 13.5 × 3.7 mm (immature), with thick chalazal base adjacent to the hilum and a slight aril-rim at the periphery of the chalaza, but absent from the hilar side. Testa 13–16 cells thick; *o.e.* as a palisade of radially elongate cells with slightly thickened outer walls, shorter and indistinct at the exostome; *mesophyll* aerenchymatous, the cells rounded; *i.e.* as a layer of small unspecialized cells. Tegmen 12–14 cells thick; *o.e.* as a narrow layer of very narrow longitudinal fibres, laterally compressed (? becoming lignified), passing abruptly into shortly clavate cells at the endostome, ending abruptly at the chalazal pore; *mesophyll* consisting of very large cells collapsing from the inner region outwards, without air-spaces; *i.e.* as a layer of narrow, thin-walled cells slightly radially elongate, with dense contents, appearing as an endothelium (? eventually crushed). Nucellus soon absorbed. Vascular supply as a short stout v.b. at the hilum, a single branch in the preraphe to the micropyle with 2–3 apical lobes at the end, and with 8–10 peripheral branches to the thickened chalaza but not entering the testa.

Follicles 3 per flower, with ventral suture from base to apex; mesocarp thin-walled, with two rings of v.b. anastomosing mainly at the apex of the follicle, the outer ring of 17–19 v.b., the inner with more numerous smaller v.b. in a plexus.

Jollydora Pierre

J. duparquetiana (Baill.) Pierre
(Congo, Brazzaville, F. Hallé n. 1480, forêt de la Djoumana, 1969; flowers, fruits and seed in alcohol.)
Figs. 146–154.

Ovary as a single carpel, thickly brown hairy, with 3 v.b. in t.s. Ovules 2, anatropous, suspended with the micropyle directed distally; o.i. 5–7 cells thick; i.i. 4–5 cells thick; nucellus scarcely massive, with thick cuticle; v.b. to the chalaza simple but with slight procambial branchings from the chalaza into the o.i.

Seeds 15–20 × 9–13 mm, covered by the pale ochre to brownish sarcotesta *c.* 1 mm thick, except at the hilum and micropyle and along the black preraphe connecting them. Testa 20–25 cells thick before development of the sarcotesta, *c.* 9–12 cells thick round the exostome and in the preraphe; *o.e.* round the exostome and in the preraphe composed of short cells with thickened brown outer walls, in a compact layer, the rest of o.e. dividing to form the sarcotesta (see below); *mesophyll* consisting of stellate-celled aerenchyma, eventually drying up and collapsing with slightly thickened yellow-brown walls, unlignified; *i.e.* with small cells as in o.e., but many cells converted into small, cuboid, stone-cells with yellow-brown lignified walls and a crystal in the lumen. Sarcotesta many cells thick, developed by periclinal division of o.e. and o.h. of the testa, consisting at maturity of large, radially elongate cells with slightly thickened, not lignified walls, filled with starch-grains, without v.b.; *o.e.* as a small-celled layer with minute orange-yellow oil-globules without starch. Tegmen 8–10 cells thick, 4 cells thick at the endostome, 10–12 cells thick round the hypostase; *o.e.* becoming a layer of longitudinal, thick-walled, pitted, lignified fibres

14–30 μ wide; *mesophyll* consisting of much enlarged aqueous cells, eventually collapsing; *i.e.* developing as an endothelium of small radially elongate cells, then collapsing and absorbed. Nucellus forming a small hypostase-like tube, eventually absorbed or collapsed. Vascular supply consisting of 4–5 branches from the chalaza ending blindly at the micropyle without anastomosis, some of the branches once or twice furcate, with a preraphe v.b. to the micropyle, giving 9–12 v.b. in t.s. of the seed. Endosperm none. Embryo consisting mainly of the two cotyledons, without starch.

Fruit *c.* 25 × 17 mm, indehiscent, reddish orange, glabrescent, somewhat flattened dorsiventrally, 1–2 seeded; exocarp composed of 4–6 layers of rather small thin-walled cells; mesocarp much thicker, aerenchymatous, the outer 5–8 cell-layers with thickened, pitted and lignified walls as a subcutaneous sclerotic layer; endocarp unspecialized; v.b. as a plexus of small strands in the outer part of the sclerotic layer of the mesocarp.

The fruit, as usual, develops more rapidly than the seed which slowly fills the large cavity. The last feature of the seed is the appearance of the sarcotesta; intermediate stages between the young seed and the mature have not been available.

Rourea Aubl.

R. minor (Gaertn.) Leenh.
(Sarawak, S 28545; alcohol-material.) Fig. 155.

Seed 27–28 × 17 × 11–12 mm, dark brown with fine subrugulose and waxy surface, compressed ellipsoid with slightly prominent micropyle, covered by the aril but eventually protruding slightly along the cleft side of the aril. Testa 33–40 cells thick, brown, massive, not lignified; *o.e.* as a layer of subcuboid cells with thickened outer wall, not radially elongate, with polygonal facets; *mesophyll* aerenchymatous with more or less stellate cells, many with brown tanniniferous contents and often in short longitudinal clusters, thin-walled, the intervening cells becoming crushed; *i.e.* unspecialized but, at the exostome, with lignified inner wall. Tegmen thin, crushed, but with a single thin layer of longitudinal, pitted and lignified fibres at the chalaza and micropyle and the adjacent parts of the seed-coat, not differentiated over the sides of the seed. Vascular bundle apparently only in the preraphe, but with short branches from the funicle to the insertion of the aril, the raphe undeveloped. Endosperm none, or a trace at the chalaza. Embryo with massive cotyledons and short included hypocotyl.

Aril 30–32 × 24 × 20 mm, –6 mm thick at the base, massive, covering the seed but cleft longitudinally on the preraphe-side, the apex shortly and broadly 3-lobed, the base widely attached round the funicle and seed-base except at the preraphe; *o.e.* composed of somewhat flattened cells with slightly thickened radial walls, the facets angular, not lignified; mesophyll pulpy, the large cells cylindric-prismatic, elongate thin-walled, with fine transverse lines on the walls, some with tannin, not lignified; v.b. none.

Fruit 35 × 24–25 × 20 mm, compressed-ellipsoid with oblique apex, apparently dehiscent; endocarp lignified.

This is *R. ovale* (Schellenb.) Leenh. which Leenhouts informs me should be reduced to *R. minor*.

CONVOLVULACEAE

Ovules anatropous, erect, unitegmic, tenuinucellate, 1–2 per loculus; integument massive, 5–40 cells thick; i.e. not as an endothelium.

Seeds 1–10 mm long, ellipsoid to subglobose, smooth or hairy, mostly grey, brown or black, more or less albuminous, exarillate. Seed-coat with multiplicative outer hypodermis forming a hypodermal palisade 1–4 cells thick, the mesophyll multiplicative or not and eventually crushed; *o.e.* as a layer of relatively large cuboid cells with polygonal or undulate facets, the outer wall often slightly thickened and finely punctate, mostly unlignified, with dark brown contents, sometimes with starch, in some cases with groups of more radially elongate cells (*Convolvulus*) or the cells projecting into hairs 1–10 mm long (*Ipomaea*); *o.h.* as a layer of short cells with thin or slightly thickened walls, lignified or not, transversely elongate in *Calystegia*, in some cases both o.h. and o.e. crushed; *palisade-layer* as a single cell-layer or, more usually, through periclinal division 2–4 cells thick, consisting of radially elongate, narrowly prismatic cells with the thickened walls lignified or not, often with a *linea lucida* in the outer layer of longer cells, with small angular facets and more or less stellate lumen in t.s., the inner cells shortest, the whole forming the mechanical layer of the seed-coat; *mesophyll* aerenchymatous, starchy, eventually crushed. Vascular bundle extending beyond the chalaza almost to the micropyle, unbranched. Endosperm nuclear, abundant or reduced to one or a few cell-layers, more or less starchy. Embryo curved or coiled; cotyledons more or less intricately folded. Figs. 156–158.

(Netolitzky 1926; Kühn 1927; Sripleng and Smith 1960; Jos 1962; Guédés 1968; v. Heel 1970b, 1971c; Kaur and Singh 1970; Govil 1971.)

Breweria, Calystegia, Convolvulus, Cuscuta, Evolvulus, Ipomaea, Jacquemontia, Merremia, Pharbitis, Rivea.

Among the unitegmic seeds of sympetalous families, the Convolvulaceous is distinguished by the special development of the three outer cell-layers of the integument. The feature occurs in *Cuscuta* and shows that there is little or no reason to make the family Cuscutaceae, and none to remove it from Convolvulaceae (Solanales) into Polemoniales, as though they were phyletic parallels. The isolated position of Convolvulaceae is discussed on p. 50.

I draw attention to the work of Guédés (1968) which

explains on the classical theory of the carpel the phyllodic syncarpy of the family.

The recent work of Govil confirms the previous investigations into the origin and construction of the hypodermal palisade of the seed. The ovular hypodermis divides periclinally to form the outer hypodermis of the seed with its small cells and the third cell-layer of the seed which may remain undivided as the single palisade-layer (*Ipomaea purpurea*) or divides again to form the 2, 3, or 4 rows of palisade-cells. He distinguishes also the structure of the seed-coat along the raphe, where the outer epidermis divides periclinally once to form a double layer of cells, and the counter-palisade of the hilum. Here, the outer epidermis divides to give 4–5 layers of isodiametric cells of which the innermost make a counter-palisade to the normal hypodermal palisade of the seed-coat.

Calystegia R.Br.

C. sepium R.Br.
(England; living material.)   Fig. 156.

Integument 12–20 cells thick in the ovule, 18–30 in the seed.

Testa; *o.e.* composed of large cuboid cells with isodiametric facets, the wall scarcely thickened; *o.h.* composed of small narrow cells elongated tangentially, mostly in transverse groups but in places longitudinally, with slightly thickened radial walls; *palisade-layer* 2–3 cells thick, the inner cells short; *mesophyll* with large thin-walled cells, becoming crushed.

Cuscuta L.

Integument 9–15 cells thick apparently not multiplicative.

Testa; *o.e.* composed of rather large cuboid cells with slightly thickened outer wall, the dark contents with much starch; *o.h.* composed of short cells with slightly lignified walls; *palisade-layer* as a single layer of prismatic cells with thick lignified and brownish walls, in places with a second inner layer of short cells. Endosperm starchy. Embryo coiled.

(Netolitzky 1926; Tiagi 1951a; Johri and Tiagi 1952.)

Evolvulus L.

E. alsinoides L.
(Ceylon; flowers and fruits in alcohol.)   Fig. 158.

Similar to *Jacquemontia* but the exotestal cells flattened and the facets with sharp re-entrant angles; palisade short.

Jacquemontia Choisy

J. paniculata (Burm.f.) Hallier f.
(Ceylon; flowers and fruits in alcohol.)   Fig. 158.

Integument 10–13 cells thick, not multiplicative except o.h.

Testa; *o.e.* composed of large cuboid cells with rather sinuous facets, the outer wall very finely perforate, slightly lignified; *o.h.* with slightly lignified walls; *palisade-layer* as a single layer of thick-walled prismatic cells, apparently not lignified; *mesophyll* slightly lignified but thin-walled and collapsing.

Ipomaea L.

Ovule 1–1.4 × 0.5–0.8 mm, rather large; integument of various thicknesses, up to 18 cells thick in the antiraphe and 40 cells thick in the raphe (*I. pes-caprae*).

Seeds 4–10 × 3–7 mm, with the seed-coat 1–2 mm thick before drying; *o.e.* with large polygonal cells with firm brown walls, in various species developing into 1(–2)-celled hairs 1–10 mm long; *o.h.* composed of small cells; *palisade-layer* single or 2–6 cells thick, the outer cells longest, prismatic, lignified or suberized or not; *mesophyll* composed of large cells, then collapsing. Raphe projecting strongly into the seed-cavity, in some cases also the antiraphe. Fig. 157.

(Woodcock 1944; Jos 1962; Misra 1963; Govil 1971.)
I. carica (L.) Sweet, *I. carnea* Jacq., *I. pes-caprae* (L.) R.Br., *I. pes-tigridis* L., *I. purpurea* Roth, *I. quamoclit* L., *I. reptans* Poir., *I. sepiaria* Koen., *I. sindica* Stapf, *I. sinuata* Orteg., *I. vitifolia* Sweet.

Pharbitis Choisy

As *Cuscuta* but with the outer hypodermis of the seed developed into a short palisade-layer as the normal third cell-layer of the seed-coat (but ? both these palisade-layers derived from the third cell-layer and the true o.h. crushed). Vascular bundle in the raphe divided into three strands united again at the chalaza and proceeding as one almost to the micropyle.

(Netolitzky 1926; Kühn 1927.)

CORIARIACEAE

Ovules anatropous, suspended, bitegmic, crassinucellate, one per carpel, with dorsal raphe; o.i. 3–4 cells thick, short; i.i. 2–3 cells thick, forming the micropyle.

Seeds small, compressed, reddish brown, exalbuminous or with a trace of endosperm, exarillate; seed-coats not multiplicative. Testa fairly persistent; *o.e.* as a hard layer of more or less cuboid cells with thickened walls (? lignified), filled with tannin; *mesophyll* and *i.e.* thin-walled, unspecialized. Tegmen crushed. Endosperm nuclear. Embryo straight.

(Mauritzon 1936d; Sharma 1968a.)
The aberrant *Coriaria*, which makes this family, has been variously and doubtfully referred to the affinity of Anacardiaceae, Dilleniaceae, Ranunculaceae, and Rutaceae. The simple exotestal structure, if correct, agrees with Ranunculaceae and excludes Dilleniaceae.

CORNACEAE

Ovules anatropous, suspended with dorsal raphe, 1 per loculus, unitegmic, crassinucellate or tenuinucellate; integument 8–15 cells thick; hypostase absent.

Seeds small to medium-size, enclosed in the endocarp or pyrenes. Seed-coat not multiplicative, thin-walled, more or less crushed; *o.e.* thin-walled, with more or less undulate facets. Vascular bundle terminating at the chalaza (*Aucuba, Cornus*). Endosperm cellular, thin-walled, starchy. Embryo small (*Aucuba*) or well-developed.

(Netolitzky 1926.)

CORYNOCARPACEAE

Ovule anatropous, suspended, solitary, bitegmic, crassinucellate; integuments short; micropyle formed by the endostome.

Seed exalbuminous. Testa 20–30 cells thick, apparently unspecialized, with branching v.b., eventually more or less crushed. Tegmen 3 cells thick, unspecialized, crushed. Endosperm nuclear. Embryo large; radicle minute.

(Netolitzky 1926; Pigott 1927.)
Possibly this is a pachychalazal seed as in Anacardiaceae.

CRASSULACEAE

As Saxifragaceae. Ovule; o.i. 2–3 cells thick; i.i. 2 cells thick. Endosperm cellular, reduced or absent in some species of *Sedum*.

(Netolitzky 1926; Subramanyam 1963, 1968.)

CROSSOSOMATACEAE

Ovules campylotropous but after fertilization amphitropous, numerous, erect, bitegmic, crassinucellate.

Seeds numerous, arillate, thinly albuminous. Testa sclerified. Tegmen; *i.e.* with fibrous thickenings on the walls. Chalaza with hypostase. Endosperm nuclear. Embryo well-developed.

(Kapil and Vani 1963.)
The family is classified in Dilleniales. The shape of the arillate seed agrees but the meagre knowledge of the seed-coat supplies no distinction from Podophyllaceae.

CRUCIFERAE

Ovules anacampylotropous, generally numerous, often very small, bitegmic, tenuinucellate; o.i. 2–3–4–5 cells thick in various genera; i.i. 3, 4, 5–7, 6–8 and 12–15 cells thick in various genera (*Bunias* 12–15, but ? postfertilization).

Seeds generally small, 0.3–4 mm long, –8 mm (Lunarieae), –17 mm (*Megacarpaea*), in some cases ribbed or winged, often more or less reticulate from the collapse of the outer testa on the indurated endotesta, exarillate, more or less exalbuminous. Testa apparently not multiplicative; *o.e.* as a continuous layer of cuboid or flattened cells, with thin or slightly thickened (? not lignified) walls, in many cases with internal mucilage, not as a palisade, with reticulate thickenings on the radial walls (*Megacarpaea*); mesophyll thin-walled, more or less collapsed, with collenchymatous thickenings (*Sinapis alba*) or rather thick-walled (*Alyssum*); *i.e.* typically as the mechanical layer of cuboid or shortly radially elongate cells with lignified thickening on the inner and radial walls (U-shaped thickening), but in many cases thin-walled (? indehiscent fruits, cf. *Crambe*) or slightly thick-walled all round the cell, or with crystals (*Barbarea, Malcolmia*). Tegmen strongly multiplicative (*Lunaria, ? Bunias*) or generally not multiplicative, thin-walled, often as a pigment layer, collapsing; *i.e.* as a lignified and more or less palisade-like layer in *Lunaria* (? *Sisymbrium*). Vascular supply only in the short raphe v.b., or with branches to the sides of the seed. Chalaza unspecialized. Endosperm nuclear, reduced in the seed to a single layer of cells. Embryo variously folded. Figs. 159–162.

(Netolitzky 1926; Berger 1964, pp. 383–404; Singh, B. 1964; Misra 1966; Vaughan 1970; Vaughan and Whitehouse 1971.)

The seeds of Cruciferae are well known through the researches of many botanists; in fact, no other family has been so thoroughly explored in this respect. Vaughan and Whitehouse have recently published an account of the mature structure in some 90 genera and 200 species. The smaller seeds are merely ovules with the integuments grown into seed-coats by enlargement of the cells with little or no multiplication. This may also be the condition in the largest (*Megacarpeae*; Vaughan and Whitehouse 1971), but the moderately large seeds of *Lunaria* are complicated by a multiplicative tegmen which, as Guignard first showed, develops an endotegmic palisade as well as an endotestal (Fig. 162). Misra's account of *Sisymbrium irio* suggests this same endotegmic structure without multiplication of the tegmen and with complete dissolution of the testa. It appears, therefore, that the Cruciferous seed may have been primitively both endotestal and endotegmic. The possibility bears on the vexed question of the affinity of this predominantly herbaceous family.

Usually Cruciferae are associated with Capparidaceae, more distantly with Dilleniaceae, Flacourtiaceae and Violaceae. Hutchinson (1959) takes Cruciferae out of this alliance and derives them from Papaverales. The seed-structure presents many difficulties in the way of either proposition. Firstly, Capparidaceae, Flacourtiaceae, Papaveraceae, and Violaceae have exotegmic fibres which have never been seen in Cruciferae; for this reason Vaughan and Whitehouse would dissociate them from Capparidaceae. Secondly, Dilleniaceae have endotestal seeds with sclerotic cells unlike those of Cruciferae and they have the tracheidal exotegmen which has not been found in the Cruciferous seed; there are numerous

differences in detail, and no one has suggested a close connection between the two families. Thirdly, the double feature of lignified internal palisades to each integument, as in *Lunaria*, is unusual. It occurs in some species of Capparidaceae (*Capparis, Cleome*), at least at the endostome in some Flacourtiaceae, and to a slight degree without lignification in Dilleniaceae, Resedaceae and, perhaps, Violaceae. If these families are referable to a common ancestor (subclass Dilleniidae, Cronquist 1968), then there is the problem of the diversification of the mechanical layer of the seed-coat (p. 32). In this, Cruciferae agree best with Dilleniaceae, while the Capparidaceous alliance must be set apart on account of the fibrous exotegmen. It is clear to me that immediate affinities between families must be dismissed and that their pachycaulous progenitors must be considered. In this light Dilleniaceae may not be so dissimilar from a multistaminate, multicarpellary and arillate Cruciferous ancestor.

CUCURBITACEAE

Ovules anatropous, bitegmic, crassinucellate; o.i. 6–8 cells thick (*Bryonia, Cephalandra, Citrullus, Cucumis, Cyclanthera, Dicoelospermum, Lagenaria*), 8–10 cells thick or ? more (*Biswarea, Cucurbita, Ecballium, Echinocystis, Edgaria, Herpetospermum, Luffa, Marah, Momordica, Trichosanthes*), 15–20 cells thick in the raphe-antiraphe; i.i. 1–4 cells thick, mostly 2–3, 4–5 (*Bryonia, Sechium*), developing before o.i.; micropyle with exostome and endostome.

Seed medium-size to large, often compressed, winged in some genera, enveloped in a false aril (placental tissue) in others, mostly more or less ellipsoid and often pitted or rugulose in a pattern on drying. Seed-coat with woody exotesta of several layers derived from o.e. (o.i.), and papery endotesta. Tegmen not multiplicative, soon disappearing. Vascular supply as a single v.b. in the endotesta continuous in the raphe-antiraphe without distinct chalazal development, but in some cases with anastomosing branches to the sides of the seed along most of its course (*Cyclanthera, Edgaria, Momordica, Trichosanthes*). Chalaza simple, unspecialized. Nucellus enlarging to fill the full-sized but immature seed, then crushed by the endosperm, often persistent at the micropyle. Endosperm nuclear, reduced to 1–2 cell-layers or absent, oily. Embryo straight; cotyledons flat.

Exotesta consisting at first of 3 cell-layers produced by two periclinal divisions of o.e. (o.i.), giving a seed-epidermis (e-layer, generally remaining one cell thick), a seed-hypodermis (e″-layer, often only 1–2 cells thick), and an innermost sclerotic layer (e′-layer, one cell thick), but in some genera the e-layer and the e″-layer multiplicative and thickening the seed-hypodermis (*Biswarea, Echinocystis, Edgaria, Herpetospermum, Marah, Momordica, Sicyos, Trichosanthes*). Seed-epidermis (e-layer) composed of cuboid or radially elongate cells, some seeds with both kinds (the shorter cells over the ridge-like thickenings of the seed-hypodermis), the outer wall thickened, the radial walls with rod-like or strap-like cellulose thickenings (in some seeds with minute crystals), not or slightly lignified, eventually forming a pulpy mucilaginous layer drying up on the ripe seed with the cells collapsing obliquely on to the woody inner tissue, reviving on wetting, without stomata. *Seed-hypodermis* (*e''*-layer) formed after the sclerotic layer, remaining one cell thick or dividing periclinally into few or many cell-layers (the number varying often in different parts of the seed to form ridges, furrows, pits, and the thickened edges), the cells more or less cuboid, becoming thick-walled, pitted and lignified, when multiplicative then often with small-celled outer layers and larger-celled inner layers, never palisade-like. *Sclerotic layer* (*e'*-layer) mostly one cell thick, in some cases two or more, the cells becoming larger than those of the seed-hypodermis, thick-walled, pitted, lignified, either radially elongate as columnar palisade-cells with lobate-undulate facets or more or less tangentially and longitudinally elongate (cuboid in t.s.) with lobate-undulate facets and ends, absent from the raphe-antiraphe at the micropylar end of the seed (thus giving the line of dehiscence on germination).

Endotesta developed from the mesophyll and i.e. of the outer integument, multiplicative mainly by periclinal divisions in o.h. (o.i.), aerenchymatous, thin-walled, often with a smaller-celled and more chloro-phyllous inner layer of cells, eventually becoming mucilaginous, drying into a papery membrane; i.e. unspecialized or the cells somewhat hourglass-shaped as those next to the exotesta. Figs. 163-172.

(Fickel 1876; Kratzer 1918; Netolitzky 1926; Singh, B. 1964; Singh, D. 1965, 1967; Singh, D. and Dathan 1969, 1971a, b; Vaughan 1970.)

The elaborate structure of the Cucurbitaceous seed has been established now for many genera. The complications lie in the thick exotesta which is derived entirely from the outer epidermis of the outer integument of the ovule after fertilization. Features which differentiate from particular cell-layers of the outer integument in other families, such as the external palisade of unlignified cells, the sclerotic mesophyll, or the internal palisade of lignified cells, arise in Cucurbitaceae from the single outer layer of this integument. The first to prove this was Fickel (1876), who also described most clearly the forms of the cells, but it has become the custom to follow the explanation and manner of description given by Kratzer (1918). The epidermal cells divide periclinally to form an outer layer and an inner or *e'*-layer. The outer cells then divide periclinally again and form the outer *e*-layer, which becomes the seed-epidermis, and the middle *e''*-layer which becomes the seed-hypodermis; the *e'*-layer becomes the sclerotic layer and separates

this multiple exotesta from the rest of the outer integument as the endotesta. Generally the *e*-layer and the *e'*-layer are not further multiplicative, whereas the *e''*-layer commonly forms two or more layers of cells. In some genera, however, both the *e*-layer and the *e''*-layer are multiplicative, and the result is a sclerotic testa of epidermal origin some 10-40 cells thick, comparable with the whole testa of Myristicaceae. The construction is so definite that Kratzer considered that no seed without this manner of development could belong to Cucurbitaceae. It seems, nevertheless, that there are, as always, exceptions; a proper seed-coat does not develop in *Sechium*, though it may have the typical inception of the three *e*-layers.

The simplest examples are afforded by *Bryonia*, *Cucumis*, *Ecballium* and *Sicyos* (Fig. 163), in which none of the three *e*-layers is multiplicative. *Cucurbita* and *Luffa* (Fig. 163) introduce multiplication in the *e''*-layer which thickens the seed-hypodermis. In the thicker seeds of *Momordica* (Fig. 167) and *Trichosanthes* (Fig. 170), both the *e*-layer and the *e''*-layer divide periclinally. Their divisions add to the seed-hypodermis and, in the places where the divisions are numerous, the ultimate cells of the *e*-layer are short and the seed-hypodermis consists of strips or patches of long radially elongate palisade-cells and of short cells. A few additions of this kind have been observed in *Echinocystis* and *Sicyos*; they occur also in *Biswarea*, *Edgaria* and *Herpetospermum*. Probably they are the reason for the very thick seed-coat of *Hodgsonia*.

The simpler seeds were examined first and have given rise to the idea that they are primitive. In seed-development, however, restriction of multiplication of cell-layers is a sign of advance. The precise structures described by Kratzer seem rather to be the final outcome of the simplification of a massive seed-coat which originally developed from an extensive seed-periderm in the manner of an overgrown seed, e.g. *Hodgsonia* with few large seeds in the fruit. *Sechium* has the solitary and very large overgrown seed without differentiation of the periderm. In other genera there follow simplifications of the seed-coat, diminution in size of the seeds, and multiplication of their numbers typical of seed-progress in so many families. *Hodgsonia*, *Sechium* and water-melons may be the extremes. The intermediate state of *Momordica* and *Trichosanthes*, which retain in placental modification the red aril, may be closer to the primitive Cucurbit-aceous seed. The vascular spines that occur on the fruits of many genera and develop beneath glandular or massive hairs, the dehiscence of the capsule with winged seeds (Fevilleae), and these arilloids point to a fleshy spinous arillate capsule as the primitive Cucurbitaceous fruit and bring the family into line with Caricaceae and, perhaps, Passifloraceae; but the Cucurbitaceous problem has been discussed on p. 38.

It is likely that genera may be defined on the microscopic structure of the seed. More research is required into genera and their species and, as the structure may vary in different parts of the seed, these will have carefully to be distinguished. The following is a very tentative key for those genera with non-alate seeds that have been studied.

Generic grouping by the microscopic structure of non-alate seeds

(1) e'-layer composed of radially elongate cells.
  (2) e-layer composed mainly or partly of radially elongate cells.
    (3) e''-layer 1 cell thick ... *Bryonia, Ecballium*
    (3) e''-layer thicker ... *Biswarea, Cephalandra, Edgaria, Herpetospermum, Luffa*
  (2) e-layer composed of cuboid cells.
    (4) e''-layer 1 cell thick ... *Sicyos*
    (4) e''-layer thicker ... *Echinocystis*, (? *Luffa*), *Marah*
(1) e'-layer composed of tangentially elongate cells, cuboid in t.s.
  (5) e''-layer composed of small cells in few layers.
    (6) e-layer composed of cuboid cells ... *Cyclanthera*
    (6) e-layer composed of radially elongate cells.
      (7) e''-layer 1–3 cells thick ... *Cucumis*
      (7) e''-layer 3–6 cells thick (or more) ... *Cucurbita*
  (5) e''-layer massive, generally more than 6 cells thick.
    (8) e-layer composed mainly or partly of radially elongate cells ... *Citrullus, Lagenaria, Momordica, Trichosanthes*
    (8) e-layer composed of more or less cuboid cells ... *Benincasa, Dicoelospermum, Melothria, Telfairia, Thladiantha*, (? *Trichosanthes*)

Cephalandra Schrad.

C. indica Naud. (*Coccinea grandis*)
(Ceylon; flowers and fruits in alcohol.)   Figs. 164, 165.
  Ovule; o.i. 5–6 cells thick; i.i. 2 cells thick; micropyle with exostome and endostome.
  Seeds 7 × 3 × 1–1.3 mm, pale brown, much compressed, smooth. Exotesta: *e-layer* composed of columnar pulpy cells, rather long at the micropyle, the thickened walls roughened on their inner faces from minute projections (caused by minute crystals on the walls), slightly lignified towards their bases in contact with the inner tissue; *e''-layer* mostly 3 cells thick, but 6–8 cells in the projections of the raphe-antiraphe, composed of small isodiametric cells with thickened, closely pitted, lignified walls but wide lumen, slightly aerenchymatous; *e'-layer* composed of shortly radially elongate cells, very thick-walled, sparsely pitted, lignified, the outer ends with undulate and interlocking outline. Endotesta *c.* 6 cells thick, thin-walled, the cells shortly elongate tangentially and transversely, the outer 1–2 layers consisting of small cells becoming slightly lignified; *i.e.* unspecialized.
  Chakravorti (1947) gave the i.i. as 3–4 cells thick and o.i. as 7–18 cells thick, and noted that periclinal divisions in o.e. (o.i.) begin in the micropylar region and proceed towards the chalaza.

Dicoelospermum C. B. Clarke

D. ritchei C. B. Clarke
Ovules 2; o.i. 6–10 cells thick; i.i. 2 cells thick.
  Seeds 5–6 × 3–4 mm, 4–4.5 mm thick, flattened at right angles to the plane of the cotyledons, pale, woody, drying up with two lateral cavities. Exotesta: *e-layer* as small thin-walled cells (? with extra periclinal divisions), but with slightly lignified and reticulately thickened walls near the micropyle; *e''-layer* divided into a thin-walled part and an inner sclerotic part, the outer part forming the succulent tissue on the sides of the seed and eventually drying up as the lateral cavities, composed of radially elongate cells, the sclerotic part composed of several layers of short, longitudinally elongate, undulate, thick-walled cells; *e'-layer* composed of thick-walled, pitted cells appearing cuboid in t.s., not as a palisade.
  (Singh, D. 1965.)

Ecballium A. Rich.

Baillon (1871) reported in *E. elaterium* (L.) A. Rich. a very small aril, evidently funicular and surrounding the hilum and micropyle, but I have been unable to see this.

Momordica L.

M. charantia L.
(Singapore, living material; Ceylon, alcohol-material.)
Figs. 166–168.
  Ovules transverse in two rows in each loculus, with the antiraphes of each pair of ovules apposed; o.i. 8 cells thick; i.i. 2 cells thick; without v.b. at the time of fertilization.
  Seeds *c.* 15 × 9 × 4 mm, drying creamy brownish, marked on the obtuse sides with paler raised areas and flat sinuous or lobed lines and patches along the raphe and antiraphe; undried seed covered in a gelatinous pellicle (*e-layer*) of varying thickness, thickest along the edges of the seed on either side of the raphe-antiraphe; hilum small. Exotesta; *e-layer* as a palisade of columnar cells varying in height, tallest along the obtuse edges of the seed, shortest in the raised areas of the dried seed, the walls becoming mucilaginous but retained by the thin cuticle, many of the cells with starch-grains, drying up as a pellicle over the irregularities of the *e''-layer*, reviving on wetting; *e''-layer* consisting of an outer zone of small contiguous cells (3–4 layers) and an inner zone of aerenchymatous tissue with substellate cells (5–11 layers), the whole becoming thick-walled and lignified at maturity; *e'-layer* as an ill-defined layer of large substellate cells, becoming very thick-walled and lignified; both *e'-* and *e''-*layers undifferentiated along the raphe-antiraphe as the line of dehiscence on germination. Endotesta consisting of an outer layer (2 cells thick) of small cells and a much thicker inner layer of large, thin-walled, substellate cells, slightly pulpy at maturity, then drying into a pellicle

without lignification; *i.e.* unspecialized. Vascular bundle in the raphe-antiraphe slightly thickened at the chalaza, giving off along its whole course small branches anastomosing into a fine network on the broad sides of the seed.

False aril enveloping the seed, scarlet, pulpy, derived from the placental tissue, as a loose bag around the seed and funicle, the wall 0.5 mm thick, composed of the compact inner layers of the loosely aerenchymatous endocarp, eventually becoming slimy, with orange carotinoid granules and abundant starch-grains in the cells.

Pericarp composed entirely of thin-walled succulent cells except the 10 longitudinal v.b. and the spongy placental tissue.

Spinous fruited variety; ovary with 8–10 external rows of dark green, soft, fleshy spines on the radii of the v.b., each spine capped by a stoma, also with short irregular papillae between the spines, pubescent with glandular hairs and scattered long hairs with basal growing region and finely verrucose mature cells; on fertilization the short spines beneath the long hairs enlarging and adding to the spinous exterior (Fig. 166).

Trichosanthes L.

T. anguina L.
(Singapore, living material.) Figs. 169, 170.

Ovules mostly set longitudinally, a few transverse; o.i. 8–10 cells thick; i.i. 1–2 cells thick.

Seeds 14–20 × 10–13 × 5 mm, as in *Momordica*, with a distinct pleurogram, enveloped in a false scarlet aril; tegmen not multiplicative, soon disappearing. Exotesta: *e-layer* as a palisade of pulpy cells of varying height, tallest along the raphe-antiraphe, on either side of the micropyle and in the central part of the pleurogram, drying into a pellicle; *e″-layer* consisting of an outer zone of varying thickness (2–10 cells or more) composed of angular contiguous cells filled with starch-grains, becoming slightly thick-walled, lignified, yellow-brown, drying into a firm spongy and outwardly papillate layer, and an inner zone (6–8 cells thick) of very thick-walled, heavily lignified cells with brown contents, forming the main woody casing of the seed; *e′-layer* consisting of cuboid–stellate cells with very thick brown lignified walls. Endotesta composed of thin-walled stellate cells, the outer cells adjoining the exotesta and those of the i.e. smaller and often hourglass-shaped, eventually drying into an unlignified pellicle. Vascular supply as in *Momordica*. Nucellus reduced to a single layer of rather elongate cells with pale olive, thinly cuticularized outer wall. Pleurogram outlined on the broad side of the seed by a wavy line open towards the hilum, with papillate central region (caused by varying thickness of *e″*-layer), delimited internally by a radial cleft in the *e″*-layer.

Pericarp composed of thin-walled tissue as in *Momordica* but with many more v.b. (present as procambial strands in the ovary).

Trichosanthes sp.
(Ceylon, Peradeniya; ripe fruits in alcohol.) Figs. 171, 172.

Seeds 8–10 × 6–6.5 × 2.5 mm, ovoid-compressed, greyish brown to black, nearly smooth, without pleurogram, embedded in the blackish green placental pulp; testa 17–23 cells thick on the sides of the seed, thicker along the obtuse edges. Exotesta: *e-layer* as shortly columnar pulpy cells, taller along the raphe-antiraphe, much elongate on the sides of the micropyle, thin-walled but with fine fibrillar thickenings (not lignified) on the radial and inner walls, the outer wall thickened and slightly lignified, at first filled with starch-grains; *e″-layer* 7–11 cells thick on the sides of the seed but much thicker on the obtuse edges of the raphe-antiraphe, composed mainly of substellate cells (radially elongate along the raphe-antiraphe), but the outer 3–4 cell-layers more or less superposed, shortly elongate tangentially and rectangular in t.s., all the cells filled with starch-grains, then thick-walled, lignified and pitted; *e′-layer* consisting of cuboid substellate, thick-walled, lignified cells. Endotesta 8–10 cells thick, thin-walled, without starch, aerenchymatous, substellate, the outer 2(–3) cell-layers composed of smaller cells becoming thinly lignified and persistent, the rest unlignified and collapsing; *i.e.* small-celled, unspecialized, more or less persistent. Tegmen disappearing except for a trace at the micropyle. Vascular bundle of the raphe-antiraphe without branches. Nucellus persistent as 2–4 cell-layers with thick external cuticle.

This is one of the more typical wild species with round red fruit and the seeds embedded in greenish black slime. They differ considerably in fruit and seed from *T. anguina* which resembles *Momordica* in these respects. The uniformly short cells of the e-layer on the sides of the seed, resulting in the absence of a pleurogram, and the unbranched v.b. are different from *T. anguina*.

CUNONIACEAE

Ovules anatropous or hemianatropous, bitegmic, crassinucellate; o.i. 1 cell thick; i.i. 2 cells thick.

Seeds smooth or pilose. Structure ? Endosperm nuclear, persistent.

(Netolitzky 1926.)

I include this family in the hope that the seeds may be studied.

CYNOCRAMBACEAE

Ovule solitary, subanatropous or campylotropous, erect, unitegmic, tenuinucellate; integument 8–12 cells thick.

Testa apparently thin-walled and becoming crushed. Endosperm nuclear. Perisperm absent.

(Woodcock 1929a; Kapil and Mohana Rao 1966b.)

These authors conclude that the family belongs in Centrospermae near to Amaranthaceae and Phytolaccaceae. Woodcock recorded cellular endosperm with starch.

DAPHNIPHYLLACEAE

Ovules anatropous, suspended, bitegmic, crassinucellate. Seed medium-size, albuminous, exarillate. Testa ? not multiplicative, unspecialized, collapsing. Tegmen multiplicative, unspecialized, eventually crushed. Endosperm oily, thin-walled. Embryo minute, straight.

Seed-structure forbids alliance of this small family with Euphorbiaceae; there is no tegmic differentiation such as happens even in drupaceous Euphorbiaceae. Hutchinson places the family in Hamamelidales and it is so similar to Buxaceae that, in view of the range in structure admitted into many large families, it is not clear why Daphnipyllaceae should be separated. The ovary, ovule and fruit resemble those of *Sarcococca*, but the seed lacks an exotestal palisade, and has a multiplicative tegmen and a minute embryo.

Daphniphyllum Bl.

D. borneense Stapf

(Borneo, Kinabalu, RSNB 761 and 5886; flowers and fruits in alcohol.) Figs. 173–176.

Ovules 4, suspended in pairs from two parietal placentas at the apex of the ovary, the ovary with a partial septum at the base; o.i. 3–4 cells thick; i.i. 4–5 cells thick; micropyle closed by the exostome.

Seed 9.5 × 6.5 mm, broadly ellipsoid, smooth. Testa not multiplicative or ? adding 1–2 layers of cells; *o.e.* and *i.e.* unspecialized. Tegmen 8–11 cells thick, unspecialized. Vascular bundle in the raphe only.

Drupe 1(–2) seeded, ripening dull red to purple-black; endocarp lignified with two layers of crossed fibres; mesocarp with scattered stone-cells (as the ovary).

D. macropodum var. *humile* (Maxim.) Rosenth.

(Sato 1972, without information on the seed.)

DATISCACEAE

Ovules anatropous, bitegmic, crassinucellate; o.i. and i.i. 2 cells thick.

Seeds small, exarillate, with little or no endosperm. Testa reduced to o.e. composed of subcuboid cells with thickened inner and radial walls. Tegmen crushed. Endosperm nuclear. Embryos straight.

(Mauritzon 1936b; Crété 1952.)

Said to be related to Begoniaceae, and the small seed agrees, but one would like to know the reasoning that connects big trees such as *Octomeles* with *Begonia*. Apart from the seed of *Datiscus*, little is known about the others.

DEGENERIACEAE

Ovules anatropous, numerous (20–32 per carpel, the majority forming seeds), bitegmic, crassinucellate; o.i. 7–8 cells thick; i.i. 3 cells thick, 4 cells at the endostome forming the micropyle.

Seeds *c.* 16 × 12 mm, ellipsoid, with bright orange-red, smooth, waxy sarcotesta, subsessile or with variously developed funicles, albuminous, exarillate. Testa multiplicative, composed of sarcotesta and woody endotesta; *o.e.* as a palisade of radially elongate cells, thin-walled but with thick cuticle; *sarcotestal mesophyll* thin-walled, with clusters of 2–10 large oil-cells opposite the ruminations, the walls of the oil-cells often breaking down to form cysts; *endotesta* composed of several layers of thin-walled cells with lignified internal fibrils (as in Magnoliaceae), the inner surface of the endotesta deeply ruminate in the endosperm through repeated periclinal divisions of the endotestal cells. Tegmen unspecialized, crushed. Chalaza with a dark-staining hypostase of crushed nucellar tissue. Vascular bundle of the raphe extended by a postchalazal branch to form a single loop round the seed from funicle to micropyle in the sarcotestal mesophyll. Endosperm cellular, oily, ruminate from the endotestal ingrowths. Embryo very small; cotyledons 3(–4).

Follicle *c.* 10.5 × 4.5 cm, rich pink to purple, falcate-oblong, fleshy without fibrous tissue but with many 'spicular cells' and v.b. in the firmer outer part, the endocarp intruding as spongy extensions between the seeds, then drying up; many-seeded, tardily dehiscent along the ventral suture after falling.

(Swamy 1949; Bhandari 1963.)

The seed of this Fijian genus *Degeneria*, which has aroused so much interest in recent years, is clearly Magnoliaceous. Sarcotesta, woody endotesta with internal lignified fibrils to the cells, and the degenerate tegmen are conclusive. The sarcotesta is covered by a palisade-like epidermal layer which evidently lacks stomata and the endosperm is more strongly ruminate than has been observed hitherto in Magnoliaceae, cf. *Michelia montana*. Lack of fibres in the testa and tegmen rule out immediate affinity with Annonaceae and Myristicaceae. It seems to me, however, from the published accounts that the seed may be in some measure pachychalazal.

As a large canopy-tree with simple, entire, obovate leaves and unicarpellary flowers, the modern species must have had a long ancestry from the pachycaul beginning.

DIDIEREACEAE

Ovary trilocular with one fertile loculus containing one basal, erect, campylotropous, bitegmic and crassinucellate ovule; o.i. and i.i. 2 cells thick.

Seeds 3–6 mm long, solitary in the indehiscent fruit, with a small funicular aril between the micropyle and the hilum; integuments not multiplicative. Testa with i.e. crushed; o.e. as a layer of short cells with thickened outer wall. Tegmen with o.e. crushed; i.e. with thickenings on the radial walls. Endosperm (? origin), persistent as a trace round the radicle. Fig. 177.

(Perrot et Guérin 1903; Schölch 1963; Johri 1967.)

This small family of cactoid habit in Madagascar is now generally assigned to Centrospermae. The ovule and seed-coats appear typically Centrospermous, and the vestigial aril is in accord.

DILLENIACEAE

Ovules numerous to solitary, reniform–campylotropous or anatropous then amphitropous after fertilization, with conical micropylar end along one side of the seed-stalk, erect, bitegmic, crassinucellate; raphe short; chalaza relatively massive; o.i. 2(–3) cells thick; i.i. 2–6 cells thick; micropyle closed by the exostome; funicle distinct, generally with a rudimentary aril-rim before anthesis; v.b. simple or branched in the chalaza.

Seeds rather small, brown or black, shaped as the ovule, with crustaceous testa, smooth or hairy with slime, generally covered by the white or red, waxy-pulpy aril. Testa not multiplicative; o.e. composed of enlarged, usually thin-walled and pulpy cells, often with tannin, prolonged into hairs over the rounded side of the seed; i.e. as a continuous layer of very thick-walled, lignified, pitted cells, more or less radially elongate but often with substellate facets, poorly developed in Tetracera. Tegmen not multiplicative, generally with the remains of o.e. and i.e., the thin-walled mesophyll crushed; o.e. composed of tangentially arranged, generally longitudinal, tracheidal cells with spiral or band-like lignified thickenings, often rather large, commonly represented only by the outer wall in the mature seed, shortened and more or less radially arranged at the endostome and the chalaza; i.e. as a layer of short, slightly thick-walled, cuboid cells, often with brown contents. Chalaza massive, often more or less inflated by the extension of the sclerotic cells of the endotesta as a cylinder several cells thick round the vascular bundle. Micropyle woody. Endosperm nuclear, oily, without starch, copious, the cell-walls slightly thickened, the outer cells in regular rows. Embryo microscopic with short cotyledons.

Aril arising as a meristematic ring at the junction of the funicle and ovule, usually enveloping the seed but often impeded at the micropyle, entire, crenate–fimbriate or deeply laciniate, without v.b., often with raphid-cells in the mesophyll. Figs. 178–186.

(Svedelius 1911; Netolitzky 1926; Gilg and Werdermann 1925; Paetow 1931; Misra 1938; Hoogland 1951, 1952; Swamy and Periasamy 1955; Rao, A. N. 1957a; Sastri 1958b; Subramanyam 1960a, b.)

The uniformity of ovule and seed in this family has been proved by investigation of Acrotrema, Davilla, Dillenia (incl. Wormia), Doliocarpus, Hibbertia, and Tetracera. The seed is distinguished by the sclerotic endotesta and the tracheidal exotegmen. The testa is limited to two cell-layers, the outer of which makes a thin sarcotesta or, as in Davilla and some species of Dillenia, it dries up. The tegmen is thicker, but largely collapses at maturity. This construction precludes Theaceous or Clusiaceous affinity and suggests derivation from Myristicaceous ancestry with modern Dilleniaceae and the non-multiplicative seed as a blind alley, not the fertile source of other families (p. 29). Concerning Paeoniaceae, see p. 26, Crossosomataceae p. 111, and Cruciferae p. 32.

Many species of Dillenia are exarillate, especially in the alliance of D. indica with indehiscent fruits, but a correlation between the presence or absence of the aril and the dehiscence of the fruit is not possible because it is not known in many species whether the fruit dehisces or not (Hoogland 1951, 1952). The indehiscent D. indica has a vestige of an aril; the indehiscent D. philippinensis has a white aril; all the dehiscent species are reported as arillate.

The suffrutescent Acrotrema with free carpels and, in a few species, with pinnate or pinnatifid leaves appears as a primitive relic of Ceylon.

Acrotrema Jack

A. arnottianum Wight

Testa with many patches of cells with tannin, causing the mottling of the seed. Tegmen with spiral thickenings on the exotegmic tracheids; i.e. with brown contents. Raphe and aril with the mature cells multinucleate.

(Swamy and Periasamy 1955.)

This seed evidently much resembles that of Dillenia suffruticosa.

A. costatum Jack

(Singapore; living material.) Figs. 178, 179.

Ovules as in Dillenia, 2–6 per carpel, mostly 3–4, developing acropetally from the carpel-base; chalazal v.b. red in the living ovule.

Seed as in Dillenia, brown with white aril; testa shortly hairy over the curved side.

Davilla Vand.

Davilla sp.

(Brazil, Manaus, Corner 99; fruits in alcohol.) Fig. 180.

Seed 1 per carpel (solitary in fruit), black, rough,

with white aril. Testa as in *Dillenia*; *o.e.* with dark brown cell-contents, not as a pulpy layer. Tegmen evidently only 2 cells thick; *o.e.* with slight and more or less scalariform thickening on the anticlinal walls; *i.e.* composed of short cells with slightly thickened brown walls. Aril with raphid cells and with rows of very small crystals in the outer walls of the epidermis.

Davilla and *Doliocarpus* have the same kind of seed according to Svedelius (1911).

Dillenia L.

D. indica L.

(Singapore; living material.)   Fig. 180.

Ovule; o.i. 2 cells thick, or 3 cells (Sastri 1958b); i.i. 4–5 cells thick; aril-rim present at the base of the chalaza, but not meristematic and not functional.

Seed 4 × 6 mm, brown. Testa; *o.e.* with the cells becoming thick-walled but not lignified, with pitted anticlinal walls, many cells developing into long, aseptate, unbranched, mucilage-hairs, drying up and becoming slightly thick-walled and pitted; *i.e.* with strongly thickened, pitted and lignified walls. Tegmen collapsing except for i.e. composed of short slightly thick-walled cells; *o.e.* apparently not tracheidal. Chalaza with thin-walled cells, the simple v.b. merely slightly dilated at the end. Funicle composed of colourless cells with slightly thickened walls, with the slight annular vestige of the aril.

D. philippinensis Rolfe

(Java, Kebun Raya, cult.; living material.)   Fig. 181.

Seeds 5 × 3 mm, brown, hairy, covered with the white entire aril. Testa; *o.e.* composed of shortly longitudinally elongate cells, developing hairs (mostly 1 per cell), the outer walls becoming finely striate, not lignified, drying up; *hairs* aseptate, obtuse, dry, with thickened walls and fine oblique striae, not lignified; *i.e.* composed of sclerotic substellate cells. Tegmen 3–4 cells thick; *o.e.* composed of longitudinal cylindric cells with longitudinal subanastomosing lignified thickenings, the inner and radial walls eventually breaking down and leaving the outer walls as a fine pellicle; *i.e.* composed of cuboid cells with slightly lignified walls. Endosperm with thin walls.

D. suffruticosa (Griff.) Martelli

(Singapore; living material.)   Figs. 182–185.

Ovules 7–10, mostly 8, per carpel; funicle short, with annular rudiment of the aril before flowering; o.i. 2 cells thick; i.i. 4–6 cells thick; both integuments thickened at the micropyle by radial elongation of the cells; nucellus with the two outer rows of cells radially elongate; chalaza massive; funicular v.b. breaking up into 4–6 short branches round the base of the nucellus and forming a slight plexus; micropyle closed by the exostome.

Seeds 3.5 × 3 × 1.7 mm, brown, subreniform, smooth, 1–9 per follicle invested by the scarlet aril. Testa; *o.e.* composed of large watery colourless and thin-walled cells with isodiametric facets but variously elongated tangentially over the curved side of the seed, the middle lamella of the anticlinal walls thickened below the cuticle and bearing small crystals; *i.e.* composed of isodiametric, lignified and pitted cells with thick yellow-brown walls. Tegmen not multiplicative; *o.e.* as tracheidal cells with spiral thickening, not or scarcely lignified; *i.e.* composed or more or less persistent cells with slightly thickened, pale brownish walls. Chalaza with a woody brown cylinder (several cells thick) round the white, thin-walled core.

Aril growing mainly by cell-division in the part next to the seed, the hypodermal cells on the outer side elongating greatly; mesophyll without air-spaces, the cell-walls becoming mucilaginous, some cells with raphids; epidermal cells small, with orange-red oil-globules.

Pericarp thin-walled, with large mucilage-canals, the outer and inner surfaces becoming somewhat collenchymatous, not lignified, without oil-cells, idioblasts or stone-cells, but with scattered raphid-cells; *i.e.* with the cells becoming substellate, with the mucilage between the arms drying up to form air-spaces, not lignified, without stomata; *mesocarp* with the inner 5–6 layers of cells aerenchymatous and developing pink cell-sap at maturity; *o.e.* with abundant vestigial hairs as pitted hair-bases without projections, without stomata.

D. triquetra (Rootb.) Gilg

Chalaza with the v.b. branching as in *D. suffruticosa*.
 (Misra 1938.)

Hibbertia Andr.

H. acicularis R.Br., H. stricta Benth.

Ovule; o.i. 2 cells thick; i.i. 3–4 cells thick.

Seed with spiral thickenings on the exotegmic cells before fertilization (*H. stricta*) or after (*H. acicularis*). Aril incomplete or rudimentary, not covering the seed. (Sastri 1958b.)

Tetracera L.

T. indica (Christm. et Panz.) Merr.

(Malaya; fruits in alcohol.)   Figs. 178, 186.

Seeds 5 × 4 × 2.5 mm, blackish brown, smooth, covered by the scarlet, deeply fimbriate aril, reniform with thick stalk-like base. Testa 2 cells thick, –4 cells near the chalaza; *o.e.* as a strong palisade 130–190 μ high, of radially elongate cells with thick brown outer wall (*c.* 15 μ), thin brownish radial walls and inner walls, with mucilaginous contents, facets isodiametric and finely granular-punctate, cuticle faintly rugulose; *i.e.* as small, coarsely pitted, lignified, sclerotic cells, but scarcely differentiated, if at all, over the convex end of the seed. Tegmen 4 cells thick, –8 cells near the chalaza; *o.e.* composed of large, tangentially arranged, tracheidal cells with longitudinal lignified bands in places dichotomous, the cells shorter at the micropyle and chalaza; *mesophyll* crushed; *i.e.* as a short and conspicuous palisade 35–50 μ high, composed of radially elongate cells with thin brown walls, with slight parallel fibrous thickenings on the radial walls, with isodiametric facets, not lignified. Vascular bundle of the funicle centric, dilated into a band just below the chalaza. Stalk of the seed composed of thin-walled cells except for the massive sclerotic tissue along the exostome. Endosperm with slightly thickened walls.

Aril 10–12 mm long, attached round the base of the seed-stalk, with the long ligulate processes free almost to the base, non-vascular, with raphid cells in the mesophyll, all the cells shortly elongate longitudinally and with pale orange oil-drops.

DIONCOPHYLLACEAE

Ovules anatropous, bitegmic, crassinucellate; micropyle formed by the endostome.

Seeds large to very large, discoid, 7–12 cm wide in *Habropetalum*, peltately attached to the long funicle. Testa –10 mm thick; structure ? Chalaza developing a transverse outgrowth in *Triphyophyllum*. Endosperm copious, ? nuclear. Embryo discoid–obconic.

(Schmid 1964.)

Unfortunately the microscopic structure of these exciting seeds is not known. Schmid relates the family with Droseraceae (minute seeds!) and Ancistrocladaceae. Hutchinson (1967) treats the family as one genus of Flacourtiaceae. Perhaps *Drosophyllum* does have a vestige of the fibrous exotegmen typical of Flacourtiaceae (p. 121).

DIPSACACEAE

Ovule anatropous, unitegmic, tenuinucellate; integument 10–15 cells thick.

Seed with persistent o.e. and o.h., the rest of the tissue crushed. Endosperm cellular, oily.

(Netolitzky 1926.)

DIPTEROCARPACEAE

Ovules 2 per loculus, anatropous, suspended, bitegmic, crassinucellate; o.i. 2–9 cells thick; i.i. 3–18 cells thick, with tegmic procambial strands in *Dipterocarpus*; micropyle often with prolonged exostome.

Seeds large, usually solitary in the fruit with 5 abortive ovules, but in some cases paired, occasionally polyembryonous, mostly exalbuminous, rarely arillate (*Upuna*, ? *Vatica*). Testa ? multiplicative, unspecialized, with a variable number of v.b. (–20 in *Vatica*); o.e. with stomata in *Dipterocarpus*. Tegmen ? multiplicative, mostly unspecialized; *o.e.* in *Dipterocarpus* as a short palisade of thick-walled, pitted, lignified cells, in *Vatica moluccana* with some of the cells sclerotic, in other genera undifferentiated; *mesophyll* with scattered, thick-walled and pitted, but not lignified, cells (*Shorea, Vateria, Vatica*), with tegmic v.b. in *Dipterocarpus spp.* Chalazal region with complicated vascularized lobes in *Dipterocarpus* and *Shorea spp.* Endosperm nuclear. Embryo large, with massive cotyledons. Figs. 187–195.

(Netolitzky 1926; Vaughan 1970.)

Little has been done on the microscopic structure of the seeds of this family since the researches of Guérin (1911). The seed-coats in most genera seem to become completely crushed at maturity and show little differentiation except in *Dipterocarpus* and *Vatica*. Nevertheless there are curious points to be investigated in the earlier stages of development, such as the vascular supply and the chalazal lobings.

The family has been variously associated with Guttiferae, Ochnales and Theales. The seed-structure forbids such alliances. The exotegmic palisade of sclerotic cells in *Dipterocarpus* and *Vatica* has the same character as in *Durio*. If truly arillate, then *Upuna* adds another resemblance. The fruit, though generally described as indehiscent, often splits into three valves on germination. The massive ovary and ovules of *Dipterocarpus* point to a massive ancestral capsule with radiating vascular bundles in the cortex and, probably, spinous excrescences beneath the peltate scales in the durian-pattern, cf. the spinous galls of various dipterocarps (Corner 1963). The geniculate petiole suggest that the lamina is the terminal leaflet of a palmate leaf. Anatomical resemblances in the phloem and petiole with Malvales have been noted by Metcalfe and Chalk (1950). But the cortical vascular bundles, the complexity of the pedicel, and the tegmic as well as testa bundles in some species of *Dipterocarpus* suggest a more primitive pachycaulous origin than other families of Malvales, Tiliales or Euphorbiales. Of all, Gonystylaceae may be the nearest approach. Geographically the critical genera *Durio, Gonystylus, Dipterocarpus, Vatica* and *Upuna* are centred in south-east Asia; they may have diverged from a common pachycaul and Mesozoic stock.

Nevertheless, *Dipterocarpus*, if retaining primitive traces in the seed, is advanced in the basipetal growth of the calyx which conceals the basipetally growing nut, in which the stylopodium appears as the capsular relic; inside there is the basipetally growing seed with basipetal proliferation of the tegmen. If this basipetal and post-fertilization growth were to be put neotenically into the floral construction before fertilization, an inferior ovary would result; and there may be epigynous advances on the dipterocarps.

Vochysiaceae, estranged in tropical America, have enough resemblance in fruit and seed to suggest derivation from the capsular ancestors of Dipterocarpaceae. The winged seed is the basipetal derivation, but it has a prolonged micropyle. Anatomical resemblances occur in the mucilage-canals and banded parenchyma. The exfoliating fruit with divergent cortical vascular bundles may be a refinement from the spinous covering.

Polyembryony is reported in *Hopea odorata* and, possibly, *Shorea argentifolia*, *S. macrophylla*, *S. parvifolia* and *S. talura* (Ghosh and Shahi 1957; Maury 1968, 1970).

Anisoptera Korth.

A. marginata Korth.
Ovule; o.i. 3–4 cells thick; i.i. 5–6 cells thick.
Seed with both integuments crushed, without differentiation.
(Guérin 1911.)

Dipterocarpus Gaertn.

Ovules rather large, with massive integuments. Testa with stomata in o.e. Tegmen; *o.e.* as a short palisade of sclerotic cells; *mesophyll* with v.b. (*D. obtusifolius*). Chalaza with complicated vascularized lobes, the cotyledons intrusive.

Endocarp with radially elongate, lignified epidermal cells, continuous or in longitudinal bands separated by thin-walled strips in the lower part of the nut.

D. alatus Roxb., *D. grandiflorus* Blanco, *D. retusus* Bl., *D. trinervis* Bl.
Ovule; o.i. 8–10 cells thick; i.i. 8–9 cells thick.
(Guérin 1911.)

D. hasseltii Bl.
(Leiden, ex Hort. Bogor. VII B 17; alcohol-material.) Fig. 187.
Ovule 1.2 mm long; o.i. 6–7 cells thick, with 8–9 longitudinal procambial strands; i.i. 6–8 cells thick, with 7 procambial strands in the chalazal part.
Ovary 3(–4) locular, without resin-canals in the axis, septate below but with parietal placentation on the edges of the septa above. Pedicel with v.b. in one ring.
(This collection may have been *D. trinervis* Bl.)

D. obtusifolius Teysm.
(Leiden, ex Hort. Bogor. VII B 62; alcohol-material.) Figs. 188–190.
Ovule *c.* 2 mm long; o.i. 7–8 cells thick, –10 cells near the chalaza, but thicker in the curved and oblique exostome, the o.e. with stomata, the mesophyll with 8–9 longitudinal procambial strands as well as the raphe v.b.; i.i. 7–10 cells thick near the endostome, thickening to 15–18 cells near the chalaza, with 9–10 longitudinal procambial strands in the chalazal third of the ovule.
Pedicel of the flower with a ring of five large concentric v.b.
(Guérin 1911.)

D. zeylanicus Thw.
(Ceylon, Peradeniya; full-sized but immature fruits without embryo in alcohol.) Figs. 191–195.
Ovule; o.i. and i.i. 4–5 cells thick (?, counted in abortive ovules.)
Seed (immature, 6–9 mm long, apparently without embryo), pyriform, descending at the rounded chalazal end into the basally growing region of the fruit, with long micropyle, exarillate. Testa 5–8 cells thick, aerenchymatous, unspecialized, carrying the post-chalazal v.b.; *o.e.* with stomata, some without guard-cells. Tegmen 9–15 cells thick, much thicker at the chalazal end, in the micropylar region separated from the testa by a mucilage-space, the nucellar lumen flattened with short diverticula; *o.e.* becoming shortly palisade-like at the micropylar end; at the chalazal end developing thick folds projecting and lobing into the nucellar cavity and almost occluding it, each lobe with a v.b. derived from the chalaza. Chalaza with an infundibuliform orange-coloured hypostase of thin-walled cells leading to the remains of the nucellus. Vascular supply as a stout bundle in the raphe, dividing at the chalaza into *c.* 9 post-chalazal bundles, 3(–4) almost reaching the micropyle, and the chalaza with inward branches to the tegmic lobes.

Pericarp thick, fibrous; inner epidermis developing from the apex of the loculus to the base into a palisade of radially elongate, prismatic, thick-walled and lignified cells; endocarp with a compact ring of v.b. and accompanying resin-canals, with oblique branches ascending outwards towards the exocarp and somewhat anastomosing; base of pericarp with the axile cylinder of v.b. from the peduncle entering the axis of the ovary and giving off the ring of endocarp v.b. Pedicel without accessory v.b.

Neither earlier nor later stages in development were available. Guérin, however, described the tegmic lobes at the chalazal end of the young seed and found that the embryo developed late and that the cotyledons grew between these folds and thrust them apart. These tegmic lobes suggest, at first sight, some apogamous embryonic growth.

Dipterocarpus sp.
(Malaya; dried material.) Fig. 195.
Testa crushed, unspecialized. Tegmen; *o.e.* as a short palisade of dark brown cells, shortly elongate radially, with thick, deeply pitted, lignified walls, not stellately expanded; *mesophyll* collapsed; *i.e.* composed of small cuboid cells with slightly thickened brownish walls.

Doona Thw.

D. nervosa Thw.
Ovule; o.i. 3 cells thick; i.i. 4 cells thick.
(Guérin 1911.)

Hopea Roxb.

H. nigra Burck, *H. odorata* Roxb., *H. pierrei* Hance
Ovule; o.i. 2–3 cells thick; i.i. 4–5 cells thick.

Seed with the integuments more or less crushed, unspecialized.

(Guérin 1911.)

Shorea Roxb.

S. cochinchinensis Pierre, *S. pinanga* Scheff., *S. scaberrima* Burck, *S. zelanica* Bl.
Ovule; o.i. 3–4 cells thick (2–3 in *Isoptera*); i.i. 4–5 cells thick (3–4 in *Isoptera*).

Testa crushed. Tegmen with most of the cells thick-walled and pitted, not lignified, not as a palisade (? *Isoptera*).

(Guérin 1911.)

S. talura Roxb.
Ovule; o.i. 5 cells thick; i.i. 2–3 cells thick; endostome projecting as the micropyle.

Seed evidently with multiplicative tegmen 4–6 cells thick (more in the lower part of the seed); tegmen apparently developing basal lobes as in *Dipterocarpus*.

(Rao 1953.)

Stemonoporus Thw.

Mature fruit of a species which I collected in Ceylon had the integuments entirely crushed.

Upuna Symington

Fruit dehiscent loculicidally into 3 valves on germination. Seed 1, with thin pale aril from the funicle, covering the red-brown testa. (Symington 1941; Ashton 1964.)

This seed needs critical investigation, for it may have the primitive structure in the family. Symington considered that the genus might come between Monotideae and Dipterocarpoideae.

Vatica L.

V. bantamensis (Hassk.) Burck, *V. lamponga* Burck, *V. moluccana* Burck
Ovule; o.i. 4 cells thick; i.i. 5 cells thick.

Testa and tegmen ? multiplicative, becoming nearly twice as thick in the seed as in the ovule (*V. bantamensis*, *V. lamponga*), persistent but compressed, with up to 20 v.b. variously orientated in the testa; tegmen with thick cellulose walls. Testa in *V. moluccana* persistent but the tegmen crushed except for some sclerotic cells as remnants of o.e., and some cells with thick cellulose walls from the mesophyll.

(Guérin 1911.)

DROSERACEAE

Ovules anatropous, bitegmic, crassinucellate or tenuinucellate; o.i. and i.i. 2 cells thick in *Drosera* (4 cells at the endostome), 2–3 cells thick in *Dionaea*; micropyle formed by the endostome.

Seeds minute, albuminous, exarillate. Testa not multiplicative (? *Aldrovanda*, *Drosophyllum*); o.e. persistent, with short cells (*Drosophyllum*) or radially elongate cells in a palisade (*Aldrovanda*, *Dionaea*, *Drosera*), the outer wall more or less thickened, with transverse thickenings on the radial walls (*Dionaea*); mesophyll unspecialized or elaborate (*Drosophyllum*), with pitted walls (*Aldrovanda*); i.e. crushed or persistent with thickened radial and inner walls (*Aldrovanda*, *Drosophyllum*) as crystal-cells. Tegmen not multiplicative, with thick-walled endostome as an operculum (*Dionaea*); o.e. crushed, but ? fibrous in *Drosophyllum*; i.e. with thickened, more or less sclerotic, radial and inner walls (*Drosera*, *Drosophyllum*), or only the inner wall thickened (*Aldrovanda*), or with mucilaginous walls (*Drosera*). Endosperm nuclear, starchy. Embryo small or microscopic (*Drosera*).

The construction of the minute Droseraceous seed is problematic. There is evidence from one or other genus that every epidermal layer of the seed-coat is potentially specialized, the exotesta as a palisade, the endotesta as thick-walled crystal-cells, the exotegmen as fibres, and the endotegmen as sclerotic cuboid cells with U-shaped thickening of the wall. The most complicated seed is that of *Drosophyllum* which lacks only a well-defined exotestal palisade. If it truly has a fibrous exotegmen, then the alliance of Droseraceae must be with Aristolochiaceae, Linaceae, Oxalidaceae and Sauvagesiaceae, if not Flacourtiaceae. Schmid (1964) favours a connection with Dioncophyllaceae but there is yet no evidence to connect the minute seeds of Droseraceae with the giant seeds of Dioncophyllaceae.

Aldrovanda Monti.

Testa 3–4 cells thick; o.e. as a black palisade without distinct cell-outlines; mesophyll with unevenly thickened, pitted walls; i.e. as a palisade of radially elongate cells with slightly thickened walls. Tegmen 2 cells thick; o.e. as a layer of colourless thin-walled cells; i.e. as a layer of brown tabular cells with thickened inner wall.

(Netolitzky 1926.)

Dionaea Ellis.

Ovules anatropous, erect, basal; o.i. and i.i. 2–3 cells thick, not multiplicative.

Seeds black, half-embedded in placental tissue. Testa; o.e. as a palisade of radially elongate cells with thick outer wall and an irregular reticulate thickening on the radial walls; mesophyll and i.e. thin-walled. Tegmen thin-walled, but the cells at the endostome thick-walled and lignified to form an operculum.

(Smith 1929.)

Possibly this seed is pachychalazal.

Drosera L.

Testa in some species larger than the tegmen and an air-space occurring between them; *o.e.* with radially elongate, thin-walled cells. Tegmen crushed (*D. burmanni*) or i.e. persistent with thickened and lignified radial and inner walls (*D. intermedia*, *D. rotundifolia*), or mucilaginous and becoming crushed (*D. indica*, *D. peltata*).

(Netolitzky 1926; Wynne 1944; Venkatasubban 1951; Patankar 1956; Favard 1963.)

Drosophyllum Link

Seeds ribbed. Testa 4 cells thick; *o.e.* composed or cuboid cells with dark brown contents, the outer wall thickened, many cells somewhat papilliform; *mesophyll* with a hypodermal layer of cells in the ribs of the seed, the cells elongated transversely to those of o.e., the inner layer of cells elongate and with dark contents; *i.e.* composed of cuboid crystal-cells with thickened radial and inner walls. Tegmen 2 cells thick; *o.e.* as a layer of longitudinally elongate, thick-walled, pitted cells with the lumen practically obliterated; *i.e.* composed of cuboid cells with slight reticulation on the walls.

(Netolitzky 1926.)

EBENACEAE

Ovules 2 per loculus, anàtropous, suspended, the short funicle overarching the adaxial micropyle, bitegmic with small nucellus or tenuinucellate; o.i. 5–7 cells thick on the sides, thicker along the raphe and antiraphe; i.i. 5–12 cells thick, soon obliterated after fertilization; nucellus at first deeply immersed, soon obliterated (? enlarging after fertilization in some spp., cf. *D. oblonga*); micropyle formed by both integuments; v.b. as a hoop round the ovule in the raphe–antiraphe, unbranched, not perichalazal.

Seeds 4–30 mm long, ovoid-compressed to sub-globose, hard, shiny, brown to black, albuminous, exarillate but in some spp. invested by pulpy endocarp. Testa more or less multiplicative; *o.e.* as a palisade of enlarged cuboid or radially elongate cells filled with mucilage, with hexagonal or gyrose facets, or composed of obliquely longitudinal fibriform cells, the walls more or less thickened and lignified, often sparsely and finely pitted on the radial and inner walls; *mesophyll* thin-walled or with scattered brown sclerotic cells, the inner layers often obliterated; *i.e.* with thickenings on the anticlinal walls, in some cases as crystal-cells, or obliterated. Tegmen with the cells enlarging, then becoming mucilaginous and obliterated, or with a lignified trace at the endostome, ? not multiplicative. Chalaza more or less enlarged. Vascular supply as in the ovule, eventually mucilaginous with spiral thickenings extensible as threads. Endosperm nuclear, horny, thick-walled, oily,

without starch, in some spp. ruminate from ingrowths of the testa. Embryo rather short; hypocotyl often relatively long and stout; cotyledons small, thin, flat.

Pericarp with nests of sclerotic cells scattered throughout, or adjacent to the endocarp, or aggregated into a peripheral layer separated from o.e. by a layer of cells with thickened cellulose walls (*D. lotos*); mesocarp with many large tannin-cells with brown contents; endocarp unspecialized or forming a jacket of radially elongate oil-pulpy cells round each seed. Figs. 196–203.

(Netolitzky 1926; Periasamy 1966.)

The Ebenaceous seed is spuriously like the Annonaceous in shape and rumination but it lacks the fibrous seed-coat, has a characteristic exotestal palisade and larger embryo, and is derived from an ovule with dorsal raphe as in Lauraceae, Buxaceae, some Theaceae and others (p. 5). It seems to be tenuinucellate with the tegmen enlarging much after fertilization, mostly by inflation of the inner layers of cells; then it becomes mucilaginous and is crushed by the endosperm except for the slender woody endostome of narrow elongate cells (Fig. 202). The study of *D. oblonga* which I made in 1943, however, seemed to show a multicellular nucellus enlarging after fertilization to take the place of the inner integument (Fig. 199); I have no means of checking this account and I may have mistaken the enlarged inner cells of the tegmen for those of the nucellus, as in *D. quaesita*.

The ovary develops into the fruit with basipetal growth (Fig. 201), and this is shown by the basipetal growth of the ovules into the seeds, the chalazal part of which thickens and may form the greater part of the seed. The basipetal growth begins in the ovary of the flower-bud and results in the secondary septa which partition the lower ends of the ovules into individual loculi (double the number of placentas). According to Periasamy (1966), the chalazal end forms most of the seed in *D. chloroxylon* and *D. tomentosa*; from this end the ruminations develop and the free parts of the integument are micropylar. Thus the seeds appear pachychalazal or perichalazal, but this requires confirmation because, in other species which I have studied, the integuments extend along the seed to the chalazal base. Periasamy also describes the ruminations as Annonaceous which is true in so far as they are infoldings of the testa, but the seeds are not perichalazal and the ruminations, if always basal and longitudinal as Periasamy has described, do not correspond with the transverse ruminations from the sides of the Annonaceous seed.

There is clearly much variation of systematic importance in the shape, size and contents of the exotestal cells, in the lignification in the mesophyll, in the nature of the endotesta, as well as in the construction of the endocarp and the arrangement of the sclerotic cells in the mesocarp.

The mesotestal construction of some species suggests

affinity with Theaceae, in which the sympetalous character is already indicated. A slight extension of the basipetal growth of the Ebenaceous ovary would result in the inferior ovary of Symplocaceae. Concerning the affinity with Sapotaceae, see p. 249.

Diospyros L.

D. chloroxylon Roxb.
Seed developed mainly from the enlarged chalazal end of the ovule, the free part of the testa confined to the conical micropylar end. Tegmen soon disappearing. Ruminations as slight ridges from the chalazal part of the seed-coat.
(Periasamy 1966.)

D. ebenum Koen.
(Ceylon, Yala; ripe fruits in alcohol.)
Testa 5–7 cell thick; *o.e.* as a palisade of thin-walled cells, the outer wall slightly thickened and brownish, with colourless rather watery contents but with tannin, facets variously elongate and gyrose; *mesophyll* thin-walled; *i.e.* as a compact layer of small cuboid cells with short flange-like thickenings on the radial walls, not lignified, without crystals. Endosperm with very thick-walled cells, not ruminate.

D. lotos L.
(Cambridge, University Botanic Garden; living material.) Figs. 196, 197.
Ovule; o.i. 5 cells thick; i.i. 7–12 cells thick.
Seeds 6 × 5 × 3 mm. Testa 7–9 cells thick; *o.e.* as a layer of variously tangentially elongate cells with subgyrose facets. Endosperm not ruminate.

D. mollis Wall.
(Ceylon, Peradeniya, cult.; ripe fruits in alcohol.) Fig. 198.
Seeds 10–11 × 7 × 4–5 mm, pale brown, not ruminate. Testa 7–11 cells thick; *o.e.* composed of thick-walled, lignified, fibriform cells, obliquely longitudinal, 120–300 × 37–45 μ (high), 11–21 μ wide; *mesophyll* thin-walled, collapsed, some cells with a large tetrahedral crystal (especially near the hilum and round the micropyle); *i.e.* indistinct, some cells with a crystal. Tegmen crushed except for the narrow sclerotic endostome and i.e. composed of small rectangular cells with slightly thickened brownish walls, not lignified except round the endostome; *endostome* wholly lignified, the outer and inner layers with thick-walled sclerotic cells, the 1–2 intervening layers of cells with thin walls.
Endocarp forming a thin pellicle round each seed, composed of a few layers of small thin-walled cells with i.e. as a layer of more or less transversely elongate cells with mucilaginous walls, not as a palisade, not lignified.

D. oblonga Wall.
(Singapore; living material.)   Figs. 199, 200.
Ovule; o.i. 5–7 cells thick; i.i. 8–10 cells thick.
Endosperm ruminate.
Unfortunately I was never able to conclude this investigation.

D. quaesita Thw.
(Ceylon, Peradeniya, cult.; immature fruits in alcohol.) Fig. 201.
Seeds 30 × 15 × 5 mm, not ruminate. Testa 6–7 cells thick in the ovule soon after fertilization, becoming 15–18 cells thick in the half-grown seed; *o.e.* becoming an irregularly folded or rugulose palisade of radially elongate cells 140–280 μ high, with elliptic–rectangular, shortly elongate or subgyrose facets, with mucilage-contents; *mesophyll* developing many crystal-cells at the micropylar end; *i.e.* composed of small cells without crystals. Tegmen 7–9 cells thick in the ovule soon after fertilization, ? not thickening but the inner cells enlarging with thickened mucilaginous walls, then crushed by the endosperm; *o.e.* unspecialized but slightly lignified especially in the long endostome with narrow elongate and scarcely thick-walled cells. Nucellar tissue slight, soon obliterated.
Pericarp very thick with many closely set nuggets of sclerotic tissue in the inner part of the mesocarp; 10–12 locular.

D. tomentosa Roxb.
As in *D. chloroxylon*, but the ruminations as longitudinal flanges from the chalazal end of the seed.
(Periasamy 1966.)

Diospyros sp.
(Brazil, Mato Grosso, Chavantina, Philcox 4178; fruits in alcohol.)   Figs. 202, 203.
Seeds 15 × 10 × 7 mm. Testa *c.* 12 cells thick; *o.e.* composed of very large cells with somewhat thickened brown walls, filled with mucilage; *mesophyll* with thick brown and partly lignified cell-walls; *i.e.* with single crystals in many cells. Tegmen persistent near the micropyle as a single layer of cuboid thin-walled cells and forming a woody endostome 2–3 cells thick. Endosperm not ruminate.
Fruit 3 × 4 cm, with 5 small sepals. Pericarp with many sclerotic cells in the outer tissue; endocarp as a palisade of cells with oily contents and slightly thickened walls, as a jacket round each seed.

Euclea L.

E. divinorum Hiern
(Kenya, D. R. D. Chetham n. 3; mature fruits in alcohol.)
Fruit 8 mm wide, 1-seeded. Seed subglobose, smooth, blackish. Testa thin, firm; *o.e.* as a layer

of cuboid cells filled with mucilage, with shortly gyrose rectangular facets (much as in *D. lotos*), the walls scarcely thickened, not lignified; *mesophyll* 5–6 cells thick, more or less crushed, not lignified; i.e. as a layer of small crystal-cells. Endosperm subruminate.

ELAEAGNACEAE

Ovule anatropous, basal, erect, bitegmic, crassinucellate; o.i. 5–6 cells thick; i.i. 3–4 cells thick.

Seed enclosed in the membranous pericarp, exalbuminous, exarillate. Testa 14–40 cells thick, strongly multiplicative; o.e. as a palisade of radially elongate, thick-walled cells with lignified middle lamella, with hexagonal facets and stellate lumina, or reduced to short cells with undulate facets; *mesophyll* aerenchymatous, unspecialized or partly thick-walled. Tegmen not or slightly multiplicative, unspecialized, soon crushed except for a trace at the micropyle. Vascular bundle only in the raphe, simple or branched. Chalaza with hypostase. Endosperm nuclear, more or less absorbed. Embryo straight; cotyledons large, thick, plano-convex; radicle short. Figs. 204–206.

(Netolitzky 1926; Wunderlich 1968.)

The seed-structure is typically Rhamnaceous. As observed by Wunderlich (1968), resemblance with Thymelaeaceae is spurious, though it is maintained by Rau and Sharma (see Johri 1967b). The range of seeds in the family has not been explored. *Elaeagnus* offers in the two species that have been studied differences in thickness of the testa and in the form of the exotestal cells.

Elaeagnus L.

E. conferta Roxb.

Testa 35–40 cells thick; o.e. as a palisade of thick-walled cells filled with tannin; *mesophyll* with patches of thick-walled cells (? lignified).

(Singh, B. 1964.)

E. latifolia L.

(Ceylon; fruit in alcohol.) Fig. 204.

Seeds 5 mm long, oblong. Testa 17–20 cells thick; o.e. as a layer of short cells, not radially elongate, with slightly thickened brown radial and outer walls, with stellate–undulate facets, not lignified; *mesophyll* ? with mucilage-spaces in the inner part. Vascular bundle spreading shortly after entry into the seed over more than half of the raphe-side of the seed.

Pericarp *c.* 4 cells thick, thin-walled, with 3 longitudinal v.b. concentrically constructed. Perianth-wall with thick outer part composed of large cells (brown contents in alcohol) and a thin inner part with small colourless cells, not lignified; i.e. with a tough felt of white stellate hairs.

Perhaps these seeds are pachychalazal.

Hippophae L.

H. rhamnoides L.

(England; living material.) Figs. 205, 206.

Testa 14–18 cells thick, more along the raphe; o.e. as a palisade. Tegmen 3–4 cells thick, 5–7 at the chalaza. Endosperm slight, round the radicle.

Pericarp with 2 longitudinal v.b. Perianth-tube with the outer layer relatively thin with small cells, the inner layer thick with large pulpy cells and 6 longitudinal v.b.

ELAEOCARPACEAE

Ovules anatropous, bitegmic, crassinucellate; o.i. 2–5 cells thick; i.i. 3–7 cells thick; chalaza rather massive, conical; hypostase present or not; micropyle formed by the thickened exostome, out of line with the endostome.

Seed massive to very small (*Muntingia*, *Pentenaea*), arillate or not, albuminous. Testa 2–16 cells thick, in the larger seeds multiplicative, unspecialized or with thick-walled exotesta or some hypodermal cells sclerotic, without v.b.; o.e. in some cases hairy; i.e. with crystal-cells (*Muntingia*, *Sloanea*). Tegmen 4–40 cells thick, multiplicative; o.e. as a layer of rather short, pitted, lignified, longitudinal fibres, in some cases unspecialized; o.h. as several layers of longitudinal fibres (*Sloanea*) or unspecialized; *mesophyll* aerenchymatous, densely sclerotic in *Aristotelia*; i.e. as a layer of cuboid cells with brown walls usually not lignified, unspecialized in *Muntingia*. Vascular supply limited to the raphe, with a chalazal plexus (*Sloanea*) or also extending into the tegmen almost to the micropyle as several discrete v.b. or as a plexus (*Elaeocarpus*); absent from the testal sides. Nucellus soon destroyed by the endosperm. Endosperm nuclear, thin-walled, oily, abundant. Embryo straight; cotyledons often rather thick.

Aril in *Echinocarpus*, *Peripentadenia*, *Sloanea* spp., arising from the raphe and chalazal end of the seed or from the chalaza only, without v.b., more or less covering the seed, red, orange or yellow, or much reduced. Figs. 207–215.

(Netolitzky 1926; Mauritzon 1934b; Venkata Rao 1953a; Smith, L. S. 1957.)

What little is known of the structure of the fruit and seed in this family confirms the view that it contains many durian-relics (Corner 1953). *Sloanea*, in the widest sense, has possibly a complete series from the spinous to the naked capsule, the arillate to the exarillate seed, and the thick multiple fibrous exotegmen to the state with a single layer of fibres in the exotegmen. The thick pericarp with radial fibro-vascular bundles, typical of the durian-fruits, is retained in the drupes of *Elaeocarpus*, the seeds in which add the unexpected feature of tegmic vascular bundles.

The family is usually placed near Tiliaceae. The fibrous exotegmen, even if the fibres may be rather short, denies this immediate affinity and points to a seed-ancestry common with Flacourtiaceae, Celastraceae, Meliaceae, Violaceae, etc. But it introduces a new point. In *Sloanea*, perhaps also in *Aristotelia*, the tegmen has many outer layers of fibres, and the evidence in *Sloanea* suggests that the single fibrous layer is a simplification of the thickly fibrous tegmen. If so, then the arillate Elaeocarpaceous seed is the most primitive in this range of families. Critical species with which to test this conclusion will be *S. paradisearum* with 16 large arillate seeds per loculus of the capsule and, possibly, *Peripentadenia* (Smith, L. S. 1957). Both these are Australian and may indicate the focus of this Asian–American or Pacific, as opposed to Atlantic, family.

The opposite is the small simple berry of *Muntingia*. The minute seeds have the shortly fibrous exotegmen as in *Elaeocarpus* and *Sloanea*, unlike the palisade exotegmen of Tiliaceae in which *Muntingia* is usually placed. On anatomical evidence, however, the genus has been associated with *Dicraspidia* (with similar but inferior berry) in Elaeocarpaceae (Metcalfe and Chalk 1950). *Dicraspidia* is placed with *Neotessmannia* (fruit unknown) in Tiliaceae tr. Neotessmannieae by Hutchinson (1967), whereas *Muntingia* is associated with *Tilia* and its entirely discordant ovary, fruit and seed. These three American genera may be the neotropical specialization of Elaeocarpaceae far removed in space and time from the origin of the family. *Muntingia* is very advanced in its neotenic ovary ready for pollination before the ovules have developed, as in Orchidaceae, and in its minute neotenic seeds. Theaceae and Rosaceae have a similar divergence in large- and small-seeded genera. The relics of Elaeocarpaceae prove its former significance as a source of evolution for fruit and seed.

There is a striking resemblance in these respects between Elaeocarpaceae and Gonystylaceae. Both have the massively constructed capsule, the raphe–chalaza aril, the shortly fibrous exotegmen and the tegmic vascular supply. Perhaps the phyletic connection of Elaeocarpaceae is with Thymelaeales as a relic of the stock from which the Malvaceous and Sterculiaceous alliance with Tiliaceae diverged. I trace their fundamental differences to the ancestral pachycaul states.

Aristotelia L'Herit.

Ovules without hypotase; o.i. 4–5 cells thick, more at the exostome; i.i. 6–7 cells thick.

Testa? not multiplicative; o.e. composed of thick-walled cells. Tegmen 20–22 cells thick (*A. racemosa*), 13–15 cells thick (*A. maqui*); ? o.e. fibrous; *mesophyll* densely sclerotic.

(Ochsenius 1899; Mauritzon 1934b.)

Elaeocarpus L.

Seeds in hard endocarps of drupes, but the seed-coats well-formed. Testa with smooth or hairy surface, not or slightly multiplicative. Tegmen highly multiplicative; *o.e.* with longitudinal fibres in some species; *mesophyll* with tegmic v.b. in some species (? not all).

This large genus invites enquiry into the microscopic structure of fruit and seed.

E. edulis Teysm. et Binn.

(Ceylon, Peradeniya, cult.; flowers and fruits in alcohol.) Figs. 207, 208.

Ovary 3-locular; ovules 5–6 per loculus, superposed in two rows, anatropous, suspended; o.i. (3–)4 cells thick; i.i. 5–6 cells thick; chalaza rather massive, subacute; nucellus large; micropyle formed by the exostome.

Seeds 1 per fruit, 17–18 × 6 × 6.5 mm, fusoid, whitish, appressedly hairy and with erect hairs at each end. Testa 5–10 cells thick, variable; *o.e.* with longitudinally elongate cells, somewhat thick-walled, especially at the micropyle and chalaza, pitted and lignified, many cells producing hairs; *mesophyll* aerenchymatous, thin-walled, eventually crushed, some of the cells of the inner layers with a single large crystal, especially near the micropyle and along the raphe; *i.e.* with small cuboid cells, thin-walled, many with a single large crystal, not lignified. *Hairs* on the testa of two kinds; single or clustered, aseptate, acute hairs becoming thick-walled and lignified, often appearing as arms of a stellate hair but with separate, if usually adjacent, bases, in clusters of 2–8; closely septate hairs in similar clusters, becoming slightly thick-walled and lignified. Tegmen 18–28 cells thick, at first massive, then most of the tissue crushed; *o.e.* as a single layer, in places incomplete, of shortly fibriform, pitted and lignified cells with moderately thick walls, longitudinal, in several layers at the chalaza; *mesophyll* large-celled, aerenchymatous, thin-walled, carrying the v.b. peripherally; *i.e.* as cuboid cells with brownish unlignified walls, eventually crushed. Chalaza as the massive woody basal and acute prominence of the seed, hairy, most of the testal tissue sclerotic as well as the multiple fibres of the tegmen. Vascular supply as a massive v.b. in the raphe, arching at the chalazal apex and dividing into *c.* 5 branches ascending in the tegmic mesophyll almost to the micropyle, 1–2 dividing into 2–3 close longitudinal branches, not anastomosing. Nucellus soon crushed. Endosperm with thin walls, the cells in radial rows, oily. Embryo straight (immature).

Fruit glabrous, crisply succulent around the fibrous, 3-angled, oblong stone; mesocarp succulent, permeated by many slender fibrovascular strands decurved towards the base of the fruit, not anastomosing, apparently unbranched; endocarp (stone) composed of long interwoven lignified fibres with the main v.b. at the angles and with many small longitudinal v.b. on the faces; cavity unilocular, the septa flattened against the sides, the endocarp and septa developing many microscopic, multicellular, and more or less peltate scales or gland-patches.

No young fruits of suitable size were available to explain the development of the fibro-vascular bundles in the mesocarp or the origin of the peltate gland-patches on the endocarp and septa. However, these radiating fibrovascular bundles seem homologous with those of the echinate capsules of *Sloanea* and indicate the durian-ancestry of the smooth drupe.

E. ganitrus Roxb., **E. robustus** Roxb.

Ovule; o.i. 3 cells thick, more at the exostome; i.i. 5–6 cells thick.

(Venkata Rao 1953a.)

E. petiolatus (Jack) Wall.

(Singapore; immature, living fruits.)

Testa 7–9 cells thick. Tegmen –40 cells thick or more, undifferentiated as yet; *i.e.* as a regular membrane of small cuboid cells with brownish walls and rectangular facets.

E. serratus L.

(Ceylon, Kandy; fruits in alcohol, immature.)

Seed as in *E. edulis* but glabrous. Testa 5–6 cells thick, 3–4 cells towards the micropyle but –13 cells at the exostome, the cells with thin brown walls, slightly lignified; *o.e.* composed of longitudinally elongate cells with patches of isodiametric cells (? original o.e. disrupted), these cells with small air-spaces at the angles; *i.e.* unspecialized, without crystals. Tegmen 28–34 cells thick; *o.e.* undifferentiated except near the micropyle and along the inside of the raphe towards the chalaza, there with longitudinally elongate, but rather short, pitted, lignified fibres, most of the o.e. more or less indistinguishable from the sclerotic sheaths of the v.b. in the close vascular network; *mesophyll* aerenchymatous, with large cells inwards, thin-walled, collapsing from the inside; *i.e.* as cuboid cells with thin brown walls and rectangular facets, not lignified. Vascular supply as in *E. edulis* but the tegmic v.b. forming a very close network in the peripheral tissue, the individual v.b. with a close sclerotic sheath.

Fruit 1-seeded (ovary 3-locular); pericarp with an outer plexus of small longitudinal v.b. near the epidermis, connected by obliquely ascending or transverse branches with the main longitudinal v.b. in the fibrous-woody stone, without decurved v.b.; stone somewhat 5-angled at the apex, but for the greater part 3-angled; septa with small clusters of gland-hairs, not as peltate discs.

Muntingia L.

M. calabura L.

(Ceylon; flowers and fruits in alcohol.)   Fig. 209.

Ovules very numerous, anatropous, mostly transverse, merely as primordia at anthesis, fertilization following full development 12–15 days after pollination, crassi-

nucellate; o.i. 2 cells thick; i.i. 3 cells thick; nucellus relatively massive; micropyle formed by the exostome.

Seeds 0.5 × 0.3 mm (excluding the mucilaginous pellicle), brown, embedded in the pulpy berry. Testa not multiplicative; *o.e.* apparently becoming entirely mucilaginous with its outer surface pressed against that of adjacent seeds; *i.e.* as a layer of crystal-cells with firm radial and inner walls, slightly thickened and lignified at the angles, the outer wall becoming mucilaginous and disappearing, some of the cells projecting in small groups. Tegmen 4 cells thick but only the o.e. persisting; *o.e.* as a compact layer of short, pitted, strongly lignified, longitudinal fibres 45–110 × 12–20 μ, shortened into subcuboid sclerotic cells at the micropyle and chalaza, the outer wall scarcely pitted, projecting into the angles of the reticulum of crystal-cells of the endotesta; *mesophyll* with the cells enlarging, then crushed; *i.e.* unspecialized. Hypostase conspicuous with dark staining cells. Raphe becoming mucilaginous, v.b. ending at the chalaza. Endosperm composed of large, thin-walled cells, oily. Embryo short, straight, the radicle longer than the cotyledons.

(Venkata Rao 1951, 1952a.)

This common little tree, well-known in elementary biological classes in the tropics, is a deceptively simple problem in angiosperm evolution (p. 60).

Sloanea L.

Ovules numerous per loculus, anatropous. Seeds rather large, brown to black, more or less arillate, with multiplicative integuments. Testa not lignified; *o.e.* as a compact layer of cuboid cells; *i.e.* as a layer of crystal-cells, in some cases incomplete. Tegmen often thicker than the testa, then the inner part crushed; outer 3–9 cell-layers transformed into short, lignified, pitted, longitudinal fibres, inner tissue thin-walled; *i.e.* unspecialized or as a discontinuous layer of small, cuboid, lignified cells (*S. alnifolia*). Vascular bundle only in the raphe, unbranched. Endosperm thin-walled, oily.

Aril developing from the raphe and chalaza (*S. javanica*) or only from the chalaza, not vascular, red or orange.

Capsule spiny or smooth.

The four species which I have been able to study show the loss of the spines on the capsule, the reduction of the aril, and the simplification of the fibrous tegmen. In *S. sigun* this layer is 6–9 cells thick; in *S. celebica* and *S. javanica* it is 5–7 cells thick, though thicker along the raphe; in *S. alnifolia* it is merely 1–3 cells thick. There may be some variation even in *S. javanica*, leading to a thinner fibrous layer.

One is apt to assume that a complicated vascular supply goes with the large size of the seed. This is not so in *Sloanea*.

S. alnifolia Mart.

(Brazil, Goiania, Corner s.n. 1968; fruits in alcohol.) Figs. 210–212.

Seeds 9 × 5 mm, ellipsoid, pale brown, smooth, covered with the brilliant orange-red pulpy aril. Testa 9–16 cells thick, composed of thin-walled aerenchyma; *o.e.* with thickened outer walls; *i.e.* with 1–2 crystals in the small thin-walled cells. Tegmen 14–17 cells thick, thin-walled except the epidermal layers, the inner part more or less collapsing; *o.e.* and *o.h.* as 1–3 layers of longitudinal fibres with thick brownish and pitted walls, in places apparently not developed; *i.e.* composed of shortly rectangular cells with thick brown pitted walls, more or less lignified, but in places with patches of thin-walled cells. Vascular bundle of the raphe forming a plexus at the chalaza.

Aril attached broadly round the chalaza, not along the raphe, entire with even or crenulate edge, composed of a colourless and thick-walled epidermis of small cells and a pulpy interior of thin-walled cells full of orange-yellow oil-drops, with numerous large thin-walled cells without v.b.

Capsule –18 × 9 mm, green, set with acicular bristles 7–9 mm long, 4-locular but by abortion unilocular with one seed, splitting into 4 lobes; exocarp densely set with thick-walled simple hairs; bristles multicellular, traversed by a single v.b., the surface set with minute rigid hairs, scattered distally; mesocarp thin-walled, traversed by many radial v.b. to the bristles, the v.b. often flanked by fibres or stone-cells; endocarp with 8 longitudinal bands of fibres (oblique and longitudinal).

S. celebica Boerl. et Koord.

(Koorders 19030; Herb. Bogor., fruits in alcohol.) Fig. 214.

Seed 10 × 5 mm, ellipsoid, black, with very short aril developed from the chalaza with a slight free border. Testa 5–6 cells thick; *o.e.* with compact cuboid cells, with thickened outer walls and hexagonal facets; *mesophyll* thin-walled; *i.e.* as a layer of small cells with a crystal. Tegmen many cells thick, but the inner layers crushed; outer 5–7 layers of cells as longitudinal, pitted, rather short fibres –200 μ long, the outermost shortest (as in *S. javanica*). Funicle short.

S. javanica (Miq.) Szysz.

(Singapore, living material, 1943; Ceylon, Peradeniya, seeds in alcohol, 1972; Indonesia, Koorders 42807, Herb. Bogor., fruits in alcohol.) Figs. 213, 214.

Seeds 20–24 × 14–15 mm, ovoid, shiny black, more or less covered by the scarlet waxy-pulpy aril. Testa 0.5 mm thick, brown with blackish outer layer, dry, *c.* 45–48 cells thick at the micropyle, 16–25 cells thick on the sides of the seed, not lignified; *o.e.* composed of short cells with strongly thickened outer walls, filled with dark brown gum; *mesophyll* with rounded aerenchy-matous cells, the outer 7–11 layers with dark brown walls, some of these outer layers with brown contents, the inner tissue becoming crushed; *i.e.* with single crystals in many cells, especially along the raphe. Tegmen 30–35 cells thick; *outer mesophyll* woody, composed of 2–3 layers of shortly tangentially elongate, thick-walled, pitted and lignified fibres or sclerotic cells 50–130 × 22–45 μ, the outermost as o.e., then with 2–3 layers of shortly fibriform but similar sclerotic cells 90–280 × 25–45 μ, the sclerotic tissue absent from the brown, thin-walled endostome and its immediate surround, or as slightly thick-walled o.e., the sclerotic tissue strongly developed –13 cells thick along the broad raphe (Ceylon material); the sclerotic tissue as a single layer of short fibres, and in 2–3 layers along the raphe (Koorders 42807); *inner mesophyll* thin-walled and crushed; *i.e.* unspecialized. Vascular supply as several anastomosing strands along the raphe, surrounded by isodiametric sclerotic cells, and as a plexus in the chalaza. Seed-stalk 5–7 mm long, magenta-pink, consisting mostly of placental tissue stripped off the loculus, without true funicle.

Aril attached along the whole broad raphe and chalaza, their outer tissue also arilloid, scarlet from large yellow oil-drops and yellowish cytoplasm, and small orange-red globules in the cells, without v.b.

Capsule splitting loculicidally into 4 valves, mostly 1(–2) seeded. Exocarp red from anthocyanin, minutely villous with simple hairs; mesocarp thick, with a thick woody layer in the inner part of each valve, with woody fibres traversing the outer part of the mesocarp radially; endocarp thin, consisting of a bony-woody layer in each loculus and a pale pink, thin-walled, inner tissue lining the loculus, this inner tissue much thickened into a bright magenta-pink pith in the sterile loculi.

S. sigun (Bl.) K. Schum.

(Java, Tjibodas; living material.) Fig. 215.

Ovules 5–6 or more per loculus, only 1–2 forming seeds; o.i. 4 cells thick; i.i. 3–4 cells thick.

Seed 9–12 × 6–7 mm, (immature but full-grown), ellipsoid, with a chalazal aril covering a quarter to one-third of the seed. Testa 7–10 cells thick; *o.e.* composed of cuboid cells with subhexagonal facets, the outer wall slightly thickened; *mesophyll* thin-walled, aerenchy-matous; *i.e.* composed of small thin-walled cells, many with a crystal. Tegmen 20–27 cells thick; outer 6–9 layers of cells converted into shortly longitudinally elongate sclerotic cells with thick, lignified, coarsely pitted walls, the inner cells progressively shorter and smaller, the most internal subglobose, becoming thick-walled from the outer layer inwards; *inner mesophyll* 13–18 cells thick, composed of large, thin-walled, unlignified, subglobose cells, aerenchymatous, eventually crushed; *i.e.* as a layer of small, unspecialized, subcuboid cells. Vascular bundle of the raphe dividing into a few short branches in the chalaza. Hypostase as a thin disc of slightly lignified cells.

Aril red, non-vascular, developing round the chalaza, with free crenulate margin, not along the raphe or round the micropyle.

Capsule with soft, shortly hairy endocarp, lignified in the outer part of the endocarp; mesocarp traversed radially by coarse branching and subanastomosing fibrovascular bundles, with one v.b. passing into each spine of the exocarp; spines with sclerotic lignified cortex round the small central v.b.

ELATINACEAE

Ovules anatropous, bitegmic, crassinucellate (? tenuinucellate); o.i. 2–3 cells thick, 4–5 cells (Netolitzky); i.i. 2–3 cells thick.

Seeds minute, straight, with reticulate surface, more or less exalbuminous, exarillate; seed-coats not multiplicative. Testa; *o.e.* thin-walled, unspecialized, with tannin; *i.e.* unspecialized and crushed or with an annular thickening on the radial walls (*Elatine*). Tegmen; *o.e.* as a layer of thick-walled, pitted, stellate–undulate cells, lignified or not, oblong in *Bergia*; *mesophyll* soon crushed (*Bergia*); *i.e.* thin-walled with tannin (*Bergia*). Endosperm nuclear, reduced to 1–5 cell-layers in *Bergia* or none in *Elatine*. Embryo straight, relatively large.

(Netolitzky 1926; Raghavan 1940; Salisbury 1967; Dathan and Singh 1970, 1971.)

The seed-structure is decisive in the classification of this family. Hutchinson refers it to Caryophyllales (exotestal seed with perisperm). Takhtajan places it in Theales near Hypericaceae and Sauvagesiaceae (Ochnaceae–Luxemburgieae). There is a difference here. Hypericaceae have stellate–undulate and lignified exotegmic cells; Sauvagesiaceae have exotegmic fibres. As the accounts of the Elatinaceous seed were not clear in this detail, I examined the seeds of *Bergia ammannioides* Roxb. (leg. Saxton 2796, Bombay; CGE) and found that these minute structures (*c.* 350 × 170 μ), like ovules with enlarged and differentiated cells, had more or less oblong, strongly stellate–undulate cells (45–70 × 35 μ) with thick, yellow-brown, finely pitted, but hardly lignified walls, very much as in *Hypericum*. This finding has been confirmed by Dathan and Singh (1971).

I conclude that Elatinaceae belong with Hypericaceae and, probably, with Bonnetiaceae for these exotegmic cells of *Bergia* agree with those of *Ploiarium*. Exceptional, perhaps, is the nucellus which is said to be massive in Elatinaceae.

ERICALES (Clethraceae, Cyrillaceae, Diapensiaceae, Empetraceae, Epacridaceae, Ericaceae, Lennoaceae, Monotropaceae, Pyrolaceae, Vacciniaceae)

Ovules anatropous (campylotropous in Diapensiaceae *pr. p.*), unitegmic (? bitegmic in *Schizocodon*, Diapensiaceae), tenuinucellate (? also crassinucellate in Cyrillaceae *pr. p.*); integument 4–10 cells thick or 2–3 cells (Monotropaceae, Pyrolaceae); i.e. as an endothelium; funicle rather long in Cyrillaceae.

Seeds small or minute, dry, in some cases alate or appendaged at each end (Monotropaceae, Pyrolaceae), albuminous, exarillate. Seed-coat apparently not multiplicative, exotestal; *o.e.* unspecialized (Cyrillaceae, Empetraceae) or generally with thick pitted and lignified inner and radial walls (with short hair-like outgrowths in *Daboecia*), or the wall thickened all round (*Andromeda*, *Gaultheria*, *Gaylussacia*), with a stout annular thickening on the radial walls (*Diapensia*), with fine reticulate thickening (Lennoaceae); *mesophyll* and *i.e.* crushed, but *o.h.* persistent and slightly thick-walled in *Daboecia*. Vascular bundle in the raphe (Clethraceae, Diapensiaceae, *Enkianthus*), generally without v.b. (Epacridaceae, Ericaceae, Pyrolaceae). Endosperm cellular in all families but nuclear in Ericaceae (except *Rhododendron*, cellular), oily and proteinaceous. Embryo relatively large, straight, but minute in Lennonaceae, Mocotropaceae and Pyrolaceae.

(Netolitzky 1926; Copeland 1953; Maheshwari, P. 1963; Ganapathy and Palser 1964; Ganapathy, Vijayaraghavan, in Johri 1967b; Stushnoff and Palser 1970.)

These families, assigned to Ericales, are so similar in their small exotegmic and unitegmic seeds that I have taken them together. Whether they may be defined on exotestal peculiarities remains for future investigation; too few genera have been sampled. The seed deteriorates, evidently, from Theaceous complexity (p. 29).

Fagerlind (1947) would place Grubbiaceae in Ericales, but I have retained it as usual in Santalales. The seed offers no difference from Ericales.

ERYTHROXYLACEAE

Ovary 1–4 locular, one loculus fertile with 1–2 ovules, the others with a sterile ovule. Ovules anatropous, suspended, bitegmic, crassinucellate; o.i. 2–5 cells thick; i.i. 7–9 cells thick, with i.e. as an endothelium; micropyle formed by the endostome; v.b. of the raphe commonly double.

Seed with thin brown papery seed-coats, tightly enclosed in the thin, but woody, often angled endocarp, albuminous, arillate in *Aneulophos*. Testa not multiplicative; *o.e.* as thin-walled cells somewhat transversely elongate; *mesophyll* and *i.e.* unspecialized, eventually crushed. Tegmen multiplicative, 8–24 cells thick; *o.e.* as a continuous or somewhat interrupted layer of narrow, longitudinally elongate, thick-walled, pitted, lignified fibres, shorter and rounded into sclerotic cells at the prominent endostome; *mesophyll* composed of large thin-walled cells, somewhat radially arranged, aqueous, not aerenchymatous, eventually crushed; *i.e.* at first as an endothelium with firm walls, eventually crushed. Chalaza unspecialized. Vascular bundle ending at the chalaza. Endosperm nuclear, thin-walled, starchy. Embryo straight, varying rather small to massive with thick flat cotyledons and little endosperm.

Ovary and fruit with a hypodermal layer of large, radially elongate cells filled with oleaginous matter (*Erythroxylon*) or starch (Rao, D. 1968); aborting loculi in some species developing into strips of endocarp separate from the fertile endocarp or joined with it. Figs. 216–219.

(Netolitzky 1926; Mauritzon 1934d; Rao, D. 1968.)

This description refers mainly to *Erythroxylon*. If the seed of *Aneulophos* is truly arillate, it should show the primitive construction of the seed-coats; according to Netolitzky it has a very thick testa and thin tegmen.

Erythroxylon L.

E. cuneatum Kurz, *E. ecarinatum* Hochr., *E. kunthianum* A. St Hil., *E. lanceum* Boj., *E. novogranatense* (Morris) Hieron.

(Rao, D. 1968.)

E. coca Lamk.
(Ceylon, Peradeniya, cult.; flowers and fruits in alcohol.) Fig. 216.

Ovule; o.i. 3–4 cells thick; i.i. 5 cells thick; v.b. of raphe double.

Testa 4 cells thick; *o.e.* composed of longitudinally elongate cells with slightly thickened, brown, outer walls; *mesophyll* and *i.e.* crushed except in the angles of the endocarp. Tegmen 18–24 cells thick; *o.e.* as a continuous jacket of lignified fibres, forming a massive endostome-plug. Endosperm thick.

Drupe with the sterile loculi separate and as a sterile strip of sclerotic tissue at the angles of the fertile endocarp.

(Mauritzon 1934d.)

E. monogynum Roxb.
(Ceylon, Yala; flowers and fruits in alcohol.) Figs. 217–219.

Ovule; o.i. 2–3 cells thick; i.i. 7–8 cells thick.

Testa; *o.e.* with the cells shortly transversely elongate. Tegmen 8–15 cells thick; *o.e.* often with undulate fibres; *i.e.* as a persistent membrane with angular–lobate cell-outlines, not lignified. Vascular bundle of the raphe simple.

Drupe with the sterile endocarps attached along one side to the fertile.

E. moonii Hochr.
(Ceylon, Peradeniya; flowers and fruits in alcohol.) Fig. 216.

Ovule; o.i. 4–5 cells thick; i.i. 7–9 cells thick; v.b.

of raphe double, the two strands widely separate in the seed.

Seed as in *E. monogynum* but with the exotegmic fibres lacking in places. Embryo massive.

Drupe without sterile loculi.

ESCALLONIACEAE

As Hydrangeaceae; ovules bitegmic (*Brexia*) or unitegmic (*Escallonia*), crassinucellate or tenuinucellate. Endosperm nuclear.

(Netolitzky 1926, p. 152.)

Corokia A. Cunn. with unitegmic ovule, placed usually in Cornaceae, is referred to Escalloniaceae by Eyde (1966).

EUCOMMIACEAE

Ovules anatropous, suspended with dorsal raphe and adaxial micropyle, unitegmic, subtenuinucellate; integument 5 cells thick.

Seed albuminous. Seed-coat membranous, apparently thin-walled and not multiplicative. Endosperm cellular. Embryo straight.

(Eckardt 1963.)

The orientation of the ovule does not fit Urticales.

EUPHORBIACEAE

Ovules anatropous, suspended, solitary (Crotonoideae) or 2 per loculus (Phyllanthoideae), bitegmic, crassinucellate; obturator present.

Seeds 1–80 mm wide, exotegmic, albuminous, exarillate or with a caruncle (some Crotonoideae); seed-coats multiplicative or not. Testa generally thin-walled, unlignified, pulpy or dry; *i.e.* often as a thin-walled palisade (Crotonoideae). Tegmen with o.e. as the mechanical layer; in Crotonoideae as a palisade of Malpighian cells; in Phyllanthoideae as a layer of longitudinal sclerotic cells, fibres, cuboid sclerotic cells or a short palisade. Vascular supply often complicated with testal and (in some Crotonoideae) tegmic bundles or a pachychalazal complex. Chalaza with a small heteropyle and in various genera of Crotonoideae with a vascular tegmic pachychalaza. Funicle generally undeveloped. Endosperm nuclear, becoming thin-walled, oily. Embryo relatively large, straight or slightly curved; radicle short; cotyledons thin, flat. Figs. 220–267.

Exceptions: unitegmic ovules (*Drypetes macrostigma*, Phyllanthoideae); corky testa (*Sapium spp.*, Crotonoideae).

(Netolitzky 1926; Mandl 1926; Wiehr 1930; Vaughan 1970.)

Crotonoideae

Ovules 1 per loculus; o.i. 3–8 cells thick; i.i. 3–25 cells thick; micropyle formed by the exostome but often occluded by the nucellar projection reaching to the obturator.

Seeds *in several genera arillate with a caruncle.* Testa usually multiplicative in the larger seeds, pulpy, then often drying up into a pellicle, often mottled with patches of pigmented cells, in some cases papillate with short aseptate hairs; *i.e. generally as a palisade of short, thin-walled, columnar cells,* not or slightly lignified, or with calcified deposits in the lumen or on the walls. *Tegmen 4–80 cells thick, generally much thicker than the testa; o.e. as a palisade of lignified Malpighian cells 0.2–3 mm long* (? more in *Joannesia*), uniform or with groups of shorter cells giving the pitted appearance in surface-view, the cells often curved on the sides of the seed with the inner end towards the micropyle; mesophyll aerenchymatous; *i.e.* unspecialized or with slightly thickened, subligneous, radial walls and slight ridge-like thickenings. Vascular supply with or without postchalazal branches in the testa; absent from the tegmen or as a peripheral plexus of tegmic v.b. from the chalaza, or as a funnel-shaped basket of slender v.b. in the tegmic pachychalaza. Figs. 220–249.

Phyllanthoideae

Ovules 2 per loculus (solitary in *Galearia, Microdesmis*), both or one only forming seeds, varying hemianatropous; o.i. 2–9 cells thick; i.i. 3–10 cells thick.

Seed exarillate, the endostome sometimes exposed. Testa 2–20 cells thick, thickly pulpy and multiplicative or dry; *i.e.* unspecialized, not as a palisade, but in sarcotestal seeds the cells of i.h. more or less radially elongate. *Tegmen not or slightly multiplicative; o.e. composed of cuboid or shortly radially elongate sclerotic cells, stellate–undulate sclerotic cells, or short or long, lignified, longitudinal fibres* (cylindric or ribbon-like); i.h. in some cases also fibrous or sclerotic; mesophyll little developed, collapsing; i.e. as cuboid cells, often with brownish walls. Vascular supply with or without postchalazal branches in the testa, without tegmic v.b. Chalaza often with a broad plate-like hypostase supplied with a funnel-shaped expansion of tracheids from the v.b. of the heteropyle, or as a vascular pachychalaza (*Glochidion*). Figs. 250–267.

In spite of recent innovations in the classification of this large and complicated family (Airy Shaw 1972), I have followed the conventional system because it is suitable for the description of the comparatively few seeds that are sufficiently known. There is a uniformity in the exotegmic palisade of Crotonoideae which allies them with Bombacaceae, Malvaceae, Sterculiaceae and Tiliaceae. The contrast is the varied condition in Phyllanthoideae which, in some genera, have the fibrous exotegmen of Celastraceae, Flacourtiaceae and Violaceae, and in other genera a diversification into various forms of sclerotic cell which might be regarded as a degenerate palisade.

These variations of Phyllanthoideae are detailed on p. 138. Clearly they are important if such families as Bischofiaceae, Hymenocardiaceae, Pandaceae and Stilaginaceae are to be distinguished, but it is equally clear that many more genera and species need investigation in order to discover whether the differences in the exotegmic cells are constant and phyletic or transitional within a natural Phyllanthoid series.

In Crotonoideae the tegmen thickens greatly; it introduces two problems. There is that of the tegmic vascular supply, which may be considered to be primitive as in Myristicaceae; its systematic occurrence hardly accords with this conclusion; it occurs in the carunculate seeds which, according to the durian-theory, have a vestigial aril and may be the more primitive; it does not occur in the sarcotestal seeds of Phyllanthoideae which should be primitive according to van der Pijl's theory, and which occur in the less reduced ovary with paired ovules. And there is the problem of the tegmic pachychalaza with its basket of fine vascular bundles in the thickened tissue between the hypostase and the exotegmen which surrounds the heteropyle (Figs. 222, 248). In mature seeds with crushed tegmen it may be impossible to distinguish between the two states. The immature seed, however, at once reveals the difference; the endotegmen in the first case can be traced from the micropyle to the region over the heteropyle (Figs. 220, 228, 235) but, in the seed with tegmic pachychalaza, it extends a much shorter way from the micropyle (Figs. 229, 248); it may be that the short free part of the tegmen can be discerned in such pachychalazal seeds (Fig. 226). The truly tegmic vasculature is peripheral whereas the tegmic pachychalaza becomes situated close to the endosperm as it encroaches into the pachychalaza.

The expression 'tegmic pachychalaza' is cumbrous but it is consistent with the other usage of these words and is preferable to 'pachy-hypostase'. The condition is not fully pachychalazal because it affects only the interior of the tegmen, not the testa, cf. Fig. 378 for the pachychalaza and Fig. 248 for the tegmic pachychalaza. It seems that the vascular tegmic pachychalaza is a complication imposed on the normal anatropous seed, similar to the true pachychalaza, and that the tegmic vasculature may, therefore, also be an innovation rather than a primitive inheritance. This problem has to be considered in connection with other genera with tegmic vasculature such as *Dipterocarpus, Elaeocarpus, Gonystylus,* and *Thymelaeaceae* (p. 21).

The caruncle is a hard round body that sits on the raphe-side of the micropyle and generally surrounds it; it has little similarity in appearance with the pulpy micropylar aril. It develops after fertilization and must not be confused with the obturator which develops before fertilization and usually dries up without enlargement as the seed develops; the obturators persist, however, in *Aporosa* (Phyllanthoideae) to supply the funicle with the levers that split the capsule and elevate the seeds. The caruncle occurs in all three seed-groups of Crotonoideae, and variously degenerates. In *Croton* and *Manihot* it may be conspicuous or absent; in *C. tiglium* it is so slight as scarcely to be noticeable. In *Acalypha* the lateral flowers of the inflorescence may have carunculate seeds but not the terminal. If *Petalostigma* and *Toxicodendron* belong with Phyllanthoideae, they are the sole examples with the caruncle in this subfamily. I regard the caruncle as a vestigial aril, homologous with the aril of Bombacaceae and Sterculiaceae; it is independent of the occurrence of a sarcotesta.

In most, if not all, Euphorbiaceae the seed reaches its full size and begins to differentiate the structure of the seed-coats while the endosperm and embryo are yet rudimentary (Fig. 227). The final state of the tegmen may be so hard that a chisel is better than a razor.

CROTONOIDEAE

For characters of ovule and seed, see p. 129.

The following three states of construction can be distinguished.

(1) Seeds with tegmic vasculature; tegmen attached to the short chalaza and supplied by v.b. from the heteropyle to form a peripheral and longitudinal network, simple, double or treble: *Aleurites, Croton, Givotia, Hevea* (?), *Jatropha, Manihot* (?), *Pimeliodendron*: (? *Dalechampia, Codiaeum, Cremophyllum*, according to Le Monnier 1872).

(2) Seeds with tegmic pachychalaza; tegmen broadly inserted on the thick tissue between the hypostase and the heteropyle, without tegmic v.b. but with a funnel-shaped basket of slender v.b. extending from the heteropyle to the base of the nucellus; free part of the tegmen limited to the micropylar end of the seed: *Cleidion, Dimorphocalyx, Gelonium, Ricinus*; (? *Acalypha, Codiaeum, Mallotus, Sphaerostylis*).

(3) Seeds of normal construction; without tegmic v.b. or tegmic pachychalaza; sarcotesta often well-developed: *Acalypha, Claoxylon, Chrozophora, Euphorbia, Hevea* (?), *Homalanthus, Joannesia, Macaranga, Mallotus* (?), *Manihot* (?), *Melanolepis, Mercurialis, Micrococca, Sapium, Sebastiana, Trewia* (?).

The vasculature of the testa varies but may be characteristic of a genus or a generic alliance. Thus, in two species of *Mallotus* (Figs. 240, 241) short vascular bundles pass from the raphe, near its origin, on either side of the micropyle and join with the ends of the postchalazal bundles, such as in *Durio. Melanolepis, Micrococca* and *Trewia* agree. In some species of *Croton* there is a single preraphe bundle (Fig. 227) and traces of such occur in other genera.

The thin-walled endotestal palisade seems to be homologous with the woody endotesta of Myristicaceae, but its cells may variously fail to elongate, e.g. in *Manihot*.

Very detailed accounts of the dried seeds of *Aleurites, Croton, Hevea, Jatropha, Manihot*, and *Ricinus* have been given by Wiehr (1930).

Acalypha L.

Ovule; o.i. 3–4 cells thick, 6–8 at the exostome, 3–6 in *A. malabarica* with 8–10 at the exostome; i.i. 4–5 cells thick, in *A. malabarica* 2–4 cells and 5–6 on the chalazal side.

Testa not multiplicative, thin-walled, with small cells; *o.e.* with tannin; *i.e.* enlarged, cuboid, as a short palisade, the radial walls with fibrous thickenings. Tegmen 7–8 cells thick, or not multiplicative (*A. ciliata, A. malabarica*); *o.e.* with curved Malpighian cells; *mesophyll* without v.b.; *i.e.* with band-like thickenings on the walls. Chalaza somewhat thickened, with a short conical or cupular plexus of fine v.b. round the small hypostase and leading to the base of the nucellus.

Caruncle present but only in seeds from lateral flowers, not on the seeds of terminal flowers (*A. alnifolia, A. brachystachya, A. indica*); epidermis formed by division of 2–3 hypodermal layers of the thickened micropylar testa.

A. alnifolia Klein, *A. brachystachya* Hornem., *A. ciliata* Forsk., *A. indica* L., *A. malabarica* Muell. Arg., *A. rhomboidea* Rafin.

(Landes 1946; Johri and Kapil 1953; Kapil 1960; Mukherjee 1964.)

Aleurites Forst.

Seeds large 15–40 × 10–30 mm, plump, without caruncle. Testa *c.* 15 cells thick (*A. cordata*) or 28–38 cells thick (*A. triloba*), white, pulpy, with postchalazal v.b. in the mesophyll; *o.e.* composed of brownish cuboid cells; *i.e.* as a palisade of narrow prismatic cells with slightly thickened or slightly lignified walls, with hexagonal facets, with calcified deposits in the cells (*A. moluccana, A. triloba*). Tegmen with very thick, hard, palisade of long, heavily lignified and pitted cells, 0.5–3.5 mm thick; *mesophyll* becoming very massive with a fine network of longitudinal v.b. towards the outer side; *i.e.* unspecialized. Chalaza relatively small. Endosperm thin–walled.

A. cordata (Thunb.) R.Br.

Seed 15 × 15 × 10 mm. Testa *c.* 15 cells thick; *i.e.* shortly prismatic, with little or no calcification. Tegmic palisade 400–550 μ thick.

(Mandl 1926; Wiehr 1930.)

A. moluccana (L.) Willd.

(Java, Bogor; flowers and immature fruits.) Fig. 220.

Ovule with o.i. and i.i. 4–6 thick; obturator massive; nucellus slightly projecting through the micropyle.

Seed 35–40 × 25–30 mm, with massive vascular testa and tegmen. Testa 28–36 cells thick, mostly pulpy, white; *o.e.* with firm walls; *i.e.* as a short palisade of columnar cells with firm walls, not lignified. Tegmen 70–85 cells thick, 12–15 cell-layers external to the ring of v.b., 60–70 cells-layers internally, most of the inner tissue crushed by the endosperm; *o.e.* as a dark brown, woody palisade 2–3 mm thick; *mesophyll* thin-walled, aerenchymatous, composed mostly of more or less radially elongate cells but with a layer (7–10 cells thick) of small cells just internal to the ring of v.b.; *i.e.* unspecialized, with slight scalariform thickening on the inner walls. Chalaza small. Vascular bundle of the raphe giving off a micropylar branch on entry to the ovule, then 1–2 lateral branches on each side of the raphe, and 2–3 postchalazal branches ramifying over the sides of the seed almost to the micropyle; tegmic v.b. derived from the short branch entering the heteropyle, dividing and anastomosing in a peripheral cylinder placed in the outer part of the tegmen.

(Wiehr 1930.)

A. triloba Forst.

(Ceylon; mature fruits in alcohol.)   Fig. 221.

Seed 27–30 × 26 × 22 mm. Testa 1–1.8 mm thick, 28–38 cells thick; *mesophyll* aerenchymatous with substellate cells, not lignified; *i.e.* as a palisade 220–250 μ high, the narrowly prismatic cells with slightly lignified walls and calcareous deposits in the lumen. Tegmic palisade 2–2.5 mm thick. Vascular supply in the testa with 5 postchalazal branches dividing over the seed and almost reaching the micropyle; tegmic vascular supply as in *A. moluccana*.

Pericarp with the thin woody cocci composed of transverse fibres; exocarp with an interrupted sclerotic layer; v.b. in 3(?–4) rings of branching strands in the endocarp, with small scattered clusters of variously elongate sclerotic cells.

(Mandl 1926.)

Chrozophora Neck.

Ovule; o.i. 3–4 cells thick, 4–5 at the exostome; i.i. 4–5 cells thick; nucellus projecting as a cylinder through the micropyle, curving on to the obturator.

Seeds with caruncle. Testa –9 cells thick, unspecialized, collapsing; *i.e.* as a palisade of shortly radially elongate cells. Tegmen –11 cells thick; *o.e.* with curving Malpighian cells, unequally elongate to give the rugulose surface of the dried seed; *i.e.* with fine spiral or annular thickenings. Vascular bundle only in the raphe.

Fruit papillate with peltate scales on the short papillae.

C. obliqua A. Juss., *C. prostrata* v. *parvifolia* Klotsch
(Kapil 1956.)

Claoxylon A. Juss.

C. indicum (Reinw.) Hassk.
(Singapore; living material.)

Seed with red sarcotesta, the cells containing pink oil-drops. Tegmic palisade black, carbonaceous, the cells unequally elongate to give the pitted surface.

Capsule without bony cocci, the wall soft, but the inner epidermis with irregular sinuous cells with thickened, pitted and lignified walls, many of these cells bearing simple hairs with similar walls.

Cleidion Bl.

C. javanicum Bl.
Collection A: (Ceylon, Yala, Jan. 1969; immature fruits in alcohol). Collection B: (Java, Tjibodas, May 1972; mature fruits in alcohol).

Collection A.   Figs. 222–225.

Seeds (immature) *c.* 12 mm wide, round, with massive vascular tegmic pachychalaza. Testa 0.4 mm thick, pulpy, many cells thick, not lignified; *o.e.* with cuboid cells, not elongate; *mesophyll* composed of large thin-walled cells; *i.e.* as a short palisade of radially elongate thin-walled cells, closely abutting on the exotegmen. Tegmen massive; *o.e.* as a massive brown palisade 0.4 mm thick, the curving cells with thick brown pitted and lignified walls, with hexagonal facets; *mesophyll c.* 55 cells thick in the pachychalazal region, thin-walled, aerenchymatous, many cells with small spherical clusters of pointed crystals, the innermost tissue collapsing; *i.e.* as a distinct layer of rectangular cells with slightly lignified walls, with slight reticulate thickening on the radial walls. Vascular bundle of the raphe prolonged into the antiraphe with 3–5 short branches; the short branch to the heteropyle forming a funnel-shaped basket of *c.* 30 fine, more or less anastomosing v.b. between the chalaza and the base of the nucellus, not entering the free part of the tegmen. Nucellar tissue with many scattered sphaerocrystals as in the tegmic mesophyll.

Fruit with 2 loculi; style 1, with 2 bifurcate stigmatic arms; seed 1 per loculus, surrounded by the bony coccus.

Collection B.   Fig. 226.

Seeds 17–18 × 12 mm, ellipsoid, with white pulpy-pellicular testa and hard black tegmen; micropyle distant from the hilum, without caruncle, with vascular tegmic pachychalaza. Testa 0.8 mm thick along the raphe and chalaza, much thinner (6–7 cells thick) over the sides of the seed, thin-walled, aerenchymatous, not lignified, colourless; *o.e.* with hexagonal facets, the outer walls slightly thickened and pitted; *i.e.* as a short thin-walled palisade. Tegmen *c.* 25 cells thick in the micropylar region; *o.e.* 250–300 μ thick, as a palisade of lignified cells with curving bases, dark brown pitted walls, and hexagonal facets; *mesophyll* thin-walled. Vascular bundle of the raphe ending at the chalaza; pachychalazal v.b. *c.* 16 as a funnel-shaped basket.

Nucellus reduced to 1–2 layers of more or less crushed cells, at first projecting into the micropyle.

Fruit 3-locular.

Though I obtained the same names for these two collections, which appear to be congeneric, I cannot believe that they represent the same species. As Thwaites noted, the fruit is always bilocular in the Ceylon trees. The ovule and seed of *Dimorphocalyx* are similar.

Codiaeum A. Juss.

C. variegatum Bl.

Ovule; o.i. 3–4 cells thick, but thicker at the exostome; i.i. 7 cells thick, but thicker at the endostome and towards the chalaza; v.b. of the raphe entering the chalaza and with branches anastomosing in the i.i.

Seed 6–6.5 × 4 mm, brown, paler mottled. Testa not multiplicative; *o.e.* composed of cuboid cells with isodiametric facets, the outer walls finely pitted, the contents brown; *i.e.* with brown cell-contents. Tegmen multiplicative; *o.e.* with curving Malpighian cells; *mesophyll* with a network of v.b. from the chalaza.

Caruncle white; cells elongate with fine spiral and crossing fibres on the walls.

(Landes 1946; Singh, R. P. 1965.)

Though described with tegmic vascular bundles, the seed may have a tegmic pachychalaza as in *Dimorphocalyx* which is placed with *Ostodes* (cf. *Cleidion*) in the subtribe Codiaeneae.

Croton L.

Seeds 5–15 mm long, carunculate or not. Testa 3–9 cells thick; *i.e.* as a short palisade of thin-walled cells with little or no calcification. Tegmen multiplicative; *palisade* 100–360 μ high; *mesophyll* with tegmic v.b.

C. argyratum Bl.

Seed 10 × 7–8 × 4–5 mm. Testa 5–6 cells thick. Tegmic palisade 275–345 μ high.

(Wiehr 1930.)

C. bonplandianum Baill.

Seed 5.5 × 3 mm, black, mottled white, the caruncle shortly 3-lobed. Testa 3 cells thick; *o.e.* as a short palisade, the radial walls with transverse bands.

(Singh, R. P. and Chopra 1970.)

C. glandulosus L.

Ovule with nucellar beak.

(Landes 1946.)

C. laevifolium Bl.

(Singapore; flowers and fruits in alcohol.) Fig. 227.

Ovule; o.i. 4–5 cells thick; i.i. 17–21 cells thick, with 9–10 procambial strands; nucellus projecting through the micropyle. Seed 8 × 6 × 5 mm, without caruncle,

brown. Testa not multiplicative, thin-walled, drying up. Tegmic palisade *c.* 140 μ thick; *mesophyll* with 20–24 peripheral anastomosing v.b. Vascular bundle in the raphe with a conspicuous branch to the micropyle. Hilum distant from the micropyle.

C. mubango Muell. Arg.

Seed 10 × 5–6 mm. Testa 5–6 cells thick. Tegmic palisade 315–360 μ high.

(Wiehr 1930.)

C. oblongum Burm. f.

(Sarawak, S 28482; fruit in alcohol.)

Seed 15 × 14 × 10 mm, with mottled testa and dark brown, very woody tegmen; *caruncle* 1.5 mm high, 3.5 mm wide, small. Testa 7–9 cells thick, apparently without v.b. except in the raphe; *o.e.* composed of short cuboid cells with dark brown contents, the outer wall distinctly thickened; *mesophyll* aerenchymatous, the cells with long arms; *i.e.* as a palisade of cylindric cells *c.* 55 μ high. Tegmen –40 cells thick (? exact thickness, the inner cells crushed); *o.e.* as a palisade 300 μ high, with curving cells, pitted and heavily lignified; *mesophyll* aerenchymatous, thin-walled, containing the network of v.b.; *i.e.* apparently unspecialized, crushed. Tegmic v.b. forming a reticulum in 2–3 layers from the heteropyle to the micropyle, the outermost layer *c.* 15 cells within the tegmic palisade, the innermost near the endotegmen.

Fruit 28 mm wide, globose, subdepressed, muricate with small, crowded, soft papillae 0.5–2.5 mm high, stellate-hairy, the papillae without v.b.; mesocarp with a double layer of interconnecting v.b.; cocci 1–1.5 mm thick, hard; trilocular with 2 empty loculi.

C. rivulosum Muell. Arg.

Seed 5 × 3 × 2.5 mm. Testa *c.* 5 cells thick. Tegmic palisade 100–165 μ high.

(Wiehr 1930.)

C. tiglium L.

(Ceylon; flowers and seeds in alcohol.)

Ovule; o.i. 4–5 cells thick, –8 in the raphe; i.i. 20–25 cells thick; nucellus penetrating the micropyle.

Seed 10 × 6 mm, with a vestige of the caruncle as a slight thickening of the exostome. Testa 5–6 cells thick; *o.e.* composed of cuboid, thin-walled cells with polygonal facets; *mesophyll* aerenchymatous; *i.e.* as a short palisade of thin-walled cells. Tegmen ? multiplicative, the inner part crushed; *o.e.* as a palisade 210–300 μ high, the cells with thick brown walls; *mesophyll* unspecialized but with v.b. in the raphe entering the heteropyle, without postchalazal branches; tegmic v.b. in the outer part as a longitudinal reticulum of 20–24 main v.b.

(Wiehr 1930.)

Croton sp.

(Ceylon, Alagala, summit; flowers and seeds in alcohol.) Fig. 228.

Ovule; o.e. 4–6 cells thick; i.i. 14–18 cells thick, with several procambial v.b.

Seed *c.* 5 × 3 mm (immature, but ? full-sized), with the caruncle surrounding the micropyle, widest on the hilar side, developing after fertilization. Testa 5–6 cells thick, unspecialized, aerenchymatous, the cells of o.e. with polygonal facets. Tegmen 30–35 cells thick; *o.e.* as a palisade of columnar cells. Vascular bundle of the raphe with a projection towards the micropyle; tegmic v.b. arising as *c.* 5 branches from the heteropyle, branching and anastomosing in the periphery of the tegmen to give *c.* 12 longitudinal v.b.

Dimorphocalyx Thw.

D. glabellus Thw.

(Ceylon, Peradeniya, Corner 197–; flowers and fruits in alcohol.) Fig. 229.

Ovule with obturator; o.i. 4–7 cells thick; i.i. 7–10 cells thick, less at the endostome, much shorter than o.i.; nucellus with a cylindrical projection into the exostome and with enlarged base as the incipient pachychalaza.

Seed 7 mm long, subglobose, plump, mottled brown, without caruncle. Testa not multiplicative, watery-pulpy then drying; *o.e.* as cuboid cells with isodiametric or shortly oblong facets; *i.e.* unspecialized. Tegmen multiplicative; *o.e.* as a palisade 100 μ high, the lignified cells with pitted walls and hexagonal facets; *i.e.* unspecialized, limited to the micropylar third of the seed. Vascular bundle in the raphe only but with two short processes on entry to the sides of the micropyle. Pachychalaza formed between the nucellar base and the heteropyle, with a cone of fine v.b. from the chalaza, forming a network midway between the exotegmen and the developing endosperm, the thin-walled tissue drying up at maturity.

The immature seeds have essentially the same construction as those of *Cleidion*. The pachychalaza is already indicated in the ovule.

Euphorbia L.

Ovule; o.i. 3–4 cells thick, –6 in *E. geniculata*, 12–14 at the micropyle in *E. dracunculoides*; i.i. 3–9 cells thick, mostly 5–7, 3–4 in *E. dulcis* and *E. thymifolia*, 4–5 in *E. hirta*, 7–8 in *E. peltata*; nucellus in some species projecting; obturator present.

Seed with small caruncle. Testa not multiplicative, thin-walled, collapsing; *o.e.* with tabular cells or shortly projecting (*E. peltata*); *i.e.* as a short palisade of cuboid cells with thickened radial and inner walls. Tegmen 4–10 cells thick, not or slightly multiplicative; *o.e.* with curving Malpighian cells; *i.e.* unspecialized or with striations on the radial walls (*E. dentata*). Vascular supply

only in the raphe, but with branches to the base of the nucellus in *E. geniculata*. Fig. 230.

E. dentata Michx., *E. dracunculoides* Lamk., *E. dulcis* L., *E. geniculata* Orteg, *E. helioscopia* L., *E. hirta* L., *E. hypericifolia* L., *E. lathyris* L., *E. microphylla* Heyne, *E. myrsinites* L., *E. peltata* Roxb., *E. pilosa* L., *E. thymifolia* Burm.

(Netolitzky 1926; Mandl 1926; Banerji and Dutt 1945; Landes 1946; Kajale 1954b; Singh, S. P. 1959; Kapil 1961; Mukherjee 1965; Singh, R. P. 1969.)

Gelonium Roxb.

G. glomerulatum Hassk.

(Java, Kebun Raya, cult.; living material.) Fig. 231.

Seed 8 × 7 × 6 mm, thinly whitish pulpy over the black stony tegmen, without caruncle. Testa *c.* 6 cells thick, 9–12 cells round the micropyle, thin-walled, unspecialized, drying up; *o.e.* composed of small tabular cells with subhexagonal facets; *i.e.* as a short thin-walled palisade. Tegmen 12–15 cells thick; *o.e.* as a rather thick palisade of dark brown curving cells; *mesophyll* composed of large thin-walled cells, drying up; *i.e.* composed of small unspecialized cuboid cells. Chalaza wide, the vascular bundle of the raphe entering without postchalazal branches and dividing in the thickened nucellar base into an umbrella of fine v.b. radiating to the broad insertion of the tegmen; no tegmic v.b. Nucellar beak at first projecting into the micropyle.

Capsule red, with fleshy mesocarp, splitting from base to apex into 3 lobes and exposing the greyish black seeds attached to the axial column.

This evidently has a tegmic pachychalaza, though it may be short. The fruit resembles that of *Aporosa* which has two seeds per loculus.

Givotia Griff.

G. rottleriformis Griff.

(Ceylon, Dambullah; immature fruits in alcohol.)

Seeds 15 × 12 mm, ellipsoid, one or two in the drupe, without caruncle. Testa 7–10 cells thick, thin-walled, pulpy, white; *o.e.* as cuboid cells with slightly thickened walls; *mesophyll* with 7 longitudinal v.b. from the chalaza in addition to the raphe v.b.; *i.e.* as a well-formed thin-walled palisade. Tegmen 30–35 cells thick; *o.e.* as a palisade 1–1.2 mm thick, with curving pitted cells, heavily lignified; *mesophyll* with *c.* 20 longitudinal anastomosing v.b. in the outer part, the inner part composed of large cells. Vascular bundle of the raphe with 7 postchalazal branches and with tegmic v.b.

The seed-structure is like that of *Aleurites*, though the two genera are widely separate in classification.

Hevea Aubl.

Seeds large. Testa 2–6 cells thick; *i.e.* not as a palisade of radially elongate cells. Tegmic palisade 400–900 μ high. Tegmen with v.b.

H. brasiliensis Muell. Arg.

Ovule; o.i. 6 cells thick, with o.e. reddish from antho-cyanin; i.i. 5–6 cells thick.

Testa 5–6 cells thick. Tegmic palisade 730–900 μ high.

(Wiehr 1930; Bouharmont 1962.)

H. guyanensis Aubl.

Testa 2 cells thick (? 8 cells, Wiehr 1930). Tegmic palisade 0.7–1 mm thick (200–600 μ, Wiehr 1930); ? without tegmic v.b.

(Mandl 1926; Wiehr 1930; Vaughan 1970.)

H. spruceana Muell. Arg.

Testa *c*. 6 cells thick. Tegmic palisade 730–900 μ high.

(Wiehr 1930.)

Homalanthus A. Juss.

H. populneus (Geisel.) Pax

(Ceylon, cult.; flowers and fruits in alcohol.) Figs. 232, 233.

Ovule; o.i. 6–7 cells thick; i.i. 4–5 cells thick.

Seeds 5–5.5 × 4 × 3 mm, without caruncle. Testa 6–7 cells thick, pulpy, the cells shortly elongate transversely; *o.e.* with isodiametric facets; *i.e.* not as a palisade. Tegmen 5–6 cells thick; *o.e.* as a palisade of finely pitted, curving Malpighian cells; *mesophyll* thin-walled, the cells transversely elongate, eventually crushed; *i.e.* forming a hyaline layer of cuboid cells with finely scalariform to subannular thickening, slightly lignified, persistent. Vascular bundle only in the raphe.

Except for the endotegmic layer, the seed resembles that of *Mercurialis*.

Jatropha L.

Seeds 7–21 mm long, carunculate. Testa 6–12 cells thick; *i.e.* as a short palisade with or without calcification. Tegmic palisade 115–400 μ high. Tegmic v.b. in a single ring placed a few cells internally from the palisade. Figs. 234–237.

J. curcas L.

Ovule; o.i. 5–8 cells thick; i.i. 18–22 cells thick, with 16–18 procambial strands more or less anastomosing from the chalaza in the periphery of the tegmen (4–6 cells internal from the o.e.); nucellus protruding through the micropyle.

Seeds 20–21 × 12 × 9 mm (16–17 × 8–10 × 7–9 mm, ? dried, Wiehr 1930), subtrigonous in t.s., black, with a small hard brownish white and minutely lobed caruncle round the micropyle. Testa 10–12 cells thick, –20 cells round the chalaza; *o.e.* as a layer of narrow columnar cells with slightly thickened and pitted radial walls, contents dark brown, not lignified, but with small patches of thin-walled unspecialized cells (drying into fine pits on the surface of the testa), without stomata;

mesophyll aerenchymatous, with latex-tubes, the cell-walls eventually lignified but scarcely thickened; *i.e.* as a short palisade of narrow, thin-walled, unlignified cells with calcified inclusions, becoming 2–3 cells deep round the chalaza. Tegmen 26–30 cells thick, eventually crushed except the palisade; *o.e.* as a palisade of straight or curving Malpighian cells with finely pitted brown walls, 150–180 μ high near the micropyle, 250–400 μ on the sides of the seed, 500–600 μ round the chalaza; *mesophyll* large-celled, aerenchymatous, the inner cells becoming radially elongate before being crushed; *i.e.* unspecialized. Vascular bundle of the raphe entering the heteropyle and forming 16–18 longitudinal tegmic v.b., anastomosing in the periphery of the tegmen, situated 8–10 cells within the palisade; no postchalazal v.b. in the testa.

Caruncle as a slight lobulate rim on the sides of the micropyle, not on the funicle or antiraphe-side of the exostome, composed of firm, unspecialized cells.

Pericarp with numerous v.b. in a network of 2(–3) circles.

(Mandl 1926; Wiehr 1930; Vaughan 1970; Singh, R. P. 1970a.)

J. glandulifera Roxb.

Seed 6 × 4 × 2–3 mm. Testa *c*. 8 cells thick; *i.e.* with little or no calcification. Tegmic palisade 130 μ high.

(Wiehr 1930.)

J. gossypiifolia L.

As *J. curcas* but without patches of thin-walled cells in the exotesta; nucellus in the ovule less protruding.

(Singh, R. P. 1970a.)

J. hastata Jacq.

Seeds 7–8 × 5 × 3 mm. Testa *c*. 6 cells thick; *i.e.* not or slightly calcified. Tegmic palisade 115 μ high.

(Wiehr 1930.)

J. integerrima Jacq.

(Mandl 1926; Kühn 1927, p. 341.)

Joannesia Vell.

J. gomesii Juss.

Testa; *o.e.* with tall palisade-like cells –250 μ high. Tegmic palisade 1–2.5 mm thick; *mesophyll* without v.b.

(Mandl 1926; Vaughan 1970.)

Mallotus Lour.

Seeds with incipient tegmic pachychalaza.

M. barbatus (Wall.) Muell. Arg.

(Singapore; living material.)   Figs. 238–240.

Ovule; o.i. 6–5 cells thick, with oblique exostome; i.i. *c*. 14 cells thick, massive; v.b. of the raphe with a branch on entry to each side of the micropyle, the postchalazal branches merely as procambial strands; chalaza with a small fountain of tracheids at the base of the relatively small nucellus; obturator with the cells full of oil-drops, as those of the conducting tissue in the style.

Testa 7–9 cells thick, thin-walled, the cells with yellowish oily contents, sarcostestal but drying up at maturity as a blackish pellicle; *i.e.* as a palisade of shortly elongate cells. Tegmen multiplicative, thin-walled except the exotegmic palisade with brown, sparsely pitted, lignified walls. Nucellus enlarging at the base into a small bulb of thin-walled cells surrounded by the tracheids of the chalaza, then crushed by the endosperm. Vascular bundle of the raphe with the two micropylar branches and 8–9 postchalazal branches curving over the sides of the seed, anastomosing sparsely and joining the micropylar branches.

(Corner 1949a, figs. 25–28.)

This is really a seed with small and, perhaps, incipient pachychalaza.

Mallotus sp.

(Ceylon, Alagala, summit; flowers and fruits in alcohol.) Figs. 241, 242.

Ovule; o.i. 4–6 cells thick, 5–7 cells on the sides, the exostome oblique; i.i. 6–8 cells thick.

Seed 5–6 × 4 mm, ovoid, blackish, without caruncle, the obturator dried but with a conspicuous stellate hair on each side. Testa 5–8 cells thick, the cells enlarging with little or no division; *o.e.* composed of cuboid cells with brown walls; *i.e.* unspecialized, closely appressed to the tegmen. Tegmen 12–14 cells thick, multiplicative by division of the mesophyll cells; *o.e.* as a palisade of Malpighian cells, curving, thick-walled but not pitted, the outer ends projecting irregularly to form the rough surface; *i.e.* unspecialized. Chalaza with a short basket of tracheids surrounding the slightly swollen base of the nucellus. Vascular supply as in *M. barbatus.*

Manihot Adans.

Seeds 12–20 mm long, carunculate or not. Testa; *i.e.* as a rather short palisade of thin-walled cells but with some cells more elongate, with calcareous deposits in the cells. Tegmic palisade 0.7–2(–3) mm thick. Tegmic v.b. present (? absent from *M. utilissima*).

M. dichotoma Ule, *M. glaziovii* Muell. Arg., *M. piauhyensis* Ule.

(Wiehr 1930.)

M. ? dulcis (J. F. Gmel.) Pax
(Ceylon; seeds in alcohol.)

Seed-structure evidently as in *Jatropha curcas* with tegmic v.b. (*c.* 8 in t.s. of the seed). Testa; *o.e.* mottled with anthocyanin cells in groups.

M. utilissima Pohl
(Singapore; living material.) Figs. 243, 244.

Seed variable in size, 8–13 mm long, oblong, date-brown, very hard, with a dry membranous silvery grey skin mottled fuscous black from blue anthocyanin cells, with caruncle. Testa 12–13 cells thick, multiplicative, thin-walled; *o.e.* composed of shortly columnar cells with ingrowing thickenings of the outer walls extending to the middle of the cells, rarely occluding the whole lumen, some of the cells with deep blue anthocyanin; *i.e.* as a palisade of thin-walled columnar cells, as a soft aqueous tissue extending into the micropylar canal. Tegmen thin-walled, collapsing except for the thick woody palisade with pitted brown walls, mostly with curving cells. Vascular supply as a single v.b. in the raphe without branches, without tegmic v.b. (Mandl 1926).

Caruncle rather massive, white, hard, 3-lobed, developed from the short preraphe and the micropyle, composed of radial rows of rather thick-walled, aqueous, colourless cells, shorter outwards to the epidermic.

(Mandl 1926; Wiehr 1930.)

This well-known species clearly needs full investigation of the seed to determine whether the tegmic v.b. are really absent.

Melanolepis Reichenb. f. et Zoll.

M. multiglandulosa Reichenb. f.
Fig. 245.

Ovule and seed as in *Mallotus barbatus.* Testa entirely pulpy and thin-walled. Tegmen becoming very thick and with a pitted surface through localized inhibition of the radial elongation of the exotegmen; *o.e.* as a black lignified palisade, the cells of varying length. Vascular supply with two micropylar branches from the raphe on entry and 5 postchalazal branches, two of these joining the micropylar branches.

Mercurialis L.

M. perennis L.
(Cambridge; living material.)

Ovule; o.i. 6–8 cells thick; i.i. 4–6 cells thick; nucellus projecting through the endostome.

Testa 9–10 cells thick, –14 round the micropyle, thin-walled; *i.e.* as a palisade of thin-walled columnar cells, extending into the pits of the tegmen. Tegmen 7–8 cells thick; *o.e.* as a palisade of curving Malpighian cells, shortened in many places to give the pitted exterior; *mesophyll* much enlarged, then crushed; *i.e.* unspecialized. Vascular supply only in the raphe. Caruncle vestigial.

(Mandl 1926; Bresinsky 1963.)

Micrococca Benth.

M. mercurialis Benth.
Fig. 246.

Ovule; o.i. and i.i. 5 cells thick; nucellus projecting.

Testa not multiplicative; *o.e.* and *o.h.* persistent, the rest crushed. Tegmen with the palisade of curving cells persistent. Vascular supply as the raphe v.b. with a short branch to the micropyle on entry.

(Nair and Abraham 1963.)

Pimeleodendron Hassk.

P. macrocarpum J.J.Sm.

(Malaya, Johore, Corner s.n. 1935; fruits in alcohol.)

Tegmen with exotegmic palisade and tegmic v.b. as in *Aleurites*.

Fruits with 14 loculi, each with a seed.

Ricinus L.

Seeds carunculate, with tegmic pachychalaza.

R. communis L.

(Ceylon; flowers and fruits in alcohol.) Figs. 247–249.

Ovule; o.i. 5–8 cells thick; i.i. 14–17 cells thick at the nucellar end (10 cells, Singh, R. P. 1954); nucellus short, conical; tegmic pachychalaza well-developed as the main mass of the ovule, already with a longitudinal cylinder of more or less anastomosing procambial strands, the tissue *c.* 15–20 cells thick external and internal to the procambial cylinder.

Seed very variable in size according to varieties, 6–18 × 5–18 × 3–8 mm (Wiehr 1930), oblong, blackish, surmounted by a short, white, firm caruncle. Testa 6–8 cells thick, as a thin pulpy covering to the seed, drying up at maturity; *o.e.* composed of cuboid cells; *mesophyll* with loose oblong cells; *i.e.* shortly columnar, the cell-walls not thickened or lignified, with small calcified deposits in the cells. Tegmen with the free portion restricted to the region of the radicle, the tegmic pachychalaza much enlarged with the intravascular part crushed by the endosperm, the extravascular part drying up except for the tegmic palisade, the vascular cylinder remaining as a pellicle round the endosperm; *o.e.* as a palisade 0.2–0.3 mm thick (140–250 μ, Wiehr 1930), dark brown to black, with closely pitted walls; *i.e.* with slight band-like thickenings on the tangential walls. Vascular supply as the raphe v.b. and the pachychalazal cylinder of *c.* 20 anastomosing bundles.

Caruncle 4 mm wide, 1.5 mm thick, composed of thin-walled cells as in the testal mesophyll, and covered by a shortly columnar epidermis, not lignified.

(Mandl 1926; Kühn 1927, who described the nucellus as united with the tegmen; Wiehr 1930; Singh, R. P. 1954; Vaughan 1970.)

Sapium P. Br.

S. indicum Willd.

(Malaya; fruits in alcohol.)

Testa 10–15 cells thick, all the cell-walls slightly thickened, widely pitted and lignified except the outer walls of o.e., more or less aerenchymatous as a corky coat to the seed. Tegmic palisade 180 μ high. Chalaza wide. Vascular supply apparently only in the raphe. Caruncle absent.

S. jamaicense Sw.

(Singapore; living material.)

Seed crimson-red from the epidermal cells of the sarcotesta, the walls red; *mesophyll* white, pulpy, the cells full of oily matter. Tegmen with uneven, warty, black, hard palisade.

S. sebiferum (L.) Roxb.

(Japan, Kyoto; ripe fruits in alcohol.)

Seeds 9 × 7 × 6 mm, without caruncle. Testa pulpy, 12–16 cells thick; *o.e.* composed of short cuboid and thin-walled cells with oily mucilaginous contents, the facets polygonal; *mesophyll* aerenchymatous, thin-walled, the outer 2–3 cell-layers with the same oily mucilaginous contents as o.e.; *i.e.* unspecialized. Tegmen with woody palisade 400–475 μ high; *mesophyll* crushed; *i.e.* as a layer of contiguous, rather long, isodiametric or shortly oblong cells with fine transverse lignified thickenings as close subanastomosing striae. Chalaza as a thick pad of tissue in the tegmen at the base of the endosperm. Vascular bundle of the raphe entering the heteropyle and forming a network of fine radiating bundles beneath the hypostase, apparently not entering the tegmen.

(Mandl 1926; Vaughan 1970.)

This seed needs to be checked in development for a tegmic pachychalaza.

Sebastiania Spreng.

S. chamaelea Muell. Arg.

Ovule with the obturator entering the micropyle; o.i. 3–4 cells thick, –6 at the exostome; i.i. 4–5 cells thick; hypostase present.

Seed *c.* 12 × 5 mm, with caruncle. Testa not multiplicative, thin-walled, soft. Tegmen 8–10 cells thick through multiplication of the mesophyll; *o.e.* as a lignified palisade; *i.e.* with small cells, becoming crushed. Vascular bundle of the raphe ending at the chalaza.

(Nair and Maitreyi 1962.)

Sphaerostylis Baill.

S. malaccensis (Hook.f.) Pax et Hoffm.

(Johore, Corner s.n. 1966; fruits in alcohol.)

Seed *c.* 6 mm long, globose, mottled, slightly hairy, without caruncle. Testa 6–7 cells thick; *o.e.* with brown cuboid cells, clusters of the cells here and there with a short hair-like papilla each; *mesophyll* aerenchymatous with substellate cells; *i.e.* composed of short, cuboid, unspecialized cells. Tegmen; *o.e.* as a brown, lignified palisade of pitted columnar cells; *mesophyll* many cells thick; *i.e.* unspecialized. Vascular supply with the raphe-bundle forming 5 postchalazal v.b. reaching almost to the micropyle, and with tegmic v.b. forming a network near the nucellus.

I could not decide whether this had tegmic v.b. or a tegmic pachychalaza because the seeds, though not yet mature, were already too hard to section.

Trewia L.

T. nudiflora L.
Ovule; o.i. and i.i. 3–9 cells thick; no nucellar beak.

T. polycarpa Benth. et Hook.
Testa with postchalazal v.b. and with short branches to the micropyle from the beginning of the raphe.
(Singh, B. 1964.)

PHYLLANTHOIDEAE

For characters of ovule and seed, see p. 130.

Concerning the structure of the exotegmen, I distinguish the following categories:

(1) Exotegmen as a short palisade of prismatic sclerotic cells: *Glochidion*, ? *Phyllanthus*.

(2) Exotegmen as a layer of more or less cuboid sclerotic cells:
(*a*) cells cuboid: *Drypetes*, ? *Putranjiva*, (*b*) cells stellate–undulate: *Actephila*.

(3) Exotegmen composed of oblong sclerotic cells, shortly tangentially elongate ($1\frac{1}{2}$–4 times as long as wide), not fibriform with pointed ends: *Breynia*, ? *Putranjiva*.

(4) Exotegmen as a layer of lignified longitudinal fibres, narrow, cylindric or ribbon-like; sometimes 1–2 hypodermal layers also fibrous:
(*a*) fibres rather short: *Cicca*, *Emblica*, *Securinega*, (*b*) fibres long; sarcotesta often well-developed: *Andrachne*, *Antidesma*, *Aporosa*, *Baccaurea* (ribbon-like fibres), *Bischofia* (ribbon-like fibres), *Bridelia*, *Galearia*, *Microdesmis*.

An evolutionary series could be read from the first to the fourth category or vice versa. Alternatively the first two may be distinguished phyletically, as simplifications of the Crotonoid exotegmen, from the last with fibrous exotegmen as in such families as Celastraceae, Flacourtiaceae and Violaceae where *Rinorea* is a match for *Baccaurea*. Then the third category becomes the problem. Dehiscent fruits occur in the first, third and fourth categories; the second and the fourth have genera with indehiscent fruits. There is no evidence in the fourth that indehiscence affects the differentiation of the exotegmen but in the second the differentiation may be imperfect in *Drypetes*, one unitegmic species of which has no exotegmic structure. The second category, therefore, may be a modification of the first and the third may be another diversification. Thus, on first inspection, Phyllanthoid seeds appear to belong in two phyletic groups, the Crotonoid and the Celastraceous–Violaceous. A Violaceous alliance has been suggested on grounds of wood-anatomy (Metcalfe and Chalk 1950). The situation is puzzling. One fact which has troubled me is the form of the leaf. The Crotonoid is built on the plan of the palmate leaf, as in the whole of the Bombacaceous–Tiliaceous alliance with its exotegmic palisade. There is no trace of this in the Phyllanthoid leaf, and many Phyllanthoid genera by contrast develop phyllomorphic ramuli of pinnate form.

Four other points need consideration. The pachychalaza of *Glochidion* may be connected with the tegmic pachychalaza of Crotonoideae, and this genus may well be critical. Then, in the fourth seed-category, the ovules of *Cicca*, *Emblica* and *Securinega pr. p.*, on developing into the seed, become hemi-anatropous; the very short raphe proceeds directly from the lateral hilum to the chalaza. This simplified form of seed may lead to the orthotropous ovule in Pandaceae (Forman 1966). The point introduces the third consideration. *Galearia* and *Microdesmis* are exceptional in Phyllanthoideae because of their solitary ovules. Forman (1966, 1971) has provided much evidence for their transfer to Pandaceae. If correct, then here is another family with fibrous exotegmen comparable in the case of *Panda* (if correctly interpreted, see p. 212) with Capparidaceae. I have retained *Galearia* and *Microdesmis* in Phyllanthoideae in order that they may be considered in connection with the other genera of this problematic subfamily, and because I am not satisfied that the seed-structure of *Panda* is sufficiently known. Lastly, if *Petalostigma* and *Toxicodendron* are arillate or carunculate, it is important to the whole problem of Euphorbiaceae that their seed-structure be made known.

Actephila Bl.

A. javanica Miq.
(Java, Hort, Bogor. XII B 15; Herb. Bogor., fruits in alcohol.)

Tegmen; *o.e.* as a woody layer of large, isodiametric or shortly longitudinally elongate, sclerotic cells with strongly stellate–undulate and pitted radial walls, with wide lumen, varying with the inner wall more or less undulate, the facets irregularly stellate–undulate.

Placed in Phyllanthinae by Airy Shaw (1972), yet the exotegmen is remarkably different from that of *Glochidion* and *Phyllanthus*, so far as known.

Andrachne L.

A. colchica Fisch. et Mey.
(Cambridge, University Botanic Garden; living material.)

Ovules 2 per loculus, with obturators; o.i. 2(–3) cells thick; i.i. 3–4 cells thick; nucellus large, projecting through the endostome.

Seed-coats not multiplicative; raphe elongate, the hilum subapical. Testa; *i.e.* not as a palisade. Tegmen; *o.e.* composed of rather short, longitudinal, lignified fibres, pitted; *i.e.* persistent, the small cuboid cells soon developing slightly thickened fuscous brown walls, not lignified; *mesophyll* enlarging, then crushed. Vascular bundle only in the raphe.

Antidesma L.

A. montanum Bl.

(Java, Hort. Bogor.; living material.)

Tegmen; *o.e.* as a single layer of longitudinal, lignified fibres.

A. menasu Miq.

(Singh, R. P. 1961.)

Aporosa Bl.

The unspectacular female flower of *Aporosa* is the embryonic forerunner, neotenic in pollination, of the elegant fruit. The swelling of the funicle may be the relic of the aril.

A. frutescens Bl.

(Singapore; living material.) Figs. 250–253.

Ovule; *o.i.* ? thickness; *i.i.* 3 cells thick; funicle short, thickened into the papillate obturator, the papillae as short cylindric projections of the epidermal cells; chalaza well-formed.

Seed *c.* 9 × 8 mm, slightly compressed, red, one per loculus, not detached. Testa 10–11 cells thick, pulpy as a sarcotesta, thin-walled, the cells filled with orange oil-globules; *o.e.* with thickened outer walls and rhombic facets; *i.e.* and *i.h.* not as palisades. Tegmen 4 cells thick; *o.e.* composed of a single layer of longitudinal, thick-walled, pitted, lignified fibres, shortened and radially orientated at the micropyle and chalaza. Chalaza broad, with an expanded plate of tracheids between the brown, thin-walled hypostase and the radially directed cells of the exotegmen. Vascular bundle only in the raphe, but with short minute branches to the sides of the seed. Embryo with thin green cotyledons; hypocotyl tetragonal in t.s., intensely purple towards the short radicle; endosperm cells near the radicle also with purple sap.

Fruit 15 mm wide, subglobose, pale green flushed purple over the apex, stigmata dark brown and sunken, splitting into 3 fleshy sappy parts falling off and leaving the three thin, pale yellowish white, bony, membranous cocci (endocarp not splitting into 6 parts); axis of the fruit persistent as a whitish fibrous core swollen at the apex by the soft dilated funicles, the funicle-cells swelling, holding the seeds horizontally and discarding the cocci; *mesocarp* thin-walled, large-celled, with some scattered and slightly thick-walled cells; *endocarp* (cocci) with oblique, thick-walled, pitted and lignified cells; *loculi* extended by basipetal growth of the ovary after fertilization and accommodating the enlarging seeds.

Aporosa sp.

(Penang; ripe fruits in alcohol.)

As *A. frutescens*; tegmic fibres 250–700 × 105–125 μ

(high), 15–26 μ (wide), in a single layer, laterally compressed and almost ribbon-like.

Baccaurea Lour.

Ovules 2 per loculus, with obturators.

Seeds large, 1 per loculus, with white, yellow, orange, red, pink, purple or blue sarcotesta, exarillate. Testa 10–20 cells thick, permeated with v.b.; *i.h.* with the cells radially elongate into large pulpy cells, often curved in various parts of the seed; *i.e.* unspecialized. Tegmen 4–7 cells thick; outer zone composed of 2–3 layers of longitudinal, thick-walled, pitted, lignified fibres, laterally compressed and appearing as a palisade in t.s., shortened into a single layer of radially elongate, sclerotic cells round the micropyle; *mesophyll* thin-walled; *i.e.* unspecialized. Chalaza broad, the tracheids spreading in a plate from the raphe v.b. and separated by a brown-celled hypostase from the endosperm, and by a thick mass of sclerotic tissue from the pulpy testa, the sclerotic tissue composed of more or less isodiametric cells but continued into the fibrous layers of the tegmen.

The tegmen and chalaza of *Baccaurea* resemble the same structures in Violaceae. The chalazal fibres do not become radial as in *Aporosa* but adjoin a mass of sclerotic cells in the mesophyll external to the spread of tracheids. In Violaceae, so far as known, this sclerotic tissue is slight and a sclerotic plate is interposed between the spread of tracheids and the endosperm where in *Baccaurea* and *Aporosa* there is the unlignified hypostase. The testa of Violaceae seems not to become so thick and pulpy as in these Phyllanthoideae. The many differences between this group and Violaceae suggest that the resemblances are the outcome of parallel evolution; it would seem impossible to distinguish between them by a transverse section of the tegmen.

B. motleyana Muell. Arg.

Figs. 254–257.

Ovule; *o.i.* 5–8 cells thick; *i.i.* 3–4 cells thick; micropyle formed by the exostome.

Seeds 13–15 × 10–11 × 5–6 mm, ellipsoid compressed, pulpy, yellowish white, .veined, sessile. Testa 10–16 cells thick, pulpy with longitudinal v.b.; *o.e.* with cuboid cells, unspecialized; *mesophyll* large-celled, aerenchymatous; *inner hypodermis* elongating into large, watery-pulpy cylindric, radiating cells; *i.e.* composed of short cells with brown contents, thin-walled, not lignified. Tegmen 4–5 cells thick, free to the chalazal base of the seed; *o.e.* developing into longitudinal fibres, often with a layer of hypodermal fibres, as a thin yellowish mechanical layer to the seed, thickened at the chalazal end, shortened into radial palisade-cells round the micropyle; *mesophyll* unspecialized, collapsing; *i.e.* as small unspecialized cells with reddish brown contents, soon collapsing. Vascular supply consisting of the raphe v.b. with 1–2 lateral v.b. near the chalaza and *c.* 7 postchalazal v.b. recurrent almost to the micropyle, unbranched or with 1–2 short branches, not anastomosing.

Baccaurea sp.

(Sarawak, S. 29458; mature fruits in alcohol.) Figs. 258, 259.

Seeds 10–12 × 8 mm, somewhat compressed. Testa 17–20 cells thick; *o.e.* as a compact layer of cuboid cells with polygonal facets, the outer walls slightly thickened; *mesophyll* aerenchymatous, the cells with short arms, then pulpy; *i.h.* as a layer of radially elongate, cylindric, thin-walled pulpy cells; *i.e.* as small cuboid brownish cells. Tegmen ? 5–7 cells thick; *outer fibre-layer* 160 μ thick, composed of 2(–3) layers of laterally compressed, thick-walled, strongly pitted, lignified fibres 170–450 μ long, 50–80 μ high, 15–25 μ wide; *mesophyll* 2–3 cells thick, with large thin-walled cells; *i.e.* composed of small brownish cells, not lignified. Chalaza broad, flat, the tissue between the hypostase and the testa thickened and wholly sclerotic except for the heteropyle; *v.b.* forming a small plexus between the sclerotic layer and the hypostase, the tracheids not entering the tegmen. Hypostase with a central pillar, composed of brown cells, not lignified. Vascular bundle of the raphe with 3–4 small branches from each side and two large postchalazal branches with 2 smaller pairs, recurrent in the testa and forming in the micropylar half a fine, rather open, reticulum.

Fruit 25 mm wide, subglobose; mesocarp 4–5 mm thick, firm, permeated by a plexus of v.b.; endocarp thin, lignified, consisting of a single layer of longitudinal fibres 40–50 μ wide, and immediately internally a sclerotic zone –150 μ thick, composed of 3–5 layers of rounded cells.

Bischofia Bl.

B. javanica Bl.

(China, Henry 1558A, and Borneo, SAN 44575; dried seeds ex Herb. Kew., det. H. K. Airy Shaw. Ceylon, leg. Corner s.n. 1972; fruits in alcohol.) Fig. 260.

Ovule; *o.i.* 2–3 cells thick; *i.i.* 3–5 cells thick; nucellar beak entering the base of the elongate endostome.

Seed 3.5 × 2 mm, blackish brown, shiny, oblong, pointed at the micropylar end, exarillate, hilum subapical. Testa 2 cells thick; *o.e.* composed of tabular cells with undulate facets and brown contents, not lignified; *i.e.* with hyaline, thin-walled rectangular cells transversely elongate, with a large crystal in each. Tegmen 4–5 cells thick; *o.e.* as a layer of longitudinal, ribbon-like, thick-walled, rather sparsely pitted, lignified fibres 100–600 μ long, 35–45 μ high, 11–14 μ wide, the tapered ends overlapping, shortened into rectangular or columnar cells at the micropyle and chalaza; *mesophyll* composed of thin-walled, flattened, transversely elongate cells in one or (?) two layers, with scattered sclerotic cells near the raphe in the chalazal half of the seed, the mesophyll

cells round the chalaza entirely sclerotic; *i.e.* as a layer of cuboid cells with lignified subscalariform thickenings on the radial walls. Vascular bundle only in the raphe, ending in the sclerotic tissue of the chalaza. Nucellus persistent as a single layer of longitudinally elongate cells. Endosperm thin-walled, oily. Embryo with short radicle and rather large flat cotyledons.

(Singh, R. P. 1962d.)

The seed has the general structure of that of *Aporosa* but the testa is thin, the exotesta is composed of flat cells with undulate outline to the facets, and the endotesta has crystals. The genus is peculiar in other ways, including the trifoliate leaf, and has been given its own family Bischofiaceae (Airy Shaw 1965, 1967). Alliance with Staphyleaceae has been suggested but, for seed-purposes, it is necessary to divide this family into two; tr. Tapiscieae has the fibrous exotegmen as in *Bischofia*, whereas the remainder of the family appears to be mesotestal with sclerotic cells as in Sapindaceae (p. 256). Wood-anatomy gives evidence for affinity with Flacourtiaceae (Metcalfe and Chalk 1950). The seed-structure clearly places *Bischofia* in the Celastraceous–Flacourtiaceous–Violaceous alliance, but whether it is a special family or one of the Phyllanthoideae is impossible to decide unless the crystal-cells of the endotesta are indicative, cf. Connaraceae, Meliaceae.

Breynia J. R. et G. Forst.

B. rhamnoides (Retz.) Muell. Arg.

(Malaya, Johore, leg. Corner 1972; ripe fruits in alcohol.)

Seed superficially as that of *Aporosa*. Testa pale orange, pulpy, with carotin oil-drops in the epidermal cells. Tegmen blackish brown, with 2–3 outer layers of shortly tangentially elongate, longitudinal, rectangular cells with dark brown, pitted, lignified walls, about 1½–4 times as long as wide, not fibriform.

As *Melanthesa thamnoides* Wt. (*Melanthesa* being a synonym of *Breynia*). R. P. Singh gives the following description which recalls that of *Glochidion*:

Ovules suborthotropous with broad oblique chalaza; *o.i.* 2–3 cells thick; *i.i.* 3 cells thick; nucellus protruding beyond the endostome.

Seeds 3.5–4.5 × 2–3 mm, ovoid with pointed micropylar end, the upper or micropylar two-thirds hard and dark brown, the lower chalazal third soft and dirty white. Testa multiplicative near the chalaza through division of *o.e.*, thin-walled, unspecialized. Tegmen; *o.e.* multiplicative into 2–5 layers of isodiametric, pitted, thick-walled, sclerotic cells, –7 cell-layers near the chalaza; *mesophyll* unspecialized; *i.e.* as a short endothelial palisade, thin-walled, with brown cell-contents.

(Singh, R. P. 1968.)

Cicca L.

C. acida L.

(Ceylon; mature fruits in alcohol.) Fig. 261.

Seed *c.* 4 × 2–2.5 × 1–1.2 mm, ellipsoid–fusiform, compressed, brown, 1(–2) per loculus, embedded in the thick woody endocarp, without caruncle; hilum median-lateral; raphe none.

Testa 2 cells thick; *o.e.* with somewhat tabular cells, facets hexagonal, thin-walled; *i.e.* crushed. Tegmen 4–5 cells thick; *o.e.* composed of more or less elongate, longitudinal, oblong or subfibriform cells with thick, finely and closely pitted, lignified walls; *mesophyll* thin-walled, crushed; *i.e.* as a layer of pale brown, thin-walled, subcuboid cells, contiguous. Chalaza conspicuous, lateral, adjacent to the hilum, with 8–10 layers of mesophyll cells (testa) thickened into oblong sclerotic cells round the end of the short v.b.

This genus is generally referred to *Phyllanthus* but it remains to be determined whether they agree in seed-structure.

Drypetes Vahl

Ovary bilocular. Ovules 2 per loculus, bitegmic or unitegmic (*D. macrostigma*), with obturators.

Seeds with thin seed-coats, albuminous, exarillate, commonly only one per loculus; exotegmen consisting of cuboid sclerotic cells.

Pericarp massive, with a sclerotic zone in the exocarp, the mesocarp with transverse radiating fibrovascular bundles, endocarp woody.

D. laevis Pax et Hoffm.
(Java, Hort. Bogor. III J 36a; Herb. Bogor., fruits in alcohol.) Fig. 262.

Seeds 22 × 13 × 6.5 mm, compressed ellipsoid, slightly curved. Testa 6–8 cells thick, not lignified; *o.e.* composed of shortly tangentially elongate, interlocking cells; *mesophyll* and *i.e.* unspecialized. Tegmen 6–8 cells thick at the chalazal end, 4–6 towards the micropyle; *o.e.* as a fairly compact layer of cuboid, pitted, sclerotic cells, not elongate; *mesophyll* thin-walled, collapsing, but more or less sclerotic in the woody endostome; *i.e.* as a firm layer of shortly tangentially elongate cells, not lignified. Vascular bundle of the raphe with 11–13 postchalazal branches ramifying over the sides of the seed almost to the micropyle, with few anastomoses; no tegmic v.b. Chalaza simple.

D. longifolia Pax et Hoffm.
(Java, Koorders 20200; Herb. Bogor., fruits in alcohol.)

Seed-structure intermediate between *D. laevis* and *D. macrostigma*, apparently bitegmic. Tegmen very thin, as a trace at the endostome with scattered sclerotic cells, over most of the seed crushed except for scattered sclerotic cells, without a fully differentiated exotegmen.

D. macrostigma J.J.Sm.
(Java, Hort. Bogor. VIII b.3; Herb. Bogor., flowers and fruits in alcohol.) Fig. 263.

Ovules unitegmic; integument 6–9 cells thick, with procambial strands; nucellar epidermis firm.

Seeds 17 × 11 × 6 mm. Testa apparently not multiplicative; *o.e.* as in *D. laevis*; *mesophyll* thin-walled or slightly thick-walled, sclerotic round the v.b.; *i.e.* unspecialized. Vascular supply as in *D. laevis*. Chalaza simple.

There is no sclerotic tissue in this seed to correspond with the exotegmen of *D. laevis*, for both ovule and seed appear to have lost the tegmen.

D. neglecta Pax et Hoffm.
(Java, Koorders 21205; Herb. Bogor., fruits in alcohol.)

As in *D. laevis*, but the fruits and seeds smaller, the exotegmic sclerotic cells smaller and in places thin-walled and unlignified.

This species has a degenerating exotegmen.

Emblica Gaertn.

E. officinalis Gaertn.
(Ceylon; fruits in alcohol.)

Seeds 2 per loculus but most loculi empty, shaped as in *Cicca* and without raphe, brown, angled, filling the loculus, exarillate. Testa 2 cells thick; *o.e.* composed of large brown thin-walled cells; *i.e.* crushed. Tegmen eventually with only *o.e.* and (?) *o.h.*, the rest of the tissue crushed; *o.e.* composed of rather large, laterally compressed, rectangular cells, less elongate than in *Cicca*, thick-walled, pitted, lignified.

(Mandl 1926.)

Galearia Zoll. et Mor.

Ovules solitary in the loculi, anatropous or (?) orthotropous in *G. maingayi*, suspended. Tegmen with 1–5 layers of longitudinal fibres. (Forman 1966, 1971.)

G. celebica Koord.
Tegmic fibres in 4–5 layers.
(Vaughan and Rest 1969.)

G. filiformis (Bl.) Boerl.
(Java, Hort. Bogor. VIII B 54; Herb. Bogor., fruits in alcohol.) Fig. 264.

Testa very thin; *o.e.* composed of tangentially elongate, longitudinal narrow cells with slightly thickened walls. Tegmen; *o.e.* composed of longitudinal fibriform cells, many converted into thick-walled lignified fibres, but the fibres in groups separated more or less by thin-walled elongate cells, both with side-arms; *mesophyll* thin-walled, ? 3–4 cells thick; *i.e.* as a compact layer of cuboid cells, not lignified.

G. maingayi Hook. f.
Ovules ? orthotropous. Tegmen with fibres in 3 layers.
(Vaughan and Rest 1969.)

Galearia sp.
Tegmen with the fibres in 1–2 layers.
   (Vaughan and Rest 1969.)

Glochidion J. R. et G. Forst.

Unfortunately I have failed to investigate thoroughly the seeds of the many common species of this genus. In cursory examination I noted a tangentially oblong facet for the exotegmic cells, as if they agreed with those of *Breynia*. In view, however, of the curved palisade-cells of the following species, *G. zeylanicum*, I think these oblong facets may have resulted from oblique sections.

G. zeylanicum A. Juss.
Fig. 265.
   Seed *c.* 6 × 4.5 mm, plano-convex. Testa 12–15 cells thick, 4–5 at the exostome, pulpy with aerenchymatous mesophyll; *o.e.* composed of cuboid cells with thick lignified brown walls, closely and finely pitted; *i.e.* unspecialized. Tegmen *c.* 6 cells thick; *o.e.* as a short palisade of sclerotic cells with lignified, finely pitted walls and hexagonal facets, the cells somewhat curving on the sides of the seed; *o.h.* with small scattered sclerotic cells, especially near the junction of the tegmen and the pachychalaza; *i.e.* as a short thin-walled palisade, then crushed, not lignified. Pachychalaza as a convex swelling occupying most of the adaxial side of the seed, pervaded by fine v.b. from the raphe, lining the nucellar base with a thin lignified tissue of small brown sclerotic cells, continued as a flange from the base of the tegmen and occluding most of the small lateral hilum. Vascular supply as a short raphe v.b. giving a short branch to the testa on one side of the seed, the main v.b. dividing up into branches radiating into the pachychalaza. Endosperm thin-walled, oily, relatively slight. Embryo curved round the pachychalaza.

Microdesmis Hook. f.

Ovules anatropous, solitary in the loculus, suspended.
   Seed apparently with the testa reduced to *o.e.* composed of cuboid cells with slightly thickened walls. Tegmen with a single layer of longitudinal, pitted, lignified, cylindric fibres, in a continuous layer (*M. casearifolia*) or interrupted with thin-walled cells (*M. magallanensis*, *M. yafungana*); *i.e.* as a layer of slightly thick-walled cuboid cells.
   (Forman 1966; Vaughan and Rest 1969.)
   This genus has been removed with *Galearia* to Pandaceae by Forman.

Phyllanthus L.

Ovules; o.i. 2–4 cells thick; i.i. 2–4 cells thick; nucellar beak extending through the micropyle; obturators absent.
   Seeds without caruncle. Testa not multiplicative, the cells enlarging, then more or less crushed, unspecialized. Tegmen not multiplicative; *o.e.* sclerotic (? cell-shape); *mesophyll* crushed, unspecialized; *i.e.* as thin-walled tannin-cells.
   (Banerji and Dutt 1945; Singh, R. P. 1962b; Bancilhon 1971.)

Putranjiva Wall.

P. roxburghii Wall.
Ovules; o.i. 4–8 cells thick, with procambial strands as branches from the raphe; i.i. 7–10 cells thick; nucellus reduced to a column of cells at the base of the embryo-sac.
   Seeds 9–10 mm long, fusiform, without caruncle. Testa thin-walled, unspecialized. Tegmen; *o.e.* as a layer of cuboid or shortly longitudinally oblong, thick-walled, pitted, sclerotic cells; *mesophyll* crushed, but with some scattered sclerotic cells. Vascular bundle with several postchalazal branches.
   (Singh, R. P. 1962c, 1970b.)

Securinega Commers.

S. leucopyrus (Willd.) Muell. Arg.
(Ceylon; fruits in alcohol.)   Fig. 266.
   Seed 2–2.3 mm long, subcompressed, brown, without raphe or caruncle. Testa 2 cells thick; *o.e.* composed of large pulpy cells with firm brown walls; *i.e.* composed of rectangular, more or less transversely elongate cells. Tegmen 3–4 cells thick; *outer fibrous tissue* composed of 2–3 layers of thick-walled, finely pitted, short, lignified fibres, rather obliquely orientated; *i.e.* composed of small, unspecialized cells with dark brown contents. Funicle composed of lignified fibres as in the tegmen.
   Berry 6–7 mm wide, round, white, pulpy; endocarp very thin, as a single layer of transversely elongate fibres; seeds 2–3.
   There is no stone in the berry, or drupe, to protect the seeds as there is in *Cicca* and *Emblica*, and the seed has the thicker fibrous tegmen.

S. suffruticosa (Pall.) Rehder
(Kenya; flowers and fruits in alcohol.)   Fig. 267.
   Ovule with short raphe; o.i. 2–3 cells thick; i.i. 3 cells thick; nucellus projecting through the endostome.
   Seeds 2–2.2 × 1.7–1.8 mm, with subapical hilum and distinct raphe, without caruncle; seed-coats not multiplicative. Testa pulpy, the inner layers more or less crushed; *i.e.* not as a palisade. Tegmen; *o.e.* composed of shortly longitudinally elongate, finely pitted, lignified fibres; *mesophyll* crushed; *i.e.* soon developed into a short palisade of brown cells with homogeneous brown contents, the walls neither pitted nor lignified.
   The form of this seed is transitional between that typical of Euphorbiaceae and the raphe-less state in *Cicca* and *Emblica*.

EUPOMATIACEAE

Eupomatia R.Br.

E. laurina R.Br.
(Queensland, C. E. Hubbard 2504; New South Wales, V. Hadley 2741; Herb. Kew. dried seeds.) Fig. 268.

Ovules 3–4 per carpel, basal, anatropous, erect, bitegmic, crassinucellate.

Seeds 5–6 × 2–3 mm, pale brownish ochre, uneven, subangular, rather coarsely but shallowly pitted, with ruminate endosperm; embryo minute.

Testa 7–11 cells thick; *o.e.* composed of flattened tabular cells with thick, unlignified, outer wall and brown contents; *mesophyll* composed of longitudinal fibres with sparsely pitted lignified walls, the fibres entering into the main ruminations but with 2–3 thin-walled hypodermal cells marking the pits over the ruminations; *i.e.* composed of thin-walled cuboid cells, more or less crushed, not lignified. Tegmen; *o.e.* composed of large cuboid cells with thin, slightly lignified walls, most with a large yellowish oil-drop; *inner tissue* (? 1–2 cell-layers) crushed. Endosperm oily, thin-walled. Ruminations as finger-like intrusions of the tegmen, the larger with fibres of the testa.

Fine details of structure could not be made out in the dried seeds, but the fibrous nature of the testa proves the Annonaceous affinity of the genus. The fibres seem to be longitudinal except where they dip into the ruminations.

With long anthers, *Eupomatia* suggests an early derivation from Annonaceous ancestry. The perigynous construction of the flower is in parallel with that of Monimiaceae, Calycanthaceae and advanced Lauraceae. Wood-structure refers *Eupomatia* to Magnoliaceae (van der Wyk and Canright 1956) where the seed would not agree.

EUPTELEACEAE

Ovules anatropous, bitegmic, 1–4 per carpel, crassinucellate; o.i. 2–5 cells thick (3–5 in *Euptelea polyandra*); i.i. 2 cells thick.

Seeds small, black, crustaceous; seed-coats not multiplicative. Testa; *o.e.* consisting of large thin-walled cells with red-brown contents giving the reticulate surface of the seed; *mesophyll*, if present, more or less sclerotic; *i.e.* sclerotic. Tegmen composed of small thin-walled cells. Endsperm cellular, oily. Embryo minute.

(Bailey and Nast 1946; Smith, A. C. 1946; Metcalfe and Chalk 1950; Endress 1970.)

The very small seeds are merely inflated ovules, yet they seem to be endotestal as in Trochodendraceae. In *Euptelea* it is the fruit, not the seed, which becomes alate. However, Bailey and Nast, Smith, Metcalfe and Chalk find little in common with *Trochodendron* and maintain the separate family. Endress considers that *Euptelea* is nearer to Schisandraceae and Cercidiphyllaceae.

FAGACEAE

Ovules anatropous, suspended, 2 per loculus, bitegmic, unitegmic in *Nothofagus*, crassinucellate; o.i. and i.i. several cells thick, but i.i. small and thinner than o.i. in *Fagus*; integument in *Nothofagus* 4–5 cells thick; nucellar tracheids in *Fagus* and *Quercus*.

Seeds medium-size to large, exalbuminous, enclosed in the nut. Testa ? not multiplicative, thin-walled, more or less persistent or crushed; *o.e.* as a layer of firm cells or somewhat enlarged. Tegmen soon crushed, unspecialized. Vascular bundles branching in the testa, the branches arising from the raphe and chalaza. Endosperm nuclear, completely absorbed or reduced to a single layer of cells (*Fagus*). Embryo large, straight; cotyledons thick or plicate.

(Netolitzky 1926; Hjelmquist 1948, 1957; Poole 1952; Kummerow y Labariz 1961; Stairs 1964; Vaughan 1970.)

The nuts of this family are 1-seeded, indehiscent capsules derived from inferior ovaries, and all but a single fertile loculus are obliterated in the course of development. The seed conforms with the cavity of the nut and has lost practically all specialization of the cells except for the copious supply of vascular bundles in the manner of overgrown or pachychalazal seeds. The tegmen has no part in the seed. The inner integument is much reduced in *Fagus* and it is evidently that which has disappeared in *Nothofagus*. In this genus, the ovules at anthesis are merely slight bulges which do not develop fully until fertilization occurs 9–10 weeks later (Poole 1952). In *Quercus* subgen. *Erythrobalanus* there is even longer delay for fertilization occurs in the spring after the year of pollination and the nut matures in the second autumn (Stairs 1964). The family, though it has been placed low in the scale of dicotyledonous evolution, is highly advanced and replete with relict features in flower, fruit, seed and leaf. Thus, biologically, there is a problem in its tremendous association with fungi, especially basidiomycetes.

FLACOURTIACEAE

The seeds of the few species that have been studied in this family fall into two groups between which there is so little resemblance that the natural unity of the family becomes suspect. The *Flacourtia*-group, or Flacourtiaceae proper, has a fibrous exotegmen as in Celastraceae. The *Hydnocarpus*-group has a very thick pachychalazal seed-coat without trace of the fibrous exotegmen, even round the micropyle where the two vestigial integuments persist; and here, where the fibrous exotegmen occurs in similar examples of pachychalazal seeds in Meliaceae, there is a thick sclerotic endotesta as in Magnoliaceae.

The genera put in Flacourtiaceae have many relics of fruit-ancestry. There are spinous epicarps, thick mesocarps with cortical and radiating fibrovascular bundles, dehiscent and indehiscent fruits, and arillate and alate seeds. It is one of the more interesting assemblages of tropical dicotyledons but obscured by intricate floral detail which renders the family almost impossible to recognize in the forest. Flowering and fruiting are usually seasonal. Thus it is not easy to obtain a full range of material for seed-investigation, but collectors should be urged to preserve in alcohol or suitable fixative what they may chance to discover.

The recent discovery of the substance mangostin in the bark of *Hydnocarpus* in contrast with mangiferin in *Flacourtia* may mark the unnatural assemblage of these genera in one family (Gunasekera and Sultanbawa 1973).

Flacourtia group

Ovules anatropous and suspended, or orthotropous, bitegmic, crassinucellate; o.i. 2–3 cells thick; i.i. 3–5 cells thick.

Seeds rather small, shaped as the ovules, arillate or not, albuminous, in some cases with projecting exostome or endostome. Testa multiplicative or not, 2–9 cells thick; *o.e.* unspecialized or with stomata; *mesophyll* thin-walled or with sclerotic cells at the micropyle and chalazal ends (*Casearia, Flacourtia*); *i.e.* unspecialized or as palisade-cells (*Oncoba*). Tegmen slightly multiplicative, 4–9 cells thick; *o.e.* as a layer of longitudinal, pitted, lignified fibres, cylindric or laterally compressed, shortened at the endostome and chalaza and often with 2–4 layers of short sclerotic cells; *mesophyll* unspecialized; *i.e.* often sclerotic at the endostome. Chalaza with thick unlignified hypostase, also with sclerotic cells (*Oncoba*). Vascular supply only as the raphe v.b. Endosperm nuclear, thick- or thin-walled, oily. Embryo with thin flat cotyledons.

Aril funicular, more or less fimbriate, with many slender v.b. (*Casearia*).

Casearia, Oncoba, with orthotropous seeds; *Flacourtia* with anatropous seeds; ? *Idesia* with subanatropous ovule. Figs. 269–276.

Until more genera have been studied, it is difficult to decide upon distinctions between Flacourtiaceous seeds and Celastraceous, Violaceous or those of other families with the fibrous exotegmen (p. 14).

Hydnocarpus group

Ovules anatropous, in *Scaphocalyx* orthotropous, bitegmic, crassinucellate; o.i. 3–6 cells thick; i.i. 3–5 cells thick.

Seed massive, often lump-like, with thick pachychalazal seed-coat 1–3 mm, 40–100 or more cell-layers thick, the cells arranged in four tissue-layers; albuminous, exarillate. Testa (round the micropyle) with the two outer layers of the pachychalazal seed-coat. Tegmen (round the micropyle) undifferentiated. Vascular bundles

branching in the soft outer layer of the seed-coat, but not in *Scaphocalyx*. Endosperm nuclear, thin-walled, oily. Embryo with thin, flat or slightly curved, cotyledons.

Pachychalazal seed-coat with the following four tissue-layers (as in *Taraktogenos*):

(1) the outer soft or fleshy layer, aerenchymatous, with scattered sclerotic cells and groups of thin-walled cells with brown tannin-contents, this layer carrying the v.b.,

(2) a sclerotic layer with (*a*) outer isodiametric cells, (*b*) more or less radially elongate cells in the middle part, (*c*) more or less tangentially elongate or gyrosé cells in the inner part,

(3) thin-walled tissue with small cells separating layers (2) and (4),

(4) thin-walled cells with brown contents and without air-spaces, as the hypostasial lining.

Hydnocarpus, Scaphocalyx, Taraktogenos, ? Kiggelaria, ? Pangium.

The relation between these four layers and the free parts of the vestigial testa and tegmen can be seen at their junction near the micropyle (Figs. 279, 280). The sclerotic layers (2*a*) and (2*b*) correspond with the inner mesophyll and i.e. of the testa, while the layer (2*c*) corresponds with the outer layers of the tegmen; the layer (4) corresponds with the endotegmen. These conclusions may be taken to suppose that testa and tegmen have fused to form the pachychalaza, but the developing seed shows that there is no fusion (Fig. 281); the chalazal region expands and differentiates tissues which correspond with those of testa and tegmen, presumably through a diversion of gene-activity into this new growth. In so far as the sclerotic layer (2) of the pachychalazal seed-coat is more or less fibrous, without internal fibrillar lignification of the cells, the construction is Annonaceous, not Magnoliaceous.

Casearia Jacq.

C. clarkei King
(Malaya, Sing. F. N. 33593; fruits in alcohol.)
Seed-structure as in *C. rugulosa*.

C. rugulosa Bl.
(Bornea. Kinabalu, RSNB 81; fruits in alcohol.) Figs. 269–271.

Seeds 6 × 4 mm, many per fruit, sessile, brown, smooth, covered with a red aril, orthotropous. Testa 7–9 cells thick, –12 cells near the chalaza, thin-walled, aerenchymatous, not lignified, without v.b.; *o.e.* composed of small angular cells with slightly thickened brown walls; exostome projecting, with sclerotic cells in i.e. and 2–3 adjacent layers of the mesophyll. Tegmen 7–9 cells thick, –10 near the chalaza; *o.e.* consisting of very long, ribbon-like, longitudinal fibres with thick, closely pitted and lignified walls, shortened into sclerotic cells at the endostome and as 2–3 layers of sclerotic cells at the chalaza; *mesophyll* composed of thin-walled cells, shortly elongate longitudinally; *i.e.* composed of rectangular cells with firm, not lignified but slightly pitted walls, sclerotic at the endostome. Endosperm copious, thick-walled. Embryo rather small, with long radicle.

Aril saccate over most of the seed, dividing into irregular laciniae distally, arising from the very short funicle all round the base of the seed, free from the testa, composed of thin-walled tissue, with many slender longitudinal v.b. in places anastomosing.

Pericarp not lignified, pervaded by mucilage-canals and a complex system of v.b. divergent from endocarp to exocarp.

C. tomentosa Roxb.
Ovules orthotropous; o.i. and i.i. 2 cells thick, but thicker at the micropyle; hypostase present.
(Narayanaswami and Sawhney 1959.)

C. tuberculata Bl.
(Java, Tjibodas, Corner s.n. 1972; living fruits.)
Seed white; aril red, fimbriate. Testa entirely thinly pulpy. Tegmen; *o.e.* as strongly lignified longitudinal fibres, ribbon-like; *endostome* projecting beyond the exostome.

Flacourtia L'Hérit.

F. indica (Burm.f.) Merr.
(Ceylon; ripe fruits in alcohol.) Fig. 272.
Seeds *c.* 4.5 × 3 mm, 1(–2) in a pyrene, flattened, pale brown, smooth, with thin firm pellicular seed-coat, exarillate. Testa 3 cells thick on the sides of the seed; *o.e.* composed of more or less tabular cells, not lignified, with scattered stomata; *mesophyll* and *i.e.* with thin-walled cells, crushed on the sides of the seed, transformed into sclerotic pitted cells at the micropylar end (–12 cells thick) and at the chalazal end (6–8 cells thick). Tegmen evidently 4 cells thick; *o.e.* as a compact layer of longitudinal, subcylindric, lignified, pitted fibres; *mesophyll* thin-walled, crushed; *i.e.* as a persistent coherent layer of cuboid cells with lignified thickenings at the angles of the radial walls. Vascular supply only in the raphe, expanding below the rather wide discoid hypostase of small brownish cells (not lignified). Endosperm thin-walled.

Hydnocarpus Gaertn.
Evidently with the same construction of the seed as in *Taraktogenos*.
(Netolitzky 1926; Vaughan 1970.)

Idesia Maxim.

I. polycarpa Maxim.
Ovules subanatropous; o.i. 3 cells thick; i.i. 4 cells thick.
Seed practically orthotropous. Testa with enlarged cells, not multiplicative, eventually crushed. Tegmen;

o.e. developing into a palisade-layer; *mesophyll* collapsed; *i.e.* composed of small cuboid cells filled with yellow brown material, very hard, but with radially elongate cells at the thick endostome. Chalaza simple.
(Mauritzon 1936b.)

If the tegmen is a palisade of radially elongate prismatic cells as in *Bixa*, according to Mauritzon, then Idesia cannot belong in Flacourtiaceae, but no tangential sections or surface-views of this palisade were given and it may be a layer of fibres.

Kiggelaria L.

K. africana L.
Ovule anatropous; o.i. and i.i. 3 cells thick; micropyle formed by the exostome.
Seed ? arillate or sarcotestal. Testa *c.* 30 cells thick, the cell-layers increased especially by cell-division of i.e. Tegmen *c.* 10 cells thick. Vascular bundle at the chalaza dividing into many branches.
(Mauritzon 1936b.)

This description suggests a seed as in *Taraktogenos*, and the aril that has been ascribed to it may be the thick pachychalazal seed-coat.

Oncoba Forsk.

O. spinosa Forsk.
(Ceylon, Peradeniya, cult.; fruits in alcohol.) Figs. 273–276.
Seeds *c.* 6 mm long, ellipsoid-compressed, blackish brown, smooth, hard, orthotropous with small projecting endostome, the base somewhat oblique, exarillate. Testa 2 cells thick, soft; *o.e.* composed of short cuboid cells or, in places, shortly radially elongate, unspecialized, the brownish walls not lignified; *i.e.* as a pellicle of columnar cells, the brownish walls for the most part unlignified but groups of cells in places with lignified middle lamella at the cell-angles, especially in their inner parts next the tegmen. Tegmen 5–7 cells thick; *o.e.* as lignified, longitudinal, ribbon-like fibres 180–500 μ long, pitted, shorter and palisade-like at the projecting endostome, also near the chalaza and thickened into a sclerotic annulus round the chalaza; *mesophyll* 3–5 cells thick, with transversely elongate cells, thin-walled, then crushed; *i.e.* composed of small cuboid cells with brown contents, the thin slightly lignified walls with slight bar-like, spiral or subannular thickenings, at the endostome developed into heavily lignified, sclerotic cells meeting those of o.e. Vascular bundle very short, slightly oblique, without branches to the sides of the seed. Hypostase flat, discoid, with firm, not thickened, dark brown walls as a plug within the sclerotic annulus of the tegmen, supported by irregular groups of sclerotic cells on the outside. Endosperm thin-walled. Embryo with the flat, thin cotyledons longer than the radicle.

Pericarp with heavily lignified exocarp and orange mealy or floury pulp embedding the seeds.

(Vaughan 1970.)

The account given by Vaughan for *O. echinata* describes the seed-coat with a thick endotesta of 6–8 layers of fibrous stone-cells and there is no mention of an endotestal palisade of pulpy cells.

Pangium Reinw.

P. edule Reinw.

(Java, Bogor; living mature fruits.)

Seeds 4.5–6 × 3–4 cm (overall), more or less angular from compression, lump-like. Testa very thick, in three layers: (1) an outer pulpy yellowish layer 2–5 mm thick (the so-called aril but actually a sarcotesta); (2) a hard woody middle layer 0.7–1.7 mm thick, composed of isodiametric sclerotic cells, but the inner cells more or less fibriform and interwoven; (3) an inner pithy layer of collapsed cells. Embryo with short radicle and thin flat cotyledons.

Whether this seed was pachychalazal as in *Taraktogenos*, I was unable to determine because the woody part of the seed-coat was too massive and hard to section. The pulpy sarcotesta passes gradually into the woody layer through a region of small cells.

Scaphocalyx Ridley

Ovules sessile, orthotropous, transverse in two whorls; o.i. and i.i. 5 or more cells thick, both with 2–4 triangular lobes at the micropyle; nucellus beaked.

Seed 10–15 × 6–8 mm, reniform, pachychalazal with minute vestigial integuments round the micropyle. Pachychalazal testa evidently as in *Taraktogenos* but without branches from the chalazal v.b.

(van Heel 1973.)

Taraktogenos Hassk.

Seeds massive, angular, with flattened sides and rounded ends, pachychalazal with thick seed-coat; micropyle near the hilum, scarcely visible. Seed-coat as described (p. 144). Vascular bundles in the seed-coat numerous, not anastomosing, derived from the chalazal end of the raphe v.b., a few from the raphe itself. Endosperm thin-walled. Embryo rather large; hypocotyl clavate; cotyledons thin, flat, cordate, acute.

T. heterophylla (Bl.) v. Sl.

(Ceylon, Peradeniya, cult.; fruits in alcohol.) Figs. 277, 278.

Seeds 2.5–2.8 × 1.7–2 cm. Seed-coat; *outer layer* 1.4–2 mm thick, comprising 25–45 cell-layers, thin-walled, aerenchymatous, carrying the v.b., with rather abrupt transition to the middle sclerotic layer; *sclerotic layer* 0.6 mm thick, comprising 12–16 cell-layers, the outer cells isodiametric, the middle cells radially and obliquely elongate, the inner cells sinuous and tangential; *colourless layer* of thin-walled cells 3–5 cells thick; *inner hypostasial layer* 5–9 cells thick, without air-spaces,

with firm thin brown walls. Vascular bundles c. 12, ascending from the chalazal network almost to the micropyle, some branched, not anastomosing; raphe v.b. with 1–2 small v.b. from each side to the seed-coat.

T. kurzii King

(Ceylon, Peradeniya, cult.; fruits in alcohol.) Figs. 279, 280.

Seeds essentially as in *T. heterophylla*. Seed-coat; *outer layer* –2 mm thick, comprising 100 or more cell-layers, brown flecked in the outer part and at the surface with clusters of thin-walled cells with brown contents, with a few sclerotic cells scattered singly or in small groups throughout the tissue but mainly near the v.b. (not in the epidermis); *woody zone* not yet lignified except in parts of the inner layer of tangential cells, the radially arranged cells formed by late anticlinal and periclinal division of the mesophyll cells adjacent to the layer of tangential cells; *brown hypostasial layer* 6–9 cells thick, compact. Vascular bundles evidently as in *T. heterophylla*. Embryo as in *T. heterophylla*.

Pericarp complicated, consisting of four layers: (1) a brown periderm cracking and causing the finely verrucose or scurfy surface, with tufts of ovary-hairs on the particles; (2) the outer cortex traversed radially by many simple or branching, sometimes anastomosing, fibres of varying size, composed of radially arranged, shortly elongate, sclerotic cells without v.b., also with a few small masses of sclerotic cells in the ground-tissue; (3) a zone of dense, more or less contiguous, large masses of sclerotic cells, heavily lignified, giving rise to the fibres on the outside, with scattered v.b. on the inner side, some v.b. traversing the zone radially; (4) a wide, thin-walled endocarp with c. 4 rings of v.b., abundant small masses of sclerotic cells (few near the inner epidermis), and small masses of thin-walled cells with brown tannin-contents (giving brown flecks as in the seed), the endocarp not generally lignified or fibrous.

(Vaughan 1970.)

At the micropyle of the immature seeds the origin of the woody layers could be discerned. The inner layer of tangential-gyrose cells is derived from the part equivalent to the mesophyll of the inner integument, where this zone is lignifying round the endostome. The endostome is surrounded by brown tissue confluent with that of the inner layer of the pachychalaza. The layer of radially elongate cells of the pachychalazal coat corresponds with the endotesta. The main mass of the testal tissue, confluent with the vascular outer layer of the pachychalaza, swells up and obliterates the course of the micropyle which is not visible at the surface.

Taraktogenos sp.

(Ceylon, Peradeniya, cult.; flowers and immature fruits in alcohol.) Figs. 281, 282.

Ovary dark brown hairy, with incipient cortical v.b.;

style short, with 4 cuneate and flattened stigmatic lobes, thinly hairy externally; placentas (3–)4, entirely parietal. Ovules numerous, anatropous, variously directed, mostly transverse; o.i. 5–6 cells thick; i.i. 3–4 cells thick, extending about half-way along the ovule; nucellus fairly massive, 4–5 cells thick on the sides, 6–8 cells at the apex; micropyle formed by the endostome.

Seed (immature); pachychalazal seed-coat 60–70 cells thick, aerenchymatous, some cells with sphaerocrystals, many cells at the chalazal end and micropylar end becoming sclerotic; *inner hypostasial layer* 4–5 cells thick, continuous with the thicker original hypostase at the chalazal end of the seed; *endostome* becoming overgrown by the exostome. Vascular bundles ? as a network in the outer layer of the pachychalaza.

Fruit ovoid subglobose, dark brown hairy, developing a thick sclerotic wall beneath the thin exocarp, the endocarp becoming lignified; mesocarp with a network of v.b.

The immature seeds showed clearly the vestigial integuments around the micropyle.

FOUQUIERACEAE

Ovules anatropous, bitegmic, tenuinucellate; o.i. 3 cells thick; i.i. 4 cells thick; micropyle formed by the long endostome.

Seeds small, flattened, with a marginal wing formed by exotestal hairs, albuminous, exarillate. Testa multiplicative, –15 cells thick; structure ? Tegmen –6 cells thick; structure ?; *i.e.* as an endothelium. Endosperm cellular, 4–6 cells thick, oily, without starch. Embryo straight.

(Netolitzky 1926; Mauritzon 1936b.)

The seeds with multiplicative integuments could serve as an ancestral state to the reduced seeds of Frankeniaceae and Tamaricaceae, with which the family is usually associated, though with oily endosperm. The unspecialized wood fits a pachycaul relic.

Dried seeds of *Fouquiera* that I examined showed little structure other than the embryo, endosperm, crushed seed-coat, and exotestal hairs with coarse, double spiral thickening on the walls.

FRANKENIACEAE

Ovules anatropous, bitegmic, thinly crassinucellate; o.i. and i.i. 2(–3) cells thick, not multiplicative in the seed; with a vestigial third integument on the antiraphe-side; funicle rather long.

Seeds small, albuminous, exarillate. Testa; *o.e.* thin-walled, developing a thick-walled papilla from each cell; *i.e.* with the cells tangentially elongate, thin-walled. Tegmen; *o.e.* crushed; *i.e.* composed of cuboid cells with slightly thickened, cuticulate, inner wall. Endosperm nuclear, starchy. Embryo straight.

(Netolitzky 1926; Walia and Kapil 1965.)

This seed with starchy endosperm and uniformly papillate exotesta seems related to that of Tamaricaceae. Walia and Kapil suggest affinity with Elatinaceae which, however, is exotegmic with Hypericaceae.

FUMARIACEAE

Ovules campylotropous, bitegmic, crassinucellate; o.i. 2–3 cells thick; i.i. 3 cells thick.

Seeds small, albuminous, arillate or not; seed-coats not multiplicative. Testa; *o.e.* as a layer of cuboid or shortly radially elongate cells with thickened outer wall; *i.e.* with fibrillar internal and more or less lignified thickening as in Papaveraceae, or the cells unspecialized and crushed. Tegmen; *o.e.* and *mesophyll* unspecialized, thin-walled, crushed, or with o.e. as longitudinally elongate cells; *i.e.* persistent, with striate or reticulate thickening on the radial and inner walls (? not lignified), or thin-walled. Vascular bundle of the raphe short, ending at the chalaza. Endosperm nuclear, oily. Embryo minute.

Aril (elaiosome) as a small, white or yellowish, outgrowth of the raphe or funicle (*Corydalis, Dicentra*), with thin-walled oily cells radially elongate into a crest, simple or lobulate (*Dicentra formosa*), attached to the raphe by small short cells.

(Meunier 1891; Netolitzky 1926; Saksena 1954; Röder 1958; Singh, D. and Negi 1962; Berg 1969.)

Adlumia Raf., *Corydalis* Vent., *Cysticapnos* Miller, *Dicentra* Bernh. *Fumaria* L.

This family, treated by some as a part of Papaveraceae, has a simplified seed-coat in which the characteristic endotestal palisade rarely occurs and the exotegmen is not fibrous. This deterioration occurs in various Papaveraceae and the simplified seeds of the two families are structurally alike. The campylotropous form is but another modification of the anatropous.

The small arillate seeds are distributed by ants. In the case of *Dicentra* with arillate and exarillate seeds, Berg considers that the arillate have evolved from the exarillate but, as he is unacquainted with durianology, such a conclusion is a guess.

GARRYACEAE

Ovules anatropous, pendulous, unitegmic, crassinucellate, 2 per loculus; integument 12–15, 15–25, or 28–30 cells thick, very massive; hypostase developed after fertilization; funicle rather long.

Seeds rather small, sarcotestal. Seed-coat with multiplicative o.e. cutting off a hypodermal layer and this layer further multiplicative or not, thin-walled, mostly becoming crushed or with 4–5 outer layers of cells persistent; *o.e.* composed of large, thin-walled, radiating cells, more or less columnar but often oblique, with wine-purple contents, drying up at maturity. Vascular supply in the raphe, terminating at the chalaza. Endosperm nuclear, starchy, horny. Embryo minute.

(Netolitzky 1926; Hallock 1930; Mosely and Beeks 1955; Kapil, R. N. and Mohana Rao 1966a.)

Hallock explained the sarcotestal character of the seed as 'an arilloid development of the outer cell-layer of the integument', and this simple integument he considered on account of its thickness to represent two connate integuments. Probably the seed is pachychalazal.

GENTIANACEAE

Ovules anatropous or more or less atropous (*Cotylanthera*, *Halenia*, *Voyria*), unitegmic, tenuinucellate; integument 2–6 cells thick; i.e. as an endothelium (Menyantheae).

Seeds small, albuminous, exarillate, mostly exotestal. Seed-coat not multiplicative; *o.e.* persistent with thickened radial and inner walls, or with spiral thickening (*Voyria*), or with thickened and pitted outer wall (*Menyanthes*); *mesophyll* and *i.e.* unspecialized, crushed, or with several rows of sclerotic cells (*Menyanthes*). Vascular bundle slight or none, but extending to the micropyle in Menyantheae. Hypostase present in *Halenia* and *Menyanthes*. Endosperm nuclear or cellular (Menyantheae), copious, oily. Embryo small.

(Netolitzky 1926; Maheswari Devi 1962; Sharma, D. R. and Upadhyay 1962; Singh, B. 1964; Vijayaraghavan and Padmanaban 1969.)

The separation of Menyanthaceae, for which the genera *Limnanthemum*, *Menyanthes* and *Villarsia* have been investigated, is favoured on embryological grounds by Maheswari Devi. Possibly in support there are the points of the endothelium, the exotesta, the vascular supply and the cellular endosperm.

GERANIACEAE

Ovules anatropous or campylotropous, erect, with short basal raphe and micropyle turned to the ovary-axis, bitegmic, crassinucellate; o.i. and i.i. 2–3 cells thick, not multiplicative (? *Biebersteinia*); exostome forming the micropyle.

Seeds small, anatropous (*Biebersteinia*) or campylotropous, exarillate, more or less exalbuminous. Testa not lignified; *o.e.* with isodiametric or shortly longitudinally elongate facets, variously enlarged (*Geranium*), with scattered stomata (*Geranium* spp.); *i.e.* generally as a layer of small crystal-cells, the outer wall often strongly thickened, or unspecialized (*Biebersteinia*, ? *Monsonia*). Tegmen; *o.e.* as a compact layer of more or less thick-walled cells with undulate–lobulate facets, not or scarcely lignified, not or scarcely radially elongate, but in *Biebersteinia* with radially elongate arms to the strongly lignified cells; *i.e.* unspecialized or with slight thickenings on the radial walls, not lignified or slightly

(*Biebersteinia*). Vascular bundle of the raphe ending at the chalaza or with a postchalazal extension (*Biebersteinia*). Hypostase brownish, more or less suberized. Nucellar cuticle persistent. Endosperm nuclear, much reduced or absent, oily.   Figs. 283–286.

(Netolitzky 1926.)

The seeds of this family have a uniform structure distinguished by the undulate outline of the mechanical cells of the exotegmen. In *Erodium*, *Monsonia* (Fig. 286) and *Pelargonium* these cells are scarcely radially elongate; in various species of *Geranium* they are shortly elongate (Figs. 284, 285); in *Biebersteinia* they have radially elongate arms which give an irregular palisade-effect in transverse section. The seed of this genus is unusually large, anatropous, with slight endosperm and an embryo that is only slightly curved. Moreover this seed has postchalazal vascular bundles and, perhaps, multiplicative integuments. It appears as the least reduced seed in the family and suggests that the Geraniaceous exotegmen has been derived from a more elaborate palisade, that of *Monsonia* representing the greatest simplification. However, the seed of the capsular *Viviania* needs investigation.

Concerning the affinity of the family with Hypericaceae, see p. 35.

Biebersteinia Steph.

B. multifida DC
(Laman 1852; herbarium material, CGE.)   Fig. 283.

Seeds 6 × 4 × 2.5 mm, subtriquetrous, subcompressed, shortly elongate, with rounded raphe, brown, dried rugulose, anatropous. Testa 5–7 cells thick, thin-walled, more or less collapsed; *o.e.* with slightly thickened outer walls, facets isodiametric; *mesophyll* with small sphaerocrystals in some cells; *i.e.* small-celled, without crystals. Tegmen; *o.e.* composed of thick-walled, strongly lignified, scarcely pitted, lobate–undulate cells with outward, radially directed arms with short unlignified, prismatic extremities, appearing as an irregular palisade; *mesophyll* ?, collapsed; *i.e.* as a pellicle of closely adherent tabular cells with slightly thickened and lignified radial and inner walls, the radial walls with short thickened bands, some cells apparently with lignified contents. Vascular bundle of the raphe with several postchalazal branches dividing to give, with the raphe v.b., *c.* 14 v.b. in t.s. of the seed.

Monsonia L.

M. angustifolia E. Mey.
(Transvaal, Burtt-Davy 17499; herbarium material, CGE.)   Fig. 286.

Testa 3 cells thick; *o.e.* with cuboid cells, larger at the obtuse end of the seed; *mesophyll* with transversely elongate cells; *i.e.* as small crystal-cells. Tegmen with the exotegmen of stellate–lobate cells 16–18 μ high, scarcely lignified.

The seeds of *Pelargonium* and *Erodium* have the same structure.

M. senegalensis Guill. et Perr.

Ovule amphitropous, erect; o.i. and i.i. 3 cells thick.

Testa 4–5 cells thick by multiplication of the mesophyll; *o.e.* with large, thin-walled cells, somewhat tangentially elongate, with tannin; *mesophyll* and *i.e.* becoming thick-walled with rounded cells. Tegmen not multiplicative, thin-walled; *i.e.* with tannin.

(Narayana and Arora 1963.)

GESNERIACEAE

Ovules anatropous, unitegmic, tenuinucellate; integument 3–4 cells thick; hypostase none.

Seeds small or minute, with hair-tufts at the ends (*Aeschynanthus*), albuminous or not, exarillate, the integument not multiplicative; *o.e.* more or less thick-walled, various; *mesophyll* crushed; *i.e.* unspecialized, persistent or not; haustorial growth of the embryo in some cases entering the seed-coat between the epidermal layers. Endosperm nuclear, but cellular in *Platystemma*. Embryo small, straight.

(Netolitzky 1926; Adatia, Sharma and Vijayaraghavan 1971.)

GLAUCIDIACEAE

Ovules anatropous, suspended, bitegmic, crassinucellate; o.i. 6–13 cells thick; i.i. 5 cells thick.

Seeds alate, 12–14 × 9–10 mm (including the wing 2–3.5 mm wide), flattened, albuminous, exarillate. Embryo large.

(Tamura 1972.)

Glaucidium Sieb. et Zucc., placed usually in Ranunculaceae, has been put into its own family by Tamura on account of the centrifugal stamens, the thick ovular integuments, the 2(–3) carpels connate at the base and ripening into follicles dehiscent along both sutures, the diploid chromosome number 20, and the presence of coumarin and other substances. He considered that Glaucidiaceae should be placed in Hypericales, but mentioned that Siebold and Zuccarini had suggested affinity with *Paeonia*. This is supported by the thick integuments as well as the centrifugal stamens. The seed-structure is not known.

GONYSTYLACEAE

Ovules anatropous, bitegmic, crassinucellate.

Seeds large, arillate, exalbuminous, with tegmic vasculature, with conical prominent exostome. Testa *c.* 6 cells thick, unspecialized, thin-walled, not lignified, without v.b. Tegmen multiplicative, –30 cells thick; *o.e.* as a layer of thick-walled sclerotic cells, often shortly longitudinally elongate, not fibriform. Vascular supply as many discrete longitudinal v.b. in the raphe, forming a plexus in the tegmen; heteropyle rather wide. Embryo with short radicle and thick, not folded, cotyledons.

Aril developed from the raphe and chalaza, partly investing the seed, thin-walled, with longitudinal branching v.b.

Gonystylus is the remnant of a line of arillate ancestry that seems to bear on the vexed relation between families with an exotegmic palisade and those with exotegmic fibres. Its customary affinity with Thymelaeaceae is supported by the seed-structure, but it has the more primitive marks of the multilocular ovary, the loculicidal capsule with massively fibrous pericarp, the numerous stamens, and the large arillate seeds with tegmic vascular bundles. Spines appear to have been lost from the capsules unless there are traces in the verrucose or rugulose fruits of some species. The exarillate, mostly unilocular, and 1-seeded fruits of Thymelaeaceae are derivative. Between the two families there is the tribe, or family, Aquilarieae, with bilocular capsules and what appears to be a curiously modified aril at the chalazal end where there is the chalazal aril in *Gonystylus* (Fig. 287); the clue to this feature may be found in the allies of *Aquilaria*, namely *Lethedon* and *Solmsia*.

Nevertheless the seed of *Gonystylus* introduces some peculiarities. There is the multiple vascular supply of the raphe, which becomes itself arilloid at maturity. Then the exotegmen has shortly tangentially elongate cells, almost exactly intermediate in form between the columnar cells of the palisade and the fibrous cells. There is the tegmic plexus of vascular bundles. The aril comes from the raphe and chalaza and is also vascular. These characters of the tegmen, aril and massive capsule relate *Gonystylus* with Elaeocarpaceae (p. 125), and thus with other families with more typically fibrous exotegmen and similar, if non-vascular, aril, such as Connaraceae, Flacourtiaceae, Meliaceae, and Violaceae. The scattered vascular bundles of the raphe suggest an expansion of the stout raphe-bundle of other large seeds, e.g. *Connarus*, *Durio*, *Myristica* or *Sterculia*, in which the single bundle is an aggregation of many small concentric bundles; the slender raphe-bundle of the small Thymelaeaceous seed is the simplification which sets the limit to the development of the ovule.

Gonystylus, in its modern tree-form, appears to be all that is left of the pachycaul ancestry that divided into Elaeocarpaceae and Thymelaeaceae.

Gonystylus T. et B.

G. forbesii Gilg

(Sumatra, Palembang, Boschpr. n. 15G P17; Herb. Bogor., fruits in alcohol.) Fig. 214.

Seeds 28–35 × 13–15 × 11–12 mm, oblong–ellipsoid, 1–2 per fruit, the raphe dilated into flanges round the long beak-like exostome, expanded along its whole length

6

and forming a cap at the chalaza. Testa *c.* 6 cells thick, thin-walled, unspecialized. Tegmen much thickened; *o.e.* as a layer of short, thick-walled, finely perforate, woody cells, varying shortly longitudinally elongate and oblique, in places subprismatic and radial; *mesophyll* thin-walled, carrying the plexus of v.b. Chalaza with wide heteropyle traversed by the tegmic v.b., the exotegmen thickened round the entry. Raphe with many v.b., narrowly attached along its course to the seed, becoming arilloid.

Aril derived from the raphe and chalaza, covering barely one-half of the seed.

Gonystylus sp.

(Sarawak, S. 29461; fruits in alcohol.)   Figs. 287, 288.

Seeds 25–28 mm long, 1 per loculus, massive, dark brown, covered on one side by the massive, rather fibrous aril, the exostome as a beak –5 mm long exsert between the basal lobes of the aril. Testa 4–6 cells thick, –10 cells at the chalaza, –15 cells at the micropyle, aerenchymatous, unlignified, without v.b., collapsing into a greyish pellicle; *o.e.* composed of thin-walled cuboid cells; *i.e.* unspecialized. Tegmen 26–32 cells thick; *o.e.* as a continuous dark brown layer of subcuboid cells, varying shortly longitudinally elongate, very thick-walled, lignified, with rather sparse fine pitting, absent from the micropylar region; *mesophyll* pervaded by a very close network of v.b. from the chalaza nearly to the micropyle, aerenchymatous, the outer cells, external to the v.b., with brownish walls, the inner cells colourless and crushed; *i.e.* unspecialized. Raphe arilloid, joined to the seed by a narrow ridge, with many longitudinal v.b. Chalaza relatively thin, with many v.b. passing through the heteropyle into the tegmen.

Aril derived from the raphe, pervaded by many longitudinal branching v.b., composed of thin-walled cells; *o.e.* with small cells.

Capsule 45 mm wide, subglobose, rough, not verrucose, 3-locular. Pericarp with an inner ring of v.b. and divergent, subdividing branches to the cortex almost reaching the uneven (? excoriating) surface, with scattered sclerotic cells and nests of sclerotic cells very abundant in the inner part of the pericarp; v.b. with a strong fibrous sheath; endocarp thin, pellicular.

GOODENIACEAE

Ovules anatropous or campylotropous, unitegmic, tenuinucellate, erect or suspended (*Catosperma, Pentaptilon*), 1–2-many per loculus; integument 6–20 cells thick, or more along the raphe.

Seeds small to medium-size, ovoid or more or less compressed, in some cases becoming campylotropous (*Dampiera*), stropiolate with slight funicular swelling in

Goodenia spp. Seed-coat reduced more or less to o.e., the inner tissue crushed or with some lignified outer tissue in the mesophyll (*Scaevola*); *o.e.* of four kinds: (1) with a palisade of short or elongate lignified cells with thick, pitted and laminate walls, with sinuous facets, in the more or less compressed seeds usually with thin-walled mucilage-cells along the raphe-antiraphe, in some cases developed into a wing (*Calogyne, Catosperma, Goodenia, Neogoodenia, Pentaptilon, Scaevola pr. p., Selliera, Velleia, Verreauxia*); (2) composed of thin-walled cells with lignified reticulate or band-like thickenings (*Dampiera, Diaspasis, Leschenaultia, Scaevola pr. p.*); (3) composed of cells with thickened radial walls (*Anthotium*); (4) composed of thin-walled and more or less compressed cells (*Brunonia*). Vascular bundle short, reaching or slightly exceeding the chalaza (*Anthotium, Brunonia, Dampiera, Leschenaultia,* double in *Dampiera*), or as a single bundle extending in the antiraphe more or less to the micropyle (*Calogyne, Catosperma, Diaspasis, Goodenia, Neogoodenia, Pentaptilon, Scaevola, Selliera, Velleia, Verreauxia*). Endosperm nuclear, oily, copious or reduced to a single layer of cells. Embryo straight or slightly curved.

(Netolitzky 1926; Brough 1927; Carolin 1966.)

Reference must be made for numerous generic and subgeneric details to the account of the seed and fruit of Goodeniaceae by Carolin. He divides the genera into two groups according to the length of the vascular bundle of the seed, and the character fits also with floral classification. *Goodenia*, with capsular fruit, has the fullest development of the seed-coat, which deteriorates in several genera with indehiscent fruits, though still displayed by *Scaevola*. Three species of *Goodenia* have a funicular strophiole as if a relic from the arillate seed (*G. barbata, G. chisholmii, G. strophiolata*). The exact form of the exotestal cells needs evaluation from surface-view. There is a resemblance between the compressed seed of *Goodenia* with mucilaginous periphery along the raphe-antiraphe and that of Cucurbitaceae.

The problematic genus *Emblingia* has been referred to Goodeniaceae but the bitegmic crassinucellate ovules, the spirally coiled embryo and, perhaps, the aril do not agree (Erdtman, Leins, Melville and Metcalfe 1969). If the seed-coats are differentiated, which seems likely in an arillate seed, they should provide the evidence for its classification.

Scaevola L.

Seeds more or less compressed; *o.e.* with thin-walled cells strengthened by thickening bands, or as thick-walled, lignified and compressed cells (*S. sericea*), both kinds of cell with polygonal facets, or as thick-walled cells with undulate facets (*S. fasciculata, S. helmsii, S. stenophylla*), the periphery of the seed without mucilage-cells. Vascular bundle reaching almost to the micropyle.

(Carolin 1966.)

S. sericea Vahl

(Ceylon; flowers and fruits in alcohol.)

Ovules 1 per loculus, anatropous, erect, basal, large, $1.7 \times 1.2 \times 0.5$ mm, obconic, slightly flattened; integument *c.* 20 cells thick on the sides, 30–40 cells thick in the raphe-antiraphe; i.e. not endothelial; vascular bundle as a procambial strand round the periphery of the ovule from hilum to micropyle.

Seeds 4–4.5×2.8–3×0.8–1 mm, plano–convex, pale brownish, enclosed in the very woody endocarp; *o.e.* as a compact layer of small tabular cells with polygonal-subrectangular facets, the wall rather strongly thickened, lignified, finely and irregularly pitted, not as a palisade; *mesophyll* composed of fairly large, thin-walled cells, the outer 5–6 layers with rather thin but lignified walls; *i.e.* crushed. Vascular bundle unbranched.

The exotestal cells and the lignified outer cells of the mesophyll seem different from what occurs in the species described by Carolin.

GROSSULARIACEAE

Ovules anatropous, bitegmic, crassinucellate; o.i. 3–5 cells thick, the cells of o.e. often well-formed with slightly thickened walls; i.i. 2 cells thick; micropyle formed by the exostome; hypostase of small brown cells, thin-walled.

Seeds rather small, sarcotestal with more or less gelatinous pulp, the integuments not multiplicative, arillate to exarillate, albuminous. Testa; *o.e.* converted into a thick palisade of radiating cells filled with mucilage, thin-walled except the somewhat thickened outer wall, not lignified; *mesophyll* composed of small hyaline cells, not lignified; *i.e.* as a compact layer of small cuboid crystal-cells with thickened and lignified radial and inner walls, 1(–3) crystals in a cell. Tegmen; *o.e.* soon crushed, but with a trace at the chalaza; *i.e.* as a layer of enlarged thin-walled cells with firm brown tanniniferous contents, not lignified. Vascular bundle only in the raphe. Chalaza unspecialized. Endosperm cellular, oily, with slightly thickened walls. Embryo small.

Aril funicular, investing the seed, somewhat lobed, or unilateral or reduced to a crenulate fleshy placental ridge, pulpy, 3–6 cells thick, not vascular. Figs. 289, 290.

(Pohl 1922; Netolitzky 1926; Yen 1936.)

Arillate seeds were described for several species of *Ribes* by Pohl; the fact is little known and was overlooked in my list of families with arils (Corner 1953). Arils cover the seeds in the slightly fleshy pericarps of *R. aureum*, *R. alpinum*, and *R. nigrum*; the aril covers half of the seed in *R. grossularia* with very fleshy pericarp; it is practically absent in *R. caucasicum*, *R. gracile*, *R. rubrum*, and *R. sanguineum*, though in these cases the prominent placenta becomes fleshy and slightly crenulate

at the base of the seed. The firm layer of the seed is the lignified, yet narrow, endotesta to which the larger-celled endotegmen is closely adnate.

The family is usually classified with Saxifragales, but placed in Cunoniales by Hutchinson. The seed-structure is similar to that of Crassulaceae and Saxifragaceae but the aril, the sarcotesta and the firm endotesta suggest a more primitive state, for which reason I have placed the family among the endotestal seeds with the arillate Dilleniaceae and Polygalaceae. Even the gooseberry with its finely vascular bristles has the evidence of derivation from a spinous capsule of arillate seeds.

HALORRHAGACEAE

Ovules anatropous, suspended with dorsal raphe, bitegmic and crassinucellate (*Gunnera*, *Halorrhagis*, *Lauremburgia*, *Myriophyllum*) or unitegmic and tenuinucellate (*Hippuris*); o.i. 2–3 cells thick; i.i. 2 cells thick; pachychalazal with very short integuments in *Myriophyllum*.

Seeds small, albuminous, exarillate, the seed-coats not multiplicative, reduced to the tabular thin-walled cells of the exotesta, the remainder of the testa and the tegmen crushed. Endosperm cellular or nuclear (*Lauremburgia*), starchy. Embryo straight.

(Netolitzky 1926; Johri 1963; Nagaraj and Nijalingappa 1967; Kapil and Bawa 1968; Bawa 1969a, b.)

The seed-coats are simplified within the small nut or drupe and convey little indication of affinity. The genera, sometimes split into Gunneraceae, Halorrhagaceae and Hippuridaceae, are relics of pachycaul marsh ancestry, similar to the Polygonaceae but with much reduced seed, fruit and flower. There is no evidence of Lythralean or Myrtalean affinity in the seed. The pachychalazal ovule of *Myriophyllum* suggests the derivation of the unitegmic ovule of *Hippuris*.

HAMAMELIDACEAE

Ovules anatropous, suspended, bitegmic, crassinucellate; o.i. evidently massive, 8–12 cells thick (*Distylium*, *Hamamelis*), but merely 2 cells thick in *Altingia*; i.i. 3 cells thick, 5 cells in *Altingia*.

Seeds mostly medium-size, oblong, smooth, brown or black, in some genera winged, albuminous, exarillate. Testa multiplicative, massive, hard, sclerotic, 13–30 cells thick, but not multiplicative in *Altingia*; *o.e.* unspecialized or more or less sclerotic, in some cases with undulate facets; *mesophyll* composed of isodiametric or shortly elongate sclerotic cells, with crystals in some cases; *i.e.* unspecialized, crushed (? with crystal-cells in *Liquidambar*). Tegmen not multiplicative, thin-walled, crushed; *i.e.* in *Altingia* with small oblong cells with thin lignified walls. Vascular bundle of the raphe ending at the chalaza. Perisperm persistent in some genera as

6-2

2-4 cell-layers (*Distylium, Hamamelis*). Endosperm nuclear or cellular (*Parrotiopsis*), oily. Embryo straight. (Netolitzky 1926; Kaul 1969.)

What little is known of these seeds indicates a massive sclerotic mesotesta as in Theaceae and some Rosaceae. The seed of *Liquidambar*, reported by Netolitzky as having a fibrous tegmen, needs re-investigation and comparison with other winged seeds such as those of *Rhodoleia* and *Exbucklandia*. The seed of *Altingia*, which I studied in Java, is much simplified with non-multiplicative integuments.

HERNANDIACEAE

Ovule solitary, pendent from the apex of the ovary, apotropous then (before or after fertilization) anatropous with dorsal raphe, bitegmic, crassinucellate.

Seed solitary, enclosed in the woody endocarp, rather large, exalbuminous, pachychalazal in *Hernandia*, exarillate. Testa thick, undifferentiated or possibly with tracheidal endotesta in *Gyrocarpus*. Tegmen soon crushed, unspecialized. Chalaza much enlarged in *Hernandia*. Nucellus becoming ruminate in *Hernandia*. Endosperm ? Embryo large; *cotyledons* plano-convex, folded or lobate and more or less connate in the ruminations (*Hernandia*). (Netolitzky 1926; Sastri 1963; Kubitzki 1969; v. Heel 1971a, b.)

As yet only the ovule and seed of *Hernandia* have been studied in any detail. The ovule is Lauraceous in shape and position, and so is the exalbuminous seed in the woody endocarp. Tracheidal cells (? endotestal) have been reported in *Gyrocarpus* (Sastri 1963). These points confirm the Lauraceous affinity which is generally recognised, but there is a Myristicaceous resemblance. The seed of *Hernandia*, enclosed in the pericarp with woody endocarpic palisade and surrounded by the cupule, resembles an arillate Myristicaceous seed. The dark exocarp resembles the testa of *Myristica* with stomata; the endocarp resembles the endotesta. The ruminations of the testa correspond with the tegmic ruminations of *Myristica* though they are not vascular and they arise as extrusions from the nucellus. It seems as if the testal characters of *Myristica* have been transferred to the pericarp in the indehiscent fruit of Hernandiaceae. The outward shift of the loci of genetic materialization recalls that in *Artocarpus*. Thus Hernandiaceae may stem from the offshoot of Myristicaceous ancestry which has led to Lauraceae.

Hernandia L.

H. peltata Meissn.
(Java; Herb. Bogor., young and mature fruits in alcohol.) Fig. 291.
Ovule becoming anatropous after fertilization; o.i. thick; i.i. 4–5 cells thick, the endostome at first with *c*. 5 lobules (v. Heel 1971a, b).

Seeds 14–16 mm wide, subglobose, with thick unspecialized testa and massive lobate–ruminate cotyledons, pachychalazal, closely invested with the woody endocarp. Testa becoming *c*. 50 cells thick, then largely crushed by ruminate outgrowths of the nucellus from the expanded chalaza; *o.e.* slightly differentiated, thin-walled, continuous; *mesophyll* aerenchymatous, with oil-cells. Tegmen only in the micropylar part of the young seed, soon crushed. Vascular bundle of the raphe dividing into 3 branches at the chalaza, not ramifying in the testa.

Pericarp; o.e. as a short palisade of subcuboid, compact cells with strongly thickened outer wall and dark brown contents, with abundant stomata; mesocarp aerenchymatous, but the outer few hypodermal cell-layers more or less sclerotic; endocarp as a thick woody palisade, 0.5 mm thick over most of the seed but –0.7 mm thick round the thin apical disc 0.3 mm thick (as the side of the funicle-base), composed of a single layer of narrow prismatic cells with thick, scarcely pitted, lignified walls, the apical disc with distinct linea-lucida. (v. Heel 1971a, b.)

The chalazal region of the ovule develops enormously to form the body of the seed. The wall of this region becomes ruminate from profuse lobings of the enlarging nucellus and the cotyledons eventually bulge into these lobings to substitute the nucellus; the endosperm is slight and ill-formed. The true ovule wall remains more or less undeveloped and is crushed at the micropylar end of the pachychalazal seed.

HIPPOCASTANACEAE

Ovules anacampylotropous, bitegmic, crassinucellate, 2 per loculus, suspended or the lower erect, becoming strongly funiculate (at least after fertilization); o.i. 8–10 cells thick; i.i. 4–6 cells thick.

Seed massive, rounded, with coriaceous brown testa, exarillate or (?) arillate in *Billia*, with a partial septum for the radicle-pocket, with large hilum, ? pachychalazal. Testa (? pachychalazal wall) strongly multiplicative; *o.e.* as a short palisade of cells with thickened outer and radial walls; *mesophyll* aerenchymatous, much thickened, thin-walled, then slightly thick-walled, carrying the v.b.; *i.e.* unspecialized. Tegmen persistent round the radicle, not multiplicative, thin-walled, unspecialized. Vascular supply as a reticulum in the testa, arising from divisions of two branches of the short stout compound v.b. of the raphe. Endosperm nuclear, soon absorbed. Embryo with thick cotyledons and short radicle, curved.

The large seeds appear to be pachychalazal and there must be caution in interpreting the structure. They are certainly overgrown seeds with highly vascular wall and the tegmen limited to the radicular pocket, as with various Sapindaceae. According to van der Pijl (1957), the stout funicle is a vestigial arillode or funicular aril, such as may occur in a more fully developed state in the American *Billia*, seeds of which have proved inprocurable. The practically exarillate seed of *Aesculus* with its tough seed-coat is considered by van der Pijl to be 'an adaptation to a life in extra-tropical regions', thus admitting the primitive and tropical nature of the arillate seed in accordance with the durian-theory, but the abundance of tough seed-coats among tropical trees, e.g. Lecythidaceae, shows that the character is an adaptation of the tropical forest suitable for extra-tropical regions.

Aesculus L.

A. hippocastanum L.
Fig. 502.

Testa (? pachychalazal wall) –84 cells thick through multiplication of the mesophyll, but only half of this thickness round the radicular pocket; *o.e.* with the palisade-cells 20–30 × 10–20 μ; *mesophyll* with (1) a thick outer layer of cells with brown walls, at first aerenchymatous, (2) a layer (*c.* 10 cells thick) of flattened cells with pigmented contents carrying the v.b., (3) a layer of colourless thin-walled cells, 4–5 cells thick, and (4) an innermost layer (*c.* 9 cells thick) of flattened pigmented cells, more or less transversely elongate. Tegmen ? with i.e. more or less thick-walled. Funicle thick, widened upwards to the base of the seed, subarilloid, with a few short branches from the v.b. in the distal part.

(Guérin 1901; Netolitzky 1926; Baehni et Bonner 1953; v.d. Pijl 1957; Vaughan 1970.)

HYDNORACEAE

Ovules orthotropous, unitegmic, tenuinucellate; integument 2–4 cells thick.

Seeds minute. Seed-coat not multiplicative; *o.e.* thin-walled; *i.e.* with thickened inner wall, pitted (*Prosopanche*) or not (*Hydnora*). Perisperm as a single layer of cells. Endosperm abundant. Embryo minute.

(Netolitzky 1926; Dastur 1921.)

HYDRANGEACEAE

Ovules anatropous, unitegmic, tenuinucellate.

Seeds small, in some cases winged, apparently simply exotestal. Endosperm cellular, abundant. Embryo straight.

(Netolitzky 1926.)

HYDROPHYLLACEAE

Ovules more or less anatropous, numerous, unitegmic, tenuinucellate; integument 6–8 cells thick (*Hydrolea*); i.e. becoming an endothelium.

Seeds, small, albuminous. Seed-coat reduced to o.e. as a layer of large, but collapsing, cells often with the inner wall bulging into the mesophyll or endosperm, the radial walls often with spiral-reticulate thickening, the contents slimy. Endosperm nuclear, the cell-walls amyloid. Embryo small, straight.

(Netolitzky 1926; Mitra 1947.)

In seed-structure the family seems near to Polemoniaceae and is usually so classified.

HYPERICACEAE

Ovules anatropous, bitegmic, tenuinucellate; o.i. 2 cells thick, 3–5 cells at the micropyle; i.i. 2–6 cells thick; micropyle formed by the exostome.

Seeds small, brown, the seed-coats not multiplicative, exarillate, in some genera alate, thinly albuminous. Testa; *o.e.* with more or less strongly enlarged cells set in longitudinal rows, with brown tannin contents and slightly thickened radial and inner walls, forming the reticulate surface of the dried seed; *i.e.* collapsed, often apparently absent. Tegmen; *o.e.* as a palisade of thick-walled, finely pitted, not or slightly lignified cells with stellate–undulate facets; *mesophyll* and *i.e.* unspecialized, crushed. Vascular supply only in the raphe, slender, the tissue round the v.b. becoming widely lacunose. Chalaza simple. Endosperm nuclear, reduced to 1–2 layers of thin-walled cells, oily. Embryo straight. Figs. 292–295.

(Netolitzky 1926.)

The small seeds of this family appear to be uniformly constructed. The characteristic layer is the exotegmen of thick-walled stellate cells as in Clusiaceae–Clusieae, Elatinaceae and Geraniaceae. The crystal-cells (endotesta) of Geraniaceae seem to be absent from Hypericaceae. The seed of *Cratoxylon* has a testal wing with circuitous and peripheral v.b. (Fig. 292); it suggests an arillate ancestor, as in Bonnetiaceae and Ixonanthaceae.

Cratoxylon Bl.

Seeds winged from the chalazal end or, in *C. arborescens*, narrowly all round. Exotegmic palisade thin (*C. arborescens*) or thick, not lignified. Vascular bundle of the raphe following the periphery of the wing and conspicuously recurved at the chalaza.

(Gogelein 1967.)

C. arborescens (Vahl) Bl.
(Malacca, leg. Griffiths; dried material, CGE.) Fig. 292.

Exotesta with large cells. Exotegmen 10–12 μ high.

C. cochinchinense (Lour.) Bl.
(Thailand, Sing. F. N. 2948; dried material, CGE.) Fig. 292.

Exotesta with rounded cells, some bulging to the exterior singly or in groups, especially at the micropylar end. Exotegmen *c.* 50 μ high, as a hard palisade, not or slightly lignified.

C. *formosum* (Jack) Dyer has a similar seed.

Haronga Thouars

H. paniculata Lodd
(Congo, Burtt-Davy 17932; dried material, CGE.)

Berry with lignified endocarp. Seeds as in *Hypericum*.

Hypericum L.

Seeds typical of the family, not winged. Figs. 293–295.

H. androsaemum L., *H. calycinum* L., *H. maculatum* Cr., *H. mysorense* Heyne. *H. patulum* Thunb., *H. perforatum* L., *H. quadrangulare* L.

(Netolitzky 1926; Crété 1936b; Rao, A. N. 1957b; Salisbury 1967.)

In *H. calycinum* and *H. patulum* the sclerotic cells of the exotegmen become bullately lobed at the outer angles and press into the spaces between the cells of the exotesta. Rao described the endotestal cells of *H. patulum* as radially elongate with thickened outer and radial walls, which would be exceptional in the family; I did not observe such cells in the material which I examined.

Vismia Vand.

Seeds as in Hypericum; exotegmen apparently not lignified.

V. guyanensis (Aubl.) Choisy
(Brazil, Manaus, Corner 65.)

Vismia sp.
(Brazil, Manaus, Corner 232.)   Fig. 292.

ICACINACEAE

Ovules anatropous, suspended from the apex of the ovary, paired, unitegmic but with free integuments at the micropyle in *Phytocrene*, tenuinucellate; integument more than 10 cells thick, with long micropyle.

Seeds ? pachychalazal, mostly albuminous. Seed-coat with the outer layers of cells elongate with thickened walls (*Gomphandra*). Endosperm nuclear. Embryo small.

(Netolitzky 1926; Mauritzon 1936c; Fagerlind 1945; Padmanabhan 1961.)

Apodytes, Desmostachys, Gomphandra, Gonocaryum, Icacina, Lasianthera, Leptaulus, Leretia, Pennantia, Phytocrene, Poraqueiba, Raphiostyles, Stemonurus, Villaresia.

ILLICIACEAE

Ovule anatropous, erect, bitegmic, crassinucellate, 1 per carpel; o.i. 3–4 cells thick; i.i. 2(–3) cells thick; chalaza small; micropyle formed by the exostome.

Seeds with hard smooth, black or brown testa, the micropyle thickened, exarillate, albuminous. Testa multiplicative, 6–11 cells thick; *o.e.* as a palisade of thick-walled lignified pitted cells with undulate facets, without stomata; *mesophyll* thin-walled or with 1–2 hypodermal layers modified into sclerotic cells as those of o.e. Tegmen disappearing or with persistent i.e. (*I. floridanum*). Raphe v.b. terminating at the unspecialized chalaza. Endosperm cellular or nuclear, copious, oily. Embryo minute.

Pericarp thin-walled but the inner epidermis becoming a palisade of columnar, thick-walled, lignified cells.

(Netolitzky 1926; Hayashi 1964; Bhandari 1971.)

The exotestal structure of the seed allies *Illicium* with Schisandraceae and Winteraceae, not with Magnoliaceae (p. 28). These exotestal families may refer back to the ancestry of Ranunculales and Berberidales, but as leptocaul trees with simple leaves and small exarillate seeds their modern representation is variously advanced.

Illicium L.

For *I. religiosum*, Buxbaum (1961) mentions a strophiole as a slight swelling between the funicle and the micropyle; for *I. verum*, he figures a slight exostomal flap on the funicular side of the micropyle, as if a vestigial aril. He relates *Illicium* with Phytolaccaceae where funicular and exostomal arils may be well-developed. These structures have not been found by others in *Illicium* but may well occur.

I. kinabaluense A. C. Smith
(Borneo, Kinabalu, RSNB 4256; material in alcohol.)
Figs. 296, 297.

Ovules; o.i. 3–4 cells thick; i.i. 2 cells thick.

Testa 6–7 cells thick. Tegmen not multiplicative but the cells enlarging, those of i.e. tangentially elongate with slightly thickened walls, eventually the whole tissue crushed. Micropyle thickened and lignified.

Illicium sp.
(Malaya, Pahang, Cameron Highlands, Corner s.n. Oct. 1966; fruits in alcohol.)   Fig. 298.

Seeds 8 × 5.5 × 5 mm, as in *I. kinabaluense*.

IXONANTHACEAE

Ovules anatropous, suspended, 2 or 1 per loculus, bitegmic.

Seeds arillate or exarillate and alate, or round and massive. Testa apparently without a palisade, reduced to o.e., or massive (*Irvingia*). Tegmen; *o.e.* as massive fibres, or without fibres (*Irvingia*). Endosperm present or much reduced (*Irvingia*).

(Netolitzky 1926; Forman 1965.)

This small family, as recognized by Forman, is a

relic of arillate ancestry. A laciniate aril covers the seed in *Phyllocosmos*, but it is variously reduced in other genera, even to a minute funicular process in *Allantospermum*; it is absent from Irvingioideae. In *Ixonanthes* and *Ochthocosmos* the arillate seed transforms into the alate. Microscopic study is needed to confirm agreement with *Ixonanthes*.

Irvingia is puzzling. It has been associated with *Desbordesia* and *Klainedoxa* in a separate subfamily or family (Irvingiaceae), but, whereas both these genera appear to have the characteristic fibrous exotegmen, it is absent from the massive seed of *Irvingia*. The testa of *Irvingia* is highly multiplicative and develops sclerotic cells throughout; the vascular bundle has postchalazal branches into this tissue. The seed may be pachychalazal in the fibrous woody endocarp. There is undoubted resemblance in seed with Simarubaceae and Balanitaceae. It is strange that such outstanding trees as *Irvingia* should be so baffling in their alliance.

It seems probable that *Ploiarium* (p. 82) belongs with *Ixonanthes*.

Allantospermum Forman

Seeds allantoid, exalbuminous, not winged, but with a small arillar process adherent to the columella of the fruit.

Cyrillopsis Kuhlm.

Seeds with a 2-lobed membranous aril.

Desbordesia v. Tiegh.

Ovule; o.i. 2 cells thick; i.i. 3 cells thick.

Testa ? not multiplicative; *o.e.* composed of large thin-walled cells; underlying cells tangentially elongate, thin-walled, brown. Tegmen; *o.e.* composed of thick-walled longitudinal fibres (? endotesta of Netolitzky 1926, p. 179).

According to Netolitzky, *Klainedoxa* Pierre has the same structure.

Irvingia Hook. f.

Ovule; o.i. 3 cells thick; i.i. 4 cells thick; funicle with obturator.

Seed large, –35 mm long, ovoid-compressed, more or less concrescent with the endocarp and placenta. Testa *c.* 18 cells thick, composed of many layers of sclerotic cells with spiral or reticulate thickening, and with numerous postchalazal v.b. in the mesophyll. Tegmen ? Endosperm more or less reduced to a single layer of cells.

(Netolitzky 1926; Vaughan 1970.)

Ixonanthes Jack

Seeds winged from the chalazal end, except in *I. icosandra* with funicular aril having 3 narrow tapered lobes.

I. beccarii Hall. f.
(Sarawak, S. 26830; dried seeds.) Fig. 299.

Seed 15–16 × 3.5–4.5 mm, flattened, winged, brown, dry. Testa ? 3–4 cells thick; *o.e.* composed of narrow longitudinal cells with convex and slightly thickened outer walls, not lignified; *mesophyll* and *i.e.* thin-walled, crushed, without crystal-cells. Tegmen; *o.e.* as a continuous layer of longitudinal fibres 170–600 × 40–70 μ, of uneven thickness, the irregularly undulate walls thickened, lignified and rather finely pitted, shortened to sclerotic cells at the chalaza and micropyle; *mesophyll* crushed; *i.e.* as tabular cells with slight ribbing on the radial walls, not lignified, more or less crushed. Chalaza with contracted, dark brown hypostase, not lignified. Vascular bundle ending at the chalaza, not entering the heteropyle. Wing of the seed formed by the extended mesophyll and o.e. of the testa, the cells aerenchymatous, with thin and slightly suberized walls. Endosperm thin-walled, oily. Embryo straight.

Ochthocosmos Benth.

Seeds flattened, winged with a funicular extension (? aril) of pale fleshy tissue.

Phyllocosmos Kl.

Seeds ellipsoid, covered with a laciniate aril (? funicular).

JUGLANDACEAE

Ovules orthotropous, erect, basal, unitegmic, crassinucellate; integument 6 cells thick (*Juglans*), 8–10 (*Pterocarya*); exostome lobulate or bilabiate.

Seeds large, contained in the endocarp, practically exalbuminous. Seed-coat highly vascularized, reduced eventually to o.e. as a layer of rather large, slightly radially elongate, thin-walled cells, with stomata (*Carya, Juglans*). Vascular bundles *c.* 20, freely branched, postchalazal. Endosperm nuclear, oily, reduced to a single layer of cells (*Carya, Juglans*). Embryo massive; cotyledons often contorted.

(Netolitzky 1926; Kühn 1927; Hjelmquist 1948, 1957; Meeuse and Houthuesen 1964; Verhoog 1968; Vaughan 1970; v. Heel and Bouman 1972.)

The tardily dehiscent capsule is practically a drupe, nut or samara. Neither this point nor the pinnate leaf has been explained by those who suppose the family to be so primitive in its ovule and ovary as to relate with Gnetales. According to v. Heel and Bouman, the integument begins as a slight outgrowth on opposite sides of the nucellus. The two outgrowths are then carried up by a basal intercalary growth which forms the cylindrical integument. This suggests a pachychalazal construction, as may occur in Anacardiaceae and Simarubaceae, and it would explain the vascular seed-coat, which was taken by Netolitzky to represent the testa with stomata and vascular bundles. The orthotropous ovule appears as a result of basipetal growth of the ovary and does not in itself appear primitive in angiosperms, e.g. Piperales, Polygonales, Proteales, Urticales.

JULIANACEAE

Ovules more or less atropous, unitegmic (? bitegmic), with long funicle as in Anacardiaceae.

Seeds exalbuminous. Seed-coat consisting of 5 layers of thin-walled cells, more or less tangentially elongate; *i.e.* with dense contents in the cuboid cells.

(Netolitzky 1926.)

KRAMERIACEAE

Ovules 2, anatropous, suspended, bitegmic.

Seed exalbuminous, enclosed in the barbed fruit, without special development of the seed-coats. Testa; *o.e.* composed of tabular cells with tannin-contents, with stomata; *mesophyll* with numerous v.b.

(Netolitzky 1926; Milby 1971.)

The absence of any of the characters of the Leguminous seed seems to prevent affinity with this family but, of course, there are simplified and overgrown seeds in it, such as *Euchresta*, even with stomata as in *Inocarpus*. Similarly there is nothing Polygalaceous in the seed.

LABIATAE

Ovules anatropous, basal, erect, solitary, unitegmic, tenuinucellate; integument 6–9 cells thick; i.e. as an endothelium or not.

Seeds enclosed in pericarp, albuminous or not, exarillate. Seed-coat reduced to o.e. with or without 1–3 layers of mesophyll, or entirely crushed; *o.e.* with cuboid or tabular cells, often with reticulate or bar-like thickenings on the inner wall, sometimes on adjacent parts of the radial walls, entirely thick-walled (*Salvia spp.*), with glandular hairs (*Teucrium*), or thin-walled, tabular, and more or less crushed; *mesophyll* crushed or with 1–2 outer layers of thick-walled cells (*Anisomeles, Pogostemon*). Vascular bundle short, ending at the chalaza. Endosperm cellular, oily.

(Netolitzky 1926; Murthi 1947; Jaitly 1966, 1969b, 1971; Vaughan 1970.)

The schizocarpic seed in this family has the seed-coat structure so much deteriorated as to be almost negligible. However, few genera have, as yet, been studied.

LACISTEMACEAE

Ovules anatropous, suspended, parietal, bitegmic, crassinucellate.

Seeds albuminous; seed-coat thinly fleshy with a crustaceous inner layer. Endosperm nuclear. Cotyledons foliaceous.

(Netolitzky 1926.)

It will be interesting to learn the nature of the crustaceous layer of the seed.

LACTORIDACEAE

Ovules anatropous, bitegmic, ? tenuinucellate; o.i. 2 cells thick; i.i. 3 cells thick, forming the micropyle.

Seeds small, albuminous. Seed-coats apparently not multiplicative, not differentiating into mechanical tissue, the cell-structure obliterated at maturity. Endosperm nuclear, oily. Embryo minute.

(Bouman 1971b.)

The illustration of Bouman suggests a simplified state of the Winteraceous seed.

LARDIZABALACEAE

Ovules more or less anatropous, campylotropous (*Lardizabala*) or atropous (*Holboellia, Parvatia*), bitegmic, crassinucellate, numerous or solitary; o.i. 3–4 cells thick, or more; i.i. 2–3 cells thick.

Seeds small to medium-size, ovoid-subcompressed to subreniform, albuminous, exarillate or with vestigial aril (*Akebia*). Testa not multiplicative or slightly (*Decaisnea*) or distinctly (*Akebia*); o.e. apparently as a palisade of narrow cells with thickened outer and radial walls and shortly oblong in surface-view, as a high palisade in *Decaisnea*, as narrow unlignified fibres in *Akebia*; *mesophyll* with brown, unlignified walls, the outer cell-layers in *Akebia* thickened into longitudinal and oblique fibres; *i.e.* unspecialized, (? with thickened radial and inner walls in *Decaisnea*). Tegmen unspecialized, not multiplicative, crushed (? *Decaisnea*). Endosperm cellular, oily, also starchy in *Decaisnea*. Embryo short.

(Netolitzky 1926; Swamy 1953b; Payne and Seago 1968.)

The few genera of this family appear as relics of former diversity. The seeds of *Akebia* and *Decaisnea* are superficially similar and are surrounded by placental pulp in the fleshy follicles, yet microscopically they are extraordinarily different. *Decaisnea* has the shortly oblong exotestal cells of Berberidaceae, as seen in surface-view, but they form actually a high palisade. In *Akebia* they are not at all radially elongate but develop into very long, colourless cells which, with the brown thick-walled outer layers of the mesophyll, form a fibrous covering to the seed reminiscent of that in Annonaceae. The fleshy follicles suggest arillate ancestry and there is a vestigial aril-patch on the short raphe of *Akebia*. Whether this occurs in *Decaisnea*, as the erect and pinnate-leafed pachycaul, is uncertain. This aril-patch could be regarded as a vestigial sarcotesta, but that would not explain its presence round the hilum and on the raphe, which are the loci of the aril in

Podophyllaceae and Melianthaceae (p. 193). The seed of *Holboellia* is said to have a fibrous coat as in *Akebia*.

The evolutionary significance of the family is discussed on p. 28.

Akebia Decne.

A. lobata

(Cambridge, University Botanic Garden; living material.) Fig. 300.

Seeds 7–8 × 5–6 × 3 mm, ovoid-compressed, more or less campylotropous, smooth, blackish brown with a soft white aril-patch round the small hilum and along the short raphe. Testa multiplicative, 11–14 cells thick, not lignified; *o.e.* as more or less longitudinal, narrow, fibre-like cells 100–500 *μ* long, shortly oblong in the raphe, with thickened and hyaline outer and radial walls, not as a palisade; *mesophyll* with 5–6 layers of brown, thick-walled fibres (8–9 on the raphe-side), the outer 1–2 layers more or less oblique, the inner longitudinal, with 5–7 inner layers of shortly oblong, aerenchymatous, thin-walled and colourless cells; *i.e.* unspecialized. Tegmen crushed, even at the endostome. Vascular bundle in the short raphe ending at the chalaza with 7–8 short branches not extending to the sides of the seed. Chalaza simple. Endosperm thin-walled, oily. Embryo 2.5 mm long, with short cotyledons.

Aril-patch composed of long, radially elongate cells with slightly thickened walls, with hexagonal facets, filled with oily cytoplasm and starch-grains, forming a soft but firm, colourless tissue adnate to the seed, with rather abrupt transition to the testa.

Follicle massive, fleshy, without sclerotic tissue or fibres; v.b. few, distant, roughly in two layers, the outer with smaller, more or less anastomosing v.b.; placental tissue soft, spongy, sweet, filling the gaps between the seeds and embedding them.

Decaisnea Hook. f. et Th.

D. insignis Hook. f. et Th.

(India, Sikkim, leg. Hook. f.; dried material, CGE.) Fig. 300.

Ovule; o.i. 3 cells thick; i.i. 4 cells thick.

Seed 11 × 7 × 3 mm, ellipsoid, compressed. Testa ? 5–6 cells thick; *o.e.* as a palisade *c.* 170 *μ* high, composed of narrow cells with thickened lignified walls, but wide lumen, the cells shortly elongate longitudinally with oblong facets 27–80 × 7–9 *μ*; *mesophyll* crushed but the outer hypodermal cells rather large; *i.e.* unspecialized. Tegmen 5–6 cells thick, thin-walled, more or less crushed; *i.e.* as a layer of cuboid cells with thick brown radial and inner walls, not lignified. Endosperm thin-walled, oily and with many starch-grains in the cells. Embryo 2 mm long; cotyledons short.

(Swamy 1953b.)

What I have described as thickened tegmen (as seen in the dried seeds) may really be the colourless inner layer of the testa, as in *Akebia*. Swamy, who was unable to investigate full-grown seeds, describes the tegmen as becoming 5–6 cells in thickness.

LAURACEAE

Ovule solitary, anatropous, pendent, with dorsal raphe and the micropyle on the placental side with overarching funicle, bitegmic, crassinucellate; o.i. 3–5 cells thick; i.i. 2–4 cells thick; micropyle closed by the endostome or by the nucellus; chalaza massive, unspecialized; embryo-sac often large and elongate.

Seed massive with thin testa, in many genera enclosed by the lignified endocarp (appearing as the exotesta), in many cases (? most) more or less pachychalazal or perichalazal. Testa 4–20 cells, thick, more or less multiplicative, most layers persistent or the inner crushed, in *Cryptocarya* entirely crushed except i.e., without stomata; *o.e.* as a layer of cuboid cells with tannin, or unspecialized, shortly radially elongate in *Cassytha* and *Cinnamomum*, pressed closely to the endocarp; *mesophyll* unspecialized, without oil-cells; *i.e.* typically composed of longitudinally and tangentially elongate, narrow tracheids with spiral or annular thickening, slightly lignified or not, 2–3 layers of tracheids at the micropyle or in the free part of the testa (*Laurus*). Tegmen not or scarcely thickening, soon crushed. Vascular bundle of the raphe commonly dividing into 2–3 branches at the chalaza and forming a network in the periphery of the pachychalazal part of the seed, undivided in *Cryptocarya* as a flattened band in the perichalaza, or with short branches from the raphe and the chalaza to the sides of the seed (*Laurus canariensis*). Chalaza not extended (*Cassytha*) or enlarged as a pachychalaza to form one- to two-thirds of the length of the seed. Nucellus soon absorbed or persistent as ruminations (*Cryptocarya*, *Ravensara*). Endosperm nucellar, absorbed by the embryo, cellular in *Cassytha*. Embryo filling the seed with thick massive cotyledons concealing the very short radicle, the young embryo at first with divergent cotyledons.

Pericarp with oil-cells and in many cases with stone-cells; endocarp commonly as a single layer of more or less columnar, thick-walled, pitted, and lignified cells with stellate–undulate facets, varying unspecialized. Figs. 301–309.

(Netolitzky 1926; Coy 1928; Bambacioni-Mezzetti 1935, 1938; Kasapligil 1951; Sastri 1952, 1958a, 1959b, 1962, 1963, 1965; Schroeder 1952; Chowdhury and Mitra 1953; Kostermans 1957; Hutchinson 1964.)

The affinity of this family with Monimiaceae is proved by the construction of the carpel, ovule, seed, and fruit (p. 26). The observations of Kasapligil and Sastri on the rudimentary margin of the carpellary primordium reveal the Lauraceous ovary as a single carpel; the conclusion is confirmed by the extra-locular ovules that may occur in *Persea* (Bambacioni-Mezzetti 1938). Basipetal growth of this carpel leads as in Monimiaceae to genera with the ovary and fruit more or less immersed in a receptacular or perianth tube. But Lauraceae have two main advances. Progressive neoteny has produced the small flowers which, as simple florets, permit aggregation into clusters or heads with more gradual display in inflorescences and diminish the loss on lack of pollination. Reduction to a single carpel permits the elaboration of the one ovule into the large and exalbuminous seed which secures rapid establishment of the seedling in the forest. Steps in both directions occur in Monimiaceae, but the combination is decisive for Lauraceae. There is no evidence in Laurales of an ancestor with follicles containing albuminous seeds, as there is for the solitary follicle of Papilionaceae. The indehiscent and one-seeded fruit of Laurales seems derived from the indehiscent follicle, as seen in Annonaceae, and is in marked contrast with the reduced and dehiscent fruits of Magnoliaceae and Myristicaceae. Whether there is any phyletic significance in the divergent cotyledons of Myristicaceae, some Monimiaceae, and the young embryos of Lauraceae requires more extended investigation for proof.

The Lauraceous seed, nevertheless, offers in the thick and multiplicative testa a more primitive condition than in Monimiaceae. It suggests divergence from the ancestral stock before the distinction of Monimiaceae. In spite of their many resemblances, the two families appear to have evolved in parallel, and this is borne out by a detail of the exotesta which has a palisade-form in *Cassytha* and *Cinnamomum*. It must refer to an arillate seed exposed in a follicle, rather than to the Magnoliaceous sarcotesta.

There is not only this least derived state of the testa in *Cassytha* but the absence of pachychalazal modification. The parasitic habit seems to be the manner of survival of this relic of early Lauraceae.

The way in which the ovule turns into the seed may serve as a check on generic alliances in Lauraceae, which are based on floral detail. *Cryptocarya*, with the seed deeply immersed in the pericarp and the perianth-tube, has the chalazal expansion limited to the perichalaza in the proximal end of the seed. *Cinnamomum* and *Litsea* have extensively pachychalazal seeds, but slight development of the perianth-tube.

Actinodaphne Nees

A. speciosa Nees
(Ceylon, Hakgala; fruits in alcohol.)   Fig. 302.

Testa; *i.e.* as 2–3 layers of cells in the micropylar region with coarsely spiral, scarcely lignified, thickenings, the inner layer of cells radially elongate, the outer 1–2 layers shortly tangentially elongate, on the sides of the seed unspecialized with shortly radially elongate cells.

Cassytha L.

Ovule; o.i. 4–5 cells thick; i.i. 2 cells thick; micropyle occluded by the nucellus.

Testa 10–11 cells thick; *o.e.* with shortly radially elongate cells; *mesophyll* thin-walled, with or without mucilage-canals; *i.e.* with spiral thickenings on the walls. Vascular supply ? not ramified. Embryo developing rapidly, taking 7–10 days to enlarge from 0.7 mm to full-size at 3 mm long.

Pericarp with pulpy mesocarp and woody endocarp; o.e. composed of slightly elongate and projecting cells; mesocarp thin-walled; endocarp consisting of i.e. as a palisade 200–500 μ high, of thick-walled, pitted, lignified, prismatic cells, the shorter-celled inner hypodermis with scarcely thickened walls.

C. filiformis L., *C. glabella* R.Br., *C. pubescens* R.Br. Fig. 301.

(Sastri 1952, 1962.)

Cinnamomum Bl.

C. iners Reinw., *C. zeylanicum* Bl.
Figs. 301, 303.

Ovule; o.i. 4–5 cells thick; i.i. 3–4 cells thick; micropyle occluded by the nucellus.

Testa 10–12 cells thick; *o.e.* composed of shortly radially elongate cells; *mesophyll* unspecialized; *i.e.* composed of longitudinally elongate cells with unlignified spiral thickening on the radial walls. Vascular bundle of the raphe dividing into 2 postchalazal branches ascending to the micropyle.

Pericarp consisting of 5 layers; (1) o.e. short-celled; (2) o.h. 4–5 cells thick, as stone-cells; (3) pulpy mesophyll 4–5 cells thick, thin-walled; (4) fibrous endomesophyll (i.h.), 4 cells thick, the fibres tangentially elongate; (5) i.e. composed of columnar cells in one layer.

(Sastri 1952, 1958a; Chowdhury and Mitra 1953.)

Cryptocarya R.Br.

C. wightiana Thw.
(Ceylon, Peradeniya; flowers and fruits in alcohol.)
Figs. 304, 305.

Ovule; o.i. 3 cells thick; i.i. 2–3 cells thick.

Testa 4–5 cells thick, thin-walled, unspecialized (except i.e.) and becoming crushed; *i.e.* composed of longitudinal tracheids with annular or closely spiral, lignified thickening, differentiating soon after fertilization, forming a pellicle over the embryo. Tegmen not multiplicative, soon crushed. Vascular bundle of the raphe not branching, becoming a flattened band in the perichalaza, with a thin brownish hypostase. Chalaza extending into a perichalaza in the lower part of the seed. Nucellus well-developed, forming a crescentic ingrowth from the perichalaza between the cotyledons, appearing as a rumination, eventually drying up.

This seed has four peculiarities: (1) the exotesta does

not form a palisade but consists of small tabular cells that are soon crushed; (2) the only part of the seed-coats to persist is the tracheidal endotesta; (3) the ovule enlarges regularly into the seed except at the chalaza which is drawn out as a crescentic band round the lower end of the seed, but it is not a pachychalaza because the integuments of the immature seed extend on its sides to the chalazal base; (4) the vascular bundle does not form a reticulum, but the function of such in this case seems to be taken over by the endotestal pellicle of tracheids.

The persistent perianth-tube, surrounding the immersed pericarp, gives to the fruit in section the appearance of a large seed in which the perianth-tube would represent the testa, the pericarp the tegmen, the testa the nucellus, and the nucellus the endosperm. The peculiar ruminations of *Ravensara* (Madagascar) may be similar to the nucellar rumination of *Cryptocarya*, but they may in fact be endocarp-folds.

Laurus L.

L. nobilis L.
Figs. 306, 307.

Seed with massive pachychalaza composing about two-thirds of its body. Testa 13–20 cells thick; *o.e.* and 1–2 hypodermal layers with tannin and ? slightly lignified; *i.e.* and *i.h.* as 2(–3) cell-layers with slight annular or reticulate thickening, slightly lignified, tangentially and longitudinally elongate but shorter and slightly radially elongate at the micropyle. Vascular bundle of the raphe breaking up at the chalaza into 3 branches, subdividing and anastomosing in the periphery of the swollen pachychalaza, absent from the free part of the testa.

(Bambacioni-Mezzetti 1935, 1941; Kasapligil 1951; Vaughan 1970.)

According to Le Monnier (1872) the vascular bundle of the raphe in *L. canariensis* has short branches to the sides of the seed throughout its length as well as from the chalaza.

Litsea Lamk.

L. sebifera Pers.
Fig. 301.

Ovule; o.i. 4–5 cells thick; i.i. 2 cells thick; micropyle formed by the endostome.

Seed and pericarp evidently as in *Cinnamomum*.
(Sastri 1958a.)

L. singapurensis Gamble
(Singapore; living material.) Figs. 308, 309.

Ovule; o.i. 4 cells thick, the radial walls of o.e. soon beginning to thicken; i.i. 3 cells thick, 2 cells near the micropyle, 4 cells near the chalaza; nucellus occluding the micropyle.

Testa 6–9 cells thick, the inner 2–3 layers collapsing;

o.e. unspecialized, the walls slightly thickened, brownish. Tegmen 4–5 cells thick, soon collapsing. Vascular bundle of the raphe breaking into 2–3 branches shortly after entry into the seed, descending to the chalaza, then ascending in the testa as a plexus ending with *c.* 5 v.b. below the micropyle.

Pericarp pink, pulpy; o.e. without stomata; mesophyll thin-walled, the inner layers crushed; i.e. as a lignified palisade.

Persea Plum.

P. americana Mill.
Testa apparently *c.* 10 cells thick, pervaded by a plexus of v.b., without specialization.

(Bambacioni-Mezzetti 1938; Schroeder 1952; Vaughan 1970.)

LECYTHIDACEAE

Ovules anatropous, bitegmic, tenuinucellate, often with conspicuous funicle; o.i. massive, 6–18 cells thick; i.i. 3–7 cells thick.

Seeds small to large, various, ovoid, smooth or hairy, nut-like with uneven and angled surface (*Bertholletia*), winged (*Couratari*), tuber-like (*Barringtonia*), almost or completely exalbuminous, exarillate or with slight funicular aril (*Gustavia*). Testa highly multiplicative, massive; *o.e.* palisade-like and lignified (*Bertholletia*, *Couroupita*), with hairs (*Couroupita*), or composed of short cells with undulate facets and thick, lignified, outer wall (*Gustavia*); *mesophyll* thin-walled, aerenchymatous, with scattered sclerotic cells (*Gustavia*) or extensively sclerotic; *i.e.* unspecialized. Tegmen ? not multiplicative, unspecialized or i.e. as an endothelium, soon crushed. Vascular supply as a single bundle in the raphe-antiraphe, or with 2 (*Bertholletia*) or several (*Barringtonia*) postchalazal branches, or dividing into several branches on entry into the seed (*Gustavia*). Chalaza distinct or not (*Couroupita*, *Gustavia*). Endosperm nuclear. Embryo various, large; with long radicle and thin plicate cotyledons (*Couratari*, *Couroupita*, *Planchonia*); with short radicle and thick cotyledons (*Napoleona*) or cotyledons shortly lobed (*Gustavia*); hypocotylar, tuber-like with minute cotyledons (*Allantoma*, *Barringtonia*, *Bertholletia*, *Careya*, *Chytroma*, *Eschweilera*, *Lecythis*). Funicle much enlarged in several genera, arilloid with complicated vascular supply in *Gustavia*. Figs. 310–314.

(Netolitzky 1926; Venkateswarlu 1952a.)

The large fruits and seeds of this family have not been accommodating to microscopists. There appears the problem common to Melastomataceae, Myrtaceae, Rosaceae and Theaceae whether the seed was primitively exostomal or mesotestal or pantestal with wholly sclerotic testa. *Gustavia* with rather thin-walled and dehiscent capsule and with stout arilloid funicle is exotestal;

Bertholletia is the sclerotic extreme both in fruit and seed. *Couroupita* with softly disorganizing pericarp is intermediate. The single seed of *Barringtonia* is overgrown and does not differentiate testal structure. If seed-structure evolved along with fruit-structure, then the exotestal seed of *Gustavia* appears the more primitive, but this is contrary to the conclusion reached in the other mesotestal families. The problem is clearly unresolved. With mesotestal seeds and centrifugal stamens, however, the family seems to have come from the same stock as Theaceae.

Bertholletia HBK

B. excelsa HBK

Seeds 3–5 cm long, triquetrous, transversely rugulose, brown, nut-like. Testa massive, woody; *o.e.* as a tall palisade, –1 mm high, of colourless, radially elongate cells with irregularly thickened and lignified walls, the linear lumen dilated at the outer end of the cell; *mesophyll* sclerotic, but the inner cells compressed. Vascular bundle of the raphe with 2(–3) postchalazal branches, producing transverse subdividing branchlets over the sides of the seed. Perisperm as several layers of compressed cells. Endosperm reduced to 2–3 layers of cells, oily, the outer layer thick-walled. Embryo hypocotylar with minute cotyledons.

(Netolitzky 1926; Vaughan 1970.)

Couroupita Aubl.

C. guyanensis Aubl.

(Ceylon, Peradeniya, cult.; flowers and fruits in alcohol.) Figs. 310–313.

Ovules; o.i. 6–9 cells thick, forming the micropyle; i.i. 3–4 cells thick, not multiplicative in the seed, but disintegrating on the crushing of the endotesta.

Seeds *c.* 12 × 10 × 5.5 mm, ellipsoid, lenticular, brown, densely comose with long, aseptate, slightly thick-walled but unlignified hairs 4–5 mm long, pervading the endocarp; hilum small; micropyle practically invisible; funicle long. Testa multiplicative to *c.* 50 cells thick, but the inner layers becoming crushed and leaving 15–20 outer cell-layers in the mature seed; *o.e.* soon palisade-like and developing the long hairs, one from each cell, the cell-body in the palisade then becoming lignified with variously shaped wide pits in the thick walls; *o.h.* consisting of 2–3 layers of small cells with reticulately thickened and lignified walls, most with a tetrahedral crystal, the outermost being smallest; *mesophyll* consisting of thin-walled, unlignified cells, aerenchymatous; *the inner cell-layers* crushed into a thin pellicle without trace of i.e. Tegmen with i.e. as an endothelium, soon crushed. Vascular bundle as a single strand round the seed almost to the micropyle, unbranched, without chalazal complication. Embryo developing late, crushing the tegmen and inner part of the testa; radicle long, curved; cotyledons inflexed, foliaceous, plicate.

Ovary 6-locular; placents axile; v.b. in an axile plexus of small bundles and another plexus of small bundles just external to the loculi, connected with the v.b. in the septa, the middle and outer parts of the ovary-wall with v.b. to the floral parts and a plexus of small v.b. in the outer cortex. Pedicel of the flower with a ring of small cortical v.b. outside the main vascular cylinder.

The flower, fruit and seed of the cannon-ball tree exemplify the massively primitive construction of the family.

Gustavia L.

Gustavia sp.

(Java, Kebun Raya, cult., ? *G. augusta*; living material.) Fig. 314.

Seed 18–20 × 10–12 mm, ellipsoid, compressed on some sides, shiny black, with a narrow white zone 0.5–1.5 mm wide at the junction with the funicle, exalbuminous, subarillate with massive funicle. Testa 30–35 cells thick; *o.e.* composed of short cells with undulate facets, the outer wall thickened, lignified and dark brown, as a tough layer; *mesophyll* mostly thin-walled, unlignified, but with scattered sclerotic cells, eventually more or less crushed, but sclerotic in the whitish zone round the hilum and micropyle; *i.e.* unspecialized, soon crushed. Tegmen apparently not multiplicative, soon crushed. Vascular bundle from the funicle (main v.b.) soon dividing on entry into the seed into 3 branches, without distinct raphe or chalaza, the branches dividing and laxly reticulate on the sides of the seed, descending to the micropyle. Embryo with short radicle; cotyledons massive, 2(–3) lobed distally.

Funicle long, massive, toughly fleshy, pale orange, subcrenulate, becoming flexed during development, then straightening on dehiscence of the capsule; *o.e.* composed of longitudinal fibriform cells with thick, pitted, lignified walls and wide lumen; *mesophyll* composed of compact cells with air-spaces between them. Vascular supply as a main central bundle with 10–12 small bundles in the cortex, sheathed by lignified fibres, derived independently from the placental plexus, with many fibres in irregular sclerotic masses among the small v.b.; at the funicle-apex most of the small v.b. disappearing into fibrous tissue, 2–3 entering the seed with the main v.b. and surrounded by two zones of sclerotic tissue continued into that of the white hilar zone.

Aril represented by the slightly dilated funicle-head, with a lobe concealing the micropyle, and with short lateral lobes on each side of the seed.

This massive arilloid funicle with short lobes at the head suggests, as in *Acacia*, that the seed of *Gustavia* was primitively more or less covered with a funicular aril. The seed is practically exotestal.

LEGNOTIDACEAE

Ovules anatropous, bitegmic, crassinucellate; o.i. 2–3 cells thick; i.i. 2–4 cells thick.

Seeds small, albuminous, exarillate, seed-coats ? not multiplicative. Testa not lignified; *o.e.* as a palisade of large cells with mucilaginous and tanniniferous contents, thin-walled except for the thick outer wall; *i.e.* composed of very small cells, thin-walled, with a crystal in each (*Gynotroches*). Tegmen; *o.e.* as a compact layer of longitudinally elongate, pitted, lignified cells, commonly dilated radially at the ends to give the uneven surface, not fibriform; *mesophyll* and *i.e.* unspecialized. Vascular bundle only in the raphe, minute. Embryo straight.

Though *Carallia*, *Gynotroches* and *Pellacalyx* are classified in Rhizophoraceae–Legnotideae, there is such a striking difference between the seeds of *Carallia* with thick testa and the small tegmic seeds of the other two genera that I prefer to maintain, at least for the present, Legnotidaceae for these two genera. *Carallia* remains in Rhizophoraceae.

The exotegmen of long sclerotic cells places Legnotidaceae in the seed-group with fibrous exotegmen. The perignyous tendency may suggest alliance with Combretaceae–Lythraceae–Punicaceae, though no sclerotic mesophyll occurs in the small seeds of Legnotidaceae, but the ribbon-like form of the sclerotic cells agrees rather with Capparidaceae–Celastraceae. Three other families, however, must be considered. Bonnetiaceae have elongate but undulate exotegmic cells; Sauvagesiaceae have narrow exotegmic fibres; Turneraceae have very oblique exotegmic palisade-cells and the Passifloraceous testa. I think that it is impossible to decide on the affinity of Legnotidaceae until the seed-structure is known for other genera of the family and, indeed, for all this complex of families with longitudinally elongate and lignified exotegmic cells. The superior ovary of *Gynotroches* does not fit Rhizophoraceae.

Gynotroches Bl.

G. axillaris Bl.
(Malaya; immature fruits in alcohol.)   Fig. 315.

Seeds 1.6–1.8 × 0.8–1 mm, brown, smooth, drying areolate, several in the berry. Testa 2(–3) cells thick; *o.e.* as a palisade of large cells with brown mucilaginous contents and thickened outer wall, with isodiametric polygonal facets; *i.e.* composed of very small thin-walled cells with a small crystal in each, often with another cell or two on the outside in the gaps between the bases of the exotestal cells. Tegmen 3–4 cells thick; *o.e.* with the sclerotic cells 80–195 × 27–46 μ (high) × 13–23 μ (wide), with thickened inner and radial walls, shorter as palisade-cells at the endostome and chalaza; *mesophyll* and *i.e.* with thin-walled, slightly lignified, shortly transversely elongate cells, eventually crushed. Hypostase very small, lignified.

Pericarp with small oil-cells in the outer hypodermis; outer and inner epidermal cells with brown mucilaginous contents as those of the exotesta.

Pellacalyx Korth.

P. axillaris Korth., *P. saccardianus* Scort.
(Indonesia; Herb. Bogor., material in alcohol.)   Fig. 316.

Seeds 1.5–1.7 × 0.7–0.8 mm, blackish brown, smooth, many in the fruit. Testa 2 cells thick; *o.e.* as a palisade of large cells with brown tanniniferous and mucilaginous contents, and strongly thickened outer wall; *i.e.* soon obliterated and apparently without crystals. Tegmen with ribbon-like sclerotic cells strongly dilated radially at the ends, shortened to a palisade at the endostome and the chalaza; *i.e.* soon obliterated. Nucellus becoming 10–11 cells thick, persistent as 2–4 layers, not as a storage tissue, the inner part crushed. Endosperm thin-walled, starchy. Embryo with short cotyledons.

The lack of crystal-cells in the endotesta, the thin tegmen, and the more uneven surface of the exotegmen may be distinctions from *Gynotroches*.

LEGUMINOSAE

Ovules anatropous or more or less campylotropous, bitegmic, crassinucellate; o.i. 2–10 cells thick; i.i. 2–3 cells thick.

Seeds small to very large, generally with hard smooth surface, arillate or not, albuminous or not. Testa generally multiplicative; *o.e.* typically as a palisade of prismatic, thick-walled Malpighian cells, with *linea lucida*, facets hexagonal, lumen linear and often substellate in t.s., not or slightly lignified, often with the outer wall mucilaginous below the cuticle, but in overgrown seeds more or less undifferentiated; *mesophyll* aerenchymatous, typically with o.h. as a layer of thick-walled hourglass-cells, often becoming thick-walled but not lignified, more or less crushed at maturity; *i.e.* unspecialized or as a layer of hourglass-cells. Tegmen not multiplicative, unspecialized, crushed in the fully grown seed. Chalaza simple. Vascular supply various. Hilum with tracheid bar in Papilionatae. Endosperm nuclear. Aril funicular, ? micropylar in *Archidendron*. Funicle often rather long or stout.   Figs. 317–344.

Mimosoideae

Ovule and seed anatropous. Testa in many cases with a heart-shaped pleurogram defined by a *linea fissura* open at the hilar end. Vascular bundle extending typically round the seed from the hilum to the micropyle in the raphe-antiraphe. Hilum small, round or shortly oblong, closed by the palisade of the testa, simple. Funicle often long and slender, in some cases with funicular aril, mostly without. Endosperm with thickened walls or absent. Embryo straight; radicle short, thick;

cotyledons with thin-walled cells, generally without starch-grains (present in *Acacia auriculaeformis*, *Pithecellobium dulce*). Figs. 317–319.

Caesalpinioideae

As Mimosoideae but without the *linea fissura*, the seed more various in shape and structure; i.e. (testa) often as hourglass-cells; funicle often much thickened and arilloid, rarely investing the seed.

Amherstieae–Cynometreae. Hilum often shortly oblong, without a palisade; cotyledons with thick-walled cells, a few species with starch-grains. Fig. 339.

Bauhinieae. Hilum generally crescentic with one or two aril-lobes (except *Cercis*); raphe usually longer than the antiraphe and the seeds obcampylotropous (except *Cercis*); vascular bundle without postchalazal extension; hourglass-cells absent. Figs. 320–322.

Caesalpinieae. Raphe in some cases with a short palisade-tube round the v.b. on entry (*Caesalpinia*, *Delonix*).

Cassieae. With slender Mimosoid funicle in *Cassia* and a pleurogram (without *linea fissura*). Figs. 324–334, 341.

Papilionatae

Ovule more or less campylotropous. Seed bean-shaped with round or usually more or less elongate hilum, without pleurogram; raphe shorter than the antiraphe; hilum typically with rim-aril, double palisade, median groove, tracheid-bar, and two recurrent v.b. (one in *Cadia*); endosperm usually absent, when present thick-walled; embryo mostly curved with long curved radicle (except Sophoreae), the cotyledons mostly with thin-walled cells and starch; funicle usually rather short and flattened. Figs. 323, 336–338, 340, 342, 344.

Swartzioideae

Intermediate between Caesalpinioideae and Papilionatae (p. 172). Figs. 335, 343.

(Netolitzky 1926; Kühn 1927; Pitot 1935a, b; Joshi 1938; Pantulu 1942, 1945, 1951; Boelcke 1946; Corner 1951; Dnyansagar 1951, 1954, 1956, 1958; Isely 1955; Steiner and Jancke 1955; Bocquet et Bersier 1959; Pal 1960; Thomson 1960; Saber, Balbaz and Awad 1962; Singh, B. 1964; Wellendorf 1964; Hallé 1965; Maheshwari, J. K. 1967; Chowdhury and Buth 1970; Deshpande and Untawale 1971; Evrard, Vieux et Kabele-Ngiefu 1971.)

Leguminosae are one of the great lines of dicotyledonous evolution. They far exceed the idea of an order or family, except within the artificial confines of nomenclature. They have progressed from doubly pinnate leaves and stipules to the simple, from polypetalous multistaminate multicarpellary and actinomorphis flowers to the sympetalous, the oligostemonous, the unicarpellary, the zygomorphic, the unisexual and the apetalous, and in fruit from the many-seeded arillate legume to one-seeded drupes, nuts, samaras, and achenes. They have not achieved the epigynous construction though most are variously and, even, extensively perigynous. Pachycaul forms are scarce and seem confined to the saplings of trees, e.g. *Enterolobium*, *Sclerolobium*. They have extended through the lowland tropics and the temperate regions with trees, climbers and herbs in greater abundance and variety than any other line of dicotyledons; yet, they have never become epiphytic, parasitic, or saprophytic, and very few are aquatic. With such manifestation it is difficult to frame a definition. Genera are placed in alliances but there are intermediates, and the whole is one of the more remarkable concatenations of the plant kingdom. It is so vast that it has still to be explored microscopically, and this vastness, replete with evolutionary side-lines and relict genera, seems to have dismayed writers on the evolution of angiosperms who see, not a mighty contingent, but a standard family with zygomorphic flower, one carpel, and rosaceous or saxifragaceous parentage. Yet there are living plants to explain how the diverse zygomorphy of *Bauhinia* and *Vicia* has arisen, how the holy *Saraca* has lost its petals and become secondarily actinomorphic, how *Parkia* has reverted to the primitiveness of *Archidendron*, how the bean has gained its hilum, how bats and moths play box and cox with birds and butterflies, how *Inocarpus* has lost most of its leguminous marks, why *Trifolium* has dentate leaflets, *Acacia* and *Lathyrus* have phyllodes, and *Bauhinia monandra* one stamen, like an orchid. The neap-tide of modern botany never uncovers its riches. From arctic circle to tropics, desert to pergola, bacteria to plough, field to mouth, and legend to science, Leguminosae invest our lives, and a feeble backwash seeps through our universities. We wait that treatise which will quicken the herbarium into the living tree of phylogeny.

Of general marks to define Leguminosae there is the microscopic structure of the seed; it is exotestal with a palisade of Malpighian cells and hypodermal with hourglass-cells. It holds better than most, as the student of seed-structure would expect. The tegmen is unspecialized and does not contribute to the seed-coat. There are no fibres or sclerotic cells in the mesophyll. In the over-grown seeds of indehiscent pods, this mark may fail. It is, however, a sharp distinction from Rosaceae and Saxifragaceae, and when the aril is added in the many-seeded pod, there is nothing so primitive in those families. Indeed, no genus of theirs has a fruit to compare with that of *Archidendron*. While it may be held that Rosaceae and Saxifragaceae are wayward products of Leguminous ancestry, I am struck with the resemblance in seed-structure between Leguminosae and the families Berberidaceae, Lardizabalaceae and Melianthaceae (p. 28).

In this account I summarize the main features of leguminous seeds and add details on critical species that have become available. The form of the ovule determines, as usual, that of the seed. These variations from the anatropous to the campylotropous have been investigated by Bocquet who concludes that the anatropous is the primitive. This accords with the advanced character of the papilionaceous hilum. Maheshwari has studied the form of the starch-grains in the embryo and finds their occurrence wide-spread in Papilionatae, occasional in Mimosoideae, and absent, so far as known, from Caesalpinioideae. My attention has been drawn by Professor A. Burckart to Boelcke's account of mimosoid seeds where he used the term *linea fissura*, in place of the unsuitable *linea sutura* of Capitaine (1912), for the fissure which delimits what I called the pleurogram. Both terms have their place. The pleurogram is the special area which marks the side of the seed, whether in Mimosoideae or *Cassia* (Figs. 318, 331), while the *linea fissura* is the boundary of the mark. The variations that may be expected in large genera have been revealed by Vassal in the case of *Acacia*, particularly in respect of the presence or absence of endosperm. *Swartzia* is such a case for which I describe here the albuminous seed of *S. madagascariensis*. Knowledge of leguminous seeds may be extensive but it is not yet comprehensive enough for finer distinctions of generic, if not tribal, rank. Nevertheless, the peculiarity of the seed of Bauhinieae (Corner 1951) is largely sustained by *Cercis* (p. 168) and *Barklya* (p. 165), which has hitherto been mistaken as papilionaceous. The removal of *Barklya* from Papilionatae tr. Cadieae and the discovery of the papilionaceous seed in *Cadia* helps to clarify the nature of the tribe which seems to be allied with Swartzioideae (p. 172).

Then there appear to be differences in the way in which the mesophyll of the testa multiplies its cell-layers (Pitot 1935a, b). In Caesalpinioideae the mesophyll is diffusely multiplicative, but in Papilionatae o.h., i.e. and the middle layer of cells may be variously engaged.

For purposes of identification, Chowdhury and Buth have recently published a critical study of the commoner pulses of India, namely *Cajanus, Cicer, Cyanopsis, Dolichos, Lathyrus, Lens, Phaseolus, Pisum, Vicia, Vigna*.

In proposing the durian-theory (Corner 1949a), I drew attention to the clear evidence in Leguminosae for the loss or degeneration of the aril. My purpose was to employ a large, natural and well-known group of plants as an instance of that comparative morphology which is necessary for the evolutionary approach. Examples may be picked out from the whole range of flowering plants to prove almost any conjecture, unfortunately too commonly; the test is the systematic evidence

within an alliance. I concluded that the arillate, fleshy and, possibly, spinous pod was the primitive leguminous fruit in which the exotestal seeds were of the Mimosoid–Caesalpinioid kind. The conclusion has been vigorously attacked by van der Pijl (1952, 1955, 1956, 1966; Corner 1952). He has paid little attention to the points which I made, particularly in regard to the vestigial aril in indehiscent fruits and the degenerate aril of the bean-hilum, except to assume that the aril in such cases is disappearing. He argues instead his thesis that the primitive leguminous seed was sarcotestal (see p. 51). The facts which he advances for this supposition in Leguminosae I find untenable. Firstly, he assumes that *Archidendron* has a sarcotesta. The genus has been monographed (de Wit 1952), and no mention is made of a sarcotesta. It does not occur in the species here described (p. 165), which is plainly exotestal. De Wit did not consider the multicarpellary gynoecium of *Archidendron* of phyletic significance, for it is little more than a multiplication of the single carpel in *Pithecellobium*, but the multifollicular fruit is surely of importance because it shows the massive post-anthesis growth of the flower on a grandly primitive scale, comparable with that of Annonaceae. The unicarpellary *Parkia*, with perfect flowers over the centre of the capitulum, develops a similar cluster of pods but, as its flowers are individually simplified, it employs an inflorescence for the effect of the fruiting flower of *Archidendron*. Therefore, that this primitive leguminous fruit should have typical exotestal seeds disproves the hypothesis of the sarcotesta. Secondly, for *Parkia*, van der Pijl again assumes a sarcotesta, and again there is none: what soft tissue adheres to the seed is endocarpic (Corner 1951). The capitulum of *Parkia* with male and perfect flowers, even with sterile petaloid flowers at the base in American species, functions as a single flower of second-order construction similar to the capitulum of Compositae; that it should fruit with a cluster of pods, functioning as those of *Archidendron*, with exotestal seeds, discredits a sarcotestal genesis. Finally there is the ambiguous case of *Inga*, in particular the Peruvian fruit of *I. feuillei*. The seeds of *Inga* are what I had called overgrown, with little or no differentiation of the testa. Some, as I have seen, have a thin black testa and rather pulpy mesophyll, comparable with some seeds of *Pithecellobium* (e.g. *P. clypearia, P. ellipticum*). In others, including *I. feuillei*, the testa is thin-walled and pressed so tightly against the endocarp pulp that it is impossible to separate the two and the embryo appears to lie naked in the loculus of the fruit (Boelcke 1946); these fruits, too, may be more or less fleshy and indehiscent. The testa in such cases appears to be more or less sarcotestal but it is certainly not structurally comparable with the sarcotesta of *Magnolia* formed on the outside of a woody endotesta in the inside of a woody dehiscent follicle, or with the unitegmic sarcotesta of *Cycas*. In stating that this is so, van der Pijl uses the term sarcotesta with that ambiguity of which he complains in the use of the word aril (p. 51). Further, in proof of the derived and overgrown nature of the seed of *I. feuillei*, Boelcke found that it is commonly polyembryonic, as may happen with the overgrown seed of the mango.

The sarcotestal theory is unnecessarily involved. It requires, though the author is not explicit, firstly an unspecialized pulpy testa, then with its hardening in the exotesta the production *de novo* of an aril as the attraction about the better protected seed; then this aril is lost as dispersal becomes independent of animals, or, if still dependent, then the fleshy and drupaceous legume appears around the seed in which the testa has reverted to its primitively unspecialized structure. No systematic evidence is adduced for these transformations. The last state may as well be that of *Inga* as the first. On the durian-theory, *Inga* has arrived at the exarillate state with unspecialized testa such as may be found in many other genera of Leguminosae, e.g. *Pithecellobium*, *Koompassia* (p. 170, with soft testa), *Bauhinia* (Corner 1951, fig. 27), *Cordyla* (p. 169), *Millettia* (Corner 1951, figs. 25, 26), *Inocarpus* (p. 170, with stomata as in *Magnolia*), *Pongamia* and *Arachis* (Vaughan 1970). If this state were primitive it would imply that these genera, or their immediate allies, had evolved independently of each other that set of genes which make the universal family character of the leguminous seed, as if apetaly, which occurs throughout the family, had evolved in every case into a corolliferous flower, or the one-seeded indehiscent fruit had become on many separate lines, not necessarily the apetalous or that with undifferentiated testa, into the many-seeded dehiscent legume. It would mean that the distinctive characters of the Leguminosae were evolved in parallel along many different generic lines, which is absurd for these characters are the primes of the ancestral heritage. I conclude that the sarcotestal theory is not founded on fact and introduces only confusion.

As red and black to the aril round the seed, so in Leguminosae there are red and black seeds. To durianologists they appear as seeds on to which the red colour of the aril has been transferred and in the end they become red seeds. They are not primitive, but I still consider that they are an outcome of the arillate seed, perhaps in modern avian circumstances. Parrots are fond of these hard seeds. With suitable beaks they crack them and eat the kernel, which is no use for the plant, but parrots are vociferous and quarrelsome, and half the seeds they crack are spluttered down to germinate, when intact they fail. I was drawn to this detail by the Mimosoid *Adenanthera pavonina* (Fig. 318). Its red seed has at the top of the short pink funicle a vestigial aril, which is a significant detail overlooked by van der Pijl in his criticism. New evidence has come forward from the papilionaceous *Bowringia* and *Leucomphalos* (Hallé

1965), in support of this idea of transference of function. There is no sarcotesta in any of these seeds, or their allies.

The Leguminous aril is funicular. *Archidendron* may be an exception. It is said to be exarillate but there is a vestigial aril-like outgrowth from the micropyle of *A. solomonense* (Fig. 319). Such minute details are easily overlooked in the examination of dried specimens and, if found, are as difficult to investigate. It is to be hoped that adequately preserved material of other species can be studied.

The hourglass-cells are situated between the palisade-cells and the more or less stellately armed cells of the mesophyll. There appear to be two form-factors, one for the columnar palisade-cells, the other for the stellate-cells, and where they meet the intermediate form of the hourglass-cell develops. In the case of the inner hourglass-cells derived from the endotestal layer, it seems that both factors must be at work but that that of the palisade is weak. Their presence may indicate that there was an endotestal palisade in the primitive Leguminous seed, as there is in the Myristicaceous.

Custom divides Caesalpinioideae into six or seven tribes, as in Bentham and Hooker's Genera Plantarum. Hutchinson (1964) has dismissed them and appointed in their place as many groups of genera which are so strangely unnatural that they can be accepted only as an artificial aid to superficial identification. *Tamarindus* becomes widely separate from *Amherstia* and grouped with *Cassia*. *Saraca* is next to *Hardwickia*. *Cercis* (Fig. 334) goes with *Koompassia* and *Sindora* (Fig. 341), while, mostly ridiculously, *Bauhinia* (Figs. 321, 322) comes between *Hymenaea* and *Trachylobium* (Fig. 339). If the customary classification is followed, it will be found that several tribes have remarkable and constant distinctions, to some of which Baillon and Thompson (1924, 1925) have drawn attention. Amherstieae, decimated by Hutchinson, have four sepals (the two posterior being connate); the receptacular tube develops on the anterior side of the flower to give the posterior ovary; the style is recurved and coiled in the bud; the funicle is often arillate or clavate. In contrast, Bauhinieae have five connate sepals, posterior receptacular tube with anterior ovary, straight style in the bud, and a particular seed (p. 162). Caesalpinieae have the central ovary and the style irregularly curved or kinked in the bud. Cassieae have an almost hypogynous flower with the style following the curve of the ovary to the posterior side. Cynometreae, however, are a mixture of Amherstieae and Caesalpinieae (? also Cassieae); the tribe is a grade of genera with paucivulate legumes. Dimorphandreae have Mimosoid links, as noted by Brenan (1967) in the most recent and extensive account of a large tropical Leguminous flora. The floral characters of the main tribes are supported by those of leaf and seed, and will surely be supported by those of the fruit when its construction is known. I conclude, indeed, that seed-structure will become the basis for the natural classification of Leguminosae.

Acacia Mill.

For details of the seed in many species of *Acacia*, particularly in regard to the *linea fissura*, the funicle, the thickness of the testa, and the presence of endosperm, there are the recent exhaustive accounts given by Vassal (1968, 1971).

A. auriculaeformis A. Cunn.
(Singapore; living material.) Fig. 317.

The development of the long funicle is shown in Fig. 317. The ovules are fixed in shallow sockets along the centre of the growing pod. It reaches full-size before the ovules enlarge into seeds, so the funicle is elongated and then coiled through its own intercalary extension and the backthrust of the enlarging seed.

Adenanthera L.

Ovules anatropous; o.i. 3 cells thick; i.i. 2 cells thick.

Seeds 6–11 mm long, ovoid-subcordate, slightly compressed, shining red or with the chalazal half black, with *linea fissura*, hanging on persistent funicles.

A. pavonina L.
(Singapore; living material.) Fig. 318.

Seeds typically mimosoid (Corner 1951).

In this species there is a minute, pink, and lobulate aril covering the micropyle and developed from the end of the funicle, which is pink from anthocyanin. There is a slight horny ridge on the testa at the entry of the raphe v.b., and the palisade in this part has no *linea lucida* and does not swell in water. Thus the seed is well sealed, until cracked by parrots or cooked by a forest fire or, perhaps, decayed after many years in the soil. Van der Pijl (1955) overlooked the detail of the aril which shows that the seed is derived from a state with black seed and red aril, cf. *Pithecellobium*. It is in no way a sarcotestal derivative.

Archidendron F. Muell.

A. solomonense Hemsl.
(Solomon Islands, RSS 2845; fruits in alcohol.) Fig. 319.

Seeds 17–20 × 11–13 mm, ellipsoid, smooth, black with blue-grey bloom, hanging on short black funicles 4–5 × 0.5–1 mm. Testa *c.* 23–30 cells thick; *o.e.* as a palisade 58–70 μ high, composed of prismatic cells, the radial walls thickened outwards with narrowed lumen, the outer wall thick, not or slightly lignified, without *linea lucida*; *mesophyll* aerenchymatous, more or less crushed, without hourglass-cells (not even in the hilum); *i.e.* unspecialized. Tegmen evidently crushed. Vascular bundle as a single broad strand in raphe and antiraphe, dilated and shortly 3–4 lobed at the termination in the antiraphe near the micropyle. Hilar region with rounded, shortly armed, more or less thick-walled, unlignified cells. Micropyle surrounded with a small, pale brown patch, 1.5–2 × 1–1.5 mm, of excrescent mesophyll-like cells with thick walls, smaller at the surface but not covered by the exotestal palisade. Endosperm none. Cotyledons shortly bilobed at the base.

Pods 1–2(–3) per flower, borne on the trunk, curved, more or less deeply constricted between the 7–9 seed-cavities, but only 1–3 seeds per pod, 15–27 mm thick, the wall 6–7 mm thick, scarlet with bright yellow interior. Mesocarp composed of large cells, more or less radially elongate with an inner ring of large v.b. giving smaller fibro-vascular branches to the outer part and forming a network of fine, mostly fibro-sclerotic, strands near the short-celled outer epidermis, many of these sclerotic cells like idioblasts, also with many scattered or clustered sclerotic cells in the general tissue; *endocarp* composed of 3–5 layers of cells transformed into lignified, obliquely crossing, narrow fibres, the inner epidermis composed of small thin-walled cells.

The small area of excrescent cells surrounding the micropyle, but devoid of testal palisade and free from the funicle, is peculiar. It suggests a vestigial exostomal aril, but only mature seeds were available for study and the tissue of this small outgrowth was too friable to make out its exact structure. No aril has been recorded for the genus (de Wit 1952), but the vivid pods suggest by analogy with *Acacia* and *Pithecellobium* that vestigial arils may occur.

The trunk of this tree, *c.* 20 m high, was covered with the brilliant pods. They may be eaten by opossums which abound in these islands, and the cockatoos may pluck off the dangling seeds.

Barklya F. Muell.

B. syringifolia F. Muell.
(Brisbane, leg. L. J. Webb; flowers and pods, dried and in alcohol.) Fig. 320.

Seeds 12–14 × 6–7 × 2 mm, rather bean-shaped but much compressed, with slightly projecting micropyle and hilar region rendering the base asymmetric, grey-brown, smooth, albuminous, ? exarillate. Testa 9–11 cells thick on the sides of the seed, thicker along the raphe; *o.e.* as a palisade of Malpighian cells with distinct *linea lucida*, not lignified, with hexagonal facets and subquadrate lumen (in t.s.); *mesophyll* thin-walled, aerenchymatous, without hourglass-cells. Tegmen crushed. Chalaza on the opposite side of the seed from the micropyle and at the same level, unspecialized. Hilum very small, slender, crescentic as in *Bauhinia*, without counterpalisade, without tracheid bar, without recurrent v.b., with the *linea lucida* about half-way up the palisade around the funicle v.b.; subhilar tissue aerenchymatous, thin-walled. Vascular bundle simple, extending from the hilum to the chalaza. Endosperm copious, thick-walled. Embryo with short curved radicle; cotyledons thin-walled.

The one species of this genus has been referred to Papilionaceae tr. Cadieae where its simple leaf and the variable aestivation of the corolla, noted by Baillon, are exceptional. I am grateful to Dr L. J. Webb who sent me in 1951 material from the tree in the Brisbane Botanic Gardens. From this I discovered that the seed agrees in all details with that of *Bauhinia* except for the curved radicle, and this seems but a slight enhancement of the condition in other species of *Bauhinia*, e.g. *B. picta* (Corner 1951, fig. 10). The much compressed seed, the perimeter of which is built mostly from the raphe, and the fine crescentic attachment of the funicle are decisive. The simple leaf, as in *Cercis*, is the terminal pinna of a pinnate leaf, from which the bifid state in *Bauhinia* has been derived (Cusset 1966). Indeed, it is not clear that *Barklya* is distinct from *Bauhinia* in its wide sense or from *Gigasiphon* in particular, for the origin of which Cusset has postulated eastern Malaysia.

Bauhinia L.

Seeds much flattened. Testa; *o.e.* as a palisade of Malpighian cells, or reduced to a layer of cuboid cells; *mesophyll* without hourglass-cells. Raphe longer than the antiraphe; *v.b.* without postchalazal extension. Hilum crescentic with 1–2 aril-lobes as narrowly adnate ridges along part of the raphe, or absent; without recurrent v.b. in the subhilar tissue, but often with a sudden flexure of the v.b. into the raphe and a slight projection into the subhilum. Funicle flattened, conspicuous, smooth or hairy and then often connate with the loop of endocarp hairs. Endosperm copious, sparse round the radicle, or none, the cells thick-walled. Embryo with broad flat cotyledons, the thin-walled cells without starch; radicle short, straight or slightly curved. Figs. 321, 322.

(Corner 1951; de Wit 1956.)

There is clearly much diversity in detail in the seeds of this large genus, or cluster of genera. One expects all transitions from the simpler states in *Barklya* and *Cercis*, which are without aril-lobes, to those in the more typical species of *Bauhinia*. The aril-lobes, or funicular branches of de Wit, are two extensions of the funicle-head along the raphe, developed after fertilization, and each connects with the mesophyll of the testa through a narrow isthmus in the exotestal palisade (Fig. 321). The lobes consist of a horny outer layer, 2–3 cells thick, composed of large watery cells with slightly thickened, contiguous walls, as in the vestigial arils of *Detarium*, *Hymenaea* and *Sindora*, and a central mass of substellate, thin-walled cells containing chlorophyll. On ripening of the seed, this central tissue dries up and detaches the lobe from the testa.

In *B. kockiana* (Fig. 322a) the aril-lobes are well developed and circumvent more than half the perimeter of the seed, but they become obscured with endocarp-hairs, as in *B. integrifolia*. In *B. acuminata* the lobes extend along most of one side of the seed (Fig. 322j). In *B. picta* and *B. fassoglensis* they are short (Corner 1951, figs. 10, 11, 12, 15), and they are even shorter in *B. tomentosa* and *B. variegata*. In *B. purpurea* one lobe is very short and the other may be missing (Corner 1951, fig. 13). In *B. integrifolia* (= *B. flammifera*) there are no aril-lobes, but the hairy funicle becomes entangled with the endocarp-hairs which extend around the narrow edge of the seed and detach from it as if they had formed an adnate hoop (Fig. 322h); de Wit, however, describes the funicle with two branches extending two-thirds of the circumference of the seed. How all these differences may be related in a comprehensive investigation of the genus which, notoriously has never been monographed, is impossible to say.

The seed of *Bauhinia*, with raphe longer than the antiraphe, is the counterpart of the Papilionaceous which extends the antiraphe or the hilum, just as the ascending imbrication of the corolla is the counterpart of the descending papilionaceous imbrication.

Cadia Forsk.

C. purpurea (Picc.) Ait.
(Ethiopia, J. J. de Wilde n. 4103, 4164; dried seeds.)
Fig. 323.

Seeds 8–9 × 5.5–6 × 3 mm, pinkish red to purplish or reddish brown, compressed, with subapical hilum. Testa; *o.e.* as a palisade of Malpighian cells, the angular facets with minutely stellate lumen, not lignified; hourglass-cells in one layer, or in 2 layers round the hilar region, slightly lignified; *mesophyll* thin-walled, but as a hard mass of thick-walled, substellate, unlignified cells round the hilum. Vascular supply consisting of a short raphe bundle and one recurrent bundle in the hilum, this median recurrent bundle evidently double at the junction with the raphe bundle. Hilum papilionaceous, small, circular, with the tough remains of the disrupted funicle; *tracheid bar* composed of lignified cells with annular-subreticulate thickenings. Endosperm copious, thick-walled.

(v.d. Maesen 1970.)

Seeds of this genus, which I have long endeavoured to study, were kindly sent to me by Dr L. J. G. van der Maesen. In his monograph of *Cadia*, he reviewed the problematic position of the genus, whether caesalpinioid or papilionaceous, and decided tentatively in favour of the first. The seed, however, is truly papilionaceous, if peculiar in the single median recurrent vascular bundle of the hilum. The two 'roots' of this bundle suggest that it arises in double. The tracheid bar also contacts the raphe bundle where it begins its course from the funicle. Hence I am forced to conclude that Cadieae are Papilionaceous and that the concept of the tribe has been perplexed by the inclusion of *Barklya* (Bauhinieae, p. 165). Nevertheless, the seeds of all genera of Cadieae need investigation.

The seed of *C. purpurea* resembles that of *Swartzia madagascariensis* (p. 172) in most features, but there is no recurrent v.b. in the hilum of *S. madagascariensis*. Yet, *S. pinnata* has one recurrent v.b., as in *C. purpurea*, but lacks the tracheid bar. One can hardly doubt that there is some significance in this resemblance, which may be explained when the seed-structure is known in other species of *Swartzia* and genera of Swartzieae.

Cassia L.

Ovules anatropous, often very numerous (–120); o.i. 3–4 cells thick; i.i. 2–3 cells thick.

Seeds 2–12 mm long, brown, generally somewhat flattened, often with filiform funicles, exarillate (? in all), albuminous, typically caesalpinioid, variously orientated with septation of the pod. Testa multiplicative, brown from the cell-contents of all tissues, in some cases from the pale brown walls; *o.e.* as a Malpighian palisade, in some species with the outer stratum developed into a pleurogram; *mesophyll* generally thick-walled, not or scarcely aerenchymatous except in the subhilar tissue, generally with hourglass-cells in o.h. (short or absent from the raphe-antiraphe); *i.e.* often with hourglass-cells. Hilum very small, closed by the palisade of the testa. Cotyledons flat, warped, or folded, with thin-walled cells containing oil and protein, without starch or with few and very small grains 1–2 μ wide. Figs. 324–333.

Septation of the pods

The development of transverse septa from the endocarp between the young seeds affects the position and shape of the mature seed. How the septum develops is shown in Figs. 326, 328. The ovary lengthens after pollination. Its internal epidermis begins to multiply and to form between the young seeds annular ingrowths that soon meet along the mid-line of the pod. The septa then appear complete but the lines of junction of the ingrowths can generally be detected microscopically, if not with the hand-lens (Fig. 327). The ingrowths form the endocarp-pith which, as a pale green watery tissue, shapes the cavities in which the young seed expands and then cushions them before, in most cases, it dries up into a white papery membrane. The manner of ingrowth is primitive and recalls the development of the cortex in parenchymatous algae with proliferating epidermis. The young pod, as usual, enlarges to its full size before there is much growth of the ovules, other than lengthening of the funicle.

As the septa grow, the hypodermis of the endocarp begins to thicken by cell-division and elongation; eventually it forms the woody, fibrous, endocarp-valve.

At the same time the centres of the septa develop radially elongate cells and differentiate into the more or less fibrous or woody plates which persist as the septa of the ripe pod after the endocarp-pith has dried up or become slimy. These plates girder the pod between the longitudinal valves and fortify the seed-cavities. Further growth of the pod introduces the following modifications:

(1) *Indehiscent pods* (subgen. *Cassia*). In *C. fistula* the endocarp-valves are thick, woody and straight, but the septal plates are thin (Fig. 324). The mesocarp is thin and dries into the black skin over the valves. The endocarp-pith disorganises into a brownish olive slime which fills the loculi round the seeds. In *C. javanica*, with similar long, black, cylindric and indehiscent pods, the endocarp-valves are thin; the septal plates are strengthened by a semi-annular thickening on each side between the edges of the seeds (Fig. 324); the endocarp-pith forms a rather corky tissue round the seeds which appear embedded individually in thick discs. In this subgenus the seed is strictly transverse and flattened in this plane at right angles to that of the raphe-antiraphe, which is longitudinal in the pod. The cotyledons are warped and oblique (de Wit 1955).

(2) In *C. spectabilis* (subgen. *Senna*) the thin endocarp-valves become arched between the seeds as seen in transmedian l.s. of the pod (Fig. 327). They are strengthened by accessory and semi-annular hoops of woody, fibrous tissue formed by division of the adjacent mesophyll-cells (Fig. 328). The endocarp-pith makes complete septa but begins to shrivel to a dry, white and papery film as the seeds enlarge. As the pod ripens, the mesophyll dries, and the accessory hoops give the transverse corrugations of the mature pod. These contain 50–70 seeds and dehisce from apex to base, in the usual basipetal manner; the distal seeds may ripen several days before the proximal. The seeds are transverse in the pod but flattened in the plane of the raphe-antiraphe which lies transversely.

(3) In *C. siamea* (subgen. *Senna*) the seeds are transverse but orientated with the plane of the raphe-antiraphe longitudinal in the pod (as in subgen. *Cassia*) and they are flattened in this same longitudinal plane (Fig. 325). The septa, as seen in transmedian l.s. of the pod, are short and the endocarp-pith, which extends longitudinally in the loculi, dries up. *C. auriculata* and *C. multijuga* are similar.

(4) In other species of subgen. *Senna* (*C. alata*, *C. hirsuta*, *C. occidentalis*, *C. tora*) the septa develop as in *C. spectabilis* but they are not strengthened by accessory hoops of fibrous mesophyll and the septal plates are usually incomplete so that a central hole appears in them as the endocarp-pith dries up. The septa are transverse in *C. alata*, slightly oblique in *C. hirsuta* and *C. occidentalis*, and strongly oblique in *C. tora*. Their seeds show the following complexities:

(i) raphe-antiraphe in the longitudinal plane of the pod.
  (*a*) cotyledons in the transverse plane, slightly folded;
        seed flattened transversely ... *C. hirsuta*
  (*a*) cotyledons in the longitudinal plane.
      (*b*) seeds not flattened; cotyledons much folded ... *C. tora*
      (*b*) seeds flattened transversely, subtetragonal in raphe-
            view; cotyledons warped ... *C. alata*
(i) raphe-antiraphe in the transverse plane; seeds flattened
    in the same plane; cotyledons in the same plane, not
    folded ... *C. occidentalis*

In the last case of *C. occidentalis* the very young seeds
rotate through 90° (Fig. 329). At this time the cortex
of the funicle, the hypostasial tissue, and the inner
epidermis of the exostome colour pink through anthocyanin; the colour soon fades in exostome and hypostase,
but persists in the funicle which is dull red at maturity.
I have not observed this coloration in other species,
but in *C. hirsuta* the chalazal tissue turns dark brown
soon after fertilization. The pink of *C. occidentalis*
recalls the coloration of the funicle in *Adenanthera*.

Pleurogram
This mark occurs in several species of subgen. *Senna*,
but its systematic value is unexplored. It is a lateral
patch completely demarcated by a line or zone, but it is
not surrounded by a *linea fissura*. The demarcation is
caused by the position of the *linea lucida* in the palisade
of the testa (Fig. 333). Over the pleurogram the *linea
lucida* lies about the middle of the palisade which has an
outer stratum comprising the thickened and diffluent
ends of the Malpighian cells without lumen, and an
inner stratum consisting of the body of these cells
with lumen. Over the rest of the seed the *linea lucida*
lies, as usual, near the surface of the palisade. The
difference is readily seen in sections mounted in water
which swells the almost structureless outer stratum.
In the dried seed this layer is contracted and the
pleurogram is slightly sunken. Its surface is often marked
with minute, oblique, and short lines caused by the
uneven thickness of the palisade; they suggest stomatal
pits, as in *Magnolia*, but the Leguminous palisade is
closed and stomata occur in the family, so far as known,
only on the undifferentiated testa of *Inocarpus*.

The shape and size of the pleurogram depends to some
extent on the flattening of the seed and is often, but by
no means always, larger when the side of the seed is
broad because the plane of the raphe-antiraphe is
transverse. The distribution of the endotestal hourglass-
cells may also be connected with the pleurogram. Its
features, however, are not obviously connected to any
mechanics of the seed. In *C. multijuga* (narrow, oblong,
pale pleurogram, and seeds orientated as in *C. siamea*),
in *C. hirsuta* and *C. tora* (both with narrow, well-
defined pleurogram) the endotestal hourglass-cells
occur only under the pleurogram, not elsewhere in the
seed. In *C. occidentalis* they occur also in the hilum. In

C. splendida, without pleurogram, they occur all through
the testa, even under the v.b. of raphe and antiraphe,
but the side of the seed is marked with the same minute
and oblique slits which occur on the pleurogram in
other species. In *C. spectabilis*, also with seeds orientated
as in *C. occidentalis*, there is a vague oblong pleurogram
and the inner hourglass-cells occur all over the testa
except for the strip along the v.b. This is the case in
C. siamea and in *C. auriculata*, both of which have the
same longitudinal flattening of the seed and slight
pleurogram, but it occurs also in *C. alata* which has the
seed-orientation of *C. hirsuta* and a well-defined pleurogram. As the seed of *C. alata* ripens, the pleurogram
turns metallic green from an evanescent pigment in the
outer stratum of the palisade.

Funicle
As the pod widens and the developing endocarp-pith
fixes the ovule in the centre, the funicle lengthens. When
the seed grows the funicle is thrust back on itself and in
species in which the endocarp-pith dries up and con-
tracts, the funicle becomes coiled at the apex. Thus the
Cassia-seed, in other ways similar to that of *Delonix*
(Corner 1951), is peculiar among Caesalpinioideae for the
long cylindrical funicle as in Mimosoideae. The mature
tissue is horny with thick-walled and contiguous cells in
the cortex, and pale yellow, except in *C. occidentalis*.

Classification
This variety in pod-structure, seed-orientation, embryo-
orientation, pleurogram, and inner hourglass-cells
needs to be worked out for all species of this large genus
before they can be satisfactorily classified. The *Cassia*-
pod is one of the most fertile. In *C. fistula* there are 100–
120 ovules in the carpel and most, if not all, become seeds
in the pod which can reach 90 cm long. This number
diminishes to 3–4 seeds in some species of subgen.
Lasiorhegma, e.g. *C. mindanensis* (de Wit 1955). It is
not clear whether the state with most ovules is primitive,
as in *Delonix*, or it is a multiplication of an intermediate
state with *c.* 20 ovules, but the smaller numbers seem to
be reductions fitting the smaller flowers or smaller
plants. The indehiscence of the large pods of subgen.
Cassia seems to result from the excessive lignification
of the septal plates.

In the related genera *Labichea* (2–3 ovules per ovary)
and *Petalostylis* (several ovules), the pods have 1–2
seeds and the funicle is dilated into a white saucer-shaped
aril in *Petalostylis* or a globose swelling in *Labichea* (Fig.
341). Possibly a vestige of an aril may occur in some
species of *Cassia*, e.g. *C. absus* (de Wit 1955).

Cercis L.

C. siliquastrum L.
(Cambridge, University Botanic Garden; living material.)
Fig. 334.

Seeds 4.5–5 × 3.7 × 1.7 mm, ellipsoid, compressed, smooth, brown, albuminous, exarillate. Testa 5–6 cells thick on the sides of the seed, 7–8 cells in the antiraphe, –12 cells in the raphe; o.e. as a palisade 70–80 μ high, thicker round the hilum, composed of slightly lignified Malpighian cells with quadrate lumina and distinct *linea lucida*; *mesophyll* thin-walled, without hourglass-cells. Tegmen crushed. Hilum very small, 0.1 mm wide, ellipsoid, the surrounding palisade thickened with the *linea lucida* across the middle of the cells, without counter-palisade; without crescentic attachment of the funicle. Chalaza unspecialized, at the distal end of the seed. Funicle well-developed, green then drying brown, slightly plicate, attached to the seed only by the centre of its discoid head, composed of loosely aerenchymatous tissue traversed by the v.b., with a compact thick-walled epidermis of short cuboid cells, not lignified. Endosperm thick-walled. Embryo straight; cotyledon-tissue thin-walled.

I studied these seeds for comparison with those of *Barklya*. In both, as in all Bauhinieae so far as known, the v.b. of the raphe stops at the chalaza and there are no hourglass-cells, but the seed of *Cercis* is anatropous without becoming obcampylotropous and the hilum is simple. That is to say, the seed of *Cercis* is less typically Bauhinioid than that of *Barklya*. The leaf of *Cercis*, with pulvinus at the apex of the petiole, is clearly the terminal pinna of a pinnate leaf, as in *Barklya* and *Gigasiphon* (Cusset 1966). Thus *Cercis*, distributed as a Tertiary relic, appears as an ally of *Barklya* with the more primitive hilum and the vestige of the funicular aril of so many Caesalpinioideae. The central ovary is another primitive mark.

Cordyla Lour.

C. africana Lour.
(Rhodesia, leg. H. D. L. Corby; seeds in alcohol.) Fig. 335.

Seeds c. 4 × 2.5 × 1.2 cm, exarillate, exalbuminous. Testa c. 20 cells thick, all the tissues with lignified walls except i.e.; *palisade* c. 130 μ high, 2(–3) cells thick, composed of radially elongate, oblong cells with more or less thickened and pitted walls, not as Malpighian cells, absent from numerous small areas especially over the v.b. and round the hilum, these areas with thinner-walled cells not in a palisade; *o.h.* as a single layer of short, ill-defined, hourglass-cells; *mesophyll* aerenchymatous, the cells with stellate arms, becoming compressed, the outer 3–5 cell-layers with brown tannin contents, these brown layers more numerous round the micropyle; *i.e.* as a discontinuous layer of short cuboid cells with slightly thickened, unlignified walls and brownish contents. Tegmen crushed. Vascular bundle of the raphe stout, dividing at the chalaza into 3 branches, dividing to give 9–11 branches descending the seed almost to the micropyle, one or two branches anastomosing but most

remaining free. Hilum without tracheid-bar. Embryo straight; radicle short, immersed in the basal lobes of the cotyledons and far removed from the micropyle; cotyledons very massive, thick, each thickly 2-lobed at the base, the lobes of different cotyledons connate in pairs below the radicle.

(Guignard 1905; Engler 1910.)

Guignard argued that *Cordyla*, placed in Swartzieae, could be included in Papilionaceae in spite of the straight embryo or radicle, for such occurs in various Sophoreae, *Arachis*, *Cicer* and *Voandzeia*. However, there is no tracheid-bar or recurrent v.b. in the hilum but, if the seed is overgrown and but partly differentiated, as would appear from the double palisade without Malpighian cells, then these Papilionaceous features may have failed to develop, cf. *Millettia* (Corner 1951). The flower is not Papilionaceous.

Erythrina L.

E. subumbrans Merr.
(Singapore; living material.) Figs. 336, 337.

Seed typically papilionaceous, brown. Testa; *o.e.* with typical palisade; *mesophyll* with o.h. as ill-defined hourglass-cells, 2–3 layers of substellate cells, and a wide inner layer of subaerenchymatous tissue carrying the reticulum of v.b. (from the raphe and recurrent v.b. of the hilum). Hilum apparently without counter-palisade; *subhilar tissue* much thickened with stellate cells, separated from the palisade by many layers of thick-walled, more or less contiguous cells; *tracheid-bar* with isodiametric cells, lignified with reticulate thickening. Rim-aril white, horny, adherent to the testa, composed of radially elongate cells with close annular thickening, not lignified, the inner tissue connected with the subhilar tissue by a narrow annulus of cells with subannular thickenings, appearing as an isthmus in t.s. Funicle composed of horny agglutinated cells with thick walls, the central tissue loose and disrupting round the v.b., but this thin-walled tissue continued as a soft band along the side of the funicle towards the suture of the pod.

I have figured this seed because the connection of the rim-aril to the subhilar tissue, across the palisade of the testa, is the same as in *Bauhinia* (Figs. 336, 337).

Euchresta Benn.

E. japonica Benth.
(Japan, Kyoto, leg. E. J. H. Corner s.n.; fruits in alcohol.) Fig. 338.

Seeds 15 × 8 mm, ellipsoid, smooth, symmetrical, exalbuminous. Testa 14–18 cells thick on the sides, c. 30 cells thick along the raphe, 7–10 cells thick along the antiraphe, thin-walled, not lignified; *o.e.* as a short palisade of cells with slightly thickened outer and radial walls, with wide lumen, and hexagonal facets shortly oblong in various directions in different parts of the seed. Tegmen crushed. Chalaza unspecialized. Vascular bundle of the raphe ending at the chalaza, without branches. Hilum 1 mm wide, small, round, unspecialized, without rim-aril, tracheid-bar or recurrent v.b. Cotyledon-tissue thin-walled, starchy.

Pod 1-seeded, drupaceous, bluish purple; o.e. composed of small cells with thick outer walls; exocarp c. 10 cells thick, composed of small cells; mesocarp composed of large oblong cells, many with slightly lignified, but thin, walls; endocarp as a thin-layer of transversely elongate sclerotic, almost fibre-like, lignified cells.

This is clearly an overgrown seed with minimal testal differentiation, hard to identify as Leguminous.

Hymenaea L. and Trachylobium Heyne

These related genera with indehiscent fruits and large seeds both possess a vestigial funicular aril similar to that of *Intsia* (Corner 1951). All three belong naturally in Amherstieae.

H. courbaril L.

(Singapore, cult.; living material.)   Fig. 339.

Seeds 2.5 × 1.6–1.8 × 1.3–1.4 cm, dark brown, exalbuminous. Testa; *o.e.* as a tall palisade marked along the middle with a dark line; *mesophyll* with hourglass-cells, a single layer of stellate cells, and a thick inner layer of tangentially elongate cells, strongly thickened along the raphe-antiraphe. Funicle thickened into a small white fleshy aril covering the hilum and micropyle, consisting of an outer, rather horny layer of large, contiguous, thin-walled cells and an inner pithy tissue of substellate cells filled with starch, eventually drying up.

T. verrucosum (Gaertn.) Oliver

(Singapore Botanic Gardens, cult.; living material.) Fig. 339.

Seeds as in *Hymenaea courbaril*, but also with a hoop-like thickening of the testa along the sides of the seed, thus giving four internal ridges in t.s.

Inocarpus Forst.

I. edulis Forst.

(Singapore, Botanic Gardens, cult.; living material.) Fig. 340.

Seed large, exalbuminous, exarillate. Testa 7–8 cells thick, thin-walled and unspecialized except for the stomata in o.e. and the v.b. Tegmen crushed. Hilum with slight tracheid-bar in young seeds but, apparently not differentiating a mature structure. Vascular supply consisting of a short raphe v.b. and two recurrent v.b. in the hilum giving off many branches to the sides of the seed. Cotyledon-cells thin-walled with abundant starch.

This enigmatic genus has been referred to Sapotaceae (Jussieu 1789), Lauraceae (Sprengel 1817), Hernandi-aceae (Blume 1825, Endlicher 1836–40), Thymelaeaceae (Endlicher 1842) and Papilionaceae–Dalbergieae (Bentham 1862). Hutchinson (1964) places it in the mixture of genera which he assigns to Cadieae and next to *Barklya* which should go with Bauhinieae (p. 165). The presence of recurrent v.b. in the hilum, from which the vascular supply of the sides of the testa is derived, as in *Erythrina* and *Mucuna* (Corner 1951), and the vestigial tracheid-bar prove that the genus is Papilionaceous, though the exact affinity is unknown. The presence of stomata in the undifferentiated testa is unique among Leguminosae. The radial symmetry of the flower with K(5) C(5) A 5 + 5 G1 is, if anything, leguminous, then mimosoid, but the almost entirely gamosepalous calyx recalls Swartzieae. *Inocarpus* may be a relic of the common ancestry of Swartzieae and Papilionaceae with secondarily actinomorphic flower comparable with the mimosoid; or it may be a primarily actinomorphic flower with reduced indehiscent pod and overgrown seed, comparable with *Aldina* (Swartzieae). Now that universities are being established in the Pacific region, botany may give more attention to this common tree that defies classification. The red gum in the trunk may be a help.

Koompassia Maingay

K. malaccensis Maingay

(Singapore; living material.)   Fig. 341.

Seeds 5 × 2 × 1 cm, flattened, pale pinkish brown, exarillate, easily detached from the short funicle. Testa soft, subcoriaceous, unspecialized, the thin-walled cells eventually collapsed. Endosperm horny, grey, thick-walled, swelling in water into a structureless jelly. Cotyledons with thin-walled cells containing oil-droplets and a few very small starch-grains.

Pod 1-seeded, indehiscent, twisted at the base; mesocarp thin, drying into powder and disappearing except for the veins; endocarp pale brown, as a tough papery capsule round the seed, without endocarp-valves.

It will be interesting to learn how this peculiar pod and overgrown seed may have been derived in Cassieae.

Ormosia Jack

Seeds papilionaceous, red or red and black with the black area at the micropylar end or along or near the raphe, exalbuminous. Testa; *o.e.* without distinct *linea lucida*, the cells without mucilaginous ends, the inner half of the cells with rather wide lumen; *mesophyll* with outer hypodermal hourglass-cells rather poorly differentiated except round the hilum in some species, the rest of the tissue composed of slightly thick-walled, substellate cells. Vascular supply consisting of the unbranched raphe-bundle and two short recurrent v.b. in the hilar tissue. Hilum shortly oblong, with a counter-palisade; *tracheid-bar* surrounded by thick-walled stellate-celled aerenchyma. Funicle rather short and wide, with the rim-aril detaching or not from the seed, or without rim-aril but the anterior edge of the funicle thickened into a horny strand and serving to suspend the seed. Embryo with short straight radicle; cotyledons with thick-walled cells, not amyloid, without starch.

This genus, comparable in its red or red and black seeds with *Erythrina*, has particular interest because, firstly, it is Caesalpinioid in the anatropous ovule and seed and in the straight embryo with thick-walled cotyledon cells, but it is Papilionaceous in the structure of the hilum. Then secondly, it relates with the rare *Arillaria robusta* with black seed covered by a pulpy red aril. *Arillaria* has been reduced to *Ormosia*, in which case *Ormosia* displays all transitions from the arillate seed to that with the rim-aril and to the exarillate, but seed-structure needs to be studied in *Arillaria*. It is to be hoped that someone in Assam, Burma or Thailand will encounter it before deforestation extinguishes it and so many other trees of exceptional interest and rarity in those countries.

O. bancana Prain
(Singapore; living material.) Fig. 342.
Seeds 8–9 mm long, 1–3 per pod, coral red, hanging. Hilum without rim-aril. Funicle with the anterior border (that towards the distal end of the pod and on the micropylar side) thickened into an undulate horny strip, detaching from the rest of the funicle and suspending the seed by adnation to one side of the hilum near the micropyle.

Though the red seeds contrast with the dark green foliage, they seem generally to drop off through wind and rain and to germinate under the tree. Most Asiatic species of *Ormosia* resemble *O. bancana* in seed.

O. henryi Prain
(Hong Kong, H.K. 1783; dried material.) Fig. 342.
Seeds 10 mm long, subcompressed, 3–8 per pod (7–11 × 3 cm), detaching completely from the rim-aril, ? not suspended. Funicle without horny anterior border; rim-aril horny, yellow, composed of radially elongate cells with thick yellowish walls; subhilar tissue with well-developed hourglass-cells in 4–5 hypodermal layers but the inner ones more or less as stellate cells.

O. macrodisca Baker
Seeds large, round, subcompressed, 1 per pod, red with black hilar area. Funicle apparently without rim-aril or horny anterior border.

Similar seeds occur in *O. fordiana* (South China) and *O. coutinhoi* (Brazil) which may have the largest seeds in the genus, measuring 3–4 × 2.5–3.5 × 1.5–2 cm.

O. semicastrata Hance
(Hong Kong, H.K. 10068; dried material.) Fig. 342.

Seed as in *O. henryi* but plumper with both narrow rim-aril, remaining attached to the seed, and horny convolute anterior border to the funicle, this acting as a pseudofunicle in suspending the seed and commonly splitting into two.

Physostigma Balf.
P. venenosum Balf.
(Berger 1964, p. 351.)

Pithecellobium Mart.
Seeds mimosoid but in many species with a rather soft testa, the palisade short and flexible; arillate or, mostly, exarillate.

P. ellipticum (Bl.) Hassk.
(Malaya; living material.)
Seeds 25 × 16 × 12 mm, ellipsoid, plump, greyish black with a waxy bloom, hanging on dried brown funicles 2–3 mm long, exarillate, without the *linea fissura*, exalbuminous. Testa with a short unlignified palisade *c.* 40 μ high, and with hourglass-cells, with a thin layer of wax on the outside of the palisade.

This is not a sarcotestal seed as van der Pijl (1955) would have us believe.

Pongamia Vent.
Vascular supply to the sides of the testa apparently derived from the recurrent v.b. of the hilum (as with *Mucuna*).
(Le Monnier 1872.)

Sindora Miq.
Seeds caesalpinioid, exalbuminous. Funicle thickened into a stout, hard, red, orange or yellow aril-like structure about as large as the seed.

S. wallichii Graham
(Singapore; living material.) Fig. 341.
Seeds 22 × 20 × 9–10 mm, ellipsoid, obtuse, compressed, blackish brown, but orange-brown before maturity, becoming transversely cracked. Testa very hard and woody; *o.e.* as a tall palisade of Malpighian cells with yellow-brown walls, traversed by a median brown line; *mesophyll* with a single layer of tall hourglass-cells in o.h., several layers of stellate-celled aerenchyma with thickened cell-walls, and internally several layers of tangentially elongate cells with slightly thickened, pitted walls; *i.e.* as a narrow layer of small, irregularly stellate cells with strongly thickened, colourless walls, as ill-defined hourglass-cells. Funicle-aril reddish brown to reddish orange, waxy, hard, traversed asymmetrically by the v.b., covering the micropyle but not embracing the seed, composed of thin- or slightly thick-walled compact cells with very oily contents, the colour in the walls of the epidermal cells (mainly the outer walls).

Hilum short, linear, consisting of a narrow strand of small thin-walled cells with the v.b. curving abruptly at one end into the seed. Cotyledons with very thick-walled cells, the walls amyloid, without starch.

Squirrels and rats eat the funicle-aril. Monkeys prefer the unripe but full-sized seed, and the prickly pod does not seem deterrent.

Sophora L.

S. tomentosa L.

(Singapore; living material.) Fig. 344.

Seeds 7 mm wide, subglobose, light brown, exalbuminous. Testa thin, except at the hilum; *o.e.* with a palisade of Malpighian cells, their inner ends rather wide and almost as hourglass-cells; *mesophyll* with or without very slightly differentiated hourglass-cells in o.h. Hilum oblong, with counter-palisade, the rim-aril slight and adnate to the testa; subhilar tissue composed mainly of thick-walled, more or less contiguous cells. Vascular bundle in the raphe, unbranched, with two short recurrent v.b. Embryo with short straight radicle; cotyledon-cells slightly thick-walled, with a few small starch-grains.

Swartzia Schreb.

The seed of *S. madagascariensis*, here described, is so different from that of *S. pinnata* (Corner 1951) that it is hard to believe they are congeneric. The first is almost papilionaceous, though albuminous and lacking recurrent v.b. in the hilum, and the second is almost caesalpiniaceous, though exalbuminous and with recurrent v.b. An error crept in, unfortunately, in my figure of *S. pinnata* where the cotyledon-tissue was labelled endosperm (Corner 1951, fig. 24A). The differences between the species are tabulated as follows:

	S. pinnata	*S. madagascariensis*
seed	large, ovoid, not compressed, not bean-like; raphe equal to antiraphe	small, compressed, subreniform, bean-like; raphe? shorter than antiraphe
aril	present	absent (at least from detached seeds)
hilum	elongate, convex, basal; counter-palisade scarcely developed; subhilar tissue lignified, bony	circular, o.8 mm wide, concave, lateral; counter-palisade well-developed subhilar tissue not or scarcely lignified
testa	thin, cracking and peeling off; without hourglass-cells	very hard and tough, not cracking; with hourglass-cells
v.b.	with short postchalazal extension; with 2 recurrent v.b. uniting into one subhilar v.b.; no tracheid-bar	? with postchalazal extension; without recurrent v.b., but with a short tracheid-bar joined to the raphe v.b.
endosperm	none	thick, thin-walled
embryo	straight; radicle short; cotyledons very thick	curved; radicle long; cotyledons scarcely thickened

These seeds, however, have some points in common. The unlignified palisade-cells of the exotesta have a stellate lumen in t.s. and they lack the *linea lucida*, except at the hilum of *S. madagascariensis*. The raphe bundle is unbranched but gives off along its sides slight processes; in *S. madagascariensis* the processes have no xylem. The raphe bundle does not reach the micropyle, as it does in Caesalpinioideae (except Bauhinieae) and Mimosoideae. The hilum has a longitudinal groove which is very fine in *S. pinnata* but more conspicuous in *S. madagascariensis* where it leads immediately to the tracheid-bar and this structure joins up with the raphe v.b. on its entry to the seed. It seems clear that the median subhilar v.b. of *S. pinnata* is the equivalent of the tracheid-bar, but it has longitudinal tracheids whereas *S. madagascariensis*, as in Papillonaceae, has transverse tracheids directed to the hilar groove. It will be interesting to learn of the seed-structure in other species of *Swartzia* of which there are more than a hundred. They may complete the range from the caesalpinioid to the papilionaceous or extend it. The seed of *Cordyla* (p. 169) is overgrown but reverts in symmetry to the caesalpinoid. Somehow the tracheid-bar, of unknown function, took its place in Papilionaceous ancestry; the difficulty is to decide what may be primitive in its inception and what is simple through deterioration. It seems that the bar was a vascular product of the raphe, perhaps with double origin which became the triple structure of bar and two recurrent vascular bundles. When, therefore, a hilum similar in structure to that of *S. pinnata* turns up in the Papilionaceous *Millettia ovalifolia*, with a bony ridge of sclerotic cells in place of the tracheid bar (Pal 1960), it may be that this structure is not primitive but a step in degeneration of the hilar structure as seen more definitely in *Millettia atropurpurea* (Corner 1951).

Concerning affinity with *Cadia*, see p. 166.

S. madagascariensis Desv.

(Rhodesia, leg. W. P. L. Sandmann; dried seeds.) Fig. 343.

Seeds 7–8.5 × 6 × 3.5 mm, ovoid-compressed, subreniform, smooth, light olivaceous fawn, albuminous, exarillate; hilum o.8 mm wide, round. Testa *c.* 9–11 cells thick on the sides of the seed, thicker along the raphe and at the hilum; *o.e.* as a palisade of Malpighian cells 80–90 μ high, with distinct *linea lucida* only at the hilum, not lignified, the lumen stellate in t.s.; *mesophyll* aerenchymatous, unspecialized except for the layer of well-formed hourglass-cells (o.h.), thick-walled in the hilar tissue with rounded cells and short arms, some of these cells slightly lignified; *i.e.* unspecialized. Tegmen crushed, but the endostome slightly lignified. Vascular bundle from the hilum encircling nearly one half of the seed, evidently as the raphe v.b., visible on the surface as a dark line with a slight purplish dilation half-way along its course and at the end (? somewhat dilated chalaza), giving off a short thick subhilar process as the tracheid bar composed of tracheids with annular and spiral or subreticulate thickenings and arranged more or less perpendicularly to the hilum, without recurrent v.b. Hilum with a lignified counter-palisade as well developed as the exotestal, remaining attached to the seed, the two palisades interrupted along the midline of the hilum by the slit leading to the tracheid-bar. Funicle with very slender v.b. and without trace of aril. Endosperm thick-walled, not starchy. Embryo curved; cotyledon-cells thin-walled.

Seeds of this African tree were kindly sent to me by Dr H. D. L. Corby (University College of Rhodesia). It is possible that they had a rim-aril which, on detachment of the seed, remained on the funicle.

LEITNERIACEAE

Ovules subanatropous, suspended, solitary, unitegmic, crassinucellate.

Seed albuminous. Testa *c.* 8 cells thick; *o.e.* thin-walled, unspecialized; *mesophyll* with the cells becoming slightly thick-walled with reticulate meshes; *i.e.* ? Tegmen ? 5–6 cells thick, obliterated except i.e. as a layer of rectangular cells with slightly thickened walls and reticulate markings. Perisperm as a few cell-layers. Endosperm nuclear, starchy. Embryo large, straight.

(Netolitzky 1926; Hjelmquist 1948.)

The starchy endosperm and the reticulate thickenings on the walls of the mesophyll-cells (testa) and of the endotegmen suggest that this family may not relate directly with other Amentiferae. Nevertheless a mesotestal derivation fits with Rosales and Hamamelidales.

LENTIBULARIACEAE

Ovules anatropous, unitegmic, tenuinucellate; integument 3–5 cells thick.

Seeds minute, 0.2–2 mm long, exalbuminous or with a trace of endosperm, exarillate. Seed-coat apparently not multiplicative, reduced to o.e. with thickened walls, often with minute external processes or cuticular spikes, or more or less papillate, the papillae in some cases barbed at the apex; *i.e.* at first endothelial. Endosperm cellular, absorbed or reduced to a single layer of cells.

(Netolitzky 1926; Kausik 1938; Khan 1954; Haccius und Hartl-Baude 1957; Casper 1966; Farooq 1964, 1965, 1966; Taylor 1964; Farooq and Siddiqui 1967.)

LIMNANTHACEAE

Ovules anatropous, solitary, erect unitegmic, tenuinucellate; integument 14–16 cells thick.

Seeds small, exalbuminous, exarillate. Seed-coat not multiplicative, unspecialized but with lignified cells at the micropylar and chalazal ends, 4–5 outer cell-layers persistent. Vascular bundle with 5–9 postchalazal branches. Endosperm nuclear. Embryo straight; radicle small.

(Netolitzky 1926; Mathur 1956; Maheshwari and Johri 1956; Johri 1963.)

Concerning the affinity of this family, see p. 35.

LINACEAE

Ovules anatropous, suspended, bitegmic, crassinucellate to tenuinucellate; o.i. 2–3 cells thick; i.i. 3–12 cells thick; micropyle formed by the endostome, or by the exostome (*Hugonia, Radiola*).

Seeds small, exarillate (? *Tirpitzia*); seed-coats multiplicative or not (*Linum, Radiola, Reinwardtia*). Testa 2–6 cells thick, multiplicative in *Hugonia*; *o.e.* often with thickened outer walls, with mucilaginous outer walls (*Linum*), or not (*Hugonia, Radiola*); *mesophyll* with sclerotic cells in *Hugonia*; *i.e.* sometimes dividing to form a mesophyll-layer, without crystals, lignified in *Hugonia*. Tegmen 4–12 cells thick; *o.e.* as a more or less compact layer of longitudinal, thick-walled, laterally compressed, pitted and regular fibres, appearing as a palisade in t.s., but largely absent from *Hugonia*; *mesophyll* unspecialized, crushed; *i.e.* at first as an endothelium, then as a layer of rectangular tannin-cells with slightly thickened and barred radial walls, not lignified. Vascular bundle ending at the chalaza. Endosperm nuclear, oily, 1–6 cells thick. Embryo straight; cotyledons flat.

(Guignard 1893; Netolitzky 1926; Crété 1937b; Narayana, L. L. 1964; Forman 1965.)

Durandea, Hugonia, Linum, Radiola, Reinwardtia, Roucheria.

A fibrous exotegmen, similar to that of Celastraceae, appears to be characteristic of Linaceae. It may be incomplete, however, in *Reinwardtia* with patches of thin-walled tissue between the bundles of thick-walled fibres. Hugonieae, with very hard endocarp to the drupe, seem to show the degeneration of this layer which is slightly lignified and widely pitted in *Roucheria* and unlignified for the greater part in *Hugonia*. Both these genera, however, have a multiplicative testa with abundant sclerotic cells in the mesophyll and a solid sclerotic mass round the ending of the vascular bundle at the chalaza. These sclerotic cells occur also in the endotesta of *Hugonia*. The simple tegmen of Lineae may, therefore, be reduced from a mesotestal construction represented by Hugonieae; it must be noted, nevertheless, that similar sclerotic cells occur in *Euonymus glandulosus* in its family of Celastraceae distinguished by the fibrous exotegmen. Possibly the seed of *Tirpitzia* will be critical (Forman 1965).

The disposition of Linaceae near Geraniaceae does not accord with seed-structure. The alliance is clearly with Malpighiaceae and, possibly, Oxalidaceae, also with mucilaginous seeds.

Hugonia L.

H. mystax L.
(Ceylon; flowers and fruits in alcohol.)　Fig. 345.

Ovules anatropous, suspended, several per loculus, bitegmic, crassinucellate; o.i. 2–3 cells thick; i.i. 3–5 cells thick; hypostase present.

Seeds 5–5.5 × 2.5 × 1.7 mm, compressed, pale brown, smooth, rather hard, exarillate but with rather prominent woody chalaza. Testa 4–6 cells thick, 2–3 at the exostome; *o.e.* with slightly thickened outer walls, little specialized, apparently with small air-spaces between the angles of the cells in the chalazal part of the seed (appearing as stomata in t.s.); *mesophyll* with small cells, many becoming subcuboid sclerotic cells, not as a continuous layer; *i.e.* composed of small cells, most becoming sclerotic, not elongate, without crystals. Tegmen *c.* 10 cells thick, thin-walled and eventually crushed; *o.e.* composed of somewhat longitudinally elongate cells, thin-walled, but with slightly thickened and lignified walls near the endostome formed of a plug of short sclerotic cells; *i.e.* composed of small cells with pale brown walls, not lignified. Vascular bundle terminating in a slight umbrella of tracheids in the chalaza. Chalaza woody, most of the tissue round the v.b. becoming sclerotic as the mesophyll-cells of the testa.

Fruit 4–5 locular, the hard stone incorporating the peripheral ring of numerous v.b.; mesocarp pulpy; exocarp with very thick-walled epidermal cells.

Linum L.

Ovule; o.i. 2 cells thick; i.i. 5–12 cells thick.

Seed with non-multiplicative integuments. Testa; *o.e.* with thick mucilaginous outer walls, the cells not or shortly radially elongate, with polygonal facets; *mesophyll* and *i.e.* crushed. Tegmen; *o.e.* with fibres –300 μ long, appearing as a palisade in t.s. Endosperm 3–6 cells thick.

(Guignard 1893; Dorasami and Gopinath 1945; Berger 1964, p. 313; Narayana, L. L. 1964; Vaughan 1970.)

L. grandiflorum Desf., *L. mysorense* Heyne, *L. rubrum* Rafin., *L. usitatissimum* L.

Radiola (Dill.) Roth

R. linoides Roth
Ovule; o.i. 2 cells thick; i.i. 4–6 cells thick through multiplication of i.e.; exostome exceeding the endostome.

Testa not mucilaginous; *o.e.* with thickened outer walls.

(Mauritzon 1934d; Crété 1937b.)

Reinwardtia Dumort.

R. trigyna Planch.
Ovule; o.i. 3 cells thick; i.i. 6–8 cells thick; integuments not multiplicative in seed.

Testa; *o.e.* with mucilaginous cells shortly tangentially elongate. Tegmen with clusters of thick-walled cells (? fibres), not as a continuous layer of thick-walled cells. Endosperm 1 cell thick.

(Narayana, L. L. 1964.)

Roucheria Planch.

R. griffithiana Planch.
(Bornea, G. D. Haviland n. 1473; dried seeds, CGE.)

As in *Hugonia* but the walls of all the cells in the testa slightly lignified, thin-walled except the abundant, scattered, sclerotic cells. Tegmen with o.e. evidently as a layer of longitudinal, regular, fibrous cells with slightly thickened, lignified walls and broad, more or less scalariform, pitting. Chalaza thickly sclerotic.

Tirpitzia H. Hallier

T. sinensis (Hemsl.) Hallier f.
According to a note from L. L. Forman, the seed is enclosed in a flattened arillate bag arising from the funicle and, possibly, serving as a wing.

(Forman 1965.)

LOASACEAE

Ovules anatropous, unitegmic, tenuinucellate; integument few to many cells thick, initially 2 cells thick then multiplicative through periclinal divisions of o.e. even to 17 cells thick; raphe v.b. undeveloped.

Seeds 0.5–5 mm long, mostly small, smooth, striate or reticulate, in a few cases winged, albuminous. Testa exotestal, not multiplicative, mostly crushed; *o.e.* with variously thickened and characteristic cells (? not lignified); *mesophyll* generally crushed; *i.e.* endothelial, then crushed. Endosperm cellular, oily, thin-walled. Embryo straight; cotyledons rather short.

(Kratzer 1918; Netolitzky 1926; Garcia 1962.)

The seeds of this family appear as neotenic enlarged ovules with an exotesta that may be characteristic of the genera. It is given as a palisade of cells with strongly thickened outer wall in *Loasa*, and as tabular polygonal cells with thick cuticle in *Eucnide*. *Blumenbachia* has fibriform cells with a peculiar stout annular thickening on the radial walls. *Cajophora* has subcuboid cells with reticulate thickening on the radial walls.

Blumenbachia Schrad.

B. insignis Schrad.
(Cambridge, University Botanic Garden; living material.) Fig. 346.

Ovule with the integument 12–15 cells thick.

Seed 2.5 × 1.5 mm, ellipsoid, pale straw brown, finely striate. Testa; *o.e.* composed of longitudinal fibre-like cells –900 × 100 μ, with strong yellowish brown thickenings on the radial walls, not pitted, not lignified; *mesophyll* thin-walled, the outer layers more or less persistent; *o.h.* with cushion-like, minutely subechinulate, thickenings on the walls, not lignified.

Cajophora Presl

Seeds small, brown, coarsely alveolate–reticulate. Testa; *o.e.* as a layer of large, radially elongate cells with stout brown walls, the radial walls with more or less reticulate brown thickenings, with hexagonal facets, not lignified; *mesophyll* crushed. Fig. 346.

I examined dried material of three species that I had collected in Bolivia around La Paz in 1947.

LOBELIACEAE

As Campanulaceae.
(Netolitzky 1926; Rosen 1932; Hewitt 1939; Cooper 1942a; Subramanyam 1949.)

LOGANIACEAE

Ovules more or less anatropous, unitegmic, tenuinucellate.

Seeds small to rather large, albuminous, exarillate. Seed-coat reduced to o.e., the inner cell-layers crushed; *o.e.* as a firm lignified layer of cuboid to radially elongate cells of varying character, hairy and with undulate facets in *Strychnos*; *mesophyll* thin-walled; *i.e.* unspecialized. Endosperm cellular, oily, with thin or thick walls, or nuclear in *Strychnos*. Embryo small and little-formed to moderately large. Figs. 347–350.

(Netolitzky 1926.)

With unitegmic and exotestal seeds Loganiaceae have little to add to the general theory unless to show the primitive status of the poorly developed embryo in sympetalous families. Details of the outer epidermal cells appear important in generic classification. The seeds of Buddleioideae have, it seems, a thin-walled and unspecialized coat, more or less crushed at maturity, but in *Polypremum* the inner epidermis persists with thickened inner wall; this may be a distinction of Buddleiaceae, though Hutchinson (1959) refers *Polypremum* to Loganiaceae.

Fagraea Thunb.

F. ceilanica Thunb.
(Ceylon, Peradeniya; fruits in alcohol.) Fig. 347.

Seeds 2.5 × 1.5 mm, ellipsoid–reniform, dark brown, covered with a thin gelatinous pellicle; hilum at one end. Testa reduced to o.e. with 2–3 layers of more or less crushed, thin-walled mesophyll-cells; *o.e.* as a palisade of shortly radially elongate cells, strongly elongate round the hilum, hard, crustaceous, with brown contents, the outer wall becoming mucilaginous and detaching the cuticle as the seed-pellicle (except at the hilum), the radial walls much thickened, lignified, with fine branching pits, but their outer ends thin-walled, the inner wall with a strong lignified reticulum of irregular meshes. Endosperm thin-walled, oily, and with aleurone crystals. Embryo small; cotyledons very short.

Fruit-peduncle with a ring of small cortical v.b. Fruit-wall with an outer and an inner ring of v.b.; ground-tissue with abundant idioblasts.

Geniostoma Forst.

G. randianum Merr. et Perry
(Bougainville Island, NGF 13593; flowers and fruits in alcohol.) Fig. 348.

Ovary bilocular. Ovules 230 × 160 μ, numerous, anatropous, half-embedded in the periphery of the thick axile placenta; integument 4–6 cells thick.

Seeds 0.8–0.9 mm long and wide, sublenticular, dark brown, minutely papillate, with the rigid funicle projecting from a slight depression on the flattened side of the seed. Seed-coat consisting of o.e. and the crushed brown inner layers of cells; *o.e.* composed of papilliform cells with brown, rigid, crustaceous and slightly lignified inner wall thickened to occlude the lumen, perforate radially by fine canals. Endosperm thin-walled, rather oily, not starchy. Embryo straight, 500 × 140 μ; cotyledons short.

Fruit capsular, ripening dull orange-buff, splitting longitudinally with the two orange-buff spongy placentas hanging out between the valves; seeds embedded in the placentas.

G. rupestre Forst.
(Solomon Islands, RSS 2694; flowers and fruits in alcohol.)

Ovules and seeds as in *G. randianum*; funicular side of the seed slightly concave and concealing the lignified head of the funicle in the depression. Seed-coat; *o.e.* with narrow remains of the cell-lumen and the thin outer wall external to the strongly lignified, finely perforate, and massive inner wall.

Fruit 12 × 11 × 10 mm, subglobose, slightly compressed laterally.

Polypremum L.

P. procumbens L.
Ovules amphitropous; integument 3–4 cells thick, –6 cells on the funicular side.

Seeds with thin-walled coat except for i.e., not multiplicative, compressed; *i.e.* developing as an endothelium, then with thickened inner wall, persistent.

(Moore 1948.)

Strychnos L.

Seeds medium-size to rather large, generally more or less orbicular–lenticular, covered with a short close appressed tomentum of unbranched hairs or glabrous. Seed-coat reduced to the lignified outer epidermis; *o.e.* consisting of somewhat radially elongate cells with thickened and pitted radial and inner walls, but with undulate facets, commonly with the outer wall lengthened into lignified hairs up to 1 mm long, in some cases with lignified rods inside the wall of the hair. Vascular bundle in the raphe short, shortly branched. Endosperm massive, thick-walled, oily. Embryo rather small but well-formed.

(Berger 1964, pp. 445–451.)

There are clearly many differences in the hairs on the seed-coat, as Netolitzky observed, and it seems that they may be critical in the classification of the species.

S. nux-vomica L.
(Ceylon; ripe fruits in alcohol.) Fig. 349.

Seeds 23 × 8 mm, thickly discoid, greyish, shortly and densely appressedly silky-hairy. Seed-coat crushed except for o.e., or with a few layers of thin-walled mesophyll-cells; *o.e.* composed of thick-walled, pitted, lignified, strongly undulate cells giving rise to one (? or several) stout hairs –1 mm long, supported internally by 6–15 longitudinal lignified rods attenuate towards the tips of the hairs and sometimes dichotomous, the inner face of the epidermal cell with bar-like or reticulate thickening. Endosperm with thick walls, oily. Embryo rather small, compressed; cotyledons thin, flat.

S. potatorum Linn. f.
(Ceylon, Yala, leg. Corner 1969; fruits in alcohol.) Fig. 350.

Seed-coat –12 cells thick, eventually all the cells more or less crushed except the outer epidermis; *o.e.* with very irregular stellate–undulate cells elongating into smooth, cylindric, obtuse hairs, deflexed on to the seed-coat, supported internally by 3–7 lignified rods, the bases of the hairs cemented together by the membranous base of the epidermal cells; *mesophyll* thin-walled, unspecialized, many of the cells with one or more crystals, especially in the inner hypodermal layer; *i.e.* unspecialized, composed of small tabular cells. Embryo with thick cotyledons.

LYTHRACEAE

Ovules anatropous, bitegmic, crassinucellate; o.i. 4–6 cells thick (*Cuphea*, *Lagerstroemia*, *Pemphis*) or 2–3 cells thick, generally thicker at the exostome; i.i. 2–3 cells thick, 3–4 at the endostome; micropyle formed by the exostome, out of line with the endostome.

Seeds minute to small and subcylindric to trigonous or pyramidal with convex base (outer end), or medium-size and winged all round (*Lafoensia*) or by extension of the chalazal end of the raphe (*Lagerstroemia*), exalbuminous, exarillate. Testa more or less multiplicative in the larger seeds, not in the smaller; *o.e.* as a palisade of rather wide cells (*Lafoensia*, *Lagerstroemia*, *Pemphis*), or composed of cuboid and often somewhat compressed cells, or cuboid with an invaginated mucilage-hair (*Ammannia*, *Cuphea*, *Lythrum*, *Peplis*); *mesophyll* thin-walled, more or less crushed in small seeds, or in two parts with an outer large-celled, thin-walled but pitted and lignified part and an inner densely sclerotic part (*Lawsonia*; *Lagerstroemia* with crystals in the cells) or wholly sclerotic (*Pemphis*), or with the sclerotic layer in the outer part (*Lafoensia*); *i.e.* as a layer of crystal-cells with thickened, pitted and lignified inner and radial walls, not pitted (*Lythrum*), or with scarcely thickened walls (*Ammannia*). Tegmen not multiplicative; *o.e.* as a continuous hard pellicle of narrow, longitudinally elongate tracheids with spiral, annular, scalariform or subreticulate thickening (*Lafoensia*, *Lagerstroemia*, *Lawsonia*, *Pemphis*), or as very narrow, thick-walled fibres without pitting (*Lythrum*, ? other herbaceous genera), the tracheids and fibres lignified, but composed of thick-walled cuboid cells in Ammannia; *i.e.* unspecialized with thin-walled oblong cells or as a layer of fibres crossing the exotegmic fibres (*Lythrum*, ? other herbaceous genera). Chalaza small, simple. Vascular bundle only in the raphe. Endosperm nuclear, absorbed. Embryo straight; cotyledons flat or convolute in the larger seeds. Figs. 351–353.

(Netolitzky 1926; Mauritzon 1934a; Joshi and Venkateswarlu 1935a, b, 1936; Venkateswarlu 1937b; Singh, B. 1964; Salisbury 1972.)

With narrowly fibrous or tracheidal exotegmic seed, Lythraceae comes in the alliance of Combretaceae, Onagraceae, Punicaceae, Sonneratiaceae (if distinct) and Trapaceae (p. 37). In all cases it is the larger seeds which bear the closest resemblance and the small seeds which show in their reduction the peculiarities. This is very evident in Lythraceae, sometimes misconstrued as a herbaceous family, for which reasons I have examined the seeds of several woody genera. The diversity which they show may indicate as many herbaceous trends in the family.

Lafoensia Vand.

L. densiflora St. Hil.
(Ceylon, Kandy, cult.; flowers and fruits in alcohol.) Fig. 351.

Ovules basally attached in the loculus, erect, numerous; o.i. 3 cells thick (–8 along the raphe); i.i. 2 cells thick, very thin; micropyle formed by the exostome.

Seeds 15–17 × 10–12 × 1 mm, elliptic-alate, compressed at right angles to the median plane, surrounded by the broad wing, with small seed-body and long micropyle, pale brown, dry. Testa multiplicative; *o.e.* as a palisade of shortly radially elongate cells with hexagonal facets, but more or less oblong over the micropylar end, shorter on the seed-body, much reduced along the raphe, with thickened outer wall, not lignified; *mesophyll* thin-walled, more or less crushed round the seed-body but with rounded lignified and finely pitted cells as an outer hypodermis, and as a hard flange (–4 cells thick) of small sclerotic cells along the junction of the seed-body with the wing; *i.e.* as a continuous layer of small sclerotic cells with a large crystal; *wing* covered by the exotestal palisade and filled with an extension of the mesophyll as rather large rounded cells with slightly thickened and lignified walls. Tegmen; *o.e.* with tracheidal fibres –130 × 5–7 μ, with spiral, annular, or subscalariform thickening; *i.e.* thin-walled, crushed. Nucellus more or less persistent as 2 layers of thin-walled cells. Embryo straight; radicle relatively long; cotyledons thin, slightly convolute.

Lagerstroemia L.

Seeds winged by extension of the distal end of the raphe; *wing* many cells thick, finally thinly lignified, without v.b. Seed-body transverse at the base of the seed, with the raphe on the side towards the apex of the capsule; testa with a densely sclerotic inner layer of crystals cells, 2–5 cells thick. Embryo with folded cotyledons.

L. flos-reginae Retz.
(Malaya, Penang; fruits in alcohol.) Fig. 352.

Seeds with the body *c.* 4 mm long, the wing 8–10 × 4–6 mm, pale brown. Testa multiplicative; *o.e.* as a short palisade of cells with hexagonal facets, slightly thickened outer walls, not lignified; *mesophyll* with an outer layer 2–4 cells thick over the seed-body, composed of large, slightly thick-walled, pitted and lignified cells, much thicker in the wing, and with an inner layer 4–5 cells thick, composed of small, irregular sclerotic cells each with a large crystal; *i.e.* as the inner layer of the sclerotic mesophyll. Tegmen; *o.e.* as narrow tracheidal fibres 7–12 μ wide, with close annular or subspiral thickening; *i.e.* with larger, shorter, thin-walled oblong cells (? not lignified).

Pericarp with 3 rings of small v.b., with mucilage-spaces and scattered stone-cells; endocarp as a thick woody layer of *c.* 20 rows of transverse and oblique fibres.

Lawsonia L.
L. inermis L.
(Malaya; flowers and fruits in alcohol.) Fig. 353.

Ovules very numerous per loculus, minute, transverse or radially directed from the placenta; o.i. 2–3 cells thick, 4–5 in the raphe; i.i. 2 cells thick; funicle short, slender; micropyle formed by the exostome.

Seeds 2.5–3 × 1.7–2 mm, pyramidal with convex base, angular-compressed, with slender pointed micropyle and minute funicle, grey-brown. Testa 4–12 cells thick, multiplicative, the thick outer part rather succulent, then drying spongy and more or less lignified; *o.e.* composed of rather compressed cells with slightly thickened outer walls and irregular facets, not or scarcely lignified; *mesophyll* with the outer part consisting of large, polyhedral cells with slightly thickened, finely pitted and lignified walls, without air-spaces, 1–8 cells thick, thinnest near the micropyle, thickest towards the chalazal end, and the inner part composed of 3–4 layers of thick-walled, densely lignified vaguely pitted, isodiametric cells, apparently without crystals; *i.e.* at first with shortly radially elongate cells, then sclerotic and forming the inner layer of the sclerotic mesophyll, ? with minute crystals in some cells. Tegmen; *o.e.* with narrow longitudinal tracheids or fibres 6–10 μ wide, with rather coarse pitting, in places subscalariform; *i.e.* composed of rather large, longitudinally oblong cells with thin, but lignified walls. Embryo with short radicle and flat cotyledons.

The seed resembles that of *Pemphis* but has a thicker testa with the two layers of mesophyll as in *Lagerstroemia*.

Pemphis J. R. et G. Forst.
P. acidula J. R. et G. Forst.
(Java, Herb. Bogor; flowers and fruits in alcohol.)

Ovule; o.i. 3–4 cells thick, 4–6 in the raphe; i.i. 2 cells thick.

Seeds *c.* 3 mm long, pyramidal, angular-compressed. Testa not multiplicative except along the raphe-antiraphe *o.e.* as cuboid cells, scarcely radially elongate, with hexagonal facets and thickened outer walls; *mesophyll* wholly sclerotic; *i.e.* sclerotic (? as crystal cells). Tegmen; *o.e.* with elongate tracheids with spiral thickening, very narrow; *i.e.* unspecialized. Embryo with short radicle and flat cotyledons.

Except for the simpler embryo, this seed is a miniature of that of *Sonneratia*.

MAGNOLIACEAE

Magnolieae

Ovules 1, 2 or many per carpel, anatropous, bitegmic, crassinucellate, shortly funiculate but soon sessile in seed-development; o.i. 4–10 cells thick; i.i. 2–3(–4) cells thick; micropyle formed by the exostome, simple or lobate (*Michelia*).

Seed 8–20 mm long, fairly massive, sessile, in dehiscent fruits becoming suspended on lignin-fibrils of the

raphe v.b. as pseudofunicles, exarillate but sarcotestal.
Testa strongly multiplicative to form a sarcotesta and
woody endotesta; *sarcotesta* composed of a pink pellicle,
2–6 cells thick, developed from o.e. (o.i.) by periclinal
division, the epidermis with stomata (*Magnolia, Man-
glietia, Michelia,* ? in all genera), and of white oily
mesophyll derived from the middle layers of o.i.,
carrying the v.b.; *endotesta* composed of 2–10 (? more)
layers of lignified cells, contiguous without air-species,
derived by periclinal division from i.e. (o.i.), the cells
more or less in radial rows and somewhat radially
elongate but not prismatic, most with a single crystal,
not thick-walled but developing an internal fibrous
lignification throughout the cell-cavity. Tegmen not
multiplicative, eventually shrivelled and crushed,
unspecialized except for participation with the endotesta
in the chalazal tube, without v.b. Chalaza massive,
surrounded by the tubular endotesta to form the
heteropyle. Vascular supply consisting of the raphe-
bundle with extension into the chalazal tube and one or
two (? more) postchalazal bundles reaching to the
micropyle. Endosperm cellular, oily, not or slightly
ruminate. Embryo small or microscopic.   Figs. 354–359.

(Gray 1858; Maneval 1914; Netolitzky 1926; Earle
1938; Hutchinson 1964; Kapil and Bhandari 1964.)

Liriodendreae

Ovules as in Magnolieae, provided with obturators as an
outgrowth of the funicle.
    Seed *c.* 5 mm long, retained in the samaroid follicle.
Testa ?
    (Guard 1943; Kaeiser and Boyce 1962.)
    I can find no detailed account of the seed-structure.

Because the massive flowers of this family have a
primitive apocarpous and hypogenous construction, it
is often assumed, though unrealistically for the theory of
evolution, that in other respects, particularly the seed,
the family must also be primitive. Vegetatively this is
manifestly untrue because the family consists mainly of
successful leptocaul trees of wide distribution with
simple leaf, suited to forest-advancement, and stipulate
buds with intermittent growth. There are, also, various
advances through simplification of the flower which
can become unicarpellary and, even, unisexual (*Kmeria*),
and of the fruit which may become a one-seeded nut
or drupe. The seed, nevertheless, when compared with
the Myristicaceous, has such primitive features as the
multiplicative testa, stomata, ligneous endotesta, massive
chalaza, copious endosperm and minute embryo. It is
advanced in the apparently functionless nature of the
tegmen, perhaps in the loss of rumination, and in the
substitution of the aril by the sarcotesta. How far any
of these matters refer to tr. Liriodendreae seems to be
unknown.
    The specialities of the seed of tr. Magnolieae are the

pink pellicle of the sarcotesta, the nature of the cells
composing the multiple endotesta, and the tubular
lignified sheath around the chalaza. This elaboration
of the heteropyle is, as in Myristicaceae, a means of
occluding the wide chalaza of the developing seed to
prevent the desiccation of the endosperm while the embryo
is growing after dispersal. The lignified cells of the
endotesta are peculiar and will need study by electron-
microscopy for elucidation which is beyond the resolution of the ordinary microscope. A system of lignified
filaments pervades each cell, as if the endoreticulum
were lignified, and even the nucleus appears to become
lignified. Swamy (1949) and Padmanabhan (1960)
refer to it as vacuolar incrustation. As discussed on p. 32,
a simpler state in Chloranthaceae, Papaveraceae and
Proteaceae connects this endotestal structure with the
endotestal and fibrous-exotegmic of Myristicaceae, and
Papaveraceae show the loss of the aril from the pachycaulous plants, which are now missing from Magnoliaceae.
    The pink pellicle and the white mesotesta, which
compose the sarcotesta, have the microscopic structure
of the aril of Myristicaceae (Figs. 359, 412, 422). Hence
I assume that, accompanying the loss of the aril, its
construction and function have been transferred to the
testa in tr. Magnolieae. *Xylopia* (Annonaceae) and
Nandina provide parallels.
    The structure of the Magnolia-seed was first explained
by Gray (1858) in the case of *M. umbrella* Desr. (*M.
tripetala* L.). It has been substantiated by Maneval
(1914) and Earle (1938), who showed that the earlier
work of Brandza on *M. macrophylla* was mistaken in
referring the woody layer to the tegmen and the tegmic
layer to the nucellus. In the testa of *M. grandiflora*,
Earle distinguished five layers: (1) a pellicle of 2–3
cell-layers; (2) a hypodermis of 2–3 layers of flattened
cells; (3) a fleshy tissue of 3–6 layers of large round oily
cells; (4) an inner hypodermis of 2 layers of small cuboid
cells; and (5) the woody endotesta of *c.* 4 layers of cells.
Kapil and Bhandari (1964) found the same construction
in *M. obovata* and *M. stellata*. Indeed, there is so much
similarity that it is not clear if there are specific or generic
distinctions in the seed-coat of Magnolieae. The most
primitive state is that of *Michelia* with multiovulate
carpels. It has been studied by Padmanabhan (1960)
and it is here illustrated from sections of living material
that I studied long ago in Singapore. In *Michelia
montana* I find that the endotesta is pitted on its outside
and that these pits correspond with slight infoldings
into the endosperm, as if there slight ruminations as in
Degeneria (p. 116).

Michelia L.

M. champaka L.
(Singapore; living material.)   Fig. 356–359.
    Ovules 6–10 (–12) per carpel, variously orientated,

angular from compression; o.i. 4 cells thick, the exostome often plicate and incised with short lobes, occasionally with one or two overlapping lobes; i.i. 2 cells thick.

Testa light vermilion, developing rapidly after fertilization, swelling round the base of the seed and obliterating the funicle; *pellicle* 4–6 cells thick, the cells prosenchymatous and elongated obliquely round the seed, with thick agglutinated walls and oily contents, the inner cells with pinkish walls, the numerous stomata sunk in elongate pits and communicating by a small airspace with the middle layer of the testa; *middle layer* composed of large colourless thin-walled cells filled with oil-globules, the innermost 1–2 cell-layers with the oily contents congealed and hardened (as intermediate to the woody endotesta); *endotesta* composed of 9–10 layers of cells, lignified with fibrous endoreticulum. Tegmen not multiplicative, the cells enlarging and then crushed by the endosperm. Vascular supply consisting of the raphe-bundle with a process to the umbrella of tracheids in the chalaza, and two postchalazal branches reaching to the micropyle, the testa in t.s. with one median and two lateral v.b.

Follicles (1–) 4–6 (–8) seeded, ripening from base to apex along the receptacle, dehiscing along the dorsal and ventral lines; mesocarp thin-walled, with scattered oil-cells and groups of sclerotic cells; endocarp 1–3 cells thick, papery-tough, composed of obliquely directed, thick-walled fibres. Placenta emitting, after fertilization, groups of cells as short green flanges of tissue in the gaps between the seeds, simulating aril-processes and containing stone-cells in the tissue.

M. nilagirica Zenker
(Ceylon; flowers and young fruits in alcohol.)
Ovules and immature seeds as in *M. champaka*.

MALPIGHIACEAE

Ovules solitary in each loculus, anatropous or sub-campylotropous, suspended, bitegmic, crassinucellate; o.i. 2–4 cells thick; i.i. 3–6 cells thick; nucellus projecting through the micropyle in some cases (*Hiptage, Stigmatophyllum*).

Seed rather small, 3–6 mm long, compressed-obconic or pyriform, more or less exalbuminous, exarillate; seed-coats not multiplicative, the inner tissues collapsing, reduced to the exotesta and fibrous exotegmen with sclerotic endotegmen or only the endotegmen. Testa unspecialized, subaerenchymatous; *o.e.* with tabular cells, more or less undulate facets, with stomata in *Tristellateia*. Tegmen; *o.e.* as a layer of longitudinal pitted fibres (*Thryallis*) or undifferentiated and merging imperceptibly into the testa; *i.e.* as a layer of more or less rectangular or shortly oblong cells, lignified with pitted radial and inner walls, firmly adherent to the

nucellus, often separating from the rest of the tegmen. Chalaza simple, without hypostase. Vascular bundle only in the raphe or dividing into 3–5 short branches in the chalaza (*Heteropteris, Tristellateia*). Nucellus enlarging greatly and filling the seed, then crushed by the endosperm and embryo. Endosperm nuclear, mostly absorbed. Embryo straight or curved; cotyledons rather thick, sometimes lobed. Figs. 360–363.

(Netolitzky 1926; Singh, B. 1964.)

Few Malpighiaceous seeds have been studied in detail. It seems clear that they are primarily tegmic seeds with a fibrous exotegmen and a sclerotic endotegmen, but one or both of these features may disappear. Seeds which are liberated from the fruits have the full construction, e.g. *Thryallis*; it degenerates in those which are retained in drupes, schizocarps or samaras. The fibrous exotegmen seems to disappear first and then the sclerotic endotegmen. In *Tristellateia* with tardily dehiscent schizocarps, there is a strong endotegmen; in the samara of *Heteropteris* it is already feeble, and *Malpighia* may be similar. It is this state with reduced endotegmen that seems to correspond with some Zygophyllaceous seeds. Doubtless other marks of deterioration or modification will be found for the fruits would repay detailed investigation. *Trichomarieae* show the loss of durian-spines; *Tristellateia* may show their modification into the flattened samara. Arillate seeds seem unknown.

The seed of Erythroxylaceae has a multiplicative tegmen with fibrous exotegmen, but there is no lignified endotegmen.

Heteropteris Kunth

H. angustifolia Griseb.
(Ceylon, Peradeniya, cult.; flowers and fruits in alcohol.)
Fig. 360.

Ovules set tangentially across the loculus, laterally attached to the axis by a short stout curved funicle; o.i. 3–4 cells thick, 2–3 at the micropyle, –5 cells at the chalaza; i.i. 3(–4) cells thick, almost indistinguishable from the o.i. in the basal part of the ovule; nucellus extending into the micropyle; chalaza with the raphe v.b. dividing into 4–5 branches; micropyle incurved, formed by the exostome.

Seeds 4–5 mm wide, pallid with brownish chalazal area; mature seed-coat reduced to a layer 16–23 μ thick, consisting of the tabular exotestal cells and the thinly lignified endotegmen. Testa with undulate facets to the o.e., not lignified. Tegmen in the lower part of the seed soon amalgamated with the testa and becoming crushed except for i.e., but in the micropylar region distinct for some time with the cells of the middle layer enlarged and shortly radially elongate; *i.e.* very soon after fertilization becoming punctate-pitted and lignified. Cotyledons thick, unequal, becoming irregularly lobed.

A section through the lower part of the ovule or seed of this species suggests that it is unitegmic because the two integuments seem to amalgamate, but their separation can be made out at the micropyle. There is no fibrous exotegmen to preserve their identity.

Thryallis L.

T. glauca (Cav.) Kuntze
(Ceylon, Peradeniya, cult.; flowers and fruits in alcohol.)
Fig. 361.

Ovule ? campylotropous; chalaza abrupt, small; micropyle formed by the exostome.

Seeds 3.5 × 3 mm, subcompressed, curved, brown. Testa 2(–3) cells thick, 6–8 cells at the micropyle; *o.e.* with thickened brown outer wall eventually lignified (Singh, B. 1964), with subundulate facets. Tegmen 5–6 cells thick; *o.e.* composed of longitudinal, coarsely pitted, lignified fibres –300 μ long, short and subcuboid at the endostome; *mesophyll* with thin-walled cells, shortly transversely elongate across the fibres, here and there with some cells in o.h. thick-walled and pitted, especially near the partial septum of the seed; *i.e.* composed of rather large cells with strongly thickened and finely pitted radial and inner walls, lignified.

(Singh, B. 1964.)

Tristellateia Thou.

T. australasiae A. Rich.
(Ceylon, Singapore; flowers and fruits in alcohol.)
Figs. 362, 363.

Ovules as in *Heteropteris*; o.i. 3–4 cells thick; i.i. 2–3 cells thick.

Seeds 5 × 3.5 mm, campylotropous-pyriform with the stout funicle to one side of the prominent micropyle, pale brown with dark brown chalazal patch. Testa; *o.e.* composed of tabular cells with undulate facets, with scattered stomata arranged singly or in groups of 2–3–4; *mesophyll* aerenchymatous, the inner layer of cells much enlarged as those of i.e., then more or less crushed. Tegmen; *o.e.* and middle cell-layer soon disappearing; *i.e.* as a firm pellicle of shortly oblong cells set transversely in the micropylar region and in various directions on the sides of the seed, with spiral annular, lignified thickening, becoming finely reticulate or scalariform on the radial and inner walls. Vascular bundle of the raphe with a few slight branches at the chalaza. Nucellus filling the full-grown seed, then crushed. Embryo with subequal cotyledons, not lobed.

The seed-structure is similar to that of *Heteropteris* but the endotegmen is better developed, the vascular supply is hardly branched at the chalaza, and the exotesta has stomata which are often curiously clustered.

MALVACEAE

Ovules campylotropous or (Hibisceae) anatropous, bitegmic, crassinucellate; o.i. 2–3 cells thick, 3–6 cells in Hibisceae; i.i. 4–8 cells thick, (? –15, Hibisceae), o.e.

often with shortly radially elongate cells; micropyle formed by the exostome, out of line with the endostome.

Seeds small to moderate size (10 mm), usually brownish or black, smooth, rough, rugose or hairy, exarillate. Testa not multiplicative; *o.e.* with more or less thickened walls and often lignified (Hibisceae), or thin-walled and collapsing, with stomata in *Gossypium* and *Hibiscus*; *hairs* simple, septate or not, lignified or not, developed from special epidermal cells often indicated in the ovule; *mesophyll* in some cases (especially Hibisceae) pigmented with tannin-cells, o.h. in some cases lignified or thick-walled; *i.e.* unspecialized or (Hibisceae) as 1–3 layers of small cells each with a crystal. Tegmen 4–28 cells thick, multiplicative in the larger seeds; *o.e.* as a palisade of Malpighian cells or shortly radially elongate, pitted cells, lignified (at least in the inner part); *mesophyll* with 1–3 or more (Hibisceae) outer layers of small, tannin-pigmented cells, becoming more or less crushed; *i.e.* composed of somewhat longitudinally elongate cells with thickened, brownish, often pitted radial walls, somewhat lignified or not. Vascular bundle in the raphe only or (Hibisceae) with postchalazal branches reaching even to the micropyle. Chalaza usually with hypostase. Endosperm nuclear, persistent but sometimes reduced to a single layer of cells, oily, starch absent or scarce (*Gossypium*).[1] Embryo often curved and with folded cotyledons. Figs. 364–367.

(Netolitzky 1926; Venkata Rao 1955.)

Hibisceae: *Abelmoschus, Gossypium, Hibiscus, Thespesia*.

Malveae: *Abutilon, Althaea, Callirhoe, Lavatera, Malachra, Malva, Malvaviscus, Modiola, Pavonia, Sida, Sidalcea*.

So far as known, Hibisceae are distinguished by the anatropous ovules with thick integuments both of which are multiplicative in the seed, by the well developed exotegmen of Malpighian cells with hexagonal facets, by the crystal-cells in the endotesta, and by the presence of resin-glands in the cotyledons. Reeves (1936) considered such ovules and seeds to be derived from the campylotropous ovules and seeds of other tribes, but as the capsular fruits of most Hibisceae are clearly antecedent to the schizocarps of the others, I consider the larger and better developed seeds of Hibisceae to be the more primitive. In fact, as in other families, Malvaceae show the diminution of the large and elaborate seed into the neotenic, ovular, and often campylotropous seed; in *Malvaviscus* even the exotegmen, so typical of the family, degenerates into cuboid cells. This view accords with the derivation of the family from ancestral Bombacaceae with larger arillate seeds. *Hampea*, placed hitherto in Bombacaceae, has been transferred to the neighbourhood of *Thespesia* by Fryxell (1968), as the sole arillate example in Malvaceae. The two families practically merge, as the durianologist would expect, in their arillate representation.

In some genera, as *Althaea* and *Lavatera*, the exotegmic palisade and the outer hypodermis of the tegmen come to resemble the exotestal palisade and the outer hypodermis of hourglass-cells of Leguminosae.

Abelmoschus Medik.

Testa rugulose; *o.e.* as thick-walled lignified cells on the ridges, as thin-walled cells in the intervals and collapsing. Tegmen 12–15 cells thick; *i.e.* in places 2 cells thick and slightly lignified. Perisperm as a thin layer.

(Singh, B. 1968.)

Abutilon Mill.

A. avicennae Gaertn.
Seed 3 mm long, cuniform, puberulous. Testa; *o.e.* with lignified unicellular hairs; *mesophyll* none; *i.e.* with lignified walls in the places below the hairs. Tegmen with exotegmic palisade 150 μ thick.

(Vaughan 1970.)

A. theophrasti L.
Ovule; o.i. 2 cells thick; i.i. 6 cells thick. Seed as in *A. avicennae*.

(Winter 1960.)

Althaea L.

A. officinalis L.
Fig. 364.
Tegmen 5–6 cells thick; *o.e.* as short Malpighian cells 50 μ long; *i.h.* as convex cells with slightly thickened walls.

(Guignard 1893.)

A. rosea (L.) Cav.
Fig. 365.
Ovule campylotropous; o.i. 2 cells thick; i.i. 4(–5) cells thick, but thicker in the raphe and at the micropyle.

Seed about twice the size of the ovule, the seed-coats not multiplicative. Testa thinly pulpy, then drying up; *o.e.* developing on the antiraphe somewhat thick-walled, aseptate, lignified hairs, longest near the micropyle. Tegmen; *o.e.* as short palisade-cells 35 μ long; *o.h.* composed of subconic cells pointing inwards, with thickened anticlinal walls, not lignified; *i.e.* as a pavement of thin-walled cells with rod-like lignified thickenings scattered on the anticlinal walls.

This species shows a minimal seed-growth. The seed-coats dry up to the palisade layer of the exotegmen in the schizocarp.

Gossypium L.

Ovules anatropous; o.i. 4–6 cells thick, some cells of o.e. as incipient hairs developing from the chalaza; i.i. 6–8 cells thick.

Seeds 7–11 mm long, black, hairy. Testa not multiplicative (*G. arborescens*) or –8 cells thick through 1–3

periclinal divisions of i.e. (*G. barbadense*, *G. herbaceum*, *G. hirsutum*); *o.e.* composed of thick-walled cuboid cells with tannin, of hair-cells, and scattered stomata (arising first near the chalaza); *mesophyll* with brown cell-contents; *hairs* aseptate, unbranched, 0.5–50 mm long, of two kinds, long 'lint hairs' arising pre-fertilization or in the ensuing 1–6 days and becoming flattened, short 'fuzz hairs' 1–5 mm long developing 7–12 days after fertilization, both kinds mixed over the seed, or the fuzz hairs in the micropylar third and the rest of the seed with lint hairs (*G. arboreum*, *G. barbadense*, *G. herbaceum*, *G. hirsutum*), the epidermal cell dilated into a foot at the base of the hair in many species. Tegmen more or less multiplicative, 8–25 cells thick; *o.e.* as a palisade of Malpighian cells 150–240 μ long, lignified; *mesophyll* with brown cell-contents in the outer 3–4 layers of cells; *i.e.* with slightly thickened, pitted, anticlinal walls. Endosperm reduced to 5–6 layers of cells, or less, at each end of the seed, oily and starchy. Embryo with folded cotyledons.   Fig. 366.

(Balls 1915; Barritt 1929; Reeves and Valle 1932; Reeves 1936; Fryxell 1964; Ramchandani, Joshi and Pundir 1966; Vaughan 1970; Chowdhury and Buth 1971.)

The fullest account of cotton-seeds is that by Ramchandani, Joshi and Pundir. Whereas the hairs are initiated before fertilization and soon after, the palisade cells of the exotegmen do not begin to lengthen until 6–12 days after fertilization and are not fully grown until the 20–24th day. They grow into hexagonal prisms by radial elongation from the inner end outwards; the inner part loses its lumen as the walls thicken and lignify; the process continues almost to the outer ends, which remain pitted, until there results an almost solid mass of woody tissue (Fig. 366). It seems that the walls thicken as inwardly growing plates of cellulose, thus giving the stellate lumen in t.s., and that these plates continue outwards to the nuclear end of the cell, while they proceed to coalesce along the central line of the cell in the same direction.

Hibiscus L.

Seeds glabrous or hairy, or the exotestal cells papillaeform.

H. cannabinus L.
Seeds 5 mm long, puberulous. Testa; *o.e.* composed of thin-walled and thick-walled cells, some pigmented, some of the thick-walled cells forming short hairs aggregated into plates; *mesophyll* none; *i.e.* composed of thin- and thick-walled cells. Tegmen; *o.e.* palisade 120 μ thick; *o.h.* composed of bottle-shaped pigmented cells.

(Vaughan 1970.)

H. esculentus L.
Seeds 4.5 mm long, similar to *H. cannabinus* but the endotesta with thickened radial and inner walls. Tegmen-palisade 165 μ thick.

(Vaughan 1970.)

Lavatera L.

L. trimestris L.
Fig. 364.

Testa 2 cells thick. Tegmen *c.* 6 cells thick; *o.e.* as a palisade –80 μ high; *i.h.* composed of conico-convex cells with strongly thickened walls.

(Guidnard 1893.)

Malvaviscus Fabr.

M. drummondii T. et G.
Ovule anatropous, becoming campylotropous; o.i. 2–3 cells thick; i.i. 6–8 cells thick; neither integument multiplicative in the seed.

Tegmen; *o.e.* as a layer of cuboid cells with much thickened, pitted, lignified walls, not as a palisade.

(Reeves 1936.)

Sidalcea A. Gray

S. malvaeflora DC
(living material.)   Fig. 367.

Ovule anatropous, then campylotropous after fertilization; o.i. 2 cells thick, non-multiplicative; i.i. 4–5 cells thick in the antiraphe, 5–7 cells thick on the sides, 7–9 cells thick in the raphe.

Tegmen 6–8 cells thick; *o.e.* as a layer of sclerotic cells, slightly elongate radially, sometimes also tangentially, the inner walls thick and strongly pitted, the outer walls slightly thickened and not pitted, the radial walls of intermediate thickness with longitudinal pits, eventually lignified, not as Malpighian cells; *mesophyll* collapsing; *i.e.* unspecialized.

Thespesia Soland.

T. populnea (L.) Soland.
Fig. 364.

Testa 7 cells thick; *o.e.* with thick-walled cells and long, soft, lignified hairs thickly developed at the angles of the seed; *mesophyll* pigmented with brown tannin-cells; *i.h.* and *i.e.* as crystal-cells. Tegmen ? 8 cells thick; *o.e.* as a palisade of Malpighian cells –200 μ high; *mesophyll* with *c.* 4 outer layers of pigmented tannin-cells.

(Reeves 1936.)

T. populneoides (Roxb.) Kostel.
Seeds covered with a dense short pubescence of erect clavate, or bulbous, hairs.

(Fosberg and Sachet 1972.)

MARCGRAVIACEAE

Ovules anatropous, bitegmic, tenuinucellate; o.i. and i.i. 2–3 cells thick; micropyle formed by the projecting endostome.

Seeds small, slightly albuminous, exarillate. Testa not or slightly multiplicative, –4 cells thick; *o.e.* composed of shortly radially and transversely elongate cells with brown contents (? lignified walls); *mesophyll* thin-walled, crushed, but some cells near the chalaza with raphids. Tegmen not multiplicative, unspecialized, crushed. Endosperm cellular, more or less absorbed, starchy in *Norantea*. Embryo relatively large; radicle short; cotyledons long.

(Netolitzky 1926; Mauritzon 1939; Swamy 1948b.)

This description is based on *Marcgravia*. *Norantea* is evidently similar. If the family belongs in Theales, then it seems nearest in seed-structure to Adinandreae.

MARTYNIACEAE

Ovules anatropous, unitegmic, tenuinucellate.

Seeds small to medium-size, thinly albuminous or exalbuminous, exarillate. Testa reduced to a subgelatinous pellicle of large, thin-walled cells (*Martynia*) or with short sclerotic cells with thickened inner and radial walls (*Proboscidea*); *mesophyll* more or less crushed; *i.e.* as small tabular cells (*Martynia*). Endosperm cellular. Embryo straight.

(Netolitzky 1926.)

MELASTOMATACEAE

Ovules anatropous or subcampylotropous, bitegmic, crassinucellate; o.i. 2–5 cells thick, 3–6 at the exostome, often with thickened raphe; i.i. 2 cells thick; micropyle formed by the exostome, often curved; hypostase generally present.

Seeds 0.3–8 mm wide (or more ?), anatropous or becoming more or less campylotropous with a partial septum forming a radicle-pocket, exalbuminous, exarillate, a few genera with thickened raphe (? vestigial aril). Testa multiplicative in the larger seeds, not in the smaller; *o.e.* varying from a palisade of radially elongate, lignified cells with hexagonal facets to a layer of cuboid cells with thickened outer wall, sometimes also the radial walls thickened, sometimes papillaeform or with undulate facets; *mesophyll* thin-walled or with small groups of sclerotic cells in o.h.; *i.e.* unspecialized or with crystal-cells (*Miconia*), generally crushed. Tegmen not multiplicative, soon crushed, or persistent at the endostome. Hilum small, simple, but wide in the larger seeds and with sclerotic subhilar tissue, in small seeds generally with a single layer of sclerotic cells between the raphe and the embryo and this layer forming an operculum on germination. Vascular bundle simple or branched (*Memecylon*). Endosperm nuclear, not forming a cellular tissue. Embryo in small seeds straight or curved, with well-developed hypocotyl and short cotyledons; in larger seeds either straight with thick cotyledons and very short radicle (*Mouriria*) or curved with long hypocotyl, radicle-pocket and thin convolute cotyledons

(*Memecylon*). Raphe arilloid in *Amphiblemma*, *Blakea*, *Calvoa*, *Gravesia*, *Miconia*, *Veprecella*. Figs. 368–373.

(Netolitzky 1926; Ziegler 1925; Etheridge and Herr 1968; Whiffin and Spencer Tomb 1972.)

Most seeds of this family are small, 0.3–1.5 mm long, and plainly exotestal. This may be true of the larger seeds of *Memecylon* and *Mouriria* with well-developed exotestal palisade, but the massive sclerotic tissue in the sub-hilum of their seeds may extend some distance along the raphe v.b., and scattered clusters of sclerotic hypodermal cells occur in *Memecylon*. Such details in seed-anatomy suggest simplification from a more extensively mesotestal construction. The small seeds, studied in *Bertolonia*, *Dissotis*, *Heeria*, *Medinilla*, *Miconia*, *Monochaetum* and *Tibouchina* by Ziegler, are clearly just adult ovules with non-multiplicative seed-coats and, as such, reductions from the larger and more complicated seeds around the base of which there are many undeveloped ovules which have formed the exotestal character of the small seeds; here, indeed, is proof of the derivative nature of these small ovule-like seeds which have precociously matured. The single layer of sclerotic cells in these small seeds may be the vestige of the massive sclerotic tissue in the subhilum of the larger seeds, now adapted to an oper-culum in the refinement of small structures, and the absence of thick-walled testa from the arilloid raphe may be the neotenic effect of the manner of develop-ment of the thick exotesta from the antiraphe-side of the large seed (cf. *Mouriria*, Fig. 372). The family has proceeded along the path of producing many small seeds from one fruit. The primitive state with many large seeds in a capsule, as in Lecythidaceae, seems extinct, but that reduced to a single large seed prospers in the small forest trees of *Memecylon* and *Mouriria*.

Details of the seed are important in the classification of the genera. The large-seeded are separated as Memecyleae, but there are great differences between the seeds of *Mouriria* and *Memecylon*. The same differences occur in the small-seeded genera and classification by this means may provide a better alignment. The value of the external surface of the small seeds has been shown by Whiffin and Spencer Tomb, who distinguish five kinds, representative of five tribes. *Mouriria* has a large seed with straight embryo without radicle-pocket, as in the small seeds of *Bertolonia*, *Blakea*, *Miconia* and *Oxyspora*, and it is noteworthy that *Miconia* has an exotesta which, if short, is exactly comparable with that of *Mouriria*. *Memecylon* has the campylotropous seed with radicle-pocket as in the small seeds of *Dissotis*, *Heeria*, *Mono-chaetum* and *Osbeckia* (Fig. 373). Thus, at least, two lines of reduction of the Melastomataceous seeds appear and within each there have been modifications of the exotesta. *Medinilla venosa*, as figured by Ziegler, has a curved embryo but no distinct radicle-pocket; it may resemble *Blakea*.

The semi-inferior ovary of various genera develops septal pockets between its outer surface and the recep-tacular tissue. The deflexed anthers fit each into a pocket (Fig. 373). According to Ziegler, the anther-primordia grow towards the base of the flower as the receptacular tube begins to form and, as they fit on to the ovary, so the conjoint tissue of ovary-base and receptacular tube extend by basipetal growth to make the pockets. Thus the syncarpous ovary becomes longitudinally septate both externally and internally.

To what extent the raphe-arils are relics of larger structures is unknown, but there is certainly funicular thickening in the seed of *Mouriria*. The complex hairs and vascular processes on the receptacular fruits, that occur in many genera, suggest durian-relics, even if on the inferior ovary.

Blakea P. Br.

B. trinervia L.

(Ceylon, Peradeniya; flowers and fruits in alcohol.) Fig. 368.

Ovules anatropous, slender, elongate, funiculate, very numerous; o.i. and i.i. 2 cells thick; hypostase minute, not becoming lignified.

Seeds 1.3 × 0.7 mm, obovoid, anatropous, pale brown with reddish brown succulent raphe, without radicle-pocket. Testa not multiplicative; *o.e.* as a lignified layer of thick-walled cells, shortly radially elongate but with irregularly sinuous anticlinal walls, not making a regular palisade, pitted, the outer walls convex to subpapillaeform and sparsely pitted; *i.e.* composed of small thin-walled cells, subpersistent. Tegmen soon crushed. Raphe composed of large cells except the tabu-lar thin-walled epidermal cells with undulate facets; the inner part in contact with the tegmen and embryo, composed of a hard layer of shortly prismatic lignified cells with much thickened and pitted anticlinal walls and the wall next to the tegmen; several hypodermal cells with sphaerocrystals. Embryo slightly curved.

Pericarp with a mass of sclerotic cells between the epidermis and v.b., some with short lobes as idioblasts, and with very thick-walled, acicular, longitudinal fibres accompanying the main v.b.; many hypodermal cells with spaerocrystals.

The form of the epidermal cells of the testa is halfway between the palisade-cells of *Mouriria* and the short cells with undulate facets of *Osbeckia*.

Melastoma L.

M. malabathricum L.

Ovule slightly curved; o.i. and i.i. 2 cells thick, 4–6 cells at the micropyle.

Seeds small, many without embryo coloured purple or black. Testa reduced to o.e. composed of cuboid cells with much thickened and pitted outer and radial walls, papillaeform.

(Subramanyam 1948a.)

Memecylon L.

Ovule; o.i. 2 cells thick, 3-5 cells at the micropyle; i.i. 2 cells thick.

Seed solitary, round, smooth, brown, campylotropous with thick sclerotic hilum and a radicle-pocket, exalbuminous. Testa multiplicative with lignified o.e. and, in some cases, sclerotic cells in the mesophyll; *i.e.* unspecialized. Tegmen soon crushed, unspecialized. Embryo with long hypocotyl-radicle and thin, much folded cotyledons.

(Subramanyam 1942.)

M. umbellatum Burm. f.
(Ceylon; fruits in alcohol.)   Figs. 369, 370.

Seeds 4-4.5 mm wide, subglobose, pale brown, surrounded at the base with *c.* 10 undeveloped ovules. Testa 0.2-0.3 mm thick; *o.e.* as a palisade of radially elongate cells with thick, coarsely pitted, lignified walls, attenuate round the base of the seed into a thin layer of shortly longitudinally elongate, thick-walled cells; *mesophyll* 8-12 cells thick (? more), thin-walled, more or less crushed, especially the inner layers, but with scattered clusters of small sclerotic cells in o.h.; *i.e.* unspecialized, crushed. Tegmen crushed. Hilum broad, closed by internal sclerotic tissue permeated by v.b., the sclerotic cells extending shortly into the radicle-septum and on the chalazal side of the seed-base. Vascular supply consisting of several v.b. entering the hilum, passing through the sclerotic tissue and spreading obliquely through the mesophyll of the testa (not along the antiraphe), anastomosing, without distinct raphe-bundle.

Pericarp with many v.b., permeated with branching, fibre-like idioblasts, without special endocarp.

Memecylon sp.
(Java, Sukabumi distr., Lengkong; fruits in alcohol.)

Seeds 7-8 mm wide, subglobose, pale brown, smooth. Testa as a thin hard shell 8-10 cells; *o.e.* as a short palisade of columnar cells with hexagonal facets, with thick lignified outer and radial walls, rather sparsely pitted; *mesophyll* and *i.e.* thin-walled, unspecialized, crushed. Tegmen crushed. Hilum thickly sclerotic. Vascular supply without distinct raphe or chalaza, dividing into several, more or less anastomosing branches soon after entry into the seed, surrounding the seed in a loose network and almost reaching the micropyle.

Mesocarp with scattered idioblasts.

Miconia Ruiz et Pav.

Seeds *c.* 1 mm long, anatropous, subcuneate, pale, with purple-red raphe. Testa not multiplicative; *o.e.* as a compact layer of cuboid or shortly radially elongate cells, with undulate facets, with thick lignified outer and radial walls, thin-walled over the raphe; *i.e.* thin-walled but lignified, composed of rectangular cells with single

crystals in some species. Tegmen as a pellicle of subundulate cells, not lignified. Nucellus (?) with the remains of crushed elongate cells with brown contents.

Ziegler reported elongate cells in the endotesta of *M. magnifica* Triana, which Netolitzky interpreted as fibres, though that much is not evident from Ziegler's description. I examined the seeds of two species (herbarium-material) but was unable to find such cells.

Mouriria Aubl.

M. guyanensis Aubl.
(Ceylon, Peradeniya; flowers and fruits in alcohol.)
Figs. 371, 372.

Ovules anatropous, erect, 3-5 per loculus; o.i. 4-5 cells thick, much thicker in the raphe; i.i. 2 cells thick; nucellus massive; hypostase not formed; micropyle curved, formed by the exostome; occasionally two nucelli in an ovule.

Seeds 6 mm wide, subglobose, smooth, brown. Testa 10-14 cells thick, multiplicative; *o.e.* as a palisade *c.* 150 μ thick, of high radially elongate prismatic cells with thick, rather sparsely pitted and lignified walls, with hexagonal facets, the inner ends of the cells thinwalled; *mesophyll* thin-walled, the cells with brownish contents, compressed and the inner layers crushed, with some sclerotic cells round the v.b. on its entry to the raphe and a few on the antiraphe-side of the exostome. Tegmen not thickening, unspecialized, crushed, but persisting as the narrow lignified endostome. Vascular bundle in the raphe only, broad, ending at the wide but unspecialized chalaza. Hilum wide, closed by a plate of sclerotic cells perforated by the v.b. Embryo consisting mostly of two thick cotyledons; radicle minute, not inserted in a radicle-pocket.

Pericarp without woody endocarp but with shortly and spicately lobate idioblasts in the middle tissue. Ovary 2-3 locular in the lower part; style with 5 longitudinal v.b. round the stylar canal.

Half-grown fruits show the cells of the exotesta beginning to elongate and mature over the antiraphe before there is any differentiation on the raphe-side (Fig. 372). The short funicle also begins to thicken at this stage by numerous periclinal divisions of its epidermis.

Osbeckia L.

Ovules anatropous with slightly curved nucellus; o.i. and i.i. 2 cells thick; nucellus small.

Seeds 0.5 mm long, brown, finely papillate, becoming subcampylotropous with a partial septum. Testa not multiplicative; *o.e.* composed of cuboid cells with undulate facets and more or less of a central papilla, with thick punctate outer wall and thickened radial walls, lignified; *i.e.* crushed. Tegmen soon crushed. Nucellus with the cells much enlarged, mostly in one cell-layer, then crushed by the embryo.

O. octandra DC

(Ceylon; flowers, fruits in alcohol.)   Fig. 373.

Tegmen (apparently) as a pellicle of thin-walled, angular cells with fine striations on the tangential walls, not lignified (? i.e.).

O. wightiana Benth.

(Subramanyam 1942.)

Oxyspora DC

O. paniculata DC

Ovule; o.i. and i.i. 2 cells thick.

Seeds developing a cylindrical projection from the antiraphe at the chalazal end. Testa with only o.e. persistent as cells with undulate facets, not radially elongate, but with convex papillaeform and slightly thickened outer wall.

(Subramanyam 1951b.)

Pternandra Jack

P. azurea (Bl.) Burk

(Java; living material.)

Ovules very numerous, minute; o.i. and i.i. 2 cells thick.

Seeds 0.8 × 0.4–0.5 mm, the seed-coats not multiplicative. Testa; *o.e.* composed of shortly longitudinally elongate cells with undulate facets, with very thick outer and anticlinal walls, lignified; *i.e.* crushed. Tegmen crushed.

MELIACEAE

Ovules anatropous, suspended or the upper ones erect, or orthotropous (*Chisocheton*), bitegmic, crassinucellate; o.i. 2–5 cells thick; i.i. 2–4 cells thick.

Seeds various, large and arillate or exarillate, with or without a sarcotesta, or massive with fleshy or corky pachychalaza, or winged from the extension of the raphe-exostome, or small exarillate and reduced in indehiscent drupaceous fruits with woody endocarp (stone) or cocci; albuminous or exalbuminous. Testa strongly multiplicative or not; *o.e.* as a short palisade-layer with thickened outer and radial walls or unspecialized, with stomata (*Swietenia*) but generally without; *i.e.* in some cases as a layer of small crystal-cells. Tegmen multiplicative or scarcely so; *o.e.* typically composed of longitudinal, lignified, pitted fibres. Pachychalaza well-developed in many cases and then forming most of the seed-coat, with a plexus of v.b. external to a more or less sclerotic layer (several cells thick) and an extensive hypostasial layer; in other cases perichalazal along most of the raphe. Endosperm nuclear, oily. Aril funicular and from the adaxial side of the exostome, but also perichalazal and from the raphe or elongate hilum, vascular or not.   Figs. 374–392.

(Netolitzky 1926.)

This is a family in which the variety in fruits and seeds far outweighs in complexity the floral details which are the basis of classification. The neotenic flowers have advanced to a static level but the fruits retain a great many stages in evolution from the fleshy arillate capsule to the dry capsule with alate seeds and the drupe. The more primitive fruits, because of their size and consistency, make poor herbarium-specimens; aril and sarcotesta become confused in the dried specimen. There is great need to overhaul fruit- and seed-structure in order to understand specific alliances within genera, and generic alliances.

Three characters of the seed are outstanding. First, the seed is primitively arillate and, if this is a funicular aril, it may be partly exostomal (*Aphanamixis*) and variously extended along the raphe, chalaza and antiraphe (*Aglaia, Lansium, Dysoxylon*). Second, the seed-structure is exotegmic with fibres; this is Celastraceous, cf. *Turraea* with Meliaceous flower and Celastraceous seed. Third, the arillate seeds are more or less massive. All these features may deteriorate. Exarillate seeds occur not only in the drupaceous genera and those with alate seeds, but also in all the large arillate genera (*Aglaia, Chisocheton, Dysoxylon, Trichilia*), in several of which the loss is associated with the development of a pachychalaza and the transference of aril-characters to form a sarcotesta; there is no rigid exotestal palisade to impede this process as there is in Leguminosae. When the pachychalaza develops, the fibrous exotegmen disappears (*Aglaia, Aphanamixis, Carapa, Chisocheton, Dysoxylon, Lansium, Sandoricum*); it also deteriorates and disappears from some drupaceous fruits with slight development of the seed-coats (*Melia–Azedirachta*) and from some alate seeds (*Cedrela, Cedrelopsis*). The anatropous seed may become orthotropous (*Chisocheton*).

The primitive Meliaceous seed appears to have had the following characters:

(1) a massive construction, fitting the forest-requirement;

(2) a red funicular aril, perhaps also exostomal;

(3) a thick testa with black or red cuticle or epidermal walls, with stomata, with thin-walled mesophyll, and with the endotesta as a layer of crystal-cells;

(4) a thick tegmen with a fibrous exotegmen of ribbon-like sclereids as in Celastraceae, and a tanniniferous endotegmen;

(5) endosperm.

Concerning the vascular supply of this seed, it is difficult to arrive at a conclusion. The massive supply in many of the large seeds is associated with the expansion of the chalaza. The large seeds of *Dysoxylon* which are not pachychalazal have only the raphe-bundle, as if this were primitive. Yet, in *Chisocheton*, there are tegmic vascular bundles. The aril is vascular only in *Aphanamixis* with a single v.b.

Whether such a primitive seed exists is not known.

It is most likely to occur among the species of *Dysoxylon* or *Turraea*. Various transformations of the Meliaceous seed are shown diagrammatically in Fig. 374. The marked basipetal tendency in these massive seeds expresses the strong basipetal growth of the ovary as it becomes the fruit (Fig. 375).

Aglaia Lour.

The definition of this genus and its distinction from *Aphanamixis*, *Lansium* and *Trichilia* are debatable (Kostermans 1966). The construction of the seed varies from arillate to exarillate, to perichalazal and pachychalazal with sarcotesta. As with *Chisocheton* and *Dysoxylon* all these points, which are much more complicated than the floral details, need to be investigated before there can be an adequate basis for classification. Among the following species which I have studied, *A. acida* is arillate and possibly pachychalazal; *A. trichostemon* is arillate and perichalazal in the origin of the aril, but possibly pachychalazal in the construction of the seed-coat; *Aglaia* sp. (Ceylon) appears to be sarcotestal. I have not, however, had an adequate series of stages in which to trace these developments.

A. acida K. et V.

(Java, Koorders 19968; Herb. Bogor., fruits in alcohol.)

Seeds rather large, ellipsoid, completely surrounded by a pulpy aril derived from the narrowly elongate hilum, covering the micropyle and forming a slit-like arillostome on the antiraphe side of the seed. Testa *c.* 12 cells thick, thin-walled, unlignified, with a close plexus of v.b. divergent from the raphe at its entry. Tegmen not seen. Embryo transverse; cotyledons thick; radicle short, enclosed, pointing to the hilum.

Seeds of *A. latifolia* K. et V. (Koorders 4698) appeared to be identical. The construction resembles that of *Lansium domesticum* but, instead of being perichalazal, it seems to be entirely pachychalazal.

A. trichostemon C.DC

(Malaya; fruits in alcohol.)　Fig. 376.

Ovary 2-celled, with 2 anatropous suspended ovules in each cell.

Seed *c.* 10 × 9 × 7.5 mm, arillate with membranous reddish orange testa and large white embryo. Seed-coat (? pachychalaza) 10–15 cells thick, remaining thin-walled, many of the outer cells (especially along the raphe) with reddish orange sap; *v.b.* as a stout bundle in the raphe and antiraphe, with subdividing branches from the antiraphe over the sides of the seed.

Aril pale yellow then more or less orange, adnate to the testa along the raphe and antiraphe but free on the sides of the seed, bilobed and free over the micropylar end, composed of large, thin-walled, radiating cells, many of the inner cells with pale orange sap; sour-sweet and slightly astringent.

Fruit 12–13 × 11–12 mm, oblong–ovoid, fawn-ferruginous scurfy with stellate hairs, hard, 1- (rarely 2-) seeded, dehiscing irregularly by a vertical or transverse split caused by the swelling of the aril; pericarp with small nests of sclerotic cells just internal to the thin green outer cortex, the white mesocarp thin-walled, the endocarp unspecialized.

This account is incomplete. The young seed, if not the ovule, seems to be perichalazal with the pulpy aril-tissue developed from the raphe-antiraphe. The aril is slightly coloured, the testa more strongly.

Aglaia sp.

(Ceylon, Yala, leg. E. J. H. Corner; ripe fruits in alcohol.) Fig. 377.

Seed brownish with white pulpy sarcotesta, exarillate. Sarcotesta (?) composed of 8–10 layers of radially elongate cells, the epidermal cells short and thin-walled; *mesophyll* of testa composed of small compact cells with thin brownish walls, connected with the sarcotesta by a layer of somewhat radially elongate, large, and slightly thick-walled cells. Tegmen evidently crushed. Vascular bundles forming a plexus from the raphe on both sides of the seed in the testal mesophyll.

Fruit *c.* 15 × 12 mm, 1(–2) seeded, apparently indehiscent; pericarp with many small nests of sclerotic cells below the outer epidermis.

It was impossible to make out in the ripe seeds whether the pulpy outer tissue was a sarcotesta or an aril as in *A. trichostemon*. If an aril, then the layer of enlarged cells with slightly thickened walls, separating the pulpy tissue from the compact mesophyll of the testa, must be the exotesta.

Aphanamixis Bl.

Seeds arillate, pachychalazal, exotegmic with a layer of longitudinal fibres, with thick woody endostome, exalbuminous. Aril with a single longitudinal v.b.

A. grandifolia Bl.

(Singapore, living material; Ceylon, alcohol-material.) Fig. 378–380.

Ovules 2 per loculus, superposed, anatropous, pendulous; o.i. 4–5 cells thick; i.i. 3 cells thick, attached along the length of the raphe.

Seeds 1–3 per fruit, the others more or less abortive though invested by small arils, 17–22 × 11–13 mm overall, invested by the red aril with white oily flesh; seed-body 12–15 × 9–14 mm, black or brownish black, smooth, shiny, with a narrow white strip on the adaxial side as the extended chalaza with aril-attachment along either side from the micropyle to the chalazal extremity, the free testa and tegmen limited to the upper micropylar quarter of the seed, the remainder surrounded by the thicker pachychalaza with a network of v.b.; ripe seeds stripping the endocarp and hanging by it from the capsules. *Testa* 18–22 cells thick; *o.e.* as a hard palisade of shortly radially elongate cells with thick brown outer and radial walls, with polygonal facets, cuticulate, not lignified; *mesophyll* composed of tangentially elongate thin-walled cells with short arms, aerenchymatous; *i.e.* unspecialized or some cells with a large crystal; *exostome* very oblique, prolonged but inconspicuous, thin-walled, appressed by the aril. *Tegmen* 5–7 cells thick; *o.e.* as a layer of longitudinal, lignified, regular fibres 230–580 × 23–30 μ with moderately thick, pitted walls and wide lumen; *mesophyll* aerenchymatous, the cells shortly transversely elongate with short arms, thin-walled; *i.e.* as a continuous layer of cuboid cells with slightly thickened brown walls, not lignified; *endostome* as a thick mass of sclerotic cells. *Pachychalaza* 30–37 cells thick, comprising about three-quarters of the seed-coat, consisting of three fairly distinct layers; *o.e.* as the exotesta; *outer layer* 10–11 cells thick, with thin, pale brown walls, aerenchymatous not lignified; *middle layer* 6–10 cells thick, composed of lignified cells with thickened pitted walls, shortly longitudinally elongate, with brown contents, slightly aerenchymatous; *inner layer* 12–15 cells thick, composed of contiguous, thin-walled, colourless cells, not lignified; *i.e.* unspecialized. Vascular bundle in the arilloid raphe stout, multiple-concentric, at the chalazal end of the seed branching to give a plexus of broad v.b. in the pachychalaza at the junction of the outer and middle layers of the tissue, not entering the free parts of the testa or tegmen.

Aril with red surfaces and white oily flesh, developed after fertilization from the dilated funicle and the adaxial side of the exostome, investing the seed as an entire cupule, covering the micropyle but often incomplete at the chalazal end on the antiraphe-side; *o.e.* with red-brown, distinctly thickened, outer walls, the cells cuboid; *i.e.* with distinctly thickened cuticle; *mesophyll* composed of rounded thin-walled cells full of oil-drops, with slight air-spaces at the angles, attached to the seed along the line of the v.b. Funicle very short but dilated to make the broad attachment of the seed. Endosperm reduced to a single, more or less crushed, layer, at first oily. Embryo with two large, practically concrescent, thick, white cotyledons, plumule and radicle very short; germination hypogeal.

Pericarp in three layers; exocarp with clusters of sclerotic cells; mesocarp thin-walled, not lignified; endocarp *c.* 130 μ thick, composed of 6–7 layers of transverse, lignified fibres; *v.b.* as scattered longitudinal bundles in the exocarp and mesocarp; latex-tubes in the mesocarp.

(Corner 1964, pl. 22; Vaughan 1970, fig. 84.)

The seed is complicated and consists of the micropylar quarter with free integuments, showing the characteristic Celastraceous structure, and the much larger pachychalazal part of Sapindaceous resemblance; this is the part illustrated by Vaughan. The seed grows mainly by extension of the pachychalaza from the lower extremity of which several vascular bundles develop upwards in the pachychalazal wall as a stout network without penetrating the free testa or tegmen. The aril develops rapidly after fertilization from the sides of the raphe-like part of the pachychalaza and from the free testa on the adaxial side between the hilum and the micropyle; the attachment of the aril is revealed as the narrow white strip on the seed after its removal. The vascular bundle of the raphe-chalaza lies in the arilloid tissue. As there is no evident funicle and part of the aril is exostomal, without the abaxial side of the exostome developing an arillostome, the aril is exactly intermediate between the funicular aril and the exostomal arilloid; it disposes of the academic distinction.

Carapa Aubl.

C. guyanensis Aubl.
(Ceylon, Singapore; seeds in alcohol.)   Fig. 381.

Seeds 3–4 × 2.5 cm, light brown, darkening on exposure to air, with wide hilum and a smooth area on one side leading to the micropyle; testa and tegmen free in an area 12–15 mm wide round the micropyle, the remainder of the seed composed of pachychalazal tissue 1–1.5 mm thick and with a close network of v.b. near the inner surface of this tissue. Testa 40–50 cells thick; *o.e.* with large cells, little projecting; *i.e.* unspecialized; *exostome* as a minute slit, unspecialized. Tegmen *c.* 30 cells thick, with smaller cells than those of the testa, unspecialized except o.e., without v.b.; *o.e.* as a more or less continuous layer of lignified, pitted fibres –550 × 15–25 μ, with wide lumen; *i.e.* with brownish cells. Pachychalazal tissue *c.* 30 cells thick, becoming entirely lignified with slightly thickened, more or less pitted walls, except for the hypostase-tissue; *o.e.* composed of large, somewhat thick-walled cells with rather shortly projecting, conical, obtuse apices, some pitted, finally crushed; *mesophyll* aerenchymatous, substellate, without special sclerotic cells; *v.b.* close to the hypostase; *hypostase* several cells thick, lining the pachychalaza, with small brown, thin-walled, contiguous cells. Embryo with very short, thick, radicle; cotyledons more or less connate. Endosperm none.

(Vaughan 1970.)

As with *Aphanamixis*, the seed has a vestigial area with fibrous exotegmen round the micropyle; most of the seed-coat, as illustrated by Vaughan, is the single massive wall of the pachychalaza. The outer sclerotic cells, described by Vaughan, are evidently the crushed papillate epidermal cells of this pachychalaza; it supplies the corky texture for the water-borne seed.

Cedrela P.Br.

C. toona Roxb.
(Ceylon, Peradeniya; ripe fruits in alcohol.)   Fig. 382.

Seeds 12–15 mm long, the seed-body 5–6 × 3–3.5 mm,

pendent, with a thin unilateral wing from the raphe-exostome; funicle very short, subapical; exarillate. Testa 4–5 cells thick, not lignified; *o.e.* without stomata; *mesophyll* composed of 2–3 layers of large, thin-walled cells, collapsing except in the raphe-antiraphe and the wing; *i.e.* unspecialized. Tegmen 4–5 cells thick, unlignified, more or less crushed, unspecialized; *endostome* as a hard conical brown plug, unlignified. Endosperm oily. Embryo straight; cotyledons flat; hypocotyl moderately long.

Pericarp with lignified endocarp valves and columella; cortex thick, with a single ring of v.b.; endocarp-valves *c.* 180 μ thick, with transverse fibres and a strand of longitudinal fibres along each edge; endocarp-lining of the columella with 3–4 layers of longitudinal fibres.

The seed is similar to that of *Swietenia* but smaller with thinner and unspecialized tissues. No exotegmic fibres or endotestal crystal-cells occur, but the chalaza seems to extend as in *Swietenia*. Compare *Chloroxylon* (Rutaceae, p. 234).

Cedrelopsis Baill.

C. grevei Baill.
Seed-coat; *o.e.* composed of thick-walled, papillate cells; *mesophyll* aerenchymatous with secretory cells, the inner tissue crushed; *i.e.* small-celled. Endosperm none.
(Netolitzky 1926.)

Chisocheton Bl.

So far as I have been able to study the seeds of this genus, I find that they differ from those of *Dysoxylon* in being orthotropous (as the ovules) with tegmic vascular bundles. The embryo is longitudinal in the seed but transverse to the fruit. There are arillate and exarillate species. The aril is chalazal or funicular, for there is no sharp distinction between these parts. As with *Dysoxylon*, there is need of extensive specific investigation.

C. divergens Bl.
(Java, Koorders 4885; Herb. Bogor., mature fruits in alcohol.) Fig. 383.
Seeds 12–15 mm wide, ellipsoid-compressed, blackish brown, orthotropous with the micropyle depressed in the middle of the convex free surface; aril funicular, thick, as a red cushion with a short flange round the base of the seed. Testa 2 cells thick, thin-walled, unspecialized, pellicular. Tegmen *c.* 20 cells thick; *o.e.* as a layer of longitudinally elongate cells with slightly thickened walls, not or scarcely lignified, not truly fibriform; *mesophyll* thin-walled, pervaded with a network of v.b. from the wide pachychalaza; *i.e.* unspecialized. Embryo straight; radicle minute; plumule with long hairs.

C. sandoricocarpus K. et V.
(Java, Koorders 28607; Herb. Bogor., fruits in alcohol.) Fig. 383.

Seeds as in *C. divergens* but subglobose with less dilated chalaza, exarillate. Testa and tegmen thick; *v.b.* as a network in the tegmen.

Cipadessa Bl.

C. baccifera (Roth) Miq.
(Java, Kebun Raya; living material.)
Ovule; o.i. 2 cells thick; i.i. 4–5 cells thick.
Testa 3 cells thick, the outer two layers thin-walled and crushed; *i.e.* composed of small cells, each with a crystal. Tegmen 6–7 cells thick, the inner layers becoming large-celled, then crushed; *o.e.* as a compact layer of narrow, subcylindric, thick-walled, pitted, longitudinal fibres, not or slightly lignified, forming a pellicle round the endosperm.
(Narayana, L. L. 1958, without information on the seed.)

The small seeds in the indehiscent fruits have the fibrous exotegmen, if poorly differentiated, of the family and the crystal-cells of the endotesta as in *Melia azedarach* (p. 190).

Dysoxylon Bl.

Ovules anatropous, suspended, sessile with broad attachment, without distinct raphe.

Seeds large, ovoid, black, with pale hilum, arillate or not, exalbuminous. Testa multiplicative, thick; *o.e.* as a layer of cuboid cells with more or less isodiametric facets, thickened brownish walls and brown contents, not lignified; *mesophyll* thin-walled; *i.e.* as a layer of small cells with rectangular facets and 1-several crystals in each. Tegmen ? not or little multiplicative; *o.e.* typically as a layer of lignified, pitted, longitudinal fibres (not ribbon-like); *mesophyll* crushed; *i.e.* composed of small thin-walled cells transversely orientated, with slightly elongate rectangular facets. Vascular bundle in the raphe stout, with branches passing obliquely over the sides of the seed towards the micropyle in the testa. Embryo longitudinal in the fruit, rarely transverse; cotyledons thick, green.

Aril thick, oily, without v.b., the cells thin-walled; typically red, orange or yellow, arising from the hilar (or raphe-) side of the seed, partly enclosing the seed or reduced to a rim or merely as a thickening of the raphe; in some species the whole testa arilloid and sarcotestal.

The fruits and seeds of this genus have great interest for the durian-theory. Unfortunately classification of the species is based on the flowers and it is extremely difficult to identify fruiting trees, though they are so much more conspicuous. In the Singapore herbarium there appear to be more unnamed species than named from the Malay Peninsula and Borneo. I made a large collection of their fruits in alcohol in Singapore with the intention of studying them in detail, but many years have elapsed and most specimens have dried out or been hydrolysed in vinegar. In the few that remain I have detected transitions from the typically arillate seed with fibrous exotegmen and hilar (or raphe-) aril to exarillate seeds and sarcotestal, but pachychalazal, seeds; these, as one would expect, lack the fibrous exotegmen and prove the derivative nature of the sarcotesta. I list the following:

(1) fully arillate (*Dysoxylon sp.*, Corner 14 Nov. 1937);

(2) reduced aril (*D. acutangulum, D. cauliflorum*);

(3) exarillate with blackish brown testa (*D. costulatum*; *Dysoxylon sp.*, Sing. F. No. 32810);

(4) exarillate and mostly pachychalazal with orange sarcotesta (*Dysoxylon sp.*, Sing. F. No. 32212, 34930);

(5) exarillate and wholly sarcotestal with red or orange-red seed-coat (*D. angustifolium*, ? *D. arborescens*).

It will be seen, also, from the illustration (Fig. 384) that there are considerable differences in the size and extension of the hilum and in the place of entry of the vascular bundle from the placenta (whether at the micropylar or the chalazal end of the seed). There can be no doubt that all the species of *Dysoxylon* need full examination of their seeds before there can be an adequate understanding and classification of them.

D. acutangulum Miq.
(Sing. F. No. 34950, 34989; fruits in alcohol.)

Seeds brown to black, with a small orange patch as the arilloid hilum on one side. Testa thick, rather pulpy, the greater part appearing to be pachychalazal. Tegmen without fibres.

D. arborescens (Bl.) Miq.
(Java, Bakh. v.d. Brink 7622; Herb. Bogor., fruits in alcohol.) Fig. 384.

Seed 17 × 8 mm, ellipsoid, exarillate, but with thick fleshy arilloid raphe (not developing a free aril). Testa thin; *o.e.* composed of cuboid cells with brown, not lignified, walls; *mesophyll* with v.b. Tegmen crushed, without a fibrous layer.

D. cauliflorum Hiern
(Java, Kebun Raya; living material.) Figs. 375, 384.

Ovule; o.i. 3–4 cells thick; i.i. 2–3 cells thick.

Seed 13–17 × 8–10 mm overall; aril orange-red, waxy, as a rather narrow flange round the hilar side of the seed, much thickened at the chalaza. Testa 6–9 cells thick; *o.e.* as a firm layer of cuboid cells; *mesophyll* thin-walled; *i.e.* with 1-several crystals in each cell. Tegmen ? not multiplicative; *o.e.* as a layer of longitudinal, lignified, cylindric fibres completely surrounding the seed except at the chalaza; *mesophyll* crushed; *i.e.* unspecialized. Vascular bundle passing directly to the chalaza without a distinct raphe, stout, giving off from the chalaza a laxly branched system of v.b. extending along the wide hilum (or placental side) of the seed towards the micropyle. Hilar tissue much thickened and extending along the adaxial side of the seed, thin-walled, oily as the aril.

Aril with thin-walled oily cells, the outer epidermal walls red.

Except for the reduced aril, this seed appears typical of *Dysoxylon*. Because of the basal position of the raphe v.b., there is a pre-raphe v.b. reminiscent of the Connaraceous seed.

D. costulatum Miq.
(Sing. F. No. 36415; Singapore, Bukit Timah Forest, as tree 334; fruits in alcohol.)

As *D. acutangulum* but the larger seed with thin testa and without an aril.

Dysoxylon sp.
(Singapore, Mandai Road, Corner s.n. 14 Nov. 1937; fruits in alcohol.) Fig. 384.

Seeds 14 × 9 mm, black, covered with the orange-red aril except on the abaxial side. Testa *c.* 12 cells thick; *o.e.* composed of cuboid cells with slightly thickened and brown outer walls; *i.e.* as a layer of small crystal-cells. Tegmen crushed but the o.e. composed of longitudinally elongate, thin-walled cells and the o.h. composed of shortly transversely elongate cells.

The seed resembles that of *D. cauliflorum* but the aril is better developed; the vascular bundle from the placenta enters the seed at the level of the micropyle and passes down a raphe-like portion to the chalaza; the hilum is a narrow strip of tissue along this raphe. Another unnamed species (Sing. F. No. 37099) has a similar seed.

Dysoxylum sp.
(Sing. F. No. 32810, Pahang, Cameron Highlands; fruits in alcohol.)

Seeds exarillate. Tegmen without fibres. Embryo more or less transverse (as in *Chisocheton*) but the seed anatropous. Capsule 1-seeded.

This seems to be near to *D. thyrsoideum*.

Dysoxylon sp.
(Sing. F. No. 32212, 34930; fruits in alcohol.)

Seeds 3.4 × 2.6 cm overall, covered by an orange pachychalazal sarcotesta 1–2 mm thick, except for the black testal region round the micropyle. Hilum small, round, white.

This tree grew in the Reservoir Jungle of Singapore. It was cut down during the Japanese occupation and seems now to be extinct in the island. It is near to *D. andamanicum* King.

Guarea L.

G. trichilioides L.
(Brazil, Belem-do-Pará, Museu Goeldi; living material.) Fig. 385.

Seeds 12 × 7.5 mm, ovoid, red, 1–4 per fruit, exarillate, with broad white hilum. Seed-coat comprising the following four layers of tissue, from the outside inwards: (1) epidermis with red cuticle, the cells rather large, 20μ high, not radially elongate, without stomata; (2) a pulpy layer 150–200μ thick, composed of 4–6 layers of rounded, thin-walled cells with yellow oil-drops; (3) a sclerenchymatous layer 200–300μ thick, composed of 2–3 layers of tracheid-like pitted cells, the outer cells longitudinally elongate, the inner cells smaller and more or less cuboid; (4) an innermost layer of thin-walled cells with starch, more or less crushed. Micropyle (? endostome) as a pale yellowish white orifice surrounded by more or less isodiametric sclerotic cells continuous with the sclerotic cell-layer of the seed-coat. Vascular bundle of the raphe dividing at the chalaza into $c.$ 5 branches subdividing to form a narrow plexus of v.b. at the junction of the wide chalazal area with the sarcotesta, but $c.$ 6 slender v.b. extending from this plexus into the seed-coat between the pulpy and sclerotic layers, the antiraphe without v.b. Chalazal area composed of sclerotic tissue, continuous with that of the seed-coat but thicker.

Capsule fawn brown, coriaceous, splitting loculicidally into 4 parts with a columella; endocarp thin, separating as a thin coccus round each seed; seeds hanging out on placental strips detached from the columella.

I studied these seeds in Brazil before I was aware of the complexity of the Meliaceous seed. Layers (1) and (2) evidently represent the sarcotesta; layers (3) and (4) represent either the tegmen with fibrous exotegmen and sclerotic mesophyll (cf. *Swietenia*) or, as seems now more probable, the tissue of a pachychalaza (as in *Aphanamixis*). Re-investigation is needed from ovule to ripe seed.

Lansium Corr.

L. domesticum Corr.
(Ceylon; ripe fruits.)   Fig. 386.

Ovules anatropous, suspended.

Seeds $c.$ 25×15 mm (overall), the brownish seed-body 13×7 mm covered with the cream-white pulpy aril arising from the short thick funicle and exostome and in a median perichalazal band along the antiraphe side of the seed; seed-coat perichalazal with a pachychalazal plate-like expansion on the antiraphe side; exalbuminous. Testa 11–15 cells thick; *o.e.* unspecialized but with thick cuticle; *mesophyll* unspecialized, subaerenchymatous, some cells with sphaerocrystals; *i.e.* composed of smaller unspecialized cells, some with a sphaerocrystal. Tegmen 4–5 cells thick, developed over the apex and on the side of the seed; *o.e.* composed of longitudinal, pitted, lignified fibres with thick walls; *mesophyll* unspecialized but with some sclerotic cells along the raphe and at the chalaza; *i.e.* unspecialized, but with sclerotic cells at the endostome and round the chalaza. Perichalaza thin-walled, with v.b. on the wider antiraphe-side. Vascular supply consisting of a raphe-bundle

(compound with several anastomosing v.b.), with a fascicle of v.b. to the micropylar end of the aril, these v.b. combining to enter the micropylar end of the perichalaza on the antiraphe; at the chalaza the v.b. combining into a stout compound bundle entering the lower end of the perichalaza and forming with the micropylar bundles the plexus of v.b. along the antiraphe.

Aril with large pulpy cells, those near the testa radially elongate; *o.e.* as smaller tabular cells.

Pericarp rather thin, with small scattered groups of sclerotic cells and numerous latex-tubes; v.b. small in 2–3 rings; endocarp $c.$ 100μ thick, collenchymatous, the fibrous cells variously directed in bundles.

This complicated seed needs full investigation from the ovule in order to determine exactly how the aril arises. In the few seeds available to me I was unable to see the accrescent surfaces of the aril. The species was the thin-skinned 'langsat' of Malaya.

Melia L.

Seed-structure associates *M. azedarach* with *Cipadessa* (fibrous exotegmen, endotestal crystal-cells) and distinguishes *M. azadirachta* and *M. dubia* (exotegmen without fibres, endotesta without crystal-cells). *M. excelsa*, transferred to *Azadirachta* by Jacobs (1961), has an unusually large seed (Corner 1939) and evidently agrees with *M. azadirachta* (*Azadirachta indica*). It is not clear to me that *Cipadessa*, *Melia* and *Azadirachta* have been satisfactorily distinguished.

M. azedarach L.
(Ceylon; flowers and fruits in alcohol.)   Fig. 387.

Ovules 2 per loculus, superposed, the upper smaller and erect, the lower suspended and fertile; o.i. (2–3) cells thick; i.i. 3–4 cells thick, with i.e. well-formed; micropyle formed by the endostome (Nair).

Seeds 3.5×1.6 mm, 1 per loculus, 1–5 per fruit, oblong, brown, smooth, exarillate, with very short funicle, each enclosed in a fibrous-woody coccus (transverse fibres). Testa 2 cells thick; *o.e.* composed of somewhat tabular thin-walled cells with polygonal facets, not lignified; *i.e.* as a layer of small crystal-cells with lignified walls. Tegmen 5–6 cells thick; *o.e.* as a compact layer of narrow longitudinal fibres -200μ long, lignified, pitted; *mesophyll* consisting of 3–4 layers of large thin-walled cells, eventually crushed; *i.e.* as a layer of shortly radially elongate cells with brown walls, not lignified. Vascular bundle only in the raphe, not perichalazal. Endosperm oily.

(Netolitzky 1926; Nair 1959b; Vaughan 1970.)

While this description agrees fairly well with that of Netolitzky, Nair gives the testa as 7–8 cells thick, and neither he nor Vaughan mentions a fibrous exotegmen; none of these authors mentions the crystal-cells of the endotesta. They may not have studied mature seeds. The species may be variable, or there may have been confusion with *M. dubia*, the fruits of which are similar.

M. azadirachta L.

Ovule: o.i. and i.i. 2–4 cells thick; endostome 5–6 cells thick, forming the micropyle.

Seeds apparently with non-multiplicative integuments. Testa; *o.e.* thick-walled; *mesophyll* composed of *c.* 4 layers of compact flattened cells (Vaughan), of 2–3 layers of thin-walled and loosely arranged cells (Nair). Tegmen unspecialized (Nair), ? with loose sclerotic cells or fibres (Vaughan). Perisperm persistent as 2–3 layers of cells (Nair).

(Nair and Kanta 1961; Vaughan 1970.)

There are discrepancies here also. The sclerotic cells figured by Vaughan suggest exotegmic fibres.

M. dubia Cav.

(Ceylon; ripe fruits in alcohol.)   Fig. 388.

Seeds 1 per loculus, in the very hard endocarp, 1–3 per fruit, *c.* 12 × 3 mm, oblong, elliptic in section, acute at the chalazal end, flattened towards the short funicle, dark brown, smooth, exarillate. Testa well-developed with conspicuous palisade (o.e.) and more or less crushed mesophyll, not lignified, *c.* 9–10 cells thick on the sides of the seed, 15–20 cells thick in the elongate exostome-raphe; *o.e.* as a palisade of prismatic cells with thickened outer and inner yellowish walls (faintly punctate–striate) and brown middle part (? thick-walled); *mesophyll* composed of small, tangentially elongate, substellate cells with brown contents; *i.e.* unspecialized. Tegmen crushed or absorbed except for a single layer of large, more or less hexagonal cells with brownish walls, not lignified; *endostome* as a small plug of crushed brown cells, unlignified. Vascular bundle only in the raphe, dividing at the chalaza into 7–9 branches shortly recurrent to the base of the endosperm, apparently perichalazal along the raphe. Hypostase not evident. Endosperm oily.

There is no fibrous exotegmen. Whether I have interpreted the exotestal palisade-cells correctly must be decided from earlier stages than were available. At first sight the exotesta seems to be composed of three layers of cells but in surface-view, as well as in section, the radial walls can be traced as continuous lines from the outside to the inner side of the complex. The nature of the yellowish thickened outer and inner ends of the cells I could not determine. A few palisade cells, particularly along the raphe, had the brown middle part of the cell clearly delimited with thick walls, but most cells had merely glairy contents.

Naregamia W. et A.

N. alata W. et A.

Ovules 2 per loculus, collateral, suspended; o.i. 2 cells thick, 3–4 at the exostome; i.i. 2–4 cells thick; micropyle

formed by the exostome; obturator present, crushed after fertilization.

Seed ? small (? 2 mm long), dark brown, with white aril from the sides of the funicle. Testa not multiplicative; *o.e.* with the cells radially elongate with papillae-form apex drawn out into a fine hair, with thick cuticle. Tegmen 5–7 cells thick; *o.e.* with the cells elongate tangentially, but radially at the micropyle; *i.e.* unspecialized. Vascular bundle only in the raphe. Aril non-vascular.

(Nair 1959a).

Exotegmic fibres are not mentioned, but it is not clear if ripe seeds were studied.

Sandoricum Cav.

S. koetjape Merr.

(Ceylon; ripe seeds in alcohol.)   Fig. 375.

Ovules 2 per loculus, suspended; o.i. 3–4 cells thick; i.i. 2 cells thick.

Seed 16–18 × 9–11 × 7–8 mm, oblong, subcompressed, faintly keeled at the micropylar end, pale brownish, exarillate. Testa 1 mm thick, 3–4 mm at the micropylar end, 1.5–2 mm thick at the chalazal end, 25–50 cells thick, not lignified except for the v.b.; *o.e.* as a layer of shortly longitudinally elongate cells, not as a palisade; *mesophyll* more or less aerenchymatous with elongate substellate cells and latex tubes, spongy, firm, thick; *i.e.* composed of small cells, unspecialized. Tegmen and endosperm without trace. Vascular bundle in the raphe stout, with 4–5 branches to each side of the seed from the lower chalazal part, the branches ascending the sides of the seed obliquely, branching and anastomosing, apparently not in the antiraphe. Embryo straight; cotyledons large, bright pink, 2-lobed at the base round the short radicle; plumule minute.

Pericarp thick, thin-walled, with scattered v.b., without sclerotic cells, the inner tissue pulpy; endocarp *c.* 50 μ thick, composed of a few layers of narrow, slightly lignified, transverse fibres.

(Juliano 1934; Nair 1958.)

It seems that this is a pachychalazal seed with little differentiation of the tissues. The pulp round the seeds is mesocarp.

Swietenia Jacqu.

Seeds numerous and biseriate in each loculus, imbricate and hanging by the thin end of the wing (raphe-exostome extension) with the enlarged seed-body towards the base of the capsule, exarillate, practically exalbuminous. Testa 12–15 cells thick on the sides of the seed-body, thicker at the basal end, much thinner in the wing, all the tissue with slightly thickened, lignified walls; *o.e.* as a regular layer of cells, mostly with pitted walls, some unpitted, not radially elongate, with frequent stomata; *mesophyll* composed of large subglobose cells with very short arms, the walls rather widely and shallowly pitted, aerenchymatous, corky; *i.e.* as a layer of small cuboid cells with 1–2 crystals in each. Tegmen 5–6 cells thick, attached along the length of the perichalaza; *o.e.* as a compact layer of obliquely longitudinal, finely pitted, lignified fibres, scarcely elongate radially; *mesophyll* 2–3 cells thick, several of the outer cells (o.h.) as shortly longitudinally elongate sclerotic cells, widely pitted, with rather thin walls (not in a continuous layer), the inner cells thin-walled, enlarged, not lignified, not aerenchymatous; *i.e.* as small cells with tannin, not lignified. Vascular bundle as a stout concentric bundle in the elongate alate raphe-exostome, curving in its descent and dividing into 3–4 bundles (more or less concentric) in the elongate perichalaza along one side of the seed, ending blindly at the lower end, without branches to the sides of the seed. Hypostase as a layer of small brown cells along the perichalaza, thin-walled. Endosperm reduced to 1–2 layers of small oily cells. Embryo with the cotyledons in the broad plane of the seed; plumule and radicle minute, immersed, appearing lateral.

S. humilis Zucc.
(Netolitzky 1926.)

S. macrophylla King
(Ceylon; ripe seeds in alcohol.) Fig. 389.
Seeds 5–9 cm long, the seed-body *c.* 3.5 × 1.5 × 0.6 cm.

Trichilia P. Browne

In this large genus there appears to be a considerable range in the construction of the seed-coat and in the nature of the aril. The species, here described from Bolivia as *Trichilia sp.*, has a fibrous exotegmen and funicular aril which becomes compounded with those of sterile seeds in adjacent locules of the fruit; I am sure no creature is perceptive enough to distinguish this arilloid as a special ecological device.

T. emetica Vahl
Seed 14–18 mm long, dark brown, with red aril (?). Seed-coat (? pachychalaza); *o.e.* composed of flattened cells with thick brown outer walls (red contents; Netolitzky); *mesophyll* in several layers, thin-walled, oily, the inner part with v.b. and nests of sclerotic cells.
(Netolitzky 1926; Vaughan 1970.)
This may be a pachychalazal seed with sarcotesta.

Trichilia sp.
(Bolivia, Santa Cruz de la Sierra, leg. E. J. H. Corner, Feb. 1948; dried.) Figs. 390, 391.
Ovules 2 per loculus, anatropous, suspended.
Seeds 10 × 5 mm, brownish black, shiny, partly covered from the raphe-side by a red aril, 1 per fruit, hanging on the short and partially detached placental

columella of the fruit; exalbuminous; hilum short, linear. Testa 4–5 cells thick, not lignified; *o.e.* as a layer of cuboid cells with polygonal facets, not radially elongate, with thickened brown outer wall, without stomata; *mesophyll* more or less crushed; *i.e.* as a layer of small crystal-cells. Tegmen 4–5 cells thick; *o.e.* as a compact layer of longitudinal, thick-walled, finely pitted, laterally compressed, and lignified fibres, interspersed with rows of 3–8 short, pitted, sclerotic cells in a double layer; *mesophyll* crushed, but o.h. evidently composed of rather large, tabular cells with thickened (not lignified) walls and brown contents. Vascular bundle of the raphe descending into the elliptic unspecialized chalaza about the middle of the adaxial side of the seed, and spreading shortly as fine v.b. at the base of the seed, without v.b. on the sides of the seed. Embryo straight; radicle short, immersed between the 4 basal lobes of the thick green cotyledons.

Aril red, rather pulpy, compound, formed from the axis of the fruit and the raphe and funicle of the one seed, together with the arils of the two sterile loculi, the whole conjoint into an elongate body with 6 longitudinal furrows, with a lobe from the raphe capping the micropyle, and the two abortive ovules embedded in it; *cortex* red, composed of 2–4 layers of thin-walled cells in radial rows, the epidermal cells with thickened outer wall, making a minutely tesselate surface; *medulla* white, composed of enlarged cells, not vascular except for the v.b. of the fruit-axis; all the cells with oily contents.

Capsule 12–13 × 6 mm, ellipsoid, with persistent calyx and short style-base, green, loculicidal into 3 recurved valves each lined with two lignified cocci, exposing the one seed with the axile aril.

In this deceptive fruit, which I first took to be Sapindaceous, the apparent aril consists of three parts; (1) the funicular aril of the one seed, (2) the funicular arils of the five sterile ovules which become embedded in the fleshy tissue, and (3) the fleshy columella with which the funicular arils are continuous. The septa in the ripe capsule are reduced. On dehiscence the axile part of the arilloid mass begins to dry. The weight of the seed pulls out the axile strand of v.b. in the columella and on this strand the whole contents of the loculi dangle, as if a single seed on its own funicle.

Turraea L.

T. pubescens Helen
(Java, Kebun Raya, XVI.I.B.15; living material). Fig. 392.
Seed 6–7 × 3.5 mm, black, shiny, with an orange aril along the raphe-side, 1–2 seeds per loculus; chalazal end with a small beak; micropyle strongly incurved at the apex of the seed and concealed. Testa 4–7 cells thick, unspecialized; *o.e.* composed of shortly columnar cells with thick outer wall and dark brown colour

(dissolving into a reddish purple solution in potash). Tegmen; *o.e.* as a thick layer of stout, ribbon-like, longitudinal fibres with thick, pitted, lignified walls; *mesophyll* and *i.e.* crushed. Vascular bundle only in the raphe. Endosperm oily.

Aril developing from the apex of the funicle and along the raphe, not from the chalaza or micropyle, not covering the micropyle; tissue compact, not laminar, becoming pulpy and slimy, not vascular.

The fruit and seed in their complexity are Celastraceous but the flower in neotenic simplification is Meliaceous.

MELIANTHACEAE

Ovules 1–3 per loculus, anatropous, erect or the lower ones transverse, bitegmic, crassinucellate; o.i. 3–7 cells thick; i.i. 2–4 cells thick; micropyle formed by the exostome (*Bersama*) or the endostome (*Melianthus*); funicle rather massive.

Seeds moderately large, ellipsoid, albuminous, with short funicular aril in *Bersama*. Testa multiplicative, with styloids (long, prismatic crystals of calcium oxalate) in the mesophyll; *o.e.* appearing as a thick-walled palisade in t.s. but consisting of longitudinally elongate, more or less fibriform cells, not lignified; *mesophyll* thin-walled, rather small-celled; *i.e.* unspecialized. Tegmen not multiplicative, unspecialised, crushed. Vascular bundles in the raphe only, but dividing into several branches over the simple chalaza (*Bersama*). Endosperm nuclear, massive, horny, amyloid (*Bersama*). Embryo straight, rather short. Figs. 393, 552.

(Netolitzky 1926; Mauritzon 1936a.)

The seeds do not have the Sapindaceous structure, but that of Lardizabalaceae (p. 156). The raphe-bundle of *Bersama* divides exactly as that of *Akebia* (Lardizabalaceae), which has a vestigial aril. *Greyia* is excluded by Hutchinson (1964) as Greyiaceae, placed in Cunoniales. Compare, also, the Leguminous seed.

Bersama Fresen

B. abyssinica Fresen ssp. *paullinioides* (Planch.) Verdc. (Kenya, Kitale, leg. Mrs Tweedie, 1972; flowers and fruits in alcohol.) Fig. 393.

Ovules anatropous, erect, solitary in each loculus, sub-basal; o.i. 5–7 cells thick, with the cells of o.e. already slightly elongate longitudinally; i.i. 2–3 cells thick, very thin; micropyle formed by the long exostome.

Seeds 12–14 × 7–8 mm (drying 8–9 × 6 mm), black, ellipsoid, with a yellow, entire, funicular aril at the base, concealing the micropyle. Testa slightly multiplicative, 15–18 cells thick, not lignified; *o.e.* composed of longitudinally elongate, narrow cells with much thickened and horny walls, especially the outer wall, (80–) 100–300 μ long, appearing as a short palisade in t.s.; *mesophyll* composed of rather small unspecialized cells, but with

numerous styloids –350 × 16–25 μ and square in t.s.; *i.e.* unspecialized, without crystals. Tegmen not multiplicative, becoming crushed. Chalaza unspecialized but broad. Vascular bundle simple, concentric, then bifurcating into two close branches, finally dividing into a plexus over the apex (chalazal tip) of the seed. Endosperm massive, horny, slightly thick-walled, with diffusely amyloid cell-contents (violet-black in iodine). Embryo relatively short; hypocotyl long; cotyledons narrow, flat.

Funicle stout, with a central core of *c.* 5 v.b. in a concentric ring, and with a small v.b. external to the main ring on the micropylar side and dividing into 3–5 strands ending blindly; epidermis similar to that of the aril; mesophyll with abundant styloids.

Aril developing late in half-grown seeds, arising as a broad annulus round the apex of the funicle, eventually more or less covering the funicle with the swollen base of the aril' and pressing round the base of the seed to obscure the micropyle, the free border crenulate and sometimes cleft along the micropylar line; *o.e.* composed of small thick-walled cells like those of the exotesta but not or scarcely elongate; mesophyll composed of large thin-walled cells with large yellow oil-drops, without v.b., with few or no styloids except at the junction with the firm funicular tissue.

Capsule 3 × 2.8 cm, splitting loculicidally into four parts; mesocarp thin-walled, pervaded radially by slender fibro-vascular bundles from the woody endocarp, the slender bundles ending blindly near the outer surface; endocarp massively wood, carrying the longitudinal v.b.

This fruit with radial fibres has all the marks of derivation from a spinous arillate capsule.

Melianthus L.

M. comosus Vahl, *M. major* L., *M. minor* L.

Testa –20 cells thick; *o.e.* as a thick-walled palisade, as thick as the rest of the seed-coat (but ? the cells longitudinally elongate); *mesophyll* collapsing, the outer few layers of cells often with styloids. Tegmen 3 cells thick, collapsing.

(Guérin 1901.)

MENISPERMACEAE

Ovules 2, more or less anatropous or campylotropous, bitegmic or unitegmic (*Cissampelos, Stephania, Tinospora*), crassinucellate, one ovule aborting after fertilization; o.i. 4–5 cells thick (*Tiliacora*), 2–3 cells (*Cocculus*) or the one integument 3–4 cells thick (*Cissampelos*); i.i. 2 cells thick.

Seeds generally curved or amphitropous, the integuments unspecialized and crushed, exarillate, albuminous or not. Testa; *o.e.* persistent in some cases as a layer of tabular cells with thin lignified walls. Endosperm nuclear, oily, ruminate with infoldings of the testa

(*Tiliacora, Tinospora*) or from infoldings or undulations of the fibrous endocarp (*Cissampelos, Cocculus, Stephania*). Embryos relatively large, often curved.

(Netolitzky 1926; Sastri 1954, 1964.)

The carpel has a ventral suture and marginal ovules. The seed in the indehiscent fruit has a much simplified coat which, in the unitegmic, may be pachychalazal. If, as most authors agree, the family is related to Berberidaceae, then the seed-coat seems to have lost all character.

MONIMIACEAE

Ovules 1 per carpel, anatropous, crassinucellate, either suspended with dorsal raphe from the apex of the loculus and bitegmic (Monimioideae), or basal erect and bitegmic (*Amborella*), or basal-erect with ventral raphe circumscribing the ovule and unitegmic (Atherospermoideae); o.i. 2–4 cells; i.i. 2–3 cells thick; the single integument 3–5 cells thick; funicle (Monimioideae) stout, with obturator; (? orthotropous ovules in *Daphnandra*).

Seed small to moderately large, not exposed, exarillate, albuminous, more or less perichalazal; integuments not multiplicative or the testa slightly, the tegmen eventually crushed; funicle short. Testa unspecialized except for i.e.; *mesophyll* in some cases with oil-cells; *i.e.* unspecialized or composed of short, lignified, tracheidal cells with spiral-reticulate or subscalariform thickening (*Hortonia, Kibara, Steganthera*). Tegmen unspecialized, the cells generally cuboid or somewhat longitudinally elongate, enlarging, then crushed. Endosperm cellular, oily, copious. Embryo very small to moderately large, straight; cotyledons appressed or divergent.

Fruit drupaceous, indehiscent, free or sunk in the enlarged receptacle or enclosed by it, such receptacles eventually dehiscing; mesocarp with stone-cells (? absent in Atherospermoideae) and oil-cells; endocarp woody with 1–20 layers of thick-walled, lobate, sclerotic cells. Figs. 394–405.

(Perkins and Gilg 1911; Netolitzky 1926; Mauritzon 1935a; Bailey and Swamy 1948, 1950; Hutchinson 1964; Rodenburg 1971.)

Affinity with Lauraceae is supported by the construction of the carpel, ovule, seed and fruit. Both families have the tubular carpel developed basipetally below the short crescentic primordium, which becomes the style and stigma, and the carpel-body lacks the ventral groove or suture. The solitary ovule in Lauraceae and Monimiaceae–Monimioideae has the dorsal raphe which arches over the micropyle, and it becomes more or less perichalazal as the seed develops basipetally in the extending fruit. The endotesta, if it differentiates, becomes tracheidal in both groups and the exostome sclerotic, though these features may be lost as in the indehiscent fruits, while the tegmen enlarges and disappears without specialization. The drupe has a woody endocarp composed of lobate-substellate cells, which may be in numerous layers in Monomiaceae but reduces to one layer in Lauraceae. In these respects Lauraceae differ mainly in the solitary carpel, the strongly multiplicative testa with extensive vascular supply, and the large embryo without endosperm. Except for this elaboration of the testa, Lauraceae appear as the more advanced derivative of a common ancestry.

Monimiaceae are a neglected family full of systematic detail which awaits evolution interpretation. The flower seems based on the primitive Magnolialean formula $P \propto A \propto G \propto$, but the floral parts are reduced and small, the floral axis is short (as in Annonaceae) and tends to become urceolate as a receptacular tube or a complete envelope in the more advanced genera. The uni-ovulate carpel supposes a multi-ovulate and follicular ancestor which has been simplified in two ways; the solitary ovule is apical and pendent to fit the basipetal growth of the carpel in Monimioideae, but it is basal and erect in the Atherospermoideae, fitting an intercalary extension of the carpel between its base and the crescentic apex. This distinction appears fundamental because the Atherospermoid ovule (as shown by *Siparuna*) has become unitegmic with encircling raphe. Presumably in the ancestral follicles, the Monimiaceous seeds were endotestal with tracheidal cells.

Neoteny has diminished the size and complexity of the floral parts but has elaborated much detail. Tepals may disappear; stamens and carpels become few, even solitary; anthers have short filaments or are sessile as in their primordial inception, and they develop crescentic, equatorial or valvate dehiscence. In compensation for the reduction of the floral parts, the receptacle widens with tangential and basipetal growth into discoid, cupular, urceolate and syconium-like forms (very much as in Moraceae), and it assumes the functions of perianth and pericarp. The primitive Annonaceous display of stalked drupes is converted into a pyriform enclosure. The seed-like drupelets of *Siparuna* are isolated in loculi by receptacular septa and the fruit resembles a Rosaceous pome. The drupelets of *Tambourissa* are embedded in sockets of the receptacle like those of *Dorstenia*, but this advanced must have been in parallel with that of *Siparuna* because *Tambourissa* has the Monimioid ovule; an analogy is *Nelumbium*. A final touch in this outward displacement of function is provided by *Hennecartia* in which, according to Perkins and Gilg, apical processes of the urceolate receptacle become stigmatic lobes in mimicry of a syncarpous, bilocular, and inferior ovary. Thus, experimenting in the conversion of a Magnolialean flower into a modern neotenic flower, Monimiaceae have ranged from an Annonaceous resemblance into Rosaceous, Saxifragaceous, Moraceous, Malvaceous and, indeed, Lauraceous effects. The existing genera offer diverse combinations of these results which must have begun in the early evolution of dicotyledonous forest, for they are demarcated geographically as well as structurally. Some main branches may have been recognized as Amborellaceae, Atherospermataceae, Austrobaileyaceae, Trimeniaceae (? with very hard testa, Rodenburg 1971), and, even, Lauraceae.

In this light of neoteny, simplification and transference of function, the seed-like drupelet of *Siparuna* finds an explanation. It expresses the outward transfer of function from seed to receptacle whereby the apocarpus fruit has been converted into a structure similar to that of an inferior syncarpous ovary. The tegmen of Monimiaceae, if physiological in the early growth of the seed, is vestigial in its adult state and seems to have been lost from the Atherospermoid ovule (or merged with the testa in a perichalazal form). In *Siparuna* its place is taken by the one integument which, in the adult seed, is crushed between the endosperm and endocarp. As the next outer layer, the endocarp takes over the mechanical function of the testa. The pulpy mesocarp resembles the sarcotesta, and the receptacular wall of the fig-like fruit acts as a dehiscent pericarp. So the drupelet resembles a Magnoliaceous or Annonaceous seed exposed on an irregularly dehiscent receptacular matrix (cf. *Xylopia*, Fig. 16). But in some species of *Siparuna* the drupelet develops a red, oily, and bilobed outgrowth which can be called a carpellary aril because it resembles in structure and function a seed-aril displaced, or transferred, to the outside of the carpel. That it is an ecological device and an arilloid (van d. Pijl 1955, 1966) is sure, but that does not imply that it is an innovation without genetic precedents. It originates, according to the durian-theory, through the transference of function of the genetic characters of the seed-aril to the outside of the drupelet, itself modified into a seed, as the pericarp characters are transferred to the receptacular wall; a comparable example is *Artocarpus* (Corner 1962). Through this process of externalization, there result a second-order receptacular pericarp, a second-order pericarp testa, a second-order carpellary aril, and a second-order testal tegmen. The arillate drupelet of *Siparuna* is layered in cell-division and differentiation much as in the arillate seed of *Carica*, the reticulations in which must be genetically akin to those of *Siparuna*. If this explanation seems strange, it is only one of the many vestigial diversifications of a unique family that has strangely embarked on lines of progress that later families of more secure establishment have standardized with the improved syncarpic construction. Critics avoid, or overlook, the morphogenetic consequences of the single integument. The short processes on the fruits of certain species of *Siparuna* are explicable in the same way, not as the genetic innovations of an ecological creation, but as the genetic transfer of the primitively exocarpic spines to the receptacle. Thus, *Siparuna* in its specialization has retained primitive durian-features that have disappeared from Monimioideae, and Atherospermoideae supply evidence of an earlier divergence

from pre-Monimiaceae than Lauraceae. It is in this light that the vessel-less *Amborella* (Monimioideae) with erect bitegmic ovule must be considered (Bailey and Swamy 1948).

Calycanthaceae have many points in common with Monimiaceae. The endotestal seed confirms this alliance and disposes of the suggestion to place the family in Rosales (Hutchinson 1964). An early divergence from the pre-Monimiaceous stock in also indicated by the presence of two ovules in the carpel (anatropous and erect, but only the basal one fertile, as in Atherospermoideae, though bitegmic) and by the strongly multiplicative testa with well-developed outer epidermis. The exalbuminous seed with large embryo parallels the Lauraceae with Monimioid ovule.

Amborella Baill.

Ovule sub-basal, anatropous, erect, bitegmic; o.i. 5–7 cells thick; i.i. 3 cells thick.

Seed 3 × 1 mm. Testa apparently unspecialized, not multiplicative; *o.e.* with dark-staining contents. Tegmen crushed. Endosperm copious. Embryo minute.

(Bailey and Swamy 1948.)

Remarkable for the lack of vessels and the erect bitegmic ovule.

Hortonia Wight

Ovule solitary, anatropous, pendulous with dorsal raphe, the short funicle arched over the micropyle with slight obturator; o.i. 2–3 cells thick; o.e. with brownish walls (alcohol-material); i.i. 2–3 cells thick; micropyle formed by the exostome.

Seed *c.* 10 × 6 mm, with membranous seed-coat enclosed in the endocarp; developing an extensive perichalaza as a pseudoraphe with the integuments attached to it; exarillate. Testa 4–5 cells thick, thicker in the perichalaza, with the sclerotic endotesta 1–2 cells thick but the testa wholly sclerotic at the exostome; *o.e.* composed of polygonal to subrectangular cells, eventually crushed; *o.h.* thin-walled, with scattered large oil-cells, becoming crushed; *mesophyll* thin-walled or the cells with more or less scalariform-reticulate thickening; *i.e.* as a continuous layer of tracheidal cells with scalariform-reticulate, lignified thickening, not elongate, in places the cells of i.h. similarly modified. Tegmen 3 cells thick, 4 cells at the endostome, the cells thin-walled, shortly longitudinally elongate, enlarging, some with oil-drops (mesophyll), eventually crushed. Nucellus with cuticle, eventually crushed but the nucellar beak at the endostome persistent with thick-walled (not lignified) cells. Perichalaza extending along the dorsal (abaxial) side of the seed to the antipodal end, carrying the single v.b., separated from the nucellus by a thin strip of hypostase-tissue with firm brown (not lignified) walls, with the seed-coats attached along the sides of the perichalaza; in *H. angustifolia* the perichalaza much shorter, the raphe with free tegmen extending for two-thirds the length of the seed. Funicle with sclerotic, tracheidal tissue, 2–3 cells thick, over the surface and continuous with the endocarp.

Drupe *c.* 18 × 10 mm, shortly stipitate, 1–10 per flower; *pericarp* composed of thin-walled tissue with scattered oil-cells and sclerotic cells; *v.b.* 13–15 in a ring, the dorsal and ventral v.b. large and opposed, the ventral v.b. recurved into the funicle; *endocarp c.* 350 μ thick, composed of 3–5 layers of very thick-walled, pitted cells with undulate-lobate anticlinal walls, lignified, derived from i.e. by anticlinal divisions, continuous with the sclerotic surface of the funicle.

H. angustifolia (Thw.) Trim.
(Ceylon; flowers and fruits in alcohol.)

H. floribunda Wight
(Ceylon; flowers and fruits in alcohol.)   Figs. 394, 395.

Kibara Endl.

K. coriacea (Bl.) Endl.
(Java; young and nearly mature fruits in alcohol.)

Testa 4–6 cells thick, −14 cells in the raphe, thin-walled, more or less crushed except the endotesta as a more or less continuous pellicle (one cell thick) of short, longitudinal or transverse, tracheids with spiral or subannular, lignified thickening. Tegmen soon crushed. Vascular bundle only in the raphe, with scattered sclerotic cells in the sheath.

Palmeria F.Muell.

Palmeria sp.
(New Guinea, NGF 12901; fruits in alcohol.)   Fig. 396.

Seed 5 × 4mm, developed from a pendulous ovule, enclosed in the thick woody endocarp. Testa 3–4 cells thick, the cells tangentially elongate, thin-walled; *o.e.* with slightly thickened brown walls. Tegmen apparently entirely crushed, without trace. Embryo small; cotyledons divergent.

Drupes 1–4, invested by the perianth-tube, smooth, black; *exocarp* 4 cells thick, the walls dark brown and slightly thickened, the cells cuboid and small; *mesocarp* with thin-walled cells packed with starch-grains, without stone-cells; *endocarp* 1.5 mm thick, very hard, composed of many layers of very thick-walled, contiguous, lignified cells. Perianth-tube splitting at maturity into 5 lobes (corresponding with the minute tepals), covered on both surfaces with short stellate hairs with 2–4 arms, internally with large clusters of stone-cells.

Siparuna Aubl.

Ovary deeply perigynous or epigynous, apocarpic with 9–15 carpels separated into individual loculi by thin confluent receptacular septa, only the stigmatic tips of the styles projecting.

Ovule unitegmic, anatropous, basal, erect with ventral raphe extending round the ovule to the micropyle; funicle short; integument 3–4 (–5) cells thick, not multiplicative, disintegrating in the seed or i.e. persistent with tangentially elongate, thin-walled cells. Embryo minute.

Drupes filling the carpellary loculus; *mesocarp* thin-walled, with oil-cells, without stone-cells; *endocarp* 7–20 cells thick, the cells lobate with thick, pitted, heavily lignified walls, often with uneven tuberculate or reticulate outer surface showing through the pellicular mesocarp, the endocarp incipient in the carpel as 5–6 layers of small cells derived from i.e. Carpellary aril developed in some species from the upper part of the carpel-body, shortly bilobed, without v.b.

Receptacle firmly pulpy or fleshy, without stone-cells, splitting irregularly at maturity, fragrant or foul-smelling, in some species with soft or thorn-like, simple or branched processes from the outside. Vascular bundles to the carpels derived from the inner descending plexus of v.b. Hairs stellate, peltate, bifid or simple.

(Heilborn 1931.)

I have been unfortunately unable to identify any of the four species which I have collected and studied.

Siparuna sp.
(Brazil, Rio de Janeiro, Corcovado, leg. E. J. H. Corner 24 Nov. 1948; material in alcohol.)   Figs. 397–399, 401.

Aril carpellary, developed from the upper part of the carpel-body at the junction with the style-base, becoming shortly bilobed on the sides of the carpel, pink to red, oily, thin-walled.

Carpels *c.* 11 per flower. Receptacle with foetid oily smell at maturity.

Siparuna sp.
(Peru, Iquitos, leg. E. J. H. Corner 1948; material in alcohol.)

Drupes 6 × 3 mm, 10–15 per flower, light grey, tuberculate from the endocarp; carpellary aril dark crimson, waxy, shortly bilobed, *c.* 2 × 4 mm.

Receptacle 12–13 × 15 mm, with a stalk 6–9 × 2 mm, fig-shaped, ripening rose-red to dark purple, with green lenticels, splitting into 5–8 irregular lobes, exposing the pink mealy inner surface with the more or less embedded drupes, smelling strongly of citronella (as the flowers, but the leaves with little or no smell).

(Corner 1949a, Fig. 8; the drupes treated as seeds.)

Siparuna sp.
(Brazil, Manaus, leg. E.J. H. Corner 1948; material in alcohol.)   Fig. 400.

Without carpellary aril.

Siparuna sp.
(Brazil, Mato Grosso, Base Camp, Royal Society Expedition, leg. E. J. H. Corner 27 Jan. 1968; material in alcohol.)   Fig. 401.

Without carpellary aril. Endocarp papillate, very

hard. Receptacle ripening red with pink inner surface, dehiscing irregularly, foul-smelling.

Steganthera Perkins

Steganthera sp.
(Solomon Islands, RSS 151; material in alcohol.) Figs 402–404.

Ovule as in *Hortonia;* o.i. 4 cells thick i.i. 3 cells thick; ? without obturator; integuments not multiplicative.

Seed 11–12 × 9 mm, subglobose, enclosed in the black drupe with yellow-ochre pedicel. Testa thin-walled; *i.e.* as a single layer of cells with spiral-reticulate thickening, not or slightly lignified. Tegmen disintegrating. Chalaza ? extended as in *Hortonia*, and perichalazal.

Pericarp with oil-cells and groups of stone-cells; endocarp as a single layer of shortly columnar cells cells with pitted, lignified walls. Hairs simple.

Steganthera sp.
(New Guinea, NGF 12904; material in alcohol.)

As the preceding but the endotesta with lignified, spiral-reticulate thickenings; endocarp with short cells.

Tambourissa Sonn.

Ovules bitegmic, anatropous, pendulous, the long funicle with an obturator, developed in individual pockets of the receptacle.

Fruit almost closed, fleshy. Drupes with thin, pulpy mesocarp and woody endocarp, without carpellary aril. Fig. 405.

(Baillon 1871, p. 302.)

MORACEAE

Ovules more or less anatropous, suspended, solitary, bitegmic, crassinucellate; o.i. 3–4 cells thick; i.i. 3 cells thick, 4–5 at the endostome.

Seeds small to large, albuminous or not, exarillate, enclosed in the woody endocarp. Testa mostly membranous, ? not multiplicative, thin-walled, often with brown walls, more or less crushed, or many cells with thickened walls and full of starch (*Prainea*); *o.e.* generally persistent as cuboid cells with rectangular facets, in some cases with slightly thickened outer wall. Tegmen crushed, unspecialized or i.e. persistent. Hypostase usually present. Vascular supply in the raphe only or with branches to the testa in large seeds. Endosperm nuclear, oily. Embryo straight or curved, small or large; cotyledons flat and thin or thick, or convolute.

(Netolitzky 1926; Johri and Konar 1956; Corner 1962; Kaur in Johri 1967b.)

The so-called seeds of this family are pyrenes, as in Urticaceae. I have failed to find any definite structure in the seed-coats which, at most, are slightly exotestal.

This large family is one of the successes of tropical forest. Though much simplified in flower and fruit, it is replete with vegetative evolution which has led to the modern tree. It proves, thereby, that the reproductive character of the family and, probably, of most of the main genera evolved in the pachycaul stage of dicotyledons (Corner 1949a, 1962). Compared with Magnoliales, the family is modern and its primitive growth-forms have not all been exterminated. Traces, even, of the ancient arillate capsule can be read in *Artocarpus* with the primitive genes acting, not in the seed, but in the fruit, much as in *Siparuna* (Monimiaceae).

MORINGACEAE

Moringa Adans.

M. oleifera Lamk.
(Ceylon; flowers and fruits in alcohol.) Figs. 406–408.

Ovules anatropous, bitegmic, crassinucellate, with distinct funicle; o.i. 5–6 cells thick; *i.i.* 3 cells thick; micropyle formed by the exostome.

Seeds trialate, the blackish brown subglobose body 10 mm wide, the broad longitudinal wings colourless, exarillate, practically exalbuminous. Testa highly multiplicative, 40–50 cells thick, for a long time thin-walled and the outer green cells with much starch, eventually thick-walled, lignified and consolidated from the middle tissue towards the unspecialized outer and inner epidermis; *mesophyll* in three layers, an outer and an inner layer each of several cells in thickness with lax criss-crossing spiral thickenings on the walls, the cells drying up to form the rather corky tissue, and a middle layer as a continuous shell (many cells thick) of very compact, heavily lignified, pitted cells shortly elongate tangentially and transversely but hardly fibriform (the cells –150 μ long, Vaughan 1970). Tegmen becoming 5 cells thick, eventually crushed without lignification, the cells enlarging; *i.e.* as a compact layer of cuboid cells with dense contents, thin-walled, then crushed. Vascular supply consisting of the raphe v.b. and two postchalazal branches extending almost to the micropyle, the three v.b. placed on the radii of the three wings. Wings developing early as extensions of the testa into the narrow triangular lumen of the young fruit, composed of thin-walled cells, unlignified, drying up into scarious membranes at maturity. Funicle elongating but more or less connate with the endocarp. Endosperm nuclear, aqueous, evanescent or persistent as a single layer of cells. Embryo with thick hemispheric cotyledons; plumule with 2 (–3) leaf primordia; radicle very short.

Ovary syncarpous with tubular style, without stigmatic lobes, the lumen hairy between the ovules; v.b. 12 in 3 circles, 10–12 in the style almost to the tip.

Pericarp developing a complicated vascular supply with an incomplete inner ring of small v.b. with inverted xylem; cortex with many close bundles of sclerotic cells, with gutter-shaped longitudinal rows of sclerotic cells in the endocarp along the three sutures; exocarp with scattered mucilage-sacs. Ovary and fruit developing by intercalary and basipetal growth.

(Netolitzky 1926; Narayana, H. S. 1962a; Vaughan 1970.)

The mesotestal structure of this seed, agreeing neither with Cruciferae nor Capparidaceae, is discussed on p. 33.

MYRICACEAE

Ovule solitary, orthotropous, more or less basal, unitegmic, crassinucellate; integument 4–7 cells thick, 3–5 (*M. gale*).

Seed small, enclosed in the drupe, exalbuminous. Testa not multiplicative, crushed except o.e. as a compact layer of more or less thick-walled cells. Vascular bundle in the raphe very short, with *c.* 9 postchalazal branches reaching almost to the micropyle. Hypostase with lignified cells. Endosperm nuclear, then cellular, ephemeral, not as a storage tissue, or persistent as a single layer of cells with oil and aleurone. Embryo straight.

(Netolitzky 1926; Kühn 1927; Hjelmqvist 1948; Håkansson 1955; Vaughan 1970.)

Though the ovary and ovule are regarded by some as primitive, the orthotropous form in the drupe is a sign of reduction cf. Urticaceae. In *M. gale* fertilization is delayed for several weeks after pollination. The seed-coat may be pachychalazal.

MYRISTICACEAE

Ovule solitary, sub-basal, sessile or shortly funiculate, anatropous and erect or suborthotropous, bitegmic, crassinucellate; o.i. and i.i. 7–10 cells thick.

Seeds 1–5 cm long, ellipsoid to subglobose, massive, brown, black, pallid or white, arillate. Testa generally multiplicative, vascular; o.e. composed of cuboid cells with isodiametric facets or shortly tangentially elongate, with thickened walls, not or slightly lignified, with stomata (*Knema, Myristica*) or without; *mesophyll* subaerenchymatous; i.h. often as a short palisade of unlignified cells or indistinct (*Gymnacranthera, Knema* spp.); i.e. as a strong palisade of prismatic cells with thick lignified walls and hexagonal facets, often with crystals in the lumen. Tegmen multiplicative, vascular, developing ruminations; o.e. as a single layer of longitudinal lignified fibres, compact or incomplete in places with thin-walled cells (*Horsfieldia irya, Myristica lowiana*) or as a single layer of short sclerotic cells or tracheidal cells (*Knema*) or unspecialized (*M. fragrans*) in all cases at the chalaza with a counter-palisade of lignified cells recurved from the endotesta; *mesophyll* thin-walled, generally with oil-cells in the inner part; i.e. unspecialized. Chalaza large, surrounded by the double tubular palisade (endotesta–exotegmen), central or pressed to the raphe-side (*Horsfieldia, Myristica*). Vascular supply of the testa derived from post-chalazal branches, that of the tegmen and its ruminations from the chalaza. Nucellus enlarging greatly, then absorbed and crushed. Endosperm nuclear, thin-walled, oily or also with starch (*Myristica*), ruminate. Ruminations developed as pegs (not plates) lobing from the chalaza and the sides of the tegmen (*Horsfieldia, Knema intermedia, Myristica*) or only from the chalaza (*Gymnacranthera, Knema*), with v.b. and oil-cells. Embryo small, 0.5–3 mm long; radicle very short; cotyledons divergent.

Aril micropylar to form an arillostome and from the base of the seed, entire or deeply laciniate, generally covering the seed, red or yellow, waxy or pulpy, corticate or not, with many v.b. Figs. 409–426.

(Voigt 1888; Netolitzky 1926; van de Wyk and Canright 1956; Sinclair 1958; Sastri 1959a, 1963; Uphof 1959; Periasamy 1961; Canright 1963; Hutchinson 1964.)

Some of the larger and the most complicated seeds of dicotyledons occur in this family. Their primitive nature is discussed on p. 55. Few species have been examined in detail and, while there is general agreement, there are differences in the form of the exotestal and exotegmic cells, in the development of the ruminations, in the position and symmetry of the chalaza with its double palisade, in the presence of stomata, and in the structure of the aril which may help in the definition of genera. *Knema* appears to be distinguished by the short sclerotic or tracheidal cells of the exotegmen and by the central, more or less columnar, chalaza; it has both kinds of rumination. *Myristica* has stomata, lateral chalaza, and lateral rumination. *Horsfieldia* has a central or lateral chalaza and lateral rumination, but lacks stomata; its ovule may also be suborthotropous. *Gymnacranthera* has the laciniate aril and stomata of *Myristica* but the central chalaza and chalazal rumination of *Knema*. Elongate exotestal cells occur in *Gymnacranthera* and *Knema pr. p.*

The construction of the exotegmen may readily be discovered in ripe seeds by scraping off the tissue next to the inner surface of the endotesta and examining it microscopically with or without the use of phloroglucin. The exotegmic fibres can be seen at the frayed edge of the broken seed-coat.

Gymnacranthera Warb.

Gymnacranthera sp.
(Sumatra, Palembang, Boschproefst. E606 and E620; Herb. Bogor., in alcohol.)

Testa 6–7 cells thick, ? not multiplicative; o.e. with the cells shortly longitudinally elongate, slightly lignified, with stomata; i.h. not as a palisade; i.e. as a rather short palisade, every cell with a stout crystal at the outer end. Tegmen as a layer of long, very thick-walled, sparsely pitted fibres. Chalaza central, tubular. Ruminations chalazal.

Horsfieldia Willd.

Chalaza lateral in *H. macrocoma* and *H. subglobosa*, central and shortly infundibuliform in *H. irya*. Stomata absent. Aril entire.

H. irya Warb.
(Sumatra, Palembang, Lambach n. 1311; Herb. Bogor., in alcohol.) Fig. 409.

Seed 15–17 mm wide, subglobose. Testa; *o.e.* with isodiametric facets; stomata none. Tegmen; *o.e.* as a single interrupted layer of finely pitted fibres, the groups of fibres separated by groups of thin-walled cells. Chalaza shortly columnar, not appressed to one side. Ruminations chalazal and lateral.

H. macrocoma Wall.
(Java, Hort. Bogor. IV-H-10; Herb. Bogor., in alcohol.)

Seed-structure as in *H. subglobosa* v. *brachiata*. Testa; *o.e.* with isodiametric facets; stomata none. Tegmen; *o.e.* as a single layer of stout longitudinal fibres. Chalaza appressed, appearing lateral. Ruminations lateral.

H. subglobosa (Miq.) Warb. v. *brachiata* (King) J. Sinclair
(Singapore; living material.) Figs. 410–412.

Seed 20 mm wide, subglobose, white, surrounded by the orange-scarlet aril with a few short involute lobes at the apex. Testa *c.* 0.6 mm thick; *o.e.* consisting of small irregular flattened colourless cells with thin or slightly thickened walls, without stomata; *outer mesophyll* 4–5 cells thick, thin-walled, colourless, with abundant small starch-grains and abundant air-spaces; *inner mesophyll* as a brownish layer 3–4 cells thick, with slightly thickened brown walls and, often, brown gummy contents, but also with patches of colourless cells, without air-spaces; *i.h.* with the cells slightly radially elongate but scarcely as a palisade except round the chalaza; *i.e.* as a palisade *c.* 250 μ thick, composed of prismatic cells with thick yellowish brown walls. Vascular bundles in the outer mesophyll (as in *Myristica fragrans*). Tegmen evidently multiplicative; *o.e.* as thick-walled, pitted fibres; *outer mesophyll* composed of brownish, more or less crushed cells; *inner mesophyll* composed of large, thin-walled cells with yellow oily-vitreous contents, forming the main mass of the tegmen; *i.e.* crushed with the nucellar remains. Vascular supply as in *Myristica fragrans*. Chalaza simpler than in *M. fragrans*. Ruminations as in *M. fragrans*. Endosperm greyish white, horny-floury, composed of large thin-walled cells with oily contents and few starch-grains (absent from the smaller peripheral cells). Embryo very small.

Aril 2–3 mm thick, consisting of four layers; *outer palisade cortex* red, 3–4 cells thick, with the epidermis composed of prismatic cells and the inner layers of shorter rounded cells with slightly thickened, toughly submucilaginous walls, with abundant orange to reddish oil-globules; *outer mesophyll* yellowish, oily, composed of rounded cells, thin-walled, most with one large oil-drop, but in places these cells arranged in two layers separated by a strip of colourless cells, this tissue becoming friable and mealy at maturity; *inner mesophyll* colourless, watery, composed of thin-walled cells often tangentially or obliquely elongate, without oil-drops, the v.b. lying in the outer part of this thick layer; *inner cortex* red, generally consisting of only the inner epidermis, the cells shortly elongate with red oil-globules and thickened red outer walls, but the cell-cavity not pointed outwards as in the outer epidermal cells.

Pericarp composed of thin-walled cells, not differentiated into layers, but with complicated idioblasts in the outer half; v.b. as in *Myristica fragrans*.

H. sylvestris (Houtt.) Warb.
(Singapore; living material, without fertile seeds.) Fig. 413.

Ovule sub-basal, suborthotropous, the micropyle obliquely erect and pointing to the carpel-midrib, subfuniculate.

Seed (sterile) pale yellow, covered by the entire, orange-scarlet, waxy-pulpy aril with intricately fimbriate-lobulate apex. Testa as in *H. subglobosa* v. *brachiata*, without an inner hypodermal palisade, without stomata. Tegmen soon shrivelled in the sterile seeds. Chalaza projecting deeply into the nucellus. Ruminations as in *Myristica fragrans*.

Aril with a tough red cortex, 5–6 cells thick, on both sides, the cell-walls slightly thickened, the cells containing orange oil-globules, without air-spaces; medulla pale yellowish white, composed of thin-walled pulpy cells with pale yellowish oily contents, abundant air-spaces; v.b. as in *M. fragrans*; without stomata.

Pericarp as in *H. subglobosa* v. *brachiata*.

Vascular supply to the flower, ovary and fruit as in *M. fragrans* but few or no v.b. of the inner ring of the ovary developed until after fertilization.

Knema Lour.

Testa; *o.e.* with isodiametric or shortly elongate facets, with or without stomata. Tegmen; *o.e.* not fibrous but with a single layer of short sclerotic or tracheidal cells. Chalaza central and columnar. Ruminations chalazal or chalazal and lateral, the lateral much reduced in some cases. Aril entire, but with lobulate or fimbriate apex.

K. intermedia Warb.
(Java, Hort. Bogor. IV-G-83; Herb. Bogor., in alcohol.)

Testa; *o.e.* with isodiametric facets, stomata none. Tegmen; *o.e.* as a continuous layer of short tracheidal cells with more or less annular thickening. Chalaza shortly columnar (as in *Horsfieldia irya*). Ruminations chalazal and lateral.

K. laurina Bl.
(Java, Dungus Iwul Nature Reserve; living material.)
Figs. 415, 416.

Seed 10 × 6 mm, obovoid. Testa 6–8 cells thick; *o.e.* as a layer of subcuboid cells with isodiametric facets, without stomata; *mesophyll* thin-walled, aerenchymatous; *i.h.* as a short palisade of radially elongate cells, thin-walled, unlignified; *i.e.* as a palisade of strongly lignified cells with thick walls and narrow stellate lumen, some with crystals towards the outer ends. Tegmen 10–12 cells thick; *o.e.* as a narrow layer of longitudinally elongate, lignified, tracheidal cells with transversely pitted walls; *o.h.* as a layer of shortly radially elongate or cuboid cells with thinly lignified walls; *mesophyll* aerenchymatous, becoming crushed; *i.e.* as a layer of small subcuboid cells, slightly lignified, becoming crushed. Chalaza more or less central, tubular. Vascular supply of both testa and tegmen as a network in the thin-walled mesophyll, originating from post-chalazal branches in the testa, and from the end of the raphe in the tubular chalaza for the tegmen; *v.b.* of the tegmen accompanied by hypodermal oil-cells. Ruminations developed only from the chalazal region as 9 ridges lobing into the endosperm, traversed by branching *v.b.*, not arising from the sides of the tegmen.

Aril *c.* 10 cells thick, lobulate only at the apex; *o.h.* with many oil-cells; *v.b.* very numerous, small, longitudinal.

Pericarp with *c.* 3 rows of *v.b.*; idioblasts closely set in the outer hypodermis as more or less radial groups.

Knema sp.
(Java, Lengkong, Sukabumi distr.; living material.)
Fig. 417.

Seed *c.* 19 × 12 mm. Testa 14–17 cells thick; *o.e.* as cuboid cells with subisodiametric facets, stomata very scattered; *mesophyll* thin-walled; *i.h.* as a short palisade of rather wide, thin-walled, columnar cells, not lignified; *i.e.* as a palisade of thick-walled, lignified prismatic cells, differentiating from the chalaza. Tegmen 12–15 cells thick; *o.e.* as short, longitudinal tracheidal cells; *o.h.* with shortly radially elongate cells, thin-walled as the mesophyll; *i.e.* as small cuboid cells with brown walls, not lignified. Chalaza central, columnar. Vascular bundles as a network in the testa; longitudinal in the ruminations of the tegmic wall and in the chalazal ruminations. Ruminations developed from the chalaza and permeating the endosperm, but also as narrow longitudinal ridges from the tegmic wall (disappearing at the micropylar end), the two sets of ruminations not anastomosing though in places almost contiguous.

Aril with many longitudinal *v.b.*

Myristica Gron.

Testa; *o.e.* with isodiametric facets and stomata. Tegmen; *o.e.* with a single layer of longitudinal fibres, incomplete with gaps of thin-walled cells, or absent (*M. fragrans, M. iners*). Chalaza displaced to one side. Ruminations chalazal and lateral. Aril laciniate.

M. fragrans Houtt.
(Singapore; living material and material in alcohol.)
Figs. 418–424.

Ovule 0.7 mm long; *o.i.* and *i.i.* 6–8 cells thick; micropyle formed by *i.i.*; *v.b.* only in the raphe.

Seeds *c.* 25 × 20 mm, ellipsoid, dark brown, longitudinally grooved from aril-lobes. Testa 1.5 mm thick, 10–12 cells thick in small seeds, –23 cells thick in large seeds, multiplicative by anticlinal and periclinal divisions of the mesophyll-cells; *o.e.* composed of cuboid cells with brown and slightly thickened walls, especially the outer, the contents colourless, with stomata at intervals of *c.* 0.3 mm, opening in the half-grown seed; *mesophyll* consisting of thin-walled, colourless cells with air-spaces, slightly pulpy before maturity, then drying up; *i.h.* as a rather short palisade of watery unlignified cells 100–150 μ long, with firm walls slightly thickened at the angles, not lignified, hexagonal facets; *i.e.* as a palisade of prismatic lignified cells 600–800 μ long, 500 μ near the micropyle, with hexagonal facets, the brown walls unevenly thickened with inward flanges giving the stellate lumen, obliquely pitted, differentiating from the chalaza, then from the micropyle, lastly over the sides of the seed. Tegmen –30 cells thick, multiplicative, becoming much thicker than the testa, eventually drying up as a brown papery layer contracting from the testa except along the raphe; *o.e.* unspecialized except as the counter-palisade at the chalaza; *o.h.* unspecialized except as a short thin-walled palisade at the chalaza; *mesophyll* in the outer part (*c.* 20 cells thick) thin-walled but slightly lignified, colourless, and in the inner part (8–10 cells thick) thin-walled with brown cell-contents and oil-cells, not lignified, developing the ruminations; *i.e.* unspecialized. Chalaza thickened as a hard ellipsoid patch flattened on the raphe-side of the seed, the raphe *v.b.* entering obliquely into the compressed chalazal tube. Vascular supply in the testa with 5–7 post-chalazal *v.b.*, differentiating into the inner part of the mesophyll, with a few slender branches from the raphe, all anastomosing in an elaborate reticulum connecting the micropylar and basal ends of the seed with the arillar *v.b.*; in the tegmen derived from *c.* 5 branches from the chalazal expansion of the raphe, extending into the inner tissue of the mesophyll but not reaching the micropyle, branching and anastomosing, with branches to the ruminations. Nucellus enlarging by cell-growth without division, then absorbed and crushed, without *v.b.* Endosperm watery-pulpy, then drying into a firm white floury consistency, oily and starchy. Ruminations formed by outgrowths of the inner brown vascular layer of the tegmen, pushing the nucellar tissue into the endosperm, but the pointed micropylar end of the endosperm without ruminations. Embryo developing very slowly, still microscopic in the fully grown but immature seed.

Aril developing after fertilization as a crescentic lobe from the exostome and as a crescentic flange round the base of the ovule, the lobe and flange becoming concrescent, then dividing into laciniae extending to the chalazal end of the seed, the laciniae becoming spaced by the growth of the seed, finally with their tips entwining over the chalazal end, white then scarlet, turning dull crimson on exposure to air; *epidermis* composed of tangentially elongate cells, prosenchymatous, pitted, not lignified, with pink contents, without stomata; *mesophyll* thick, the cell-walls slightly thickened, subgelatinous, without air-spaces, many cells enlarging and containing a large yellow oil-drop, the others with finely disperse, red, oily pigment; *v.b.* numerous, derived from the funicle and from the inner plexus of v.b. in the pericarp, anastomosing at the base of the seed, then becoming dispersed singly into the aril-laciniae but subdividing again with little anastomosis in the widening laciniae.

Vascular supply of flower and fruit; *pedicel* with a ring of *c.* 21 v.b. giving off v.b. to the perianth, these subdividing to give 6–10 v.b. to each perianth-lobe, the axial ring of v.b. then forming an inner ring of smaller v.b. to supply the ovary with two rings of v.b., these two uniting in the style and giving a single v.b. to each stigmatic lobe; *funicular v.b.* derived from both of the larger v.b. of the carpel-margins at the base of the ovary; at the time of flowering, the inner ovarian ring of v.b. lignified only in the parts distal to the ovule, lignification proceeding basipetally with the growth of the fruit; *pericarp* supplied after fertilization with 3–4 additional rings of v.b. between the inner and outer ovarian rings, giving 5–6 rings of v.b., but with anastomosing branches; *idioblasts* in the outer part of the ovary-wall and as a narrow subepidermal layer in the pericarp.

The structure of the nutmeg-seed has been wrongly described. Periasamy, following Voigt and Netolitzky, described it as having (1) a free inner integument limited to the micropylar half of the ovule, the chalazal half consisting of the nucellar base directly in contact with the outer integument; (2) this nucellar base enlarged after fertilization to accommodate the endosperm, the micropylar half of the ovule forming little more than the endostome; (3) ruminations formed from the expanded pachychalaza. Accordingly other authors refer to the ruminating tegmen of the seed as perisperm. The account which I give was drawn up in 1943 and I have checked it with material collected in Ceylon in 1968. Sastri's account of the ovule agrees with mine. What I find in the nutmeg accords with the construction of the seed in other members of the family.

The aril is both exostomal and funicular. If the two lobes did not become concrescent, the nutmeg would have both an aril and an arillode, as may happen in Clusiaceae. The distinctions, of course, are not profound.

M. iners Bl.
(Sarawak, S. 26862; seed in alcohol.) Fig. 425.
Seed 48 × 21 mm, generally constructed as in *M. fragrans*. Testa 9–10 cells thick (excluding v.b.), stomata small and sparse; *i.h.* as a dark line of short, columnar cells with brown contents; *i.e.* with the palisade −250 μ thick. Tegmen *c.* 25 cells thick, −40 cells at the chalaza, the inner vascular tissue with a layer of oil-cells separated from the i.e. by 2–3 layers of small brown cells, the layer of oil-cells extending into the ruminations. Chalaza more extended. Aril with 2–4 v.b. in the laciniae; *mesophyll* with a hypodermal layer of oil-cells (2–3 deep).

M. lowiana King
(Indonesia, Bangka, Grashoff n. 114; Herb. Bogor., in alcohol.) Fig. 409.
Seeds *c.* 42 × 23 mm; aril laciniate. Testa; *o.e.* with isodiametric facets and stomata. Tegmen; *o.e.* as an interrupted layer of lignified fibres. Chalaza appressed laterally. Ruminations lateral.

Pycnanthus Warb.
P. kombo Warb.
Seeds 20 mm long, ovoid, dark brown. Testa few cells thick; *i.h.* as a short palisade of cuboid cells; *i.e.* as a lignified palisade 150 μ thick, with crystals in the cell-lumina. Tegmen; *o.e.* as fibres −200 μ long; *mesophyll* thin-walled, with oil-cells.
(Vaughan 1970.)

Virola Aubl.
V. sebifera Aubl.
(Brazil, Mato Grosso, leg. Corner 29 Jan. 1968; seeds in alcohol.) Fig. 426.
Seeds 7 mm wide, subglobose (apparently all sterile); aril laciniate, scarlet, the lobes joined only at the base. Testa 6–7 cells thick, but thicker at the chalaza; *o.e.* with thickened cell-walls; *mesophyll* thin-walled, with brown cell-contents, aerenchymatous, rather pulpy; *i.h.* as a short palisade, thin-walled, cell-contents brown; *i.e.* as a lignified palisade *c.* 140 μ thick, with brownish walls and numerous small crystals in the cell-lumina. Tegmen much thickened; *o.e.* as a layer of longitudinal fibres, thick-walled, lignified, pitted, 30–40 μ wide. Chalaza projecting as a conical peg into the tegmen, with double palisade, that of the tegmen passing abruptly into the fibrous exotegmen. Vascular supply as in *Myristica fragrans*. Aril with cortex and medulla; cortex 2–4 cells thick, as a firm envelope of tangentially elongate cells with thick walls, not lignified, without stomata; medulla composed of pulpy cells with slightly thickened walls and oily contents; laciniae each with one v.b.

Virola sp.
Seeds 15 mm wide, subglobose, dark brown to black. Testa; *i.e.* as a palisade 100 μ thick, with crystals in the

cells. Tegmen; *o.e.* with lignified fibres –1 mm long, 40 μ wide.

(Vaughan 1970.)

MYRSINACEAE (including Theophrastaceae)

Ovules anatropous or subcampylotropous, erect, bitegmic or unitegmic (*Aegiceras*), numerous, often embedded in the placenta; o.i. and i.i. 2–3 cells thick.

Seeds small to medium-size, commonly few or one per fruit, albuminous, exarillate. Testa multiplicative (? in all), unspecialized or with thickwalled o.e. Tegmen ? multiplicative, eventually crushed without specialization. Endosperm nuclear, oily, in some cases ruminate, or absent (*Aegiceras*). Embryo straight or slightly curved.

The indehiscent fruit seems to have induced the degradation of the seed-coat in this family and it is too much simplified for any sure conclusion. Floral structure, it has been suggested, associates the family with Primulaceae but, if Primulaceae are Centrospermous in alliance with Caryophyllaceae, there is no evidence in the seeds of Myrsinaceae. The primitive pachycaul habit of *Clavija* and *Theophrastus* may relate the family with Rhamnaceous or Sapotaceous origins, as suggested by Hutchinson's arrangement of the families. The thickening of the tegmen before it is obliterated is an Ebenaceous feature.

Aegiceras Gaertn.

Ovules unitegmic. Seed-coat thin, with two outer layers of thin-walled cells and three layers of brown cells. Endosperm absorbed.

Possibly this is a pachychalazal seed.

Ardisia Sw.

A. willisii Mez.

(Ceylon, Peradeniya; flowers and immature fruits in alcohol.)

Ovules many, embedded in the placenta; o.i. 2–3 cells thick; i.i. 5–6 cells thick.

Seed 1 per fruit. Testa (immature) 2–4 cells thick; *o.e.* composed of cuboid cells. Tegmen –9 cells thick, with much cell-division, unspecialized, without endothelium.

Embelia Burm. f.

Testa composed of thin-walled cells; *o.e.* with scattered stomata. Tegmen consisting of 2–4 layers of cells with firm walls and brown contents.

(Netolitzky 1926.)

Jacquinia L.

J. ruscifolia Jacq.

Testa *c.* 10 cells thick; *o.e.* composed of flattened cells with uniformly thickened walls; *o.h.* composed of flattened cells with thick anticlinal walls; *mesophyll* composed of small thin-walled cells, many in the outer 1–2 layers with rhombic crystals or spherical clusters of crystals, more scattered in the inner cells; *i.e.* unspecialized.

(Mennechet 1902.)

MYRTACEAE

Ovules anatropous to campylotropous, bitegmic or unitegmic (*Eugenia spp.*), crassinucellate; o.i. 2–4 cells thick, 6–8 cells in *Arillastrum*; i.i. 2–4 cells thick; micropyle formed by the exostome.

Seeds small to large, 0.5–20 mm long or more, anatropous, campylotropous or hemitropous, smooth, rough, ribbed, round to ellipsoid or oblong, in some cases alate, almost or quite exalbuminous, exarillate. Testa multiplicative or not, developing thick sclerotic tissue or the exotesta with more or less thickened and lignified outer wall and the mesophyll thin-walled; *i.e.* unspecialized, or as a crystal-layer with thin walls or with more or less strongly thickened and lignified inner walls (*Arillastrum, Eucalyptus spp.*), or developing radially elongate sclerotic cells at the micropyle (*Decaspermum, Rhodamnia, Rhodomyrtus*). Tegmen not multiplicative, mostly unspecialized and crushed, in some cases i.e. persisting as a layer of tannin-cells, or with slight unlignified thickenings (*Psidium*). Vascular supply generally simple in the raphe only, but variously extended in *Eugenia*. Endosperm nuclear, generally absorbed or persisting as a cell-layer near the micropyle. Embryo various, straight or curved, with thin or thick cotyledons, or hypocotylar. Figs. 427–432.

(Netolitzky 1926.)

The seeds of *Eucalyptus* have been investigated in great detail. Few other genera have been studied and in most only one or two species. It seems that a primitively anatropous seed with multiplicative and extensively sclerotic testa, comparable with the Rosaceous, Rutaceous and Theaceous, has undergone varying specialization and reduction into hemitropous and campylotropous forms with non-multiplicative and greatly simplified testa; the smallest seeds appear as enlarged ovules.

Seed-structure supports the division of the family into capsular and baccate genera. The baccate, such as *Decaspermum, Myrtus, Psidium, Rhodamnia,* and *Rhodomyrtus*, have the pantestal construction with generalized sclerotic tissue except for a few thin-walled inner layers of cells. In the non-multiplicative *Rhodomyrtus* this layer is one cell thick as the inner epidermis, and the arrangement suggests that in *Myrtus* and *Psidium* the inner epidermis forms 3–5 layers of thin-walled cells. In *Decaspermum* and *Rhodamnia*, 2–3 layers of thin-walled cells are so formed and the sclerotic part of the testa is derived from the outer epidermis.

In the capsular genera the testa is generally not multiplicative. The outer epidermis is thick-walled and protective, the mesophyll thin-walled, and the inner epidermis a layer of crystal-cells. Transitions from the more complicated seed of the baccate genera to that of the capsular are found in *Eucalyptus*; some species have the former, most have the simpler exotestal construction with modifications in the form of the seed, but others have the testa merely two cells thick as the simplified seeds of *Callistemon*, *Darwinia* and *Kunzea*, which are no more than enlarged and differentiated ovules. A peculiarity occurs in *Arillastrum*, some species of *Eucalyptus* (sect. *Macrantherae* ser. *Eudesmieae*) and to a lesser extent in *Metrosideros*, where the inner epidermis develops strongly thickened and lignified inner walls and becomes a palisade-like endotesta of crystal-cells.

Eugenia introduces a large and, evidently, overgrown seed with practically unspecialized testa, but scattered sclerotic cells may occur to indicate derivation from the massively sclerotic seed of the baccate kind. However, several species of *Eugenia* are unitegmic and the seed-coat may be pachychalazal.

Yet another peculiarity, which may be of wide occurrence, is found at the micropyle in *Decaspermum*, *Myrtus*, *Psidium* and *Rhodomyrtus*. Here the thin-walled tissue of the inner part of the testa becomes sclerotic, even with radially elongate cells in *Decaspermum* and *Rhodomyrtus*, and forms an inner plug to the micropyle, particularly noticeable in the ripe seeds of *Myrtus* and *Psidium*. Thus, the functions of the outer and inner layers of the testa are exchanged, as if the seed were once endotestal.

Most Myrtaceae have numerous ovules in the ovary. Some of these may be variously abortive with reduced nucellus and others, either unfertilized or inhibited, remain as small sterile structures (ovulodes) in the fruit, as in Melastomataceae. These sterile ovules may develop a simple form of the adult seed-coat with sclerotic outer and inner epidermis, much as might happen in *Rhodomyrtus*. There is evidence in *Eucalyptus* that even these structures may have taxonomic value.

I conclude that primitively the Myrtaceous seed was pantestal with extensively sclerotic tissue throughout and that the other kinds of seed have been derived by simplification. This means that the baccate fruits now have the primitive seed of the family, whereas one would expect the capsular, as antecedents of the baccate, to have such a seed. The dry capsules of Myrtaceae, however, are much derived and, on durian principles, have evolved from the larger fleshy capsules which closed to make the baccate fruits. The many genera of this large family must hold much information about the evolution of fruit and seed.

Angophora Cav.

A. floribunda (Sm.) Sweet
Ovules hemianatropous; o.i. and i.i. 4 cells thick.

Seeds 4 × 3 mm. Testa not multiplicative; *o.e.* with thin-walled, shortly radially elongate cells filled with tannin; *mesophyll* thin-walled; *i.e.* with a crystal in each cell, thin-walled. Tegmen not multiplicative, thin-walled, crushed.

(Prakash 1969d.)

The seed-structure resembles that of *Eucalyptus* ser. *Clavigerae*.

Arillastrum Panch.

A. gummiferum Panch.
Ovules numerous, campylotropous; o.i. 6–8 cells thick; i.i. 2 cells thick.

Seeds 4 × 3 mm, brownish black, usually 1 per loculus with 6–7 abortive ovules at the base, commonly 1 per capsule. Testa not multiplicative; *o.e.* composed of cells shortly elongate tangentially with slightly thickened outer wall and thin collapsed radial walls; *mesophyll* thin-walled, more or less collapsed; *i.e.* as a palisade-like layer of radially elongate cells with a crystal in each and the inner wall heavily thickened and lignified, practically obliterating the lumen. Tegmen collapsed. Embryo with long hypocotyl, very short radicle enclosed in a sheath, surrounded by the folded cotyledons.

(Netolitzky 1926, as *Spermolepis*; Dawson 1970a.)

Such an endotesta occurs in some species of *Eucalyptus* (sect. *Macrantherae* ser. *Eudesmieae*; Gauba and Pryor 1959.)

Callistemon R.Br.

C. citrinus (Curt.) Skeels
Ovules anatropous, numerous; o.i. and i.i. 2 cells thick.

Seeds 0.45 × 0.08 mm, linear. Testa not multiplicative; *o.e.* thin-walled with tannin; *i.e.* collapsed. Tegmen with o.e. collapsed and i.e. thin-walled with tannin.

(Prakash 1969b.)

Darwinia Rudge

Ovules anatropous, erect, attached to a basal placenta, 4 ovules (*D. micropetala*) or 2 (*D. fascicularis*); o.i. and i.i. 2 cells thick.

Seeds 1.2 × 0.5 mm, oblong, anatropous. Testa not multiplicative; *o.e.* composed of subcuboid cells with slightly thickened outer wall, filled with tannin; *i.e.* collapsed. Tegmen with o.e. crushed and i.e. as thin-walled tannin-cells. Embryo consisting of hypocotyl with minute radicle and cotyledons.

(Prakash 1969c.)

Decaspermum J. R. et G. Forst.

D. fruticosum J. R. et G. Forst.
(Sarawak, pr. Kuching, leg. Corner 1972; flowers and fruits in alcohol.) Fig. 427.
Ovules 7–8 per ovary, campylotropous, erect; o.i. 2(–3) cells thick, with the cell of o.e. large; i.i. 2 cells thick.

Seeds 3.5–4 × 2.5 mm, curved, cuneate in t.s., dark brown, covered in a thin layer (1–2 cells thick) of the fibrous endocarp. Testa multiplicative, becoming 4–6 cells thick through divisions in both cell-layers of o.i.; *exotesta* 2–3 cells thick, with thickened, lignified, closely pitted walls but wide lumen, the outermost cells strongly radially elongate in groups mainly along the sides of the seed and at the chalazal end, the inner cells short; *mesophyll* and *endotesta* composed of 2–3 layers of small radially compressed cells with thin but slightly lignified walls, without crystals, but the cells of the endotesta strongly radially elongate and lignified round the micropyle (as in *Rhodomyrtus*). Tegmen not multiplicative, unspecialized, crushed. Nucellus with the outer small cells long persistent. Endosperm reduced to a single layer of small cells. Embryo with long curved hypocotyl and short cotyledons with reflexed tips.

The seed is constructed essentially as in *Rhodomyrtus* but it is less campylotropous and the exotesta is clearly multiplicative to form the outer 2–3 sclerotic layers of cells, which I did not observe in *Rhodomyrtus*. Compare, also, *Rhodamnia*.

Eucalyptus L'Herit.

Ovules numerous, anatropous to more or less campylotropous, transverse; o.i. 2–4 cells thick; i.i. 2 cells thick.

Seeds small, anatropous, hemitropous or campylotropous, smooth, ribbed, or winged. Testa multiplicative or not, either more or less sclerotic throughout or with sclerotic o.e., thin-walled mesophyll (if present), and i.e. as a continuous or interrupted layer of crystal-cells with thin or thickened (not lignified) inner wall. Tegmen not multiplicative or in some species becoming 3–4 cells thick in places, becoming crushed but in some species with thin suberized walls. Vascular bundle single in the raphe, very short in hemitropous seeds and often dividing into fine branches round or slightly exceeding the chalaza. Endosperm absent or reduced to a single layer of cells at the micropyle.

(Netolitzky 1926; Gauba and Pryor 1958, 1959, 1961; Pryor and Johnson 1971.)

The seeds of many species have been studied in great detail by Gauba and Pryor. The differences are generally of serial rank taxonomically and show the value of the seed in classification. It appears that the second class of seed with thin-walled mesophyll has been derived from that with sclerotic mesophyll and that the seeds show varying reduction to the non-multiplicative testa with merely two layers of cells, as the outer more or less sclerotic epidermis and the inner crystal-bearing epidermis. This state with thin-walled mesophyll characterises sect. *Renanthereae* in which the series show varying forms of outer epidermal cells, some with a mucilage-thickening within the outer wall, some without cuticle, varying degrees of the persistence of the tegmen, and varying shapes of the seed from the anatropous to the hemitropous, but these shapes have yet to be correlated with those of the ovules. As ribs or wings develop

on the seeds, so the shapes of the outer epidermal cells vary in different parts of the seed, which renders description laborious, but it seems that there is no regular palisade of radially elongate cells. Thus the *Eucalyptus*-seed conforms generally with the Myrtaceous indication of the pantestal character. However, the sterile ovules may develop a regular exotestal palisade with or without sclerotic mesophyll. As Gauba and Pryor note, these ovulodes may have taxonomic significance.

Eugenia L.

Ovules unitegmic or bitegmic, anatropous; o.i. 2–3 cells thick; i.i. 2 cells thick; single integument 3–5 cells thick (*E. caryophylloides*, *E. cumini*. *E. fruticosa*, *E. malaccensis*, *E. paniculata*).

Seeds small to large, few or one per fruit. Testa thin, papery, composed of thin-walled tissue more or less pervaded by v.b., becoming crushed, but with scattered sclerotic cells in *E. cumini*. Embryo with thick cotyledons; several species polyembryonic with nucellar embryos (*E. cumini*, *E. hookeri*, *E. jambolana*, *E. jambos*, *E. malaccensis*, *E. caryophyllifolium*).

(Netolitzky 1926; Tiwary 1926; Narayanaswami and Roy 1960b; Roy and Sahai 1962.)

I use *Eugenia* in the wide sense, lest botany be decimated. Schmid (1972) assigns a smooth seed-coat and, usually, fused cotyledons to *Eugenia* in contrast with the rough seed-coat and usually distinct cotyledons of *Syzygium*. This seed-coat, however, needs analysis and may at times refer to endocarp.

Kunzea Rchb.

K. capitata Rchb.
As in *Callistemon*. Seed 0.7 × 0.4 mm. Testa; *o.e.* with enlarged, thin-walled cells.

(Prakash 1969a.)

Metrosideros Banks

Ovules anatropous, almost linear, numerous, ascending; o.i. and i.i. 2 cells thick; nucellus 2 cells thick.

Seeds 2–5 × 0.4–0.6 mm, linear, few fertile. Testa not multiplicative; *o.e.* with the cells flattened tangentially, the outer walls more or less strongly thickened and lignified, the lumen even more or less occluded; *i.e.* with similar but smaller cells with slightly thickened and lignified inner walls. Tegmen more or less crushed. Embryo straight; hypocotyl equal to or shorter than the cotyledons.

(Dawson 1970b.)

Myrtus L.

M. communis L.
Ovule anatropous; o.i. 2 cells thick, 3–4 at the micropyle; i.i. 2 cells thick.

Seeds 3–3.5 × 2–2.5 × 1.5–1.7 mm, campylotropous,

pale ochraceous, woody. Testa multiplicative, becoming densely sclerotic, with an endotestal plug at the micropyle. Tegmen crushed. Embryo curved, slender; cotyledons nearly as long as the hypocotyl.

(Netolitzky 1926; Roy 1962.)

Psidium L.

Ovules numerous, anatropous, transverse; o.i. 2–4(–5) cells thick; i.i. 2 cells thick; micropyle formed by the exostome.

Seeds 3–5 × 3–4 × 2–2.5 mm, reniform-campylotropous, pale brown, very hard. Testa multiplicative, 15–26 cells thick or more, becoming wholly sclerotic with thick-walled, closely pitted, lignified cells, with wide lumen, without an epidermal palisade, but with a thin-walled inner layer 2–6 cells thick and eventually more or less crushed; *mesophyll* with the sclerotic cells shortly elongate tangentially in various directions; *i.e.* at first with a small crystal in each cell; *exostome* with the inner tissue of thin-walled cells transforming into a massive, very hard, micropylar plug of sclerotic cells. Tegmen not multiplicative, the cells elongating longitudinally, those of i.e. developing slight spiral thickening but not lignified, eventually collapsing, or i.e. persistent. Nucellus persistent as 2–3 layers of cells but eventually crushed. Hypostase brownish, thin-walled, slight. Embryo with long hypocotyl and very small, often reflexed, thin cotyledons.

Pericarp with many small and large clusters of sclerotic cells, and with shortly lobed idioblasts.

(Netolitzky 1926; Narayanaswami and Roy 1960a.)

P. cattleyanum Sabine, *P. cujavillus* Brum., *P. guajava* Raddi.
Fig. 428–431.

The thick testa slowly becomes sclerotic from the inner part outwards and the last part to lignify and mature is the sclerotic plug at the micropyle; it simulates an endostome.

Rhodamnia Jack

R. cinerea Jack
(Singapore, leg. Corner 1972; fruits in alcohol.)

The material which I collected was unfortunately immature. Flowering is strictly periodic, though it happens several times a year, and a succession of stages must be collected; small fruits on trees with nearly mature fruits are not young but with merely one or two seeds. A fully illustrated account would help much in understanding the group of Myrtaceae to which *Rhodamnia*, *Rhodomyrtus* and *Decaspermum* belong.

The seed has the general construction of *Decaspermum* but it is plumper; the woody part of the testa forms an even surface without the prominently elongated cells of *Decaspermum*; this woody layer is 2–3 cells thick along the antiraphe, 3–5 cells thick on the sides of the seed and

at the chalazal end, and sharply delimited from the soft tissue of the raphe; the nucellus persists as several layers of small cells until the seed is full-sized and then, only, the embryo begins to develop rapidly; the endotegmic cells enlarge into conspicuous, longitudinally oblong cells with firm, not lignified, yellowish brown walls.

Rhodomyrtus DC

R. tomentosa (Ait.) Wight
(Ceylon; flowers and fruits in alcohol.) Fig. 432.

Ovules numerous, campylotropous, transverse; o.i. 3–4 cells thick, the cells of i.e. small; i.i. 2 cells thick, the cells elongated tangentially and longitudinally.

Seeds 2–2.2 × 0.8–1 mm, campylotropous, laterally compressed, brown, papillose. Testa not multiplicative but all the cells (except those of i.e.) enlarging greatly and becoming thick-walled, closely and coarsely pitted, lignified with wide lumen; o.e. composed of the largest cells with bulging outer wall; *mesophyll* with the inner row of cells least enlarged, absent from the base of the seed; *i.e.* composed of thin-walled cuboid cells without crystals, but strongly radially elongate round the micropyle and becoming slightly thick-walled, lignified and pitted. Tegmen soon crushed into a brown membrane. Raphe short. Chalaza simple, with thin brown hypostase. Partial septum with 5–7 layers of sclerotic cells over the apex, fewer towards the base of the seed. Embryo with long curved hypocotyl and small cotyledons.

Compare *Decaspermum* with larger, fewer seeds in the fruit.

NANDINACEAE

The position of this very small family with endotegmic seed is discussed on p. 28. The seed could be described as sarcotestal with lignified endotegmen but the sarcotesta collapses at maturity and does not function in seed-dispersal.

Nandina domestica Thunb.
(Java, Tjibodas, cultivated; flowers and fruits in alcohol.) Fig. 433.

Ovary with two transverse anatropous ovules, laterally attached one above the other; o.i. 7–9 cells thick; i.i. 3(–4) cells thick, with o.e. as a distinct palisade, i.e. as a shorter palisade, the middle layer of cells more or less tangentially arranged.

Seeds 5–6 mm long and wide, 2.5–3 (–3.5) mm thick, plano-convex, subcircular, whitish with brown inner layer, 1 per fruit, occasionally 2, transverse, exarillate, albuminous; funicle rather long. Testa 9–12 cells thick, wholly thin-walled, becoming rather pulpy, not lignified; i.e. and in places i.h., composed of rather small cells each with a small crystal. Tegmen not multiplicative, or the middle layer of cells dividing once periclinally in places, eventually crushed except i.e.; o.e. with the cells enlarging, thin-walled, then crushed; *i.e.* with the cells enlarging and becoming lignified, thickened slightly on the radial walls, strongly on the inner walls, sparsely pitted, with large rhombic facets. Nucellus crushed. Chalaza small, simple. Vascular supply only as the raphe-bundle. Endosperm oily, composed of small cells with distinctly thickened walls. Embryo minute.

(Netolitzky 1926.)

The ovary of *Nandina* is said to have a single ovule. I found in the Javanese material that two were frequent, though the fruits were generally 1-seeded. Netolitzky described the exotesta as thick-walled, which I did not observe.

NEPENTHACEAE

Ovules anatropous, bitegmic, crassinucellate; integuments a few cells thick.

Seeds 3–25 mm long, slender, filiform. Testa reduced to o.e. with thick outer walls to the cells and irregular thickenings on the radial and inner walls. Tegmen crushed. Endosperm starchy. Embryo minute but well-formed, straight.

(Netolitzky 1926; Kühl 1933.)

NYCTAGINACEAE

Ovules anatropous or campylotropous, erect, basal, solitary, bitegmic or unitegmic (*Abronia, Boerhaavia*), crassinucellate; o.i. 2–4 cells thick, 5–7 cells thick(*Mirabilis*), and the single integument 3–6 cells thick; i.i. 2 cells thick; micropyle formed by the endostome; funicle short.

Seeds generally small (except *Pisonia*), shortly oblong, with or without perisperm, exarillate, integuments not multiplicative (except in *Pisonia*). Testa not lignified; *o.e.* as a persistent layer of cells with thick outer wall or uniformly thick-walled (*Mirabilis*), with brownish contents; *mesophyll* aerenchymatous (when present), crushed; *i.e.* unspecialized and crushed or persistent with thickened walls (*Mirabilis*). Tegmen; *o.e.* unspecialized, crushed; *i.e.* unspecialized and crushed or persistent with finely striate anticlinal walls (*Mirabilis*), with yellow-brown contents (*Pisonia*). Vascular bundle with two postchalazal branches reaching almost to the micropyle (*Mirabilis*), expanded in the stout raphe (*Pisonia*). Perisperm starchy, abundant (Mirabileae, Pisonieae) or scant. Endosperm nuclear, scant. Embryo curved or straight (Pisonieae).

(Netolitzky 1926; Woodcock 1929b; Heimerl 1934; Venkateswarlu 1948; B. Singh 1964; Stemmerik 1964; Wunderlich 1968.)

The seed agrees generally in structure with that of other Centrospermae in the alliance of which the family is usually classified; it forbids alliance with Thymelaeales where it has been transferred by Hutchinson. Heimerl

described three kinds of seed: (1) that of Mirabileae with the testa more or less adnate to the pericarp; (2) that of Boldeae, Colignonieae and Leucastereae (*Reichenbachia*) with distinct testa; (3) that of Pisonieae with the testa more or less adnate to the pericarp but with straight embryo. The second kind resembles that of Phytolaccaceae, but the distinctions need microscopic evaluation. The single integument of *Abronia* and *Boerhaavia* appears to be the outer integument.

The exception to these general statements occurs in *Pisonia* the ovule of which develops terminally on the floral apex (Venkateswarlu 1948). It is supplied with a large vascular bundle compounded from the remains of the procambial strands at the floral apex, and it seems that with this source of nourishment the extraordinarily large and overgrown seeds of the alliance of *P. longirostris* have been evolved (Fig. 434). The testa in these seeds is strongly multiplicative along the massive and ridge-like raphe which carries the expanded v.b., but it does not differentiate a particular structure. It is free from the true pericarp (ovary-wall), though this is more or less crushed into a pellicle, and both are separate from the receptacular or perianth-tube which makes the wall of the fruit (anthocarp). Such seeds, the dispersal of which is not known, suggest overgrown products similar to the seeds of *Aegiceras* (Myrsinaceae) or *Avicennia* (Verbenaceae), but practically nothing is known about the development of the seeds in other species of *Pisonia* and those with large fruits and seeds are inland trees, not coastal nor necessarily riverine.

Takhtajan considers the family advanced in entomophily from Phytolaccaceae. This applies more to the alliance of *Mirabilis* and scarcely to the trees of *Pisonia*. The seed has the more massive construction of Phytolaccaceae but its structure is simplified, as Netolitzky observed, in the indehiscent fruit, and it is exarillate.

Pisonia Plum.

P. longirostris Teysm. et Binn.
(New Britain, NGF 13789; Solomon Islands, RSS 6172; flowers and mature fruits in alcohol.) Fig. 434.

Ovule erect, solitary; o.i. 3–4 cells thick; i.i. 2–3 cells thick.

Seed 6–10 cm × 7–8 mm, fusiform, brownish, with soft thin seed-coat. Testa strongly multiplicative on the raphe-side, little or not at all along the antiraphe, *c.* 16 cells thick in the part adjoining the raphe, thin-walled, not lignified; o.e. as subcuboid cells with subrectangular facets, many with brown contents, without stomata; *mesophyll* aerenchymatous, many cells with brown contents, especially in the raphe and adjacent tissue; *i.e.* thin-walled, unspecialized. Tegmen crushed, apparently not multiplicative. Raphe developing into a prominent ridge projecting between the arms of the cotyledons (as seen in t.s.); *v.b.* flattening into a strand with numerous small xylem-groups and flexed into the ridge of the raphe, not branched over the sides of the seed. Perisperm and endosperm as a mucilaginous distintegrated tissue. Embryo large; radicle-hypocotyl *c.* 1omm long; cotyledons pinkish purple, unequal, thick, with convolute lamina and two basal lobes, the outer encircling the small and inner.

No young stages in the growth of the fruit have been available. How the seed is distributed from the strange and apparently 'viviparous' plants is not known. The fruits or anthocarps, which may reach 55 cm long, and hang in bunches, are illustrated by Stemmerck. The collection from New Britain had shorter seeds, –6 cm long, than that from the Solomon Islands, 7–10 cm long, and the female flowers had five staminodes, those of the Solomons four.

NYMPHAEACEAE

Ovules anatropous, bitegmic, crassinucellate; o.i. 2–6 cells thick (? more in *Euryale*); i.i. 2 cells thick.

Seeds small to medium-size or large (*Nelumbo*), arillate (*Euryale, Nymphaea, Victoria*) or not, with abundant perisperm and smaller amount of endosperm, with minute embryo, but exalbuminous with large embryo in *Nelumbo*. Testa ? not multiplicative; *o.e.* as a palisade of cuboid or shortly columnar cells with thickened, more or less lignified walls, varying stellate–undulate (*Brasenia, Nymphaea, Victoria*), in several genera thinner round the micropyle to form an operculum on germination, or unspecialized (*Barclaya* with hairs, *Nelumbo*); *mesophyll* and *i.e.* unspecialized. Tegmen not multiplicative, unspecialized, crushed. Vascular bundle in the raphe only or dividing into two branches with short processes to the sides of the seed (*Cabomba*) but complicated in *Nelumbo*. Aril white, thin, saccate, non-vascular, with stellate cells in the mucilaginous mesophyll, developed from the funicle, investing the whole seed, present as a rim on the funicle before fertilization, or developing as 4 separate outgrowths becoming concrescent (*Euryale,* ? *Victoria*), serving as an air-float. Perisperm starchy. Endosperm cellular, but nuclear in *Nelumbo.*

Cabomboideae

Seeds operculate, exarillate.

Cabomba: o.i. 2 cells thick. Testa; *o.e.* with the cells each forming a conical thick-walled papilla. Raphe v.b. dividing into 2 branches at the chalaza.

Brasenia: o.i. 3–4 cells thick. Testa; *o.e.* with very thick outer walls to the somewhat stellate–undulate cells.

Barclayoideae

Seeds without aril or operculum. Testa unspecialized, but some epidermal cells forming brittle hairs. *Barclaya.*

Nupharoideae

Seeds operculate, exarillate. Ovule; o.i. 4–6 cells thick. Testa; *o.e.* composed of cuboid cells with thick, lignified, pitted walls and hexagonal facets; *mesophyll* with somewhat thickened walls, then crushed.

Nymphaeoideae

Seeds arillate and operculate. Testa; *o.e.* as a short palisade of thick-walled, lignified, pitted cells, in some cases with short hairs, the facets stellate–undulate in *Nymphaea* and *Victoria*, hexagonal with a central papilla in *Euryale* but substellate in the fossil *Euryale* and *Eo-euryale.*

Nelumbonoideae

Ovules solitary, apical, suspended. Seeds large, exarillate, exalbuminous. Testa thin, unspecialized in the woody pericarp. Funicle with 3 v.b. descending to the chalaza, then breaking up into a network with short branches to the testa (? pachychalazal). Endosperm nuclear. Embryo green; radicle abortive; cotyledons large, thick.

(Netolitzky 1926, Kühn 1927, p. 338; Miki 1960; Khanna 1964b, 1965, 1967; Hotta 1966.)

A primitive position is assigned to Winteraceae because of their lack of vessels, their apocarpous flowers, and their albuminous seeds with small embryo. They are advanced, however, as leptocaul trees with small leaves and exarillate seeds. Nymphaeaceae offer the same primitive marks and, also, the pachycaul spinous habit with big leaves and the arillate perispermous seeds. *Cabomba* is the aquatic leptocaul which, according to Hutchinson, connects with *Ceratophyllum.* Then the arillate Nymphaeaceous seed is not much different from that of Podophyllaceae. I conclude that Nymphaeaceae are in many ways more primitive than Winteraceae and closer to the common origin of Ranunculales (including Berberidales and Illiciales, p. 27). The aquatic habitat borders the swampy riverine forest which seems to have been the home of angiosperms (Corner 1964).

A good account of the fossil seeds of the family is given by Miki (1960). According to Khanna (1965), the saccate aril of *Euryale* develops as a basipetal outgrowth uniting four separate small outgrowths that arise from the funicle before fertilization; they persist as four lobes at the top of the mature aril. Probably the aril of *Victoria* with 4–5 short apical lobes develops in the same way, (Fig. 435).

NYSSACEAE

Ovules unitegmic (? also bitegmic), ? crassinucellate. Testa very thin, ? unspecialized. Endosperm thin. Embryo relatively large.

(Netolitzky 1926; Eyde 1963; Dilcher and McQuade 1967; Tandon and Herr 1971.)

Presumably, from the silence concerning the seed-coat, it consists of thin-walled cells which are more or less obliterated as the endosperm and embryo enlarge. Thus the family may fit Araliales or Cornales in a primitive way, if the record of bitegmic ovules is correct.

OCHNACEAE

Ovules solitary, basal, anatropous or subcampylotropous, erect, bitegmic at the micropyle but more or less extensively unitegmic, massive, tenuinucellate, subsessile.

Seeds moderately large, enclosed in the lignified endocarp of the drupe, exarillate, exalbuminous. Seed-coat evidently as a highly vascularized pachychalaza, thin-walled, multiplicative, but o.e. remaining as a membrane. Endosperm nuclear, soon absorbed. Embryo various, straight with thick cotyledons and short radicle, or curved with linear cotyledons (*Brackenridgea*).   Figs. 436–442.

(Netolitzky 1926; Farron 1963.)

The description refers to subfamily Ochnoideae. Van Tieghem noted that the seed of *Luxemburgia* had a layer of thick-walled and lignified cells corresponding with what might have been the inner epidermis of the testa. This detail, together with the small and truly bitegmic seeds in a dry capsule, is so different from the state in Ochnoideae that I investigated such seeds as were available (*Euthemis*, *Luxemburgia* and *Sauvagesia*), and found that they had a typically fibrous exotegmen and a testa composed of merely two cell-layers differentiated, nevertheless into a large-celled exotesta and an endotesta of small crystal-cells. I have treated these genera as the separate family Sauvagesiaceae of Lindley because they clearly have no close affinity with Ochnoideae; they belong, rather, with Violaceae as given by Bentham and Hooker. With this exclusion, the definition of Ochnaceae changes from one of much uncertainty to a fairly close unit that resembles Simarubaceae in ovary, fruit and seed. It seems that the pachychalazal tendency in the Simarubaceous genera *Picrasma* and *Quassia* has been promoted in Ochnaceae to give the vestigial micropylar integuments. *Lophira*, however, with bilocular capsule needs investigation, though it appears that the seeds also lack testal differentiation.

The variety in form of the embryo on which van Tieghem dwelt has been largely resolved by Farron.

Lophira Banks

L. alata Banks
Testa reduced to *c.* 3 layers of unspecialized cells, compressed between the pericarp and embryo.

(Vaughan 1970.)

Ochna L.

The structure of the ovary has been described in detail for *O. multiflora* and *O. squarrosa* by Baum (1951) and V. S. Rao and Gupta (1957).

O. kirkii Oliv.
(Ceylon, Peradeniya, cult.; flowers and fruits in alcohol.) Figs. 436, 437.

Ovule with very short free integuments only at the micropyle; o.i. 3–4 cells thick; i.i. 2(–3) cells thick; ovule-wall 8–10 cells thick; vascular bundle dividing into 2–3 branches shortly beyond the beginning of the raphe, without distinct chalaza.

Seeds 8 × 6 mm, developing slowly in the enlarged cavity of the drupe; seed-coat 14–18 cells thick (excluding v.b.), strongly aerenchymatous, then crushed, with a close network of v.b.; o.e. remaining as a brown membrane, at first as a slight palisade.

Ovary with 9 carpels and free central hollow style surrounding the slightly conical vestige of the stem-apex.

Ouratea Aubl.

Ouratea sp.
(Brazil, Manaus, Corner 137; fruits in alcohol.)   Figs. 438–442.

Ovules campylotropous; integument 14–17 cells thick, aerenchymatous, many cells with sphaerocrystals; v.b. in the raphe very short, massive, giving many fine procambial strands to the outer part of the integument.

Seeds 8.5 × 6 mm, enclosed in the woody endocarp of the drupe. Seed-coat not multiplicative, thin-walled, aerenchymatous, pervaded by a fine reticulum of v.b., eventually the tissue crushed except for a few outer layers of cells, then drying up; o.e. composed of shortly elongate cells with thin walls. Embryo straight.

Ovary with 10 free carpels spaced widely round the upper margin of the obconic receptacle, each carpel protected by a large incurved umbo of the receptacle; style central, persistent, traversed by 10 v.b. and 10 inner bundles of fibres, the base of the style irregularly lacunar between these inner fibrous bundles; receptacle with a minute, more or less 10-lobed space just below the stylar base (? containing the stem-apex).

Drupes 10–11 × 8 mm, black, seated in the more or less expanded red receptacle split longitudinally. Exocarp as a thick palisade of prismatic cells with the walls slightly thickened distally, with thick cuticle, without stomata. Mesocarp pulpy, aerenchymatous, pervaded by a network of numerous fine v.b. Endocarp composed of 6–8 rows of pitted lignified fibres, longitudinal or oblique, but i.e. composed of transverse fibres.

The flowers of this bizarre fruit were not available. The ovarian receptacle forms a cone, as in *Nelumbium* but with the carpels peripheral under the incurved margin. It shows well the extravagant receptacular evolution distinctive of the family; it is a receptacle formed, not from the perianth as in *Rosa*, but from the intercalary growth of a gynophore.

OLACACEAE

Ovules anatropous, bitegmic (*Cathedra*, *Chaunochiton*, *Coula*, *Heisteria*, *Ximenia*), or unitegmic (*Olax wightiana*, *Strombosia*), or ategmic (*Aptandra*, *Olax imbricata*, *Schoepfia*); o.i. and i.i. 5–6 cells thick (*Coula*); single integument rather massive.

Seed-coat crushed, unspecialized. Endosperm cellular, helobial in *Olax*, copious, oily, also starchy (*Coula*). Embryo small or well-developed.

(Netolitzky 1926; Agarwal 1963, 1964; Johri 1963.)

This family and Opiliaceae have the opening stages of the disappearance of the integuments and seed-coat accomplished in Santalales.

OLEACEAE

Ovules anatropous to amphitropous, suspended with ventral raphe (*Chionanthus*, *Forsythia*, *Ligustrum*, *Olea*, *Phillyrea*, *Schrebera*) or with dorsal raphe (*Fontanesia*, *Fraxinus*, *Jasminum pr. p.*, *Syringa*) or erect (*Jasminum pr. p.*, *Nyctanthes*); integument few to many cells thick, –20 cells in *Nyctanthes*; hypostase present.

Seeds small, albuminous, exarilate. Seed-coat exotestal; *o.e.* with the cells more or less in a palisade, with thickened outer wall, in some cases of uneven length and the longer cells projecting in groups as ridges or circles; *mesophyll* thin-walled, compressed; *i.e.* at first as an endothelium, more or less persistent with thickened walls, in some cases with enlarged cells projecting into the endosperm. Vascular supply as the raphe-bundle ending at the chalaza, with a small recurrent v.b. in the raphe (*Syringa*), or continued nearly to the micropyle (*Fraxinus*, *Olea*, *Syringa*), or with 2 or more postchalazal v.b. (5–6 in *Nyctanthes*) simple or branching (*Chionanthus*, *Ligustrum*, *Olea*). Endosperm cellular, thick-walled, without starch. Embryo straight.

(Netolitzky 1926; Kühn 1927; p. 344; Andersson 1931; Kapil and Vani 1967; Vaughan 1970.)

ONAGRACEAE

Ovules anatropous, bitegmic, crassinucellate; o.i. 2 cells thick, 3–5 in *Fuchsia* and *Oenothera*; i.i. 2(–3) cells thick; micropyle formed by the exostome.

Seeds small, angular, corky, or somewhat alate, exarillate, exalbuminous. Testa not multiplicative; *o.e.* composed of rather large subcuboid cells with slightly thickened and often pitted outer wall, not lignified, not as a palisade of radially elongate cells, in some cases with the cell papilliform or prolonged into a hair (*Epilobium*); *mesophyll* aerenchymatous, crushed or with some sclerotic cells (*Oenothera*); *i.e.* as a layer of crystal-cells with the inner and radial walls more or less strongly thickened and lignified, stellately lobed in *Jussieua*. Tegmen not multiplicative; *o.e.* as a compact narrow layer of longitudinal lignified fibres, (3 layers of fibres in *Gongylocarpus*); *i.e.* thin-walled, crushed, or with the cells shortly longitudinally elongate with fine subreticulate thickening on the inner wall, not lignified. Vascular bundle of the raphe ending at the chalaza, with raphids in the chalazal region (*Boisduvalia*). Endosperm nuclear, absorbed. Embryo straight. Figs. 443, 444.

(Netolitzky 1926.)

Onagraceous seeds differ in no essential way from Lythraceous. The families are so close that one wonders at the pedantry that separates them and assimilates Caesalpiniaceae or Moraceae.

Gongylocarpus Cham. et Schlecht.

This genus presents the odd combination of a multiple fibrous tegmen in *G. fruticulosus* and an unspecialized seed-coat in *G. rubricaulis*.

G. fruticulosus (Benth.) T. S. Brandegee

Testa 2 cells thick; *o.e.* with thin-walled oblong cells; *i.e.* with thickened radial and inner walls and several crystals in the lumen. Tegmen ? 5 cells thick; outer 3 cell-layers as longitudinal tracheidal fibres with spiral thickening.

G. rubricaulis Cham. et Schlecht.

Testa and tegmen thin-walled, without endotestal crystal-cells or tegmic fibres.

(Carlquist and Raven 1966.)

Jussieua L.

Ovules; o.i. 2 cells thick; i.i. 2(–3) cells thick.

Seeds minute, 0.7–1 mm long, ellipsoid, brownish ochraceous, with rather prominent raphe. Testa not multiplicative; *o.e.* with the enlarged cells transversely elongate, often with subundulate walls, longitudinal in the raphe, the outer wall finely perforate, not lignified; *i.e.* composed of lignified, stellately lobed cells with thickened inner and radial walls, filled with small crystals. Tegmen not multiplicative; *o.e.* as a compact layer of radially flattened, thick-walled, pitted and lignified fibres; *i.e.* with shortly longitudinally elongate cells, collapsing, the inner wall with fine punctate subreticulate markings, not lignified.

J. peruviana L.

(Ceylon; flowers and fruits in alcohol.)   Fig. 443.

Seed 0.7–0.8 × 0.4–0.5 mm. Tegmen with clavate, pitted, lignified cells at the endostome; *o.e.* with fibres 11–18 μ wide, 5 μ deep, with fine crossed lines on the inner wall. Raphe with a few raphid-cells.

J. repens L.

(Khan 1942.)

Oenothera Spach

Ovule; o.i. 3–5 cells thick; i.i. 2 cells thick.

Seeds 1–2 × 0.6–1 mm, brown, corky, angled or with

3–4 short longitudinal wings. Testa not multiplicative or slightly at the angles (–7 cells deep), becoming winged through shrinkage of the mesophyll; *o.e.* composed of cuboid cells with hexagonal facets, the outer wall slightly thickened and finely pitted, not lignified; *mesophyll* with some thick-walled sclerotic cells, especially at the angles of the seed, the remainder of the tissue thin-walled and collapsing; *i.e.* composed of lignified cells containing several crystals, the inner and radial walls much thickened as a firm palisade. Tegmen not multiplicative; *o.e.* as a layer of narrow, slender, pitted, lignified and longitudinal fibres; *i.e.* with longitudinally elongate, thin-walled cells, crushed; *endostome* with clavate, pitted, lignified cells from o.e. (as in *Jussieua*). Fig. 444.

OPILIACEAE

Ovules orthotropous, unitegmic, tenuinucellate, ategmic in *Agonandra*.

Seeds albuminous; seed-coat thin-walled, crushed. Endosperm cellular.

(Netolitzky 1926; Ram, M. in Johri 1967b.)
Compare Olacaceae.

OROBANCHACEAE

Ovules anatropous or hemianatropous, unitegmic, tenuinucellate; integument 4–6 cells thick; 2–3 cells in *Aeginetia*.

Seeds minute, exotestal, albuminous. Seed-coat apparently not multiplicative; o.e. as tabular or cuboidal cells with dark brown spiral, annular or reticulate lignification of the radial and inner walls, with isodiametric facets; *mesophyll* and *i.e.* more or less crushed, but the mesophyll-cells with fine reticulate thickening in *Lathraea*. Endosperm cellular, oily or also starchy, starchy in *Cistanche*. Embryo minute, subglobose, undifferentiated.

(Netolitzky 1926; Tiagi 1951b, 1952a, 1952b; Kadry 1955; Singh, B. 1964.)

OXALIDACEAE

Ovules anatropous, suspended, bitegmic, tenuinucellate; o.i. 3–5 cells thick; i.i. 3–6 cells thick, becoming multiplicative in *Averrhoa* but not in small seeds (*Biophytum*, *Oxalis*).

Seeds small to medium-size, albuminous or not, exarillate except in *Dapania*. Testa becoming mucilaginous at the surface, in *Biophytum* and *Oxalis* eventually disrupting and leaving the seed with exotegmic covering; *o.e.* as a short palisade or unspecialized; *mesophyll* developing radiating cells in *Biophytum* and a multiplicative network of small cells in *Averrhoa*; *i.e.* generally as a layer of crystal-cells, small or lengthening

into a short palisade with thick unlignified inner walls. Tegmen becoming undulate in *Biophytum* and *Oxalis*; *o.e.* as a layer of straight, longitudinal, lignified fibres (*Oxalis*) or as unlignified tracheidal fibres (*Averrhoa*), unspecialized and crushed in *Biophytum*; *o.h.* as a layer of transverse fibres (*Oxalis spp.*); *i.e.* unspecialized. Endosperm nuclear, oily, subruminate (*Biophytum*, *Oxalis*). Figs. 445–447.

Aril bilabiate in *Dapania*.

(Netolitzky 1926; Veldkamp 1967.)

This family is placed customarily in Geraniales where the fibrous exotegmen renders it exceptional. *Averrhoa*, transferred to Rutales as Averrhoaceae (Hutchinson 1959), proves the point because its seed-coat is typically Oxalidaceous. As there is no evidence for the transformation of the stellate exotegmic cells into fibres or the reverse, I take the view that Oxalidaceae belong with the other families with fibrous exotegmen which distinguishes them as a great alliance of dicotyledons (p. 14). The small seeds of *Oxalis* may suggest affinity with Linaceae but the large fruit and more complicated seeds of *Averrhoa*, beyond the scope of that family, bear comparison with those of Caricaceae. A primitively massive and fleshy capsule is indicated and in their early pachycaul ancestry, at least, a pinnate leaf; for the palmate leaf of Caricaceae is surely the derivative, as it is in Araliaceae, Moraceae or Rosaceae. The families have diverged in floral neoteny and vegetative features but the seeds of *Averrhoa* and *Carica* have retained a certain structure that is distinctive. Both have a strongly multiplicative testa; the mesotesta of Caricaceae is derived from the outer hypodermis whereas that in *Averrhoa* comes from the inner hypodermis, and *Averrhoa* lacks the multiplicative exotesta of *Carica*. In both the testa becomes mucilaginous, the pellicle of *Carica* being a massive state of that in *Oxalis*. Both have the exotegmen developed into narrow lignified fibres, crossed by similar fibres in the hypodermis, and the crystal-cells of the endotesta add another detail of similarity that suggests, on the whole, consanguinity. The inner layer of tegmic fibres may be missing in *Averrhoa* as it is in *Jacaratia* (Caricaceae), and both layers seem to be absent from the small seeds of *Biophytum*. Yet it must be noted that the tracheidal form of the exotegmic fibres in *Averrhoa* may relate with those of Lythrales (p. 37). As a focus of affinity, the Oxalidaceous seed has considerable interest.

An aril was ascribed erroneously to *Oxalis*, but its presence in *Dapania*, as a yellowish white covering of the seed attached along the raphe, has been affirmed. *Averrhoa bilimbi* is a durianological relic.

Averrhoa L.

A. bilimbi L.

(Ceylon; flowers and fruits in alcohol.)   Figs. 445, 446.

Ovule; o.i. 4–5 cells thick, –7 near the chalaza; i.i. 4–5(–6) cells thick, i.e. as an endothelium; v.b. in the raphe with a short postchalazal bifurcation; micropyle formed by the exostome.

Seeds 6 × 4.5 × 2.5 mm, pale brown to dark brown, ellipsoid, compressed, with firm but apparently largely unlignified coat, exarillate. Testa 10–13 cells thick, –25 cells along the lines of the ruminations, thicker at the chalaza; *o.e.* as a palisade of shortly elongate cells, the outer wall with thick cuticle and mucilaginous inner layer, in places (as seen in surface-view) with small colourless excretions covering 1–4 cells; *mesophyll* aerenchymatous, thin-walled, with abundant starch, with smaller cells in the inner tissue and with fine radiating wedges of small-celled tissue as ruminations developed from i.h., the cells with thin brown unlignified walls, appearing as dark anastomosing streaks in tangential section, the ruminations associated towards the micropyle with infoldings of i.e., especially around the thick oblique exostome; *i.e.* as a layer of small crystal-cells, each with a crystal, with strongly thickened but unlignified inner wall, forming a thin brown hard layer in the ripe seed. Tegmen 5–6 cells thick; *o.e.* with the cells soon elongating into narrow fibres with unlignified spiral or annular thickenings; *mesophyll* with the cells enlarging, then crushed by the endosperm; *i.e.* unspecialized, at first as a short endothelium, eventually crushed; *endostome* with two layers of short, reticulately pitted, lignified cells. Vascular bundle of the raphe with 4 postchalazal branches descending towards the micropyle, one reaching the radicular part of the seed, not anastomosing. Nucellar cuticle more or less persistent. Endosperm oily, thin-walled. Embryo with short radicle and thin flat cotyledons.

Pericarp fleshy-pulpy without lignified tissue except in the xylem of the 2–3 rows of v.b.

A. carambola L.
(Java; ripe seed in alcohol.)   Fig. 447

Seeds 10 × 5 × 2.5 mm, blackish brown, with thin slimy pellicle (exotesta), as in *A. bilimbi* but larger; *endotesta* as a palisade of prismatic cells 38–45 μ high, with the inner two-thirds formed by the strongly thickened, brown, unlignified but hard, inner wall, the cells of uneven height, but the i.h. apparently without ruminations; *exotegmen* with the narrow tracheidal fibres becoming agglutinated to the strong endotesta, with spiral thickening, not lignified.

Biophytum DC

B. sensitivum (L.) DC
Ovule; o.i. and i.i. 3 cells thick, not multiplicative in the seed; exostome prolonged; funicle free from o.i.

Seeds with mucilaginous disrupting testa as in *Oxalis*. Testa; *o.e.* with cuboid cells, the outer wall slightly thickened; *mesophyll* composed of radially elongate, thin-walled cells of varying length fitting the undulations of the subruminate endosperm; *i.e.* composed of small cells (? without crystals). Tegmen with the outer two cell-layers crushed; *i.e.* at first as an endothelium, persisting as a tannin-layer. Endosperm slightly ruminate from transverse undulations of the testa and tegmen.

(Mauritzon 1934d; Thathachar 1942.)

It seems that the fibrous exotegmen has been lost in this genus.

Oxalis L.

Ovule; o.i. 3–4 cells thick, the cells of o.e. with thick outer wall; i.i. 3–4 cells thick; the integuments not multiplicative in the seed.

Seeds with the mucilaginous o.e. and mesophyll of the testa swelling and disrupting to eject the seed covered by the tegmen and partly by the endotesta. Testa; *i.e.* as small crystal-cells (crystals only in the depressions of the tegmen in *O. acetosella*), the cells on the ridges of the tegmen with thickened inner and radial walls. Tegmen with o.e. and o.h. developed into straight fibres, the longitudinal fibres of o.e. crossed by those of o.h., or only with the exotegmic fibres, the rest of the tissue crushed; developing transverse folds (*O. corniculata* and its allies) or longitudinal folds (*O. acetosella* and its allies), causing the endosperm to appear ruminate; *i.e.* at first as an endothelium.

(Netolitzky 1926; Herr and Dowd 1968.)

PAEONIACEAE

Ovules anatropous, bitegmic, crassinucellate; o.i. massive, many cells thick; i.i. *c.* 3 cells thick; nucellus absorbed by the embryo-sac before pollination; micropyle formed by the exostome.

Seeds medium-size, fleshy, red then purple-black from anthocyanin in the exotesta, often with fleshy red funicle and slight aril-rim, albuminous. Testa massive, many cells thick; *o.e.* as a palisade of columnar aqueous cells with strongly thickened outer walls forming a tough pellicle; *o.h.* usually as a palisade of thick-walled, lignified, sparsely pitted, columnar cells with isodiametric facets, usually with thin outer walls, shorter than the exotestal palisade and in some species absent; *mesophyll* aerenchymatous, thin-walled, the colourless cells filled with starch, the inner layers becoming crushed; *i.e.* unspecialized. Tegmen soon crushed. Vascular bundle of the raphe ending at the chalaza or with several postchalazal branches forming more or less of a reticulum in the testa. Chalaza simple, turning brown in the ripe seed, with hypostase. Endosperm nuclear, oily, white. Embryo minute.

Funicle short, thick fleshy, red from anthocyanin in the o.e., expanded round the apex in several species into a slight, crenulate, dark red aril.   Figs. 448–450.

(Camp and Hubbard 1963a; Sastri 1969; Sawada 1971.)

For a long time I have supposed that Paeony and *Dillenia* were related (Corner 1946). The general construction of the flower and fruit, the vascular anatomy and, in particular, the centrifugal development of the stamens are points in common. *Paeonia* would then be a survival of the Dilleniaceous pachycaul stock with pinnate leaves, anatropous ovule and seed, and massive testa. The object was to realise that *Paeonia* was not closely related to Ranunculaceae. However the structure of the testa with outer hypodermal palisade is peculiar. It might have reduced to the thin Dilleniaceous testa (p. 26) or it may be the relic of the primitive state which lead to Cucurbitaceae and Convolvulaceae (p. 50). The systematic position of the genus now turns, as the seed-anatomist would expect, upon the structure of the seed. It has recently been found that even the centrifugal androecium may be doubtful (Sawada 1971, for *P. japonica*).

Species of *Paeonia* vary in details of the seed. In *P. delavayi* the hypodermal palisade seems but slightly developed, the cells being short and scarcely thick-walled. *P. arietina* lacks the postchalazal vascular bundles. These are examples of curtailed development.

The slight aril was investigated by Camp and Hubbard and compared not with that of *Dillenia*, for they were not durianologists, but with that of *Myristica*.

See Glaucidiaceae.

PANDACEAE

Ovules orthotropous, pendent, solitary, bitegmic.

Seeds enclosed in thick endocarp. Seed-coat; *o.e.* with cuboid, somewhat thick-walled cells; *mesophyll* with *c.* 10 layers of longitudinal, cylindric, pitted, lignified fibres and an inner layer of rounded cells with thickened inner wall. Endosperm oily. Embryo straight; cotyledons thin, flat.

(Forman 1966; Vaughan and Rest 1969.)

That this seed has a fibrous seed-coat is clear but it is not certain from the description given by Vaughan and Rest for *Panda* whether the fibres occur in the testa or the tegmen. If *Galearia* and *Microdesmis*, hitherto referred to Euphorbiaceae–Phyllanthoideae (p. 138), are allies of *Panda* as Forman maintains, then the seed of *Panda* must have a multiple exotegmen of longitudinal fibres, such as found in *Capparis*.

PAPAVERACEAE

Ovules anatropous or more or less campylotropous (*Hypecoum, Papaver, Roemeria*), bitegmic, crassinucellate; o.i. 2–8 cells thick; i.i. 3 cells thick; micropyle formed by the exostome; 3 cuticular layers present.

Seeds medium-size to small (4–7 mm long, *Bocconia, Dendromecon*), usually brown or black, often with reticulate or rugose surface on small seeds, albuminous, arillate or more commonly exarillate; seed-coats not multiplicative. Testa; *o.e.* generally composed of enlarged cells with much thickened, radially striate or punctate–perforate, outer wall in genera with thin testa, the cells rather small with less thickened outer wall in genera with thick testa, with internal projections in the cells (*Dendromecon*), with fine internal reticulum on the wall (*Eschscholtzia*), mostly unlignified, with stomata in *Argemone* and *Eschscholtzia*; *mesophyll* unspecialized, more or less collapsed; *i.e.* as a layer of cuboid or columnar thin-walled cells with crystals and a dense brown network of internal cellulose or lignified fibrils. Tegmen; *o.e.* as a layer of narrow longitudinal fibres with thick unlignified walls (absent in the indehiscent fruits of *Platystemon*, but absent also from *Argemone, Chelidonium, Dicranostigma, Eschscholtzia, Glaucium ?, Roemeria, Sanguinaria*); *mesophyll* unspecialized, crushed, but as a layer of transverse fibres in *Hypecoum*; *i.e.* generally as a persistent layer of somewhat enlarged cells with fine spiral, annular or bar-like thickenings, at least on the radial walls, not lignified, often with dark contents. Chalaza simple, with hypostase. Vascular bundle of the raphe without postchalazal extension. Nucellus often persistent as a single layer of effete cells. Endosperm nuclear, oily, with thin or slightly thickened walls. Embryo small to minute.

Aril exostomal and funicular (*Dendromecon*), funicular and from the adjacent part of the raphe (*Bocconia*) or from the raphe only (*Cathcartia, Chelidonium, Macleaya, Meconopsis, Sanguinaria*). Figs. 451–456.

(Meunier 1891; Netolitzky 1926; Röder 1958.)

The microscopic structure of the seed has been studied in many genera of Papaveraceae, and it will become a main feature of their classification. Knowledge rests on the foundation so well laid by Meunier. The seed has seven marks:

(1) The integuments are not multiplicative. A certain amount of anticlinal division of the cells may occur but it seems that the small seeds are simply enlarged and differentiated ovules.

(2) The exotesta is more or less of a mechanical layer on account of its thick outer wall.

(3) The endotesta is a layer of thin-walled crystal-cells, often palisade-like, with internal fibrillar reticulation as in Magnoliaceae, but not as a multiple layer.

(4) The exotegmen is a layer of narrow longitudinal fibres, apparently unlignified, cf. Aristolochiaceae, Linaceae, Lythraceae, Malpighiaceae, Oxalidaceae (p. 15).

(5) The endotegmen is a layer of more or less tracheidal cells, cf. Rutaceae, Vitaceae.

(6) The copious endosperm is oily.

(7) The embryo is small, even microscopic.

This unusual mixture of characters singles out the Papaveraceous seed. Marks 2, 6, and 7 fit Ranunculaceae.

Mark 1, also Ranunculaceous, indicates neotenic advance. Marks 3 and 4 define the seed more exactly and are not Ranunculaceous or, for that matter, Cruciferous. Mark 5 is evidently the vestige of the general endotegmic character of the primitive seed (p. 48). Mark 4 relates Papaveraceae with other families with fibrous exotegmen, as a trace of Myristicaceous ancestry. The combination of mark 3 (endotesta with fibrillar crystal-cells) and mark 4 (fibrous exotegmen) links Papaveraceae with Chloranthaceae and Proteaceae. Mark 3 is Magnoliaceous. Thus the Papaveraceous seed combines Magnoliaceous and Ranunculaceous features but is on the line of fibrous exotegmic evolution from the Myristicalean source. Such a phyletic connection must be viewed in the long history of pachycaul beginnings, a trace of which persists in Papaveraceae.

The largest seed of the family is that of *Bocconia*. It is arillate; the testa is thick; the exotegmic fibres are well-developed; the habit of this 'tree-poppy' is typically pachycaulous; but its flowers are reduced, in parallel with the Chloranthaceous. If a large multi-ovulate ovary were restored and the panicle returned to a few massive flowers, there would result a capsule of arillate seeds resembling that of Clusieae. *Dendromecon* has similar seeds but its habit seems to be leptocaul and, whereas *Bocconia* has the thick striate outer wall of the exotesta that characterizes so many Papaveraceous herbs, *Dendromecon* lacks the feature and has peculiar inward projections in these cells (Figs. 451, 453). These two genera appear as diverse relics of ancestral Papaveraceae, and most of the resultant herbs with small exarillate seeds appear to have come from the line of *Bocconia*. The aril, as usual, is lost or in the process becomes a small oily strophiole or elaiosome for ants. The testa is simplified in periclinal divisions from the inner epidermis of the developing ovule until, in place of five or six such divisions, there is one or none and the testa has merely three or two cell-layers that mature simply by cell-enlargement and differentiation. In *Platystemon*, with indehiscent fruit, there is little differentiation; in *Hypecoum*, with schizocarps, the exotesta is scarcely differentiated. In various genera the exotegmen is undifferentiated. Thus the Papaveraceous seed becomes neotenic, small, simplified, and in effect merely ovular. *Fumaria* is the parallel with *Corydalis* in the allied *Fumariaceae*. Accompanying this diminution in size there is an increase in number of the ovules and seeds per capsule, and *Papaver* abounds in its brief way. For prime characters one must look to the massive *Bocconia*. A similar history must apply to Ranunculaceae, but they have lost the tree-pachycaul and the aril unless, perhaps, in the elaiosome of *Helleborus*. The alliance of the two families must extend far back into the seed-evolution of dicotyledons during their pachycaul phase when the flower was differentiating from the fruit and the seed was experimenting.

Argemone L.

Ovules anatropous, subglobose; o.i. 2 cells thick; i.i. 3 cells thick; integuments not multiplicative.

Seeds small, exarillate. Testa; *o.e.* composed of large, low cells with isodiametric facets, with thickened and striated outer wall, with stomata near the funicle; *i.e.* as a short palisade, 45–100 μ high, of thin-walled cells with internal cellulose fibrils between the crystals. Tegmen with o.e. and the middle layer crushed; *i.e.* with fine close thickenings on the walls (Meunier) or also collapsed (Röder, Sachar). Nucellus persistent as a single layer of cells. Fig. 451*d*.

A. hunemannii Otto et Dietr., *A. mexicana* L., *A. platyceras* Link

(Meunier 1891; Sachar 1955; Röder 1958; Chopra and Rai 1958; Vaughan 1970.)

According to Chopra and Rai, colchicine-treatment of the ovule prevents differentiation of the endotesta.

Bocconia L.

B. frutescens L.

(Colombia, in 1947, and Java, Tjibodas, in 1972; living material.) Figs. 452–454.

Ovule solitary, anatropous, erect, basally attached; o.i. 5–6 cells thick; i.i. 3 cells thick, already differentiated into the three layers; funicle rather long; aril slightly indicated.

Seed 7 × 3 mm, ellipsoid, black, surrounded in the hilar half by the red, crenulate, funicular aril, with the micropyle free from the aril; funicle 3 mm long, dark brown, scarious. Testa crustaceous, smooth, 6–7 cells thick, not or slightly multiplicative; *o.e.* composed of cuboid, then somewhat flattened, cells with hexagonal facets, the outer wall brown, slightly thickened, with fine granular thickenings on the inner surface, not perforate (? radially striate), not lignified; *mesophyll* thin-walled, more or less crushed; *i.e.* as an uneven palisade of radially elongate prismatic cells with fibriform-reticulate and lignified internal thickening, with small crystals, as a hard layer. Tegmen not multiplicative; *o.e.* as narrow longitudinal fibriform cells with thickened slightly pitted, unlignified walls; *mesophyll* composed of large thin-walled cells, eventually collapsing; *i.e.* as a layer of tabular cells with hexagonal facets, the radial walls thickened and pitted, not lignified. Vascular bundle of the raphe ending at the chalaza; with hypostase. Nucellus reduced to a single layer of radially elongate hyaline cells. Endosperm oily, thin-walled. Embryo minute.

Aril developing from the junction of the funicle and raphe, enveloping with two lobes the micropyle and most of the funicle, not vascular, watery-pulpy, plicate, red; epidermal cells radially elongate but shortened towards the attachment, filled with orange-red oil-droplets, thin-walled.

Pericarp-valves thickened by the radially elongate, thin-walled, epidermal cells.

(Meunier 1891.)

Chelidonium L.

C. majus L.

Fig. 455.

Ovule 350 μ long, anatropous; o.i. 2 cells thick; i.i. 3 cells thick; integuments not multiplicative.

Seeds 1.5 mm long, brown, with white raphe-aril (elaiosome). Testa; *o.e.* composed of large, low cells with very thick, biconvex, outer walls finely striate on the inner thickened side, with angular iso-diametric facets; *i.e.* as a short palisade of thin-walled cells filled with crystals. Tegmen undifferentiated, thin-walled, eventually crushed; *i.e.* possibly with fine striations on the radial walls. Endosperm rather thick-walled.

Aril with large, thin-walled cells containing oil and starch.

(Meunier 1891; Crété 1937a; Szemes 1943; Röder 1958.)

Meunier figures the exotegmen as fibrous, but I was unable to detect this. The aril is a vestige of that of *Bocconia*, developed only on the lower part of the raphe. Apart from this outgrowth, the seed seems to grow mainly by enlargement and differentiation of the cells of the ovule.

Dendromecon Benth.

Ovules anatropous; o.i. 8 cells thick, at first 4 cells thick, then increasing by periclinal division of i.e.; i.i. 3 cells thick; funicle short, with long hairs, degenerating after fertilization.

Seeds 4–4.5 mm long, subglobose, dark purple-brown, the surface finely alveolar, arillate (? all species). Testa not multiplicative; *o.e.* composed of large cells with slightly thickened walls, with internal papillae from the radial and inner walls; *mesophyll* 6–7 cells thick, composed of narrow, tangentially elongate, thin-walled cells, lignified at maturity, with brownish contents; *i.e.* as a short palisade of thin-walled cells with small crystals and a copious internal reticulum of brownish red cellulose fibrils. Tegmen; *o.e.* as a layer of unlignified fibres; middle layer crushed; *i.e.* composed of subcuboid cells with reticulate thickenings as stripes and ridges, unlignified. Chalaza closed by three layers of cells similar to those of the endotesta, but of irregular shape.

Aril funicular and exostomal, relatively massive, 1.5–2 mm long, yellowish white; outer two cell-layers with oil, the inner with starch.

D. harfordii Kellogg, *D. rigida* Benth. Fig. 451a.

(Berg 1966, 1967).

The seed is ant-distributed in the manner of reduced arils.

Dicranostigma Hook.f. et Th.

Testa; *o.e.* composed of large flat cells with thick, punctate-striate, outer wall; *i.e.* as a palisade of thin-walled cells with crystals. Tegmen collapsed.

D. franchetianum (Prain) Fedde

(Röder 1958.)

Eschscholtzia Cham.

E. californica Cham.

Ovule anatropous; o.i. 5–6 cells thick; i.i. 3 cells thick.

Testa; o.e. composed of moderately wide cells, some projecting in small groups, without a thickened outer wall, but the whole wall strengthened internally with lax irregular fibrillar reticulum, with scattered stomata; mesophyll thin-walled, crushed; *i.e.* as a layer of cuboid cells with crystals and internal fibrils. Tegmen; *o.e.* as a fibrous layer; *i.e.* with fine spiral thickenings on the cell-walls. Cotyledons deeply bifid.

(Meunier 1891; Röder 1958; Sachar and Mohan Ram 1958; Berg 1967.)

Glaucium Adans.

G. flavum Crantz.

Ovule anatropous; o.i. 2 cells thick; i.i. 3 cells thick; integuments not multiplicative.

Testa; *o.e.* composed of low convex cells with thickened and striated outer wall, with angular facets; *i.e.* as a palisade of thin-walled cells with small crystals among the internal cellulose fibrils. Tegmen; *o.e.* as a layer of very narrow fibres (Meunier), unspecialized (Röder); *i.e.* with fine close spiral thickenings on the walls. Nucellus persistent as a single cell-layer.

(Meunier 1891; Röder 1958.)

Hunnemannia Sweet.

Seed 1.2 mm long, ovoid, exarillate. Testa 9–10 cells thick; *o.e.* composed of rather small cuboid cells with slightly thickened outer wall; *mesophyll* thin-walled, collapsing; *i.e.* with one large crystal in each cell (? immature). Tegmen unspecialized (? immature).

(Meunier 1891.)

Hypecoum L.

Ovules subcampylotropous; o.i. 2 cells thick; i.i. 3 cells thick. Testa; *o.e.* as small thin-walled cells with slight cuticle; *i.e.* as thin-walled cuboid cells each with one large crystal and with internal fibrils. Tegmen; *o.e.* as a layer of longitudinal fibres; *middle layer* composed of transverse fibres; *i.e.* as a layer of cuboid cells with finely striate walls.

H. procumbens L.

(Meunier 1891; Röder 1958.)

Röder's account seems to refer to old seeds with more or less eroded testa. Meunier remarked on the double fibres of the tegmen, and said that the seed of *Chiazospermum lactylorum* was similar.

Macleaya Reichb.

Seeds 1.5–2 mm long, pale to blackish brown, ovoid; aril small, yellow, pulpy, along the lower part of the raphe. Testa 2–3 cells thick; *o.e.* composed of large obconic cells with the obtuse apex projecting into the endotesta, the much thickened outer wall radially striate on the inner face; *i.e.* as a palisade of thin-walled cells with cellulose fibrils among the many small crystals. Tegmen; *o.e.* as a layer of slightly lignified longitudinal fibres; *i.e.* with fine thickenings on the radial walls, the inner wall often with slight reticulate thickening.

M. cordata (Willd.) R.Br., *M. microcarpa* (Maxim.) Fedde. Figs. 451*b*, 456.
(Meunier 1891; Röder 1958.)

Papaver L.

Ovules subcampylotropous; *o.i.* 2 cells thick; *i.i.* 3 cells thick; integuments not multiplicative.

Testa; *o.e.* composed of large low cells with slightly thickened outer wall, with stellate–undulate facets; *i.e.* as a layer of small cells with numerous crystals, not as a palisade. Tegmen; *o.e.* as a compact layer of unlignified fibres –1 mm long; *mesophyll* composed of enlarged cells, thin-walled, collapsing; *i.e.* composed of transversely elongate cells with very fine bar-like thickenings on the radial walls.

(Meunier 1891; Fedde 1936; Röder 1958; Vaughan 1970.)

When crushed in water, the blackish exotestal walls in some species dissolve.

Platystemon Benth.

Ovules; *o.i.* and *i.i.* 3 cells thick. Testa and tegmen typical of the family but with small and slightly differentiated cells.

'La persistance du péricarpe autour de la graine de *Platystemon californicum* explique le peu d'épaisseur de son spermoderme … et la faiblesse relative de ses éléments constitutifs, conformément à la loi générale de substitution physiologiques des organes' (Meunier 1891).

Roemeria Medik.

Seeds small, reniform. Testa; *o.e.* composed of narrow cells; *i.e.* with dark contents, not as a palisade. Tegmen unspecialized, crushed.

R. refracta (Stev.) DC, *R. rhoeadiflora* Boiss.
(Röder 1958.)

Romneya Harv.

Testa; *o.e.* composed of somewhat enlarged, shortly radially elongate cells with thickened, striate, lignified outer and radial walls, some cells projecting strongly; *i.e.* composed of cuboid cells with crystals. Tegmen; *o.e.* composed of narrow fibres; *mesophyll* collapsed; *i.e.* thin-walled.

R. trichocalyx Eastwood
(Röder 1958.)

Sanguinaria L.

Ovules anatropous; *o.i.* 6–7 cells thick; *i.i.* 3 cells thick; integuments not multiplicative.

Seeds 3 mm long, black, shiny, with slightly swollen, subcrenate, greyish raphe-aril. Testa; *o.e.* as rather small low cells with thickened outer wall (? not striate); *mesophyll* thin-walled, collapsing; *i.e.* as a palisade of crystal-cells with internal fibrils. Tegmen unspecialized, crushed.

Aril as a thick twisted thread extending along the raphe as a narrow extension of the exotesta. Fig. 451*c*.

(Meunier 1891; Shaw 1904; Rao, A. N. and Shamanna 1963.)

PARNASSIACEAE

As in Saxifragaceae but tenuinucellate and with nuclear endosperm; *o.i.* 2–3 cells thick; *i.i.* 3–4 cells thick.

Seeds developing air-spaces in the chalazal region between the integuments (as in *Drosera*). Tegmen increasing to 6 cells thick in *Parnassia nubicola* Endosperm reduced to a single layer of oil-cells.

(Netolitzky 1926; Sharma, V. K. 1968*b*.)

Affinity with Droseraceae is often suggested but the endosperm in Droseraceae is said to be starchy and the seed-coat of *Drosophyllum* indicates a more complicated structure than has been seen in Parnassiaceae.

PASSIFLORACEAE

Ovules anatropous or orthotropous, bitegmic, crassinucellate; *o.i.* 2–4 cells thick; *i.i.* 3 cells thick, developing before *o.i.*; funicle rather long.

Seeds small to medium-size, mostly arillate, winged in *Hollrungia*, often with pitted surface on drying from the uneven exotegmen; seed-coats not multiplicative or the testa with multiplicative mesophyll; albuminous or not; funicle rather long. Testa thinly pulpy, often becoming mucilaginous; *o.e.* unspecialized or with the outer wall thickened; *mesophyll* unspecialized and becoming crushed or multiplicative to form trabeculae of cells bridging air-spaces (*Adenia*); *i.e.* as a short palisade of radially elongate cells, thin-walled, more prominent in the depressions of the tegmen. Tegmen; *o.e.* as a lignified palisade of columnar cells with thick pitted walls, often projecting unevenly into the testa and nucellus and then the palisade-cells of varying length; *mesophyll* unspecialized, crushed; *i.e.* as a thin plate of flattened cells, thin-walled or with lignified thickenings. Chalaza often umbonate, without special structure. Vascular bundle of the raphe ending at the chalaza by entering the heteropyle with a small spread of tracheids at the base of the nucellus. Endosperm nuclear, oily (?), persistent or absorbed. Embryo relatively large.

Aril arising from the end of the funicle, in some cases as a slight rim before fertilization, more or less investing the seed, thinly pulpy-mucilaginous, colourless, without v.b.   Figs. 457–459.

(Kratzer 1918; Netolitzky 1926; Dathan and Singh 1969a.)

Most species of this family appear to be arillate. Among genera cited as primitive by Hutchinson (1967), *Smeathmannia* is arillate but no aril is mentioned for *Soyauxia* (capsular) or *Barteria* (indehiscent fruit).

The position of the family among those with exotegmic palisade to the seed is discussed on p. 38.

Adenia Forsk.

A. acuminata Bl.
(Malaya, Sing. Field N. 33850; fruits in alcohol.)

Seeds covered by the aril as in *Passiflora*. Exotegmen rugoso-subtuberculate on the outside, with discrete woody papillae projecting into the mesotegmen on the inside, composed of a palisade of radially elongate cells with pitted, lignified walls. Heteropyle as a rather long, narrow canal traversed by the slender v.b.

A. venenata Forsk.
Testa with trabeculate mesophyll.
(Kratzer 1918.)

Paropsia Noronh.

P. obscura O. Hoffm.
As *Passiflora*.   Fig. 457c.
(Kratzer 1918.)

Passiflora L.

There appears to be much uniformity in the seeds of this genus. In some species the testa may be two cells thick and in others three. Specific analysis with heights of the palisade-cells is needed.

P. calcarata Mast., *P. cuprea* L., *P. edulis* Sims, *P. foetida* L., *P. hirsuta* L., *P. mollissima* HBK, *P. quadrangularis* L., *P. suberosa* L., *P. trifasciata* Lem.

(Kratzer 1918; Raju 1956a; Singh, D. 1962a; Dathan and Singh 1969a.)

P. edulis Sims
(Ceylon; flowers and fruits in alcohol.) Figs. 458, 459.

Ovule anatropous; o.i. and i.i. 3 cells thick; chalaza prominent as a hump; nucellus projecting through the endostome; funicle with slight aril-rim.

Seeds invested by the aril. Testa becoming a mucilaginous pellicle; *o.e.* subcuboid, rather flattened cells

with slightly thickened outer wall; *mesophyll* composed of somewhat longitudinally elongate cells, narrow, thin-walled, eventually crushed on the sides of the seed; *i.e.* as a palisade of thin-walled prismatic cells elongate in the depressions of the tegmen, cuboid over the ridges of the tegmen. Tegmen; *o.e.* as a lignified palisade of prismatic cells with reticulately thickened, pitted walls, forming in places conical projections into the nucellus-endosperm, these projections mostly corresponding with depressions on the surface; *middle layer* of cells very soon crushed; *i.e.* as a sheet of thin-walled, small, rather flattened cells with a conical lignified projection from the centre of the inner wall into the cell-cavity. Chalaza fairly massive but without complicated structure, the chalazal tube through the tegmen filled with small thin-walled cells, the v.b. expanding into shortly radiating tracheids at the base of the nucellus.

Aril funicular, saccate, composed of unlignified pulpy cells, 12–15 cells thick at the base, thinning to 2 cells at the margin, without v.b.

PEDALIACEAE

Ovules anatropous, suspended, unitegmic, tenuinucellate; integument 7–8 cells thick or more.

Seeds small to medium-size, exalbuminous or thinly albuminous, exarillate, alate in *Uncarina*. Testa multiplicative, –16 cells thick in *Sesamum*, exotestal; *o.e.* with thick-walled lignified cells, palisade-like in *Sesamum* with groups of more elongate cells giving the ridges on the seed, with oxalate crystals in the outer part (*S. indicum*) or the inner part (*S. radiatum*), with cuboid cells in *Harpagophytum* with a strong annular thickening on the radial walls and finer thickenings leading from it to the tangential walls, with thin-walled cells in *Trapella*; *mesophyll* more or less crushed, the cells with single crystals in *Sesamum*; *i.e.* at first as an endothelium. Hypostase present. Endosperm cellular, oily, reduced to 2–5 cell-layers in *Sesamum*. Embryo straight.

(Netolitzky 1926; Singh, S. P. 1960; Ihlenfeldt and Straka 1962; Johri 1967b; Vaughan 1970.)

The hooks and spines of the pericarp in *Harpagophytum* and *Uncarina* develop like durian-processes from the ovary-wall in connection with stellate scales (Ihlenfeldt and Straka 1962).

PHRYMACEAE

Ovules solitary, basal, suborthotropous, erect, unitegmic, tenuinucellate.

Seed-coat papery, formed of the collapsed outer cells of the integument. Endosperm cellular, reducing to 2 cell-layers in the seed.

(Cooper 1941.)

PHYTOLACCACEAE

Ovules anacampylotropous, bitegmic, erect, crassi-nucellate; o.i. 3–5 cells thick; i.i. 2 cells thick, 3–4 near the chalaza or micropyle; endostome forming the micropyle.

Seeds small, smooth, hairy in *Rivina* from endocarp-hairs, round, reniform or oblong, mostly not arillate but covered with a red aril (*Stegnosperma*) or yellow aril (*Barbeuia*), with perisperm; seed-coats not multiplicative, reducing more or less to the exotesta. Testa; *o.e.* as a palisade of radially elongate cells with dark brown contents, more or less thick-walled and with calcium-silicate in the walls (*Phytolacca*), distinctly thick-walled in *Rivina*; *mesophyll* crushed; *i.e.* with brown cell-contents, crushed. Tegmen; *o.e.* unspecialized, crushed; *i.e.* with slight bar-thickenings on the radial walls (*Rivina*) or none (*Phytolacca*), with brown contents. Perisperm starchy. Endosperm nuclear.

(Woodcock 1925; Netolitzky 1926; Kajale 1954a.)

The exotestal structure of the seed is typically Centrospermous. The aril seems to be exceptionally well-developed in the two capsular genera *Barbeuia* and *Stegnosperma*, but it is absent from the drupaceous. Vestigial arils, however, may occur, possibly in *Rivina*. Though *Stegnosperma* is separated in its own family in Pittosporales by Hutchinson (1959), the seed-structure does not agree with what is known of that order. Buxbaum (1961) compares the family with Illiciaceae and regards it as the most primitive of Centrospermae.

The related Gyrostemonaceae is of interest in this respect because its fruits and seeds seem even more primitive, though nothing is known of the seed-structure (Heimerl 1934). The seeds appear all to be arillate and the fruits are mostly, if tardily, dehiscent though those of *Tersonia* and *Cylindrocarpus* are indehiscent and have very small, evidently vestigial, arils. The aril of *Gyrostemon* was described as funicular and exostomal (Baillon 1871). The small fruit of *Tersonia* is thickly set with durian-scales, already incipient in the ovary. It is difficult to see how one can approach Centrospermous ancestry without accommodating this family.

PIPERACEAE

Ovule solitary, orthotropous, erect, bitegmic and basal in *Piper*, unitegmic and sub-basal in *Peperomia*, crassi-nucellate; o.i. and i.i. 3–5 cells thick in *Piper* or o.i. merely 2 cells thick in the micropylar part, o.i. shorter than i.i.; integument of *Peperomia* 2 cells thick.

Seeds 1–6 mm long, ovoid, enclosed in the drupe, exarillate, with copious perisperm and small endosperm; seed-coats not multiplicative. Testa (in *Piper*) soon crushed without trace or as a crushed pigmented layer at the base of the seed, (but ? with persistent, thick-walled o.e. or i.e. in some species). Tegmen (in *Piper*)

distinct round the endosperm at the micropylar end, more or less crushed into a firm brown layer with traces of o.e. and i.e. over the body of the seed; *o.e.* consisting at the micropylar end of large subcuboid or shortly longitudinally oblong cells with thick brown walls (? slightly lignified), eventually more or less crushed, over the body of the seed similar but not radially elongate, the oblong cells generally crushed; *mesophyll* consisting of small cells with thin brown walls, soon crushed or moderately persistent at the micropylar end; *i.e.* as a palisade of shortly radially elongate cells with thin or somewhat thickened brown walls (? lignified), the inner ends of the cells irregularly sinuate from lobate intrusions from the thick cuticle of the perisperm, fairly persistent at the micropylar end, over the body of the seed with shortly longitudinally oblong cells and eventually more or less crushed. Seed-coat (*Peperomia*) with both cell-layers thick-walled, brown, or only the inner layer, with minute crenulate-lobate projections of the perisperm-cuticle (as in *Piper*). Chalaza unspecialized but with discoidal hypostase of small brown cells. Vascular supply only as the very short bundle at the base of the seed. Perisperm filling most of the seed, the cells thin-walled, filled with starch-grains, but many also as large oil-cells, the cuticle very thick and minutely crenulate–lobate into the endotegmen, especially in the micropylar region. Endosperm small, micropylar, nuclear and starchy in *Piper*, cellular and oily-proteinaceous in *Peperomia*. Embryo microscopic, 50–250 μ long. Figs. 460, 461.

(Netolitzky 1926; Fagerlind 1940; Murty 1959a, b; Yoshida 1960b; Kanta 1963.)

The cells of the outer integument in *Piper* are relatively large and thin-walled at anthesis. It appears that, generally, they enlarge no more but are very soon crushed between the tegmen and the pericarp. The tegmen forms the thin brown membrane round the seed. The membrane consists over the greater part of the seed of the outer and inner epidermal layers with brown walls, the cells shortly longitudinally oblong, and the exotegmic walls are slightly thickened; but there is practically no trace of cell-cavities at maturity. Over the micropylar end of the seed, where presumably there is less of a squeeze, the cells of the exotegmen are much larger with very thick walls, a trace of mesophyll persists, and the endotegmic cells, though with thinner walls, form a short palisade; yet, even this construction becomes more or less crushed at maturity. *Peperomia* seems to have lost entirely the outer integument, and the two layers of cells in the seed-coat differentiate much as the two epidermal layers of the tegmen in *Piper*. The epidermal cells of the perisperm thrust the thick cuticle into the cells of the endotegmen and cause the crenulate appearance in section. It is by no means clear how thoroughly the mature tegmen has been studied in *Piper*.

In *Piper*, the outer integument may be 2 cells thick

and the inner 3 cells thick (*P. adunca, P. gaudichaudianum, P. medium, P. peltatum, P. umbellatum*), or o.i. 2–3 cells thick and i.i. 4 cells thick (*P. futokazura*), or both integuments are 3 cells thick (*P. longum*) or 3–5 cells thick (*P. nigrum, P. zeylanicum*), but in the last two the outer integument thins to two cells towards the micropyle.

The seeds of Piperaceae and Saururaceae are uniquely tegmic. They can be described as both endotegmic and exotegmic, though the exotegmic palisade of *Piper* with wide cell-lumina and brown walls is unlike that of other exotegmic families with a palisade (p. 13). The position of Piperales is discussed on p. 43. I list the order as endotegmic in order to distinguish it from other exotegmic families.

PITTOSPORACEAE

Pittosporum Banks

Ovules anatropous (? campylotropous), parietal, unitegmic, tenuinucellate; funicle distinct.

Seeds 2–7 mm long, subglobose, reniform, or irregular from compression, orange, brown, red or black, coated with white, yellow or red glutinous resin, exarillate, albuminous. Seed-coat 6–12 cells thick (at least), unspecialized, not lignified, aerenchymatous, the inner tissue becoming more or less crushed; *o.e.* as a compact layer of cuboid cells with strongly thickened outer wall, not lignified. Vascular bundle very short, ending in a slight swelling at the top of the funicle, without prolonged raphe; *hypostase* scarcely developed. Endosperm nuclear, oily, with more or less thickened walls. Embryo very small; cotyledons 2–5.   Figs. 462, 463.

P. molluccanum (Lamk.) Miq., *P. resiniferum* Hemsl., *P. tobira* Ait.

(Gowda 1951; Bakker and v. Steenis 1957.)

There appears to be little difference in seed-structure between the species. By contrast there are considerable differences in the structure of the pericarp (Fig. 463). In *P. resiniferum* the very large resin-canals are accompanied for the greater part by vascular bundles with little or no fibrous sheath. In the pericarp of *P. molluccanum* there appear to be no resin-canals and the conspicuous fibro-vascular bundles branch out into the outer pericarp. In *P. tobira* there are many minute resin-canals and a great complexity of fibro-vascular bundles. These differences seem to imply in Pittosporaceae the evolution of an arillate capsule with resin-canals, yet without complex sclerenchyma, into berries on the one hand and into a dry capsule on the other hand. The numerous seeds evidently prevent the drupaceous fruit. The resinous gum in which the seeds of *Pittosporum* are immersed, as if in diffluent aril-material, appears to be exuded into the fruit-cavity from the endocarp.

The structure of the seed does not support the alliance with Dilleniaceae proposed by Hutchinson (1967).

PLANTAGINACEAE

Ovules more or less anatropous, erect, or anacampylotropous, unitegmic, tenuinucellate; integument several cells thick.

Seeds small, albuminous, exarillate. Testa reduced to o.e. with tangentially elongate cells, forming mucilage in *Plantago*, the rest of the tissue crushed or absorbed by the endosperm; *i.e.* ? persistent in *Plantago lanceolata*. Endosperm cellular. Embryo straight.

(Netolitzky 1926; Cooper 1942b; Singh, B. 1964.)

Alliance with the unitegmic Polemoniaceae and Scrophulariaceae accords better with the simple seed-structure than with the bitegmic Primulaceae. Netolitzky remarked on the resemblance of the seeds with those of *Veronica*.

PLATANACEAE

Ovules more or less orthotropous, bitegmic, crassinucellate; o.i. 3–4 cells thick; i.i. 3–4 cells thick but more at the micropyle.

Seeds very small with thin, crushed seed-coat inside the achene. Endosperm thin. Embryo linear.

(Netolitzky 1926.)

It is hard to believe that fruit and seed provide no clue to the affinity of this family.

PLUMBAGINACEAE

Ovule anatropous, suspended, bitegmic, solitary, subbasal, carried on a slender funicle arching over the micropyle, crassinucellate; o.i. 2–3 cells thick; i.i. 2 cells thick; micropyle formed by the endostome.

Seeds small; seed-coats not multiplicative; only the exotesta persisting. Perisperm absent. Endosperm nuclear, starchy, or more or less absent. Embryo straight.

(Netolitzky 1926; Wunderlich 1968.)

The ovule, as in Cactaceae, and much reduced seed-coat support the classification of the family in Centrospermae.

PODOPHYLLACEAE

Ovules anatropous or subcampylotropous (*Epimedium, Vancouveria*), numerous on the adaxial side of the ovary or reduced to one, bitegmic, crassinucellate; o.i. 4–8 cells thick; i.i. 2–3 cells thick; aril as a slight rim at the apex of the funicle before fertilization (*Epimedium, Vancouveria*).

Seeds 1–7 mm long, anatropous or reniform, arillate or not, albuminous. Testa not or scarcely multiplicative; *o.e.* appearing in t.s. as a palisade of more or less thick-walled, slightly lignified cells, but the cells elongate longitudinally with oblong facets (*Jeffersonia, Podophyllum*) or fibriform with strongly thickened outer walls (*Epimedium, Vancouveria*), ? cuboid in *Leontice*; *i.e.*

as in Berberidaceae or thin-walled (*Leontice, Podophyllum, Vancouveria, ? Epimedium*). Tegmen not multiplicative, crushed or with o.e. more or less persistent as thin-walled, longitudinally elongate cells, shortened and subclavate round the endostome (*Vancouveria*), not lignified. Chalaza unspecialized. Hypostase small. Endosperm nuclear, oily. Embryo minute.

Aril developing from the apex of the funicle and to some extent along the raphe, not from the micropyle, developing as folds of the epidermis (*Epimedium, Vancouveria*) or fimbriate (*Jeffersonia*), non-vascular white, but as a thicker placental aril covering the seeds in *Podophyllum*. Fig. 464.

Podophyllaceae are separated from Berberidales by Hutchinson and placed in Ranales, but the longitudinally elongate cells of the exotesta and unspecialized cells of the endotesta agree with Berberidaceae and Lardizabalaceae. The alliance seems close with Lardizabalaceae for both families have, in addition, fleshy follicles with placental intrusions among the seeds and arils. Unfortunately several accounts of the seed in this family, as well as in others, are variously deficient, especially in the description of the exotestal cells. However, the removal of *Hydrastis* to Ranunculaceae, or Ranales, is justified by the seed-character for its exotestal cells are prismatic as in Ranunculaceae and they are filled with blue-black contents, suggestive of *Aquilegia* (Pohl 1894, cited by Netolitzky). It is likely that the distinction between simply oblong facets to the exotestal cells and fibriform will re-arrange the genera of Lardizabalaceae and Podophyllaceae in more natural alliance. Compare the endotegmic Nandinaceae.

Epimedium L.

As *Vancouveria* but with small seeds 1–2 mm long.
(Pfeiffer 1891; Netolitzky 1926.)

Jeffersonia Bart.

Aril deeply fimbriate into hair-like sections 2 cells thick, the basal cells thick-walled and pitted but not lignified, the hairs with elongate cells.
(Pfeiffer 1891.)

Leontice L.

Testa; o.e. with cuboid cells, thick-walled (? not longitudinally elongate); i.e. thin-walled, crushed. Vascular bundle of the raphe with postchalazal extension.
(Netolitzky 1926.)
This seed seems rather exceptional in Podophyllaceae.

Podophyllum L.

P. emodi Wall.
(Cambridge, University Botanic Garden; living material.)
Ovule; o.i. 7–8 cells thick; i.i. 2–3 cells thick.
Seed 7 × 5 mm, anatropous, subcompressed, dark brown, overgrown by the pulpy white placenta. Testa

not multiplicative; o.e. composed of rather thick-walled, slightly lignified cells, facets shortly oblong longitudinally, appearing as a palisade in t.s.; i.e. composed of small cuboid thin-walled cells, more or less separated. Endosperm oily but with erythrodextrin plastids in many cells, especially the more peripheral.

The very short funicle becomes fleshy along with the placental tissue which invests each seed individually with a more or less complete coat. The aril appears to have been transferred almost entirely to the placental tissue.

P. peltatum L.
(Clark 1923.)

Vancouveria C. Morr. et Decne

V. hexandra (Hook.) C. Morr. et Decne
(Cambridge, University Botanic Garden; living material.)
Fig. 464.
Ovules subcampylotropous; o.i. 4 cells thick; i.i. 2 cells thick; aril as a slight rim at the apex of the funicle before fertilization.

Seeds 3–3.5 mm long, brown, curved, half-covered by a white, fleshy, plicate aril attached along the raphe. Testa 4–5 cells thick; o.e. as a firm layer of thick-walled lignified cells much elongate longitudinally, fibriform, −600 × 16–20 μ, appearing as a palisade 45–80 μ high in t.s. with very thick outer walls and conical lumen; mesophyll and i.e. unspecialized, the cells of i.e. cuboid and separated in places. Tegmen not multiplicative, unspecialized but o.e. round the endostome with radially elongate, slightly thick-walled cells, not lignified. Vascular bundle ending at the chalaza. Hypostase brown. Endosperm oily, without starch.

Aril arising rather extensively along the raphe by anticlinal division of the epidermal cells causing the epidermis to buckle from the raphe and become thrown into numerous folds composed of the cuboid, rather thin-walled epidermal cells with oil-drops; the hypodermal layer of the raphe also extending into the folds (Berg).
(Berg 1972.)

PODOSTEMACEAE

Ovules anatropous, bitegmic, tenuinucellate; o.i. 2–4 cells thick, forming the micropyle; i.i. 2 cells thick, developing after o.i.

Seeds minute, subglobose or ellipsoid, dark, exalbuminous, exarillate; integuments not multiplicative. Testa; o.e. as a firm layer of cuboid cells with very thick thick pectic outer wall and reduced lumen, becoming entirely mucilaginous (at least in *Anastrophea, Inversodicraea, Tristicha*); mesophyll and i.e. unspecialized, more or less crushed. Tegmen with the outer wall of o.e. and the inner wall of i.e. thickened and lignified (*Indotristicha, Muraea, Tristicha*) or only the inner wall of i.e. (*Anastrophea, Inversodicraea*). Endosperm not formed, but a false embryo-sac from the nucellus.

(Netolitzky 1926; Hammond 1937; Chopra and Mukkada 1967; Jager-Zürn 1967.)

Anastrophea, Indotristicha, Inversodicraea, Muraea, Podostemon, Tristicha.

The affinity of this family is usually ascribed to Crassulaceae and Saxifragaceae in or near Rosales, but the distinctly tegmic seed-coat is not in agreement. It is the character of Piperaceae and Saururaceae, with which line of dicotyledonous descent the family may belong (p. 44). Compare, however, *Heuchera* under Saxifragaceae.

POLEMONIACEAE

Ovules anatropous, solitary or several per loculus, unitegmic, tenuinucellate; integument 7–10 cells thick (*Gilia*), 15–20 (*Polemonium*), with i.e. becoming an endothelium.

Seeds small, albuminous, exarillate, alate in *Cobaea*. Seed-coat ? not multiplicative, commonly reduced to o.e. with the rest of the tissue crushed; o.e. various, as a layer of cuboid cells (*Gilia*) or flattened cells (*Phlox*), or the cells more or less separated and cylindric with spiral or annular, cellulose thickening on the inner and radial walls at least in the proximal part of the cell (*Cobaea* with separated cells, *Collomia, Gilia spp., Loeselia*), the cells commonly containing mucilage and swelling in water to extrude the mucilage; *mesophyll* unspecialized, crushed, or persistent with thinly lignified and reticulately thickened walls (*Cobaea*); *i.e.* often as a pigment-layer, the cells flattened or somewhat ventricose with slightly thickened radial walls, with bar-like thickening. Vascular bundle of the raphe short, not reaching the chalaza. Endosperm nuclear, oily, with little or no starch. Embryo straight or slightly curved.

(Netolitzky 1926; Sunder Rao 1940; Kapil, Rustagi and Venkataraman 1969.)

POLYGALACEAE

Ovules anatropous, suspended, bitegmic, crassinucellate; o.i. 2–5 cells thick, mostly 2; i.i. 2–3 cells thick, mostly 2; micropyle with rather long exostome.

Seeds small to medium-size, arillate or not, albuminous or not. Testa multiplicative into a few cell-layers or not, smooth or hairy; o.e. unspecialized, or with some subsclerotic cells, in some cases with stomata, but palisade-like in *Badiera* and *Securidaca*; *mesophyll* subaerenchymatous, thin-walled or with thick pitted walls (*Polygala* sect. *Ligustrina*, ? lignified); *i.e.* typically as a palisade of Malpighian cells with thick, yellow or brown walls, not or slightly lignified, but becoming hard or crustaceous, generally with a small crystal at the outer end of each cell, the cells varying short, subcuboid

or subconic (*Polygala* sect. *Hebeclada, Bredemeyera, Hualania*) or undeveloped (*Monnina, Moutabea, Securidaca, Xanthophyllum*). Tegmen not multiplicative or possibly forming a middle layer of cells, unspecialized, finally crushed; o.e. and middle cell-layer enlarging, then collapsing. Vascular bundle ending at the chalaza or in some cases (*Polygala spp.*) extended as a few spiral tracheids along the antiraphe, unbranched. Chalaza simple, slightly thickened, with a firm mass of small cells as an unlignified hypostase or with slight lignification in the central part; often with thin remains of the nucellus. Endosperm nuclear, present in the mature seed or more or less absorbed, oily, with aleurone grains.

Aril micropylar or from the end of the funicle at the junction with the testa, generally more or less 3-lobed, reducing to a slight excrescence, or largely replaced by hairs; tissue pulpy or horny, without v.b.

(Chodat et Rodrigue 1893; Netolitzky 1926; Mauritzon 1936a; Mukherjee 1957; Rao, A. N. 1964; Adema 1966; v. Steenis 1968.)

This small family with 1–2 seeds in the fruit bears evidence of derivation from the many-seeded arillate capsule. Nine of the eighteen genera have arillate seeds, though the aril may be reduced or absent in various species, and the other genera have indehiscent fruits in four of which the seed-coat is itself degenerate and fails to develop the family character (*Monnina, Moutabea, Securidaca, Xanthophyllum*). The aril, generally described as micropylar or called a strophiole or caruncle, may have such an extensive junction round the funicle and testa that it is impossible to decide exactly on its point of origin. Thus *Polygala* has both the true or funicular aril and, in its smaller seeds, the false or micropylar aril of the old quibble in morphology. It seems that, as the seed becomes neotenic and smaller (Fig. 468, where the young seed of *P. pulchra* is compared with the mature seed of *P. vulgaris*), the funicular part of the aril is lost and the micropylar part persists until lost in its turn, as in the neotenic seed of *Salomonia*.

The seed-structure of the family was studied very extensively and carefully by Chodat and Rodrigue. Nevertheless most genera need further exploration and illustration. Those precise authors emphasized the endotestal character and found that in certain genera or sections, as *Bredemeyera*, the palisade-cells were so short as scarcely to be enlarged beyond the embryonic and ovular state. The crystal which usually forms in them appears in the inner half of the cell and, as the wall thickens from the inner end outwards, so the crystal shifts to the outer end of the radially elongate cell where the nucleus is; intermediate states of elongation are found as the final states of the endotesta in different species. The nature of the hard and horny wall of these cells needs investigation. The larger seeds have the multiplicative testa, the taller palisade, and relics of more extensive vascularization. The fibro-vascular bundle of the raphe terminates in a fan of tracheids in the chalaza but, in some species, a few tracheids continue in the antiraphe (the pseudoraphe of Chodat and Rodrigue 1893, p. 458) almost to the micropyle. From such vestiges of complexity, the family shows well how the simple structure of small seeds has been derived.

The family has been attached to Geraniales, more particularly to Malpighiaceae, to Violales, and on grounds of anatomy to Anacardiaceae. The strongly endotestal structure of the seed forbids such alliances. The structure of both aril and testa point to an ancestry with Myristicaceae, whence Polygalaceae specialized, doubtless during the pachycaul phase, in the small fruit with fewer small seeds of decreasing complexity until the herbaceous form was possible. *Xanthophyllum*, according to Hutchinson (1967), has the least advanced floral structure, but it is not necessarily primitive for this reason; it is a genus of advanced leptocaul trees with many common species and its advanced and indehiscent fruits have large, possibly overgrown, seeds without aril or seed-coat differentiation. Anatomically the wood of *Xanthophyllum* is exceptional to the level of family distinction (Metcalfe and Chalk 1950). The genus is not primitive but a living success in tropical forest and its ancestry must be viewed in this regard. It may be the remainder of an early pachycaul offshoot from the stock of Myristicaceae and Polygalaceae. Its seed-structure needs much more extensive investigation than was possible for Chodat, Rodrigue and Mauritzon.

Whether there is affinity with Vochysiaceae must await the study of mature seeds of that family, and the comparison of the exfoliating fruit with that of *Xanthophyllum*.

The endotestal structure brings Cruciferae and Dilleniaceae into consideration.

The work of Chodat and Rodrigue may be summarized as follows:

(1) Testa 2(–3) cells thick; endotesta as a well-developed palisade:
Acanthocladus Kl.; *Bredemeyera* sect. *Hualania* Phil. (short palisade); *Polygala* sect. *Brachytophis* (high palisade), *Gymnospora, Orthopolygala; Salomonia* Lour.
(2) Testa 3–7 (or more) cells thick;
(*a*) palisade high:
Bredemeyera sect. *Comesperma* Labill. (*pr. p.*), *Mundtia* HBK., *Muraltia* DC; *Polygala* sect. *Chamaebuxus, Hebeclada,* and *Ligustrina* (with thick-walled mesophyll).
(*b*) Palisade low, the cells cuboid or obconic:
Badiera DC (exotesta with radially elongate cells); *Bredemeyera* Willd.; *Polygala* sect. *Hebecarpa.*
(3) Testa undifferentiated:
Monnina Ruiz et Pav.; *Moutabea* Aubl.; *Securidaca* L.; *Xanthophyllum* Roxb.

Concerning *Mundtia*, I find some doubt. Chodat and Rodrigue described the large seed as arillate and remarked upon the well-differentiated testa in the drupe. Hutchinson referred *Mundtia* to *Nylandtia* Dumort. and gave the seed as not strophiolate.

Polygala L.

P. butyracea Heck.
Seed 5 mm long, brown, hairy in places, with a caruncle. Testa; *o.e.* with thick cuticle and thickened radial walls; *mesophyll* none; *i.e.* as a thick-walled palisade 65 μ high, with a crystal at the outer end of each cell. Endosperm oily.
(Vaughan 1970.)

P. pulchra (Hassk.) Chod.
(Borneo, Kinabalu, RSNB 2847; flowers and fruits in alcohol.) Figs. 465–468.
Ovule; o.i. 4–5 cells thick; i.i. 3 cells thick.

Seeds 5 mm long and wide, round, black, with orange-red pulpy aril; funicle well-developed, the swelling aril bursting the capsule. Testa 5–8 cells thick; *o.e.* with distinct cuticle, a few scattered cells over the chalaza with thick brown walls, with small stomata scattered round the chalaza and beside the raphe; *mesophyll* unspecialized, with slightly thickened unlignified walls; *i.e.* as an uneven palisade 130–240 μ high, with slightly lignified, hard, horny to crustaceous cells slightly curved on the sides of the seed, without crystals, at the chalaza slightly incurved to join the exotegmen. Hypostase not lignified.

Aril with two large lateral lobes, covering most of the seed, and a small median lobe derived from the end of the exostome; composed of large pulpy cells with small-celled epidermis.

P. vulgaris L.
Fig. 468.
Seed 3 × 1.5 mm, ellipsoid, black, finely silky hairy; aril white, horny, micropylar with 3 recurved lobes. Testa 3 cells thick on the sides of the seed, 4–5 cells thick in the raphe and antiraphe; *o.e.* with rather thick-walled cells, many producing a single, unbranched, acicular hair –300 × 8–11 μ (appressed on the attached seed in the fruit), with thick hyaline walls, aseptate, not lignified; *i.e.* as a palisade *c.* 80 μ high, with thick brown walls, horny-crustaceous, not lignified, with a small crystal in the outer dilated end of the lumen. Tegmen 3 cells thick, the outer 1–2 cell-layers enlarging, then collapsed. Necullus persisting as a single layer of cells. Endosperm thin-walled, oily.

Aril firm, with a compact palisade-like epidermis of thick-walled cells, the outer wall strongly thickened, not pulpy.

POLYGONACEAE

Ovule orthotropous, solitary, basal, bitegmic or unitegmic through connation, crassinucellate; o.i. 2–4 cells thick; i.i. 2 cells thick; hypostase present.

Seeds small, albuminous, exarillate. Testa multiplicative (?) or not, reduced to o.e. with little specialization or with reticulate thickening (*Rumex*). Tegmen not multiplicative, crushed. Chalaza much widened with saucer-shaped hypostase (*Coccoloba*) or small. Endosperm nuclear, starchy, slightly ruminate by ingrowths of the testa (*Coccoloba*). Embryo rather small, straight.

(Netolitzky 1926; Periasamy 1964a; Neubauer 1971.)

So little is known of the seeds of this family that it is impossible to decide whether they are truly exotestal or merely undeveloped through enclosure in the simplified and indehiscent syncarpous fruit. Compare Halorrhagaceae. Illecebraceae, placed in Polygonales by Hutchinson, has a campylotropous ovule and seed as in Caryophyllaceae.

PORTULACACEAE

Ovules anacampylotropous, bitegmic, crassinucellate; o.i. and i.i. 2 cells thick, not multiplicative in the seed.

Seeds small, reniform, albuminous, with strophiole (? aril) in *Claytonia* and *Talinum*, with a chalazal air-gap between the integuments in *Portulaca*. Testa; o.e. as a layer of cuboid cells with polygonal and isodiametric facets, with thickened and lignified outer wall, finely pitted (*Claytonia*), the cells papilliform (*Montia*, *Portulaca*, *Talinum*) or with a short hair-like process (*Calandrinia*); i.e. crushed, unspecialized, or with crystals (*Portulaca*). Tegmen unspecialized, crushed, or i.e. with slight bar-thickenings on the radial walls (*Montia*, *Portulaca*). Perisperm present, starchy. Endosperm nuclear. Embryo curved.

(Netolitzky 1926; Woodcock 1926; Kajale 1942; Haccius 1954.)

PRIMULACEAE

Ovules more or less anatropous, bitegmic, or campylotropous and unitegmic (*Cyclamen*, ? *Douglasia*), tenuinucellate; o.i. 2 cells thick; i.i. 3 cells thick, 2 in *Samolus*; the single integument 6–10 cells thick in *Douglasia*, (in *Cyclamen* ? 2 cells thick with pachychalazal base to the ovule, or 6–8 cells thick and tenuinucellate; Woodcock 1933); integuments not multiplicative in seed.

Seeds small, generally albuminous, exarillate or with a small aril (*Primula*). Testa with only o.e. persistent, cuticulate, with large and often projecting cells, the walls thickened and brown, in some cases (*Primula spp.*) with reticulate thickenings on the inner walls, or some cells not enlarging (*Cyclamen*). Tegmen crushed except i.e. as a layer of crystal-cells with thin or thickened walls,

the crystals solitary in the cell or several (*Hottonia*, *Lubinia*, *Lysimachia*, *Samolus*), or absent (*Androsace*, *Primula spp.*, *Soldanella*). Endosperm nuclear, oily, with thin walls, or with thick, pitted, amyloid walls. Embryo straight.

Aril as a funicular elaiosome in *Primula*.

(Netolitzky 1926; Woodcock 1933; Bresinsky 1963; Subramanyam and Narayana 1968.)

This family has been related with Caryophyllaceae, Plantaginaceae, Plumbaginaceae, Ebenaceae, Myrsinaceae and Theophrastaceae. The structure of the small seeds, with a trace of aril, agrees with the Centrospermous, and this alliance is supported by the affinity with Plumbaginaceae (Wunderlich 1968). Affinity with Ebenaceae would imply the untwisting of the Ebenaceous ovule, its neotenic simplification into a small seed, its multiplication in number per loculus, and extreme simplification of the fruit, for which there is no evidence in Ebenales. The endotegmic layer of crystals seems to be characteristic.

The construction in *Cyclamen* is not clear. According to Woodcock (1933), there is a single integument 2 cells thick, the inner layer of which is a crystal-layer with the radial and inner walls of the cells thickened and lignified. This suggests that only the inner integument remains, but the illustration which he gives suggests that the ovule and seed may be pachychalazal.

PROTEACEAE

Ovules numerous or commonly reduced to 4, 2 or 1 per carpel, anatropous, hemianatropous, orthotropous, erect, bitegmic, crassinucellate; o.i. 2–4 cells thick, generally 2; i.i. 3–5 cells thick, developing before o.i. and forming the more or less elongate micropyle.

Seeds small to large (*Macadamia*), commonly winged in follicles, with thin seed-coat (massive in *Macadamia*), exalbuminous (except *Bellendena*), exarillate. Testa multiplicative (Grevilleae 8–9 cells thick, *Bellendena* 5–6 cells thick, *Macadamia* 3–4 mm thick) or commonly not multiplicative; o.e. composed of cuboid or shortly radially elongate cells, the outer wall in some convex and papilliform, thin-walled or with thickened and pitted outer wall (*Grevillea*), with longitudinal fibres (*Hakea*), or more or less crushed, often with tannin; mesophyll thin-walled, crushed, or thickly sclerotic (*Macadamia*); i.e. as a layer of crystal-cells with one or several crystals in each, the cell-cavity pervaded with cellulose fibrils, with thickened (? lignified) inner and radial walls, or thin-walled with or without crystals or with tannin, the crystal-cells in some cases in a double layer at the chalaza, or with several such layers (*Macadamia*). Tegmen multiplicative (6–7 layers in *Bellendena* and *Lomatia*, 7–8 layers in *Embothrium*, 8–9 layers in Oriteae, –10 layers in *Stirlingia*) or commonly not multiplicative; consisting of three outer layers of thick-walled lignified fibres with the middle layer crossing the outer and inner layers of longitudinal fibres (*Grevillea*), or with several fibrous layers (*Hakea*), but in indehiscent fruits apparently undifferentiated and more or less crushed. Alate seeds with the wing developed as an extension of the testa (micropylar in *Stenocarpus*) the layer of crystal-cells absent in some cases (*Hakea, Xylomelum*). Vascular bundle of the raphe ending at the chalaza, with a short pre-raphe branch to the micropyle (*Grevillea spp.*), in some alate seeds flexed above the chalaza (*Embothrium, Lomatia*), ? with tegmic bundles in *Helicia*. Perisperm occasionally persistent as a thin layer (*Grevillea*). Endosperm nuclear, generally absorbed, persistent and starchy in *Bellendena* (? oily without starch in *Persoonia*). Embryo straight.

(Netolitzky 1926; Brough 1933; Kausik 1939, 1940a; Venkata Rao 1960–1970.)

Netolitzky's account of the seed in follicular Proteaceae gives the tegmen as fibrous and the endotesta as a layer of crystal-cells with internal fibrils. Thus he compared the seed with that of Papaveraceae and Aristolochiaceae. Subsequent investigators have not mentioned these features. One suspects that, with the employment of the microtome, the hard mature seeds may not have been sectioned. In Persoonioideae with indehiscent and drupaceous or nut-like fruits the structure deteriorates, as Netolitzky observed; both seed-coats remain thin-walled and are crushed except for the layer of crystal-cells (*Persoonia*).

In an extensive series of papers on the classification, microscopic structure, and evolution of the Proteaceae, Venkata Rao has added much information on the ovule and young seed, from which it is clear that both testa and tegmen may be multiplicative. He has described various states of the tegmen; in Persoonieae the outer epidermis persists as thin-walled cuboid cells, but palisade-like and some cells with crystals in *Embothrium*; in Oriteae the outer hypodermis forms a layer of thick-walled cells (1–2 cells deep, ? fibres). He arrives at two debatable conclusions. He considers that follicular fruits have been evolved from the indehiscent, which is contrary to the general understanding of fruit-evolution and, as many of these fruits have winged seeds, they suggest derivation from arillate follicles (Corner 1954). Then he considers that the orthotropous ovule has evolved into the anatropous, but this is not clear when its manner of growth is considered, for the shape of the ovule depends on the manner of growth of the carpel. When few ovules are formed and the carpel enlarges its cavity by basipetal growth below the attachment of the ovules, they grow downwards and become orthotropous; when the cavity is enlarged above the attachment, the ovules extend distally and become anatropous; the hemianatropous ovule has intermediate status. Thus, *Embothrium* has 12–16 anatropous ovules in two imbricating rows along the carpellary margins. As the carpel is diminished,

the few erect ovules become more or less basal and, as carpel-growth becomes mainly basipetal, they are transformed through the hemianatropous into the orthotropous form as seen in other families with such basipetal growth (Meliaceae, Palmae).

The combination of endotestal crystal-cells with internal fibrillar thickening and exotegmic fibres is found in Papaveraceae, Proteaceae and Chloranthaceae. A more intrinsic sign of affinity could hardly be expected (p. 32). The Proteaceous ancestor, according to the durian-theory, would have been a pinnate-leafed pachycaul with arillate follicles derived from an apocarpous ovary, and the albuminous seeds would have had this particular modification of the seed-coat midway between the Myristicaceous and the Magnoliaceous. Such an ancestor agrees with a similar postulate for Papaveraceae, e.g. *Bocconia*, and would serve for Chloranthaceae. The families would have diverged in the pachycaul phase of experimentation in angiosperm methods of reproduction, and parted physiologically and geographically. Affinity has been suggested with Thymelaeaceae (Hutchinson 1967), Elaeagnaceae (Takhtajan 1969), and both (Cronquist 1968), but seed-structure gives no support.

Helicia Lour.

H. fuscotomentosa Suesseng.
(Borneo, Kinabalu, leg. E. J. H. Corner 1964; immature fruits in alcohol.)

Ovules 2, basal, anatropous, erect, one abortive.

Testa 10–14 cells thick along the raphe but attenuate over the side of the seed with loose, transversely elongate, and narrowly substellate cells, thin-walled, apparently crushed along the antiraphe. Tegmen c. 30 cells thick, the outer 10–12 layers as brown cells, the remainder colourless, thin-walled, becoming crushed from the micropylar end towards the chalaza. Vascular bundle of the raphe stout, dividing into several branches over the diffuse chalaza and extending into the tegmen as a fine reticulum almost to the micropyle.

The material was insufficient for thorough study and I am not satisfied about the tegmic plexus of v.b. The testa becomes so thin over most of the seed that its inner limit is practically impossible to decide. The seed may be pachychalazal.

Macadamia F.v.M.

Ovules 2, orthotropous, pendent. Seeds 1–2 in the more or less indehiscent fruit, globose or hemispheric (if paired), with thin membranous testa or thick woody testa.

M. ternifolia F.v.M.
Ovule; o.i. and i.i. 3–4 cells thick, both shorter than the projecting nucellus.

Seeds 20–22 mm wide, with very hard testa 2–5 mm

thick, developed as a pachychalaza with very short free integuments in the micropylar region. Testa; *o.e.* as a layer of tabular cells; *mesophyll* composed of a mass of brownish sclerotic cells with embedded v.b., and with the inner 3 or more layers as rather thin-walled pitted crystal-cells.

(Sleumer 1955; Vaughan 1970; Venkata Rao 1970.)

Vaughan has described the mature seed, Venkata Rao the developing seed, but it is hard to make ends meet. The brown inner zone of crushed cells, described by Vaughan as lining part of the testa, may be the remains of the tegmen. Sleumer describes *M. hildebrandii* Steen. with a very thin testa; thus it appears to be not pachychalazal. Other species may show how it relates in seed with *M. ternifolia*.

PUNICACEAE

Ovules very numerous, anatropous, bitegmic, crassinucellate; o.i. 3–4 cells thick, forming the micropyle; i.i. 2 cells thick.

Seeds 10 × 7 mm, with watery translucent sarcotesta and sclerotic mesotesta, exarillate, exalbuminous; funicle long. Testa multiplicative; *o.e.* as a palisade of very long, thin-walled, columnar, pulpy cells 0.6–3 mm long, with isodiametric hexagonal facets; *mesophyll* composed of many layers of more or less sclerotic cells, the inner layers with the cells shortly radially elongate; i.e. unspecialized. Tegmen not multiplicative, persistent, demarcated with cuticular layers; *o.e.* as a compact layer of narrow longitudinal tracheids with spiral-annular lignified thickening; *i.e.* with thin-walled, longitudinally elongate cells. Vascular bundle terminating at the simple chalaza. Endosperm nuclear, absorbed. Embryo straight; cotyledons contort.　Fig. 469.

(Netolitzky 1926.)

The seed of the pomegranate agrees in all essentials with those of Combretaceae, Lythraceae, Onagraceae and Sonneratiaceae. The long funicle, the palisade-like exotesta, the sclerotic mesotesta and, particularly, the tracheidal exotegmen are the main characters. That of the pomegranate seems the most typical of the alliance.

RAFFLESIACEAE

Ovules anatropous or orthotropous (*Cytinus*), bitegmic or unitegmic, tenuinucellate; o.i. as a single layer of cell but 2 cells thick at the exostome (*Apodanthes*), or much reduced (*Cytinus*), or absent (*Mitrastemon*, *Rafflesia*); i.i. 2–3 cells thick.

Seeds minute, 0.5–1 mm long; seed-coats not multiplicative. Testa as a single layer of pulpy cells (*Apodanthes*, in some species with sparse hairs), or lacking (*Cytinus*, *Rafflesia*). Tegmen with the inner layer crushed and unspecialized, or thick-walled (*Mitrastemon*, *Rafflesia*); *o.e.* as a layer of sclerotic cuboid cells with

strongly pitted walls (*Apodanthes*), or as a layer of cells with finely pitted, much thickened, inner and radial walls (*Mitrastemon*, *Rafflesia*), with flattened cells (*Cytinus*). Chalaza woody, prominent (*Rafflesia*). Endosperm nuclear, oily. Embryo microscopic, with large cells in *Apodanthes*.

(Netolitzky 1926; Winkler 1927; Watanabe 1933; Kummerow 1962; Rutherford 1970.)

Doubtless these seeds need closer investigation though they have been well-illustrated by the older authors, cited by Netolitzky, and by Rutherford. Absence of a fibrous exotegmen excludes the supposed affinity of the family with Aristolochiaceae (p. 40). *Apodanthes*, to which *Pilostyles* has been reduced, is exotegmic, but *Mitrastemon* and *Rafflesia* are also endotegmic. This construction relates the seeds to those of Piperaceae, Saururaceae, and Podostemaceae, which may be thought a curious alliance, but if all can be derived from a Magnolialean beginning, the inter-relation is not impossible. It suggests a pachycaul line that has failed in vegetational dominance but discovered some of the most remarkable ways of existence.

RANUNCULACEAE

Ovules anatropous, transverse or in uniovulate carpels erect or suspended, bitegmic, crassinucellate, unitegmic in *Anemone*, *Ceratocephalus*, *Clematis*, *Delphinium*, *Helleborus pr. p.*, *Myosurus* and *Ranunculus*, tenuinucellate in *Anemone*, *Clematis*, *Helleborus* and *Ranunculus pr. p.*; o.i. 2–10 cells thick; i.i. 2–3 cells thick.

Seeds small, 1–5 mm long, albuminous, exarillate but with swollen raphe-elaiosome in *Helleborus* and swollen chalaza in *Caltha*. Testa not or slightly multiplicative; *o.e.* composed of cuboid, shortly palisade-like or tabular cells, the palisade-cells sometimes projecting in clusters, the outer wall thickened, in some cases lignified (*Hydrastis*), with or without hairs; *mesophyll* crushed; *i.e.* unspecialized or with fine bar-like thickenings on the radial walls (*Trollius*). Tegmen not multiplicative, generally more or less crushed; *o.e.* unspecialized or somewhat thick-walled (*Helleborus*); *i.e.* often with fine thickenings on the radial walls. Vascular bundle of the raphe ending at the chalaza or extending to the micropyle (*Anemone spp.*, ? pachychalazal). Chalaza usually with a small hypostase. Endosperm nuclear, oily with or without starch, copious. Embryo minute. Figs. 470–472.

(Netolitzky 1926; Bersier 1960; Bhandari 1969.)

The better developed seeds occur in genera with dehiscent fruits (Helleboreae). The testa of those with indehiscent fruits is more or less unspecialized, and the single integument in *Anemone* may be entirely crushed. Reduction of the integuments and their substitution by an intercalary integument to form a pachychalaza seems to occur as in Rosaceae. Thus *Actaea* and *Eranthis* have very short integuments and the chalazal region forms most of the seed-coat.

For two reasons Ranunculaceae are placed with Magnoliaceae near the beginning of angiosperm evolution. The apocarpous flower has a conical axis and the albuminous seed has a minute embryo. The two families differ profoundly in other respects, including the structure of the seed. Such divergence proves the previous existence of more ancestral angiosperms which I identify with the primitive pachycaul phase during which flower and seed evolved into the states which modern orders display. The history of Ranunculaceae is that of simplification for herbaceous existence and must be read in parallel with that of Papaveraceae (p. 27). The history of Magnoliaceae is that of simplification and engineering into leptocaul trees. This is the divergence that has probably taken place in many subsequent families from Rosaceae and Leguminosae to Apocynaceae and Compositae, Urticales and Malvales. It displays the potentialities of the pachycaul. They can be read in the more highly evolved families because their evolution has permitted the persistence of modernized pachycauls. The history of the more primitive families, from which the primitive pachycauls have disappeared, must be inferred from these analogies. Herbs and trees, just as syncarps and achenes, are polyphyletic; and the seed-structure introduces the primitive criterion of affinity.

Actaea L.

A. spicata L.

Ovule transverse, with thick chalaza and very short free integuments; o.i. 4–5 cells thick; i.i. 2 cells thick.

Seed-coat 6 cells thick, becoming crushed except for o.e. composed of hard cuboid cells. (? pachychalazal.)

(Jalan 1964.)

Adonis L.

Ovules suspended, incurved with adaxial funicle, one fertile ovule only (*A. chrysocyathus*) or four sterile and rudimentary ovules superposed on the basal fertile ovule (*A. aestivalis, A. annua, A. flammea*); o.i. 4–6 cells thick; i.i. 2 cells thick (4–5 near the micropyle).

Testa 7–8 cells thick; o.e. unspecialized but the cells tangentially elongate. Tegmen disappearing.

(Bhandari 1967.)

Anemone L.

Ovules unitegmic, tenuinucellar, suspended and incurved with adaxial micropyle, some species with 2 or more sterile ovules in addition to the fertile ovule; integument 5–6 cells thick.

Seed-coat 6–10 cells thick, unspecialized, becoming crushed, the endosperm abutting on the endocarp. Vascular bundle extending to the micropyle in *A. narcissiflora, A. nemorosa* and *A. ranunculoides*.

(Kühn 1927, p. 342; Bhandari 1969.)

Aquilegia L.

A. vulgaris L.

Fig. 470.

Ovule; o.i. 6–8(–10) cells thick; i.i. 2–3 cells thick.

Testa not multiplicative, blue-green when half-mature from a pigment in the thickened outer walls of the exotesta, then black; o.e. as a palisade of cells with much thickened outer wall with conical lumen, with small crystals on the anticlinal walls at the surface; *mesophyll* becoming crushed and dried. Tegmen soon disappearing or i.e. subpersistent.

Caltha L.

C. palustris L.

Ovule crassinucellate; o.i. 4–5 cells thick; i.i. 2 cells thick.

Seed with the cells in all layers enlarging, then crushed except the exotesta composed of cuboid cells with slightly thickened outer wall and cuticle, contents with tannin.

(Kapil and Jalan 1962.)

Ceratocephalus Moench.

C. falcatus Pers.

Ovules anatropous, erect, unitegmic, crassinucellate; integument 4 cells thick.

Seed-coat not multiplicative, the outer two cell-layers persistent, thin-walled; o.e. with slightly thickened outer wall, not as a palisade. Endosperm nuclear.

(Bhandari and Asnani 1968.)

Helleborus L.

Ovules bitegmic, in some species unitegmic.

Seeds 4–5.5 × 3–4 mm, ovoid, brown, with prominent raphe-elaiosome in some species, sparsely or very minutely hairy. Testa 8–10 cells thick (*H. foetidus*), 6–7 cells thick (*H. corsicus, H. niger*); o.e. as a short palisade with strongly thickened outer wall, not lignified, more elongate round the micropyle; *hairs* minute, subclavate, issuing singly from many of the epidermal cells. Tegmen 3(–4) cells thick, eventually crushed; o.e. with somewhat thick-walled cells, not lignified, more evident at the endostome. Endosperm oily and with starch. Figs. 471, 472.

(Netolitzky 1926; Singh, B. 1964.)

Nigella L.

N. damascena L.

Ovule; o.i. 4–5 cells thick; i.i. 2 cells thick.

Testa rugulose; o.e. with thick outer wall to the cells, some with a conical projection, others subclavate and projecting in groups; *mesophyll* with 3–4 cell-layers persistent. Tegmen crushed except for i.e.

(Vijayaraghavan and Marwah 1969.)

I have found that in this species the outer integument is 7–10 cells thick and that 1–2 outer hypodermal cell-layers have a yellow sap which gives the bright yellow colour to the ovules.

Ranunculus L.

Ovules unitegmic. Testa not multiplicative, more or less unspecialized and crushed.

(Singh, B. 1936.)

Thalictrum L.

T. javanicum Bl.

Ovule; o.i. 7–8 cells thick; i.i. 2–3 cells thick.

Seed with only the exotesta and endotegmen persistent, thin-walled, with tannin.

(Vijayaraghavan and Bhandari 1970.)

RESEDACEAE

Ovules anacampylotropous, bitegmic, crassinucellate or tenuinucellate; o.i. 2 cells thick, or 3 cells at the micropyle; i.i. 2–4 cells thick, the cells of o.e. often shortly elongate longitudinally.

Seeds 0.5–3 mm long, more or less reniform, smooth, verrucose or rugulose, often with a caruncle (aril?) at the hilum. Testa not multiplicative or slightly (*Reseda odorata*); o.e. unspecialized or in some cases the cells more or less papillate, with thick outer wall or cuticle, without stomata; i.e. composed of small, thin-walled, crystal-cells (absent in *R. luteola*). Tegmen slightly multiplicative or not, 3–6 cells thick; o.e. composed of shortly longitudinally elongate fibres, pitted, lignified, appearing as a palisade in t.s.; *mesophyll* thin-walled, collapsing; i.e. composed of cuboid cells with slightly thickened and shallowly pitted anticlinal walls, not lignified, eventually with dark contents. Vascular bundle terminating at the chalaza. Chalaza with a small septum of sclerotic cells between the brown hypostase and the end of the v.b. Endosperm nuclear. Fig. 473.

(Netolitzky 1926; Singh, D. and Gupta 1967.)

The Resedaceous seed is essentially Violaceous. The so-called third integument, which has been reported, is the close jacket of the endotegmen around the endosperm.

Oligomeris Cambess.

O. linifolia (Vahl) Maebr.

Seeds black, shiny, smooth. Testa; o.e. with thick cuticle, the cells thin-walled, unspecialized, with dark contents; i.e. cells with dark contents. Tegmen; o.e. composed of uniformly elongate, sclerotic cells (without the recurved ends of *Reseda*); *middle layer* composed of small, thin-walled cells.

(Singh, D. and Gupta 1967.)

Reseda L.

Testa; o.e. with thick or thin outer walls to the cells. Tegmen; o.e. with outwardly projecting ends to the sclerotic fibres causing the rough surface of the dried seed. Fig. 473.

R. alba L.

Testa 2 cells thick; o.e. with the cells shortly papillate, thin-walled. Tegmen 4 cells thick; i.e. with large cells.

(Guignard 1893.)

R. lutea L.

Testa 2 cells thick; o.e. with slightly thickened outer walls. Tegmen; i.e. with flattened cells.

(Hennig 1930.)

R. luteola L.

Testa; o.e. with a reticular thickening on the cell-walls, then the outer wall very thick and almost obliterating the lumen. Tegmen; i.e. without crystals.

(Netolitzky 1926; Crété 1936a.)

R. odorata L.

Testa 2–4 cells thick, multiplicative over the furrows of the exotegmen through division of i.e. into 3 cell-layers, the whole tissue thin-walled and drying up at maturity. Tegmen 4–6 cells thick, multiplicative through division of the middle layer into 2–4 cell-layers; *mesophyll* composed of small thin-walled cells.

(Singh, D. and Gupta 1967.)

RHAMNACEAE

Ovules anatropous, solitary, basal erect, bitegmic, crassinucellate; o.i. 4–8 cells thick; i.i. 3 cells thick.

Seeds rather small to medium-size, arillate or exarillate, albuminous or not. Testa generally multiplicative, firm, smooth; o.e. as a palisade of thick-walled prismatic cells, with hexagonal facets, but the lumen minutely stellate at the outer end, the base often thin-walled, the wall thickening from the outer end inwards, lignified or not, often with a *linea lucida* separating the lignified end of the cell, the cells short with stellate lumen in *Rhamnus*, unspecialized in *Ventilago*; *mesophyll* aerenchymatous, more or less crushed, with scattered hypodermal sclerotic cells in *Ventilago*; i.e. unspecialized. Tegmen not or slightly multiplicative, soon more or less crushed; i.e. composed of cuboid cells with rectangular or hexagonal facets, with barred or pitted radial walls, not lignified. Vascular bundle in the raphe only or with postchalazal extensions almost to the micropyle (*Ventilago*, *Zizyphus*). Endosperm nuclear, oily. Embryo straight. Figs. 474–478.

(Netolitzky 1926; Srinivasachar 1940; Dolcher 1941; Kajale 1944; Arora 1953.)

The affinity of this exotestal family is discussed on p. 36.

Colletia Comm.

Colletia sp.

(Bolivia, Titicaca, leg. E. J. H. Corner Dec. 1947; dried material.) Fig. 474.

Seeds 4 × 2.5 mm, brown then black, smooth, shiny, ellipsoid and subtriquetrous. Testa ? 8–10 cells thick, the inner part crushed; *o.e.* as a palisade 150 μ thick, composed of prismatic cells with thick brownish walls and brown lumen, with hexagonal facets, the lumen minutely stellate in t.s. at the outer end, the inner ends of the cells thin-walled, lignified only near the outer end. Tegmen crushed; *i.e.* apparently composed of large, tabular cells with hexagonal facets, not pitted or lignified.

Ventilago Gaertn.

V. madraspatana Gaertn.
(Ceylon; fruits in alcohol.)
Testa and tegmen unspecialized, crushed. Embryo with very short radicle; cotyledons thick, orbicular. Pericarp toughly fibrous.

V. malaccensis Ridley
(Penang; ripe fruits in alcohol.) Fig. 475.
Seed 6–7 mm wide, globose with a wide discoid chalazal area. Testa ? 10 cells thick; *o.e.* as a firm layer of cuboid cells with slightly thickened outer wall; *mesophyll* thin-walled, in places crushed, with small scattered sclerotic cells in the o.h., more numerous in the subhilar tissue; *i.e.* unspecialized. Tegmen crushed. Chalaza expanded into a disc 3.5 mm wide, with a plexus of tracheids separated from the crushed nucellus by a thin brown hypostase. Vascular bundle in the raphe-antiraphe.

Zizyphus L.

Z. mauritiana Lamk.
(Ceylon; flowers and fruits in alcohol.) Figs. 476–478.
Ovule; o.i. 4–5 cells thick; i.i. 3 cells thick.
Seeds 7–8 × 5–6 × 2 mm, compressed, with firm, smooth, light brown testa, embedded in the thick woody endocarp of the fruit. Testa 8–10 cells thick; *o.e.* as a compact short palisade of prismatic cells, lignified with reduced lumen; *mesophyll* with large cells, eventually more or less crushed and desiccated but some cells with lignified, though unthickened, walls; *i.e.* with small cells unspecialized. Tegmen 5–6 cells thick in the broad plane of the seed, 3–4 cells thick in the central part around the enlarged nucellus, eventually crushed except i.e. composed of rectangular cells with slightly thickened walls and slight scalariform thickenings on the radial walls, not lignified. Nucellus enlarging greatly, 15–25 cells thick, then crushed. Vascular bundle of the raphe with postchalazal extension nearly to the micropyle.
This seed is similar to that of *Z. jujuba*, studied by Srinivasachar and Kajale who found thicker integuments in the ovule (o.i. 9–12 cells thick in the micropylar end, 5–8 cells at the chalazal end; i.i. 1–2 cells thick at the micropylar end, 4–5 at the chalazal

end). Perisperm was said to persist as a storage tissue, but it is not clear if Srinivasachar studied mature seeds.

Z. rotundifolia Lamk.
Ovule; o.i. 4–5 cells thick; i.i. 3 cells thick.
Testa 9 cells thick. Tegmen 12 cells thick. Vascular bundle extending in the raphe-antiraphe.
(Arora 1953.)

RHIZOPHORACEAE (excluding Legnotidaceae, p. 161)

Ovules anatropous, bitegmic, crassinucellate.
Seeds of medium size with thick testa and distinct exotesta (*Carallia*) to large with unspecialized pellicular testa (*Anisophyllea, Rhizophora,* etc.), albuminous or exalbuminous, arillate (*Cassipourea, Crossostylis, Weihea*), or exarillate, winged in *Macarisia*. Embryos straight or curved (*Carallia*), with normal or reduced cotyledons and then more or less hypocotylar (*Anisophyllea, Carallia, Rhizophora* etc., ? *Poga*).
(Netolitzky 1926; Vaughan 1970.)
I have extracted *Gynotroches* and *Pellacalyx* as Legnotidaceae, though usually assigned to Rhizophoraceae; their seeds are exotegmic with elongate, longitudinal, sclerotic cells. How many other non-mangrove genera should also be removed must await the microscopic study of their seeds; if they are truly Rhizophoraceous, then the family offers arillate, alate, and exarillate seeds with all apparent intermediates to the viviparous mangroves and the large 'drupes' of *Anisophyllea*, which resemble those of *Barringtonia*.
Crossostylis, recently revised by Ding Hou (1968), suggests Legnotidaceae.

Carallia Roxb.

C. brachiata (Lour.) Merr.
(Java, Herb. Bogor.; fruits in alcohol.)
Seed curved, campylotropous. Testa unlignified, thick; *o.e.* as a palisade of large cells with red-brown tannin contents; *mesophyll* thick, unspecialized, many cells with sphaerocrystals; *i.e.* unspecialized. Tegmen without trace. Embryo hypocotylar, curved, with two very short, but distinct, cotyledons.

Poga Pierre

P. oleosa Pierre
Seed 10–20 mm long, ovoid, dark brown, exalbuminous. Testa *c.* 12–15 cells thick, with v.b.; *o.e.* composed of cuboid cells with slightly thickened, lignified radial and inner walls; *mesophyll* thin-walled. Embryo ? with fused cotyledons.
(Vaughan 1970.)

ROSACEAE

Ovules anatropous, transverse, suspended or erect, many, few, paired or solitary, often some aborted, bitegmic or unitegmic, crassinucellate; o.i. 3–14 cells thick; i.i. 2–6 cells thick; micropyle closed by the exostome or the endostome, or with projecting nucellus (*Rhodotypos*); hypostase present; obturator present; funicle short or none.

Seeds small to fairly large, generally with non-multiplicative seed-coats (except *Agrimonia*, ? *Eriobotrya*, ? *Prunus spp.*), exarillate, mostly albuminous. Testa with thickly sclerotic mesotesta (Pomeae, ? *Quillaja*), but in most genera more or less unspecialized and thin-walled; *o.e.* as a short palisade (Pomeae *pr. p.*), composed of rounded projecting and pitted cells (Pruneae), subcuboid with slightly thickened outer walls, or more or less tabular and then often with spiral or reticulate thickening of the wall (Potentilleae, Poterieae, Roseae), with stomata (*Cowania, Purshia*), in many cases with the outer wall becoming mucilaginous; *i.e.* unspecialized. Tegmen unspecialized; *i.e.* generally persistent. Chalaza and hypostase with suberized or lignified cells in some cases. Vascular bundle of the raphe ending at the chalaza, in some cases with a tracheidal plexus, with postchalazal branches (Pomeae *pr. p.*, Pruneae). Endosperm nuclear, varying 1–20 cells thick in different genera, thin-walled.

(Péchoutre 1902; Juel 1918; Netolitzky 1926; Kühn 1927; Hjelmquist 1962; Sterling 1964–66; Vaughan 1970.)

The seeds in this varied family offer no striking microscopic structure. Most appear to have non-multiplicative seed-coats and to be variously reduced and simplified with little differentiation of the cell-layers, in accordance with the prevailing indehiscence of the fruit. Therefore one should look to the tr. Quillajeae with follicles and winged seed, indicative of arillate ancestry, for the diagnostic structure of the family. What little is known of their seeds, when taken together with the well-developed testa of Pomeae, suggests that this structure was mesotestal with sclerotic mesophyll, if not also sclerotic exotesta. There is no specialization of any particular cell-layer such as distinguishes the Leguminous seed. This lack of definition may have favoured the unitegmic and pachychalazal tendency predominant and, evidently, of independent evolution in many tribes, even genera, of the family. The point was first explained by Péchoutre, through whose extensive and meticulous investigations the seed-structure has come to be known. It was elaborated by Juel and has recently been interpreted in the more comprehensive light of carpel-development in a long series of papers by Sterling. With progressively indehiscent fruit and basipetal growth of the ovary, leading to its inferior position, the ovule rather than the seed becomes the focus of enquiry and, possibly, no family is so varied in this respect.

In many tribes and various genera the ovule is more or less unitegmic. According to Péchoutre, the two integuments arise in different ways and, by detecting this difference, he concludes that they become connate in the unitegmic ovule or one or other integument is suppressed. The outer integument starts from hypodermal cell-divisions in the ovular primordium; the inner integument is epidermal in origin. When the two are connate, the single integument has as many cell-layers as the sum of those in the two free integuments of related species or genera, e.g. *Agrimonia, Rosa, Rubus, Sanguisorba, Spiraea pr. p.* The inner integument is suppressed in *Alchemilla, Fragaria, Geum* and *Potentilla*. The outer integument is suppressed, at least on the antiraphe-side, in *Poterium*. In other cases the ovule has the intermediate construction with two rudimentary, or vestigial, integuments at the micropyle raised on a single massive wall intercalated round the nucellus. This construction seems to show how the single thick integument of the first instance has arisen through intercalary and basipetal growth, as with all floral tubes; the pachychalazal seed is the result. The point is taken up by Sterling who finds that the more or less unitegmic state can be correlated with the absence of a sutural opening in the carpel; that is the intercalary growth makes both unitegmic ovule and tubular carpel below the original tip with free integuments or free carpellary margins. Sterling finds that the unitegmic ovule occurs in a much wider range of Pomoid and Prunoid genera than Péchoutre was able to establish. That it is 'polyphyletic' is shown by its systematic occurrence. Thus *Geum* is the unitegmic example in bitegmic Dryadeae; *Physocarpus* and *Stephanandra* are unitegmic examples in bitegmic Niellieae; many genera of Pomeae and Pruneae have both bitegmic and unitegmic ovules. The predominantly unitegmic Spiraeeae have *Aruncus* as the genus with free or fused integuments, and in Quillajeae *Exochorda* has free integuments while they are free or fused in *Quillaja*. Exactly how these differences affect the construction of the seed is not clear except in the cases which Péchoutre followed through in development.

Rosaceae divide into woody and mainly herbaceous or scandent groups. To the woody belong Chrysobalaneae (now usually treated as a family), Neuradieae, Niellieae, Osmaroneae, Pomeae, Pruneae, and Quillajeae; the seed is commonly mesotestal in some degree. To the herbaceous and scandent belong, Dryadeae, Kerrieae, Potentilleae, Poterieae, Rhodotypeae, Roseae, Rubeae, and Spiraeeae, which are unitegmic with much simplified, commonly unspecialized, seed-structure. Both groups indicate a pachycaul ancestry whence the main Chrysobalanceous, Pomaceous, Prunaceous and Rosaceous lines have evolved, and the whole concept of Rosaceae may be divided into several families of Rosales, from which Leguminosae must be excluded (p. 35). The mesotestal seed of Rosaceae fits with those of Hamamelidaceae, Theaceae and Myrtaceae, not with the precise exotestal seed of Leguminosae. *Archidendron* no more fits a Rosaceous tribe than *Chrysobalanus* a Leguminous.

Chrysobalaneae

Chrysobalanus L.
Vascular supply with postchalazal branches.
(Netolitzky 1926.)

Licania rigida Benth.
Seed-coat consisting of 5–6 layers of pigmented cells with v.b., and 3–4 inner layers of colourless cells. Endosperm 1 cell thick.
(Vaughan 1970.)

Dryadeae

Ovule solitary, erect, bitegmic or unitegmic (*Geum*). Testa with stomata in *Cowania* and *Purshia*.

Dryas octopetala L.
Ovule; o.i. 4 cells thick; i.i. 3 cells thick, with projecting endostome.
Testa with persistent o.e. and o.h., the cells of o.e. with uniformly thickened walls. Tegmen with persistent i.e. Endosperm 7–8 cells thick.

Geum urbanum L.
Ovule unitegmic with i.i. suppressed; o.i. 4 cells thick.
Testa with all cell-layers persistent, thin-walled, those of i.e. somewhat enlarged. Endosperm 1 cell thick.

Kerrieae

Ovule solitary, suspended, shortly bitegmic with a single intercalary integument covering the nucellus; o.i. 4 cells thick; i.i. 2 cells thick. Seed ?

Kerria japonica DC

Neuradeae

Ovule solitary, suspended.
Testa 3–5 cells thick, crushed except the small cells of i.e. Tegmen 8–10 cells thick; o.e. with thick walls, as a mechanical layer; *mesophyll* thin-walled; *i.e.* with small tannin-cells.
Grielum L., *Neurada* L.
(Netolitzky 1926.)
This seems to be a tegmic seed and may justify the family Neuradaceae as something quite different from Rosaceae.

Osmaronieae

Ovules 2, collateral, bitegmic, suspended.

Pomeae (Pyreae)

Ovules 2, collateral, erect, bitegmic (*Cydonia, Docynia, Pyrus*), with free integuments at the micropyle (*Cratae-gus*), bitegmic to unitegmic (*Amelanchier, Chamaemeles, Chaenomeles, Cotoneaster, Dichotomanthes, Eriobotrya, Hesperomeles, Malus, Mespilus, Osteomeles, Photinia, Pyracantha, Sorbus*); o.i. 5–14 cells thick; i.i. 3–6 cells thick.

Testa firm, brown; o.e. with thick-walled cells, often radially elongate, the inner layer of the wall suberized, the outer wall more or less mucilaginous; *mesophyll* with the outer layers thick-walled and lignified, or thin-walled (*Cotoneaster, Crataegus, Mespilus*); *i.e.* crushed. Tegmen more or less crushed except the persistent small-celled i.e. Vascular bundle of the raphe branching in the chalaza but without postchalazal extensions. Endosperm 2–15 cells thick.

As pointed out by Péchoutre, the simpler structure of the testa (5 cells thick) in *Crataegus* and its allies accompanies the thick lignified endocarp.

Amelanchier canadensis Torr. et Gray
Ovule; o.i. 8–9 cells thick; i.i. 4–5 cells thick.
Testa; *o.e.* with the cells radially elongate as a short palisade with strongly thickened outer wall; *mesophyll* with *c.* 5 layers of thick-walled cells. Tegmen with i.h. and i.e. persistent, thin-walled. Endosperm reduced to 3 cell-layers.

Chaenomeles japonica (Thunb.) Lindl.
As *Cydonia* but o.e. (testa) scarcely mucilaginous, the cells scarcely radially elongate.

Cotoneaster vulgaris Lindl.
As in *Crataegus*; testa reduced to o.e. and o.h. at maturity.

Crataegus oxyacantha L.
Ovule; o.i. 5 cells thick; i.i. 3 cells thick.
Testa not lignified; *o.e.* as short cells with thick mucilaginous outer wall, not radially elongate; *mesophyll* with the outer two cell-layers crushed, the inner two thin-walled. Tegmen crushed except i.e. with thin brown walls. Endosperm –15 cells thick.

Cydonia oblonga Mill.
Ovules many, biseriate, horizontal with the micropyle on the side towards the base of the ovary; o.i. 8–9 cells thick; i.i. 5 cells thick.
Testa; *o.e.* as a palisade of radially elongate cells with very thick mucilaginous outer wall; *mesophyll* composed of 4 layers of lignified cells, the inner layers crushed. Tegmen with the outer layers crushed and the inner two layers and i.e. persistent with thin brown walls. Endosperm 4–5 cells thick.

Eriobotrya japonica Lindl.
Seed-coat 20–30 cells thick, thin-walled, with v.b. (? overgrown or pachychalazal).
(Netolitzky 1926.)

Malus apiosa Steud.
Ovule; o.i. 11–12 cells thick; i.i. 4–5 cells thick; endostome projecting and in contact with the obturator.

Seed as in *Pyrus* but the exotesta less strongly radially elongate.

Mespilus germanica L.
Ovule and seed-coats as in *Crataegus*.

Photinia glabra Decne
Ovule; o.i. 5 cells thick; i.i. 3 cells thick.

Testa; *o.e.* with short cells, the outer mucilaginous wall strongly thickened; *mesophyll* sclerotic; *i.e.* crushed. Tegmen crushed except for i.e. Endosperm 3 cells thick.

Pyrus communis L.
Ovule; o.i. 12–14 cells thick, the outer 5–6 layers of the mesophyll having shortly longitudinally elongate cells; i.i. 5–6 cells thick; micropyle formed by the endostome.

Testa; *o.e.* as a palisade of shortly radially elongate cells with thick outer stratified wall becoming mucilaginous, not lignified; *mesophyll* composed of sclerotic cells shortly longitudinally elongate, the inner 1–2 layers crushed; *i.e.* crushed. Tegmen *o.e.* more or less crushed; *mesophyll* with 2–3 layers of enlarged cells with thin brown walls; *i.e.* with small cells, with thin brown walls. Endosperm 2 cells thick.

Sorbus aucuparia L.
Ovule; o.i. 7–8 cells thick; i.i. 4–5 cells thick.

Testa; *o.e.* becoming mucilaginous and disappearing; *mesophyll* with 4 layers of sclerotic cells, the inner layers crushed. Tegmen as in *Pyrus*. Endosperm 3 cells thick.

Potentilleae

Ovule solitary, suspended, unitegmic with i.i. suppressed. Endosperm 1 cell thick.

Fragaria vesca L.
Integument 4 cells thick. Testa with o.e. and i.e. persistent, thin-walled.

Potentilla verna L.
As *Fragaria* but the testa with only o.e. persistent.

Poterieae

Ovules solitary, suspended, unitegmic with connate integuments (*Agrimonia, Sanguisorba*), or with i.i. suppressed (*Alchemilla*) or with o.i. suppressed along the antiraphe (*Poterium*). Seed-coat not lignified. Endosperm 1 cell thick.

Agrimonia eupatoria L.
Integument 8 cells thick (4 + 4). Testa 12 cells thick, very closely pressed against the pericarp; outer 4–6 cell-layers crushed, the middle 3–4 cell-layers persistent and thin-walled, the inner 4 cell-layers crushed; *i.e.* persistent. Vascular bundle branching in the chalaza but the branches not extending over the sides of the seed.

As Péchoutre remarked, this peculiar seed is comparable with a caryopsis.

Alchemilla vulgaris L.
Ovule suborthotropous; integument 4 cells thick, representing o.i. (hypodermal in origin). Testa crushed except for the persistent o.e. with slightly cutinized outer walls.

Poterium sanguisorba L.
Integument 6 cells thick (4 + 2), but on the antiraphe-side only 2 cells thick as the i.i., the o.i. on this side suppressed as a slight bulge round the chalaza.

Seed-coat as in *Sanguisorba*, but o.e. with irregular cuticular thickening.

Sanguisorba tenuifolia L.
Integument 6 cells thick (4 + 2). Seed-coat thin, membranous, with o.e., o.h. and i.e. persistent, the intervening cells crushed.

Pruneae

Ovules 2, collateral, suspended, bitegmic with free integuments generally only at the micropyle, mostly unitegmic.

Seeds mostly pachychalazal. Testa; *o.e.* composed of rounded, enlarged, thick-walled, pitted cells, cutinized, often separated more or less by thin-walled cells or unspecialized cells; *mesophyll* and *i.e.* more or less crushed. Tegmen crushed, but i.e. persistent in some species. Vascular bundle of the raphe with 3–10 postchalazal branches reaching almost to the micropyle or variously shorter

Prunus amygdalus Batsch
Ovule; o.i. 8 cells thick; i.i. 6 cells thick. Testa; *o.e.* with the sclerotic cells projecting and papilliform, 140 × 100 μ, the wall –10 μ thick, truncated and pitted in the lower half. Tegmen with persistent i.e. Endosperm 1 cell thick.

P. armeniaca L.
Ovule; o.i. 7 cells thick; i.i. 6 cells thick. Testa; *o.e.* with sclerotic cells pitted all round, –70 × 40–50 μ. Tegmen entirely crushed. Endosperm 1 cell thick.

P. cerasus L.
Testa; *o.e.* with sclerotic cells 40 × 40–70 μ, often angular, the outer half not pitted.

P. domestica L.
Testa; *o.e.* with sclerotic cells 40–50 μ wide, the smooth outer half of the cell-wall thicker than the pitted inner half; *o.h.* also with sclerotic cells.

P. javanica (T. et B.) Miq.

(Borneo, Kinabalu, RSNB 4251; ripe fruits in alcohol.)

Testa *c.* 1 mm thick, with 26–30 layers of cells, thin-walled, unspecialized except for the peripheral v.b. and thin bands of cells with brown contents in the outer part of the tissue adjoining the o.e. in places and forming most of the testa round the micropyle; *o.e.* composed of small cells with slightly thickened walls, not lignified, the facets isodiametric, without stomata; *mesophyll* with the inner colourless tissue becoming crushed; *i.e.* unspecialized, crushed. Tegmen becoming crushed. Vascular bundle of the raphe stout, dividing progressively at the chalaza into 5–6 branches themselves dividing over the sides of the seed to give *c.* 20 v.b., with three reaching almost to the micropyle.

Possibly a pachychalazal seed.

P. juliana DC

Ovule with short o.i. (6–8 cells thick) and short i.i. (3–4 cells thick). Seed evidently pachychalazal, without tegmen. Endosperm of varying thickness –15 cells.

P. persica Batsch

Ovule; o.i. 6 cells thick; i.i. 3 cells thick. Testa; *o.e.* with sclerotic cells 80–160 × 90 μ, set in rows, the smooth outer part of the wall 12 μ thick, the pitted inner half of the wall 8 μ thick. Tegmen with persistent i.e. Endosperm 1 cell thick.

P. spinosa L.

Ovule with short o.i. (5 cells thick) and short i.i. (3 cells thick). Seed evidently pachychalazal. Endosperm of varying thickness, –12 cells thick.

Quillajeae

Ovules numerous and biseriate, or 2 suspended and collateral, or erect, bitegmic (*Exochorda*) or bitegmic and unitegmic (*Quillaja*).

Seeds more or less winged or apiculate.

Exochorda grandiflora Lindl.

Testa 5 cells thick, thin-walled; *i.e.* crushed. Tegmen 4 cells thick, becoming crushed except i.e. Endosperm 1 cell thick.

Qillaja saponaria Molina

Ovules with both integuments several cells thick. Testa with the three outer layers of cells somewhat thick-walled, the inner layers crushed. Tegmen crushed. Endosperm 1 cell thick.

Rhodotypeae

Rhodotypos kerrioides Sieb. et Zucc.

Ovules 2, collateral, suspended, bitegmic with o.i. and i.i. both 4 cells thick, very soon becoming unitegmic

(Péchoutre), entirely unitegmic (Juel); nucellus projecting strongly.

Seed-coat; *o.e.* as a layer of thick-walled mucilage-cells; *mesophyll* crushed; *i.e.* persistent, thin-walled. Endosperm 18–20 cells thick.

Roseae

Ovules solitary, suspended, unitegmic with connate integuments.

Rosa myriacantha DC

Integument 8 cells thick (4 + 4). Seed-coat with persistent, thin-walled o.e. and i.e., the mesophyll crushed. Endosperm 1 cell thick.

Rubeae

Ovules 2, collateral, suspended, unitegmic with connate integuments.

Rubus fruticosus L., *R. idaeus* L.

Integument 6 cells thick (2 + 4). Seed-coat with o.e., o.h., i.h. and i.e. persistent, thin-walled, the middle layer crushed. Endosperm 6 cells thick.

(Topham 1970.)

Spiraeeae

Ovules numerous, biseriate, transverse with the micropyle towards the apex of the ovary, or 2 collateral and suspended, mostly unitegmic, bitegmic and unitegmic in *Aruncus*; o.i. 3 cells thick; i.i. 2 cells thick; single integument 5 cells thick.

Testa; *o.e.* with enlarged cells, the outer walls thickened and cutinized; *mesophyll* and *i.e.* crushed. Endosperm 1 cell thick.

RUBIACEAE

Ovules anatropous, unitegmic, crassinucellate to tenuinucellate, ategmic in *Houstonia*; integument massive, 10–14 cells thick (*Rubia*), 7–9 cells (*Hydrophylax*), several cells thick (*Galium*), or merely 1–2 cells thick (*Callipeltis, Dentella, Oldenlandia, Sherardia, Vaillantia*).

Seeds fairly large, small or minute, round, ellipsoid, reniform, compressed or alate, albuminous, exarillate or strophiolate (Spermacoceae), covered by placental outgrowths (*Gardenia, Randia*), or with slight outgrowths (*Tarenna*). Seed-coat ? not multiplicative, generally reduced to o.e. through the enlargement of the endosperm, or with a few layers of unspecialized mesophyll-cells persisting (thick-walled in *Hydrophylax*), becoming pachychalazal in some cases with ruminate endosperm (*Psychotria, Randia, Tarenna*); *o.e.* thin-walled or with variously thickened and lignified walls, the walls thickened all round, or only the outer wall thickened, or the inner and radial walls thickened (*Gardenia, Randia*), with irregular gyrose or bar-like thickenings (*Guet-*

tarda, *Randia*), with reticulate thickening (*Cinchona*), or with longitudinally elongate sclerotic cells (*Coffea*); *mesophyll* developing ruminations by anticlinal divisions of o.e. folding the tissue (*Psychotria, Randia, Tarenna*). Vascular supply simple in the raphe only, or absent (*Cinchona*), accompanied by raphids (Galieae), or the v.b. extended towards the micropyle in pachychalazal seeds, or with branches to the sides of the seed from the chalaza and raphe (*Coffea*). Endosperm nuclear, but cellular in *Ophiorrhiza* and *Tarenna*, with thin or thick walls, oily but in some cases at first starchy before the thickening of the cell-walls. Embryo minute to relatively large.

(Netolitzky 1926; Kühn 1927; Fagerlind 1937; Mendes 1941; Raghavan and Rangaswamy 1941; Raghavan and Srinivasa 1941; Ganapathy 1956; Coolhaas, de Fluiter und Koenig 1960; Farooq 1960; Periasamy and Parameswaran 1962, 1965; Periasamy 1964b; Singh, B. 1964; Siddiqui and Siddiqui 1968a, b; Hayden and Dwyer 1969; Vaughan 1970.)

In this large family there is need of extensive investigation into the generic and specific variations of the integumental thickness, the exotestal structure, the vascular supply, the pachychalazal development with ruminations, and the placental outgrowths which may render the seeds brilliantly arilloid. The development of the nucellus has been studied in detail by Fagerlind.

RUTACEAE

Ovules anatropous, suspended, bitegmic, crassinucellate; o.i. 3–6 cells thick; i.i. 2–5 cells thick.

Seeds small to fairly large, dry, thinly pulpy, or slimy, some hairy, mostly exarillate, a few arillate (*Eriostemon* spp., *Pilocarpus, Xanthoxylum* spp.), albuminous or not. Testa more or less multiplicative; *o.e.* generally as a palisade of cells but also composed of fibrous, longitudinally elongate cells (*Citrus* and allied Aurantioideae); *mesophyll* aerenchymatous, in some genera becoming sclerotic or with scattered stone-cells or fibres; *i.e.* unspecialized. Tegmen generally not or scarcely multiplicative, but distinctly so in *Dictamnus*, typically with o.e. and, in some cases, i.e. as short tracheids with spiral or annular lignification, in a few cases also the middle layers, or undifferentiated. Endosperm nuclear, oily. Embryo various, straight or curved; cotyledons thin or thick, folded or not.

Aril funicular (? *Xanthoxylum*). Figs. 479–499.

(Netolitzky 1926; Mauritzon 1935c; Singh, B. 1964.)

As in other large and varied families, knowledge of seed-structure is tantalisingly inadequate. Rutaceous characters are the predominance of the testa as the mechanical layer, both exotestal and mesotestal, the lack of differentiation in the endotesta, and especially the tracheidal construction of the tegmen. There appear to be five main kinds of seed-structure, with several lesser

variations. The differences are mostly testal and range from the elaborate structure in *Xanthoxylum* and *Limonia* to the simple fibrous exotesta of *Citrus* and the undifferentiated state of *Glycosmis*. Such diversity shows how much there is to be learnt concerning generic and specific alliances from the seed-coat. The first theoretical problem is to discover the primitive construction. Derivatives may then be formulated and compared with those of other families.

The problem is introduced by the fruit. It is in general either capsular (Rutoideae) or baccate (Aurantioideae, Toddalioideae). In both the seeds may be reduced to one. The more primitive seed-structure may, therefore, be found among many-seeded fruits. In Aurantioideae, such would be the alliance of *Limonia* (group 3 in the following classification). In Rutoideae, represented mainly by group 1 in this classification, there is more uniformity and there occur such primitive marks as the apocarpous ovary, arillate vestiges, and arillate derivatives as the winged seed (*Terminthodia*), and of course the dehiscent fruit. From what is known of these, the black seeds of *Xanthoxylum* offer the most generalized construction. This selection means that the primitive Rutaceous seed had a sclerotic or thick-walled mesotesta covered by a palisade-like exotesta, and a thick, though transient, tegmen of tracheidal cells. Such seeds are albuminous and have a simple embryo. Derived seeds become exalbuminous and the embryo may be variously complicated with folded cotyledons (Cusparieae, Aurantioideae, *Micromelum*). *Aegle* and *Limonia*, with elaborate compound hairs, are intermediate to the generally fibrous seed-coat of Aurantioideae.

The primitive Rutaceous fruit appears in durian-form as an arillate capsule or set of follicles with glandular and spicate pericarp. The excrescences that occur on many fruits, as *Calodendrum, Citrus, Esenbeckia, Metrodorea* and *Toddaliopsis*, as well as the complex hairs of *Dictamnus*, indicate that the primitive pericarp was more elaborate than now. Loss of tracheidal tegmen is found in *Cusparia* where tracheids occur only at the chalazal end and in *Calodendron* with slight spiral thickening. *Cassimiroa* and *Triphasia* develop a more or less fibrous mesotesta, suggestive of the Annonaceous. It will be interesting to learn in this light the seed-structure of *Flindersia* which has passed from Rutaceae to Meliaceae. *Chlorozylon*, transferred from Meliaceae, has an alate seed similar to that of *Cedrela* and equally undifferentiated, but it may have a trace of the Meliaceous fibrous exotegmen; its position on grounds of seed-structure is not yet clear.

Classification of Rutaceous seed-coats

(1) Mesotesta thick, hard, with isodiametric, often thick-walled, pitted cells with dark contents, lignified or not; exotesta not mucilaginous. Tegmen with lignified spiral thickening on all cells or those of o.e. and i.e. or only one

of these layers. Chalaza often (? always) broad and conspicuous. Seeds often (? always) slightly curved with short raphe.

(a) Exotesta with the cells radially elongate, at least in groups, as thin-walled palisade-cells, giving the pitted mesotesta ... *Fagara, Xanthoxylum*

(b) Exotesta with thick-walled palisade cells ... *Calodendrum*

(c) Exotesta composed of cuboid cells. Testa lignified ...
Evodia, Pilocarpus, Toddalia, ? Cusparia
(All these are variations of the primitive construction)

(2) Testa fibrous throughout or with fibrous exotestal cells. Tegmen not lignified, soon crushed.

(a) Testa fibrous throughout; outer fibres longitudinal, inner transverse ... *Casimiroa*

(b) Exotesta fibrous, with mucilaginous outer wall, or with a few mesophyll fibres ... *Atalantia, Citrus, Murraya exotica, Triphasia*

In this case cell-division in the testa is mainly to produce new cell-walls in line with the long axis of the ovule. *Casimiroa* seems to be the mesotestal state which, through *Triphasia* with few mesotestal fibres, becomes limited to the exotesta as in *Citrus*.

(3) As (2) but the exotestal cells only shortly elongate tangentially or substellate, developing compound mucilage-hairs; mesotesta with scattered fibres ... *Aegle, Limonia*

(4) Exotesta as a palisade, not or scarcely lignified; mesotesta unspecialized, thin-walled.

(a) Palisade-cells conical, projecting. Tegmen unspecialized. Seed slightly curved, albuminous ... *Boeninghausenia, Ptelea, Ruta*

(b) Palisade-cells in an uneven layer, the outer part of the cells very thick-walled. Tegmen with tracheidal o.e. and i.e. Seed nearly straight, albuminous ... *Dictamnus*

(c) Palisade-cells much elongate with mucilaginous outer wall. Tegmen unspecialized. Seed straight, exalbuminous ... *Murraya glenei*

This is a heterogenous group, part capsular and part baccate (*Murraya*), in which the family-characters of thick-walled mesophyll and tracheidal tegmen are disappearing. *Ptelea* is peculiar in the long funicle and the absence of a raphe or, that is, in the free body of the ovule. The four genera belong in different alliances and all tend to the following unspecialized state.

(5) Testa undifferentiated, not or scarcely lignified. Tegmen unspecialized. ? *Chloroxylon, Glycosmis*

The lack of differentiation seems to have lead to postchalazal branching of the vascular bundle in *Glycosmis*. Though classified near to *Triphasia*, it is strikingly different in seed-coat.

Aegle Correa

A. marmelos Correa
Ovule; o.i. 4–5 cells thick; i.i. 3–4 cells thick.
Testa 8–12 cells thick, all the cells tangentially elongate; o.e. composed of shortly longitudinally elongate, pitted, lignified cells, often with undulate facets,

developing fascicles of closely septate hairs (as in *Limonia*); *mesophyll* thin-walled, the outermost layer with a crystal in each cell, 2–3 inner layers and i.e. with a crystal in each cell and somewhat thickened lignified walls. Tegmen ? thickening, crushed except i.e. as a layer of small tannin-cells.
(Johri and Ahuja 1957.)

Atalantia Correa

A. monophylla DC
(Ceylon, leg. E. J. H. Corner, 1969; fruits in alcohol.) Fig. 479.
Ovule anatropous, erect.
Seed *c.* 11 × 5–6 mm, the hard pale shell with mucilaginous surface, exalbuminous. Testa 12–20 cells thick, or more, finally 5–7 cells thick through crushing of the inner tissue, thin-walled, aerenchymatous, with numerous small crystal-cells scattered through the tissue; o.e. with elongate, longitudinal or oblique, thick-walled, lignified, finely pitted fibriform cells, the outer wall strongly mucilaginous, at first forming an uneven rugose surface but stretched and flattened on development of the embryo; *i.e.* unspecialized, crushed. Tegmen 3–5 cells thick, soon crushed. Nucellus massive, slowly invaded by the endosperm. Chalaza at first small, then expanding greatly to make most of the upper half of the seed, with a continuous thin cupular network of short tracheids; without v.b. on the sides of the seed. Embryo straight; radicle short; cotyledons thick.

This is clearly a pachychalazal seed and, perhaps, there are many more such kinds in the family.

Boeninghausenia Rchb.

B. albiflora (Hook.) Rchb.
(Java, Tjibodas; living material.)
Seeds essentially as in *Ruta*. Testa 3–4 cells thick; o.e. with brown, slightly lignified outer walls, the surface rough with minute particles (? crystals), the cells protruding in irregular groups, some even with a short process; *mesophyll* thin-walled, more or less crushed, not lignified; *i.e.* unspecialized. Tegmen crushed.

Calodendrum Thunb.

C. capense Thunb.
(Kenya, leg. D. J. Mabberley; alcohol material.)
Ovules anatropous, superposed, 2 per loculus, more or less transverse; o.i. 7–9 cells thick; i.i. 2–3 cells thick; nucellus massive.
Seeds 23 × 16 mm, black, ellipsoid subtriquetrous, exarillate. Testa 0.6–1 mm thick, massive, very hard, yet scarcely lignified, composed of many cell-layers; o.e. as a palisade *c.* 110 μ high, of cylindric cells with firm brown thickened walls, as a compact dry layer; *mesophyll* with thick-walled cells, brown, but the narrow inner layer (carrying v.b.) colourless, thin-walled and more or

less crushed; *i.e.*? Tegmen reduced to a firm pellicle with slight spiral lignification in i.e.

Ovary minutely papillate with short processes tipped by a minute, globose, multicellular gland and containing a large oil-space; carpels free at the apex.

Fruit muriculate-verrucose with fibrous v.b. radiating into the processes.

The fruit recalls that of *Flindersia*, now transferred to Meliaceae.

Chloroxylon DC

C. swietenia DC

(Ceylon; mature seeds, dried.)

Testa 5–7 cells thick; *o.e.* as a layer of short cells with subrectangular or shortly oblong facets, with the outer wall slightly thickened, not as a palisade, without stomata; *mesophyll* with slightly thickened, shallowly pitted, thinly lignified walls, aerenchymatous, extending into the wing; *i.e.* composed of small, thinly lignified cells, many with crystalloid masses (not sharply angular crystals). Tegmen crushed, evidently few cells thick; *o.e.* with the cells much elongate longitudinally and, in parts of the seed, with slightly thickened and pitted walls, not lignified, not tracheidal. Vascular bundle very short, from the placenta to the chalaza at the base of the seed. Nucellus as a thin layer of crushed cells. Endosperm reduced to a single layer of cells. Embryo straight; radicle very short; cotyledons thick.

This genus has been transferred from Meliaceae to Rutaceae but, from the seed-structure, it is Meliaceous and similar to *Cedrela*. My material was inadequate yet it showed clearly vestigial exotegmic fibres.

Citrus L.

Ovules anatropous; o.i. 5–6 cells thick; i.i. 2–5 cells thick.

Seeds 8–15 mm long, pallid ochraceous or whitish, slimy, pyriform, often with a raphe-ridge and micropylar beak, exarillate. Testa not or slightly multiplicative; *o.e.* composed of longitudinal, thick-walled, pitted and lignified fibres 0.5–2 mm long, with mucilaginous external wall and irregular lobulate processes from the cells into the mucilage, the fibres appearing as a palisade in t.s.; *mesophyll* aerenchymatous, pale, not lignified; *i.e.* unspecialized or as tannin-cells, the i.h. often with crystals. Tegmen not multiplicative or –10 cells thick near the chalaza, thin-walled, unspecialized. Hypostase often extensive over the chalazal end of the seed, composed of small brown cells. Vascular bundle in the raphe stout, dividing into a fine network in the chalaza. Endosperm very thin or none. Embryo straight, with thick cotyledons, often polyembryonous. Fig. 480.

(Netolitzky 1926; Vaughan 1970.)

Owing to the difficulty in specific and varietal identification, it is not easy to list specific characters for the seeds of the cultivated species. Vaughan gives the follow-ing details for the height of the fibres as seen in t.s.: 80–160 μ in *C. aurantia* L., *C. aurantifolia* Swingle, *C. paradisi* Macf., *C. reticulata* Blanco, *C. sinensis* Osbeck; 130–218 μ high in *C. limon* Burm.f.

Cusparia Humb.

Cusparia sp.

(Brazil, Manaus, Corner 133; flowers and fruits in alcohol.) Figs. 481, 482.

Ovules 2 per carpel, superposed, anatropous; o.i. 3–4 cells thick; i.i. 2–3 cells thick; nucellus large.

Seeds 11 × 5.5 mm, brown, smooth, oblong, round at the distal end, attenuate to the proximal end, with thin leathery seed-coat, exalbuminous; hilum lateral. Testa 8–10 cells thick near the chalaza, 3–5 cells over the greater part of the seed, aerenchymatous, not lignified; *o.e.* as a compact layer of short cells, the anticlinal walls with fine irregular thickenings at the outer ends on to the outer walls; *i.e.* unspecialized, the cells becoming isolated. Tegmen 2–3 cells thick, all the layers in the chalazal part of the seed developing spiral and irregularly scalariform lignified thickenings on the walls, not in the micropylar part with thin testa, eventually crushed except round the chalaza. Nucellus crushed except near the chalaza. Embryo curved; cotyledons folded and covering the long radicle.

Ovary with 5 carpels, connate in the style and at the base, the style caducous after flowering, the ovary-wall with simple or bicornuate appressed brown hairs.

Fruit consisting of 5 radiating follicles, each with bivalved sclerotic endocarp, with *c.* 40 sclerotic v.b. and very small, apparently simple, oil-glands.

The seeds seem to have vestigial periclinal division in the endotesta.

Dictamnus L.

D. albus L.

(Cambridge, University Botanic Garden; living material.) Fig. 483.

Ovules 3–4 per carpel; o.i. 4–5 cells thick; i.i. 4–5(–6) cells thick.

Seeds 5 mm long, ovoid, black, with prominent raphe, straight, albuminous. Testa 6–9 cells thick; *o.e.* as a short palisade of prismatic cells with very thick radial and outer walls, becoming brown with slight, if any, lignification; *mesophyll* and *i.e.* unspecialized, not lignified; *i.e.* unspecialized, not lignified. Tegmen 8–11 cells thick, eventually crushed except o.e. and i.e.; *o.e.* as short longitudinal tracheids with narrow oblique pitting, the wall slightly thickened and lignified; *mesophyll* with large thin-walled cells, aerenchymatous, collapsing; *i.e.* as a short palisade of prismatic cells with oblique spiral or annular thickening, slightly thick-walled, lignified. Nucellus at first large, then crushed. Vascular bundle of the raphe spreading over the wide but unspecialized hypostase. Embryo straight or the short radicle slightly curved; cotyledons thick.

Glycosmis Correa

G. pentaphylla (Retz.) Correa

(Ceylon, Peradeniya; fruits in alcohol.) Fig. 484.

Seeds 7 × 5 mm, ovoid, brown, smooth, sessile. Testa 9–15 cells thick, aerenchymatous, thin-walled, not lignified, the cells filled with brown contents; o.e. composed of cuboid cells; i.e. unspecialized. Tegmen obliterated. Vascular bundle of the raphe with c. 6 postchalazal branches extending half to three-quarters the length of the seed, dividing once or twice and anastomosing, also with 1–2 branches from the raphe near the chalaza. Endosperm none. Embryo with short radicle, thick straight cotyledons bifid round the radicle, and hairy plumule.

Limonia L.

L. acidissima L.

(Ceylon; fruits in alcohol.) Figs. 485, 486.

Seeds 5–6 × 2.5 mm, hairy, slimy, embedded in the swollen placentas. Testa 5–7 cells thick; o.e. composed of shortly longitudinally elongate, thick-walled, lignified, stellate-undulate cells with irregular short processes or bosses to the exterior, many of these cells elongating into fascicles of hairs; o.h. with many cells containing large crystals; mesophyll thin-walled, with scattered thick-walled, pitted, and lignified stone-cells and scattered, lignified, longitudinal fibres, both the sclerotic cells and the fibres in the outer part of the mesophyll but not as continuous layers, the fibres often shortly lobate; i.e. composed of cuboid cells with firm, slightly lignified walls and very faint radial striation on the radial walls; hairs as individual outgrowths of epidermal cells fascicled into unbranched compound hairs –1.5 mm long, penetrating between the enlarged placental cells and covered with mucilage, the individual hairs mostly aseptate, often with short irregular lobes, with thick lignified walls, thinner and unlignified distally. Tegmen 2–4 cells thick, thin-walled, becoming crushed. Chalaza simple, not expanded. Endosperm more or less completely absorbed. Embryo straight; radicle short; cotyledons thick.

Murraya L.

The two species that I have been able to study have remarkably different seed-coats. M. exotica agrees with Citrus. M. glenei has the palisade-construction of Rutoideae, but if its palisade-cells became oblique and longitudinal, there would result the fibriform exotesta.

M. exotica L.

(Java, Bogor, cult.; living material.)

Testa 6–8 cells thick; o.e. composed of longitudinal, thick-walled, pitted, fibriform cells, not lignified, with many fascicles of similar cells excrescent as compound hairs; mesophyll thin-walled. Tegmen thin-walled, crushed. Endosperm reduced to a single layer of cells.

M. glenei Thw.

(Ceylon; flowers and fruits in alcohol.) Figs. 487–489.

Ovules 2–3 per loculus, anatropous with short funicle; o.i. 5–6 cells thick; i.i. 3 cells thick; hypostase as a plate of thicker-walled cells.

Seeds 4.5 × 3 mm, with a subgelatinous rugulose pellicle and short white funicle, exarillate, thinly albuminous. Testa 16–21 cells thick, the inner layers becoming crushed; o.e. with the cells elongate radially to form an uneven palisade of straight or flexuous cells with slightly thickened, slightly lignified walls, obliquely and sparsely pitted, the outer wall becoming gelatinous to form the mucilaginous pellicle, continuous round the seed except at the hilum and micropyle; mesophyll composed of isodiametric cells becoming slightly lignified; i.e. and 1–2 inner hypodermal layers as small crystal-cells, thin-walled, not lignified. Tegmen –9 cells thick near the chalaza but mostly not thickening, becoming crushed, thin-walled, the cells at first elongating transversely. Chalaza with a thin brown disc as the hypostase. Vascular bundle ending at the chalaza. Endosperm 2–3 cells thick, thin-walled, oily. Embryo straight; radicle short; cotyledons massive, thick.

Fruit c. 25 × 20 mm, 3–5 shouldered, 3–5 locular, 0–1(–2) seeds in each loculus. Pericarp oily-glandular, thin-walled ground-tissue with a fine network of v.b. between the larger oil-glands; endocarp proliferating minute papillae and warts on the outer side of the loculi, eventually i.e. as a layer of sclerotic cells.

Pilocarpus Vahl

Ovules anatropous, 2 per loculus, suspended, often oblique; o.i. 4–5 cells thick; i.i. 2–3 cells thick.

Seed, black, ovoid, slightly compressed and bean-shaped, exarillate or with vestigial aril (P. racemosus), exalbuminous. Testa not or slightly multiplicative; o.e. as a layer of cuboid cells with much thickened, brown, outer wall, not lignified; mesophyll aerenchymatous, the cells with globose body and short arms, the outer layers with brown contents, the walls not or slightly thickened, ? sclerotic and lignified in some species; i.e. unspecialized or with slightly thickened inner walls, not lignified. Tegmen not multiplicative; o.e. as a layer of short tangential tracheids, variously orientated but radiating at the endostome, with spiral or annular lignified thickening; inner two layers unspecialized thin-walled, eventually the middle layer of cells often much crushed. Chalaza extensive on the adaxial side of the seed. Vascular bundle of the raphe dividing into 6–8 branches at the chalaza but not extending beyond.

(Netolitzky 1926, for P. pinnatifolius and P. trachylophus.)

P. racemosus Vahl

(Ceylon, Peradeniya, cult.; fruits in alcohol.) Figs. 490, 491.

Seeds 7×4.5–5×3.5–4.5 mm; funicle short, slightly dilated over the micropyle with a thin flange of rather horny tissue as a vestigial aril. Testa 7–8 cells thick, slightly multiplicative.

Aril with tissue similar to that of the funicle and the surrounding unlignified part of the endocarp, the epidermal cells often with a small papilla.

Fruit consisting of one fertile carpel, the other 3–4 undeveloped; endocarp hard, lignified, in two valves, but not lignified over the axile part of the loculus.

A similar funicular aril is reported in *Eriostemon* (Wilson 1970).

Ptelea L.

P. trifoliata L.

(England, cult.; living material.) Fig. 492.

Ovules 2 per loculus, superposed, suspended, with long funicle; o.i. 5–6 cells thick, both o.e. and o.h. demarcated with slightly opaque walls; i.i. 4(–5) cells thick; micropyle formed by the exostome.

Seeds 5×2.5 mm, black, ovoid with slightly curved, pale micropyle, with a slight longitudinal ridge towards the chalaza formed by the adnate funicle, exarillate, albuminous. Testa 7–9 cells thick, but the inner layers crushed; o.e. as a palisade of cells with thick brown radial walls and the outer wall generally extended into a short obtuse thick-walled papilla, forming a brittle carbonaceous layer, not lignified; o.h. persistent as rounded cells with brownish walls; *mesophyll* and *i.e.* unspecialized, thin-walled, crushed. Tegmen not multiplicative, becoming crushed, unspecialized, but the cells of i.e. enlarging in the immature seed, not lignified. Chalaza short, with thin hypostase. Vascular bundle ending at the chalaza. Funicle decurved, closely applied to the base of the seed, covered with the same papillate and carbonaceous epidermis as the testa.

Ruta L.

R. graveolens L.

(Cambridge, University Botanic Garden; living material.) Fig. 493.

Ovules anatropous, several per carpel, suspended; o.i. 3–4 cells thick; i.i. 2–3 cells thick.

Seeds 3×1.3 mm, slightly curved, dark brown to blackish, verruculose, exarillate, albuminous. Testa 7–9 cells thick, no part lignified; o.e. forming a palisade of obtusely conical cells with thick brown outer walls, becoming aggregated in small groups to form the warts, covered with a thin subgelatinous pellicle; *mesophyll* crushed except for the outer 1–2 cell-layers, unspecialized but the outer cells with granules staining with saffranin as in the cells of o.e.; *i.e.* unspecialized. Tegmen not multiplicative but the cells enlarging, soon

crushed, not lignified (but a layer of cells with finely reticulate walls, Netolitzky). Vascular bundle ending at the simple chalaza. Endosperm oily. Embryo slightly curved; radicle well-developed; cotyledons ligulate.

Toddalia Juss.

T. asiatica (L.) Lamk.

(Ceylon; flowers and fruits in alcohol.) Figs. 494, 495.

Ovules 2 per loculus, anatropous, suspended, with very short funicle; o.i. 4 cells thick; i.i. 2 cells thick.

Seeds $5 \times 3 \times 2.5$ mm, subcampylotropous, brown then black, smooth, 1–5 per fruit, 1 per loculus, exarillate. Testa 20–26 cells thick, in three layers; o.e. as a palisade of short thin-walled cells, not lignified; *outer mesophyll* 10–12 cells thick, composed of thick-walled, coarsely pitted, lignified, contiguous and angular cells; *inner mesophyll* 10–16 cells thick with the cells more or less in radial rows, somewhat thick-walled, not lignified, unspecialized, aerenchymatous; *i.e.* unspecialized. Tegmen 2–3 cells thick, thin-walled but with fine crossed spiral lignification; o.e. composed of tangentially elongate cells; *middle layer* and *i.e.* composed of rather large, shortly radially elongate cells. Vascular bundle of the raphe often curved over the chalaza, with a small vascular plexus round the chalaza and two very short branches to the antiraphe from the chalazal flexure. Endosperm ? (fully mature seeds unavailable).

Triphasia Lour.

T. aurantiola Lour.

(Ceylon, Peradeniya, cult.; flowers and fruits in alcohol.) Figs. 496, 497.

Ovules anatropous, suspended, solitary, sessile; o.e. 4–5 cells thick; i.e. 2–3 cells thick; micropyle formed by the exostome.

Seeds 6 mm wide, subglobose, whitish, with very thin scarious-pellicular testa; funicle very short; exalbuminous. Testa 7–10 cells thick, most of the tissue crushed at maturity except o.e. and some mesotestal fibres; o.e. as a layer of fibres 250–600×11–$16\,\mu$, subcylindric, longitudinal, with fusiform interlocking ends, thick-walled, pitted, lignified; *mesophyll* with larger shorter fibres, transverse or oblique developed from some cells especially round the chalaza, not as a compact or continuous layer, few or none on the sides of the seed, the intervening tissue thin-walled but becoming slightly lignified; *i.e.* unspecialized. Tegmen 3–4 cells thick, thin-walled, not lignified, soon collapsing; o.e. with short radially elongate cells; *mesophyll* with tangentially elongate cells; *i.e.* unspecialized. Vascular bundle ending at the chalaza with a slightly lignified hypostase. Embryo with very short radicle and thick cotyledons.

Xanthoxylum L.

X. simulans Hance
(Cambridge, University Botanic Garden; living material.) Figs. 498, 499.

Ovules anatropous, 2 per carpel, collateral, suspended; o.i. 3 cells thick, the cells of o.e. large; i.i. 2 cells thick.

Seeds 4.5 × 4 mm, ovoid black, dry, 1 per follicle, with small hilum, exarillate. Testa 11–20 cells thick depending on the extent of the exotestal palisade, no part lignified; *o.e.* as a palisade of prismatic cells with dark walls and cuticle, thin-walled at the base and separated, short in places, elongate in others to give the pitting on the dried seed; *outer mesophyll* 3–10 cells thick, composed of thick-walled, pitted, contiguous, more or less isodiametric cells with dark brown contents; *inner mesophyll* 7–9 cells thick, composed of cells with less thickened, pitted walls, with very dark brown contents, developed mostly from i.e. and in radial rows, eventually much compressed; *i.e.* unspecialized. Tegmen composed of 2–3 layers of enlarged cells, all with fine spiral lignified thickening, the spirals often crossed, eventually crushed. Vascular bundle ending at the chalaza with slight hypostase. Embryos 1–2 per seed.

SALICACEAE

Ovules anatropous, numerous, bitegmic with short i.i. (*Populus spp.*) or unitegmic (*Populus spp.*, *Salix*), crassinucellate; o.i. 4–6 cells thick; i.i. 2–3 cells thick; integument in *Salix* 3 cells thick.

Seeds minute, plumose at the base from placental hairs, without funicle. Testa not multiplicative, reduced to o.e. as a compact layer with thickened outer and radial walls. Hairs aseptate. Endosperm nuclear, oily, much reduced or absent from the seed. Embryo straight.

(Netolitzky 1926; Takeda 1936; Hjelmquist 1948.)

The small plumose seeds have the same structure as those of Tamaricales except that the hairs arise from the placenta instead of the testa itself. Affinity with exotegmic Flacourtiaceae and Violaceae is ruled out (p. 43). Alliance with other Amentiferae of Rosalean or Hamamelidalean descent is now generally discredited. In *Populus* the inner integument is reduced or absent and, thus, the single integument of *Salix*, which forms a seed-coat like that of *Populus*, must be the outer integument.

SALVADORACEAE

Ovules anatropous, erect, bitegmic, crassinucellate, solitary at the base of the loculus.

Seeds exalbuminous, exarillate. Endosperm nuclear (*Salvadora*).

(Maheshwari Devi 1969.)

This family is generally referred to Celastrales but it appears to lack the Celastraceous fibrous exotegmen.

The seed of *Azima* is merely exotestal without lignification.

Azima Lamk.

A. tetracantha Lamk.
(Ceylon; flowers and fruits in alcohol.) Fig. 500.

Ovary bilocular. Ovule solitary, erect; o.i. *c.* 10 cells thick, with o.e. as a short palisade, the other cells very small; i.i. 3–5 cells thick; micropyle formed by the exostome.

Seeds 5 × 2.2 mm, compressed in the plane of the raphe, brownish, submucilaginous. Testa 9–13 cells thick; *o.e.* as a palisade 70–90 μ high, composed of radially elongate cells with hexagonal facets, with thin cuticle and distinctly thickened outer wall, the radial and inner walls less thickened but becoming mucilaginous, with minute crystals at the outer ends of the radial walls; *mesophyll* composed of small unspecialized cells, slightly lignified at the hilum and round the micropyle. Tegmen joined to the raphe throughout its length, not multiplicative, unspecialized but the exotegmic cells longitudinally elongate (not as fibres), eventually crushed. Vascular supply only in the raphe. Embryo straight, the well-formed radicle concealed by basal lobes of the thick flat cotyledons.

SANTALALES

Ovules anatropous to more or less orthotropous or practically absent (Loranthaceae), unitegmic (Grubbiaceae, Santalaceae *pr. p.*), or ategmic (Myzodendraceae, Santalaceae *pr. p.*), tenuinucellate.

Seeds minute to medium-size, enclosed in endocarp with the seed-coat crushed or absent, but persistent in Grubbiaceae. Endosperm cellular or helobial (*Mida*, *Santalum*), persistent, even chlorophyllous (Viscoideae). Embryo relatively large to microscopic (Balanophoraceae).

(Netolitzky 1926; Fagerlind 1947, 1959; Paliwal 1956; Bhatnagar and Agarwal 1961; Johri 1967; Bhatnagar and Sabharwal 1969; Raj 1970; Geesink 1972; Hansen 1972.)

I have treated these families of Santalales together because they have practically no seed-coat. Pteridosperms and gymnosperms reveal the loss of the gametophytes which has led to the pollen-tube and the embryo-sac retained in the integumented megasporangium or nucellus. By analogy a similar failure and transformation of the gametophytes is supposed to have occurred in the bitegmic ancestry of angiosperms. Olacales and Santalales show the deterioration of this angiosperm heritage until the egg-cells appear in an undifferentiated placenta to form ategmic and anucellar endosperm and embryo. It is an astounding story. Flower and fruit remain but the seed goes. The endosperm persists when it fails in so many other advanced orders of flowering plants, and unimpeded by integuments or hypostase it extends with bizarre haustoria and becomes, even, chlorophyllous (*Viscum*). Here is proof of the angiosperm theory of reproduction. *Balanophora* has the minute embryo of *Magnolia*, though they are at the opposite ends of dicotyledonous classification.

Fagerlind (1947) would refer Grubbiaceae to Ericales. Possibly Olacales and Santalales are connected with Ericales from Thealean ancestry.

SAPINDACEAE

Ovules anatropous or campylotropous (at least after fertilization), bitegmic, crassinucellate, mostly solitary and erect, in some genera 2 per loculus with the upper ovule suspended or erect, a few genera with several ovules per loculus (*Magonia*, *Xanthoceras*) and then the ovules transverse with the upper tending to be suspended and the lower erect; funicle very short, thick, or none; o.i. 4–12 cells thick; i.i. 2–8 cells thick, generally thinner than o.i.; micropyle formed by the exostome or endostome; obturator frequent as a rounded placental or funicular outgrowth over or under the micropyle.

Seeds medium-size to large, rounded or ellipsoid, generally brown or black, alate in *Magonia*, mostly with crustaceous testa, arillate, sarcotestal or without such features, exalbuminous; developing after fertilization, through campylotropous growth, a more or less pronounced partial septum between the micropyle and chalaza and serving as the radicle-pocket (? in all Nephelieae); funicle short or practically none. Testa multiplicative, 8–70 cells thick, usually very much thicker than the tegmen, usually not lignified; o.e. typically as a palisade of columnar cells with thick brown walls, 30–900 μ high, but the cells stellate–prismatic in *Cardiospermum*, *Paullinia* and *Urvillea*, flattened with subundulate facets in *Allophylus*, *Appania* and *Magonia*, in a few cases hairy (*Arfeuillea*, *Harpullia spp.*); mesophyll usually with compact outer tissue, the cells in radial rows, and aerenchymatous inner tissue, the outer tissue in some cases becoming thick-walled (*Koelreuteria*, *Xanthoceras*), with a middle layer of crossed fibres (*Magonia*), pulpy in *Nephelium*; i.e. unspecialized. Tegmen 3–12 cells thick, not or slightly multiplicative, usually unspecialized and more or less crushed; o.e. developing into lignified fibres (*Alectryon*), or the cells shortly longitudinally elongate but not lignified (*Arfeuillea*, *Dodonaea*); endostome sclerotic in *Cupania* and *Harpullia*. Chalaza small (*Alectryon*, *Dodonaea*), distinctly enlarged (*Guioa*, *Sapindus*), extensively pachychalazal (*Allophylus*, *Cardiospermum*, *Cupania*, ? *Euphoria*, *Harpullia*, *Paullinia*), elongate transversely (*Nephelium*); hypostase small or extended in the pachychalaza as thin-walled brownish cells in several layers adjoining the nucellus. Vascular supply as a single v.b. or as a ring of v.b. in the funicle; v.b. of the raphe dividing soon after entry into the seed into two branches subdividing over the sides of the seed

(*Dodonaea*, *Guioa*, *Paullinia*) or dividing into several branches ramifying in the pachychalaza close to the expanded hypostase (*Allophylus*, *Cardiospermum*, *Cupania*, *Euphoria*, *Harpullia*, *Paullinia*, *Sapindus*) absent from the antiraphe and, in pachychalazal seeds, from the free part of the testa. Hilum round or linear (*Sapindus*), closed by a mass of sclerotic cells (*Alectryon*, *Allophylus*, *Arfeuillea*, *Cupania*, *Euphoria*, *Koelreuteria*, *Paullinia*, *Sapindus*), by a plate-like palisade of sclerotic cells (*Cardiospermum*) or of unlignified cells with *linea lucida* (*Dodonaea*, *Harpullia spp.*); much thickened with lignified micropylar canal in many cases. Nucellus enlarging greatly before absorption by the endosperm. Endosperm nuclear, eventually absorbed and crushed. Embryo generally curved but practically straight in many Nephelieae; cotyledons thin and spirally inrolled (*Alectryon pr. p.*, Cossigneae, Dodoneae, Koelreuterieae), thick and curved in most genera, or thick and straight.

Aril variously developed, usually more or less cupular and entire, but lobate–tesselate in *Alectryon*; funicular-exostomal in *Cupania*, *Euphoria*, *Guioa*, *Litchi*, *Paullinia*; funicular in *Cardiospermum*, *Dodonaea*, *Harpullia zanguebarica*, ? *Pometia*; funicular–raphe in *Alectryon*; funicular–placental in *Trigonachras*; more or less sarcotestal in *Harpullia spp.*, and *Paullinia*, or ? completely sarcotestal in *Nephelium*; usually without v.b., but highly vascular in *Alectryon*; often with a distinct cortex of radially elongate cells (except Paullinieae); usually oily, white, yellowish or red.   Figs. 501–527.

(Guérin 1901; Netolitzky 1926; Radlkofer 1934; Mauritzon 1936a; v.d. Pijl 1957.)

This family is generally considered from the multitude of detail which attends the small and uniform flowers. To the field-botanist it is one of the great elements of tropical floras, extremely varied in habit, leaf, vegetative anatomy, fruit and seed. In this last respect it has become the subject of conflicting theories. There are arillate, exarillate and sarcotestal seeds. According to the durian-theory, which I maintain, the arillate was the primitive from which the other states have been derived. According to van der Pijl (1957) the sarcotestal was the primitive, as a direct descendent from that of *Magnolia* and *Cycas* according to zoochorous harmony in the forest. This view, which I had also entertained and discarded because it does not fit the facts of the family, rests on an imaginary phyletic basis (p. 52); it encounters so many obstructions of its own fabrication that van der Pijl is finally obliged 'to admit that in the Sapindaceae this character [sarcotesta] may have arisen anew'. He meets disaster, however, because he fails to understand that the sarcotesta in Sapindaceae, as in Connaraceae, develops on the extended chalazal or pachychalazal base of the seed and is not therefore a primary homologue of that of *Magnolia*, but a secondary resemblance.

The Sapindaceous sarcotesta was described by Radlkofer as an adnate aril. This is to me correct because it is the pulpy outer layer of the pachychalaza continuous with the tissue of the free part of the aril, just as the inner layer of the pachychalaza is continuous with the tissue of the free testa; the characters of both aril and testa differentiate in, and are transferred to, the extended chalaza which, of course, has no integuments. When this sarcotesta is removed, it leaves a pale patch around the hilum, and this patch van der Pijl calls the pseudo-hilum. According to his theory, removal of the sarcotesta from the seed of magnolia would leave the endotesta as a pseudohilum covering the seed, which is an absurd confusion between testa and chalaza. Then, because of the diverse structural origins of the pulpy tissue on Sapindaceous seeds, van der Pijl coins the term 'arilloid' for general use where lack of exact knowledge prevents precise description, but this is merely the general botanical use of the term 'aril' which Baillon had urged (p. 23). Therefore, when van der Pijl has to speak of vestigial arilloids in Sapindaceae, he arrives at the conclusion of the durian-theory. *Paullinia*, which he chose for the exemplification of his theory, has species with a free aril, others with an adnate aril or sarcotesta, and others lacking both features, and it exemplifies the fact that all these states of the seed may occur within a genus. This generic fact leads to the systematic evidence which he has avoided. As a detail, however, I mention the case of *Castanella* Spruce, based on the very spiny capsules, but reduced by Radlkofer to *Paullinia*. Such capsules, on the durian-theory, are primitive compared with the thornless, and the species of *Castanella* have, as the durianologist would expect, a well-developed aril (Fig. 526). In the related genera with indehiscent fruits, namely *Serjania* and *Urvillea*, the seed lacks aril and sarcotesta.

The systematic evidence concerns the occurrence of the aril, the sarcotesta and the palisade-like exotesta in the genera and tribes. Firstly the aril is conspicuous in genera with capsular fruits, e.g. *Alectryon*, *Cupania*, *Guioa*, *Harpullia* and *Paullinia*, where the fruits themselves are often conspicuous with red and yellow colours. Then, the aril is absent from many of their allies, even specific allies within a genus, which have simplified and papyraceous capsules or indehiscent fruits. Thirdly the aril is absent from many large tribes with indehiscent fruits, namely Aphanieae, Lepisantheae, Melicocceae, Sapindeae, Schleichereae, Thinouieae and Thouinieae. Thus the aril appears as a prime character of the dehiscent fruits of the family, to be lost in the indehiscent fruits; if it were a secondary character, then it must have been evolved independently in these tribes and genera of Cupanieae, Harpullieae, Nephelieae and Paullinieae. This might seem possible, but the aril is associated with the hard, brown or black testal epidermis. If the Sapindaceous seed were primitively and entirely sarcotestal, as van der Pijl contends, then all these tribes, if not all their genera, must have evolved this testa independently.

Yet, it is the family-character of the seed which allies Sapindaceae with other exotestal families associated in Sapindales. A family-character is a prime and not a feature that is being evolved by some tribes and genera and not by others; for instance, the simple leaf, the entire leaflet, the winged petiole or rachis, the shrubby or the climbing habit, though frequent in various tribes and genera, are not prime family-characters. The aril goes primitively with the black or brown testa characteristic of the family; the sarcotesta comes in as a specialization of the chalaza where this testal character is missing. Nevertheless, van der Pijl endeavours to confirm his view by a message intended to refute the possibility that a testa may become sarcotestal; he writes 'a reversal of a hard testa with specialized palisade tissue to a parenchymatous [thin-walled] testa seems improbable' (v.d. Pijl 1966, p. 614). It is a mistake. It is exactly what happens in *Alectryon* and *Harpullia*. In Cupanieae the sarcotesta has a palisade-epidermis as the testa; in Paullinieae it has an epidermis of more or less stellate-undulate cells, as in the testa of the tribe. It is what happens in overgrown seeds of other families (Corner 1951). It can happen very simply by the retention of the testa, in part or wholly, in its thin-walled and embryonic state, as may be seen in the short transitional region between the aril and the testa in many seeds. What does involve progressive evolution is the conversion of thin-walled and unspecialized cells into thick-walled palisade-cells. The sarcotestal theory requires this and van der Pijl gives no explanation whence these new factors have come. The evolutionary origin of the exotestal palisade, like that of the exotegmen or indeed of the aril, is not a problem of modern scale but dates from the primitive angiosperm seed. The durian-theory of the Sapindaceous seed, tested with this opposition, is improved, which is important because there is so little direct evolutionary thought in the study of angiosperms.

Too few species of this large family have yet been investigated to understand the seed-evolution in the detail of genera and species; even, perhaps, to assign the family to a major alliance in seed-structure. The exotestal palisade of imperfect Malpighian cells, some with a *linea lucida* (*Sapindus*), suggests a connection with Leguminous ancestry. The curved form of the seed with the partial septum that makes the radicle-pocket, indicates a divergence. A few species raise the doubt whether Sapindaceae are simply exotestal or have descended from an arillate exotestal seed with fibrous exotegmen such as occurs in Celastraceae and, especially, Meliaceae. Thus, in the strongly arillate seed of *Alectryon*, the exotegmen develops lignified fibres which become spaced in a discontinuous layer, as if the feature were vestigial (Fig. 505). The endostome is sclerotic in *Harpullia zanguebarica* and slightly so in *Cupania sp.* and *Koelreuteria*. In *Dodonaea* and, possibly, *Arfeuillea* the exotegmic cells are longitudinally elongate, if not strongly, but they are not sclerotic. In general, the non-multiplicative tegmen seems vestigial for it soon degenerates in the growth of the seed, and its function is taken over by the pachychalaza. Thus the Sapindaceous seed differs from the strongly exotegmic Capparidaceous seed of similar form. The Meliaceous seed shows how the fibrous exotegmen may degenerate and the pachychalaza extend.

Evidence of widespread reduction in the family comes also from the small flowers and the one or few-seeded fruits, most of which are reduced to one ovule per loculus. The capsules may become winged and schizocarpic or indehiscent. They may be unilocular through abortion of the other two loculi, and the long axis of the fruit may remain undeveloped while the loculi enlarge transversely to give the almost gynobasic fruits characteristic of many genera (Nephelieae). In this reduction from a many-seeded capsule the few remaining seeds may have become enlarged and more elaborate. The expanding chalaza may be an instance; it thickens the hilar tissue, causes the long micropylar canal, and dilates the seed-cavity to accommodate the large embryo, but its operation is not simple. It may involve a uniform expansion or, as in *Paullinia pinnata* (Fig. 525) it may affect only the sides of the seed. In *Nephelium* its action seems to be confined to the testa (p. 246). As a subsidiary peculiarity it may well show generic or tribal differences with its independent origin in different Sapindaceous lines of evolution.

The original Sapindaceous embryo may have had spirally coiled cotyledons and, through pachychalazal evolution with its enlarged seed-cavity, the thick and more or less transverse cotyledons may have been derived. Certainly the convolute embryo persists in genera with two or more seeds per loculus, and the study of these genera with, perhaps, less modified seeds may help to solve the problem of the primitive seed-structure of the family. The large thick hilar base of the seed, filled with sclerotic tissue, may have come through the pachychalazal growth from the thinner construction sealed by a simple palisade of cells, as in *Dodonaea* and *Harpullia*. The complex vasculature of the seed seems also to be connected with the pachychalaza, for the testa (or free outer integument) is not generally vascularized; exceptions are *Dodonaea* and *Guioa*, and it is these exceptions which in all ways seem to stultify any clear idea of evolution of this seed in detail.

Nephelieae are an example of this perplexity. *Euphoria*, *Litchi* and *Pometia* have succulent free arils in their indehiscent, unicarpellary fruits. In *Euphoria* and *Litchi* the fruits are slightly knobbed or verrucose, suggesting derivation from a spinous capsule, but that of *Pometia* is smooth. Thus the primitive aril accompanies the advanced fruit. In *Nephelium* and, possibly, *Xerospermum* the succulence is retained not as a free aril but as an adnate aril or sarcotesta; the seeds have lost the exotestal palisade, except in a minute area round the micropyle; and, yet, the fruit shows well the primitive spinous character, even to vestigial lines of dehiscence, and the simplification to the warts of *Euphoria* and *Litchi*. In *Alectryon* and, possibly, its allies *Heterodendron* and *Podonephelium* the aril is extraordinarily lobulate and highly vascular; on swelling, it disrupts the smooth, gynobasic carpel transversely. If this large aril is primitive, then it accompanies an advanced fruit that has lost the normal suture of dehiscence. In *Pappea* and *Stadmannia*, placed near *Alectryon*, the aril is said to be simpler, though its details are unknown, and the carpel dehisces normally. Then, as further complexities and enigmas, the seeds of *Euphoria* and its allies appear to be generally pachychalazal. In *Nephelium* and, perhaps, in *Xerospermum*, as already mentioned, the chalazal expansion seems to affect only the testa and to convert the seed almost into an orthotropous ovule with the micropyle thrust so far from the hilum as to appear subapical. In *Alectryon* it is the raphe which is so greatly extended; as a thick sclerotic layer it makes up half of the seed-coat, embeds the vascular bundle of the raphe-funicle, and has the aril attached to it without participation of the exostome or the hilar region of the funicle. In spite of all these peculiarities, whether primitive or advanced, *Alectryon* has the spirally coiled embryo and the fibrous exotegmen. To accommodate the characters of Nephelieae one must suppose them to be an early off-shoot of Sapindaceous ancestry. There are still many hundreds of Sapindaceous seeds to be examined, and when this has been completed there may be some firmer knowledge of seed-evolution not only in the family but in dicotyledons generally.

There is clear evidence in Celastraceae and Meliaceae that the family-character of the fibrous exotegmen has been lost in some genera. Accordingly I argue from *Alectryon* that this feature has disappeared very widely not only from Sapindaceae but from other families associated in Sapindales as Aceraceae, Hippocastanaceae, Sabiaceae and Staphyleaceae. The fibrous exotegmen occurs in Staphyleaceae–Tapisceae with large chalaza, and, if this tribe is rightly classified, the feature confirms the conclusion from *Alectryon*. Sapindales show extensively the simplification of the seed-coat into plain exotestal seeds and overgrown seeds which happens in other advanced and varied families, e.g. Rosaceae, Myrtaceae, and Melastomataceae. It is this simplification which renders so difficult the assessment of exotestal families devoid of other details of the seed-coat.

Alectryon Gaertn.

Alectryon sp.

(New Guinea, Morobe district, leg. E. J. H. Corner; fruits in spirit.) Figs. 503–505.

Seeds 6 mm long, ellipsoid subglobose, black, smooth, projecting from the pyxidate follicle, more or less covered by the unilateral, tesselate-lobate, fleshy, crimson aril derived from the raphe and funicle; anacampylotropous with a partial septum, covered by the tegmen, making the radicle-pocket. Testa *c.* 20 cells thick, very hard, developed over the antiraphe half of the seed; *o.e.* as a thick palisade 200–250 μ high, of columnar prismatic cells with thick, pitted, yellow-brown walls, not lignified, with frequent immersed cells containing 3–6 large crystals (? short rows of cells); *mesophyll* with a few outer layers of cells with thickened brownish walls and crystaline contents, the remainder thin-walled and more or less crushed; *i.e.* unspecialized; *micropyle* directed distally, minute, between the palisade of the testa and the adjoining sclerotic layer of the raphe-side of the seed. Tegmen 6–7 cells thick, –12 at the endostome, as a thin brownish papery layer surrounding the whole embryo except at the minute chalaza; *o.e.* as a discontinuous layer of longitudinal, fibriform, thick-walled, lignified cells 120–650 × 8–23 μ, often shortly lobed or spicate, appearing as idioblasts; *mesophyll* and *i.e.* unspecialized; *endostome* not sclerotic, the lignified fibres stopping short of the slightly dilated endostome. Raphe-side of the seed (external to the tegmen) *c.* 20–30 cells thick; inner part, *c.* 15 cells thick, as the mesophyll of the testa; outer part 100–150 μ thick, composed of 7–15 layers of isodiametric sclerotic cells, heavily lignified, forming a thick hard layer fitted closely to the palisade of the testa at the exostome and at the hilar end of the seed, perforated minutely by the raphe v.b. Hilum disrupted from the placenta by the elongation of the aril, closed by the sclerotic layer of the raphe. Vascular bundle of the raphe extending along the side of the seed with branches to the aril, and with a minute branch piercing the sclerotic tissue of the raphe to enter the soft inner tissue to the chalaza. Endosperm almost entirely consumed. Embryo curved; radicle rather slender; cotyledons thick.

Aril 16–18 mm long on expansion, attached along the raphe and round the short funicle, firmly pulpy, lobulate with tesselate surface, supplied with numerous fine v.b. from the raphe; cells large, thin-walled; epidermis composed of large cells, not as a palisade; without an arillostome, not developing from the exostome and not round the hilum.

Pericarp with scattered and clustered sclerotic cells in the mesocarp and forming a more or less continuous layer in the endocarp, without idioblasts; v.b. ? in one circle (in t.s.).

With only ripe seed for study it was not possible to make out all details. The testa and the sclerotic tissue of the raphe-side of the seed are very hard. The aril is highly compressed in the unopened fruit; its cells enlarge, burst the pericarp transversely, as it appears, and stretch both the very short funicle, disrupting it from the hilum, and the lower part of the raphe; thus the black seed is projected like a jack-in-the-box.

Actually, as the fruiting carpel is elongated transversely, the rupture is in the transmedian longitudinal plane. This is an important point which bears on the seed of *Nephelium* also developed in a transversely elongated follicle. If the black testal side of the seed of *Alectryon* were reduced to a minute area by the micropyle and the body of the seed were formed by the expansion of the raphe-side from the cupular form in *Alectryon* into a cylinder with sclerotic tissue round the v.b. and an adnate aril on the outside, there would result the sarcotestal seed of *Nephelium* with straight embryo and distal micropyle. Compare Fig. 503 and Fig. 523. This does not imply that *Alectryon* has a wholly primitive seed for the family, but that it has the capacity to engender the seeds of the family. Other primitive features are the vascular aril, the thick palisade, and the absence of a pachychalaza.

Perhaps the most remarkable detail is the transformation of the exotegmen into long idioblast-like fibres. They do not form a continuous layer and it would seem that they are separated by thin-walled cells in the developing seed, as in *Myristica* (Fig. 409) or *Galearia* (Fig. 264). These fibres separate the tegmen from the testa all round the seed and on both sides of the partial septum which makes the radicle-pocket, but they do not enter the endostome; if this cap were sclerotic, as in *Harpullia zanguebarica*, the seed would be wholly exotegmic.

Allophylus L.

A. cobbe L.

(Ceylon, Hakgala; flowers and fruits in alcohol.) Fig. 506.

Ovule anatropous, becoming campylotropous after fertilization; o.i. 8–10 cells thick; i.i. 4–5 cells thick; hypostase conspicuous, brown; obturator slight or none.

Seeds 6–7 mm wide, subglobose, with thin, rather crustaceous coat, enclosed in the sclerotic fibrous endocarp, exarillate. Testa 8–20 cells thick, thinnest along the antiraphe, thickened into the pachychalaza on the raphe-side, 6–8 cells thick at the micropyle, thin-walled, not lignified; *o.e.* composed of flattened, thin cells with irregular undulate facets, not as a palisade; *mesophyll* aerenchymatous, loose; *i.e.* unspecialized. Tegmen 4–6 cells thick, thin-walled, the cells transversely elongate, contiguous; *i.e.* composed of narrow cells shortly longitudinally elongate, colourless. Chalaza greatly expanded, forming most of the seed-coat except along the antiraphe. Vascular bundle of the raphe dividing into several branches extending over the pachychalaza. Hilum with a thick ring of sclerotic tissue. Endosperm with a trace lining the pachychalaza. Embryo curved; cotyledons thick, folded.

Pericarp dull orange-red, thinly pulpy, with rather thick sclerotic endocarp, indehiscent.

Aphania Bl.

A. senegalensis Radlk.

Seed-coat thin, exarillate. Testa composed of many layers of flattened thin-walled cells.

(Netolitzky 1926.)

Arfeuillea Pierre

A. arborescens Pierre
Ovules 2 per loculus.

Seeds round, black, shortly hairy, 1 per loculus, exarillate but seated on a somewhat fleshy funicle, either erect or suspended with the other seed aborted. Testa; *o.e.* with simple, straight, 1-celled hairs. Hilum with a mass of sclerotic cells, the innermost as lax branching sclereids extending on to the sides of the seed but there apparently unlignified.

(Netolitzky 1926; Renard 1913.)
Possibly there is a fibrous exotegmen.

Cardiospermum L.

C. halicacabum L.
(Ceylon; young and mature fruits in alcohol.) Figs. 507–509.

Ovule subcampylotropous, distinctly campylotropous after fertilization; o.i. 9–10 cells thick; i.i. 5–7 cells thick; funicle short, thick; obturator slight; micropyle formed by the endostome, then by the exostome in the seed.

Seeds 4 mm, subglobose, black, smooth, with white heart-shaped and finely pruinose hilum, seated on the expanded head of the funicle as a slight whitish aril, breaking off from this aril across the hilar stratum. Testa *c.* 20 cells thick, multiplicative with prolonged periclinal divisions in the outer hypodermis; *o.e.* as a palisade *c.* 70 μ high, composed of stellately lobed cells, elongate radially, with thick brown walls, not lignified, with a *linea lucida* separating the outer ends of the cells as a hyaline superficial layer *c.* 10 μ thick; *mesophyll* thin-walled, the inner 3–4 layers of cells colourless, the outer cells brownish; *i.e.* unspecialized. Tegmen not multiplicative, thin-walled, unspecialized. Chalaza extending into a pachychalaza 25–30 cells thick, forming most of the seed, the free testa and tegmen limited to the radicle-end of the seed and the part between the micropyle and the chalaza; *hypostase* scarcely developed. Vascular bundle of the raphe dividing into two at the chalaza, then subdividing and anastomosing in the pachychalaza. Hilum closed by a single layer of thick-walled, prismatic cells with lignified angles and, mostly, with an oblong or tetrahedral crystal, forming a plate perforated only by the minute raphe v.b. Embryo with sigmoid cotyledons.

Aril developing after fertilization as a shallow expansion of the funicle-head, just covering the micropyle (? slightly exostomal), composed of a cortex of thick-walled cells (2–3 cells thick) and loose aerenchymatous, thin-walled inner cells; attached to the hilar plate by a layer (4–6 cells thick) of small, thin-walled, contiguous cells, many in the middle layers with a single sphaero-crystal, this tissue drying up in the mature seed and causing the separation of the seed with the hilar plate from the funicle and dry aril.

(Netolitzky 1926; v.d. Pijl 1957; Kadry 1960; Nair and Joseph 1960.)

I have studied this well-known seed carefully because previous accounts are unsatisfactory. The palisade-cells of the testa are prisms on a lobate base and have the stellate–undulate outline of the facet so different from the hexagonal outline of most Sapindaceae; the feature may by typical of Paullinieae. Most of the seed-coat is formed by the expanded chalaza. This pachychalaza has the same structure as the testa except that it carries the vascular bundles. The seed is separated from the funicle by a plate of hexagonal prismatic cells, shorter than those of the seed-epidermis, colourless, with lignified angles, and containing a crystal. The head of the funicle expands into a vestigial aril which never has a free border over the sides of the seed; it has the typical structure of many funicular arils and it detaches from the ripe seed through the drying up of the small-celled tissue on the funicular side of the hilar plate, and remains on the funicle as a minute watch-glass. The seed resembles that of *Dodonaea* except that it is pachychalazal, has the stellate epidermal cells, lacks a *linea lucida* in the hilar plate, and detaches from the aril.

The funicle is interpreted by van der Pijl as an extension of the placenta or endocarp because it has idioblasts as in the carpel-wall, but I found no idioblasts in the short stalk beneath the aril-disc, and I agree with Kadry that the stalk is a funicle carrying the v.b. detached from the placental strand. Then van der Pijl interprets the aril as an arillode; that is as an outgrowth of the testa, but I found no evidence for this. He emphasizes the difference in structure between this 'arillode' and the testa (? pachychalaza) without explanation; and it is not clear how an aril should develop from a pachychalaza. He remarks that the 'arillode' persists but functions merely as an abscission-layer. This is not true because the abscission-layer is the small-celled tissue below the hilar plate, whereas the aril is the larger-celled tissue with cortex of thick-walled cells that remains attached to the funicle. As a rudimentary and vestigial funicular aril, similar to that of *Dodonaea*, it suggests a funicular aril in Sapindaceous ancestry, similar to the Leguminous aril, and the hilar plate recalls the counter-palisade of the Papilionaceous hilum, where the vestigial aril remains on the seed as with *Dodonaea*. *Alectryon* has the funicular aril.

Cupania L.

Cupania sp.
(Brazil, Manaus, Corner n. 254, Oct. 1948; young and mature fruits in alcohol.) Figs. 510, 511.

Ovules 1 per loculus, campylotropous, erect, shortly funiculate; o.i. 6–7 cells thick; i.i. 3–4 cells thick.

Seeds 10–11 × 7 mm, ellipsoid, smooth, black, more than half covered by the waxy white aril arising from the funicle and exostome, the seed-coat mostly pachychalazal. Testa 15–22 cells thick, thinnest along the antiraphe, not lignified; *o.e.* as a palisade of prismatic cells with thick, stoutly pitted, brown walls; *mesophyll* aerenchymatous, then slightly thick-walled, finally more or less crushed; *i.e.* unspecialized; *exostome* converted into the arillostome. Tegmen 5–6 cells thick, soon crushed except at the endostome with slightly lignified, subsclerotic exotegmic cells; *i.e.* unspecialized. Pachychalazal wall similar to that of the free parts of the testa but carrying *c.* 12 v.b. near its inner side, at first continuous with the large cells of the much enlarged nucellus. Partial septum short. Hilum closed by a thick plate of isodiametric sclerotic cells, many cells thick, perforated by the exostome and the raphe v.b. Endosperm absorbed. Embryo with straight thick cotyledons, longitudinal, and short abaxial marginal radicle.

Aril consisting of a medulla of longitudinal cells and a cortex of radially elongate, thin-walled epidermal cells with oily contents, the cortex developing late in the growth of the seed from very small epidermal cells.

The aril has the same manner of origin and structure as in *Guioa*, the seed of which is scarcely pachychalazal.

Dodonaea L.

D. viscosa L.
Fig. 512.

Ovules 2 per loculus, anacampylotropous, the upper erect, the lower suspended; o.i. 8–10 cells thick; i.i. 3–4 cells thick; funicle short, thick; obturator as a slight flap.

Seeds 6 mm long and wide, subcompressed, blackish, smooth, with slight radicle-ridge, not pachychalazal; hilum closed by the adnate vestigial aril. Testa 23–30 cells thick, not lignified; *o.e.* as a palisade of short, radially elongate cells with thick brown walls and brown contents, prismatic, with hexagonal facets, with a *linea lucida* separating the thin colourless outer stratum; *mesophyll* with the outer cells more or less in radial rows from hypodermal divisions, brownish, the inner part aerenchymatous, the loosely arranged cells with short arms; *i.e.* unspecialized. Tegmen 4–6 cells thick, not lignified, extending all round the seed; *o.e.* and *o.h.* with the cells shortly longitudinally elongate; *mesophyll* and *i.e.* unspecialized or the cells of i.e. enlarged slightly. Chalaza small, simple. Embryo with rather long radicle and thin, spirally coiled cotyledons.

Aril derived from the head of the very short funicle, not investing the seed, persistent as a minute cushion in the hilum; separated from the seed-base by a plate of elongated, slightly thick-walled, not lignified cells forming a palisade with a *linea lucida* along the mid-line,

perforated by the small v.b.; cortex of the aril 2–3 cells thick, with thickened cellulose walls; inner tissue loose, aerenchymatous, adjoining the hilar plate with 3–4 layers of small thin-walled contiguous cells.

(Mauritzon 1936a.)

The hilum has a remarkable resemblance to the Papilionaceous. Possibly other species of *Dodonaea* will show a better developed aril and lignified fibres in the exotegmen. Apart from these deficiencies, the seed of *Dodonaea* seems one of the more primitive in the family.

Euphoria Comm.

Ovules anatropous, erect, becoming more or less campylotropous after fertilization; integuments ?

Seed with smooth, dark brown, hard and crustaceous coat; surrounded by the thick white pulpy aril developed from the junction of the ovule and the funicle and from the antiraphe of the exostome, without v.b.; exalbuminous. Embryo straight, formed almost entirely by the thick, almost inseparable cotyledons; radicle as a slight papilla near the micropyle; plumule not evident.

E. longana Lamk
(Ceylon, leg. E. J. H. Corner 1972; fruits in alcohol.)

Ovule with a prominent and vascular projection from the funicle below the micropyle (? obturator).

Seed with the testa 35–40 cells thick; *mesophyll* with a plexus of v.b.; tegmen crushed.

Aril late in development, rudimentary in seeds 5 mm long, only 2.5 mm long in seeds 16 mm long, but eventually covering the seed with overlapping lobes at the top.

(v.d. Pijl 1957.)

E. malaiensis (Griff.) Radlk.
(Malaya, leg. T. G. Whitmore; ripe fruits in alcohol.)
Fig. 513.

Seeds 12 mm wide, subglobose. Testa *c.* 300 μ thick (? 25–35 cells thick excluding v.b.), not lignified, closely adnate to the cotyledons; *o.e.* as a hard, not lignified, palisade 20–25 μ high, with radially elongate prismatic cells with very thick, dark brown, pitted walls; *mesophyll* more or less crushed, permeated by a close network of v.b., without obvious raphe, the v.b. apparently in a single layer, with scattered sclerotic cells mostly round the v.b.; *i.e.* unspecialized; *exostome* thickened, sclerotic. Tegmen apparently crushed. Hilum broad, with many v.b., surrounded by the aril-attachment except at the raphe-part of the micropyle, with the short thick funicle separated from the seed-coat by a more or less continuous, uneven, layer (5–10 cells thick) of heavily lignified, sclerotic cells.

Aril 2–3.5 mm thick, with 2–3(–4) connivent and overlapping lobes at the apex.

Pericarp *c.* 1 mm thick at the base, 0.3–0.5 mm thick over the distal end of the fruit, with many clusters of sclerotic cells in the outer part; v.b. in 2–3 circles, as seen in t.s., anastomosing; endocarp thin, fibrous, not lignified; superficial processes slight, elongate, scarcely developed.

Guioa Cav.

G. pubescens Radlk.
(Singapore; living material.)   Figs. 514–518.

Ovules anatropous, campylotropous after fertilization, 1 per loculus, erect; o.i. 4–6 cells thick; i.i. 2–3 cells thick; micropyle formed by the exostome; funicle very short and thick.

Seeds 6.5 × 6 × 4 mm, subglobose, subcompressed, white, then ripening through red to red-brown and finally shiny black, covered for about three-quarters or almost entirely by the thin, orange, pulpy aril, its micropylar process functioning as a funicle on dehiscence of the fruit; hilum subcircular, white. Testa multiplicative; o.e. composed of shortly columnar cells with thick, dark brown walls, colourless when immature but readily oxidizing red to black on cutting; mesophyll thick, the cells slightly pulpy in the immature seed, eventually drying up with slightly thickened brownish walls; i.e. unspecialized. Tegmen not multiplicative, crushed. Chalaza distinctly enlarged but not pachychalazal. Vascular bundle of the raphe dividing shortly into a few branches, subdividing and in places anastomosing, absent from the antiraphe-side of the seed. Endosperm absorbed. Embryo curved with thick green cotyledons, oily without starch.

Aril arising very soon after fertilization as an outgrowth round the junction of the funicle with the ovule and extending round the micropyle, starting from the raphe-side of the ovule; the micropylar appendage developing from the exostome on its antiraphe-side, extending to the base of the ovarian loculus as a long narrow tongue, eventually thrown into folds by intercalary extension; consisting of a medulla (3–4 cells thick) of elongate colourless cells with mucilaginous walls, and an epidermis of columnar cells with firmer, slightly thickened walls and rich orange, oily contents, the epidermal cells short and cuboid until ripening of the aril, then lengthening radially and assisting the dehiscence of the fruit, the immature epidermis pale greenish yellow from chloroplasts.

According to van der Pijl (1957) some species of Guioa have a pulpy endocarp and he refers to an endocarp-aril for the genus. He overlooked my account (Corner 1949a). I doubt any such peculiarity. The seed of Mischocarpus sumatranus hangs on a similar but red aril.

Harpullia Roxb.

Ovules 1–2 per loculus, suspended, epitropous. Seeds arillate, sarcotestal or merely with minute funicular aril.

H. zanguebarica Radlk.
(Ceylon, Peradeniya, cult.; nearly ripe fruits in alcohol.) Fig. 519.

Seeds 7–8 mm long, ellipsoid, finely villous with short hairs, shortly pachychalazal; aril funicular, minute, adnate to the hilum, enclosing the minute micropyle. Testa 25–30 cells thick, c. 20 cells thick along the antiraphe; o.e. as a short unlignified palisade of cells with rather strongly thickened outer walls, many of the cells elongate into stiff, simple, thick-walled, aseptate, acute, colourless hairs 100–140 μ long; mesophyll and i.e. unspecialized. Tegmen 6–8 cells thick, unspecialized, surrounding the seed except over the pachychalaza; endostome with the exotegmic cells more or less radially elongate, sclerotic, the cells round the base of the endostome somewhat longitudinally elongate (? fully mature seeds with exotegmic fibres). Pachychalaza only on the raphe-side, not extensive. Vascular bundle of the raphe dividing into several branches variously extending over the seed in the mesophyll of the testa, absent from the antiraphe. Hilum closed by a palisade of narrow elongated cells as in Dodonaea, not sclerotic.

Aril constructed as in Dodonaea.

Harpullia sp.
(Solomon Isl., Guadalcanal, Corner s.n. 27 July 1965; fruits in alcohol.)   Fig. 520.

Ovules solitary, suspended.

Seeds 18–20 × 13–15 mm, ellipsoid, campylotropous with small radicle-pocket, pale brown, covered except at the extreme base by the reddish pink fleshy aril, shortly sarcotestal round the hilum. Testa becoming 35–45 cells thick, then the inner tissue crushed and 15–20 cells thick at maturity; o.e. composed of unusually small, cuboid, colourless cells with scarcely thickened walls, not lignified; mesophyll aerenchymatous with the cells tangentially elongate, developing slightly thickened and brownish walls, not or scarcely lignified; i.e. crushed. Tegmen entirely crushed, ephemeral, 6–7 cells thick over the partial septum, ? at first merely 2 cells thick over the sides and antiraphe of the seed but soon disappearing. Nucellus massive, then crushed. Chalaza unspecialized. Vascular supply as a short stout funicular bundle to the chalaza, branching there into c. 12 bundles over the sides of the seed in the testal mesophyll. Endosperm watery, evanescent. Embryo curved, with thick cotyledons, one folded.

Aril rather thick, micropylar with thick arillostome, and funicular, descending over the seed-body, not lobed, reddish pink with white flesh, thin-walled, without v.b.; o.e. composed of cuboid cells; inner tissue with rows of larger cells.

Fruit as a 2–3-lobed capsule, dull orange-yellow with red inner surface, 1–2 seeds per capsule.

It was not possible to identify this species. It was a small unbranched forest shrub, 1½ metres high with stem

I cm thick, but at the time of fruiting with a terminal panicle it had become entirely leafless. I illustrate it, however, because it shows what an exciting variety of seeds must exist in this genus. Those of this species have the fullest development of the aril, comparable with that of *Guioa* with its erect or inverted ovules, and it has the least development consequently of the pachychalaza and sarcotesta, as befits the durian-theory. The following species is clearly more derived in its testal limitation by the expanded sarcotestal pachychalaza. *H. zanguebarica* must represent another line of evolution in the genus in which the seed has become practically exarillate without sarcotesta. The species of the Solomon Islands appears to have truly testal vascular bundles.

Harpullia sp.

(New Guinea, Busu River, pr. Lae, leg. E. J. H. Corner Sept. 1960; nearly ripe fruits in alcohol.) Fig. 521.

Ovules I per loculus; o.i. 10–12 cells thick; i.i. 5–6 cells thick.

Seeds 12–13 mm long, ellipsoid, black, glabrous, covered on the raphe-side with the orange-yellow sarcotesta with short free aril-rim, with fleshy slit-like arillostome; seed-coat extensively pachychalazal, unlignified except for the xylem of v.b.; with slight partial septum. Testa 12–25 cells thick, thinnest towards the base of the pulpy exostome; o.e. as a short palisade of columnar cells with somewhat thickened, brown radial walls and thick outer wall; *mesophyll* unspecialized, with some outer hypodermal divisions of the cells; i.e. unspecialized. Tegmen not multiplicative, unspecialized, soon crushed. Pachychalaza extending over most of the seed except the black antiraphe-side. Vascular bundle of the raphe dividing shortly into several branches extending to the exostome and over the pachychalaza, absent from the free part of the testa. Hilum without palisade or sclerotic tissue. Endosperm absorbed. Embryo with short radicle (appearing lateral) and thick transverse cotyledons.

Aril without v.b., constructed of large cells, developed from the exostome on all sides and, as the sarcotesta, from the pachychalaza; epidermis composed of small cells with thickened walls, as a short palisade; arillostome thickly fleshy and developing the micropyle into a long curved slit.

Capsule orange-red to scarlet, 2-locular.

The pachychalazal development of the seed is shown in Fig. 521. The sarcotesta, as a modification of the outer tissue of the pachychalaza, is combined with the micropylar aril and has no morphological agreement with that of *Magnolia*, as the sarcotestal theory of van der Pijl would suppose. The effect is to limit the black testa to the antiraphe side of the seed and to give the red dehisced fruit the typical appearance of the arillate capsule. The epidermal cells of the aril may eventually elongate as in *Cupania*.

Koelreuteria Laxm.

K. paniculata Laxm.

Fig. 502.

Ovules 2 per loculus, one forming a seed; o.i. 10–12 cells thick; i.i. 4–5 cells thick.

Seed round, black, hard, exarillate, not pachychalazal. Testa 20–24 cells thick, about ten times thicker than the tegmen; o.e. as a palisade of narrow cells, thick-walled; *mesophyll* becoming thick-walled in the outer part, (lignified ?); i.e. unspecialized. Tegmen not multiplicative, eventually crushed; o.e. as flattened thick-walled cells; i.e. with shortly columnar projections against the mesophyll and making air-spaces; *endostome* more or less sclerotic. Hilum closed by a sclerotic mass of tissue.

(Guérin 1901.)

It seemed to me in the examination of some seeds that the tegmen may be slightly multiplicative and become 8–10 cells thick. I did not detect exotegmic fibres.

Litchi Sonn.

Ovule and seed evidently as in *Euphoria*. Ovule somewhat campylotropous. Aril late in development, exostomal and funicular.

Pericarp tubercules formed by proliferation of the hypodermal cells.

(Banerji and Chowdhuri 1944; van der Pijl 1957.)

Banerji and Chanduri described the aril as developing from the obturator, but van der Pijl revealed this mistake.

Magonia St. Hil.

M. glabrata St. Hil.

Seeds flattened, alate, 6–8 per loculus, exarillate. Seed-coat –18 cells thick; o.e. as flattened thick-walled cells; *mesophyll* with an outer layer (7–8 cells thick) of slime-cells with mucilaginous walls, developed by periclinal division from o.e., a middle layer (6–8 cells thick) of crossed fibres with thick, pitted, lignified walls, and an inner thin-walled layer; wing of the seed with thin-walled o.e. and two layers of crossed fibres separated from o.e. by rather thick-walled, crushed tissue, without slime-cells.

(Netolitzky 1926.)

It is not easy to see how this construction fits that which is general in Sapindaceous seeds.

Nephelium L.

Seeds large, sarcotestal, ellipsoid, subcompressed, with the micropyle displaced towards the distal end of the seed; inner seed-coat crustaceous or coriaceous, pale brownish to yellowish, with many longitudinal anastomosing v.b. Embryo straight or slightly curved; radicle very short, apparently without radicle-pocket.

Fruits indehiscent or with vestigial longitudinal suture on desiccation of the pericarp; spines flexible or short and thick, with many small longitudinal fibrovascular bundles.

N. lappaceum L.

(Ceylon, Malaya; flowers and fruits in alcohol.)   Figs. 522–524.

Ovules 1 per loculus, anacampylotropous, erect; o.i. 5–7 cells thick on the antiraphe-side, thicker towards the base; i.i. 6–8 cells thick; micropyle formed by both integuments; obturator as a crescentic flange on the lower side of the ovule; procambial strands entering o.i. from the funicle; chalaza with thin-walled, brown cells in the hypostase.

Mature seeds 20–26 × 16–19 mm (larger in other varieties), covered with white or yellowish adnate pulp 2–3 mm thick, but with a small naked brown patch 2 mm wide round the micropyle at the distal end of the seed, the patch often hidden by the bulging sarcotesta. Sarcotesta; *o.e.* composed of small, thin-walled cells with hexagonal facets; *hypodermis* composed of larger cells with short arms; *mesophyll* composed of radial rows of very elongate pulpy thin-walled cells, 1–3 cells in a row, attached to the inner testa by a narrow layer of short, slightly thick-walled cells; sarcotesta developing late in young seeds when over 15 mm long. Inner testa 0.4–0.5 mm thick, pallid yellowish, crustaceous consisting of thin-walled ground-tissue with a network of v.b. more or less in three rows (or cylinders), the outer row of v.b. separated from the inner by an irregular sclerotic layer 60–90 μ thick, the sclerotic layer ending at the periphery of the micropylar region, many of the cells of the ground-tissue with a large crystal. Micropylar region composed of isodiametric cells in more or less radial rows 40–70 cells long, with rather thick walls, brownish towards the surface as an even epidermal layer (not a palisade) forming a hard disc, not lignified, without v.b.; micropylar canal obliterated. Tegmen obliterated. Hilum not sclerotic but with many v.b. branching and anastomosing between the stout funicle and the inner testa.

Immature seed up to 12 mm long, without evident embryo, appearing more or less erect and straightened through the enlargement of the carpel, with the micropyle far removed from the basal chalaza. Testa much thickened –50 cells or more, with a plexus of many small longitudinal v.b. arranged in an outer cylinder and a more or less double inner cylinder, without a prominent raphe-bundle. Tegmen apparently not multiplicative, the inner cells enlarging. Obturator persistent, slightly enlarged as a thin crescentic membrane round the abaxial side of the top of the funicle. Micropylar end of the testa developing radial rows of cells over the top of the seed, not on the sides.

Pericarp with complicated arrangement of v.b., with slime-canals in the mesocarp and endocarp. Spines formed as knobs on the ovary, developing basipetally with many fine fibrovascular bundles bifurcating but, apparently, not anastomosing.

(Van der Pijl 1957.)

The very special nature of this seed is shown by the reduced state of the ovary and limitation of the fruit in which it is produced, and by its extremely complicated vasculature. The carpels have one ovule and, commonly, only one carpel develops into a fruit, occasionally two, very rarely three. This fruit elongates transversely to the long axis of the ovary into a kind of drupe, for there is no particular endocarp. The ovule extends into the oblong cavity so formed and grows in an intercalary manner so that the micropyle is taken away from the placenta near which it started, and the seed is rendered almost orthotropous in a transverse direction. No radicle-pocket is formed; the embryo is practically straight; the radicle is very short. There is no single raphe-bundle but about twelve v.b. pass from the placenta into the funicle and then expand into the three cylinders of longitudinal v.b. in the testa. They stop short of the apical cap of radial rows of cells in the testa, where there is a vestige of the normal testal construction of the Sapindaceous seed. The epidermal and hypodermal tissue over the rest of the young seed eventually enlarge their cells at a late stage of development into the tissue of the sarcotesta. Thus the seed lacks all the main features of the family except for the absence of endosperm and the small exotestal patch round the micropyle. This patch, when traced in the growth of the ovule into the young seed, is the antiraphe of the ovule; it is the striated patch shown in Fig. 522; though highly multiplicative and, perhaps, overgrown in its restricted area, it is not vascular. This patch, figured by van der Pijl, is not the part which forms the sarcotesta, as his description suggests. Though millions of people, including hundreds of botanists, have eaten these seeds, this tell-tale patch has escaped their notice. The body of the seed is developed as a cylindrical pachychalaza lined by the ephemeral tegmen; it is, as usual, the pachychalazal exterior which becomes the sarcotesta. What I have called the obturator seems rather to be a true aril, similar to that of *Pometia*; when the young seed is merely 7 mm long (Fig. 522), this structure has almost mature cells without sign of further means of growth, and it is obliterated as soon as the sarcotesta begins to swell.

I am unable to follow van der Pijl in his fallacious argument that the sarcotestal seed of *Nephelium* with its very peculiar construction is primitive in Sapindaceae and to be related directly through the sarcotesta of *Magnolia*, with its different construction, to the gymnospermous sarcotesta. Rather it is one of the most elaborate and advanced seeds in the family. It can be explained by pursuing the consequences of converting the seed of *Alectryon* into a wholly pachychalazal seed except for the micropylar patch, for this is on the antiraphe side exactly as the restricted testa of *Alectryon* or

Harpullia, (Fig. 524). A study of other species of *Nephelium* and those of *Xerospermum* with transverse embryo will help in explanation.

In the wild species of *Nephelium* and in the less selected forms of *N. lappaceum* in cultivation, the pulp is inseparable from the firm inner layer of the testa. Selection for separable pulp has produced the 'Penang Rambutan' in which the outer thinly sclerotic layer of the inner testa breaks off with the pulp or the layer of small cells just external to this sclerotic layer breaks down to free the pulp, for both states exist. This selection has improved the eating and the sweetness but lost the wild flavour.

Paullinia L.

Seeds becoming campylotropous with slight radicle-pocket. Testa black or brown, more or less reduced by the sarcotestal chalaza, the aril with distinct or vestigial free lobes. Testa; *o.e.* composed of a palisade of thick-walled lignified cells with stellate facets; *mesophyll* with crystal-cells. Tegmen crushed.

Aril, when present, exostomal and funicular but scarcely developed on the antiraphe-side of the micropyle.

This large genus with similar exotesta to *Cardiospermum* and *Urvillea* shows well the simplification of the Sapindaceous capsule and seed according to the durian-theory (p. 238). The pachychalazal expansion seems to have taken two directions. In one, such as *P. riparia* (Fig. 526) it expands to make all of the seed-coat except the apex and micropylar or antiraphe-side. In the other the expansion is limited to the sides of the seed and the normal Sapindaceous testa develops as a broad band along the whole of the antiraphe (*P. pinnata* Fig. 525). Whether there are simply arillate species without pachychalaza, as in *Guioa*, is not clear.

P. cupania Kunth

Testa; *o.e.* as a layer of columnar cells with thickened outer ends and more or less fusiform lumen (? stellate cells with radially elongate body); *mesophyll* slightly thick-walled, often with crystals, the inner cell-layers thin-walled and crushed, but round the micropyle with thick-walled and pitted cells (? endostome). Vascular bundles branching from the hilum.

(Netolitzky 1926; v.d. Pijl 1957.)

P. pinnata L.

(Surinam, leg. E. J. H. Corner 1947; dried material.) Fig. 525.

Seeds 12 × 5–7 mm, subcylindric, slightly compressed laterally, with black arcuate antiraphe, the rest of the surface covered by the adnate aril with slight free crenulate edge; micropyle basal, as a very short arillostome at the junction with the short brown funicle 1–1.5 × 1 mm; hilum rather small, round. Testa 12–15 cells thick on the antiraphe; *o.e.* as a compact layer of thick-walled, lignified cells 25–35 μ high, with stellate facets and brown contents, limited to the antiraphe-side of the seed; *mesophyll* crushed, many cells with crystals, thickened at the hilum and aerenchymatous with shortly armed cells, the outer layers with thick lignified walls; *i.e.* crushed. Tegmen crushed, forming a short radicle-pocket; *endostome* long, not lignified. Sides of the seed with thicker mesophyll, the outer cells lignified and continuous with those of the hilum but forming a thinner sclerotic layer connecting with the fleshy tissue of the adnate aril. Hilum without a palisade. Vascular bundle of the funicle dividing at the base of the seed into two short branches, subdividing on the sides of the seed in the pachychalaza (? with short branches towards the antiraphe). Embryo transverse.

Aril funicular and exostomal, scarcely developed on the antiraphe-side of the micropyle, without v.b.

P. riparia (Spruce) Radlk.

(Brazil, Spruce 3883; dried material, CGE.) Fig. 526.

Capsule densely set with hard simple thorns –9 mm long. Seeds 2–3 per capsule, 11–12 × 8 mm, with blackish testa over the apical third and along the antiraphe-side, the remainder of the seed sarcotestal. Testa with stellate-celled *o.e.* Aril with distinct free limb except on the antiraphe-side.

Pometia Forst.

P. pinnata Forst.

Fig. 513.

Seed 2 cm long, subcylindric, obtuse, brown, solitary in the indehiscent fruit. Testa firm, rather woody-coriaceous, pervaded by v.b., with a radicle-pocket; *o.e.* as short, thick-walled cuboid cells.

Aril covering the seed or leaving the apex free, closely attached round the base of the seed, pulpy, sweet, pale yellow.

(v.d. Pijl 1957; Jacobs 1962.)

The aril is funicular but whether it is also micropylar is not clear. In ripe seeds of f. *tomentosa* (Bl.) Jacobs (*P. eximia* Thw.) I found the aril to be so closely pressed to the testa as to seem adnate until examined microscopically, when the exotesta was clearly distinguishable except at the base of the seed where they were concrescent. But, the inner side of the aril, pressed against the testa, was hard and composed of many long, lignified, scarcely pitted, longitudinal fibres, often set in bundles. The outer tissue of the aril was pulpy. I have not seen such fibres in other arils.

Sapindus L.

S. mukurossi Gaertn.

(Japan, Narra, leg. E. J. H. Corner 1966; ripe fruits in alcohol.) Fig. 513.

Seeds 14–15 mm wide, subglobose, hard, dark brown, exarillate, distinctly but not extensively pachychalazal. Testa 60–70 cells thick, not lignified; *o.e.* as a palisade of hexagonal prisms with somewhat thickened walls, 600–900 μ high round the lower part of the seed and with a *linea lucida* towards the outer ends of the cells, *c.* 100 μ high over the upper part of the seed and without a *linea lucida*; *mesophyll* in the outer half consisting of rows of slightly radially elongate cells, in the inner half aerenchymatous with subglobose cells with short arms; *i.e.* unspecialized. Tegmen 9–12 cells thick, unspecialized; *endostome* not sclerotic. Chalaza dilated into the vascular pachychalaza, with the same structure as the testa but an extensive hypostase substituted for the tegmen. Vascular bundle of the raphe dividing at the chalaza into numerous branches anastomosing in the inner tissue of the pachychalaza, the free testa without v.b. Hilum massive, developing a thick mass of sclerotic cells perforated by the micropyle and the raphe v.b. Subhilar tissue of the funicle with abundant lignified fibres, appearing as idioblasts, especially round the v.b.

Pericarp with many large mucilage-cells in the mesocarp; exocarp and mesocarp not lignified; endocarp as a thin layer of slightly lignified fibres, 1–4 fibres thick.

For *S. saponarius*, the raphe-bundle is said to divide at the chalaza into two branches, one to each side of the seed (Le Monnier 1872); the seed may be more extensively pachychalazal than in *S. mukurossi*.

For *S. trifoliatus*, Netolitzky described a fibrous layer as the innermost layer of the testa; it may be the fibrous exotegmen, as in *Alectryon*.

Trigonachras Radlk.

T. acuta Radlk.
(Singapore; living material.)   Fig. 527.

Seed *c.* 14 × 9 mm, ovoid, black, shiny, attached at the base to the thick, fleshy, orange, waxy-pulpy, 2–3-shouldered aril derived from the thickened funicle and placenta. Testa; *o.e.* as a palisade of thick-walled columnar cells, the walls becoming dark brown; *mesophyll* with slightly thickened walls, drying up.

Capsule 3-shouldered, with thick pericarp; endocarp as two woody cocci per loculus; mesocarp rather pulpy, with patches of sclerotic tissue round the cocci.

The short funicle swells after fertilization to develop the two main shoulders of the aril, to which the adjacent placenta contributes. *Toechima*, mentioned by van der Pijl (1957), appears to have a similar funicular aril.

Xanthoceras Bunge

X. sorbifolium Bunge
Fig. 502.

Ovules 7–8 per loculus, anatropous, transverse but the upper ones inclined to become suspended and the lower ones to be erect, after fertilization becoming campylotropous; o.i. 7–8 cells thick; i.i. 4–5 cells thick.

Seeds one to several per loculus, rounded subcompressed, hard, black, exarillate (? a trace of a funicular aril in young seeds). Testa multiplicative, *c.* 30 times as thick as the tegmen; *o.e.* as a short palisade of thick-walled cells; *mesophyll* composed of thick-walled tissue with strands of thin-walled tissue. Tegmen 8–11 cells thick, the outer 4–5 cell-layers thin-walled, the inner 4–6 cell-layers thick-walled and dark; *i.e.* unspecialized.

(Guérin 1901.)

SAPOTACEAE

Ovules anatropous to suborthotropous, erect, solitary, basal, apical, or lateral and axile, with a short slender funicle or sessile with a broad attachment, unitegmic, tenuinucellate; integument thick; hypostase none.

Seeds medium-size to large, 5–50 mm long, ellipsoid, mostly subcompressed, light brown to black, hard, shiny, exarillate, more or less albuminous; *hilum* pale or white, small (*Antranella, Sarcosperma*), linear and almost as long as the seed (*Achras*), or variously expanded to form a third to two-thirds of the seed. Seed-coat multiplicative in most cell-layers, including o.e.; outer part (8–25 cells thick or more) forming a heavily lignified, sclerotic case of compressed, isodiametric pitted cells; inner part (of similar thickness) thin-walled, aerenchymatous, pallid or brownish, carrying the v.b., eventually more or less crushed; *o.e.* composed of small lignified cells of varying height, not as a palisade; *i.e.* crushed, unspecialized. Chalaza indistinct. Vascular supply with several postchalazal branches along the antiraphe and over the sides of the seed especially towards the hilum, free or anastomosing. Endosperm nuclear, thin-walled or with thick, often amyloid walls, not ruminate, oily. Embryo massive; radicle short; cotyledons thick or thin. Figs. 528, 529.

(Netolitzky 1926; Chesnais 1943; Baehni 1965; Bhatnagar and Gupta 1970; Vaughan 1970.)

Sapotaceous seeds are often easy to recognize from the hard shiny coat with conspicuous hilum. They have been used for the primary classification of the genera (Baehni 1965), but their study has been largely superficial. Actually the genera have been revised so often and with such conflicting results that it is difficult to know what may be their limits or their relationships. The massively sclerotic outer part of the integument seems to correspond with the sclerotic mesotesta of the bitegmic Theaceous seed, and it may be that the inner integument has been lost; there is no evidence that the seed is pachychalazal. The ovule appears to lose its funicle in the more advanced genera and, in becoming sessile, to develop a broad attachment which results in the linear or expanded hilum (Chesnais 1943). Whether this process has been monophyletic or polyphyletic may be discovered from the anatomy of the seed. The outer tissue of the hilum is not lignified; the inner tissue may be lignified as a continuation of the sclerotic part of the seed-coat or it is thick-walled but unlignified (*Chrysophyllum sp.*, Fig. 528). The vascular supply may also be important, whether chalazal or from the hilum as well. Le Monnier (1872) figured a peculiar state for *Lucuma mammea* in which two postchalazal branches, one on each side of the hilum, gave off numerous short transverse branchlets towards the hilum.

Sapotaceae are usually classified with Ebenaceae. The Ebenaceous ovule is bitegmic and suspended with internal micropyle in contrast with the erect unitegmic Sapotaceous ovule. The Ebenaceous seed specializes in a well-developed exotesta, the Sapotaceous in a mesotesta. The Ebenaceous fruit is strongly basipetal; the Sapotaceous fruit seems to grow uniformly. In tree-form and foliar venation the families differ. If they have had a common ancestor of Ebenalean distinction, it would appear to have been as an offshoot of ancestral Theaceae. However for the unitegmic Sapotaceae an affinity with ancestral Actinidiaceae with large seeds must be considered.

SARRACENIACEAE

Ovules anatropous, unitegmic (*Heliamphora, Sarracenia*), or ? bitegmic (*Darlingtonia*), tenuinucellate.

Seed small, thinly winged (*Heliamphora, Sarracenia*), with aseptate hairs (*Darlingtonia*), albuminous, exarillate. Seed-coat; *o.e.* with thickened radial and inner walls, the rest of the tissue crushed. Endosperm cellular, oily. Embryo minute.

(Netolitzky 1926.)

The position of the family is discussed on p. 40.

SAURURACEAE

Ovules hemianatropous to orthotropous, erect, bitegmic, crassinucellate or subtenuinucellate (*Houttuynia*), 2–4 per carpel or 6–11 per placenta in the syncarpous ovaries; o.i. 2(–3) cells thick, forming the micropyle; i.i. 3–4 cells thick.

Seeds small, 1–2.5 mm long, seed-coats not multiplicative. Testa thin-walled, more or less crushed. Tegmen with o.e. and i.e. becoming very thick-walled, the middle cell-layer disappearing. Vascular bundle short, without postchalazal branches. Perisperm abundant. Endosperm cellular or helobial, small. Embryo minute.

(Netolitzky 1926; Quibell 1941; Raju 1961; Yoshida 1961.)

The seeds are tegmic with perisperm, much as in Piperaceae (p. 43). I found in *Houttuynia cordata* Thunb., with material collected at Tjibodas in Java (April 1972), that the exotesta was composed of large cells, shortly longitudinally elongate, with thin brownish walls; the exotegmic cells were similarly elongate and

were crushed in the mature seed except for the thick, but not lignified, outer walls; the endotestal cells were rather large, with brownish contents, thinly lignified, and projecting slightly into the perisperm, but eventually crushed. Thus, the firm layer of the seed-coat was practically the outer wall of the exotegmen.

SAUVAGESIACEAE

Ovules anatropous, numerous, parietal, bitegmic; o.i. ? 2 cells thick; i.i. ? 3–4 cells thick.

Seeds minute or small (–3 mm long in *Euthemis*) or filiform, in some cases more or less alate, albuminous, exarillate; seed-coats apparently not multiplicative. Testa with o.e. large-celled, and i.e. with small crystal-cells. Tegmen; *o.e.* as a compact layer of longitudinal, pitted, lignified, narrow fibres; *i.e.* as tabular cells with brown contents, not lignified. Endosperm with aleurone grains. Embryo small, straight. Fig. 530.

This family contains that part of Ochnaceae called Albuminosae or Sauvagesioideae. It is distinguished by the syncarpous ovary, the capsular or baccate fruit, and the small bitegmic seed with fibrous exotegmen and proteinaceous endosperm. Why floral detail should have caused the two groups of Ochnaceae to be combined in one family I have not analysed, but it is clear from the fruit and seed that Ochnaceae belong with Simarubaceae, and Sauvagesiaceae with Violaceae, as placed by Bentham and Hooker (*Gen. Plant.* vol. 1, 1862, 114–121).

The observation of van Tieghem, recorded by Netolitzky, that there was a layer of lignified cells in the seed-coat of *Luxemburgia*, lead me to investigate the following species, for which only dried material was available. The seed-structure is consistent even in the pyrenes of *Euthemis*. Compare Legnotidaceae.

Euthemis leucocarpa Jack
(C. Hose, Borneo, 1894; CGE.)

Seed 3 mm long; microscopic structure as in *Luxemburgia*; testa with o.e. as flattened cells.

Luxemburgia ciliosa Gardn.
(Gardner, Rio de Janeiro; CGE.) Fig. 530.

Seeds 2 × 0.8 mm, dark brown, fusiform, minutely papillate. Testa 2 cells thick, not lignified, more or less detached at the ends of the seeds; *o.e.* consisting of large, bulging cells with brown contents and distinct cuticle; *i.e.* composed of small crystal-cells. Tegmen ? 3–4 cells thick; *o.e.* as a compact layer of lignified, pitted, longitudinal fibres –250 × 7–8 μ, with fusiform ends; *mesophyll* collapsed; *i.e.* as tabular cells with brown contents, not lignified.

Sauvagesia erecta Lindl.
Seeds 0.5 mm wide, subglobose, blackish, hard, coarsely reticulate. Testa; *o.e.* composed of relatively large cells

(*c.* 10 cells from base to apex of the seed), with the outer wall collapsing but with finely subreticulate thickenings along the parts bordering the radial walls; *i.e.* with rather large crystal-cells. Tegmen as in *Luxemburgia*.

SAXIFRAGACEAE

Ovules anatropous, bitegmic or unitegmic (*Saxifraga hieracifolia*, *S. pallida*), crassinucellate; o.i. and i.i. 2 cells thick, not multiplicative in the seed.

Seeds small, albuminous, exarillate. Testa reduced to o.e. as a layer of cuboid cells, in some cases radially elongate to form ridges on the seed, or the cells papilliform, with thickened walls. Tegmen crushed or persistent in the micropylar part; *i.e.* thick-walled and persistent in *Heuchera*. Endosperm cellular, nuclear, or helobial, oily. Embryo minute.

(Netolitzky 1926; Wiggins 1959; Singh, B. 1964; Saxena 1969.)

The ridges on some seeds, formed as in Crassulaceae by radial elongation of the exotestal cells along certain lines, suggest the vestige of a uniformly palisade-like exotesta. Though considered, as Saxifragales, to be a centre of diversification for other families, the seed-structure indicates reduction in the family itself (p. 36). The thick-walled endotegmen of *Heuchera* raises the question of the affinity of Podostemaceae (p. 219).

SCHISANDRACEAE

Ovules anatropous, to campylotropous, bitegmic, crassinucellate, 2–3 per carpel; o.i. 5–7 cells thick, 10–11 at the micropyle and along the raphe; i.i. 2–3 cells thick, more at the micropyle; chalaza unspecialized; micropyle formed by the endostome and exostome, or mainly by the exostome (*S. chinensis*).

Seeds rather small, reniform, yellow to brown (*Schisandra*) or greyish brown (*Kadsura*), smooth, albuminous, exarillate. Testa not (or slightly ?) multiplicative, without v.b.; *o.e.* as a palisade of columnar, thick-walled, lignified, prismatic cells with branched pits; *o.h.* as 2–3 layers of cuboid or tangentially elongate, thick-walled, lignified cells with simple pits; *i.h.* as 1–2 layers of small thin-walled cells; *i.e.* as a layer of large, shortly radially elongate, thick-walled cells with simple pits. Tegmen not multiplicative, unspecialized, eventually crushed, but the cells at the micropyle thick-walled as those of the exotesta. Endosperm cellular, with starch and oil. Embryo minute.

Pericarp 10–12 cells thick in the flower, 15–16 cells thick in the fruit, unspecialized except i.h. with 1–2 layers of thin-walled cells greatly elongate radially, not lignified, fleshy-pulpy at maturity. Fig. 531.

(Netolitzky 1926; Berger 1952; Yoshida 1962; Kapil and Jalan 1964.)

This description refers to *Schisandra chinensis* C. Koch

and *S, grandiflora* Hook.f. et Th. According to Berger, *Kadsura japonica* differs in the short cells of the exotesta with thickened outer and radial walls and in the thin-walled, more or less crushed, mesophyll of the testa; it is an exotestal seed.

The Schisandraceous seed resembles that of *Illicium* but it may become campylotropous (*S. grandiflora*). The exotestal cells of *Illicium* are stellate–undulate whereas in Schisandraceae one is led to believe that they have hexagonal facets. Hypodermal sclerotic cells occur in 2–3 layers in *Schisandra* (though not mentioned for *S. chinensis*), and scattered sclerotic cells occur in the testal mesophyll of some species, but not all, of *Illicium*. The testa is multiplicative in *Illicium* but not in *Schisandra*. There is a sclerotic endotesta in *Schisandra*, which seems not to occur in *Kadsura*, and is absent from *Illicium*. The endocarp of *Illicium* has a strongly lignified palisade, absent from the baccate carpels of *Schisandra*. With such variations there is little definite upon which to decide the relative position of the two families. *Schisandra* is more primitive in the thick testa, the palisade of the endotesta, and the absence of lignified endocarp, but it is advanced in the campylotropous form of the seed and, perhaps, in its small size. The two families have been compared in detail by Kapil and Jalan. If close, their modern representation is fragmentary. *Schisandra* has the elongating Magnoliaceous receptacle and, perhaps, the relic of the endotesta. *Illicium* has the whorl of carpels as in *Liriodendron*; it is reduced to the solitary ovule in the dry follicle. Both families may be relics of the mesotestal ancestry of Hamamelidales, Rosales, Myrtales and Theales.

SCROPHULARIACEAE

Ovules anatropous, hemianatropous. or campylotropous, unitegmic, tenuinucellate; integument commonly 3–7 cells thick, 7–10 in *Antirrhinum*, 12–14 in *Ellisiophyllum*.

Seeds small to minute, in some cases alate, albuminous, exarillate. Seed-coat typically exotestal, multiplicative or not, with a firm outer hypodermal layer in *Limosella*, *Paulownia* (lignified) and ? *Stemodia*; with a thick-walled inner epidermis in *Angelonia*, *Celsia*, *Collinsia*, *Gerardia*, *Torenia* and *Vandelia*; *o.e.* with variously thickened and lignified cells, commonly with the inner and radial walls thickened, in some cases projecting in groups to form ridges (*Antirrhinum*), as mucilage-cells (*Ellisiophyllum*), tangentially elongate with reticulate thickening on the inner and radial walls (*Paulownia*), multiplicative in *Anticharis* but finally exotestal, thin-walled and disintegrating in seeds with thick-walled o.h. or i.e.; *mesophyll* generally crushed, but forming a thick seed-coat with reticulate or spiral thickening of the cell-walls in *Phygelius*; *i.e.* as an endothelium, in some cases persistent and thick-walled. Vascular bundle limited to the raphe or absent from smaller seeds. Endosperm cellular (? partly nuclear in *Paulownia*), oily or with starch (*Mimulus*), sometimes pitted on the outside from impressions of the seed-coat, the chalazal part of the endosperm forming an elaiosome in *Lathraea* and *Melampyrum*.

(Neotolitzky 1926; Svensson 1928; Millsaps 1936; Krishna Iyengar 1937, 1939, 1940, 1942; Woodcock 1939; Srinath 1940; Srinivasan 1940; Guilford and Fisk 1951; Yamazaki 1953, 1957; Tiagi 1956a; Hartl 1960; Bresinsky 1963; Singh, B. 1964; Arekal 1964; Schrock and Palser 1967; Arekal, Rajeshwari and Swamy 1971; Tsan-Iang Chuang and Heckard, 1972.)

Alonsoa, Angelonia, Anticharis, Antirrhinum, Calceolaria, Celsia, Centranthera, Collinsia, Cordylanthus, Deinostema, Digitalis, Dopatrium, Ellisiophyllum, Euphrasia, Gerardia, Gratiola, Herpestis, Limnophila, Limosella, Linaria, Melampyrum, Mimulus, Paulownia, Pedicularis, Phygelius, Ramphicarpus, Rhinanthus, Scoparia, Scrophularia, Sopubia, Stemodia, Striga, Torenia, Tozzia, Vandelia, Verbascum, Veronica.

In spite of this list of some 37 genera that have been studied from the point of view of one or two species, it is clear from the variety in construction of the exotestal cells and the hypodermis that there is much yet to be learnt about the family. The intermediate status of *Paulownia* (Millsaps 1936) shows that this variation must be interpreted in the light of Bignoniaceous seeds. Indeed, this list of mainly temperate genera points to the need for research into tropical seeds.

SCYPHOSTEGIACEAE

The interpretation of the one species of this family, *Scyphostegia borneensis*, has been beset with error. The ovules have been mistaken for flowers, the seeds for carpels, and the ovary for a floral receptacle or inflorescence. Monimiaceous and Moraceous affinities have been invoked. Swamy (1953c), van Steenis (1957), and van Heel (1967) have corrected these points, though still laboured by Croizat (1960). Affinity hovers between Celastraceae and Flacourtiaceae (Metcalfe 1956). Both families have the fibrous exotegmen which is the character of the seed of *Scyphostegia*; it allies them with Violaceae and *Scyphostegia* has the layer of sclerotic cells interposed between the hypostase and the termination of the vascular bundle in the chalaza which is a feature of Violaceae.

To explain the position of the many ovules over the dilated base of the ovary, which is without septa, Swamy inferred that the ancestral state might have had parietal placentation and that the ovules had, as it were, slipped from these placentas to the floor of the ovary. The suggestion does not explain the centrifugal development of the ovules from the vestigial floral apex or their uniform orientation with the micropyle directed towards the outside of the ovary (Fig. 534). Van Heel's re-

searches have discovered that the ovary begins with a circle of 9–13 subcrescentic carpel-primordia around the floral apex. They are lifted on a syncarpous ovary which is intercalated as a tube with basipetal growth. The carpel-primordia become the stigmata and the clefts between them become the arms of the stylar canal which lead through the short thick style into the top of the ovarian cavity (Fig. 535). Taken together with the oddities of the inflorescence and the male flower, these peculiarities may indicate a very early ancestral experiment in developing a syncarpous ovary without septa or placentas. The carpel-primordia, which become the stigmatic lobes on the style, retain the complicated and vestigial vasculature (Fig. 535).

The seed resembles strongly in shape and construction that of the Celastraceous *Sarawakodendron* (Fig. 94–96). It is small: its integuments are not multiplicative; and it lacks the basal aril-filaments. The aril corresponds with the exostomal and funicular cushion of *Sarawakodendron*. The main peculiarity of the seed of *Scyphostegia* is the intercalary extension of the endostome and exostome to form in conjunction with the raphe a pseudo-funicle (Fig. 532). The resemblance with *Sarawakodendron* is heightened by the distichous leaves, the closely bracteate inflorescences, the three stamens with extrorse anthers, the oblong urceolate capsule of *Sarawakodendron* with the seeds disposed over the almost ridge-less interior, and the occurrence of both monotypic genera in the northern part of Borneo (cf. Ding Hou 1967, fig. 1, and van Steenis 1957, fig. 1). Now, the ovules of other Celastraceae may be borne on the base of the ovary (Andersson 1931, fig. 1F, for *Celastrus scandens*). Dilation of the floral apex to accommodate the numerous carpel-primordia of *Scyphostegia* might well result in the centrifugal development of the ovules, comparable with the centrifugal development of stamens in other families. In its multicarpellary syncarpous ovary *Scyphostegia* is more primitive than *Sarawakodendron*, but its seed is reduced. *Sarawakodendron* has the reduced ovary, but the more primitive seed. This elongate and doubly arillate seed may have been the ancestral form in the exotegmic alliance that later converted the many follicles into capsules and diversified the construction of the syncarpous ovary into Celastraceae, Flacourtiaceae and, if it can be distinguished from the former as a precursorial relic, Scyphostegiaceae.

The immediate affinity between *Scyphostegia* and *Sarawakodendron* can be tested in the field. *Scyphostegia* is a monopodial tree with the leaves spirally arranged on the 3-angled main stems (4-angled, van Steenis 1957), though distichous on the side-branches, and these side-branches develop in the Annonaceous manner almost at right-angles to the median plane of the subtending leaf.

(Swamy 1953c; Metcalfe 1956; v. Steenis 1957; Croizat 1960; v. Heel 1967, 1970a.)

Scyphostegia borneensis Stapf
(Borneo, Kinabalu, RSNB 1609; flowers and fruits in alcohol.) Figs. 532–535.

Ovules anatropous, erect, numerous over the floor of the ovary-cavity, bitegmic, crassinucellate, with short thick funicle; o.i. 2–3 cells thick, thicker in the raphe; i.i. 3–4 cells thick; micropyle formed by the exostome, often lobulate.

Seeds 10–12 mm long (overall), arillate at the base, albuminous; seed-body 1.3–1.5 mm wide, brown, oblong, curved, thinly and minutely hairy, attached by the elongate raphe-micropyle as a stalk (pseudo-funicle) to the short funicle; seed-coats not multiplicative. Testa not lignified; o.e. composed of rather large cells with firm walls, shortly radially elongate, some produced into the narrowly conical, rather thick-walled, aseptate, unlignified hairs; *mesophyll* crushed; i.e. composed of small cuboid cells with dark brown tannin-contents, thin-walled. Tegmen; o.e. has a hard, lignified layer of longitudinal fibres 200–400 μ long, subcylindric, with thick, yellow-brown, finely pitted walls, the cells becoming radially elongate at the endostome and fibrous; *mesophyll* composed of 2 layers of shortly transversely elongate, thin-walled cells, eventually crushed; i.e. similar to the endotesta. Nucellus long persistent as a thin layer, eventually crushed. Vascular bundle of the raphe ending in a few short branches in the thick chalaza; not penetrating the hypostase of small cells with tannin-contents, but sealed off by a single layer of small sclerotic cells. Endosperm oily, the cells with slightly thickened walls. Embryo straight; cotyledons ligulate.

Aril white, pinkish downwards, firmly fleshy, not vascular, 3–4 lobed, rugulose, developed from the exostome and funicle-apex, the region of attachment elongating to give a short, fleshy, rugulose pseudo-funicle traversed by the exostome, the endostome limited to the part of the pseudo-funicle above the aril-attachment.

At first sight the seed appears to have a long thick funicle with short funicular aril. Actually most of this stalk is the pseudo-funicle made by the elongation of the micropylar part of the seed and consisting of two sections. The distal free part, inserted into the aril-cup, is a hard brown stalk stiffened by the fibres of the thick endostome, which appears as a solid mass of xylem in t.s. (Fig. 533). The proximal part is the extended exostome and raphe with the aril attached at its further end. The micropyle-opening is near the base of the pseudo-funicle which has the same texture as the aril and signifies the very short, thick, arilloid, true funicle.

SCYTOPETALACEAE

Ovules anatropous, unitegmic, ? crassinucellate.

Seeds albuminous, exarillate. Seed-coat c. 10 cells thick; o.e. with large cells, the outer wall projecting and almost hair-like (*Oubanguia*), developing hairs along the raphe (*Brazzeia, Rhaptopetalum*), or completely hairy (*Pierrina*); i.e. with radially elongate cells, ? endothelial. Vascular bundle with postchalazal extension almost to micropyle and with branches to the seed-coat, entering the ruminations in *Rhaptopetalum*. Endosperm simple or ruminate (*Rhaptopetalum, Scytopetalum*).

(Netolitzky 1926.)

If the family is allied with Elaeocarpaceae or Tiliaceae, it will need to be demonstrated from these tantalizing seeds.

SELAGINACEAE

Ovules anatropous, suspended, solitary, unitegmic, tenuinucellate; integument 5–7 cells thick.

Seeds small, albuminous, exarillate. Seed-coat not multiplicative, mesophyll more or less crushed; o.e. with reticulate thickening of the walls (*Lagotis*), or thin-walled and tangentially elongate (*Hebenstreitia*); i.e. at first endothelial, then crushed. Endosperm cellular.

(Junell 1961; Rau and Sharma 1962; Singh, B. 1964; Kapil and Masand 1964.)

Junell regards the family as a tribe of Scrophulariaceae. *Lagotis* is now placed in Veroniceae.

SIMAROUBACEAE

Ovules generally solitary, some times 2 or more per loculus, anatropous and suspended, amphitropous and suberect (*Picrasma*), bitegmic, crassinucellate, unitegmic and campylotropous (*Suriana*); o.i. 3–10 cells thick; i.i. 2–6 cells thick, the single integument of *Suriana* 6–7 cells thick; micropyle mostly formed by the endostome.

Seeds small to fairly large, enclosed in woody endocarp with little differentiation, albuminous or not, exarillate, pachychalazal in *Quassia*, less so in *Picrasma*. Testa more or less multiplicative (? *Picrasma*), eventually crushed and pellicular; o.e. composed of thin-walled tabular cells, in some cases aerenchymatous with short arms (*Brucea, Picrasma*); *mesophyll* unspecialized, but with scattered idioblasts in *Simarouba*; i.e. unspecialized or slightly lignified (*Brucea, Picrasma, Simarouba*). Tegmen not multiplicative, unspecialized, soon crushed. Nucellus much enlarging, long persistent, eventually crushed. Hypostase present. Vascular bundle only in the short raphe, with a fan-shaped network in the chalaza of *Picrasma* and *Quassia*, or with fine branches from the raphe to the testa in *Simarouba*. Endosperm nuclear, persistent and oily (*Brucea*) or reduced to a single layer of cells (*Simarouba, Suriana*), or entirely absorbed. Embryo straight or curved; radicle short; cotyledons thick, often plano-convex. Figs. 536–543.

(Netolitzky 1926.)

Though placed usually in Rutales, there is no evidence of this alliance in seed-structure unless the Simaroubaceous is entirely degenerate. Thus, if the seed of the

Rutaceous *Cusparia* (p. 234) lost the spiral or scalariform thickening of the tegmen, it would resemble the Simaroubaceous. Traces of this construction should be looked for in those genera of the family said to have dehiscent fruits.

On the other hand, the apocarpous ovary and the pachycaul habit with long pinnate leaves, well shown in *Eurycoma* (Nooteboom 1962, fig. 6), indicate a primitive status. The fruit of *Picrasma* resembles that of various Ochnaceae, and it must be considered whether the pachychalazal tendency in such Simarubaceae as *Picrasma* and *Quassia* may not have been accentuated to give that of Ochnaceae in which the seed-coat also lacks special character. That is, Simaroubaceae may be connected with the ancestry of Ochnaceae.

Concerning *Irvingia*, the seed of which has not been satisfactorily described (p. 155), it has been removed from Simaroubaceae to its own family and it has been transferred to Ixonanthaceae, where I have retained it for convenience. Its seed-structure fits Simaroubaceae better than Ixonanthaceae, but is peculiar in the sclerotic mesophyll of the testa. Compare, also, Balanitaceae. If *Irvingia* and *Balanites* are Simaroubaceous, they supply a mesotestal construction which would relate the family with Hamamelidales, Rosales, Myrtales, and Theales.

Brucea J. F. Mill.

B. javanica (L.) Merr.
(Ceylon; flowers and fruits in alcohol.)   Figs. 536, 537.

Ovule hemianatropous, subapical; o.i. 3 cells thick, somewhat lobate round the longer endostome; i.i. 2–3 cells thick, –4 cells at the endostome, exceeding the exostome and often splayed out against the carpel-wall.

Seeds 5×3 mm. Testa 7–9 cells thick, thin-walled, somewhat aerenchymatous, eventually crushed.; *o.e.* composed of thin-walled tabular cells, the walls eventually slightly lignified; *mesophyll* unspecialized; *i.e.* with slightly thickened, lignified and obscurely pitted, radial and inner walls. Tegmen not multiplicative, soon disintegrating. Vascular bundle of the raphe undivided in the chalaza. Endosperm 6–7 cells thick, oily.

Ovary with 4(–5) carpels and between them the remains of the floral axis as a short column with 4(–5) wings or flanges; styles relatively long, free, grooved at the base; carpel-body intercalated as a basipetal tube below the carpel-primordia (styles) and without ventral furrow or suture, with *c.* 10 v.b.

According to Nair and Sukumaran (1960), both integuments of the ovule are 4–6 cells thick and the endotegmen persists as a layer of slightly thick-walled cells with dark-staining contents.

Picrasma Bl.

P. javanica Bl.
(Solomon Islands, leg. E. J. H. Corner 1965; fruits in alcohol.)   Fig. 538.

Ovules solitary, basal, more or less erect with the micropyle adaxial.

Seeds 7×9–10×6.5 mm, wider than long, obtuse, pallid, with a slightly thickened extensive chalazal patch covering most of the abaxial side of the seed, sessile. Testa 3–5 cells thick, aerenchymatous, thin-walled; *o.e.* composed of somewhat tabular cells with short arms making air-spaces, without stomata, not as a palisade; *i.e.* becoming slightly lignified on the radial and inner walls but not thick-walled. Tegmen crushed. Chalazal region expanded, with *c.* 7 cell-layers external to the reticulate fan of v.b. derived from the short raphe, separated from the nucellus by a thin plate of hypostase with 1–2 layers of thin brown cells. Endosperm none. Embryo more or less transverse.

Ovary with 4(–7) free carpels with slight ventral furrow, the styles connate, the stigmata free.

Fruit consisting of 1–4 (? more) drupes derived each from a carpel, separated by the expansion of the receptacle, with the styles broken off near the base and the withered stylar column often long persistent on an undeveloped carpel; persistent petals and the receptacle red. Drupes $10 \times 12 \times 9$ mm, ripening red then black; exocarp somewhat pulpy; endocarp densely sclerotic, simulating a woody seed-coat; hilum filled with pale sclerotic tissue.

Quassia L.

Q. indica (Gaertn.) Nooteboom
(Ceylon, Peradeniya, cult.; Solomon Islands, RSS 2773, wild; flowers and fruits in alcohol.)   Figs. 539–542.

Ovules solitary, occasionally a second one more or less abortive, subapical; o.i. 7–10 cells thick, developing after the i.i. (6–8 cells thick); micropyle formed by the endostome; nucellus consisting of an outer zone (3–4 cells thick) of small cells and a massive core of larger cells.

Seeds $20 \times 18 \times 15$ mm, shrivelling much on drying of the fruit, rather bean-shaped, pale brown. Testa 7–10 cells thick, rather soft, drying papery, irregularly thickened (20–30 cells thick) and subcerebriform round the dilated chalaza (*c.* 50 cells thick); *o.e.* unspecialized; *mesophyll* aerenchymatous, thin-walled; *i.e.* unspecialized. Tegmen not or scarcely multiplicative, unspecialized, collapsing, but the mesophyll-cells near the chalaza at first enlarging greatly. Chalaza dilating after fertilisation into a massive pachychalaza forming the basal half of the seed, separated from the upper half by a slight annular thickening. Vascular bundle of the raphe dividing in the pachychalaza into a fan of fine branches, not entering the integuments. Nucellus enlarging greatly, eventually crushed by the endosperm except for the long persistent small-celled outer layer, the inner cells enlarging radially into an aqueous tissue before dissolution. Endosperm extending over the dome of the nucellus between the small-celled nucellar epidermis and the aqueous core, eventually absorbed by the embryo. Cotyledons subequal.

Ovary with 4 free carpels, the styles connate in a single column with a fine stylar canal, without stigmatic arms; carpel-primordia at first free, then closely adnate and forming the stylar-column by intercalary growth; floral stem-apex persistent between the carpel-bases and covered with simple and glandular hairs.

Simarouba Aubl.

Simarouba sp.
(Ceylon, Peradeniya, cult.; ripe fruits in alcohol.) Fig. 543.

Testa unspecialized, 7–8 cells thick on the flattened sides, the cells more or less collapsed; *o.e.* composed of tabular cells; *mesophyll* with scattered lignified idioblasts, most near the v.b.; *i.e.* (? testa) composed of cuboid cells with lignified pitted thickening at the base of the radial walls at the junctions with the inner wall. Tegmen apparently crushed. Vascular bundle of the raphe with fine branches to the side of the seed. Cotyledons, thick, flat, the cells with oily contents. Seed set transversely in the samara.

S. glauca DC
Testa unspecialized; *o.e.* with brown cells. Endosperm as a single layer of aleurone-cells.
(Vaughan 1970.)

Suriana L.

S. maritima L.
Ovules anatropous, bitegmic, crassinucellate.

Seeds 2.5 mm high, campylotropous, papillate, yellowish white. Hypostase well developed. Embryo curved.
(Gutzwiller 1961.)

As a result of her detailed researches, which did not include the microscopic structure of the seed, Gutzwiller decided that *Suriana* belonged to Geraniales, as a family Surianaceae to be placed somewhere near to Connaraceae and Sapindaceae. Cronquist associates *Suriana* with *Stylobasium* (Chrysobalanaceae) in Stylobasiaceae, which he places in Sapindales next to Connaraceae. Takhtajan keeps *Suriana* doubtfully in Simarubaceae and places Stylobasiaceae with doubt next to it. Hutchinson retains the genera in Simarubaceae and Chrysobalanaceae. Perhaps the seed will provide a mark.

SOLANACEAE

Ovules anatropous to campylotropous, numerous, small, unitegmic, tenuinucellate; integument relatively massive, varying 6–10 to 20–30 cells thick (20–30 in *Lycopersicum*); i.e. becoming an endothelium after fertilization.

Seeds small, often flattened and discoid, or subquadrate, mostly albuminous, exarillate, in some cases embedded in placental pulp (*Lycopersicum*, *Solanum pr. p.*). Seed-coat multiplicative or not, generally reduced to o.e. and i.e. with the mesophyll crushed; *o.e.* as a compact layer of cells with more or less undulate or stellate facets, either with more or less strongly thickened, in some cases lignified, inner and radial walls (*Atropa*, *Browallia*, *Cestrum*, *Lycium*, *Mandragora*, *Nicandra*, *Nicotiana*, *Petunia*, *Solanum pr.p.*, *Withania*), or the cells developed radially with the wall-thickenings continued from the inner wall radially along the angles of the cells, but remaining thin-walled distally, the thin-walled parts becoming mucilaginous and leaving the more or less reticulately thickened bases of the cells exposed with their thickened angles as hairs (*Lycopersicum*, *Solanum pr. p.*), or the epidermal cells much flattened and unspecialized (*Nicotiana*, *Nierembergia*, *Petunia*); *mesophyll* thin-walled, crushed, or the outer cells with reticulate thickenings (*Cestrum*, *Datura*); *i.e.* generally persistent as a layer of small compressed rectangular cells with rectangular of subundulate facets, the radial walls in some cases lignified and thickened. Vascular bundle short, in many cases not reaching the chalaza, or with a few short postchalazal branches (*Datura*). Endosperm nuclear, cellular or helobial, oily or starchy (*Browallia*). Embryos straight, curved, or coiled, with flat or folded cotyledons. Figs. 544, 545.
(Netolitzky 1926; Smith, O. 1935; Magtang 1936; Dnyansagar and Cooper 1960; Karuna Mohan 1968; Saxena and Singh 1969; Vaughan 1970.)

Clearly there are many differences in detail between the genera, even within the main genus *Solanum* which shows no clear distinction in seed-structure from *Lycopersicum*. The seed is exotestal and in the development of the spurious hairs from the thickenings in the exotestal cells there is great resemblance with the hairy seed-coat of *Strychnos* (Loganiaceae). In both cases the rigid parts of the hairs are the thickenings along the anticlinal walls of the cells. In *Strychnos* these cells are prolonged into free hairs which retain the thin outer wall. In *Solanum* and *Lycopersicum* the cells elongate into a thin-walled palisade with undulate facets, as in *Strychnos*, but the thickenings are freed by the dissolution of the thin walls. Whether the Solanaceous seed may be derived from the Loganiaceous and the hairless Solanaceous seed-coat may be derived from the hairy needs for an answer much more investigation of *Solanum*, in particular its pachycaul species. There is certainly no indication that the Solanaceous seed is connected with the more complicated construction of Convolvulaceae.

Cestrum L.

C. aurantiacum Lindl.
(Ceylon, Hakgala; flowers and fruits in alcohol.) Fig. 544.
Ovules with the integument 8–10 cells thick.

Seeds 1–3 per loculus, 4–5.5 × 2.5–3 × 2 mm, plano-convex, subfusiform or subtruncate at one end (2 seeds superposed), dark brown, finely and irregularly sub-reticulate. Seed-coat not multiplicative; *o.e.* as a compact layer of enlarged cells with polygonal facets, some with scarcely thickened and unlignified walls, others usually in groups with strongly thickened, closely pitted, lignified and brown radial walls, the inner wall with wide reticulate thickening, the outer wall with thinner and more or less unlignified reticulate thickening; *o.h.* and, commonly, the next cell-layer composed of small cells with widely pitted, lignified walls; *mesophyll* thin-walled, crushed, forming a brown layer; *i.e.* eventually crushed. Vascular bundle only in the very short raphe. Endosperm thin-walled, not starchy. Embryo almost straight; hypocotyl longer than the flat cotyledons.

Withania Pauq.

Withania sp.
(Kenya, D. R. D. Chetham n. 1, 1970; mature fruits in alcohol.) Fig. 545.

Seeds 2–2.3 × 2 × 0.8 mm, pale ochraceous, flattened, minutely areolate or reticulate. Seed-coat; *o.e.* composed of large stellate–undulate cells with thin outer wall but strongly thickened radial and inner walls, often with irregular obtuse projections from the inner wall into the lumen, the middle lamella slightly lignified; *mesophyll* reduced to 1–2 layers of small cells with thin, but lignified, walls; *i.e.* similar to the mesophyll-cells.

SONNERATIACEAE

Ovules anatropous, very numerous, bitegmic, crassinucellate; o.i. and i.i. 2 cells thick (*Duabanga*).

Seeds filiform with an appendage at each end (*Duabanga*), or hard, thick, ellipsoid, somewhat compressed and campylotropous (*Sonneratia*), exalbuminous, exarillate. Testa multiplicative (*Sonneratia*) or not (*Duabanga*); *o.e.* unspecialized; *mesophyll* densely sclerotic (*Sonneratia*); *i.e.* with crystal-cells (*Duabanga*), sclerotic (*Sonneratia*). Tegmen not multiplicative; *o.e.* composed of short, narrow tracheids with spiral thickening (*Sonneratia*), or with pitted, lignified fibres (*Duabanga*); *i.e.* unspecialized. Endosperm nuclear, absorbed. Embryo minute and straight with flat cotyledons (*Duabanga*) or larger, curved, and with short contort cotyledons (*Sonneratia*). Figs. 546, 547.

It is customary now to maintain this family with the two genera. They differ considerably in their seeds. Those of *Duabanga* agree with Lythraceae, as does the habit of the trees, the capsules, and the embryological details (Venkateswarlu 1937a). Those of *Sonneratia* with tracheidal exotegmen, unspecialized endotesta, and contort cotyledons suggest Punicaceae. But the seed of *Pemphis* (Lythraceae) also resembles that of *Sonneratia*. It is by no means clear that these three families have

been adequately distinguished and the genera rightly parcelled out to them. Probably there is only one.

The small seed of *Duabanga* is merely an enlarged and differentiated ovule.

Duabanga Buch.-Ham.

Ovules cylindric subfusiform, very numerous, sessile; o.i. and i.i. 2 cells thick, 3 cells in the exostome exceeding the endostome.

Seeds 4–5.5 mm long, 0.5 mm wide, filiform, chaffy, brownish, the small seed-body with a chalazal and a micropylar appendage of the testa, each about one-third the length of the seed; seed-coats not multiplicative. Testa not lignified; *o.e.* composed of large cells, often somewhat longitudinally elongate, with thick outer wall; *i.e.* as small crystal-cells becoming more or less obliterated on the antiraphe-side of the seed. Tegmen; *o.e.* as a compact, rather hard, layer of lignified, coarsely pitted, longitudinal fibres; *i.e.* thin-walled, with longitudinal cells, crushed. Vascular bundle in the long raphe slender. Embryo straight; radicle short.

D. grandiflora (Roxb.) Walp.
Seed 4 mm long. Testa; *i.e.* in places divided periclinally to form a mesophyll-layer. Endostome composed of enlarged tracheidal cells with spiral or reticulate thickening.
(Venkateswarlu 1937a, as *D. sonneratioides*.)

D. taylori Jayaweera
(Ceylon, Peradeniya, cult.; flowers and fruit in alcohol.) Fig. 546.

Seeds 5–5.5 × 0.5 mm, most of the seeds sterile. Tegmen; *o.e.* composed of fibres 20–25 μ wide, coarsely pitted, not tracheidal.
(Jayaweera 1967.)

Sonneratia Linn. f.

S. caseolaris (L.) Engl.
(Java, Herb. Bogor; fruits in alcohol.) Fig. 547.
Ovules with distinct funicles.

Seeds *c.* 6 mm long and wide, 3 mm thick, anatropous, laterally compressed. Testa highly multiplicative; *o.e.* as a layer of short cells, rectangular in t.s., with dark contents, thin-walled, becoming friable; *mesophyll* very thick, wholly sclerotic except round v.b., the inner part densely sclerotic and heavily lignified, the outer part scarcely lignified; *i.e.* consisting of small sclerotic cells. Tegmen apparently not multiplicative, *c.* 3 cells thick; *o.e.* and in places *o.h.* as small tracheidal cells shortly elongate, with fine and more or less annular thickening; *mesophyll* and *i.e.* thin-walled.

S. apetala Buch.-Ham.
Ovule with o.i. and i.i. 2 cells thick; funicle long.
(Venkateswarlu 1937a.)

SPHENOCLEACEAE

Ovules anatropous, unitegmic, tenuinucellate.

Seeds minute, exalbuminous. Testa with only o.e. persistent as a layer of cuboid cells with fibrous thickenings on the inner wall; *i.e.* at first endothelial. Endosperm cellular. Embryo straight.

(Kausik and Subramanyam 1946; Airy Shaw 1948; Subramanyam 1950b; Maheshwari, P. 1963.)

The ovule and seed support the customary position of *Sphenoclea* with or near to Campanulaceae rather than with Phytolaccaceae (bitegmic, crassinucellate), as advocated by Airy Shaw.

STACHYURACEAE

Ovules numerous, transverse, anatropous, bitegmic, crassinucellate; o.i. 3–5 cells thick; i.i. 3–4 cells thick; exostome exceeding the endostome; funicle with incipient aril on the raphe-side.

Seeds small, arillate, albuminous. Testa not multiplicative, or adding a hypodermal layer, becoming wholly lignified; *o.e.* with undulate anticlinal walls. Tegmen not multiplicative, crushed. Endosperm oily, thin-walled. Embryo rather small, straight; radicle shorter than the flat cotyledons. Aril derived from the exostome and funicle. Figs. 548–550.

(Mauritzon 1963b.)

Stachyurus is placed in Hamamelidales by Hutchinson (1967), next to Buxaceae. The wholly sclerotic testa agrees and the arillate seeds give a primitive character indicative of the winged seeds in some Hamamelidaceae. This alliance fits, also, with Theaceae (p. 29). Takhtajan (1969) placed Stachyuraceae in Violales near to Flacourtiaceae, but the absence of the fibrous exotegmen forbids the alliance. The question of Buxaceae is more problematic. *Buxus* is simply exotestal, but *Sarcococca* has a more sclerotic testal mesophyll (p. 84).

Stachyurus praecox Sieb. et Zucc.
(Cambridge, University Botanic Garden; living material.) Figs. 548–550.

Ovary with parietal placentation above the lower axile placentation.

Seeds 2–2.3 × 1–1.2 mm, slightly curved; funicle short, pulpy. Testa 4–6 cells thick; outer 2(–3) layers of cells shortly radially elongate with sinuous anticlinal walls, closely pitted, lignified, derived by periclinal division of o.e.; inner 2(–3) layers of cells smaller, not so thick-walled or so closely pitted, somewhat tangentially elongate, the outer two rows with a crystal in each cell, lignified. Tegmen collapsing, but o.e. developing into rather large, tangentially elongate cells, not lignified.

STACKHOUSIACEAE

Ovules anatropous, erect, basal, with dorsal or ventral raphe, 1 per loculus, bitegmic, tenuinucellate; i.i. soon absorbed or crushed before anthesis.

Seeds albuminous, exarillate. Endosperm nuclear, oily. Embryo straight, large; cotyledons rather thick.

(Netolitzky 1926.)

Stackhousia Sm.

S. linariaefolia A. Cunn.
Ovule; o.i. 7–9 cells thick; i.i. 2–3 cells thick, evanescent.

Testa 9–11 cells thick; *o.e.* composed of tangentially elongate tannin-cells, unequally developed in different parts of the seed; *mesophyll* thin-walled; *i.e.* unspecialized, as small rectangular cells. Vascular bundle extending in the antiraphe to the micropyle, unbranched.

(Mauritzon 1936c; Narang 1953; Maheshwari, P. 1963.)

STAPHYLEACEAE

Ovules anatropous, numerous, transverse or obliquely erect, bitegmic, crassinucellate.

Seeds moderately large, anatropous or subcampylotropous (without a partial septum), exarillate, albuminous. Testa multiplicative; *o.e.* as a palisade of thick-walled cells; *mesophyll* thick, sclerotic; *i.e.* unspecialized. Tegmen more or less multiplicative, eventually crushed. Endosperm nuclear, copious, oily. Embryo straight; radicle short; cotyledons thick, plano-convex. Figs. 551–555.

(Netolitzky 1926; v.d. Linden 1960.)

I have excluded tr. Tapiscieae from this definition because it is either wrongly in Staphyleaceae or it represents a primitive element; the tribe is described on p. 257. The remainder of the family is usually placed near to Sapindaceae. The seeds are not so campylotropous and appear not to be pachychalazal. If these are primitive marks, then it is remarkable that the Staphyleaceous seed, as I have defined it, has no indication of the fibrous exotegmen as in the primitive Sapindaceous seed of *Alectryon* (p. 240). Recently van Steenis (in v.d. Linden 1960) has re-affirmed the suggestion that Staphyleaceae are related with Cunoniaceae. Unfortunately little seems to be known of the seeds of that family. If, however, Tapiscieae represent this primitive state with fibrous exotegmen then there is a parallel in Elaeocarpaceae where *Aristotelia* shows the sclerotic tegmen.

Euscaphis Sieb. et Zucc.

E. Japonica (Thunb.) Pax
(Japan, Kyushyu, Tagawa and Konta n. 74; dried seeds, herb. Kew.) Fig. 551.

Seeds 5.5 × 3.5 mm, subcampylotropous, 1–3 in a follicle, sessile, black, smooth. Testa thick, heavily lignified; *o.e.* as a palisade of large, radially elongate cells with hexagonal and finely striated facets, with thin, finely pitted, lignified, brown, radial walls, pulpy then drying into a loosely fibrous layer; *mesophyll* 0.5 mm thick, composed of many layers of thick-walled, finely pitted, contiguous, lignified cells, more or less isodiametric, large in the more internal tissues, not fibrous; *i.e.* ? Tegmen unspecialized, thin, more or less crushed. Vascular bundle in the raphe stout, evidently with some side-branches (? from raphe or chalaza).

The seeds are described by Ohwi (*Flora of Japan*, 1965) as nearly globose, 5 mm wide, and enclosed in a thin aril. I did not see this aril which may refer to the pulpy exotesta.

Staphylea L.

Ovules anatropous; o.i. 6 cells thick; i.i. 3–4 cells thick.

Testa: *o.e.* as a short palisade with thickened and pitted brown walls (thin-walled in *S. trifoliata*), with brown contents; *mesophyll* many layers thick, the outer cells very thick-walled, the innermost thin-walled and crushed; *i.e.* unspecialized, the cells somewhat tangentially elongate, then crushed. Tegmen increasing to 9–10 cell-layers, then crushed; *i.e.* with very small cells, then crushed. Vascular bundle on entry into the raphe dividing into two subdividing branches extending to the sides of the seed and with short subdividing postchalazal branches .Fig. 552.

S. pinnata L., *S. trifoliata* L.
(Le Monnier 1872; Guérin 1901; Netolitzky 1926.)

Turpinia Vent.

T. grandis v.d. Linden
(Borneo, Endert n. 4669; dried seeds, herb. Kew.) Fig. 553.

Seeds 6–7 × 3–3.5 mm, curved, compressed, campylotropous, light brown, smooth, hard, 1–3 per loculus (10–18 per fruit); hilum relatively large, elongate on the narrow basal ridge. Testa forming the hard woody seed-coat 0.6–0.8 mm thick; *o.e.* composed of rather flattened cells with thin, undulate, anticlinal walls and thickened, lignified, outer wall pitted with rather crowded fine pores 0.5–1 μ wide; *mesophyll* consisting of many layers of very thick-walled substellate cells, lignified, closely and finely pitted, small towards the more or less crushed i.e. Vascular bundles evidently spreading from the chalaza and forming a plexus in the inner layer of the testal mesophyll.

T. pomifera DC
(Ceylon, leg. E. J. H. Corner 1972; fruits in alcohol.)

Seeds 9–10 mm long and wide, 5 mm thick, dark brown, essentially as in *T. grandis*; *exotesta* with cuboid cells and hexagonal facets, the wall lignified and pitted all round: *mesophyll* densely sclerotic.

STAPHYLEACEAE-TAPISCIEAE

Ovules solitary, erect, anatropous, bitegmic, crassinucellate.

Seeds subglobose, exarillate. Testa 5–8 cells thick, unspecialized or (in *Huertea*) with sclerotic cells in the mesophyll; *o.e.* more or less unspecialized. Tegmen; *o.e.* as a layer of longitudinal, lignified, pitted fibres, cylindrical (*Huertea*) or laterally compressed (*Tapiscia*); *i.e.* with cuboid cells, the wall brown with slight pitting on the radial walls, not lignified. Chalaza strongly bullate, pervaded by tracheids from the end of the raphe-bundle, without postchalazal v.b. Hypostase composed of thin brown cells. Endosperm oily, reniform in l.s. Embryo rather small; radicle as long as the flat cotyledons.

The seeds of *Huertea* and *Tapiscia* with fibrous exotegmen and bullate chalaza recall those of Euphorbiaceae–Phyllanthoideae, such as *Glochidion* or *Securinega*, but the ovule is solitary and erect. The tribe relates to the other families with fibrous exotegmen (p. 14) but whether it entrains the remainder of Staphyleaceae awaits investigation. The importance of these seeds was brought to my attention by my colleague H. K. Airy Shaw, who supplied the material.

Huertea cubensis Grisebach
(Cuba, Jack and Rowe n. 7453; dried seeds, herb. Kew.) Fig. 554.

Seeds 8.5 mm long, subglobose, brownish, smooth; hilum rather small, round, the micropyle scarcely pointed; seed-coat rather thin. Testa 5–7 cells thick, –18 cells round the chalaza, thin-walled but with clusters of lignified, slightly thick-walled, rather coarsely pitted cells in the mesophyll; *o.e.* with subtabular, thin-walled, unspecialized cells; *i.e.* unspecialized. Tegmen; *o.e.* as a single layer of longitudinal, contiguous fibres –600 μ long, 50–75 μ deep, 24–48 μ wide, not or slightly flattened, thick-walled, lignified but with wide lumen, rather coarsely pitted, brownish; *mesophyll* thin-walled, collapsed; *i.e.* composed of rectangular cells with slightly pitted radial walls, not lignified. Vascular bundle spreading into fine rows of tracheids in the bullate chalaza, separated from the endosperm by a dark brown layer of agglutinated cells forming the hypostase, many of these cells thick-walled and lignified. Raphe with hard sclerotic tissue of lignified pitted cells surrounding the v.b. except on the inner side towards the endosperm. Endosperm somewhat thick-walled, oily. Embryo with rather long radicle; cotyledons flat.

Tapiscia sinensis Oliver
(Szechuan, Henry 8990; Hupeh, Wilson 108; dried seeds, herb. Kew.) Fig. 555.

Seeds 6–7 × 5 mm, plump; micropyle acute at the base of the loculus, the adjacent hilum small; chalazal area slightly depressed. Testa 6–8 cells thick, thin-walled, sarcotestal; *o.e.* composed of cuboid cells, without stomata; *mesophyll* with dense contents in some cells, giving the minutely mottled appearance of the seed; *i.e.* with small cells with brown walls, the adjacent layer or two of the mesophyll with slight bar-thickenings on the walls. Tegmen; *o.e.* composed of compact longitudinal fibres 150–250 × 65–80 μ, 16–24 μ wide, laterally compressed and appearing as a palisade in t.s., very thick-walled, heavily lignified and pitted, yellow-brown, shortened and palisade-like at the micropyle and chalaza; *mesophyll* collapsed; *i.e.* as a pellicle of rectangular cells with brownish walls and thickenings on the radial walls. Chalaza bullate, filled with thin-walled tissue, pervaded by bundles of tracheids from the expanding v.b. of the raphe, separated from the endosperm by the hypostase of small, dark brown cells, surrounded by a thick annulus of thick-walled cells as the base of the exotegmen. Endosperm thin-walled, oily. Embryo with relatively long radicle and small thin cotyledons.

Fruit 1-seeded, without sclerotic tissue, apparently indehiscent.

The seed-structure resembles that of *Bischofia*.

STERCULIACEAE

Ovules anatropous, more or less erect (? also transverse), bitegmic, crassinucellate; o.i. 2–5 cells thick; i.i. 2–6 cells thick, developing later than o.i., in some cases not covering the nucellus (*Byttneria, Guazuma, Helicteres*); neither integument covering the nucellus in *Helicteres*; micropyle formed by the exostome, out of line with the endostome, chalaza often umbonate or shortly winged.

Seeds rather small to large, sessile, exarillate or with a vestigial or well-developed aril (*Chiranthodendron, Commersonia, Hannafordia, Leptonychia, Rulingia, Seringia, Sterculia*), alate in a few genera (*Eriolaena,* Helmiopsideae, *Pterospermum, Pterygota, Reevesia*). Testa 2–40 cells thick, generally multiplicative; *o.e.* with the cells as a short pallisade or tabular, in the alate seeds longitudinally elongate, with stomata and peltate scales in *Scaphium*, minute hairs in *Kleinhovia*; *mesophyll* of various kinds, often with an outer layer of dark brown cells or with mucilage-sacs (*Firmiana, Guazuma, Scaphium, Theobroma*), in some cases slightly pulpy (*Sterculia spp.*), with lignified stellate cells (*Pterygota*), carrying the integumentary v.b. when present; *i.e.* unspecialized or as cuboid cells with thickened walls, as crystal-cells in *Firmiana, Guazuma,* and *Sterculia spp.*; seed-wing formed by the exotesta and mesophyll (*Pterygota*). Tegmen 6–65 cells thick, multiplicatives, thicker than the testa (except *Sterculia spp, Theobroma*), the inner layers eventually collapsing; *o.e.* as a layer of long or short Malpighian cells, lignified (except *Waltheria*), but as longitudinal fibres in the exceptional

Leptonychia; *mesophyll* often an outer layer with brown cell-contents, with large mucilage-sacs in *Firmiana* and *Scaphium*; *i.e.* mostly as cuboid or shortly oblong cells with slightly thickened and pitted walls, lignified or not, massively thickened in *Sterculia spp.* Vascular bundle only in the raphe in the smaller seeds, in most of the larger with postchalazal and accessory funicular branches; tegmen without v.b. Chalaza often umbonate in small seeds, or the umbo developing into the wing of alate seeds, substituted by a false chalaza in *Sterculia*. Endosperm nuclear, oily or starchy, or both, much reduced in *Theobroma*, absent from *Cola*. Embryo straight, generally with flat cotyledons, but with many variations; inverted in *Sterculia*.

Aril micropylar (*Sterculia*), between the funicle and the micropyle (*Commersonia*), funicular in other cases, on the antiraphe in the aberrant *Leptonychia*. Figs. 556–574.

(Netolitzky 1926; Venkata Rao 1950–53.)

Though the classification of Sterculiaceous genera is concerned mainly with the flower, the family has much to teach about the evolution of the fruit and seed. Some of the most massive, yet primitively constructed, follicles in existence are found in *Sterculia*, separately stalked from the apocarpous flower. In *Firmiana* and *Scaphium* the follicles are as large but thin and leafy, and their enrolled edges are so slightly connivent that they expand as soon as the fruit begins to form and the seeds, which are well supplied with stomata in *Scaphium*, develop in a naked gymnospermous manner; a style, however, is interposed for pollination. If such fruits were found immature and fossilized, theorists would pen abundance of speculation on the primitive angiosperm. As it is, attention is given to the highly modified Winteraceae and the advanced flower has concealed the exceptional nature of these most primitive fruits.

Durian-relics occur in spinous fruits (*Commersonia, Guazuma*) and arils in several genera. Modifications of the arillate seed occur in the alate seeds of *Pterospermum* and *Pterygota*, the red follicles of which resemble those of *Sterculia* but are strongly fibrous. There is much generic variety to be investigated for there is surely a span between *Sterculia* and *Melochia*, or *Guazuma* and *Theobroma*, if such different fruits belong in the same family. Yet there seems to be no absolute criterion to distinguish the Sterculiaceous seed from the Bombacaceous, Malvaceous, Tiliaceous or Euphorbiaceous (Crotonoideae). Sterculiaceae may be a mixture of derivatives from the ancestry of all these families. Thus *Leptonychia* reveals the unexpected. This great assemblage of families, still struggling with the primitive heritage of fruit and seed, shows the enormous difference between the polypetalous arborescent dicotyledons and the standardized sympetalous families, such as Compositae, Labiatae or Boraginaceae, which have reduced this heritage to the unitegmic seed in an achene. Only the artifice of taxonomic nomenclature could render them commensurate.

The embryo of *Sterculia* is inverted. The radicle points to the chalaza which seems generally to lack a heteropyle though the exotegmic palisade is here thinner (Fig. 568). A false chalaza develops at the micropylar end of the seed, at the beginning of what should be the raphe. The exotegmic palisade is here lacking over a circular area through which dips a short column of non-vascular tissue continuous with the sheath of the main funicular vascular bundle. There is no raphe-bundle because the funicular bundle breaks up in the micropylar region into a ring of smaller bundles which descend the sides of the seed without reaching the chalazal end; thus there is no true chalaza. The effect is to transpose the physiological implications of the chalaza to the micropylar end of the seed, thus inverting the normal construction, cf. *Pterygota* (Fig. 564). The seed of *Sterculia* short-circuits the circuitous raphe and this accomplishment may be a reason for the success of the archaic follicles. With over 100 species possessed of this fruit, the pantropic *Sterculia* is a successful genus of forest trees. It is exceeded in number of species only by the capsular *Dombeya* (200 species) and *Hermannia* (120 species).

Sterculia shows also the degeneration of the Malpighian layer which, as the mechanical exotegmen, is the prime feature of the seed in this great Malvalean alliance of families. In. *S. foetida* the layer is 500 μ high, in *S. macrophylla* 150 μ, in *S. lanceolata* 80 μ, and in *S. rubiginosa* it is 30 μ high and so slightly developed that there is only the strong thickening of the inner wall as the inception of the Malpighian cell. The trend explains the short exotegmic layer in the massive seed of *Theobroma*, paralleled by *Pachira* in Bombacaceae, and it explains a similar reduction in the short palisade-layers of *Helicteres*, *Melochia*, and *Waltheria* which resembles that of Tiliaceae and Malvaceae. If this were an upward trend leading to the evolution of the Malpighian cell, there would be the absurd conclusion that this had taken place not only in *Sterculia*, but independently in the other families of the alliance, that the more advanced fruits of Malvaceae had the primitive exotegmen, and that primitive follicles as of *Firmiana* and *Scaphium* with their naked seeds had the advanced exotegmen. Thus the small seeds of *Helicteres* and *Melochia* with relatively simple construction, more like the little structures of herbaceous plants, are evidence of the general reduction in size and complexity of the dicotyledonous seed. But a study of the fully arillate seeds of *Sterculia* is required.

The aril of *Sterculia* is micropylar, as in *Coelostegia* (Bombacaceae) and the caruncle of Euphorbiaceae, but in *Commersonia* it arises between the funicle and the micropyle and is distinct from both. In the exceptional *Leptonychia* it arises in complete independence on the antiraphe. In other genera, yet to be examined critically,

it appears to be funicular. These differences are explicable as transferences of function or of differentiation in gene-activity, as the formation of the seed is altered, and no more require special treatment as special creations (v.d. Pijl, 1955, 1966) than the transference of the chalazal site in *Sterculia*. This genus provides the primitive fruit, the primitive arillate seed, the degeneration of the aril in the carunculate and exarillate seed, and the thin friable 'sarcotesta' as the simplified testa, not the primitive feature of van der Pijl.

The seed of *Theobroma* suggests overgrowth with minimal differentiation of tissues. The development of the zygote is suppressed for 50 days after pollination, when the fruit and seed are forming, and the seed has no apparent influence as a producer of auxins on the development of the fruit (Nichols 1964, 1965). *Heritiera* may be comparable.

Certain differences in structure between the naked seeds of *Firmiana* and *Scaphium* may mark a dichotomy in the family. The smaller seeds of *Firmiana*, borne along the edges of the follicle, have the simpler structure similar to that of exarillate species of *Sterculia*, but with normal embryo. The very thick-walled and lignified exotesta of *Firmiana* crumples into the reticulum on the surface of the seed as the mesotesta shrinks on drying. The exotesta of *Scaphium* has copious stomata and peltate scales and the walls of the cells are scarcely thickened. The seed of *Scaphium* is much larger, solitary and basal; it has much thicker seed-coats with very large mucilage-sacs and a double network of vascular bundles in the testa; the endotesta lacks crystal-cells. It may be argued that the basal position of the seed with stout vascular supply directly from the base of the follicle enables the massive structure to be developed, but equally massive seeds are formed in large numbers in the follicles of *Pterygota* and *Sterculia*; with the follicle acting as a wing in *Scaphium*, it may retain one primitively massive seed.

The vascular supply in the large seeds of *Scaphium* and *Sterculia* is unusually complex. A close ring of vascular bundles in the funicle sends numerous anastomizing bundles over the sides of the seed, much as in *Durio* and *Myristica*, but without branches to the aril. The contrast is *Pterygota* with prolonged raphe bundle and testal branches which arise only towards the chalazal end (Fig. 565). In other genera the arrangement simplifies to the single, undivided raphe-bundle, so familiar in small economical seeds; indeed, this simplification may occur in *Sterculia*, cf. *S. macrophylla*, and in *Firmiana*. *Fremontia* has a vestige of the postchalazal branches.

The rather small seeds of *Abroma*, *Byttneria*, *Dombeya*, *Helicteres*, *Melochia*, *Pentapetes*, *Waltheria* and, perhaps, *Guazuma* resemble Tiliaceous seeds in the thin testa with the exotestal cells inclined to elongate longitudinally and in the short exotegmic palisade with shortly oblong cell-facets. Whether these details separate a Tiliaceous alliance from that of *Firmiana*, *Scaphium*, *Pterygota*, *Sterculia* and *Theobroma* needs more enquiry, but may explain the difficulty in separating Tiliaceae from Sterculiaceae. Attention is directed to the detailed work of Venkata Rao (1950–53).

The seeds of Huaceae (*Afrostyrax*, *Hua*) seem to offer no distinction from those of Sterculiaceae (Baas 1972).

Abroma Jacq.

A. augusta L.
Ovule; o.i. 2 cells thick; i.i. 5 cells thick; chalaza umbonate.
Testa not multiplicative. Tegmen 8–10 cells thick. (Venkata Rao 1953b.)

Byttneria Loefl.

B. herbacea Roxb.
Ovule; o.i. 2 cells thick; i.i. 3 cells thick, covering only half of the large nucellus.
Testa 3–4 cells thick. Tegmen 8—10 cells thick. (Venkata Rao 1953b.)

Dombeya Cav.

D. mastersii Hook., *D. spectabilis* Bojer.
(Venkata Rao 1953b.)

Firmiana Marsigli

Ovules 2–4–6 on the margins of the free carpels, soon exposed in the young fruit.

Seeds ellipsoid, sessile, spaced along the edges of the open follicles, exarillate, glabrous, green then brownish, drying reticulate from the collapse of the testal mesophyll. Endosperm with abundant starch and oil.

F. colorata Roxb.
Ovule; o.i. 3 cells thick; i.i. 4 cells thick.
Testa 10–12 cells thick. Tegmen 25–30 cells thick. Chalaza with 6–9 postchalazal branches of the raphe-bundle, reaching almost to the micropyle, undivided or with few branches.
(Venkata Rao 1953b.)

F. simplex (L.) W. F. Wight
(Japan, Kyoto, leg. E. J. H. Corner 1966; dried seeds.)
Seeds 8 × 6.5 mm, 2 on each edge of the follicle. Testa 16–20 cells thick; *o.e.* composed of rather small cells, shortly radially elongate, with lignified walls, the outer wall very strongly thickened, the lumen short and basal, more distinctly elongate and palisade-like round the hilum, facets isodiametric and angular, without stomata; *mesophyll* collapsed, evidently with mucilage-sacs, pervaded with the loose network of v.b. mainly in the edges of the reticulum on the dried seed, with a few small scattered sclerotic cells especially in the o.h. round the micropyle; *i.e.* and *i.h.* (1–2 layers) composed

of small cuboid cells with slightly thickened and lignified walls, every cell with a single crystal, the inner wall of i.e. distinctly thickened. Tegmen *c.* 16 cells thick; *o.e.* as a palisade of lignified Malpighian cells 140–160 μ high, continuous except for the heteropyle, prismatic, without pits except at the extreme outer ends; *mesophyll* much collapsed, evidently with mucilage-sacs, without v.b., thin-walled; *i.e.* as a compact layer of cuboid cells with strongly pitted radial walls, but scarcely lignified. Vascular bundle of the raphe rather small, forming a small plexus in the chalaza and the postchalazal branches making the extensive plexus in the testa. Hypostase small, discoid, composed of small, thin-walled brown cells, not lignified.

Fremontia Torr.

F. californica Torr.
(Cambridge, University Botanic Garden; living material.) Fig. 556.
Ovule; o.i. 3 cells thick; i.i. 6–8 cells thick; chalaza umbonate.
Seeds 6–7 × 3.5–4 mm, exarillate. Testa 4–5 cells thick, soft, thin-walled, unspecialized; *i.e.* with the cells shortly radially elongate. Tegmen not multiplicative; *o.e.* as a palisade *c.* 180 μ high; *mesophyll* collapsing; *i.e.* composed of cuboid cells with thickened and pitted radial walls, not or very slightly lignified. Vascular bundle of the raphe branched at the chalaza into 3–4 short arms, not descending the sides of the seed. Chalaza with small heteropyle and hypostase of dark brown cells with thin suberised walls. Endosperm oily.

Guazuma Mill.

Capsules shortly and stoutly muricate to softly setose, many-seeded.

G. tomentosa Kunth
Ovule; i.i. covering three-quarters of the nucellus; chalaza umbonate.
(Venkata Rao 1953b.)

Guazuma sp.
(Mexico, leg. E. J. H. Corner 1947; dried fruits.) Fig. 557.
Seeds 4 mm long, plump, subtrigonous, black, covered by a thin greyish pellicle swelling and mucilaginous when wet. Testa *c.* 7–9 cells thick, not lignified, the cell-walls mucilaginous; *o.e.* apparently as a thin layer of narrow cells shortly longitudinally elongate; *o.h.* with many cells much inflated (? mucilage-sacs); *mesophyll* hyaline, collapsing except for the clusters of cells with dark brown contents extending from i.e. as masses 2–6 cells thick; *i.e.* as a layer of cuboid cells, each with a crystal, but interrupted by the clusters of brown cells. Tegmen ?7–8 cells thick, the inner hyaline mesophyll crushed; *o.e.* as a palisade of lignified Malpighian cells

c. 130 μ high, the outer ends slightly longitudinally elongate and strongly pitted, not lignified; *i.e.* as a persistent layer of large cuboid cells with brown walls, not lignified but the radial walls slightly thickened and finely striate-pitted. Vascular bundle apparently only in the raphe. Endosperm oily. Embryo with long radicle and thin, folded cotyledons.

Fruit 40–55 × 20–25 mm, cylindric, woody, green, apparently indehiscent, muriculate with short pyramidal spines 2–3 mm high, with wider base, 5-locular with many seeds in each cavity; spines with a large central woody strand and 3–5 smaller peripheral strands.

The elaborate testa of this species may be characteristic of the genus.

Helicteres L.

H. isora L.
Ovule; o.i. 2 cells thick; i.i. 2 cells thick; both integuments covering less than half of the large nucellus; chalaza umbonate.

Testa not multiplicative. Tegmen 8–10 cells thick. Chalaza with lignified hypostase.

(Venkata Rao 1953b.)

H. sacarolha St. Hil.
(Brazil, Mato Grosso, pr. Chavantina, leg. E. J. H. Corner 1968; fruits in alcohol.) Figs. 558, 559.

Seeds 3 × 1.7 mm (full-size but immature). Testa 5–6 cells thick on the sides of the seed, 3 cells thick along the antiraphe, 10–12 cells thick round the chalaza; *o.e.* composed of longitudinally elongate cells with slightly thickened walls; *mesophyll* with small patches of cells with brown walls, most cells with abundant starch; *i.e.* composed of rather large cuboid cells with slightly thickened brownish walls. Tegmen 7–9 cells thick on the flat sides of the seed, 10–12 cells thick at the angles, in four layers; *o.e.* as a short palisade of prismatic cells, slightly longitudinally elongate; *mesophyll* with an outer layer 4–5 cells thick, composed of small rounded cells filled with starch, aerenchymatous, and an inner layer, *c.* 3 cells thick, of large watery cells, radially elongate though rather obliquely, thin-walled, without starch, collapsing; *i.e.* as a layer of compact subcuboid cells shortly longitudinally elongate, with pale brown thickened walls and finely striate radial walls.

The construction of this seed is remarkably Tiliaceous.

Kleinhovia L.

K. hospita L.
Ovule; o.i. 3 cells thick; i.i. 6–10 cells thick: chalaza umbonate.

Testa 4 cells thick; *o.e.* with short hair-like spicular outgrowths from the cells. Tegmen 10–12 cells thick. Vascular bundles in the testa anastomosing in the micropylar end of the seed.

(Venkata Rao, 1953b.)

Leptonychia Turcz.

Placed in Tiliaceae by some botanists and in Sterculiaceae by others, though the difference between the families is practically negligible (Hutchinson 1967), *Leptonychia* is a misfit. The strangely arillate seed has a fibrous exotegmen which is neither Sterculiaceous nor Tiliaceous. The parietal placentation may favour Tiliaceae, for it is recorded in *Sicrea*, but it would follow incipient syncarpy of apocarpous Sterculiaceae. The starchy endosperm is Sterculiaceous. The seed, however, has these peculiarities:

(1) The exotegmen consists of short fibres, laterally compressed, as in Celastraceae, Elaeocarpaceae and Euphorbiacaea–Phyllanthoideae, though not inconceivably an exaggeration of the slight tangential elongation observed in Tiliaceae, cf. Gonystylaceae.

(2) The micropyle is widely separated from the hilum by a long preraphe, as in Connaraceae.

(3) The vascular supply has no postchalazal extension but recurrent branches from the raphe along the preraphe to connect with the site of the aril, somewhat reminiscent of the preraphe-bundle in Connaraceae and of the funicular complex in *Sterculia*.

(4) The aril is isolated on the antiraphe.

Leptonychia suggests in these particulars a relic of the common ancestry of Elaeocarpaceae (cf. *Sloanea*), Sterculiaceae, Tiliaceae, and Violaceae.

L. glabra Turcz.
(Malaya, leg. E. J. H. Corner; living material.) Figs. 560–562.

Ovules bitegmic, crassinucellate, funiculate, anatropous, suspended in two rows from the stout parietal placentas (convergent but not concrescent), giving 4–6 ovules per loculus; o.i. and i.i. ? thickness; exostome thickened to form the micropyle.

Seed 10–12 × 7–8 mm, subcompressed or, if 2 in a loculus, with flattened apposed ends, black, shiny, more than half covered by the waxy red aril; hilum lateral on the exposed side of the seed opposite to the aril, as a dark brown elliptic patch; micropyle separated widely from the hilum by a preraphe and placed at the opposite end of the seed from the chalaza; funicle practically none, but with a pseudo-funicle as the raphe v.b. stripped off the placenta and suspending the seed. Testa *c.* 7 cells thick; *o.e.* composed of short cells, shortly transversely elongate, with strongly thickened brown outer wall and brown gummy contents, without stomata; *mesophyll* 4–5 cells thick, aerenchymatous, with slightly thickened brownish walls and brownish contents, more or less crushed, with frequent mucilage-canals; *i.e.* unspecialized. Tegmen 10–11 cells thick; *o.e.* composed of shortly longitudinally elongate, laterally compressed, fibriform cells, thick-walled, lignified, pitted, appearing as a palisade of narrow cells in t.s., the walls yellowish, the lumen with brown contents, the cells shorter and radially elongate at the micropyle and chalaza, continuous over the hilum but interrupted at the chalaza; *mesophyll* as in the testa, more or less crushed; *i.e.* unspecialized. Vascular supply consisting of the raphe-bundle ending in a small plexus at the chalaza, and two preraphe bundles passing to the micropyle and uniting in a small plexus on the antiraphe at the attachment of the aril, one of the preraphe bundles usually dividing into two; tegmic v.b. absent. Endosperm thin-walled, with abundant starch. Embryo with rather short radicle and thin flat cotyledons.

Aril waxy, scarlet, attached to the testa over a circular patch on the antiraphe opposite to the hilum, free from both chalaza and micropyle, with two large lateral lobes spreading over the sides of the seed and a median peaked lobe over the chalaza, not covering the micropyle, the pellicular surfaces waxy and red, the interior pulp yellow and oily; epidermal cells with yellow oil-globules and thickened, orange-red outer wall; inner tissue with yellow oil-globules, thin-walled, without v.b.; arising very soon after fertilization as a minute disc on the antiraphe.

Capsule 25×18 mm, 3–5-locular, 1–2 seeds in each loculus but generally with 1–2 empty loculi in 3-locular fruits, or 2–3 empty in 5-locular fruits, the abortive ovules covered by abortive arils, the surface of the capsule light green and velvety; loculicidal with the valves shrinking back on the fruit-stalk.

Pericarp wholly thin-walled except for the vascular bundles and i.e., with many mucilage-canals especially in the inner layers; v.b. in the inner tissue, with short radiating fibrous processes supporting the lacunar mucilage-tissue; o.e. with short stellate hairs, the arms consisting of single thick-walled cells; i.e. consisting of narrow cells elongated into short woody fibres with pitted walls, mostly elongate transversely to the long axis of the capsule.

Melochia L.

M. corchorifolia L.

(Ceylon; flowers and fruits in alcohol.)   Fig. 563.

Ovule; o.i. 2 cells thick; i.i. 4 cells thick, 6–7 near the umbonate chalaza.

Seeds $2.5–2.8 \times 1.5$ mm, convex, subtriquetrous, grey-brown to blackish. Testa not multiplicative; *o.e.* with the cells longitudinally elongate; *i.e.* with subcuboid cells, the walls pale brown. Tegmen 8–10 cells thick, most of the mesophyll collapsing; *o.e.* as a palisade of shortly longitudinally elongate prismatic cells, lignified, the radial walls with inwardly projecting and more or less anastomosing ribs; *o.h.* as a layer of large cells, at first cuboid and hyaline, then strongly convex with dark brown contents, subpersistent; *i.e.* with rectangular cells, each with a crystal, persistent. Perisperm as a single layer of hyaline cells. Endosperm starchy.

(Venkata Rao 1951.)

Pentapetes L.

P. phoenicea L.

Ovules; o.i. 3 cells thick; i.i. 4 cells thick; chalaza umbonate.

Testa 3–5 cells thick. Tegmen 6–8 cells thick; *i.e.* with the cells round the micropyle elongate with lignified walls and reticulate thickening.

(Venkata Rao 1953b.)

Pterygota Schott and Endl.

P. alata (Roxb.) R.Br.

(Java, Kebun Raya, April 1972; living material.)   Figs. 564–566.

Seeds 6–8 cm long, alate, creamy white and finely rugulose-papillate, ripening brownish and leathery, many per follicle, filling the loculus, very closely packed, albuminous; seed-body *c.* $25 \times 15 \times 7$–8 mm, compressed with obtuse edges; wing –3 cm wide, rather thick; funicle thick, short, with a slight lip (? vestigial aril) above the base and with the micropyle leading to it. Testa 10–16 cells thick; *o.e.* composed of cuboid cells with rounded projecting outer wall and subhexagonal facets, many (? all) becoming rather thick-walled and pitted (? not lignified), without stomata, forming the uneven surface; *mesophyll* aerenchymatous, spongy, the stellate cells with narrow body and arms, many (? all) becoming thick-walled and lignified, finally drying pithy round the seed-body and in the wing; *i.e.* with small, thin-walled, unspecialized cells. Tegmen 20–26 cells thick; *o.e.* as a palisade of Malpighian cells *c.* 250μ high; *o.h.* with thin-walled cells often shortly radially elongate; *mesophyll* thin-walled, aerenchymatous, the cells with large body and slight arms; *i.e.* as a pellicle of small cells adherent to the endosperm, with pitted and lignified radial walls, facets subhexagonal. Chalaza as a wide plug of undifferentiated tissue, composed of elongate thin-walled cells leading from the chalazal plexus of v.b. to the brown meniscus of the outer hypostase. Hypostase in two parts; outer part consisting of thin-walled lignified outer cells and brown-walled unlignified inner cells; inner part consisting of thin-walled lignified cells forming a small plug at the base of the nucellus. Vascular supply as a stout bundle in the raphe soon producing 2–3 small accessory bundles, arching widely over the chalaza, breaking into *c.* 8 bundles descending and branching over the sides of the seed in the testal mesophyll, the ultimate branchings of the descending v.b. forming a double layer in the micropylar end of the testa and terminating round the base of the micropylar canal (not forming a vascular ring such as in *Sterculia*); vascular bundles to the wing arising as 11–12 branches from the chalazal arch of the raphe, lying in the median plane of the spongy-pithy mesophyll of the wing, but peripheral round the thick raphe-edge, extending along the tissue of the wing, branching with little anastomosis and ending with fine free branchlets round the periphery (without a marginal v.b.). Micropylar canal long, more or less occluded, not lignified, the surrounding aerenchymatous mesophyll eventually lignified. Funicular tissue thin-walled (? eventually thinly lignified); aril-lip not structurally different from the funicle. Embryo normally orientated.

The structure of these massive and complicated seeds was investigated from a fully grown, but immature follicle (body 12 × 10 cm, stalk 4 cm), which the wind had dislodged. It seems likely that all the testal mesophyll and outer epidermis would have become lignified into corky, spongy or pithy tissue. The wing, as a chalazal extension, consists mainly of this pithy aerenchymatous mesophyll and is profusely vascularized. Such massive seeds reveal how the structure of the small seeds of herbaceous Sterculiaceae have been simplified into the standard neotenic seed, exarillate with the small raphe ending at the simple chalaza. A slight crescentic groove, leading to the micropyle, seems to indicate a vestigial aril.

Pterospermum Schreb.

P. heyneanum Wall.
Ovule; o.i. 3 cells thick; i.i. 4–5 cells thick; chalaza with a long point instead of the usual umbo.

Testa 6–12 cells thick. Tegmen 8–10 cells thick. Wing developed from the pointed chalaza. Vascular supply with numerous anastomosing postchalazal branches.

(Venkata Rao 1953b.)

Scaphium Schott et Endl.

Ovules 2, basal, anatropous, erect, one becoming a seed. Carpels 2–5, free, involute with the margins connivent but not conjoint, soon opened by the developing seed. Testa with stomata and microscopic peltate scales.

S. macropodum Beumée
(Sumatra, Palembang, cult. in Hort, Bogor. IV.I.126; alcohol-material.)   Fig. 567.
Ovule; o.i. 4–5 cells thick, the cells of o.e. filled with dark brown matter; i.i. 3 cells thick. Carpels 2.

Seeds 18 × 13 mm (apparently full-sized, with endosperm and embryo, but immature), very soon exposed at the base of the enlarging and expanded follicle. Testa 32–40 cells thick; *o.e.* composed of small cells, slightly radially elongate, the outer wall thickened, with isodiametric, angular facets, not lignified, with many microscopic peltate scales and abundant stomata; *mesophyll* more or less aerenchymatous, with many small and large mucilage-sacs in two rows, but a single row at the chalaza; *i.e.* unspecialized. Tegmen 40–50 cells thick; *o.e.* as a palisade of radially elongate, thick-walled, lignified, Malpighian cells, not visibly pitted except at the extreme outer end, the inner ends more in-

tensely lignified; *mesophyll* aerenchymatous, the cells more or less stellate, filled with starch-grains, with abundant large mucilage sacs; *i.e.* as a compact layer of subcuboid, thick-walled cells with bar-like thickenings on the radial walls, lignified. Chalaza wide, unspecialized, the palisade of the exotegmen leaving a wide gap, without conspicuous hypostase. Vascular bundles in the testa only, forming a double network with inter-connections, arising from the ring of v.b. in the funicle; the outer network of small v.b. lying between the two layers of mucilage-sacs, the inner network of larger v.b. mostly internal to the inner layer of mucilage-sacs, the raphe v.b. slightly enlarged but not conspicuous, forming a plexus at the chalaza without entering the heteropyle; in the funicle with several medullary v.b. inside the compact ring of numerous peripheral v.b. Embryo with short radicle at the micropylar end of the seed.

Sterculia L.

Follicles usually pink or red. Seeds large, sessile, attached to the edges of the dehisced follicles, black or brown, in a few species extensively arillate, in other with a minute aril as a micropylar caruncle, more often exarillate. Vascular supply of the testa various. Embryo inverted, the radicle pointing to the abortive chalaza (? generally without heteropyle); endosperm attached to a false chalaza and hypostase near the beginning of the raphe; cotyledons flat, thin, to moderately thick. Endosperm oily or both oily and starchy.

The four species which have been examined differ considerably in the vascular supply to the testa and in the development of the exotegmic palisade. Clearly many more species need investigation before the seed-structure can be appreciated. The origin of the inverted embryo, which is normally directed in the allied *Firmiana*, is a problem.

S. foetida L.
(Ceylon; fruits in alcohol.)   Figs. 568–570.
Seeds 25 × 14 mm, purple-black, ellipsoid–cylindric, obtuse; aril 3–4 mm wide, pulvinate-crenulate, waxy, orange-yellow. Testa 0.9–1 mm thick before drying up, 11–16 cells thick, many of the cells with brown contents, without starch, not lignified; *o.e.* composed of cuboid cells with thickened outer wall; *mesophyll* in two aerenchymatous layers, the outer layer (4–6 cells thick) consisting of short cells tangentially elongate, with dark contents and carrying the v.b., the inner layer (5–7 cells thick) consisting of large cells elongate radially with brown contents, without starch; *i.e.* as a single layer of small cuboid colourless thin-walled cells. Tegmen *c.* 1.5 mm thick, 50–65 cells thick (at maximum development), becoming crushed by the endosperm and finally drying up with only the palisade-layer remaining; *o.e.* as a compact palisade of Malpighian cells *c.* 500 μ high, lignified except for the thin-walled outer end (70–

$100\,\mu$ long, ? immature), with oblique pitting and at the basal part with short longitudinal striations; *o.h.* as a single layer of hyaline substellate cells, shortly radially elongate, not as regular hourglass-cells, the walls slightly thickened; *mesophyll* composed of two layers roughly equal in thickness, the outer layer consisting of substellate cells with brown contents, the inner layer consisting of large, rounder, hyaline cells smaller towards the inside of the seed, both layers collapsing, without starch; *i.e.* as a single layer of small cuboid cells with longitudinal lignified thickenings on the radial walls forming small elongate pits. Chalaza unspecialized, the exotegmic palisade continuous over the chalazal end of the seed though thinner than on the sides; *mesophyll* of tegmen with a minute pore-like hypostase leading to the radicle; without heteropyle. Vascular supply in the outer mesophyll of the testa, consisting of *c.* 25 v.b. arising in the funicle, extending to the chalazal end of the seed, branching rather sparingly to form narrow elongate meshes, uniting into 6–8 fine slender vascular ends at the chalaza; derived in the funicle from two lateral branches of the funicular v.b. passing to the sides of the arillostome, uniting into a crescent round the antiraphe side, then giving off the longitudinal v.b., the main funicle-bundle dividing into 6–7 longitudinal v.b. on the raphe-side without a special raphe-bundle. Endosperm thin-walled, oily, without starch. Embryo with short stout radicle and thin flat cotyledons.

Aril without v.b., the cells firm with thickened walls, the outer cell-layers with yellow oily contents, the inner colourless with starch-grains; arising from the exostome.

(Venkata Rao 1953b; Vaughan 1970.)

S. lanceolata Cav.
(Ceylon, Peradeniya, cult.; fruits in alcohol.) Figs. 571, 572.
Seeds 16×10 mm, much as in *S. foetida* but smaller; aril very small and rudimentary. Testa 15–22 cells thick; *mesophyll* without radially elongate cells; *i.e.* small-celled, not lignified, many of the cells with a crystal. Tegmen 18–26 cells thick; palisade 70–$80\,\mu$ high, the cells with stellate lumen in t.s.; *i.e.* with very thick lignified walls. Vascular supply derived as in *S. foetida* from the main bundle of the funicle, the branches spreading out round the arillostome but not in a continuous ring.

S. macrophylla Vent.
(Singapore Botanic Gardens; living material.) Figs. 573, 574.
Seeds 18–23×11–12.5 mm, ellipsoid, black with shiny and minutely cellular-papillate surface; hilum rather wide, irregularly elliptic, not sunken; funicle none. Testa; *o.e.* as a compact layer of short cells with strongly thickened, convex outer walls and brown gummy contents; *mesophylla* in two layers, the outer layer (3–4 cells

thick) thin-walled but wholly agglutinated in the ripe seed into a dark brown horny lamella, the inner layer broad, white, aerenchymatous, the thin-walled cells stellate, with granular contents, some with brown gummy contents giving the brown flecks in the dry pithy testa (not pulpy or sarcotestal), but the cells at the junction with the aril tracheid-like with more or less scalariform thickening. Tegmen; *o.e.* as a layer of Malpighian cells 120–$150\,\mu$ high, the outer third of the cells thinner-walled with light yellow granular cell-contents, the inner two-thirds colourless and almost solid, the layer disrupted in a circular gap on the raphe-side of the endostome as in *S. foetida*; *mesophyll* as a fairly thick dark brown layer of somewhat tangentially elongate cells with slightly thickened brown walls, the outer cells substellate, the inner cells collapsing. Chalaza minute. Vascular bundle single in the raphe without branches. Endosperm thin-walled, oily.

Aril 1–1.5 mm wide, as a rather thick, horny, sub-discoid, crenulate patch on the antiraphe-side of the exostome, orange-yellow with white inner tissue; cortex 2–3 cells thick, composed of cells with slightly thickened walls and orange-yellow oily-smeary cytoplasm; inner tissue consisting of thick-walled, pitted, colourless cells with somewhat diffluent middle lamella, horny, not pulpy, without v.b.

The mature seeds hang down from the edges of the follicles often for several days before dropping off. The testa then dries, cracks and flakes off from the tough brown exotegmen. The simple vascular supply of the seed is noteworthy.

S. rubiginosa Vent
(Singapore Botanic Gardens; living material.) Fig. 573.
Seeds *c.* 13×8 cmm, dark purple, almost black, when immature brown with the v.b. of the testa showing as pale lines, with white pulpy sarcotesta 9.5–0.7 mm thick, exarillate, without funicle. Testa; *o.e.* as a compact layer of short cells with strongly thickened walls, especially the outer walls, with dark brown gummy contents; *mesophyll* with *c.* 3 outer cell-layers thin-walled and with dark brown gummy contents, the rest of the tissue aerenchymatous with the more or less stellate cells filled with starch-grains, thin-walled, becoming pulpy at maturity; *i.e.* as a layer of small compact cells without starch. Tegmen 7–9 cells thick; *o.e.* as a short palisade *c.* $30\,\mu$ high, the cells with strongly thickened inner walls and dark brown gummy contents, forming the endostome but disrupted in a circular gap by the raphe-side of the endostome; *mesophyll* composed of cells with slightly thickened, pale brown walls, collapsing, thickened at the hilum into aerenchymatous tissue; *i.e.* as a layer of small cells with thickened brown radial and inner walls, with subreticulate thickening on the radial walls. Vascular supply evidently as in *S. foetida*, the

fine longitudinal bundles in the testa rather distantly anastomosing, without a distinct raphe-bundle. Endosperm thin-walled, starchy.

Theobroma L.

T. cacao L.

Ovule; o.i. and i.i. 3–4 cells thick.

Seeds 15–30 mm long, reddish brown to dark brown, obovoid with the hilum at the broad end, the chalaza at the pointed end, exarillate. Testa massive, –25 cells thick (? more); *o.e.* unspecialized, the outer wall slightly thickened; *o.h.* consisting of radially elongate cells, some enlarged into mucilage-sacs and then usually in clusters –1 mm wide; *mesophyll* aerenchymatous, permeated with a network of v.b. from the chalaza almost to the micropyle. Tegmen much thinner than the testa, ? 8–9 cells thick; *o.e.* as a narrow interrupted layer of blocks of thick-walled, lignified cuboid cells 15–25 μ wide, with thin outer wall and pitted inner wall, spaced by unlignified cells; *i.e.* unspecialized. Vascular bundle of the raphe with *c.* 15 postchalazal branches reaching the micropyle, (? branching and anastomosing). Endosperm reduced to a single layer of cells, as a silvery skin round the embryo. Embryo chocolate-brown or purplish; cotyledons thick, folded, with puberulous surface, the hairs simple, cylindric, obtuse closely septate.

(Cheeseman 1927; Kühn 1927; Winton and Winton 1939; Berger 1964; Nichols 1964; Vaughan 1970.)

The description of this well-known seed is yet unsatisfactory. The exotegmic palisade seems to be degenerate (as in *Sterculia rubiginosa*).

Waltheria L.

W. indica L.

Ovule; o.i. 2 cells thick; i.i. 6–7 cells thick; chalaza umbonate.

Testa not multiplicative. Tegmen 10–12 cells thick; *o.e.* with short, unlignified palisade-cells.

(Venkata Rao 1950.)

STYLIDIACEAE

Ovules anatropous, unitegmic, tenuinucellate; integument 5–6 cells thick.

Seeds small, albuminous, exarillate. Seed-coat ? not multiplicative; *o.e.* persistent, sclerosed; *mesophyll* becoming crushed; *i.e.* endothelial, then crushed. Endosperm cellular. Embryo minute.

(Netolitzky 1926; Subramanyam 1950a, 1951a.)

STYRACACEAE

Ovules anatropous, bitegmic (*Styrax*) or unitegmic (*Halesia*), tenuinucellate.

Seeds albuminous, exarillate. Testa at first thick, with postchalazal v.b., then crushed. Vascular bundle of the raphe with 2 or more branches at the chalaza, subdividing over the sides of the seed, some almost reaching the micropyle. Endosperm cellular, oily, copious. Embryo straight or slightly curved.

(Netolitzky 1926; Kühn 1927 p. 333.)

The seeds of capsular genera need investigation in this family to eludicate the structure.

SYMPLOCACEAE

Ovules anatropous, unitegmic, tenuinucellate.

Seeds albuminous, exarillate, enclosed in endocarp. Seed-coat very thin, apparently unspecialized, crushed. Vascular bundle of the raphe continued to the micropyle or with 2 postchalazal branches subdividing over the sides of the seed.

(Netolitzky 1926; Kühn 1927, p. 333.)

TAMARICACEAE

Ovules anatropous, erect, bitegmic, more or less tenuinucellate; o.i. and i.i. 2 cells thick (*Myricaria, Tamarix*), 3 cells (*Reaumuria*), not multiplicative in the seed; micropyle formed by the endostome.

Seeds small, hairy or with a chalazal hair-tuft, more or less exalbuminous, exarillate. Testa; *o.e.* composed of longitudinally elongate cells, more or less thick-walled and pitted; *i.e.* composed of small crystal-cells; *hairs* produced from most of the epidermal cells (*Reaumuria*) or developed from the chalazal end and septate, more or less thick-walled along one side (? hygroscopic). Tegmen crushed. Endosperm cellular in *Tamarix tetrandra*, nuclear in other cases, absorbed or as a few layers of cells with starch (*Hololachna, Reaumuria*). Embryo straight.

(Netolitzky 1926; Mauritzon 1936b; Paroli 1940; Johri and Kak 1954.)

Tamaricales are placed near Capparidales and Violales but this is not supported by the simple exotestal structure of the seed. Compare Salicaceae.

THEACEAE

Ovules anatropous or anacampylotropous, transverse, erect or suspended, bitegmic, tenuinucellate; o.i. 3–10 cells thick; i.i. 3–11 cells thick; hypostase present.

Seeds small to medium-size, 1–20 mm long, rounded to flattened and alate, anatropous or becoming campylotropous after fertilization, exarillate, albuminous or not. Testa multiplicative or not, generally with sclerotic mesophyll supplied with crystal-cells; *o.e.* composed of thin-walled cells, not as a palisade, or the cells set in a palisade with lignified inner wall and adjacent parts of the radial walls (Adinandreae *pr. p.*), or as a multiple sarcotesta (*Anneslea*); *i.e.* unspecialized or as lignified cells like those of the mesophyll. Tegmen not multiplicative, unspecialized, becoming crushed. Endosperm nuclear, more or less persistent or absorbed, oily. Embryo various, straight or curved; hypocotyl often long, some seeds hypocotylar (*Ternstroemia pr. p.*); cotyledons narrow to broad and thick, or convolute. Figs. 575–598.

(Netolitzky 1926.)

While it is clear that the seed of Theaceae (excluding Bonnetiaceae, p. 82) is mesotestal with sclerotic cells and that the larger seeds have a thin-walled exotesta or sarcotesta, there is still much to be discovered about generic detail and interrelations. Flower-structure may define the family fairly closely, but the ovules, seeds and fruits show that the genera are fragments of more varied ancestry. Several classifications have been proposed, particularly for Camellioideae (Melchior 1925; Airy Shaw 1936; Keng 1960, 1962), but none fits satisfactorily seed-structure or ovule-orientation. This last feature supplies three, or four, criteria which need to be related:

(1) Ovules transverse or suspended with adaxial raphe: *Camellia* (? ovules always transverse), *Piquetia* (sometimes reduced to *Camellia;* ovules suspended), *Adinandra, Eurya,? Visnea.*
(2) Ovules erect with adaxial raphe: *Pyrenaria, Schima, Stewartia, Tutcheria,? Franklinia.*
(3) Ovules suspended with abaxial raphe: *Anneslea, Gordonia, Stereocarpus, Ternstroemia,? Laplaceae,? Visnea.*

The winged seeds of *Gordonia* and *Schima* appear, then, analogous and this conclusion is supported by wing-structure. In *Schima* it is an extension of the outer epidermis of the testa by periclinal division which multiplies the epidermis into the rows of lignified and unlignified cells as they occur also in the sides of the seed-body. In *Gordonia* the wing is an extension of the mesophyll covered by the single-layered epidermis. The winged seed has in general a close relation with the arillate and appears as one of its immediate derivatives in the change from animal to wind distribution (Corner 1954). *Gordonia* and *Schima* seem, therefore, to be parallel derivatives of two extinct arillate stocks, one with descending ovules (*Ternstroemia*-kind in *Gordonia*) and one with ascending ovules (*Pyrenaria*-kind in *Schima*). Ovule and seed point to Theaceous ancestry from an arillate capsule in the upper part of which the transverse seeds tend to descend and in the lower part to ascend, cf. Sapindaceae. These tendencies imply differences in the intercalary growth of the ovary.

With regard to seed-structure, the following categories must be distinguished:

(1) Seeds mostly small, 1–3 mm long (4–8 mm in *Adinandra spp.*), numerous, campylotropous, thinly albuminous; testa thin, not multiplicative, 2–3 cells thick (4–5 in *Visnea*), with o.e. composed of large cells; embryo curved, radicle long, cotyledons subterete; ovules parietal-axile.
(a) Exotesta thin-walled ... *Adinandra*

(b) Exotestal cells with thick, finely pitted, lignified, inner wall.
  (i) anticlinal walls shortly thickened ... *Eurya, ? Cleyera, Freziera*
  (ii) anticlinal walls extensively thickened ... *Visnea*
(2) Seeds large, 4–20 mm wide or long, relatively massive or winged; testa more than 3 cells thick, multiplicative; placentation axil.
(a) Seeds with thick woody testa 0.5–1 mm and a reticulum of testal v.b., not winged, exalbuminous.
  (i) cotyledons thick, straight; radicle short ... *Camellia*
  (ii) cotyledons thin, much folded; radicle long ... *Pyrenaria, Tutcheria*
(b) Seeds with thinly woody testa 0.3 mm thick or less, winged, more or less albuminous; v.b. simple; cotyledons thin, flat; radicle rather long.
  (i) seeds subcampylotropous, more or less winged all round; embryo with undulate cotyledons with unequal base, radicle oblique; wing developed from the exotesta ... *Schima*
  (ii) seeds straight, wing developed as an extension of the mesotesta; embryo straight with flat cotyledons.
    (x) wing unilateral; seeds flattened laterally ... *Gordonia, ? Laplacea*
    (y) winged all round; seed flattened dorsiventrally ... *Stewartia*
(c) Seeds campylotropous, not winged, thinly albuminous; testa rather strongly woody 0.2–0.5 mm thick, with red papillate sarcotesta; v.b. simple or bifurcate, axile; embryo curved with long radicle and flat cotyledons, or entirely hypocotylous ... *Anneslea, Ternstroemia*

The first group separated by seed-structure is tr. Adinandreae and it is supported by the pulpy berry and the placentation which is parietal in the upper part of the ovary and axile in the lower. This state represents the incompletely syncarpous substitution of apocarpy, as syncarpy proceeds more extensively on the outer (dorsal) side of the carpels than on the inner (ventral) edges. Therefore the tribe cannot be placed in the subfamilies Ternstroemioideae or Camellioideae. Nevertheless the seed of *Visnea* is in some ways intermediate with those of both *Camellia* and *Ternstroemia*. The seed of *Adinandra* appears as a simplification of that of *Anneslea*. That of *Eurya* closely resembles the unitegmic seed of *Saurauia* which, thus, appears to have lost the inner integument. The seed of *Eurya*, however, has 2–3 layers of lignified cells in the testal mesophyll as a reduction, evidently, from the thick mesotesta found in *Anneslea, Camellia* and *Ternstroemia*. Clearly the larger seeds which occur in some species of *Adinandra* need investigation and comparison with those of *Anneslea* and *Visnea*. While, therefore, the seeds of Adinandreae are reduced and neotenic, as inflated ovules, the ovary-structure is relatively primitive. The tribe seems to be an early offshoot of Theaceous ancestry, connected perhaps with Actinidiaceae. There is no agreement in testal structure with Dilleniaceae.

The group 2*a*, in the key, represents Camellioideae and it is the part of the family which relates with the mesotestal seeds of Hamamelidaceae, Rosaceae and Myrtaceae, though, as crassinucellate families, they are more primitive. Their unitegmic derivative may be Sapotaceae.

The group 2c, as Ternstroemioideae, has the ovule of Lauraceae and the derivative may be Ebenaceae, even Symplocaceae in view of the epigynous tendency in *Anneslea*. And, as mentioned, the mesotestal and sarcotestal seed would give by neoteny that of Adinandreae.

The group 2b suggests the central ancestry of Camellioideae and Ternstroemioideae. No arillate seeds are known in Theaceae, but the winged seed is the indication. Thus it appears that Theaceae have been a centre of diversification from a mesotestal, crassinucellate, and arillate ancestor (cf. Stachyuraceae) to the exarillate, tenuinucellate condition and, finally, leading to the unitegmic through loss of the tegmen. The winged seeds of Vochysiaceae need comparison.

Adinandra Jack

A. impressa Kobuski, *A. verrucosa* Stapf
(Borneo, Kinabalu, leg. E. J. H. Corner; flowers and fruits in alcohol.)   Figs. 575–579.

Ovules; o.i. and i.i. 3 cells thick.

Seeds 1.4 mm long (*A. verrucosa*), 1.9 mm long (*A. impressa*), campylotropous, laterally compressed, 0.5–0.6 mm thick, dark brown, slimy. Testa 3 cells thick; *o.e.* with the cells enlarged, thin-walled and filled with red-brown tannin material, the outer wall becoming mucilaginous; *mesophyll* thick-walled, lignified, most cells with a crystal, not fibriform; *i.e.* crushed. Tegmen 3 cells thick, the cells enlarging except the small-celled endotegmen with red-brown tannin-contents, eventually crushed. Endosperm thin-walled oily, without starch, the cells at first very large as in *Camellia*. Embryo curved; cotyledons linear.

Ovary with 5 loculi in the lower part, the upper with 5 partite-peltate parietal placentas; style with a fine hollow; stigma minutely 5-lobed, puberulous.

Berry with masses of stone-cells external to the network of fine cortical v.b.; placental cells much elongate and becoming pulpy like the whole ground-tissue; without idioblasts; peduncle without cortical v.b.

(Kobuski 1947.)

While these two species are quite distinct in the forest, there is some overlap with the description of *A. quinquepartita* Kobuski which seems not to have been clearly distinguished if, indeed, it is a separate species. The stamens develop as in *Saurauia* subcentrifugally; that is the later stamens to form are inserted laterally between the first.

The seed should be compared with that of *Anneslea*.

Anneslea Wall

Ovules several per loculus, ? campylotropous, suspended, axile, from the top of the loculus, with adaxial micropyle.

Seeds 7–10 × 5 × 4 mm, oblong, subcompressed laterally, campylotropous with rounded end, a slight median groove on each side, red, minutely papillate, exarillate, albuminous. Testa multiplicative; *o.e.* composed of thin-walled cells, not lignified, divided periclinally to form 2–7 layers of cells more or less radially arranged, with a minutely papillate surface from projecting groups of cells, forming a red sarcotesta; *mesophyll* with a thick outer sclerotic layer (7–14 cells thick) of lignified, pitted, isodiametric or shortly oblong cells, not fibriform, the outermost cells often with a crystal and thin outer wall, and a thinner inner layer (3–7 cells thick) of thin-walled, not lignified cells; *i.e.* composed of small unspecialized cells; *exostome* densely sclerotic. Tegmen ? 3–5 cells thick, ? 8–9 at the endostome, unspecialized but the endostome sclerotic. Vascular bundle of the raphe short, soon dividing into two branches descending the seed on the sides of the median septum (*A. crassipes*), ? undivided in *A. fragrans*. Endosperm thin-walled, oily. Embryo curved; radicle long; cotyledons flat.

Berry inferior, irregularly dehiscent. Peduncle without cortical v.b. (Wallich 1830.)

Though Wallich described and figured a red aril distinct from the seed, I find a sarcotesta of exotestal origin as in *Ternstroemia* with which the seed generally agrees.

A. crassipes Hook.f.

(Malaya, Pahang, Ulu Kali, 1300 m; full-grown but immature fruits in alcohol.)   Fig. 580.

Exotesta 3–7 cells thick; sclerotic mesotesta 7–9 cells thick. Vascular bundle of the raphe divided into two branches descending in the median septum of the seed.

I have long hunted this genus. In September 1972, when my former student Dr E. Soepadmo, of the University of Malaya, took me along the new road to the hill-station at Ulu Kali, we were lucky to find these fallen fruits. The seeds confirmed the observations which I had made on the old duplicates of Wallich's specimens of *A. fragrans*; yet there are some differences in the thickness of the testal layers and, perhaps, in the vascular supply.

A. fragrans Wall.
(Wallich 598; dried material, CGE.)   Fig. 581.
Sarcotesta 2(? –4) cells thick, the cells with reddish contents; sclerotic mesotesta 10–14 cells thick. Vascular bundle in the median septum undivided.

Camellia L.

C. sinensis (L.) O.K.
(Ceylon, cult.; flowers and fruits in alcohol.)   Figs. 582–584.

Ovules 4 per loculus in the 3-locular ovary, sessile, anatropous, rather large, 0.6 mm long; o.i. 9–10 cells thick; i.i. 9–11 cells thick, forming a stout endostome;

nucellus reduced to a single layer of cells round the large embryo-sac; hypostase small.

Seeds 9–12 mm wide, round or somewhat compressed, smooth, brown, sessile, exarillate. Testa massive, 30–40 cells thick; *o.e.* composed of small thin-walled cells, not or slightly lignified, with small undulate facets, the outer ends of the radial walls forming minutely vesicular partitions; *mesophyll* in two layers, (1) an outer layer (15–20 cells thick) composed of thick-walled, angular, contiguous, isodiametric cells with heavily lignified, pitted walls, many with a crystal in the lumen, and (2) an inner layer of thin-walled cells separated from the outer by the stratum of v.b.; *i.e.* with slightly lignified inner walls to the small cells. Tegmen not multiplicative, unspecialized, eventually crushed. Vascular bundle of the raphe with postchalazal branches forming a network of small v.b. in the testal mesophyll. Endosperm developing into a soft jelly-like mass of large cells filling the interior of the full-sized seed with an alveolar-reticulate pulp, then completely absorbed. Embryo remaining microscopic for a long time, finally growing into the soft endosperm of the full-sized seed; radicle short, embedded in the bases of the large cotyledons.

Pericarp with a double ring of v.b., as seen in t.s. Peduncle without cortical v.b.

(Sethi 1965.)

C. sasanqua Thunb.

Seed 17.5 mm wide, plano-convex, orange-brown. Testa with a thick outer layer of very thick-walled sclerotic cells and an inner layer of thinner-walled sclerotic cells, the inner layer carrying the v.b.

(Vaughan, 1970.)

Eurya Thunb.

E. trichocarpa Korth.
(Borneo, Kinabalu, RSNB 793; flowers and fruits in alcohol.)　Figs. 585, 586.

Ovules with short funicles, anatropous; o.i. 3–4 cells thick; i.i. 3–4 cells thick.

Seeds *c.* 1 mm long, campylotropous, compressed, dark brown, with slimy surface, exarillate. Testa not multiplicative; *o.e.* composed of large cells with thickened, finely pitted, and lignified inner wall and proximal parts of the radial walls, containing dark red-brown sap; *mesophyll* and *i.e.* consisting of thick-walled, pitted, and lignified cells with a large crystal in the lumen. Tegmen not multiplicative, soon crushed. Endosperm thin-walled, oily, at first coarsely reticulate as in *Camellia*.

Ovary 3-locular in the lower part, with 3 partite–peltate placentas in the upper part.

Pericarp with nests of sclerotic cells (not idioblasts) as in *Adinandra*.

I find essentially the same construction in *E. chinensis* ssp. *ceylanica*, collected at Hakgala in Ceylon.

Gordonia Ellis.

G. obtusa Wall.
(India, Nilghiri Hills, leg. Hook.f.; dried material, CGE.) Fig. 587.

Seeds –20 mm long, light brown, the wing derived from the raphe and composing about half the length of the seed, the seed-body 1.2–1.4 mm wide and somewhat compressed; hilum very narrow, slit-like at the base of the wing; micropyle just below the hilum. Testa; *o.e.* with thin-walled tabular collapsed cells; *mesophyll* composed of 7–8 layers of more or less isodiametric cells with thickened, lignified and pitted walls but wide lumen, the inner cells smaller; *endotesta* as 2–3 layers of small cells shortly longitudinally elongate, with slightly thickened walls; wing developed from the mesophyll and o.e., the mesophyll-cells rather thin-walled, often widely pitted, lignified, aerenchymatous. Embryo straight, with thin flat cotyledons (? immature).

Pyrenaria Bl.

P. acuminata Planch.
(Singapore, Botanic Gardens; living material.)　Fig. 588.

Seeds 11–14 × 4–7 mm, very variable in shape, dark brown, the micropyle pointing to the base of the fruit; hilum broad, oblong. Testa as in *Camellia*, very hard. Endosperm reduced to a thin layer, oily. Embryo pale greenish yellow, intensifying on exposure to the air.

Schima Reinw.

S. brevifolia (Hook.f.) Stapf
(Borneo, Kinabalu, RSNB 4360; flowers and fruits in alcohol.)　Figs. 589–591.

Ovules 3 per loculus in the 5-locular ovary, anatropous, erect, axile; o.i. 4–6 cells thick, some of the cells of o.e. with brown contents and often slightly enlarged; i.i. 4–6 cells thick; nucellar wall 2 cells thick, soon absorbed by the embryo-sac.

Seeds 7 × 4.5 × 1.5 mm, compressed, slightly winged all round but mainly at each end, brown. Testa consisting of an outer large-celled layer and an inner fibrous layer; *outer layer* 2 cells thick on the flat sides of the seed, many cells thick over the wing, some of the cells with slightly thickened, finely pitted, lignified walls especially in the wing with radial rows of lignified cells separated by the unlignified; *fibrous layer* 4–6 cells thick, composed of tangentially elongated, transverse or oblique, lignified cells, finely and rather sparsely pitted. Tegmen not multiplicative, crushed; *i.e.* at first with dark contents in the small cells. Vascular bundle of the raphe short, undivided. Endosperm thin-walled, oily. Embryo curved with asymmetric, slightly folded cotyledons; radicle long.

Fruit round, woody, dehiscent. Peduncle with cortical v.b. and idioblasts.

Unfortunately no young fruits were available to show how the outer layer of the testa developed; presumably it is by periclinal divisions of the exotestal layer of cells.

Stewartia L.

S. koreana Rehd.

(Arnold Arboretum, leg. G. DeWolf; dried seeds.) Fig. 592.

Seeds 7 × 4 × 1.2 mm, compressed ovate, rather narrowly winged all round, blackish brown. Testa 7–9 cells thick; wing developed from the sclerotic mesophyll; *o.e.* composed of subcuboid or shortly longitudinally elongate cells with slightly thickened and lignified outer wall, as a conspicuous epidermis; *mesophyll* 5–7 cells thick, densely sclerotic, the cells with more or less undulate outlines, many containing a crystal, those of o.h. small; *i.e.* as a layer of thin-walled subcuboid cells, not lignified. Tegmen ? 3 cells thick, becoming crushed, not lignified; *i.e.* persistent round the chalaza, the cells with brownish wall, subcuboid. Vascular bundle only in the raphe. Chalaza simple, surrounded by sclerotic tissue, with central brown hypostase. Endosperm copious, thin-walled, oily. Embryo straight; hypocotyl long; cotyledons suborbicular, rather short, flat.

Ternstroemia Mutis

T. bancana Miq.

(Singapore, Bukit Timah; living material.) Fig. 597.

Seeds 15 mm long, campylotropous, compressed, covered with carmine-red powder of epidermal papillae. Endosperm none. Embryo curved, hypocotylous with minute cotyledons.

T. lowii Stapf

(Borneo, Kinabalu, RSNB 4400; flower-buds and nearly mature fruits in alcohol.) Figs. 593–596.

Ovules anatropous to subcampylotropous, suspended from the apex of the loculus with the raphe abaxial; o.i. 6–9 cells thick; i.i. 3–4 cells thick.

Seeds 5–6 × 3.5–4.5 × 2 mm, compressed, often plano–convex, pallid whitish (immature), minutely red papillate in places along the edges, the seed-coat hard, woody, exarillate but the funicle somewhat thickened. Testa 14–24 cells thick on the sides of the seed; *o.e.* composed of thin-walled tabular cells, but taller on the free edges of the seed and there producing the papilliform hairs, not lignified, thinly pulpy; *o.h.* composed of small cuboid thick-walled lignified cells with a crystal in the lumen; *mesophyll* composed of larger, variously oblong but not fibriform, lignified cells without crystals, composing the main mass of the seed-coat, the inner 4–8 cell-layers consisting of thin-walled, not lignified, tangentially elongate cells, eventually crushed. Tegmen 3–4 cells thick, unspecialized, the cells enlarging, then crushed. Hairs papilliform, composed of 2–4 connate, thin-walled

outgrowths of adjacent epidermal cells of the testa, varying solitary, transversely septate, the finely alveolar-reticulate cytoplasm with diffuse orange-red pigment. Vascular bundle along the centre of the seed, unbranched. Endosperm thin-walled, oily rather scant. Embryo slender, curved; radicle-hypocotyl long; cotyledons subterete.

Ovary 2-locular with short style and shortly bilobed stigma; placentas 2, apical, axile, lobed into stout funicles.

Pericarp with clusters of stone-cells, many of the cells with short processes as incipient idioblasts; cortical v.b. in 2 rows; mesocarp developing lacunae; endocarp as a thin lignified layer round the loculus. Peduncle without cortical v.b.

Stamens numerous, apparently centrifugal in development, the inner ones being larger than the outer in the bud.

Tutcheria Dunn.

Tutcheria sp.

(Borneo, Kinabalu, RSNB 8023; fruits in alcohol.)

Seeds 15–17 × 10–11 × 8 mm, dark brown, 2 per loculus, superposed with flattened adjacent surfaces, axile, erect, with the micropyle pointing to the base of the fruit as in *Pyrenaria*; hilum oblong, large. Testa as in *Camellia*, very massive and hard, with a reticulum of v.b. Endosperm at first large-celled as in *Camellia*.

Berry as in *Pyrenaria*. Pericarp with 2 rings of small cortical v.b. as well as the inner ring of larger v.b. next to the loculi, with idioblasts but no clusters of sclerotic cells.

Fruit and seed seem indistinguishable from those of *Pyrenaria*.

Visnea Linn. f.

V. mocanera Linn. f.

(Madeira, N. H. Mason 355; dried material, CGE.) Fig. 598.

Seeds 2.5–3.5 × 1.8–2.3 mm, reddish brown, sub-compressed, 1–2 per loculus. Testa firm coriaceous; *o.e.* as a palisade of cells varying much in height in different parts of the seed, with dark red-brown walls and contents, the inner wall and the greater part of the radial walls thickened, lignified, and strongly pitted, the outer wall thickened and red-brown but not pitted; *mesophyll* 4–5 cells thick, 10–12 cells thick round the micropyle, the cells thick-walled, lignified, isodiametric, rather small, most with a crystal in the lumen; *i.e.*?

According to Melchior, the ovules (2–3 per loculus) are anatropous and suspended with the micropyle pointing outwards and upwards, but from the position of the seeds in the fruit it seemed to me that their arrangement was the same as in *Anneslea* and *Ternstroemia*. The structure of the seed resembles that of *Eurya* but with more extensively lignified cells in the exotesta.

THYMELAEACEAE (including Aquilariaceae)

Ovules anatropous, suspended, bitegmic, crassinucellate; o.i. 3–4 cells thick (*Aquilaria, Cryptadenia, Drapetes, Gnidia, Lachnaea, Ovidia, Passerina, Pimelea, Struthiola, Thymelaea, Wikstroemia indica*), 4–5 cells thick (*Dais, Daphnopsis, Edgeworthia, Peddiea*), 5–6 cells thick (*Chymococca, Daphne, Dirca, Lagetta, Octolepis, Phaleria, Wikstroemia canescens*); i.i. generally with the same number of cell-layers as o.i. in these genera or 1–2 more layers, 9–10 cells thick in *Daphne cannabina*, 4–6 in *Passerina*; micropyle usually formed by the endostome, strongly incurved towards the placenta in *Cryptadenia, Passerina* and *Pimelea*; obturator generally present.

Seeds mostly with non-multiplicative integuments. Testa becoming 4–5 cells thick in *Thymelaea*, –8 cells thick in *Ovidia andina* and *Goodallia guianensis*, –12 cells thick in *Aquilaria sinensis*, thin-walled, unspecialized; o.e. not as a palisade of elongate cells, somewhat thick-walled in *Daphne, Diarthron* and *Dirca*, hairy in *Aquilaria* and *Gyrinopsis*; mesophyll often crushed; i.e. with some enlarged and thick-walled cells in *Ovidia* or with bar-thickenings on the radial walls in *Daphne*. Tegmen becoming 4–5 cells thick in *Thymelaea* by division of the mesophyll-cells, 12–25 cells thick in *Phaleria* (pachy-chalazal), 10–12 cells thick in *Peddiea africana*; o.e. as a palisade of thick-walled, pitted, lignified, columnar cells with brown walls, the lumen often much reduced, with more or less hexagonal facets, straight or curved, of various heights in different species, apparently degenerate with the cells shortly tangentially elongate in *Synaptolepis* and absent from *Craterosiphon*; mesophyll thin-walled, becoming crushed, but 1–2 inner hypodermal layers with fibrous thickenings in *Lasiosiphon* and *Thymelaea*; i.e. as a layer of short, subcuboid, thinly lignified cells with more or less reticulate thickening, making a pellicle round the embryo or endosperm, but with scattered lignified cells in *Dicranolepis*, unlignified in *Octolepis*. Vascular supply generally only as the raphe-bundle, but in some cases also nucellar (*Linostoma*). Chalaza with shortly columnar or discoid hypostase well-developed and suberized, but somewhat pachy-chalazal with thin-walled cells in *Phaleria*. Nucellus filling the full-grown but immature seed, with a peripheral network of tracheids in *Craterosiphon, Dicranolepis* and *Synaptolepis*, with v.b. in *Linostoma*, crushed in the mature seed or with a few layers persistent. Endosperm nuclear, commonly absent from the seed or reduced to a trace, but copious in *Lachnaea* and *Pimelea*.

Aril ? in *Aquilaria, Gyrinopsis, Lethedon, Solmsia*. Figs. 599–601.

(Guérin 1916; Netolitzky 1926; Fuchs 1938; Kausik 1940b; Venkateswarlu 1945, 1947; Corner 1949a.)

Knowledge of the seed-structure of this family has been greatly advanced by the meticulous work of Guérin. The general construction, as he remarked, is extremely uniform though varying in detail among the species and genera. The seed is exotegmic with a well-marked palisade of lignified cells, as in Euphorbiaceae, and often with a lignified and pellicular endotegmen. It seems that in most species the integuments have become non-multiplicative, that the characteristic exotegmen may degenerate in *Craterosiphon* and *Synaptolepis*, and that the endotegmen may have degenerated in *Aquilaria, Dicranolepis* and *Octolepis*. Most peculiar is the tracheid-network in the periphery of the nucellus in *Craterosiphon, Dicranolepis* and *Synaptolepis*, which Guérin compared with the nucellar tracheids of gymnosperms, but these are the genera which in other respects show degeneration of the distinctive characters of the seed. Nevertheless in this category must come also *Linostoma* with nucellar vascular bundles (Fig. 601). Nucellar tracheids have not been reported for the seeds of the capsular genera (Aquilarieae) which, in retaining some evidence of the arillate seed, are more primitive. The exceptions, as described by Guérin, may be summarized as follows:

Craterosiphon peculiar in the absence of the lignified exotegmen and in the presence of nucellar tracheids.

Dicranolepis with only the radial walls of the exotegmen thickened; endotegmen with scattered lignified cells; nucellus with tracheids.

Octolepis with unthickened endotegmic cells.

Synaptolepis with the exotegmen represented as scattered lignified cells, thick-walled and pitted, but shortly tangentially elongate, radially compressed; with nucellar tracheids.

Aquilaria Lamk.

A. malaccensis Lamk.
(Malaya, Singapore; ripe fruits in alcohol.)   Fig. 599.

Seeds 12–15 mm long overall, 1 per loculus, consisting of a dark grey conoid body 5–5.5 mm wide, covered with short stiff brown hairs, a rather fleshy white spathulate and acute basal appendage with brown-hairy sides and back (adaxial side), and a short chalazal neck joining body and appendage; seeds eventually hanging upside down by the raphe v.b. from the dehisced fruits. Testa ? 3–4 cells thick, thin-walled, mostly crushed; o.e. with cuboid cells, many producing the aseptate tapering hairs with dark brown walls. Tegmen ? thickness, the inner layers thin-walled and crushed; o.e. as a dark brown palisade of thick-walled, lignified, prismatic cells. Raphe separate from the body of the seed at the chalazal end, entering the appendage about half way along its length; v.b. curving up to the chalaza, apparently without tegmentary branches. Appendage composed of longitudinally elongate, soft cells, but the epidermis firm under the hairy parts as on the testa. Endosperm reduced to a membrane. Embryo with thick cotyledons.

This seed resembles that of *A. beccariana* v. Tiegh., which I figured as *Gyrinopsis* (Corner 1949a); the two genera are now merged. *A. malaccensis* has a much shorter neck; the hairs are longer, and the appendage is somewhat different. Young seeds have not been available but it seems that the appendage is a peculiar development of the chalazal end of the seed, as a chalazal aril.

Linostoma Wall.

L. pauciflorum Griff.

(Sarawak, S. 26860; fruits in alcohol.) Figs. 600, 601.

Seeds 6.5 × 4.5 mm, ellipsoid in the sclerotic pericarp, exarillate, exalbuminous. Testa 4–5 cells thick, –10 cells in the raphe, thin-walled, unspecialized, not lignified. Tegmen *c.* 7 cells thick, –10 cells at the chalazal end, the endostome projecting; *o.e.* as a short layer of cuboid or oblong cells with thickened, lignified radial walls and elongate narrow pits, the facets shortly oblong; *mesophyll* composed of large cells, thin-walled, more or less collapsing; *i.e.* as a compact layer of cuboid cells with slightly thickened and lignified radial walls. Vascular bundle in the raphe entering the wide heteropyle and giving off 10–12 small nucellar bundles lying near the periphery of the nucellus and almost reaching the micropyle, simple or branched once or twice, long persistent in the seed. Embryo with small radicle and thick straight cotyledons.

Pericarp with sclerotic masses in the mesocarp; endocarp as a single layer of longitudinal lignified fibres.

Phaleria Jack

P. macrocarpa (Scheff.) Boerl.

(Java, Kebun Raya, cult.; living material.)

Seeds 16–18 × 13–15 × 8–10 mm, 1–2 per fruit, brown, ovoid, anatropous, the lower third made by the pachychalaza. Testa 6–7 cells thick on the sides of the seed, 12–15 cells thick round the micropyle, thin-walled, pulpy, then drying up, not lignified; *o.e.* with subrectangular facets. Tegmen 12–16 cells thick; *o.e.* as a palisade 50 μ high, with somewhat curving radiating cells with thick brown lignified walls and hexagonal isodiametric facets, prolonged in the endostome; *o.h.* composed in the micropylar region of subglobose cells with thickened, brown, slightly lignified walls; *mesophyll* thin-walled, more or less crushed; *i.e.* as a white pellicle of rhomboid cells, slightly lignified, with lax subannular to subreticulate thickenings on the outer walls. Chalaza dilated to form the lower third of the seed but for the most part covered by the exotegmen; dilated part 30–40 cells thick, pervaded by fans of tracheids from the end of the raphe.

This seed raises the question to what extent the nucellar tracheids reported in other genera may be connected with a pachychalaza.

TILIACEAE

Ovules anatropous, transverse, suspended or erect, bitegmic, crassinucellate; *o.i.* 2–4 cells thick; *i.i.* 2–5 cells thick; micropyle formed by the exostome, generally not in line with the endostome.

Seeds small to medium-size, the integuments generally multiplicative; chalaza often conspicuous; albuminous, exarillate. Testa 3–8 cells thick, mostly unspecialized and drying into a membrane; *o.e.* composed of longitudinally elongate cells with thin or thick and pitted walls, even lignified, but short in *Clappertonia* and *Jarandersonia*; *mesophyll* aerenchymatous, more or less collapsing; *i.e.* unspecialized, composed of small cells, mostly with a crystal in each cell, (sometimes also in i.h., *Grewia*). Tegmen 5–20 cells thick; *o.e.* as a palisade of radially elongate, thick-walled, lignified cells shortly longitudinally elongate with oblong facets, not fibriform, the outer part of the cell with rather wide stellate lumen and not or scarcely lignified, but the palisade deteriorating in *Grewia* spp; *mesophyll* aerenchymatous, unspecialized, eventually more or less crushed; *i.e.* unspecialized (*Tilia*) or as cuboid cells with brownish walls, the radial walls slightly thickened and pitted, not lignified, with a small crystal in each cell (*Clappertonia*). Chalaza thickened, often with brown cells; heteropyle small; hypostase rather thick, in some cases with a minute columella. Vascular supply in the raphe only, in some cases forming an umbrella of fine radiating fibrils of tracheids in the thickened chalaza (*Grewia*, *Triumfetta*). Endosperm nuclear, thin or copious, thin-walled, oily. Embryo straight or with oblique radicle (*Grewia*). Figs. 602–610.

(Netolitzky 1926.)

The crystal-cells of the endotesta and the shortly oblong facets of the palisade-cells of the exotegmen seem to distinguish Tiliaceae. The tangential elongation of the exotestal cells is also a fairly general feature. The characters are retained even though the seed is enclosed in hard endocarp. They must be compared with similar features in various genera of Sterculiaceae which may, perhaps, find a better position in Tiliaceae (p. 259).

Clappertonia Meisn.

C. ficifolia (Willd.) Decne

(Ceylon, Peradeniya, cult.; flowers and fruits in alcohol.)

Ovules minute; *o.i.* and *i.i.* 2 cells thick.

Seeds small, auriculiform, compressed, brown; seedcoats multiplicative. Testa 5–6 cells thick; *o.e.* with cuboid to shortly oblong cells, the facets isodiametric to oblong and sinuous, with a strongly thickened, pitted and lignified band around the radial walls, the inner wall thin, slightly pitted, lignified; *mesophyll* thin-walled; *i.e.* composed of small cells, many with a single crystal. Tegmen 5–7 cells thick; *o.e.* as a typical palisade, lignified in the inner half, with shortly oblong hexagonal facets; *i.e.* with slight bar-thickenings on the radial walls.

Corchorus L.

C. capsularis L., *C. olitorius* L.
Testa 3 cells thick. Tegmen 4–5 cells thick, with typical palisade.
(Iyer, Sulbha and Swaminathan 1961.)

C. acutangulus L.
Ovule; o.i. 2 cells thick; i.i. 3 cells thick, covering only a third of the nucellus at the time of fertilization.
Testa 3 cells thick, thin-walled. Tegmen 8–9 cells thick.
(Venkata Rao and Sambasiva Rao 1952.)

Entelea R.Br.

Testa with crystal-cells in o.e., not in i.e.
(Netolitzky 1926.)

Grewia L.

Seeds 1–3 in a pyrene, flattened, elliptic, with straight raphe, slight chalazal umbo, and rounded antiraphe, pale brown, smooth, slippery. Testa 3–6 cells thick on the sides of the seed; *o.e.* composed of longitudinal, thick-walled, pitted, lignified cells, often with the outer wall strongly thickened; *mesophyll* aerenchymatous, collapsing; *i.e.* and often *i.h.* composed of small compact cuboid cells with strongly thickened, pitted, but not or slightly lignified walls, many with 1–3 crystals, but in some species with scarcely thickened walls. Tegmen 6–12 cells thick; *o.e.* as a palisade of very thick-walled, lignified, prismatic cells, radially elongate to various lengths in different species, but almost undifferentiated in *G. polygama*, the distal part of the cell with stellate lumen and not or slightly lignified, but with a small lignified external cap; *mesophyll* aerenchymatous, collapsing; *i.e.* composed of cuboid cells with slightly thickened brown walls, with elliptic pits on the radial walls, in some cases with brown contents, not lignified, differentiating soon after fertilization. Vascular supply in the raphe only, or producing an umbrella of fine branches in the thickened chalaza (*G. glandulosa*), not entering the heteropyle. Chalaza and hypostase thickened with rows of thin-walled, unlignified, brown cells. Nucellus generally persisting as 1–3 rows of cells, at least in the chalazal half of the seed. Endosperm thin-walled, oily. Embryo straight or with oblique radicle (*G. acuminata*); radicle short; cotyledons flat.

Though the large pyrenes in the fruit have very thick and hard walls, yet the seeds within have well-differentiated seed-coats with the tough fibrous and lignified exotesta and the exotegmic palisade. This structure is well shown by *G. glandulosa* (Fig. 604*d*). Yet, it may deteriorate. In *G. acuminata* (Fig. 602) the exotegmic palisade has only scattered lignified cells, and in *G. polygama* with small fruits (Fig. 604*a*) the testa is not multiplicative and the exotegmic palisade is short, though

lignified. Another point that needs more comparative study among the many species of the genus is the fibrous structure of the pericarp.

G. acuminata
(Sarawak, S. 26751; fruits in alcohol.) Fig. 602.
Ovary 2-locular or incompletely 4-locular. Ovules 4 in the ovary, transverse, one generally sterile; o.i. (2–) 3 cells thick; i.i. 3–4 cells thick, incomplete with wide endostome occupied by the projecting nucellus.
Seeds 7 × 5.7 × 2.5–3 mm, obliquely ascending in the pyrene with the micropyle towards the base on the outer side. Testa 4 cells thick on the sides of the seed; *o.e.* with a few scattered, lignified, pitted fibres among the thin-walled cells; *i.e.* apparently without crystals. Tegmen –10 cells thick, –16 near the chalaza; *o.e.* with the palisade *c.* 35 μ high; *i.e.* as cuboid cells with firm, slightly thickened, pitted walls. Chalaza with woody hypostase. Vascular bundle of the raphe with simple end at the chalaza. Embryo with oblique radicle.
Fruit with 4 pyrenes, one abortive; mesocarp with peripheral plexus of fibrovascular bundles and a small strand of longitudinal fibrovascular bundles in the intervals between the pyrenes; slime-canals scattered but abundant in the lacunose tissue round the pyrenes.

G. glandulosa Vahl
(Kenya, leg. D. R. D. Chetham n. 8; fruits in alcohol.) Figs. 603–605.
Seeds 7 × 5 × 2 mm. Testa 5–6 cells thick; *o.e.* with very thick outer walls; *i.e.* and *i.h.* with strongly thickened, pitted and slightly lignified walls. Tegmen 10–12 cells thick; *i.e.* with brown contents in the small cells. Vascular bundle ending with an umbrella of fine branches in the chalaza (Fig. 605).
Fruit 23–28 mm wide. Pericarp with an outer, uneven, small-celled, sclerotic layer 3–5 cells thick; the broad thin-walled mesophyll traversed by many v.b., separated from the pyrene by a finely lacunose layer of mucilage-canals.

G. ? holstii Burr.
(Kenya, leg. D. R. D. Chetham n. 4; fruits in alcohol.) Fig. 604.
Seeds as in *G. glandulosa* but with thinner testa (3 cells thick) and tegmen (6–8 cells thick). Testa without thick-walled i.h. and i.e. Vascular bundle without the chalazal umbrella.
Fruit slightly 2-shouldered, subglabrous; pyrenes 2–4, each with 1–2–3 seeds.

G. orientalis L.
(Ceylon, leg. E. J. H. Corner; fruits in alcohol.) Figs. 603, 604.
Testa 4–5 cells thick. Tegmen 6–8 cells thick. Vascular bundle without chalazal umbrella.

Fruit round; pyrenes 2–4, each with 2(–3) seeds. Pericarp with a single ring of v.b. in the mesocarp.

G. polygama Roxb.

(Ceylon, leg. E. J. H. Corner; fruits in alcohol.) Fig. 603, 604.

Testa 3 cells thick; *i.e.* and *i.h.* slightly thick-walled. Tegmen 6–8 cells thick; *o.e.* as a scarcely differentiated layer of small, cuboid, slightly thick-walled, lignified cells, becoming shortly columnar towards the chalaza.

Fruit 6–7 mm wide, round; pyrenes 2–3–4, each with 2–3–4 small seeds.

Jarandersonia Kosterm.

J. paludosa Kosterm.

(Sarawak, S. 25415; flowers and fruit in alcohol.) Fig. 606.

Ovules 2 per loculus, anatropous, erect; o.i. 2–3 cells thick, the cells of o.e. brownish; i.i. 3–4 cells thick; micropyle formed by the exostome; hypostase scarcely developed in the ovule, with slight columella.

Seeds *c.* 4 mm wide, ovoid, plump, pale brown, apparently exarillate. Testa 5–8 cells thick; *o.e.* composed of small polygonal cells with brownish walls, the outer walls slightly thickened; *mesophyll* aerenchymatous, rather loose; *i.e.* unspecialized. Tegmen 17–20 cells thick, massive; *o.e.* as a layer of small cells slightly elongate longitudinally (? becoming a palisade layer); *mesophyll* composed of round, thin-walled cells, often in radial rows, aerenchymatous; *i.e.* as a compact layer of rather small cuboid cells with firm brown walls, not lignified. Nucellus massive, filling most of the immature seed, the outer cells at the nucellar base becoming slightly lignified but thin-walled. Chalaza massive with a conspicuous pad of vascular tissue at the end of the raphe, the surrounding tissue thin-walled; hypostase thin-walled, massive. Vascular supply only as the raphe-bundle.

Fruit 5-lobed, covered with blunt woody spines 0.5–1 mm long. Exocarp compact, separated from the mesocarp by a thick layer of longitudinal anastomosing fibro-vascular bundles with branches into the spines. Mesocarp permeated with irregular slime-canals. Endocarp thin, dense, lignified, as two valves to each loculus. Spines developing singly beneath the large stellate hairs on the ovary, set rather thinly with smaller, often pedicellate, stellate hairs, with a stout fibrovascular strand composed of anastomosing strands of phloem, xylem and fibres, not organized as one v.b., with a few minute slime-canals external to the fibrovascular strand.

This fruit is one of the better formed durian-relics in the family. The specimen available to me had but one immature seed.

Sparmannia Linn. f.

Seed-structure as in *Corchorus* and *Tilia*.

(Netolitzky 1926.)

Tilia L.

T. europea L.

Figs. 607–609.

Ovule erect; o.i. 3–4 cells thick; i.i. 4–5 cells thick; micropyle distant from the funicle.

Testa 6–7 cells thick, thin-walled, drying up; *o.e.* with the cells longitudinally elongate, the thin cuticle finely striate; *mesophyll* with scattered crystal-cells, especially in the 1–2 cell-layers next to i.e.; *i.e.* composed of longitudinally elongate, thin-walled cells, mostly with a small crystal, eventually crushed. Tegmen 8–11 cells thick; *o.e.* as a typical palisade; *i.e.* unspecialized, the cells longitudinally elongate.

Triumfetta L.

T. bartramia L.

(Ceylon; flowers and fruits in alcohol.) Fig. 610.

Ovule anatropous, suspended from the apex of the loculus, solitary; o.i. 2 cells thick; i.i. 3 cells thick, covering three-quarters of the nucellus at the time of fertilization.

Seeds 2.5 × 1.3 mm, somewhat compressed, smooth, brown, with shrunken chalaza, exarillate. Testa 3 cells thick through division of i.e. but –10 cells thick at the micropyle; *o.e.* composed of longitudinally elongate cells with brownish and scarcely thickened walls; *meso-phyll* composed of small cells, enlarged at the free angles of the seed; *i.e.* composed of small compressed cells, thin-walled, unlignified, with a crystal in most. Tegmen *c.* 5 cells thick (8–9 cells; Venkata Rao and Sambasiva Rao), multiplicative through division of the middle layer of cells in the ovule; *o.e.* as a palisade of radially elongate, thick-walled cells shortly longitudinally elongate, the outer part of the cells coarsely pitted and scarcely lignified, the inner part with much reduced lumen and strongly lignified in the primary layer of the wall, not pitted; *mesophyll* aerenchymatous, eventually more or less crushed; *i.e.* not lignified, the cuboid cells with brown walls, the radial walls slightly thickened and pitted. Vascular bundle terminating at the chalaza with an umbrella of short rows of fine tracheids. Chalaza thick, unlignified; hypostase conspicuous, with a columella, unlignified; heteropyle minute.

Fruit 4 mm wide, subglobose, echinate with hooked bristles, stellate-hairy; endocarp thickly fibrous, with a fibrovascular extension into each bristle, not organized into pyrenes; septa more or less crushed.

(Venkata Rao and Sambasiva Rao 1952.)

I have illustrated this genus because the fruit is a durian transformed on Tiliaceous lines into a small burr; it is a miniature of the strange exaggerations in Pedaliaceae. The bristles cover the ovary and terminate in a large hair which becomes hooked, evidently through the basipetal upgrowth of the ovary and of the bristle itself. After fertilization, the bristle continues its basipetal growth and its centre becomes the fibrovascular strand. A few gland-hairs occur along the bristle, as on the ovary. If these structures occurred at the base of the ovary, they would attract attention as staminodes or carpel-relics, but they are clearly refinements of durian-spines; the large terminal hair may be one arm of a stellate hair. In maintaining the ancestral characters of the durian-fruit, *Triumfetta* is a parallel to the Tiliaceous genera which have gone in for pyrenes.

TOVARIACEAE

Ovules campylotropous, numerous, bitegmic, crassi-nucellate; o.i. 2 cells thick; i.i. 3 cells thick.

Seeds small, numerous, albuminous, exarillate; seed-coats not multiplicative. Testa thin-walled. Tegmen; *o.e.* with the cells enlarged as a strong palisade at the endostome and along the raphe, with reticulately thickened walls, not over the rest of the seed. Endosperm nuclear.

(Mauritzon 1935b; *Tovaria pendula*, immature.)

The mature structure of the tegmen, which appears not to be known, will be critical for the assignment of the family to Capparidales with fibrous exotegmen.

TRAPACEAE

Ovules solitary, anatropous, suspended, bitegmic, crassinucellate; o.i. 5–7 cells thick, 15–19 at the chalazal end; i.i. 2 cells thick, 5–6 at the endostome; nucellus projecting as a slender beak through the micropyle.

Seed medium-size, enclosed in the nut-like fruit, exalbuminous, exarillate. Testa evidently multiplicative and consisting of three tissues; *outer region* consisting of 4–6 layers of thin-walled rectangular cells; *middle region* with *c.* 6 layers of suberised cells; *inner region* similar to the outer, all the cell-walls with fine spiral or reticulate cellulose thickenings; eventually these tissues crushed and the testa reduced to o.e. with thick walls; *i.e.* composed of small irregular cells with tannin. Tegmen not multiplicative; *o.e.* composed of narrow, elongate, longitudinal tracheids with spiral thickenings on the walls; *i.e.* thin-walled, crushed. Vascular bundle of the raphe terminating at the chalaza but with numerous branches along its course to the sides of the seed, apparently in the outer region of the testa. Hypostase 8–10 cells thick. Endosperm nuclear, evanescent. Embryo with one cotyledon vestigial.

(Netolitzky 1926; Maheshwari, P. 1963; Ram 1956.)

Alliance with Lythraceae is supported by the tracheidal exotegmen and more or less sclerotic testa. Possibly the seed is in some measure pachychalazal. Later accounts omit most of the features of the seed-coat described by Netolitzky. Maheshwari considered that on embryological grounds *Trapa* was not related with Onagraceae.

TREMANDRACEAE

Ovules anatropous, suspended, bitegmic, crassinucellate; o.i. thin, 4 cells thick in *Tetratheca*; i.i. thick, 15 cells thick in *Tetratheca*, with the outer cell-layers composed of tangentially elongate cells, the inner smaller and isodiametric, with i.e. as an endothelium; chalaza with a short process.

Seeds small, with a chalazal appendage (? aril), glabrous or hairy, albuminous. Seed-coat with thin-walled, tangentially elongate, outer cell-layers and internally a strongly sclerotic layer. Endosperm copious. Embryo small or microscopic.

(Netolitzky 1926; Mauritzon 1936b.)

The position and construction of the sclerotic layer in the seed-coat will be critical for the classification of the family.

TRIGONIACEAE

Ovules anatropous, suspended, bitegmic, crassinucellate; o.i. 2 cells thick (*Trigonia*), o.e. with large cells; i.i. 5–6 cells thick, the cells tangentially elongate.

Seeds small, with long hairs, albuminous. Embryo (?) transverse.

(Netolitzky 1926; Mauritzon 1936a.)

The structure of the mature seed is not known. If, as the elongate cells of the tegmen suggest, there are exotegmic fibres, the family cannot be near to Polygalaceae.

TROCHODENDRACEAE

Ovules anatropous, numerous, bitegmic, crassinucellate; o.i. 3 cells thick, or thicker at the exostome; i.i. 2 cells thick; micropyle formed by the exostome.

Seeds small, subalate, albuminous, exarillate; integuments not multiplicative. Testa; *o.e.* unspecialized, the cells tangentially elongate; *i.e.* slightly thick-walled (*Tetracentron*) or with much thickened radial and inner walls (*Trochodendron*), cuticularized. Tegmen; *o.e.* composed of small, tangentially elongate, thick-walled cells; *i.e.* thin-walled. Chalaza extended into the short wing, with the vascular bundle curved or hooked into the nucellar base. Endosperm cellular, thin-walled, oily. Embryo very small. Fig. 611.

(v. Tieghem 1900; Netolitzky 1926; Bailey, I. W. and Nast 1945b; Smith, A. C. 1945; Croizat 1947b; Canright 1953; Hara and Kanai 1964.)

Van Tieghem's description of the seeds as mesotestal, with sclerosed middle layer of the testa, has been corrected by Bailey and Nast. The small seeds, as enlarged ovules, are endotestal and exotegmic, as a vestige of Magnolialean or Myristicaceous features. The subalate chalazal extension with curving vascular bundle suggests the alate modification of an arillate seed, as in Ixonanthaceae, and there are many comparable instances such as *Cercidiphyllum* and *Cratoxylon*. In form, size and relatively simple structure the seed is advanced, but it may be primitive in the microscopic embryo.

The homoxylous wood without vessels was first described by van Tieghem. It has been taken to indicate affinity with Winteraceae, the seeds of which are exotestal. Smith, Bailey and Nast conclude that the family has no close affinity with Magnoliales. Cronquist and Thorne place it in the Hamamelidalean complex with mesotestal, but simplifying, seeds. Hutchinson refers *Trochodendron* with *Euptelea* (sclerotic mesotesta and endotesta) to Magnoliales and *Tetracentron* to Hamamelidales near *Platanus* (seed-coats unspecialized, crushed, as in *Cercidiphyllum*). Canright emphasizes the tricolpate pollen of Trochodendraceae in contrast with the monocolpate pollen of Magnoliales.

The fact remains that the seeds of *Euptelea*, *Tetracentron* and *Trochodendron* are remarkably similar. They summate mesotestal, endotestal and exotegmic mechanical tissues and, with minute embryo, they may point to the early ancestry of Hamamelidaceae.

TROPAEOLACEAE

Ovules anatropous, suspended, solitary, bitegmic, tenuinucellate.

Seeds evidently pachychalazal, the free parts of the integuments not multiplicative, exalbuminous, exarillate. Seed-coat; *o.e.* thin-walled; *mesophyll* with 5–8 outer cell-layers composed of thin-walled cells, the 4–8 inner cell-layers becoming suberized with reddish yellow contents (apparently as an extended hypostase). Vascular bundle dividing into a network in the expanded chalaza. Endosperm nuclear. Embryo straight.

(Netolitzky 1926; Kühn 1927; Walker, R. I. 1947.)
Concerning the affinity of the family, see p. 35.

TURNERACEAE

Ovules anatropous, transverse, bitegmic, crassinucellate; o.i. 2 cells thick; i.i. 3 cells thick.

Seeds small, drying reticulate from the exotegmen, arillate, with distinct funicle; seed-coats not multiplicative. Testa unspecialized. Tegmen; *o.e.* as a layer of finely pitted, sclerotic cells, shortly longitudinally elongate, the ends raised to form the ridges of the reticulum, as a palisade of oblique cells. Endosperm nuclear.

Aril funicular, arising round the apex of the funicle except on the micropylar side, becoming unilateral, reaching more than half-way along the seed, white, thinly pulpy, entire or lacerate; in *Stapfiella* shortly cupular; in *Mathurina* transformed into a pappus of long hairs.

(Netolitzky 1926.)
The affinity with Passifloraceae is discussed on p. 38.

Turnera L.

T. ulmifolia L.
(Ceylon; flowers and fruits in alcohol.)   Fig. 612.

Ovule with slight arillar swelling at the top of the funicle; o.i. 2 cells thick; i.i. 3 cells thick through periclinal division of i.e.; micropyle formed by the exostome.

Seeds 2.5–2.7 × 1–1.2 mm, ellipsoid subclavate, pale brown, arillate on one side. Testa soft, thin, drying on to the reticulate tegmen; *o.e.* composed of shortly longitudinally elongate cells with somewhat thickened walls, not lignified; *i.e.* composed of large cells in the meshes of the exotegmic reticulum and small cells along the ridges, thin-walled. Tegmen; *o.e.* appearing as a palisade in t.s., but the cells shortly longitudinally elongate with divergent ends, radially elongate at the micropyle and round the chalaza; *mesophyll* colourless, small-celled, crushed; *i.e.* composed of cuboid cells with thin walls and brown tannin-contents, not lignified. Vascular bundle in the raphe only, not expanding at the chalaza. Hypostase slight, thin-walled. Embryo straight. Endosperm thin-walled.

Aril appearing unilateral, covering most of the seed except along the micropylar side, 4–5 cells thick at the base, 2–3 cells at the edge; *o.e.* with thick cuticle, thin-walled as the internal cells.

(Raju 1956b; Vijayaraghavan and Kaur, 1967.)

ULMACEAE

Ovules anatropous, suspended, paired or solitary (*Holoptelea*), bitegmic, crassinucellate; o.i. and i.i. 4 cells thick (*Ulmus*), in *Holoptelea* o.i. 6 cells thick and i.i. 4 cells thick; micropyle formed generally by the endostome.

Seeds small, exalbuminous, enclosed in the indehiscent fruit; seed-coats not multiplicative. Testa apparently unspecialised and crushed in *Ulmus*, but o.e. with very thick walls and brown contents in *Holoptelea*. Tegmen crushed. Hypostase well-developed. Endosperm nuclear.

(Netolitzky 1926; Capoor 1937.)

This family, less advanced in flower than Moraceae and presumably an earlier branch of Urticales, has in its tropical genus *Holoptelea* evidence of the exotestal construction of the seed, though this in itself is little indication of affinity.

UMBELLIFERAE

Ovule anatropous, suspended, solitary, unitegmic, tenuinucellate.

Seeds contained in the schizocarps. Seed-coat not multiplicative, thin-walled and more or less crushed, or with o.e. persistent but unspecialized. Endosperm nuclear, oily with thick cell-walls. Embryo small.

(Netolitzky 1926; Baumann 1946; Gupta 1964; Gupta and Gupta 1964; Sehgal 1965.)

URTICACEAE

Ovules more or less orthotropous, erect, basal or sub-basal, solitary, bitegmic, crassinucellate; o.i. and i.i. thin, often only 2 cells thick, o.i. often shorter than i.i., and the micropyle formed by the endostome.

Seeds small, albuminous, enclosed in the endocarp. Testa not multiplicative, thin-walled, more or less crushed. Tegmen unspecialized, crushed. Hypostase lignified. Endosperm nuclear, oily. Embryo straight.

(Netolitzky 1926.)

There is evidence in the tribe Conocephaloideae, commonly but erroneously referred to Moraceae (Corner 1962), that the orthotropous ovule and small seed of Urticaceae have been derived from the anatropous ovule and larger seed of Moraceae. With the intermediacy of Conocephaloideae there is no other distinction between the two families; yet, *Artocarpus* and *Ficus* are more distinct from *Morus* than it is from Urticaceae. No definite seed-coat structure has been found in the family, which is one of the end-products of dicotyledonous evolution.

VALERIANACEAE

Ovules anatropous, solitary, suspended, unitegmic, tenuinucellate.

Seed small, enclosed in the indehiscent fruit, practically exalbuminous. Seed-coat apparently unspecialized, crushed, or with o.e. persistent. Endosperm cellular, reduced to a single layer of cells.

(Netolitzky 1926.)

VERBENACEAE

Ovules more or less anatropous to suborthotropous (*Avicennia*, *Vitex*), unitegmic, tenuinucellate; integument 6–10 cells thick, not covering the nucellus in *Avicennia*; i.e. as an endothelium.

Seeds small to fairly large (*Avicennia*); seed-coat ? not multiplicative; o.e. thin-walled or with lignified reticulate thickening of the outer wall (*Tectona*, in *Lippia* only round the micropyle and chalaza); *mesophyll* thin-walled and crushed, or with 1–3 layers of cells with reticulately thickened and lignified walls (*Lippia*, *Tectona*, *Vitex*). Endosperm cellular, much reduced in mature seeds or absent.

(Netolitzky 1926; Pal, N. 1951; Maheshwari, J. K. 1954; Padmanabhan 1960.)

VIOLACEAE

Ovules anatropous, bitegmic, crassinucellate; o.i. 2–4 cells thick; i.i. 3 cells thick; hypostase present, in some cases double (*Hybanthus*, *Viola tricolor*), the upper at the base of the embryo-sac and the lower at the base of the nucellus.

Seeds small to medium-size, in some cases alate, arillate in *Allexis, Fusispermum, Hybanthus, Rinorea pr. p., Viola pr. p.*; seed-coats generally not multiplicative (? *Rinorea*). Testa; o.e. composed of tabular or radially elongate cells, the outer wall thin to strongly thickened or the wall uniformly thickened (*Hybanthus*), in several cases with stomata at the chalaza (*Rinorea*, *Viola*) and with scattered, pitted, sclerotic cells (*Viola*); *mesophyll* thin-walled and collapsing, or thick-walled (*Hybanthus*); i.e. as a layer of small crystal-cells with thin or slightly thickened and lignified walls. Tegmen with an outer layer of longitudinal, pitted, lignified fibres, more or less flattened laterally, in 2–4 layers in the larger seeds (*Rinorea*), simplifying to a single layer as o.e. in small seeds (*Hybanthus*, *Viola*), then appearing as a palisade in t.s.; *mesophyll* composed of large cells, even slightly lignified, but collapsing; i.e. as a sheet of small cells with thin or slightly thickened walls and brown contents, cuboid to transversely elongate. Vascular supply in the raphe only, in large seeds ramifying in the expanded chalaza but not entering the testa, raphe-bundle surrounded with sclerotic cells in *Rinorea*. Chalaza often rather large and flattened, with a plate of sclerotic cells (*Rinorea*, *Viola*). Endosperm nuclear, oily. Embryo straight.

Aril as a small oily outgrowth of the micropyle or from the base of the raphe (*Hybanthus*, *Viola spp.*), as a small woody annulus round the apex of the funicle (*Rinorea*). Figs. 613–615.

(Netolitzky 1926.)

The structure of the seed-coats is known mainly for the herbaceous genera *Hybanthus* and *Viola*. It needs to be investigated in the larger seeds of the main tropical genus *Rinorea*, for it is clear that the small seeds are simplifications of the larger. Thus, *Viola* has an exotegmen made from a single layer of short flattened fibres, representing the outer epidermis, but in *Rinorea* this layer is several cells thick and composed of much longer, even sinuous, fibres; whether it results from multiplication of the outer epidermis is not known. Then, *Rinorea* may have a wide and complex chalaza with stomata and sclerotic mesophyll, such as is found in a reduced state in *Viola*. *Rinorea* presents, also, capsules set with complicated processes, simple or much branched, which give the appearance of a gall, but as vascular outgrowths they seem homologous with durian-spines. The aril, which is clearly vestigial ('seeds ... rarely arillate', Hutchinson 1967), needs investigation for it appears to arise from various regions at the base of the seed; being small it may well have been overlooked in tropical species.

See the note on *Baccaurea* (p. 139).

Hybanthus Jacq.

H. suffruticosus Baill.

Testa; o.e. with the cell-walls much thickened and laminate all round; *mesophyll* small-celled, thick-walled, persistent (Singh, D. 1963); *i.e.* thin-walled, without crystals. Chalaza without a sclerotic septum; hypostase double.

Aril from the base of the raphe.

(Raju 1958; Singh, D. 1963; as *Ionidium suffruticosum.*)

Rinorea Aubl.

Testa or tegmen evidently multiplicative to a small extent in some species. Tegmen with a hard fibrous outer shell, 2–5 cells thick, composed of flattened, ribbon-like, longitudinal and very thick-walled fibres.

Aril evidently vestigial in some cases, absent in others.

R. anguifera (Lour.) O.K.

(Malaya, Sing. F. n. 21784; dried seeds.) Fig. 613.

Seeds 6–7 mm wide, subglobose, not mottled, the broad ellipsoid chalaza (4–5 × 3 mm) slightly depressed; base of seed slightly swollen round the funicle (? as a rudimentary aril). Testa apparently 2 cells thick, very thin except at the chalaza; *o.e.* composed of tabular thin-walled cells; *i.e.* composed of small cells with slightly lignified walls, each with a large crystal; at the chalaza thicker, o.e. with frequent stomata, hypodermal cells (2–3 layers) rounded or ellipsoid, sclerotic, pitted. Tegmen ? 8 cells thick, collapsed except the outer fibrous layer and i.e.; *fibre-layer* 120–160 μ thick, composed of 3-4-5 layers of longitudinal, flattened, sinuous, pitted and lignified fibres –500 μ long, shortened and 2–3 cells thick at the micropyle; *i.e.* composed of small cuboid cells, thin-walled with brown contents. Vascular bundle of the raphe surrounded by a sheath of sclerotic cells. Chalaza flattened and dilated, pervaded by the subdivided raphe v.b.; chalazal plate 200–300 μ thick, between the v.b. and the endosperm, composed of 4–6 layers of sclerotic cells apparently adjoining the fibrous layer of the tegmen.

Aril-rim free from the funicle, composed of rounded sclerotic cells.

Capsule densely set with variously branched, villous processes –14 mm long, the stalks of the processes flattened, with 2–3 fibrovascular bundles.

R. bengalensis (Wall.) O.K.

(Borneo, Sing. F. n. 18655; dried seeds.)

Seeds 5–6 mm wide, round, mottled from brown patches of the testa, exarillate. Testa ? 3–4 cells thick; *o.e.* composed of tabular cells, some in patches with brown contents, with scattered stomata, more abundant at the chalaza; *mesophyll* collapsed; *i.e.* as a layer of small crystal-cells with slightly lignified walls, more or less embedded in the fibrous surface of the exotegmen. Tegmen 6–9 cells thick; *fibrous layer* 90–130 μ high, composed of 2–3 layers of fibres –300 μ long, –60 μ high, 25–38 μ wide; *mesophyll* collapsed; *i.e.* composed of closely adherent, small, thin-walled cells with brown

contents, not lignified. Chalazal plate composed of 5–7 layers of sclerotic cells.

Capsule smooth.

Viola L.

Testa; *o.e.* with reticulately thickened, sclerotic cells and stomata near the chalaza in several species. Tegmen with a single layer of fibres (2–3 layers of sclerotic cells in *V. arenaria*, *V. elatior*, and *V. pinnata*; Netolitzky 1926). Chalaza separated from the endosperm by a sclerotic plate composed of 2–5 layers of more or less isodiametric sclerotic cells adjoining the exotegmen. Aril (elaiosome) exostomal, funicular or from the base of the raphe.

V. arenaria DC

Testa; *i.e.* without crystals (Netolitzky 1926).

V. cinerea Boiss.

Testa; *o.e.* with the cells radiately elongate, the outer wall strongly thickened and laminate; *mesophyll* pigmented.

(Singh, D. and Gupta 1967.)

V. odorata L.

Figs. 614, 615.

Ovule; *o.i.* 3–4 cells thick.

Testa; *o.e.* with somewhat flattened cells, the walls scarcely thickened, but with scattered sclerotic cells with reticulately thickened walls and a few stomata near the chalaza; *mesophyll* 1–2 cells thick, developing fine annular thickenings as tracheidal cells. Micropyle with sclerotic endostome; *exostome* continuous with the large-celled tissue of the funicle, but with i.e. and i.h. of the testa developed as tracheidal cells making an inner sheath to the exostome.

Funicle with very large epidermal cells, the walls somewhat thickened, continued along the surface of the raphe for half its length, without aril.

V. tricolor L.

Hypostase double. Aril developed from micropylar tissue.

(Singh, D. 1963.)

VITACEAE

Ovules anatropous, erect, bitegmic, crassinucellate; *o.i.* 4–7 cells thick; *i.i.* 2–3 cells thick, rarely 1 cell thick (*Vitis vinifera*), usually 3–4 cells thick at the micropyle; endostome forming the micropyle in *Vitis*, the exostome in *Leea*.

Seeds medium-size, 5–12 mm long, obovoid, brown to blackish, with thin sarcotesta, thick lignified endotesta, horny endosperm generally ruminate with two main folds one from each side of the raphe, with persistent thin tegmen, and small to microscopic embryo; in some cases the seed more or less perichalazal. Testa usually multiplicative, in two layers; *outer layer* 5–15 cells thick, as a sarcotesta with thin-walled cells, more or less collapsing at maturity, often with many of the mesophyll-cells containing raphid-bundles or occasionally as scattered sclerotic cells sometimes in clusters especially near the raphe-bundle, outer epidermal cells cuboid to oblong with firm brownish outer wall; *inner layer* as a woody endotesta 2–6 cells thick or more near the base of the raphe, derived from i.e. (o.i.) by periclinal division with the cells of i.e. first becoming radially elongate as a thin-walled palisade, mature cells thick-walled, finely pitted, lignified, subprismatic, usually with one large or several smaller crystals in each cell filling the reduced lumen. Tegmen 3–4 cells thick; *o.e.* and *o.h.* (if tegmen 4 cells thick), consisting of a compact layer of short, tangentially elongate, lignified tracheids with spiral thickening (double and crossed in *Vitis*); *mesophyll* thin-walled, unspecialized; *i.e.* consisting of cuboid or tangentially oblong cells with thickened brownish walls, not or scarcely lignified, the radial walls with fine pits (*Leea*) or large pits. Chalaza simple, displaced in development to the antiraphe-side of the seed (*Vitis*) or more or less completely perichalazal (*Cissus, Cyphostemma, Leea*), with a hypostase of small cells with firm brownish walls. Vascular bundle single, terminating at the chalaza (*Vitis*) or elongate round the seed in the perichalaza almost to the micropyle (*Cissus, Cyphostemma, Leea*), for the most part embedded in thick-walled, lignified ground-tissue between the flanking parts of the endotesta, the base of the raphe heavily lignified. Micropyle with strong endotesta, formed by the exostome. Nucellus at first filling almost the whole of the full-sized, but immature, seed, thin-walled, rarely with raphid-cells. Endosperm nuclear, becoming thick-walled at least in the peripheral part, substituting all the nucellus, oily. Embryo developing late, varying almost microscopic to small with flat, thin cotyledons and short radicle.

Ruminations arising from the base of the seed at the entry of the vascular bundle, consisting of longitudinal plates of testal tissue lined by the tegmic, ascending nearly to the apex of the seed, the more complex seeds with 1–3 smaller ruminations on each side; testa often with raphid cells, and i.e. in some cases (*Cissus, Leea*) developed over the ruminations as a single layer of oblong sclerotic cells or short fibres, or such cells scattered (*Cissus*), not becoming stratified as in the endotesta of the seed-wall. Figs. 616–632.

(Berlese 1892; Netolitzky 1926; Suessenguth 1953; Miki 1956; Nair and Nambisan 1957; Nair and Parasuraman 1962; Periasamy 1962a; Nair and Bajaj 1966.)

The unexpected resemblance of the Vitaceous seed to the Magnoliaceous and Rutaceous is discussed on p. 36. Two genera, *Leea* and *Cissus*, have also in common with Annonaceae the perichalazal seed (Periasamy

1962a). This point, noted by Berlese (1892), proves the affinity of the two genera, to which *Cyphostemma* can be added, disposes of the idea that Leeaceae should be a separate family, but does not imply affinity with Annonaceae. The ovule, as well as the seed, in Annonaceae is perichalazal, but the feature develops only after fertilization in Vitaceae. Then asymmetrical growth, mostly along the raphe, of the anatropous ovule shifts the chalaza to the antiraphe side, as in obcampylotropous seeds (Fig. 632). The chalaza may also lengthen into a short ridge attached to a corresponding lengthening of the nucellar base. Periasamy finds a sequence from *Cayratia* with short chalaza, through *Ampelocissus* with slightly elongate chalaza, to *Tetrastigma* with distinctly elongate chalaza and so to the perichalazal *Cissus* and *Leea* (Fig. 616). He adds *Vitis* as the link between *Tetrastigma* and *Cissus*, but this needs verification.

Without decisive evidence, the evolutionary course of such a series may be read in either direction, both of which have grave implications. Hence I am led to a third possibility. Firstly, it may be assumed that the Vitaceous seed, primitively anatropous and erect in the manner of Magnoliaceous fruits with few seeds, became asymmetric through obcampylotropous growth and eventually perichalazal. Alternatively, this perichalazal state was primitive as in Annonaceae, and was converted into the normal state with free integuments and punctiform chalaza. Against this it can be argued at once that it implies a similar derivation for all anatropous and symmetrical seeds, which seems most improbable. As an instance of intercalary basipetal growth, it is the perichalazal construction which appears as the advanced. Against the first hypothesis it can be argued equally forcefully that it is extremely improbable that two such different genera as *Cissus* and *Leea* evolved in parallel an identical perichalazal state. In habit and pinnate leaf, *Leea* is more primitive than other genera of Vitaceae. It looks, therefore, as if *Leea* inherited the perichalaza from proto-Vitaceae, comparable with a proto-Annonaceous state with a post-fertilization perichalaza.

Hence I take the view that the perichalazal state of *Leea* was evolved from the normal and symmetrical anatropous seed in the course of the pachychalazal ancestry of Vitaceae; that, as steps in this seed-evolution progressed, so modern climbing genera diverged with successive degrees of perichalazal construction; and that *Cissus*, with the full construction, was the last and closest to modern *Leea* which itself is partly climbing. The progress was analogous to that leading to the Annonaceous seed, but has not resulted in the neotenic evolution of the perichalazal ovule. Other analogies occur in *Cryptocarya* (Lauraceae) and *Swietenia* (Meliaceae), and one must assume a polyphyletic origin for the perichalaza just as with the pachychalaza (p. 5). There are many profound differences between the Vitaceous and the Annonaceous seed-coats, yet in pursuit of Annonaceous ancestry one must suppose a normal anatropous ovule and arillate seed which could refer as well to Vitaceous ancestry and lead from a proto-Illiciaceous stock (pp. 26, 37).

Parthenocarpic fruits have often been observed in Vitaceae and, though lacking endosperm and embryo, the seeds are normal in size and construction.

Ampelocissus Planch.

Chalaza shifted to the antiraphe-side of the seed and shortly extended but not perichalazal. Ruminations as two broad low ridges. Endotesta 1–2 cells thick. Raphid-cells in the exotesta and ruminations.

A. latifolia Planch., *A. martini* Planch., *A. tomentosa* Planch.

(Berlese 1892; Periasamy 1962a.)

Ampelopsis L. C. Rich.

A. bipinnata Michx., *A. heterophylla* Sieb. et Zucc., *A. orientalis* (Lamk.) Planch.

(Berlese 1892.)

Cayratia Juss.

Raphe strongly thickened with two slight ridge-like ruminations. Chalaza shifted to the antiraphe side of the seed but not perichalazal. Vascular bundle dividing into 2–4 branches in the shortly extended chalaza. Endotesta 2 cells thick. Raphid-cells in the exotesta and thickened raphe.

C. carnosa Wall., *C. pedata* Vahl. Fig. 616.

(Periasamy 1962a.)

Cissus L.

Chalaza extended as a perichalaza round the greater part of the seed. Ruminations 2, prominent. Exotesta 7–10 cells thick. Endotesta 2 cells thick (*C. quadrangularis*, *Cissus sp.*) or 4–6 cells thick (*C. pallida*, *C. vitiginea*). Raphid-cells in the exotesta, ruminations and nucellus (*C. quadrangularis*), or only in the ruminations (*C. pallida*, *C. vitiginea*). Figs. 616–623.

C. acida L., *C. baudiniana* (Muell.) Planch., *C. connivens* Lamk., *C. pallida* Steud., *C. populnea* Guill. et Perr., *C. quadrangularis* L., *C. rotundifolia* Vahl, *C. vitiginea* L.

(Berlese 1892; Periasamy 1962a.)

C. pallida, C. vitiginea

Exotesta 7–10 cells thick, with an outer large-celled layer (3–4 cells thick) and an inner tanniniferous layer (4–6 cells thick). Endotesta 4–6 cells thick. Raphid-cells only in the ruminations.

C. quadrangularis
Figs. 617–622.

Seed 5 × 4 mm, one per berry. Endotesta ? not ligni-fied over the ruminations. Vascular bundle of the perichalaza with a sheath of sclerotic cells embedded in the endotesta. Nucellus with raphids in many of the peripheral cells.

Cissus sp.

(Ceylon, Dambulla; fruits in alcohol.) Fig. 623.

Seeds 6.5 × 5 mm, blackish brown, as in *C. quadrangularis* but the endotesta with scattered lignified cells towards the micropylar bases of the ruminations.

Cyphostemma (Planch.) Alston

C. setosum (Wall.) Alston
Ovule; o.i. 6–7 cells thick; i.i. 3 cells thick.

Testa becoming *c.* 20 cells thick; *mesophyll* with scattered raphid-cells; *endotesta* 4–5 cells thick, the cells cuboid. Perichalaza extensive as in *Cissus*. Ruminations 2, with 2 lateral processes to the side of the seed.

(Nair and Bajaj 1966.)

Leea L.

Chalaza extended as a perichalaza round most of the seed. Ruminations 2 as large plates from the short raphe, and 1–3 short ridges from each side of the seed, with the edges of the perichalaza themselves prominent and ridge-like. Endotesta 2–4 cells thick, extended as a single layer of lignified cells over the ruminations. Tegmen 3–4 cells thick, the middle layer or layers much enlarged in the developing seed. Vascular bundle of the perichalaza not embedded in endotestal tissue, without sclerotic sheath. Raphid cells mostly absent from the seed. Figs. 624–630.

L. aspera Edgew., *L. sambucina* Willd.

(Nair and Nambisan 1957; Periasamy 1962a.)

L. sambucina
Figs. 624, 625.

Ovule; o.i. 4–5 cells thick; i.i. 2–3 cells thick.

Seeds 4 × 3–3.5 × 2.5–3 mm, black. Exotesta 4–6 cells thick; endotesta 4 cells thick. Ruminations 4. Raphids absent.

Leea sp.

(Solomon Islands, RSS 6037, referred to *L. indica*, but doubtful; fruits in alcohol.) Figs. 626–630.

Seeds 10–11 × 6–7 mm, blackish brown. Exotesta 5–7 cells thick, but thicker near the perichalaza and with a few raphid-cells in this region; endotesta 2–3 cells thick. Tegmen 3–4 cells thick. Embryo apparently minute.

Berry 2–3 cm wide, subglobose, ripening ochre-yellow to red. Pericarp pulpy with many large raphid-cells; endocarp very thin, as a single cell-layer.

Parthenocissus Planch.

P. himalayana (Walp.) Planch., *P. quinquefolia* (Lamk.) Planch., *P. tricuspidata* (Sieb. et Zucc.) Planch.

(Berlese 1892.)

Tetrastigma Planch.

Chalaza shifted to the antiraphe-side of the seed and somewhat elongate. Ruminations as 2 broad ridges, with or without smaller ridges. Endotesta 3–4 cells thick. Raphid-cells in the exotesta and ruminations.

T. lanceolarium Wall., *T. pergamaceum* (Zoll.) Planch. Fig. 616.

(Berlese 1892; Perisamy 1962a.)

Vitis L.

V. labrusca L., *V. pallida* W. et A., *V. riparia* Michx., *V. rupestris* Scheele, *V. vinifera* L., *V. vulpina* Torr. et Gray.

(Berlese 1892; Nair and Parasuraman 1962; Periasamy 1962a; Vaughan 1970.)

V. vinifera
(England, cult.; living material, 'black Hamburg'.) Figs. 631, 632.

Seeds 5×3.5 mm, dark brown, obovoid with acute base. Exotesta 5–6 cells thick, with scattered raphid-cells; endotesta 2–3 cells thick (4 cells, Perisamy 1962a). Tegmen 3 cells thick. Chalaza shifted to the antiraphe-side of the seed, not perichalazal (or ? extended as a perichalaza over the apex and antiraphe-side of the seed, Periasamy 1962a). Ruminations 2, with raphid-cells.

VOCHYSIACEAE

Ovules anatropous, suspended, bitegmic, crassinucellate, one to many per loculus.

Seeds medium-size, often alate, albuminous or not, exarillate. Endosperm ? Embryo straight; cotyledons flat, plicate or convolute.

(Netolitzky 1926; Mauritzon 1936a; Stafleu 1952.)

The few observations on the microscopic structure of the ovule and seed in this family are so incomplete that its status in this respect cannot be decided. What is known suggests comparison with Dipterocarpaceae, Lythraceae, and Polygalaceae. The fruit of some genera has a massive wall with radiating fibrovascular bundles and exfoliating surface. Along with the alate seeds it suggests derivation from a spinous arillate capsule as in *Durio* or *Sloanea*. The wing in *Salvertea*, *Qualea*, and *Vochysia* is said to be made up of long testal hairs, but of this I saw no evidence in immature fruits of two species of *Qualea* which I have studied.

Qualea sp.
(Brazil, Mato Grosso, Chavantina, leg. E. J. H. Corner 1968; flowers and immature fruits in alcohol.) Figs. 633–638.

Ovules *c*. 6 per loculus, of unequal length, suspended with short thick funicles attached longitudinally for a considerable distance along the placenta, the ovule-body subcrescentic; o.i. 2–3 cells thick, but thicker near the chalaza; i.i. 2 cells thick, 3–4 cells round the chalaza; micropyle formed by the free, elongate, tubular part of o.i. with expanded and subundulate exostome.

Seed (immature), lengthening and retaining the shape of the ovule, the funicle broadening as a wing. Testa becoming 4–7 cells thick in the lower part of the ovule, mainly by periclinal division of i.e. and i.h.; o.e. composed of cuboid cells; i.e. composed of small cells, most with a few small crystals. Tegmen becoming 7–8 cells thick in the lower part of the ovule, 8–10 cells round the chalaza, mainly through periclinal division of o.e., the tissues undifferentiated; i.e. as a single layer of cells or in places 2 cells thick, the cells elongated radially and tangentially at right angles to the long axis of the ovule, appearing as a palisade in l.s. of the ovule, with yellow-brown opalescent contents. Nucellus developing a hypostase-like cushion indented in the centre by the antipodal end of the embryo-sac.

Ovary 3-locular with 3 axile placentas, externally covered with long and short, aseptate, thick-walled hairs with brown contents, some of the hairs finely papillate, the short ones undulate; ovary-wall in three layers, (1) a narrow outer cortex with large cells, without v.b., covered by the piliferous epidermis, (2) a broad middle cortex traversed more or less radially by slender fibrovascular bundles and procambial strands derived from the ring of small v.b., and (3) the inner cortex without v.b., many of the cells with a crystal or crystalline mass. Style triangular in t.s.

Fruit soon after fertilization thickening the outer cortex by periclinal divisions into minute irregular excrescences, then sloughing off from the middle cortex, the half-grown fruit becoming glabrous and showing the six longitudinal furrows of the exposed mesocarp; fibrovascular bundles permeating the mesocarp radially, with stout pitted fibres and a few strands of protoxylem with spiral thickening, the ends of these v.b. covered by short curved tracheid-cells and reaching almost to the surface of the mesocarp, frequently branched.

WINTERACEAE

Ovules anatropous, bitegmic, crassinucellate, several per carpel; o.i. 3–4 cells thick, 4–5 near the raphe; i.i. 2–3 cells thick; chalaza small; micropyle formed by the endostome, but the exostome equalling it after fertilization in *Belliolum*, *Drimys* and *Zygogynum*; funicles short.

Seeds rather small, 2–5 mm long, blackish brown, hard, smooth, often angular, exarillate, albuminous. Testa not or slightly multiplicative; o.e. as a lignified palisade of prismatic cells; *mesophyll* thin-walled or somewhat thick-walled (*Pseudowintera*); i.e. unspecialized, crushed. Tegmen not multiplicative (? *Zygogynum*), becoming crushed or with i.e. as a layer of thick-walled, tangentially elongate cells (*Belliolum*). Chalaza simple or thickened and making the seed subcampylotropous (*Drimys*). Nucellus disappearing or persistent (*Belliolum*). Endosperm cellular, thin-walled, oily. Embryo minute. Most cells of the seed-coats becoming tanniniferous after fertilization.

Fruiting carpels as berries with oil-cells and, often, stone-cells, but without special endocarp. Figs. 639–646.

(Netolitzky 1926; Bailey and Nast 1943, 1945a; Swamy 1952; Bhandari 1963; Sampson 1963; Tucker and Gifford 1966a, b; Vink 1970.)

This small family of vessel-less trees, inhabiting very humid forests, is generally considered to be related with Magnoliaceae but the exotestal seed-structure prevents any close affinity and refers it to the alliance of Illiciaceae, Schisandraceae and Ranunculaceae. Its modern representatives, as leptocaul trees with simple leaves and exarillate seeds in baccate fruits, are much derived. If primitively without vessels, the family must refer to a very primitive vessel-less pachycaul similar, perhaps, to a terrestrial water-lily; and this raises the question whether the simple exotestal seed-structure is the primitive state for the dicotyledon and angiosperm. Such a view, however, is impossible to reconcile with the evidence from the seeds of other dicotyledons which indicate a multiplicative testa and tegmen, as well as an aril, at the outset. It seems that Winteraceae have lost these features, as well as the mechanical function of the endotesta. Thus, it may be that the tegmen of *Zygogynum* thickens slightly and in *Belliolum* both exotegmen and endotegmen become more or less thick-walled, especially at the endostome. That the aril can be lost from exotestal seeds is shown by Nymphaeaceae and Podophyllaceae.

In floral construction there is a detail that requires more investigation. In *Drimys* the androecium is centrifugal (Vink 1970), but in *Pseudowintera* it is centripetal (Sampson 1963).

Belliolum v. Tiegh.

B. haplopus (B. L. Burtt) A. C. Smith
(Solomon Islands, RSS 41; flowers and fruits in alcohol.) Figs. 639–641.

Ovules numerous; o.i. 3–4 cells thick, –5 cells near the raphe; i.i. (2–) 3 cells thick; nucellus rather small; chalaza small; micropyle formed by the endostome and the exostome; funicle very short or none.

Seeds 3–4 × 2.5 × 2 mm, subcompressed, reddish black. Testa 4–6 cells thick, with some periclinal division in the mesophyll, unspecialized and aerenchymatous except o.e., eventually crushed; *o.e.* as a palisade of much elongated, lignified, brown cells with fine pits, the palisade extending round the raphe and on the short funicle. Tegmen not thickening, the cells elongating tangentially, eventually crushed except i.e. composed of thick-walled cells, and in places also some thick-walled cells in o.e., but the cells of the endostome with indurated brown walls.

Nucellus persistent as a microscopic layer. Chalaza not enlarged. Vascular bundle only in the raphe.

Baccate carpels red; mesocarp pulpy with oil-cells and abundant stone-cells, without special endocarp; exocarp eventually with a few scattered stomata.

Carpels clavate, compressed; cavity obliquely longitudinal; stigma linear, papillate, apical, as a compressed crescent with the arms approximated; placentas 2, apical–parietal, opposed; the young carpel developing as a basipetal tube (without stomata) below the crescentic primordium persistent as the stigma; ovules developed in basipetal succession, supplied with v.b. from the vascular ring shortly below the stigma.

Drimys J. R. et G. Forster

D. piperita Hook.f.
(Borneo, Kinabalu, RSNB 719, 755; flowers and fruits in alcohol.) Figs. 642–645.

Ovules several in each carpel, arranged transversely in two alternating rows; o.i. (3–) 4 cells thick, o.e. already with the appearance of a short palisade; i.i. 3 (–4) cells thick; micropyle formed by the endostome then covered by the exostome in the seed.

Seeds 2–2.3 × 1.2–1.4 × 0.7–0.8 mm, blackish brown, shiny, subcampylotropous, subcompressed, with a small curved apiculus at the base (representing the indurate funicle). Testa 4 cells thick, aerenchymatous with abundant large oil-cells in the mesophyll; *o.e.* as a palisade of shortly radially elongate cells with lignified, pitted, dark brown walls; *i.e.* unspecialized, collapsing into a dry brown layer. Tegmen unmodified, gradually crushed, but the cells of the endostome with slightly thickened and lignified walls. Raphe thickening into a conspicuous pad of thin-walled aerenchymatous tissue with oil-cells, projecting into the endosperm and causing it to appear curved. Chalaza developing a hypostase of small cells with slightly thickened and lignified walls. Vascular bundle only in the raphe. Nucellus crushed.

Baccate carpels purple-black, thin-walled, with oil-cells but without stone-cells or endocarp-fibres; epidermis with sunken stomata, also abundant on the flowering carpels.

Male flower with the stamens opening centrifugally.
(Tucker and Gifford 1966a, b.)

Pseudowintera Dandy

P. colorata (Raoul) Dandy
Fig. 646.

Ovules 8–10 per carpel, as in *Belliolum*; o.i. 3 cells thick; i.i. 2 cells thick, 3–4 at the endostome forming the micropyle.

Seeds 5 × 3.5 mm, angular, black. Testa apparently not multiplicative; *o.e.* as a palisade of thick-walled cells; *mesophyll* somewhat thick-walled, not lignified; *i.e.* unspecialized, crushed. Tegmen unspecialized, crushed.

Fruits with 2–3 seeds. Pericarp 16–17 cells thick (carpel-wall at anthesis *c.* 12 cells thick), with abundant oil-cells and stone-cells but no special endocarp; innermost tissue of the pericarp growing as plates of cells between the seeds, making the cavity septate.

(Bhandari 1963; Sampson 1963.)

Zygogynum Baill.

Ovule; o.i. 3 cells thick, 4 cells at the chalazal region, with tannin-contents; i.i. 2 cells thick, 3 cells towards the chalaza.

Seed with the exostome equalling the endostome after fertilization, all the cells of the' seed-coats containing tannin. Tegmen apparently multiplicative by 2–3 layers of cells, with radial elongation of the cells at the endostome.

(Swamy 1952.)

ZYGOPHYLLACEAE

Ovules anatropous, transverse with the micropyle uppermost or suspended, bitegmic, crassinucellate; o.i. 2–3 cells thick, 6–7 in *Seetzenia*; i.i. 2–3(–4) cells thick; funicle distinct; micropyle often prolonged.

Seeds small to medium-size, exarillate, albuminous or not; seed-coats generally not multiplicative (except *Guiacum, Seetzenia*); *o.e.* as a palisade of thin-walled cells elongate radially with slightly thickened and lignified walls in *Peganum*, the cells disjunct in *Zygophyllum*, tabular and crushed in *Tribulus*; *mesophyll* unspecialized; *i.e.* mostly as short crystal-cells, with lignified bands on the walls in *Tribulus*, large and radially elongate without crystals in *Peganum*. Tegmen with o.e. and mesophyll crushed; *i.e.* as narrow longitudinal lignified rectangular cells in *Peganum, Seetzenia* and *Tribulus*. Vascular bundle only in the raphe. Chalaza unspecialized. Endosperm nuclear, oily, slightly thick-walled (*Peganum*), with distinctly thickened walls (*Guiacum, Zygophyllum*), ruminate (*Guiacum*). Embryo straight or curved; cotyledons flat, in exalbuminous seeds thick. Fig. 647.

(Netolitzky 1926; Engler 1931.)

There is considerable variety in the detail of the seed-coats in this family. Genera with indehiscent fruits or with endocarp cocci seem to have a more or less degraded structure, but there are differences also in the capsular genera. Thus the capsular *Peganum* has an exotestal palisade, a less distinct and thin-walled endotestal palisade, and a lignified, almost fibriform endotegmen. In contrast, *Guiacum* with larger seeds has only the exotestal palisade. Then among genera with indehiscent fruits or with endocarp cocci, the lignified subfibriform endotegmen occurs in *Seetzenia* and *Tribulus* and the endotesta consists of short crystal-cells which are lignified in *Tribulus*. Such crystal-cells occur in *Guiacum* but not in *Peganum*. The following distinctions can be made:

(1) Endotesta as crystal-cells with thickened walls.
  (*a*) Endotegmen with longitudinal, pitted, lignified cells ... *Seetzenia, Tribulus*
  (*b*) Endotegmen not lignified ... *Guiacum, Zygophyllum* (? *Larrea*, ? *Porlieria*)
(2) Endotesta without crystal-cells.
  (*a*) Endotegmen with longitudinal, pitted, lignified cells ... *Peganum*
  (*b*) Endotegmen without lignified cells ... *Kallstroemia* (? *Fagonia*)

On the assumption that lignification degenerates in the seed-coats in the indehiscent fruits, the primitive seed of the family would appear to have been exotestal with a palisade, endotegmic with lignified fibriform cells, and supplied with crystal-cells in the endotesta, possibly as a lignified endotesta. The relation of this construction to the Malpighiaceous is discussed on p. 18.

Fagonia L.

F. cretica L.
Ovules 2 per loculus; o.i. and i.i. 2–3 cells thick, not multiplicative in the seed; micropyle formed by the endostome.

Testa; *o.e.* with the cells enlarged, radially elongate, many enlarging further and projecting as clavate ends, finally mucifying, apparently not thick-walled; *mesophyll* and *i.e.* unspecialized, or i.e. as crystal-cells (Engler 1931). Tegmen; *o.e.* persistent, thin-walled; *i.e.* at first as an endothelium, then crushed as the mesophyll.

(Nair and Gupta 1961.)

Guiacum L.

G. officinale L.
(West Indies, Trinidad; dried fruits.)
Seeds 10 × 5 × 3.5 mm, pale straw colour, soft but drying hard, ovoid compressed, 1 per loculus, exarillate. Testa 9–12 cells thick, multiplicative with regular anticlinal walls, not lignified; *o.e.* as a compact palisade of shortly radially elongate cells with thick outer wall; *mesophyll* composed of rather small, somewhat mucilaginous cells, aerenchymatous but without stellate arms; *i.e.* as a continuous layer of small crystal-cells. Tegmen collapsed, neither lignified nor fibrous. Nucellus present as a cuticular lamella. Vascular bundle thickened on entry near the micropyle, terminating at the chalaza. Ruminations as finger-like processes of the inner tissue of the testa and of the tegmen. Endosperm thick-walled. Embryo straight; hypocotyl rather short; cotyledons broad.

According to Engler (1931) *Larrea* and *Porlieria* have the same kind of seed-structure.

Kallstroemia Scop.

K. maxima (L.) Torr. et Gray
(Jamaica, leg. K. R. Sporne; dried material CGE.)
Seeds in schizocarps as in *Tribulus* and of the same

shape. Seed-coat reduced to a single layer (exotesta) of small cuboid cells with thin, slightly lignified walls, without recognizable inner tissue.

Peganum L.

P. harmala L.
(Cambridge, University Botanic Garden; living material.) Fig. 647.

Ovule; o.i. 3–4 cells thick, 7–9 at the exostome, o.e. with large brown cells; i.i. 5–7 cells thick (2–3, Kapil and Ahlwalia), exceeded by o.i.; funicle rather long.

Seeds 3.5–4 mm long, subconic, angled, brown; seed-coats not multiplicative. Testa; *o.e.* as a palisade of large cells, shortly radially elongate, with thick cuticle, the outer walls and the outer ends of the radial walls distinctly thickened, brown and slightly lignified in the middle lamella; *mesophyll* with small thin-walled cells slightly suberized or lignified, eventually crushed; *i.e.* as a palisade of thin-walled cells shortly radially elongate, not as long as the cells of o.e., without crystals. Tegmen soon crushed except i.e., composed of narrow, longitudinally elongate, lignified cells, finely pitted. Endosperm oily, rather thick-walled.

(Netolitzky 1926; Kapil and Ahlwalia 1963; Johri 1963.)

The lignified endotegmen appears to have been overlooked by earlier investigators. Kapil and Ahlwalia figure a small ligulate aril from the raphe-side of the funicle, developed after fertilization, but I could not find this.

Seetzenia R.Br.

S. orientalis Decne
Ovule; o.i. 6–7 cells thick; i.i. 3 cells thick.

Testa 6–7 cells thick or more through division of the inner cell-layers; *o.e.* as a palisade of radially elongate, thin-walled cells their ends more or less contiguous; *mesophyll* thin-walled; *i.e.* as a layer of crystal-cells, shortly radially elongate, with thick (? not lignified) brown radial and inner walls. Tegmen not multiplicative; *o.e.* and *mesophyll* crushed; *i.e.* with the cells becoming tangentially elongate lignified and finely pitted.

(Narayana and Prakasa Rao 1963.)

The seed-coat is similar to that of *Tribulus* except for the exotestal palisade.

Tribulus L.

T. terrestris L.
(Ceylon; flowers and fruits in alcohol.) Fig. 647.

Ovules transverse, the micropyle on the stylar side of the funicle; o.i. 3(–4) cells thick; i.i. 3 cells thick;

exostome long, crushed on the development of the seed.

Seeds small, pale brown, 2–3 superposed in each schizocarp; seed-coats not multiplicative. Testa; *o.e.* and 1(–2) mesophyll-layers thin-walled, the cells enlarging much then crushed against the endocarp; *i.e.* as a thin sheet of closely adherent cells with lignified radial and inner walls, marked with thickened bands, not elongate, the cells at first with a small crystal. Tegmen; *o.e.* and *mesophyll* crushed; *i.e.* at first as a short endothelium, then compressed, the cells longitudinally elongate, persisting as a sub-lignified, finely punctate layer closely adherent to the endotesta.

Netolitzky interpreted the layer of lignified cells with semi-annular thickening as the exotegmen, but the outer layers of the tegmen are soon crushed in the growth of the seed and it seems to me that the layer is the endotesta as shown in Fig. 647. Unfortunately critical stages with incipient thickening were not available, but the interpretation agrees with the structure of *Seetzenia*.

Zygophyllum L.

Z. fabago L.
Ovule; o.i. 2 cells thick; i.i. 3–4 cells thick; the integuments not multiplicative in the seed; micropyle formed by the endostome.

Testa; *o.e.* as a palisade of radially elongate cells with chloroplasts the outer walls thickened; *i.e.* as a layer of small thin-walled crystal-cells. Tegmen thin-walled, crushed; *i.e.* at first as a short endothelium.

(Mauritzon 1934d; Masand 1964.)

Z. subtrijugum Cam.
(USSR, leg. A. Dosagoba 30 July 1955; dried material, CGE.) Fig. 647.

Seed 5 × 2 × 0.8 mm, oblong, flattened, fuscous, minutely whitish papillate-villous. Testa ? 2 cells thick; *o.e.* as large, isolated, projecting cells, cylindric with the walls strengthened by variously ramified thickenings, not lignified; *i.e.* as a layer of crystal-cells with thick brown radial and inner walls, not lignified. Tegmen apparently reduced to a layer of narrow, longitudinally elongate cells with brown walls, neither pitted nor lignified. Endosperm thick-walled, oily.

According to Engler (1931) such projecting exotestal cells with spiral or dendritic thickenings of the walls occur in *Z. album*, *Z. coccineum*, *Z. flexuosum*, *Z. latialatum*, *Z. microcarpum*, and *Z. sessilifolium*, and also in the genera *Augea* and *Bulnesia*.

References

Adatia, R. D. and S. V. Gavde (1962) Embryology of the Celastraceae. In *Plant Embryology, a Symposium*, pp. 1–11. CSIR, Delhi.

Adatia, R. D., Y. B. Sharma and M. R. Vijayaraghavan (1971) Studies in Gesneriaceae: 1. Morphology and embryology of *Platystemma violoides* Wall. *Bot. Notiser* **124**, 25–38.

Adema, F. (1966) A review of the herbaceous species of *Polygala* in Malesia. *Blumea* **14**, 253–276.

Agarwal, S. (1963) Morphological and embryological studies in the family Olacaceae: 1, *Olax* L. *Phytomorph.* **13**, 185–196.

(1964) Morphological and embryological studies in the family Olacaceae: 2, *Strombosia* Blume. *Phytomorph.* **13**, 348–356.

Airy Shaw, H. K. (1934) Notes on the genus *Schima* and on the classification of the Theaceae–Camellioideae. *Kew Bull.* (1936), pp. 496–499.

(1948) Sphenocleaceae. *Flora Malesiana* ser. 1, **4**, 27–28.

(1965) Diagnoses of new families, new names etc. *Kew Bull.* **18**, 249–273.

(1967) Notes on the genus *Bischofia* Bl. (Bischofiaceae). *Kew Bull.* **21**, 327–329.

(1972) The Euphorbiaceae of Siam. *Kew Bull.* **26**, 191–363.

Andersson, A. (1931) Studien ueber die Embryologie der Familien Celastraceae, Oleaceae und Apocynaceae. *Acta Univ. lund.* N.F. **27**, 1–112.

Archibald, E. E. A. (1939) The development of the ovule and seed of jointed cactus (*Opuntia aurantiaca* Lindley). *S. Afr. J. Sci.* **36**, 195–211.

Arekal, G. D. (1964) Contribution to the embryology of *Gerardia pedicularia* L. (Scrophulariaceae). *J. Indian bot. Soc.* **43**, 409–423.

Arekal, G. D., S. Rajeshwari and S. N. R. Swamy (1971) Contribution to the embryology of *Scoparia dulcis* L. *Bot. Notiser* **124**, 237–248.

Armour, Helen M. (1906) On the morphology of *Chloranthus*. *New Phytol.* **5**, 49–55.

Arora, N. (1953) The embryology of *Zizyphus rotundifolia* Lamk. *Phytomorph.* **3**, 88–98.

Artschwager, E. (1927) Development of flowers and seed in the sugar beet. *J. agric. Res.* **34**, 1–25.

Ashton, P. S. (1964) *Manual of the Dipterocarp trees of Brunei State*. Oxford University Press.

Baas, P. (1972) Anatomical contributions to plant anatomy: 2, The affinities of *Hua* Pierre and *Afrostyrax* Perkins et Gilg. *Blumea* **20**, 161–192.

Baehni, C. (1965) Mémoire sur les Sapotacées: 3, Inventaire des genres. *Boissiera* **11**, 1–262.

Baehni, C. et C. E. B. Bonner (1953) Les faisceaux vasculaires dans l'ovaire de l'*Aesculus parviflora*. *Candollea* **14**, 85–91.

Bailey, I. W. and C. G. Nast (1943) The comparative morphology of the Winteraceae: 2, Carpels. *J. Arnold Arb.* **24**, 472–481.

(1945a) ibid. 7, Summary and conclusions. *J. Arnold Arb.* **26**, 37–47.

(1945b) Morphology and relationships of *Trochodendron* and *Tetracentron*: 2, Inflorescence, flower and fruit. *J. Arnold Arb.* **26**, 267–276.

(1946) Morphology of *Euptelea* and comparison with *Trochodendron*. *J. Arnold Arb.* **27**, 186–192.

Bailey, I. W. and B. G. L. Swamy (1948) *Amborella trichopoda* Baill. A new morphological type of vesselless dicotyledon. *J. Arnold Arb.* **29**, 245–254.

(1950) Morphology and relationships of Monimiaceae. *J. Arnold Arb.* **31**, 372.

Baillon, H. (1871) Sur une nouvelle forme d'ovules. *Adansonia* **10**, 157.

Bakker, K. and C. G. G. J. van Steenis (1957) Pittosporaceae. *Flora Malesiana* ser. 1, **5**, 345–362.

Bakshi, T. S. (1952) Floral morphology and embryology of *Psilostachys sericea* Hook. f. *Phytomorph.* **2**, 151–161.

Balls, W. L. (1915) *The development and properties of raw cotton*. A. & C. Black, London.

Bambacioni-Mezzetti, V. (1935) Ricerche morfologiche sulle Lauraceae. Lo sviluppo dell'ovulo e dei sacchi pollinici nel *Laurus nobilis* L. *Ann. Bot. Roma* **21**, 186–204.

(1938) Gimnoovulae in *Persea gratissima* Gaertn. e considerazioni sulla monomera del pistillo di questa pianta. *Ann. Bot. Roma* **21**, 503–509.

(1941) Ricerche morfologiche sulle Lauraceae. Embriologia della *Umbellularia californica* e del *Laurus canariensis*. *Ann. Bot. Roma* **22**, 1–18.

Bancilhon, L. (1971) Contribution à l'étude taxonomique du genre *Phyllanthus* (Euphorbiacées). *Boissiera* **18**, 6–81.

Banerji, I. and K. L. Chanduri (1944) A contribution to the life-history of *Litchi chinensis. Proc. Indian Acad. Sci.* B 19, 19–27.

Banerji, I. and M. K. Dutt (1945) The development of the female gametophyte in some members of the Euphorbiaceae. *Proc. Indian Acad. Sci.* B 20, 51–60.

Barritt, N. W. (1929) The structure of the seed-coat in *Gossypium. Ann. Bot. Lond.* 43, 483–489.

Bate-Smith, E. C. (1972) Chemistry and phylogeny of the angiosperms. *Nature, Lond.* 236, 353–354.

Bate-Smith, E. C. and P. Ribereau-Gayon (1959) Leucoanthocyanins in seeds. *Qualitas Pl. Mater. veg.* 5, 189–198.

Baum, H. (1951) Die frucht von *Ochna multiflora. Ost. bot. Z.* 98, 383–394.

Baumann, M. G. (1946) *Myodocarpus* und die Phylogenie der Umbelliferen-Frucht. *Ber. Schweiz. bot. Ges.* 56, 13–112.

Bawa, S. B. (1969a) Embryological studies on the Haloragidaceae: 2, *Laurembergia brevipes* Schindl. *Proc. nat. Inst. Sci. India* B 35, 273–290.

(1969a) Embryological studies on the Haloragidaceae: 3, *Myriophyllum intermedium* DC. *Beitr. Biol. Pfl.* 45, 447–464.

Beltran, I. C. (1970) Embryology of *Isotoma petraea. Aust. J. Bot.* 18, 213–221.

Berg, R. Y. (1966) Seed-dispersal of *Dendromecon*: its ecological, evolutionary and taxonomic significance. *Am. J. Bot.* 53, 61–73.

(1967) Megagametogenesis and seed-development in *Dendromecon rigida* (Papaveraceae). *Phytomorph.* 17, 223–233.

(1969) Adaptation and evolution in *Dicentra* (Fumariaceae) with special reference to seed, fruit and dispersal-mechanism. *Nytt Mag. Bot.* 16, 49–75.

(1972) Dispersal ecology of *Vancouveria* (Berberidaceae). *Am. J. Bot.* 59, 109–122.

Berger, F. (1952) *Handbuch der Drogenkunde* Bd. 3. Wihelm Maudrich, Vienna.

(1964) *ibid.* Bd. 6.

Berggren, Greta (1963) Is the ovule type of importance for the water-absorption of the ripe seed? *Svensk. bot. Tidskr.* 57, 377–395.

Berlese, A. N. (1892) Studi sulla forma, struttura e svilluppo del seme nelle Ampelidee. *Malpighia* 6, 293–324, 482–536.

Bersier, J.-D. (1960) L'ovule anatrope: Ranunculaceae. *Ber. Schweiz. bot. Ges.* 70, 171–176.

Bhandari, N. N. (1963) Embryology of *Pseudowintera colorata. Phytomorph.* 13, 302–316.

(1967) Studies in the family Ranunculaceae: 9, Embryology of *Adonis* Dill. ex Linn. *Phytomorph.* 16, 578–587.

(1969) Studies in the family Ranunculaceae: 10, Embryology of *Anemone* L. *Phytomorph.* 18, 487–497.

(1971) Embryology of the Magnoliales and comments on their relationships. *J. Arnold Arb.* 52, 1–39, 285–304.

Bhandari, N. N. and S. Asnani (1968) Studies in the family Ranunculaceae: 6, Morphology and embryology of *Ceratocephalus falcatus* Pers. *Beitr. Biol. Pfl.* 45, 271–290.

Bhargava, H. R. (1932) Contributions to the morphology of *Boerhaavia repanda. J. Indian bot. Soc.* 11, 303–326.

(1936) The life-history of *Chenopodium album* Linn. *Proc. Indian Acad. Sci.* 4, 179–200.

Bhatnagar, S. P. and S. Agarwal (1961) Morphological and embryological studies in the family Santalaceae: 6, *Thesium* L. *Phytomorph.* 11, 273–282.

Bhatnagar, S. P. and M. Gupta (1970) Morphology and embryology of *Mimusops elengi* L. *Bot. Notiser* 123, 455–473.

Bhatnagar, S. P. and S. Puri (1970) Morphology and embryology of *Justicia betonica* Linn. *Ost. bot. Z.* 118, 55–71.

Bhatnagar, S. P. and G. Sabharwal (1969) Morphology and embryology of *Iodina rhombifolia* Hook. et Arn. *Beitr. Biol. Pfl.* 45, 465–479.

Biswas, I. (1957) Embryological studies in *Demia extensa* Br. *J. Indian bot. Soc.* 36, 207–222.

Blakelock, R. A. (1951) A synopsis of the genus *Euonymus* L. *Kew Bull.* pp. 210–290.

Bocquet, G. (1960) The campylotropous ovule. *Phytomorph.* 9, 222–227.

Bocquet, G. et J. D. Bersier (1959) Les formes d'ovules chez les Légumineuses. *Acta Soc. Helvet. Sci. nat.* 158–159.

Boelcke, O. (1946) Estudi morfologico de las semillas de Leguminosas Mimosoideas y Caesalpinioideas de interes agronomico en la Argentina. *Darwiniana* 7, 240–321.

de Boer, R. and F. Bouman (1972) Integumentary studies in the Polycarpicae. *Acta bot. neerl.* 21, 617–629.

Bouharmont, J. (1962) Fécondation de l'ovule et développement de la graine après croisement et autopollination chez *Hevea brasiliensis. Cellule* 62, 119–130.

Bouman, F. (1971a) The application of tegumentary studies to taxonomic and phylogenetic problems. *Ber. dt. bot. Ges.* 84, 169–177.

(1971b) Integumentary studies in the Polycarpicae: 1. Lactoridaceae. *Acta bot. neerl.* 20, 565–569.

Boursnell, J. G. (1950) The symbiotic seed borne fungus in the Cistaceae. *Ann. Bot. Lond.* n.s. 14, 217–243.

Brandza, G. (1908) La germination des Hypéricacées et des Guttifères. *Annls Sci. nat.* (*Bot.*) ser. 9, 8, 221–300.

Brenan, J. P. M. (1967) *Leguminosae* (Part 2). Flora of Tropical East Africa. Royal Botanical Gardens, Kew.

Bresinsky, A. (1963) Bau, Entwickelungsgeschichte und Inhaltsstoffe der Elaiosomen. *Biblthca bot.* no. 126.

Brofferio, Ida (1930) Osservazioni sullo sviluppo della Calycanthaceae. *Ann. Bot. Roma* 18, 387–394.

Brough, P. (1927) Studies in the Goodeniaceae: 1. The life-history of *Dampiera stricta* (R.Br.). *Proc. Linn. Soc. N.S.W.* 52, 471–498.

— (1933) The life-history of *Grevillea robusta* Cunn. *Proc. Linn. Soc. N.S.W.* 58, 33–73.

Buell, Katherine M. (1952) Development morphology in *Dianthus*: 1. Structure of the pistil and seed development. *Am. J. Bot.* 39, 194–210.

Buxbaum, F. (1958) Morphologie der Kakteen: 4. Samen. In H. Kranz, F. Buxbaum und W. Andrew, Die Kakteen, *Lieferung* 9, 79–97.

— (1961) Vorläufige Untersuchungen über Umfang, systematische Stellung und Gliederung der Caryophyllales (Centrospermae). *Beitr. Biol. Pfl.* 36, 1–56.

— (1968) Die Entwickelungslinien der tribus Cereeae Britt. et Rose emend. F. Buxbaum (Cactaceae–Cactoideae). *Beitr. Biol. Pfl.* 44, 215–276, 389–433.

Camp, W. H. and M. M. Hubbard (1963a) Vascular supply and structure of the ovule and aril in Peony and of the aril in nutmeg. *Am. J. Bot.* 50, 174–178.

— (1963b) On the origins of the ovule and cupule in Lyginopterid Pteridosperms. *Am. J. Bot.* 50, 235–243.

Canright, J. E. (1953) The comparative morphology and relationships of Magnoliaceae: 2, Significance of pollen. *Phytomorph.* 3, 355–365.

— (1960) Comparative morphology and relationships of Magnoliaceae: 3, *Am. J. Bot.* 47, 145–155.

— (1963) Contribution of pollen morphology to the phylogeny of some Ranalean families. *Grana Palynologica* 4, 64–72.

Capitaine, L. (1912) *Les graines des Legumineuses.* Paris.

Capoor, S. P. (1937) The life-history of *Holoptelea integrifolia* Planch. (Ulmaceae). *Beih. bot. Zbl.* 57, Abt. A, 233–249.

Carlquist, S. and P. H. Raven (1966) The systematics and anatomy of *Gongylocarpus* (Onagraceae). *Am. J. Bot.* 53, 378–390.

Carolin, R. C. (1966) Seeds and fruits of the Goodeniaceae. *Proc. Linn. Soc. N.S.W.* 91, 58–83.

Casper, S. Jost (1966) Monographie der Gattung *Pinguicula. Biblthea bot.* no. 127.

Chakravorti, A. K. (1947) The development of the female gametophyte and seed of *Coccinia indica* W. et A. *J. Indian bot. Soc.* 26, 95–104.

Cheesman, E. E. (1927) Fertilization and embryogeny in *Theobroma cacao* L. *Ann. Bot. Lond.* 41, 107–126.

Chesnais, F. (1943) Sur la formation de la cicatrice de la graine chez les Sapotacées. *Bull. Soc. bot. Fr.* 90, 177–181.

Chitaley, S. D. and S. A. Paradkar (1972) *Rodeites Sahni* re-investigated. *Bot. J. Linn. Soc.* 65, 109–117.

Chodat, R. et A. Rodrigue (1893) Recherches sur la structure du tégument seminal des Polygalacées. *Bull. Herb. Boissier* 1, 197–202, 450–463, 517–541, 571–583.

Chopra, R. N. and H. Kaur (1965) Embryology of *Bixa orellana* Linn. *Phytomorph.* 15, 211–214.

Chopra, R. N. and A. J. Mukkada (1967) Gametogenesis and pseudo-embryo sac in *Indotristicha ramosissima* (Wight) van Royen. *Phytomorph.* 16, 182–188.

Chopra, R. N. and K. S. Rai (1958) Response of ovules of *Argemone mexicana* L. to colchicine treatment in vivo. *Phytomorph.* 8, 107–113.

Chowdhury, K. A. and G. M. Buth (1971) Cotton seeds from the Neolithic in Egyptian Nubia and the origin of Old World cotton. *Biol. J. Linn. Soc.* 3, 303–312.

Chowdhury, J. K. and J. N. Mitra (1953) Abnormal tricotyledonous embryo and the morphological structure of normal fruit and seed of *Cinnamomum camphora* F. Nees. *Sci. Cult.* 19, 159–160.

Clark, L. (1923) The embryogeny of *Podophyllum peltatum.* Minn. *Stud. Pl. Sci.* 4, 111–124.

Coolhaas, C., H. J. de Fluiter and H. P. Koenig (1960) *Kaffee. Tropische und subtropische Weltwirtschaftspflanzen* vol. 3, part 2. Ferdinand Enke Verlag, Stuttgart.

Cooper, D. C. (1941) Macrosporogenesis and the development of the seed of *Phryma leptostachya. Am. J. Bot.* 28, 755–761.

Cooper, G. O. (1942a) Microsporogenesis and development of seed in *Lobelia cardinalis. Bot. Gaz.* 104, 72–81.

— (1942b) Development of the ovule and the formation of the seed in *Plantago lanceolata. Am. J. Bot.* 29, 577–581.

Copeland, H. F. (1953) Observations on the Cyrillaceae particularly on the reproductive structures of the North American species. *Phytomorph.* 3, 405–411.

— (1955) The reproductive structures of *Pistacia chinensis* (Anacardiaceae). *Phytomorph.* 5, 440–449.

— (1959) The reproductive structures of *Schinus molle* (Anacardiaceae). *Madroño* 15, 14–25.

— (1962) Observations on the reproductive structures of *Anacardium occidentale. Phytomorph.* 11, 315–325.

(1964) Structural notes on Hollies (*Ilex aquifolium* and *I. cornutum*, family Aquifoliaceae). *Phytomorph.* **13**, 455–464.

(1967) Morphology and embryology of *Euonymus japonica. Phytomorph.* **16**, 326–333.

de Cordemoy, H. J. (1911) Contribution à l'étude de la structure du fruit et de la graine des Clusiacées. *Annls Mus. col. Mars.* ser. 2, 9, 1–22.

Corner, E. J. H. (1939) Notes on the systematy and distribution of Malayan phanerogams 3. *Gdns' Bull. Straits Settl.* **10**, 239–329.

(1952) Durians and dogma. *Indones. J. nat. Sci.* **108**, 141–145.

(1946) Centrifugal stamens. *J. Arnold Arb.* **27**, 423–437.

(1949a) The durian theory or the origin of the modern tree. *Ann. Bot. Lond.* n.s. **13**, 367–414.

(1949b) The Annonaceous seed and its four integuments. *New Phytol.* **48**, 332–364.

(1951) The leguminous seed. *Phytomorph.* **1**, 1–34.

(1953–54) The durian theory extended: 1. *Phytomorph.* **3**, 465–476; 2. *Phytomorph.* **4**, 152–165; 3. *Phytomorph.* **4**, 263–274.

(1958) Transference of function. *J. Linn. Soc.* (*Bot.*) **56**, 33–40.

(1962) The classification of Moraceae. *Gdns' Bull. Singapore* **19**, 187–252.

(1963) A Dipterocarp clue to the biochemistry of durianology. *Ann. Bot. Lond.* n.s. **27**, 339–341.

(1964) *Life of Plants.* Weidenfeld and Nicolson, London.

(1967) *Ficus* in the Solomon Islands. *Phil. Trans. R. Soc.* B **253**, 23–159.

(1970a) *Ficus* subgen. *Ficus.* Two rare and primitive pachycaul species. *Phil. Trans. R. Soc.* B **259**, 353–382.

(1970b) *Ficus* subgen. *Pharmacosycea* with reference to the species of New Caledonia. *Phil. Trans. R. Soc.* B **259**, 383–433.

Coy, G. V. (1928) Morphology of *Sassafras* in relation to phylogeny of Angiosperms. *Bot. Gaz.* **86**, 148–171.

Crété, P. (1936a) Développement et structure du tégument séminal chez le *Reseda luteola* L. *Bull. Soc. bot. Fr.* **83**, 43–46.

(1936b) Transformation de l'ovule en graine chez l'*Androsaemum officinale* All. *Bull. Soc. bot. Fr.* **83**, 654–657.

(1937a) Etude sur la strophiole du *Chelidonium majus* L. *Bull. Soc. bot. Fr.* **84**, 196–199.

(1937b) Developpement et structure du tégument séminal chez le *Radiola linoides* Roth. *Bull. Soc. bot. Fr.* **84**, 655–659.

(1944) Recherches anatomiques sur la séminogenèse de l'*Actinidia chinensis* Planch. *Bull. Soc. bot. Fr.* **91**, 153–160.

(1952) Contribution à l'étude embryologique des Datiscacées. *Bull. Soc. bot. Fr.* **99**, 152–156.

Croizat, L. (1947a) A study in the Celastraceae. *Lilloa* **13**, 31–43.

(1947b) *Trochodendron, Tetracentron,* and their meaning in phylogeny. *Bull. Torrey bot. Club* **74**, 60–76.

(1960) *Principia Botanica,* Wheldon and Wesley, Hitchin, England.

Cronquist, A. (1968) *The evolution and classification of flowering plants.* Houghton Mifflin, Boston.

Cusset, G. (1966) Essai d'une taxinomie foliaire dans la tribu des Bauhineae. *Adansonia* n.s. 6, 251–280.

Dastur, R. H. (1921) Notes on the development of the ovule, embryo-sac and embryo of *Hydnora africana* Thunb. *Trans. R. Soc. S. Afr.* **10**, 27–31.

Dathan, A. S. R. and D. Singh (1969a) Development of female gametophyte and seed of *Tacsonia mollissima* HBK. *Proc. 56th Indian Sci. Congr.* Part 3 (4), pp. 389–390.

(1969b) Female gametophyte and seed of *Carica candamarensis* Hook. f. *Proc. 56th Indian Sci. Congr.* Part 3 (4), pp. 394–395.

(1970) Development and structure of female gametophyte and seed of *Bergia odorata* Edgew. *Proc. 57th Indian Sci. Congr.* Part 3 (4), pp. 254–255.

(1971) Embryology and seed-development in *Bergia* L. *J. Indian bot. Soc.* **50**, 362–370.

Davis, Gwenda L. (1961) The life-history of *Podolepis jaceoides* (Sims) Voss.: 2, Megasporogenesis, female gametophyte and embryogeny. *Phytomorph.* **11**, 206–219.

Dawson, J. W. (1970a) Pacific capsular Myrtaceae: 1. Reproductive morphology of *Arillastrum gummiferum* Panch. ex Baillon (New Caledonia). *Blumea* **18**, 431–440.

(1970b) Pacific capsular Myrtaceae: 2, 3. The *Metrosideros* complex. *Blumea* **18**, 441–452.

Delay, Cécile et G. Mangenot (1960) Le développement de la graine chez *Allanblackia floribunda* Oliv. *Annls Sci. nat.* (*Bot.*) ser. 12, 1, 387–440.

Deshpande, P. K. (1962a) Fertilization and development of endosperm, embryo and fruit in *Flaveria repanda* Lag. *J. Indian bot. Soc.* **41**, 505–509.

(1962b) Contribution to the embryology of *Caesulia axillaris* Roxb. *J. Indian bot. Soc.* **41**, 540–550.

(1964a) A contribution to the embryology of *Bidens biternata* (Lour.) Merr. et Sherff. *J. Indian bot. Soc.* **43**, 149–157.

(1964b) A contribution to the life-history of *Volutarella ramosa* Roxb. *J. Indian bot. Soc.* **43**, 141–148.

Deshpande, P. K. and A. G. Untawale (1971) Development of seed and fruit in *Indigofera enneaphylla* L. *Bot. Gaz.* **132**, 96–102.

Dickison, W. C. (1971) Anatomical studies in the Connaraceae: 1. Carpels. *J. Elisha Mitchell sci. Soc.* **87**, 77–80.

Dilcher, D. L. and J. F. McQuade (1967) A morphological study of *Nyssa* endocarps from Eocene deposits in western Tennessee. *Bull. Torrey bot. Club* **94**, 35–40.

Ding Hou (1962) Celastraceae. *Flora Malesiana* ser. 1, 6, 227–291.

(1967) *Sarawakodendron*, a new genus of Celastraceae. *Blumea* **15**, 139–143.

(1968) *Crossostylis* in the Solomon Islands and the New Hebrides (Rhizophoraceae). *Blumea* **16**, 129–132.

Dittrich, M. (1968) Morphologische Untersuchungen an den Früchten der subtribus Cardueae–Centaureinae (Compositae). *Willdenowia* **5**, 67–107.

Dnyansagar, V. R. (1951) Embryological studies in Leguminosae: 2, A contribution to the embryology of *Mimosa hamata*. *J. Indian bot. Soc.* **30**, 100–107.

(1954) Embryological studies in the Leguminosae: 7, *Neptunia triquetra* Benth. and *Prosopis spicigera* Linn. *J. Indian bot. Soc.* **33**, 247–253.

(1956) Embryological studies in the Leguminose: 5, *Prosopis spicigera* and *Desmanthus virgatus*. *Bot. Gaz.* **118**, 180–186.

(1958) Embryological studies in the Leguminosae: 8, *Acacia auriculaeformis* A. Cunn., *Adenanthera pavonina* Linn., *Calliandra hematocephala* Hassk. and *Calliandra grandiflora* Benth. *Lloydia* **21**, 1–25.

Dnyansagar, V. R. and D. C. Cooper (1960) Development of the seed of *Solanum phureja*. *Am. J. Bot.* **47**, 176–186.

Dnyansagar, V. R. and R. S. Malkhede (1962) Development of the seed in *Trianthema portulacastrum* Linn. *Proc. 49th Indian Sci. Congr.* Part 3, p. 266.

Dolcher, T. (1941) Ricerche embriologiche sulla familia delle Rhamnaceae. *Nuovo G. bot. ital.* **54**, 1–26.

Domke, W. (1934) Untersuchungen ueber die systematische und geographische Gliederung der Thymelaeaceen. *Biblthca bot.* no. 111.

Dorasami, L. S. and D. M. Gopinath (1945) An embryological study of *Linum mysorense* Heyne. *Proc. Indian Acad. Sci.* B **22**, 6–9.

Earle, T. T. (1938) Origin of the seed-coats in *Magnolia*. *Am. J. Bot.* **25**, 221.

Eckardt, T. (1963) Some observations on the morphology and embryology of *Eucommia ulmoides* Oliv. *J. Indian bot. Soc.* **42A**, 27–34.

(1967) Blütenbau and Blütentwickelung von *Dysphania myriocephala* Benth. *Bot. Jb.* **86**, 20–37.

Endress, P. (1970) Gesichtspunkte zur Systematischen Stellung der Eupteleaceen (Magnoliales). *Ber. Schweiz. bot. Ges.* **79**, 229–278.

Engleman, E. M. (1960) Ovule and seed development in certain cacti. *Am. J. Bot.* **47**, 460–467.

Engler, A. (1910) *Die Vegetation der Erde* vol. 9 (1, 1) p. 169.

(1931) *Zygophyllaceae. Die Naturlichen Pflanzenfamilien*, 2nd edn, vol. 19a, pp. 144–184.

Erdtman, G., P. Leins, R. Melville and C. R. Metcalfe (1969) On the relationships of *Emblingia*. *Bot. J. Linn. Soc.* **62**, 169–186.

Etheridge, A. L. and J. M. Herr Jr. (1968) The development of the ovule and megagametophyte in *Rhexia marcana*. *Can. J. Bot.* **46**, 133–139.

Evrard, C., A. Vieux et C. Kabele-Ngiefu (1971) Relations entre les corps gras des graines de Legumineuses et la classification morphologique. *Mitt. bot. Staatssamml. München* **10**, 201–204.

Eyde, R. H. (1963) Morphological and paleobotanical studies of the Nyssaceae. *J. Arnold Arb.* **44**, 1–59, 328–376.

(1966) Systematic anatomy of the flower and fruit of *Corokia*. *Am. J. Bot.* **53**, 833–847.

(1968) Flowers, fruits and phylogeny of Alangiaceae. *J. Arnold Arb.* **49**, 167–192.

Fagerlind, F. (1937) Embryologische und Bestaubungsexperimentelle Studien in der familie Rubiaceae nebst Bermerkungen über einige Polyploiditatsprobleme. *Acta Horti Bergiani* **11**, 195–470.

(1940) Die Entwickelung des Embryosacks bei *Peperomia pellucida* Kunth. *Ark. Bot.* **29A**, 1–15.

(1941) Der Bau der Samenanlage und des Makrogametophyten bei *Quisqualis indica*. *Bot. Notiser* pp. 217–222.

(1945) Bau des Gynöceums, der Samenanlage und des Embryosackes bei einigen Repräsentäten der Familie Icacinaceae. *Svensk Bot. Tidskr.* **39**, 346–364.

(1947) Die Systematische Stellung der Familie Grubbiaceae. *Svensk Bot. Tidskr.* **41**, 315–320.

(1959) Development and structure of the flower and gametophytes in the genus *Exocarpos*. *Svensk Bot. Tidskr.* **53**, 257–282.

Fahn, A. (1967) *Plant Anatomy*. Pergamon Press, Oxford.

Farooq, M. (1960) The embryology of *Galium asperifolium* Wall. *J. Indian bot. Soc.* **39**, 171–175.

(1964) Studies in the Lentibulariaceae: 1, The embryology of *Utricularia stellaris* Linn. f. var. *inflexa* Clarke. *Proc. nat. Inst. Sci. India* B **30**, 262–299.

(1965) Studies in the Lentibulariaceae: 3, The embryology of *Utricularia uliginosa* Vahl. *Phytomorph.* **15**, 122–131.

(1966) Studies in the Lentibulariaceae: 4, The embryology of *Utricularia striata* Sm. *J. Indian bot. Soc.* **45**, 1–13.

Farooq, M. and S. A. Siddiqui (1967) Studies in the Lentibulariaceae: 6, The embryology of *Utricularia stellaris* Linn. *J. Indian bot. Soc.* **46**, 31–44.

Farron, C. (1963) Contribution à la taxonomie des Ourateae Engl. (Ochnacées). *Ber. Schweiz. bot. Ges.* **72**, 196–217.

Favard, Anna (1963) Contributions à l'étude histologique de la croissance du développement des *Drosera*. *Annls Sci. nat.* (*Bot.*) ser. 12, **4**, 265–538.

Fedde, F. (1936) *Papaveraceae*. In Engler's *Die Natürlichen Pflanzenfamilien*, vol. 17b.

Fickel, J. F. (1876) Ueber dei Anatomie und Entwickelungsgeschichte der Samenschalen einiger Cucurbitaceen. *Bot. Ztg* **49**, 770–776, 784–792, t. xi.

Forman, L. L. (1965) A new genus of Ixonanthaceae with notes on the family. *Kew Bull.* **19**, 517–526.

(1966) The re-instatement of *Galearia* Zoll. et Mor. and *Microdesmis* Hook. f. in the Pandaceae. *Kew Bull.* **20**, 309.

(1971) A synopsis of *Galearia* Zoll. et Mor. (Pandaceae). *Kew Bull.* **26**, 153–165.

Fosberg, F. R. and M. H. Sachet (1972) *Thespesia populnea* (L.) Solander ex Correa and *Thespesia populneoides* (Roxb.) Kotel. (Malvaceae). *Smithson. Contr. Bot.* **7**, 1–13.

Foster, L. T. (1943) Morphological and cytological studies on *Carica papaya*. *Bot. Gaz.* **105**, 116–126.

Fryxell, P. A. (1964) Morphology of the base of seed-hairs of *Gossypium*. *Bot. Gaz.* **125**, 108–114.

(1968) A redefinition of the tribe Gossypeae. *Bot. Gaz.* **129**, 296–308.

Fuchs, A. (1938) Beiträge zur Embryologie der Thymelaeaceae. *Ost. bot. Z.* **87**, 1–41.

Ganapathy, P. M. (1956) Floral morphology and embryology of *Hydrophylax maritima* Linn. f. *J. Madras Univ.* B **26**, 263–275.

Ganapathy, P. S. and Barbara F. Palser (1964) Studies of floral morphology in the Ericales: 7, Embryology in the Phyllodoceae. *Bot. Gaz.* **125**, 280–297.

Garcia, V. (1962) Embryological studies on the Loasaceae. In *Plant Embryology, a Symposium*, pp. 157–161. CSIR, Delhi.

Gauba, E. and L. D. Pryor (1958) Seed coat anatomy and taxonomy in *Eucalyptus*, 1. *Proc. Linn. Soc. N.S.W.* **83**, 20–32.

(1959) ibid., 2. *Proc. Linn. Soc. N.S.W.* **84**, 278–291.

(1961) ibid., 3. *Proc. Linn. Soc. N.S.W.* **86**, 96–111.

Geesink, R. (1972) A new species of *Langsdorffia* from New Guinea (Balanophoraceae). *Acta bot. neerl.* **21**, 102–106.

Ghatak, J. (1956) A contribution to the life-history of *Oroxylum indicum* Vent. *Proc. Indian Acad. Sci.* B **43**, 72–86.

Ghosh, S. S. and R. Shahi (1957) A case of polyembryony in *Shorea robusta* Gaertn. *Sci. Cult.* **23**, 254–256.

Gilg, E. und E. Werdermann (1925) *Dilleniaceae*. In Engler's *Die Natürlichen Pflanzenfamilien*, vol. 21, pp. 7–36.

Gogelein, A. J. F. (1967) A revision of the genus *Cratoxylon* Bl. (Guttiferae). *Blumea* **15**, 453–475.

Gottsberger, G. (1970) Beiträge zur Biologie von Annonaceen-Blüten. *Ost. bot. Z.* **118**, 237–239.

Govil, C. M. (1971) Morphological studies in the family Convolvulaceae: 1, Development and structure of the seed-coat. *J. Indian bot. Soc.* **50**, 32–38.

Govinda, H. C. (1951) Studies in the embryology of some members of the Bignoniaceae. *Proc. Indian Acad. Sci.* B **32**, 164–178.

Gowda, M. (1951) The genus *Pittosporum* in the Sino-Indian region. *J. Arnold Arb.* **32**, 263–343.

Gray, A. (1858) A short exposition of the structure of the ovule and seed-coats of Magnolia. *J. Linn. Soc. Bot.* **2**, 106–110.

Grundwag, M. and A. Fahn (1969) The relation of embryology to the low seed set in *Pistacia vera* (Anacardiaceae). *Phytomorph.* **19**, 225–235.

Guard, A. (1943) The development of the seed of *Liriodendron tulipifera* L. *Proc. Indiana Acad. Sci.* **53**, 75–77.

Guédés, M. (1968) Le carpelle et le gynécée de *Merremia angustifolia* Hall. (Convolvulaceae). *Cellule* **67**, 139–161.

Guérin, P. (1901) Tégument séminal de quelques Sapindacées. *J. Bot. Paris* **15**, 336–362.

(1911) Recherches sur la structure . . . de la graine des Diptérocarpacées. *Bull. Soc. bot. Fr.* **58**, 9–17, 39–48, 82–89.

(1916) Reliquiae Treubianae I. Recherches sur la structure anatomique de l'ovule et de la graine des Thymeleacées. *Annls Jard. bot. Buitenz.* ser. 2, **14**, 3–35.

Guignard, L. (1893) Recherches sur le développement de la graine. *J. Bot. Paris* **7**, 100–153.

(1905) Quelques observations sur le *Cordyla africana*. *J. Bot. Paris* **19**, 109–124.

Guilford, V. B. and Emma L. Fisk (1951) Megasporogenesis and seed-development in *Mimulus tigrinus* and *Torenia fournieri*. *Bull. Torrey bot. Club* **79**, 6–24.

Gunasekera, S. P. and N. U. Sultanbawa (1973) Mangostin from the barks of *Hydnocarpus*. *Phytochem.* **12**, 232.

Gupta, S. C. (1964) The embryology of *Coriandrum sativum* L. and *Foeniculum vulgare* Mill. *Phytomorph.* **14**, 530–547.

Gupta, S. C. and M. Gupta (1964) Embryological investigations on *Bupleurum tenue* Buch. Ham. ex Don. *Beitr. Biol. Pfl.* **40**, 301–323.

Gutzwiller, Marie-Anne (1961) Die phylogenetische Stellung von *Suriana maritima* L. *Bot. Jb.* **81**, 1–49.

Haccius, B. (1954) Embryologische und histogenetische Studien an 'monokotylen Dikotylen': 1, *Claytonia virginica* L. *Ost. bot. Z.* **101**, 285–303.

Haccius, B. und E. Hartl-Baude (1957) Embryologische und histogenetische Studien an 'monokotylen Dikotylen': 2, *Pinguicula vulgaris* L. und *Pinguicula alpina* L. *Ost. bot. Z.* **103**, 567–587.

Håkansson, A. (1955) Endosperm formation in *Myrica gale* L. *Bot. Notiser* **108**, 6–16.

Hallé, N. (1965) Presence de graines bicolores chez *Leucomphalos capparideus* Benth. ex Planch. *Webbia* **19**, 847–853.

Hallock, F. A. (1930) The relationship of *Garrya*. *Ann. Bot. Lond.* **44**, 771–812.

Hammond, B. L. (1937) Development of *Podostemon ceratophyllum*. *Bull. Torrey bot. Club* **64**, 17–36.

Hansen, B. (1972) The genus *Balanophora* J. R. and G. Forster. A taxonomic monograph. *Dansk bot. Ark.* **28**, 1–188.

Hara, H. and H. Kanai (1964) *Tetracentron* in Nepal. *J. Jap. Bot.* **39**, 1.

Harris, T. M. (1951) The relationships of Caytoniales. *Phytomorph.* **1**, 29–39.

Hartl, D. (1960) Das alveolierte Endosperm bei Scrophulariaceen, seine Enstehung, Anatomie und taxonomische Bedeutung. *Beitr. Biol. Pfl.* **35**, 95–110.

Hayashi, Y. (1964) The comparative embryology of the Magnoliaceae in relation to the systematic consideration of the family. *Sci. Rep. Tôhoku Univ.* ser. IV (Biol.) **31**, 29–44.

Hayden, Sister M. Victoria and J. D. Dwyer (1969) Seed morphology in the tribe Morindeae (Rubiaceae). *Bull. Torrey bot. Club* **96**, 704–710.

van Heel, W. A. (1967) Anatomical and ontogenetic investigations on the morphology of the flowers and fruit of *Scyphostegia borneensis* Stapf (Scyphostegiaceae). *Blumea* **15**, 107–125.

(1970a) Distally lobed integuments in some angiosperm ovules. *Blumea* **18**, 67–70.

(1970b) Some unusual tropical labyrinth seeds. *Proc. K. ned. Akad. Wet.* C **73**, 288–301.

(1971a) The distally lobed inner integument of *Hernandia peltata* Meissn. in DC (Hernandiaceae). *Blumea* **19**, 147–148.

(1971b) The labyrinth seed of *Hernandia peltata* Meissn. in DC. *Proc. K. ned. Akad. Wet* C **74**, 46–51.

(1971c) Notes on some tropical labyrinth seeds. *Blumea* **19**, 109–111.

(1973) Flowers and fruits in Flacourtiaceae: 1. *Scaphocalyx spathacea* Ridley. *Blumea* **21**, 259–279.

van Heel, W. A. and F. Bouman (1972) Note on the early development of the integument in some Juglandaceae. *Blumea* **20**, 155–159.

Heilborn, O. (1931) Studies on the taxonomy, geographical distribution and embryology of the genus *Siparuna* Aubl. *Svensk Bot. Tidskr.* **25**, 202–228.

Heimerl, A. (1934) *Nyctaginaceae*. In Engler's *Die Natürlichen Pflanzenfamilien*, vol. 16c, pp. 86–134.

Hennig, L. (1930) Beiträge zur Kenntnis der Resedaceenblüte und Frucht. *Planta* **9**, 506–563.

Herr, J. M., Jr., (1959) The development of the ovule and megagametophyte in the genus *Ilex*. *J. Elisha Mitchell sci. Soc.* **75**, 107–128.

(1961) Endosperm development and associated ovule modifications in the genus *Ilex*. *J. Elisha Mitchell sci. Soc.* **77**, 26–32.

Herr, J. M., Jr. and M. L. Dowd (1968) Development of the ovule and megagametophyte in *Oxalis corniculata* L. *Phytomorph.* **18**, 43–53.

Hewitt, W. C. (1939) Seed development of *Lobelia amoena*. *J. Elisha Mitchell sci. Soc.* **55**, 63–82.

Hindmarsh, G. J. (1966) An embryological study of five species of *Bassia* All. (Chenopodiaceae). *Proc. Linn. Soc. N.S.W.* **90**, 274–289.

Hintzinger, A. (1927) Uber die Ablösung der Samen von der Placenta. *Sber. Akad. Wiss. Wien: Math.-nat. Kl.* **136** (1), 257–279.

Hjelmquist, H. (1948) Studies in the floral morphology and phylogeny of the Amentiferae. *Bot. Notiser* Suppl. 2, part 1, pp. 1–171.

(1957) Some notes on the endosperm and embryo development in Fagales and related orders. *Bot. Notiser* **110**, 173–195.

(1962) The embryo-sac development of some *Cotoneaster* species. *Bot. Notiser* **115**, 208–236.

Hoogland, R. D. (1951) Dilleniaceae. *Flora Malesiána* ser. 1, 4, 141–174.

(1952) A revision of the genus *Dillenia*. E. Ijdo N.V., Leiden.

Hotta, M. (1966) Notes on Bornean plants, 1. *Acta phytotax. geobot.* **22**, 1–10.

Hutchinson, J. (1959) *Families of flowering plants*, vol. 1, 2nd edn, Oxford University Press.

(1964) *The genera of flowering plants*: 1, *Dicotyledons*. Oxford University Press.

(1967) *The genera of flowering plants*: 2, *Angiospermae*. Oxford University Press.

Ihlenfeldt, H. -D. (1959) Uber die Entwickelung der Blüte und den Bau der Frucht von *Carotophora skiatophytoides* Leistn. (Ficoidaceae) – Formenkreis. *Z. Bot.* **47**, 290–304.

Ihlenfeldt, H. -D. and H. Straka (1962) Uber die Morphologie und Entwickelungsgeschichte der Früchte von *Uncarina* (Baill.) Stapf (Pedaliaceae). *Z. Bot.* **50**, 154–168.

Isely, Duane (1955) Observations on seeds of Leguminosae: Mimosoideae and Caesalpinioideae. *Proc. Iowa Acad. Sci.* **62**, 142–149.

Iyer, R. D., K. Sulbha and M. S. Swaminathan (1961) Fertilization and seed development in crosses between *Corchorus olitorius* and *C. capsularis*. *Indian J. Genet. Pl. Breed.* **21**, 191–200.

Jacobs, M. (1961) The generic identity of *Melia excelsa* Jack. *Gdns' Bull. Singapore* **18**, 71–75.

(1962) *Pometia* (Sapindaceae) a study in variability. *Reinwardtia* **6**, 109–144.

Jäger-Zürn, I. (1967) Embryologische Untersuchungen an vier Podostemaceen. *Ost. bot. Z.* **114**, 20–45.

Jaitly, S. C. (1966) Development of seeds and fruits in *Anisomeles indica* (Linn.) O. Kze. *Phytomorph.* **16**, 430–436.

(1969a) Morphology and embryology of *Justicia betonica* Linn. In Johri, Kapil and Rashid (ed.), *Symposium on morphology, anatomy and embryology of land plants*, pp. 62–63. University of Delhi.

(1969b) Structure and development of seed and fruit in certain members of the Labiatae. In Johri, Kapil and Rashid, ibid. pp. 64–65.

(1971) Studies in the Labiatae: 1. Structure and development of seed and fruit in two species of *Salvia*. *J. Indian bot. Soc.* **50**, 182–188.

Jalan, S. (1964) Studies in the family Ranunculaceae: 4, The embryology of *Actaea spicata* Linn. *Phytomorph.* **13**, 339–347.

Jayaweera, D. M. A. (1967) The genus *Duabanga*. *J. Arnold Arb.* **48**, 89–100.

Johri, B. M. (1935) The gametophytes of *Berberis nepalensis* Spreng. *Proc. Indian Acad. Sci.* **1**, 640–649.

(1963) Embryology and taxonomy. In P. Maheshwari (ed.), *Recent advances in the embryology of Angiosperms*, pp. 395–444. University of Delhi.

(1967a) Angiosperm embryology and taxonomy. Symposium on newer trends in taxonomy. *Bull. nat. Inst. Sci. India* B **34**, 263–268.

(1967b) *Seminar on comparative embryology of Angiosperms*. University of Delhi.

Johri, B. M. and M. R. Ahuja (1957) A contribution to the floral morphology and embryology of *Aegle marmelos* Correa. *Phytomorph.* **7**, 10–24.

Johri, B. M. and D. Kak (1954) The embryology of *Tamarix* Linn. *Phytomorph.* **4**, 230–247.

Johri, B. M. and R. N. Kapil (1953) Contribution to the morphology and life history of *Acalypha indica* Linn. *Phytomorph.* **3**, 137–151.

Johri, B. M. and R. N. Konar (1956) The floral morphology and embryology of *Ficus religiosa* Linn. *Phytomorph.* **6**, 97–111.

Johri, B. M. and H. Singh (1959) The morphology, embryology systematic position of *Elytraria acaulis* (Linn. f.) Lindau. *Bot. Notiser* **112**, 227–251.

Johri, B. M. and B. Tiagi (1952) Floral morphology and seed formation in *Cuscuta reflexa* Roxb. *Phytomorph.* **2**, 162–180.

Johri, B. M. and L. K. Vasil (1956) The embryology of *Ehretia laevis* Roxb. *Phytomorph.* **6**, 134–143.

Jos, J. S. (1962) The structure and development of seeds in Convolvulaceae, *Ipomaea* species. *Proc. 49th Indian Sci. Congr.* Part 3, p. 268.

Joshi, A. C. (1938) A note on the morphology of the gynoecium, ovule and embryo-sac of *Psoralea corylifolia* L. *J. Indian bot. Soc.* **17**, 169–172.

Joshi, A. C. and J. Venkateswarlu (1935a) Embryological studies in the Lythraceae: 1. *Lawsonia inermis* Linn. *Proc. Indian Acad. Sci.* B **2**, 481–493.

(1935b) ibid. 2, *Lagerstroemia* Linn. *Proc. Indian Acad. Sci.* B **2**, 523–534.

(1936) ibid. 3, *Lagerstroemia* Linn. *Proc. Indian Acad. Sci.* B **3**, 377–400.

Joshi, A. C., A. M. Wadhwani and B. M. Johri (1967) Morphological and embryological studies of *Gossypium* L. *Proc. nat. Inst. Sci. India* B **33**, 37–93.

Juel, H. O. (1918) Beiträge zur Blütenanatomie und zur Systematik der Rosaceen. *K. Sv. Vetensk. Handl.* **58** (5), 1–81.

Juliano, J. B. (1934) Studies on the morphology of the Meliaceae: I. *Sandoricum koetjape* (Burm. f.) Merrill. *Philipp. Agric.* **23**, 11–35.

Junell, S. (1931) Die Entwickelungsgeschichte von *Circaeaster agrestis*. *Svensk bot. Tidskr.* **25**, 238–270.

(1961) Ovarian morphology and taxonomical position of Selagineae. *Svensk bot. Tidskr.* **55**, 168–192.

Kadry, A. E. R. (1955) The development of endosperm and embryo in *Cistanche tinctoria* (Forsk.) G. Beck. *Bot. Notiser* **108**, 231–243.

(1960) The seed of *Cardiospermum halicacabum* L. A criticism *Acta bot. neerl.* **9**, 330–332.

Kaeiser, Margaret and S. G. Boyce (1962) Embryology of *Liriodendron tulipifera* L. *Phytomorph.* **12**, 103–109.

Kajale, L. B. (1942) A contribution to the embryology of the genus *Portulaca*. *J. Indian bot. Soc.* **21**, 1–19.

(1944) A contribution to the life-history of *Zizyphus jujuba* Lamk. *Proc. nat. Inst. Sci. India* **10**, 387–391.

(1954a) A contribution to the embryology of the Phytolaccaceae: 2, *Rivina humilis* Linn. and *Phytolacca dioica* Linn. *J. Indian bot. Soc.* **33**, 206–225.

(1954b) Fertilization and the development of embryo and seed in *Euphorbia hirta* Linn. *Proc. nat. Inst. Sci. India* **20**, 353–360.

Kajale, L. B. and G. V. Rao (1943) Pollen and embryo-sac of two Euphorbiaceae. *J. Indian bot. Soc.* **22**, 224–236.

Kanta, K. (1963) Morphology and embryology of *Piper nigrum* L. *Phytomorph.* **12**, 207–22.

Kapil, R. N. (1956a) Development of embryo-sac and endosperm in *Chrozophora rottleri* A. Juss. *Bot. Gaz.* **117**, 242–247.

(1956b) A further contribution to the morphology and life-history of *Chrozophora* Neck. *Phytomorph.* 6, 278–288.

(1960) Embryology of *Acalypha* Linn. *Phytomorph.* 10, 174–184.

(1961) Some embryological aspects of *Euphorbia dulcis* L. *Phytomorph.* 11, 24–36.

Kapil, R. N. and K. Ahlwalia (1963) Embryology of *Peganum harmala* Linn. *Phytomorph.* 13, 127–140.

Kapil, R. N. and S. B. Bawa (1968) Embryological studies on the Haloragidaceae I. *Haloragis colensoi* Skottsb. *Bot. Notiser* 121, 11–28.

Kapil, R. N. and S. Bela Sethi (1963a) Gametogenesis and seed development in *Ainsliaea aptera* DC. *Phytomorph.* 12, 222–234.

(1963b) Development of seed in *Tridax trilobata* Hemsl. *Phytomorph.* 12, 235–239.

Kapil, R. N. and N. N. Bhandari (1964) Morphology and embryology of *Magnolia. Proc. nat. Inst. Sci. India* 30, 245–262.

Kapil, R. N. and S. Jalan (1962) Studies in the family Ranunculaceae: 1. The embryology of *Caltha palustris* L. In *Plant Embryology, a Symposium,* pp. 205–214. CSIR, Delhi.

(1964) *Schisandra* Michaux – its embryology and systematic position. *Bot. Notiser* 117, 285–306.

Kapil, R. N. and P. Maheshwari (1965) Embryology of *Helianthemum vulgare* Gaertn. *Phytomorph.* 14, 547–557.

Kapil, R. N. and P. Masand (1964) Embryology of *Hebenstreitia integrifolia* Linn. *Proc. nat. Inst. Sci. India* B 30, 99–113.

Kapil, R. N. and P. R. Mohana Rao (1966a) Studies of the Garryaceae: 2, Embryology and systematic position of *Garrya* Douglas ex Lindley. *Phytomorph.* 16, 565–578.

(1966b) Embryology and systematic position of *Theligonum* Linn. *Proc. nat. Inst. Sci. India* B 32, 218–232.

Kapil, R. N. and N. Prakash (1969) Embryology of *Cereus jamacaru* and *Ferocactus wislizeni* and comments on the systematic position of the Cactaceae. *Bot. Notiser* 122, 409–426.

Kapil, R. N., P. N. Rustagi and R. Venkataraman (1969) A contribution to the embryology of Polemoniaceae. *Phytomorph.* 18, 403–412.

Kapil, R. N. and R. S. Vani (1963) Embryology and systematic position of *Crossosoma californicum* Nutt. *Curr. Sci.* pp. 493–495.

(1967) *Nyctanthes arbor-tristis* Linn.: embryology and relationships. *Phytomorph.* 16, 553–563.

Kapil, R. N. and I. K. Vasil (1963) The ovule. In P. Maheshwari (ed.), *Recent advances in the embryology of Angiosperms,* pp. 41–67. University of Delhi.

Kapil, R. N. and M. R. Vijayaraghavan (1965) Embryology of *Pentaphragma horsfieldii* (Miq.) Airy Shaw with a discussion on the systematic position of the genus. *Phytomorph.* 15, 93–102.

Kaplan, D. R. (1970) Seed development in *Dowringia. Phytomorph.* 19, 253–278.

Karuna Mohan, K. (1968) Morphological studies in Solanaceae: 2, Morphology, development and structure of seed of *Browallia demissa* Linn. *Proc. nat. Inst. Sci. India* B 34, 142–148.

Kasapligil, B. (1951) Morphological and ontogenetic studies of *Umbellularia californica* Nutt. and *Laurus nobilis* L. *Univ. Calif. Publs Bot.* 25, 115–240.

Kaul, U. (1969) Endosperm in *Parrotiopsis jacquemontiana. Phytomorph.* 19, 197–199.

Kaur, H. (1969) Embryological investigations on *Bixa orellana* Linn. *Proc. nat. Inst. Sci. India* B 35, 487–506.

Kaur, H. K. and R. P. Singh (1970) Structure and development of seeds in three *Ipomaea* species. *J. Indian bot. Soc.* 49, 168–174.

Kausik, S. B. (1938) Pollen-development and seed-formation in *Utricularia coerulea* L. *Beih. bot. Zbl.* 58A, 365–378.

(1939) Studies in Proteaceae: 3, Embryology of *Grevillea banksii* R. Br. *Ann. Bot. Lond.* n.s. 3, 815–824.

(1940a) Studies in Proteaceae: 4, Structure and development of the ovule of *Hakea saligna* Knight. *Ann. Bot. Lond.* n.s. 4, 73–80.

(1940b) Structure and development of the ovule and embryo-sac of *Lasiosiphon eriocephalus* Decne. *Proc. nat. Inst. Sci. India* 6, 117–132.

Kausik, S. B. and K. Subramanyam (1946) A contribution to the life-history of *Sphenoclea zeylanica* Gaertn. *Proc. Indian Acad. Sci.* B 23, 274–278.

(1947) Embryology of *Cephalostigma schimperi. Bot. Gaz.* 109, 85–90.

Kelkar, S. S. (1958a) Embryology of *Rhus mysurensis·* Heyne. *J. Indian bot. Soc.* 37, 114–122.

(1958b) A contribution to the embryology of *Lannea coromandelica* (Houtt.) Merr. *J. Univ. Bombay* 26, 152–169.

Keng, H. (1960) A brief note on the classification of the subfamily Camellioideae. *American Camellia Yearbook,* pp. 5–8.

(1962) Comparative morphological studies in Theaceae. *Univ. Calif. Publs Bot.* 33 (4), 269–383.

Khan, P. S. H. (1970) Structure and development of seed coat and fruit wall in *Catharanthus pusillus* (Murr.) G. Don. *Proc. nat. Acad. Sci. India* B 40, 21–24.

Khan, R. (1942) A contribution to the embryology of *Jussieua repens* Linn. *J. Indian bot. Soc.* 21, 267–282.

(1954) A contribution to the embryology of *Utricularia flexuosa* Vahl. *Phytomorph.* 4, 80–117.

Khanna, P. (1964a) Embryology of *Mertensia*. *J. Indian bot. Soc.* **43**, 192–202.

(1964b) Morphological and embryological studies in Nymphaeaceae: 1, *Euryale ferox* Salisb. *Proc. Indian Acad. Sci.* B **59**, 237–243.

(1965) ibid. 2, *Brasenia schreberi* Gmel. and *Nelumbo necifera* Gaertn. *Aust. J. Bot.* **13**, 379–387.

(1967) ibid. 3, *Victoria cruziana* D'Orb. and *Nymphaea stellata* Willd. *Bot. Mag. Tokyo* **80**, 305–312.

Kobuski, C. E. (1947) Studies in Theaceae: 15, A review of the genus *Adinandra*. *J. Arnold Arb.* **28**, 1–98.

Kostermans, A. J. G. (1956) The genus *Cullenia* Wight (Bombacaceae). *Reinwardtia* **4** (1), 69–74.

(1957) Lauraceae. *Reinwardtia* **4** (2), 193–256.

(1958) The genus *Durio* Adans. (Bombacaceae). *Reinwardtia* **4** (3), 47–153.

(1962) Miscellaneous botanical notes 3. *Reinwardtia* **6**, 155–187.

(1966) A monograph of *Aglaia* sect. *Lansium* Kosterm. (Meliaceae). *Reinwardtia* **7**, 221–282.

Kostermans, A. J. G. H. and W. Soegeng Reksodihardjo (1958) *A monograph of the genus* Durio *Adans. (Bombacaceae). Part* 1, *Bornean species*. Comm. For. Res. Inst. Indon. no. 61.

Kratzer, J. (1918) Die verwandschaftliche Beziehungen der Cucurbitaceen auf Grund ihrer Samenentwickelung. *Flora* **110**, 274–343.

Krishna Iyengar, C. V. (1937) Development of the embryo-sac and endosperm-haustoria in some members of Scrophulariaceae 1. *J. Indian bot. Soc.* **16**, 99–109.

(1939) ibid. 2, 3, 4. *J. Indian bot. Soc.* **18**, 13–20, 35–42, 179–189.

(1940) Structure and development of seed in *Sopubia trifida* Ham. *J. Indian bot. Soc.* **19**, 251–261.

(1942) Development of seed and its nutritional mechanism in Scrophulariaceae 1. *Proc. nat. Inst. Sci. India* **8**, 249–261.

Kubitzki, K. (1969) Monographie der Hernandiaceen. *Bot. Jb.* **89**, 78–148.

Kühl, R. (1933) Vergleichend-entwickelungeschichtliche Untersuchungen an der Insectivore *Nepenthes*. *Beih. bot. Zbl.* **51**, 311–334.

Kühn, Gertrud (1927) Beiträge zur Kenntnis der intraseminalen Leitbündel bei den Angiospermen. *Bot. Jb.* **61**, 325–379.

Kummerow, J. (1962) *Pilostyles berterii* Guill., eine wenig bekannte Rafflesiacee in Mittelchile. *Z. Bot.* **50**, 321–337.

Kummerow, J. y C. Labariz (1961) Estudios sobre el fruto y la semilla de *Nothofagus alpina* (Poepp. y Endl.) Krasser. *Phyton, B. Aires* **17**, 205–210.

Landes, M. (1946) Seed development in *Acalypha rhomboidea* and some other Euphorbiaceae. *Am. J. Bot.* **33**, 562–568.

de Lanesson, J. (1876), in H. Baillon, *Dictionnaire de Botanique* vol. I, p. 258.

Lange, A. H. (1961) Effect of the sarcotesta on germination of *Carica papaya*. *Bot. Gaz.* **122**, 305–311.

Leemann, A. (1927) Contribution à l'étude de l'*Asarum europaeum* L. *Bull. Soc. bot. Fr.* **19**, 92–173.

Leenhouts, P. W. (1958) Connaraceae. *Flora Malesiana* ser. 1, **5**, 495–541.

van der Linden, B. L. (1960) Staphyleaceae. *Flora Malesiana* ser. I, **6**, 49–59.

Long, A. G. (1966) Some lower Carboniferous fructifications from Berwickshire, together with a theoretical account of the evolution of ovules, cupules and carpels. *Trans. R. Soc. Edinburgh* **66**, 345–374.

van der Maesen, L. J. G. (1970) Primitiae Africanae VIII. A revision of the genus *Cadia* Forsk. *Acta bot. neerl.* **19**, 227–248.

Magtang, M. V. (1936) Floral biology and morphology of the egg-plant. *Philipp. Agric.* **25**, 30–53.

Maheshwari, J. K. (1954) Floral morphology and embryology of *Lippia nodiflora* Rich. *Phytomorph.* **4**, 217–230.

(1967) Starch grains of Leguminous seed. *Phyton* **12**, 191–199.

Maheshwari, P. (1963) *Embryology in relation to taxonomy*. Vistas in botany vol. 4, ed. W. B. Turrill, pp. 55–97.

Maheshwari, P. and R. N. Chopra (1955) The structure and development of the ovule and seed of *Opuntia dillenii* Haw. *Phytomorph.* **5**, 112–122.

Maheshwari, P. and B. M. Johri (1956) The morphology and embryology of *Floerkia proserpinoides* Willd. with a discussion on the sytematic position of the family Limnanthaceae. *Bot. Mag. Tokyo* **69**, 410–423.

Maheshwari, P. and V. Negi (1955) The embryology of *Dipteracanthus patulus* (Jacq.) Nees. *Phytomorph.* **5**, 456–472.

Maheshwari Devi, H. (1962) Embryological studies in Gentianaceae. 1, Gentianoideae; 2, Menyanthoideae. *Proc. 49th Indian Sci. Congr.* Part 3, pp. 268–269.

(1969) Embryology and systematic position of Salvadoraceae. In Johri, Kapil and Rashid (ed.), *Symposium on the morphology, anatomy and embryology of land plants*, pp. 72–73. University of Delhi.

(1971) Embryology of Apocynaceae 1. Plumiereae. *J. Indian bot. Soc.* **50**, 74–85.

Mandl, K. (1926) Beitrag zur Kenntnis der Anatomie der Samen mehreren Euphorbiaceen-Arten. *Ost. bot. Z.* **75**, 1–17.

Maneval, W. E. (1914) The development of *Magnolia* and *Liriodendron*. *Bot. Gaz.* **57**, 1–31.

Markgraf, F. (1971) Florae Malesianae praecursores LI. Apocynaceae L. *Blumea* 19, 149–166.

Masand, P. (1964) Embryology of *Zygophyllum fabago* Linn. *Phytomorph.* 13, 293–302.

Masand, P. and R. N. Kapil (1966) Nutrition of the embryo-sac. A morphological approach. *Phytomorph.* 16, 158–174.

Mathur, N. (1956) The embryology of *Limnanthes*. *Phytomorph.* 6, 41–51.

Mauritzon, J. (1934a) Zur Embryologie einiger Lythraceen. *Acta Horti gothoburg.* 9, 1–21.

— (1934b) Zur Embryologie der Elaeocarpaceae. *Ark. Bot.* 26A, no. 10, pp. 1–8.

— (1934c) Ein Beitrag zur Embryologie der Phytolaccaceen und Cactaceen. *Bot. Notiser* 111–135.

— (1934d) Zur Embryologie einiger Gruinales. *Svensk bot. Tidskr.* 28, 84–102.

— (1935a) Zur Embryologie von *Peumus boldus*. *Bot. Arch.* 11.

— (1935b) Die Embryologie einiger Capparidaceen sowie von *Tovaria pendula*. *Ark. Bot.* 26A, hefte 4, no. 15, pp. 1–14.

— (1935c) Ueber die Embryologie einiger Rutaceen. *Svensk bot. Tidskr.* 29, 319–347.

— (1936a) Zur Embryologie und systematischen Abgrenzung der Reihen Terebinthales und Celastrales. *Bot. Notiser* pp. 161–212.

— (1936b) Zur Embryologie einiger Parietales-Familien. *Svensk bot. Tidskr.* 30, 104.

— (1936c) Embryologische Angaben über Stackhousiaceae, Hippocrateaceae und Icacinaceae. *Svensk. bot. Tidskr.* 30, 541–550.

— (1936d) Zur Embryologie der Berberidaceen. *Acta Horti gothoburg.* 11, 1–18.

— (1939) Ueber die Embryologie von *Marcgravia*. *Bot. Notiser.* 92, 249–255.

Maury, G. (1968) Germinations anormales chez les Dipterocarpacées de Malaisie. *Bull. Soc. Hist. nat. Toulouse* 104, 187–202.

— (1970) Différents types de polyembryonie chez quelques Dipterocarpacées asiatiques. *Bull. Soc. Hist. nat. Toulouse* 106, 282–288.

Mayr, B. (1969) Ontogenetische Studien an Myrtales-Blüten. *Bot. Jb.* 89, 210–271.

Meeuse, A. D. J. and J. Houthuesen (1964) The gynoecium of *Engelhardia spicata* (Juglandaceae) and its phylogenetic significance. *Acta bot. neerl.* 13, 352–366.

Melchior, H. (1925) *Theaceae*. In Engler's *Die Natürlichen Pflanzenfamilien*, 2nd edn, vol. 21, pp. 109–154.

Melchior, H. und W. Schultze-Motel (1959) *Canellaceae*. In Engler's *Die Natürlichen Pflanzenfamilien*, 2nd edn, vol. 17a, part 2, pp. 221–224.

Mendes, A. J. T. (1941) Cytological observations in *Coffea*: 6, Embryo and endosperm development in *Coffea arabica* L. *Am. J. Bot.* 28, 784–789.

Mennechet, L. A. (1902) Sur le fruit de *Jacquinia ruscifolia*. *J. Bot. Paris* 16, 349–357.

Metcalfe, C. R. (1956) *Scyphostegia borneensis* Stapf. Anatomy of stem and leaf in relation to its taxonomic position. *Reinwardtia* 4, 99–104.

Metcalfe, C. R. and L. Chalk (1950) *Anatomy of the Dicotyledons*. 2 vol. Clarendon Press, Oxford.

Meunier, A. (1891) Les téguments sèminaux des Papavéracées. *Cellule* 7, 375–412.

Miki, S. (1956) Seed remains of Vitaceae in Japan. *J. Inst. Polytechnics Osaka*, ser. D, 7, 247–271.

— (1960) Nymphaeaceae remains in Japan, with new fossil genus *Eoeuryale*. *J. Inst. Polytechnics Osaka*, ser. D, 11, 63–78.

Milby, T. H. (1971) Floral anatomy of *Krameria lanceolata*. *Am. J. Bot.* 58, 569–576.

Millsaps, Veronica (1936) The structure and development of the seed of *Paulownia tomentosa* Steud. *J. Elisha Mitchell sci. Soc.* 52, 56–75.

— (1940) Structure and development of the seed of *Cynoglossum amabile* Stapf et Drumm. *J. Elisha Mitchell sci. Soc.* 56, 140–164.

Misra, B. N. (1963) Germination of seeds of *Ipomaea crassicaulis* (Benth.) Robinson. *J. Indian bot. Soc.* 42, 358–366.

Misra, K. C. (1938) Vascular supply of the pedicel and ovule in species of Dilleniaceae. *Nature Lond.* 141, 204.

Misra, R. C. (1966) Morphological studies on *Sisymbrium irio* Linn. *J. Indian bot. Soc.* 45, 14–23.

Mitra, J. N. (1947) A contribution to the life-history of *Hydrolea zeylanica* Vahl. *J. Indian bot. Soc.* 26, 51–61.

Mohan Ram, H. Y. (1960) Post-fertilization studies in the ovule of *Ruellia tuberosa* Linn. *Lloydia* 23, 21–27.

— (1962) Post-fertilization development of the ovule in *Barleria cristata* Linn. *J. Indian bot. Soc.* 41, 288–296.

Mohan Ram, H. Y. and P. Masand (1963) The embryology of *Nelsonia campestris* R. Br. *Phytomorph.* 13, 82–91.

Mohan Ram, H. Y. and R. Nath (1964) The morphology and embryology of *Cannabis sativa* Linn. *Phytomorph.* 14, 414–429.

Le Monnier, G. (1872) Recherches sur la nervation de la graine. *Annls Sci. nat.* (*Bot.*) ser. 5, 16, 232–305.

Moore, R. J. (1946) Investigations on rubber-bearing plants III. Development of normal and aborting seeds in *Asclepias syriaca* L. *Can. J. Res.* 24, sect. C, 55–65.

— (1948) Cytotaxonomic studies in the Loganiaceae II. Embryology of *Polypremum procumbens* L. *Am. J. Bot.* 35, 404–410.

Mosely, M. F., Jr., and R. M. Beeks (1955) Studies of the Garryaceae: 1, The comparative morphology and phylogeny. *Phytomorph.* 5, 314–346.

Mukherjee, P. K. (1957) Embryology of *Polygala*. *Bull. bot. Soc. Coll. Sci. Nagpur* 11, 58.

— (1964) Further contributions to the embryology of the genus *Acalypha* L. *Proc. nat. Inst. Sci. India* B 34, 129–141.

— (1965) Contribution to the embryology of *Euphorbia peltata* Roxb. *Proc. nat. Inst. Sci. India* B 35, 327–337.

Mullan, D. P. (1936) On the seed structure and germination of *Acanthus ilicifolius* Linn. *J. Indian bot. Soc.* 15, 143–147.

Murthi, S. N. (1947) Studies in the Labiatae: 4, Contribution to the morphology of *Orthosiphon stamineus* Benth. *J. Indian bot. Soc.* 26, 87–94.

Murty, Y. S. (1959a) Studies in the order Piperales 5. *Proc. Indian Acad. Sci.* 49, 52–65.

— (1959b) ibid. 6. *Proc. Indian Acad. Sci.* 49, 82–85.

Nagaraj, M. and B. H. M. Nijalingappa (1967) Embryological studies in *Myriophyllum intermedium* DC. *Proc. Indian Acad. Sci.* B 65, 210–220.

Nair, N. C. (1958) Studies on Meliaceae: 3, Floral morphology and embryology of *Sandoricum indicum* Cav. *Phyton* 10, 145–151.

— (1959a) Studies on Meliaceae: 1, Floral morphology and embryology of *Naregamia alata* W. et A. *J. Indian bot. Soc.* 38, 353–366.

— (1959b) Studies on Meliaceae: 2, Floral morphology and embryology of *Melia azedarach* Linn. – a reinvestigation. *J. Indian bot. Soc.* 38, 367–378.

Nair, N. C. and V. Abraham (1963) A contribution to the morphology and embryology of *Micrococca mercurialis* Benth. *J. Indian bot. Soc.* 42, 583–593.

Nair, N. C. and Y. P. S. Bajaj (1966) Floral morphology and embryology of *Cyphostemma setosum* (Wall.) Alston and a discussion on the taxonomic position of the genus *Cyphostemma* (Planch.) Alston. *J. Indian bot. Soc.* 45, 103–115.

Nair, N. C. and I. Gupta (1961) A contribution to the floral morphology and embryology of *Fagonia cretica* Linn. *J. Indian bot. Soc.* 41, 635–640.

Nair, N. C. and R. K. Jain (1956) Floral morphology and embryology of *Balanites roxburghii* Planch. *Lloydia* 19, 269–279.

Nair, N. C. and T. Joseph (1960) Morphology and embryology of *Cardiospermum halicacabum* Linn. *J. Indian bot. Soc.* 39, 176–194.

Nair, N. C. and K. Kanta (1961) Studies in Meliaceae: 4, Floral morphology and embryology of *Azadirachta indica* A. Juss. – a re-investigation. *J. Indian bot. Soc.* 40, 382–396.

Nair, N. C. and N. Maitreyi (1962) Morphology and embryology of *Sebastiana chamaelea* Muell. *Bot. Gaz.* 124, 58–68.

Nair, N. C. and P. N. N. Nambisan (1957) Contributions to the floral morphology and embryology of *Leea sambucina* Willd. *Bot. Notiser* 110, 160–172.

Nair, N. C. and K. R. Narayanan (1961) Studies on the Aristolochiaceae; 2, Contribution to the embryology of *Bragantia wallichii. Lloydia* 24, 199–203.

Nair, N. C. and V. Parasuraman. (1962) Contribution to the embryology of *Vitis pallida* W. et A. *Phyton* B. Aires 18, 157–164.

Nair, N. C. and N. P. Sukumaran (1960) Floral morphology and embryology of *Brucea amarissima*. *Bot. Gaz.* 121, 175–185.

Narang, N. (1953) The life-history of *Stackhousia linariaefolia* A. Cunn. with a discussion on its systematic position. *Phytomorph.* 3, 485–493.

Narayana, H. S. (1962a) Postfertilization study on *Moringa oleifera* Lamk. – a reinvestigation. *Phytomorph.* 12, 65–69.

— (1962b) Studies in the Capparidaceae: 1, The embryology of *Capparis decidua* (Forsk.) Pax. *Phytomorph.* 12, 167–177.

— (1962c) Seed-structure in the Aizoaceae. In Maheshwari, Johri and Vasil (ed.), *Proceedings of the Summer School of Botany*, pp. 220–230. University of Delhi.

— (1965) Studies in Capparidaceae: 2, Floral morphology and embryology of *Cadaba indica* Lamk. and *Crataeva nurvala* Buch. Ham. *Phytomorph.* 15, 158–175.

Narayana, H. S. and P. K. Arora (1963) The embryology of *Monsonia senegalensis* Guill. et Perr. *Am. Midl. Nat.* 70, 309–318.

Narayana, H. S. and K. Jain (1962) A contribution to the embryology of *Limeum indicum. Lloydia* 25, 100–108.

Narayana, H. S. and C. G. Prakasa Rao (1963) Floral morphology and embryology of *Seetzenia orientalis* Decne. *Phytomorph.* 13, 197–205.

Narayana, L. L. (1958) Floral anatomy and embryology of *Cipadessa baccifera* Miq. *J. Indian bot. Soc.* 37, 147–154.

— (1960) Studies in Burseraceae, 2. *J. Indian bot. Soc.* 29, 402–409.

— (1963) Contributions to the embryology of Balsaminaceae, 1. *J. Indian bot. Soc.* 42, 102–109.

— (1964) A contribution to the floral anatomy and embryology of Linaceae. *J. Indian bot. Soc.* 43, 343–357.

Narayana, L. L. and M. Sayeeduddin (1959) A study of the gametophytes in *Impatiens leschenaultii* Wall. *J. Indian bot. Soc.* 38, 391–397.

Narayanaswami, S. and S. K. Roy (1960a) Embryology of the genus *Psidium. J. Indian bot. Soc.* 39, 35–45.

— (1960b) Embryo sac development and polyembryony in *Syzygium cumini* (Linn.) Skeels. *Bot. Notiser* 113, 273–284.

Narayanaswami, S. and S. Sawhney (1959) Micro-sporogenesis and embryo sac development in *Casearia tomentosa* Roxb. *Phyton, B. Aires* 13, 133–144.

Netolitzky, F. (1926) *Anatomie der Angiospermen Samen*. Linsbauer *Handb. Pfl. Anat.* vol. 10.

Neubauer, B. F. (1971) The development of the achene of *Polygonum pennsylvanicum. Am. J. Bot.* 58, 655–664.

Neumann, M. (1935) Die Entwickelung des Pollens, der Samenanlage und des Embryosackes von *Pereskia amapola* v. *argentina. Ost. bot. Z.* 84, 1–30.

Nichols, R. (1964) Studies of fruit-development of cacao (*Theobroma cacao*) in relation to cherelle wilt, 1. *Ann. Bot. Lond.* n.s. 28, 618–635.

(1965) ibid. 2. *Ann. Bot. Lond.* n.s. 29, 181–196, 197–203.

Nooteboom, H. P. (1962) Simaroubaceae. *Flora Malesiana* ser. 1, 6, 193–226.

Ochsenius, C. (1899) Ueber Maqui. *Bot. Zbl.* 38, 721–727.

Orr, M. Y. (1921a) The structure of the ovular integuments and the development of the testa in *Cleome* and *Isomeris Not. R. bot. Gdn Edinb.* 12, 243–248.

(1921b) The occurrence of tracheal tissue enveloping the embryo in certain Capparidaceae. *Not. R. bot. Gdn Edinb.* 12, 249–257.

(1921c) Observations on the structure of the seed in Capparidaceae and Resedaceae. *Not. R. bot. Gdn. Edinb.* 12, 259–260.

Padhye, M. D. (1962) Solanad type of embryo development in *Gomphrena celosioides* Mart., a member of the Amaranthaceae. *J. Indian bot. Soc.* 41, 52–63.

Padmanabhan, D. (1960) The embryology of *Avicennia officinalis* I, Floral morphology and gametophytes. *Proc. Indian Acad. Sci.* 52, 131–145.

(1961) A contribution to the embryology of *Gomphandra polymorpha. Proc. nat. Inst. Sci. India* B 27, 389–398.

Paetow, W. (1931) Embryologische Untersuchungen an Taccaceen, Meliaceen und Dilleniaceen. *Planta* 14, 441–470.

Pal, N. (1951) Studies in the embryology of some Verbenaceae. *J. Indian bot. Soc.* 30, 59–74.

(1960) Development of the seed of *Millettia ovalifolia. Bot. Gaz.* 122, 130–137.

Pal, P. (1963) Comparative studies in four species of *Heliotropium* L. *Proc. nat. Inst. Sci. India* B 29, 1–36.

Paliwal, R. L. (1956) Morphological and embryological studies in some Santalaceae. *Agra Univ. J. Res.* (*Sci.*) 5, 193–284.

Pantulu, J. V. (1942) A contribution to the life-history of *Desmodium gangeticum* DC. *J. Indian bot. Soc.* 21, 137–144.

(1945) Studies in Caesalpiniaceae: 1, A contribution to the embryology of the genus *Cassia. J. Indian bot. Soc.* 24, 10–24.

(1951) ibid. 2, Development of the endosperm and embryo in *Cassia occidentalis* L. *J. Indian bot. Soc.* 30, 95–99.

Parameswaran, N. (1961) Ruminate endosperm in the Canellaceae. *Curr. Sci.* 30, 344–345.

(1962) Floral morphology and embryology in some taxa of the Canellaceae. *Proc. Indian Acad. Sci.* B 55, 167–182.

Paroli, V. (1940) Contributo allo studio embryologico dell Tamaricee. *Ann. Bot. Roma* 22, 1–18.

Patankar, T. B. (1956) Further contribution to the embryology of *Drosera burmanni* Vahl. *Proc. Indian Acad. Sci.* B 43, 161–171.

Pax, F. und K. Hoffmann (1936) Capparidaceae. In Engler's *Die Naturlichen Pflanzenfamilien* 2nd edn, vol. 17b, 146–223.

Payne, W. W. and J. L. Seago (1968) The open conduplicate carpel of *Akebia quinata* (Berberidales: Lardizabalaceae). *Am. J. Bot.* 55, 575–581.

Pearson, Norma L. (1948) Observations on seed and seed hair growth in *Asclepias syriaca* L. *Am. J. Bot.* 35, 27–36.

Péchoutre, F. (1902) Contribution à l'étude du développement de l'ovule et de la graine des Rosacées. *Annls Sci. nat.* (*Bot.*) ser. 8, 16, 1–158.

Periasamy, K. (1961) Studies on seeds with ruminate endosperm: 1, Morphology of ruminating tissues in *Myristica fragrans. J. Madras Univ.* B 31, 53–58.

(1962a) ibid. 2, Development of rumination in the Vitaceae. *Proc. Indian Acad. Sci.* B 56, 13–26.

(1962b) The ruminate endosperm: development and types of rumination. In *Plant Embryology, a Symposium*, pp. 62–74. CSIR, Delhi.

(1963) Studies on seeds with ruminate endosperm: 3, Development of rumination in certain members of Apocynaceae. *Proc. Indian Acad. Sci.* B 58, 325–332.

(1964a) ibid. 4, Development of rumination in *Coccoloba uvifera. J. Indian bot. Soc.* 43, 543–548.

(1964b) ibid. 5, Seed development and rumination in two genera of Rubiaceae. *Proc. Indian Acad. Sci.* B 60, 351–360.

(1966) ibid. 6, Rumination in the Araliaceae, Aristolochiaceae, Caprifoliaceae and Ebenaceae. *Proc. Indian Acad. Sci.* B 64, 127–134.

Periasamy, K. and N. Parameswaran (1962) Extra-ovular outgrowths in Rubiaceae. *Curr. Sci.* 31, 300–301.

(1965) A contribution to the floral morphology and embryology of *Tarenna asiatica. Beitr. Biol. Pfl.* 41, 123–138.

Periasamy, K. and B. G. L. Swamy (1961) Studies in the Annonaceae 2. *J. Indian bot. Soc.* 40, 206–216.

Perkins, J. und E. Gilg (1911) Monimiaceae. *Das Pflanzenreich* 4, 101, Suppl.

Perrot, E. et P. Guérin (1903) Les *Didierea* de Madagascar. *J. Bot. Paris* 17, 233–251.

Pfeiffer, A. (1891) Die Arillargebilde der Pflanzensamen. *Bot. Jb.* 13, 492–540.

Phatak, V. G. and K. B. Ambegaokar (1961) Embryological studies in Acanthaceae: 4, Development of embryo-sac and seed formation in *Haplanthus tentaculatus* Nees. *J. Indian bot. Soc.* 40, 525–534.

Pigott, Ellen (1927) Observations on *Corynocarpus laevigata* Forst. *Trans. Proc. N.Z. Inst.* 58, 52–71.

van der Pijl, L. (1952) Ecological variations on the theme pod. *Indones. J. nat. Sci.* 1, 6–12.

— (1955) Sarcotesta, aril, pulpa and the evolution of the angiosperm fruit. *Proc. K. ned. Akad. Wet.* C 58, 154–161, 307–312.

— (1956) Classification of the Leguminous fruits according to their ecological and morphological properties. *Proc. K. ned. Akad. Wet.* C 59, 301–313.

— (1957) On the arilloids of *Nephelium*, *Euphoria*, *Litchi* and *Aesculus*, and the seeds of Sapindaceae in general. *Acta bot. neerl.* 6, 618–641.

— (1966) Ecological aspects of fruit evolution. *Proc. K. ned. Akad Wet.* C 69, 597–640.

Pitot, A. (1935a) Sur certaines particularités du tégument de la graine dans deux espèces de *Canavalia*. *Bull. Soc. bot. Fr.* 82, 379–381.

— (1935b) Le développement du tégument des graines de Legumineuses. *Bull. Soc. bot. Fr.* 83, 307–308, 311–314.

Planchon, J. E. et Triana (1860–62) Mémoire sur la famille des Guttifères. *Annls Sci. nat.* (*Bot.*) ser. 4, 13, 306–376; 14, 226–367: 15, 240–319; 16, 263–308.

Pohl, F. (1922) Zur Kenntnis unserer Beerenfrüchte. *Beih. bot. Zbl.* 29 (1), 206–221.

Poole, A. L. (1952) The development of *Nothofagus* seed. *Trans. Proc. R.S.N.Z.* 80, 207–212.

Popovici, R. (1947) Dr. Fritz Netolitzky. *Bul. Grăd. bot. Mus. bot. Univ. Cluj* 27 (1–4), 87–92.

Prakash, N. (1967a) Aizoaceae: a study of its embryology and systematics. *Bot. Notiser* 120, 305–323.

— (1967b) Life-history of *Tetragonia tetragonioides* (Pall.) O. Kuntze. *Aust. J. Bot.* 15, 413–424.

— (1969a) The embryology of *Kunzea capitata* Reichb. *Aust. J. Bot.* 17, 97–106.

— (1969b) Some aspects of the life-history of *Callistemon citrinus* (Curt.) Skeels. *Aust. J. Bot.* 17, 107–117.

— (1969c) Reproductive development in two species of *Darwinia* Rudge (Myrtaceae). *Aust. J. Bot.* 17, 215–217.

— (1969d) A contribution to the life-history of *Angophora floribunda* (Sm.) Sweet (Myrtaceae). *Aust. J. Bot.* 17, 457–469.

Pryor, L. D. and L. A. S. Johnson (1971) *A classification of the Eucalypts.* The Australian National University.

Puri, V. and M. L. Garg (1953) A contribution to the anatomy of the sporocarp of *Marsilea minuta* L. with a discussion of the nature of the sporocarp in the Marsiliaceae. *Phytomorph.* 3, 190–209.

Puri, V. and B. Singh (1935) Studies in the family Amarenthaceae: 1, Life-history of *Digera arvensis* Forsk. *Proc. Indian Acad. Sci.* 1, 893–908.

Quibell, C. H. (1941) Floral anatomy and morphology of *Anemopsis californica*. *Bot. Gaz.* 102, 749–758.

Radlkofer, L. (1934) Sapindaceae. *Das Pflanzenreich* 4.

Raghavan, T. S. (1937) Studies in the Capparidaceae: 1. The life-history of *Cleome chelidonii* Linn. f. *J. Linn. Soc. Bot.* 51, 43–72.

— (1940) A contribution to the life-history of *Bergia capensis* L. *J. Indian bot. Soc.* 19, 283–291.

Raghavan, T. S. and K. Rangaswamy (1941) Studies in the family Rubiaceae 1. *J. Indian bot. Soc.* 20, 341–356.

Raghavan, T. S. and Srinivasa (1941) ibid. 2. *Proc. Indian Acad. Sci.* 14, 412–426.

Raj, B. (1970) Morphological and embryological studies in the family Santalaceae. *Beitr. Biol. Pfl.* 49, 193–207.

Raju, M. S. V. (1958) Seed development and fruit dehiscence in *Ionidium suffruticosum* Ging. *Phytomorph.* 8, 218–224.

— (1961) Morphological anatomy of the Saururaceae: 1, Floral anatomy and embryology. *Ann. Mo. bot. Gdn* 48, 107–124.

— (1956a) Embryology of Passifloraceae 1. Gametogenesis and seed development of *Passiflora calcarata* Mast. *J. Indian bot. Soc.* 35, 126–138.

— (1956b) Development of embryo and seed-coat in *Turnera ulmifolia* L. v. *angustifolia* Willd. *Bot. Notiser* 109, 308–312.

Ram, Manasi (1956) Floral morphology and embryology of *Trapa bispinosa* Roxb. with a discussion on the systematic position of the genus. *Phytomorph.* 6, 312–323.

Ramchandani, S. P. C. Joshi and N. S. Pundir (1966) Seed development in *Gossypium* Linn. *Indian Cotton J.* 20, 97–106.

Rao, A. N. (1953) Embryology of *Shorea talura* Roxb. *Phytomorph.* 3, 476–484.

— (1957a) A contribution to the embryology of Dilleniaceae. *Proc. Iowa Acad. Sci.* 64, 172–176.

— (1957b) The embryology of *Hypericum patulum* Thunb. and *H. mysorense* Heyne. *Phytomorph.* 7, 36–45.

— (1964) An embryological study of *Salomonia cantoniensis* Lour. *New Phytol.* 63, 281–288.

Rao, A. N. and S. Shamanna (1963) Rudimentary aril in *Sanguinaria canadensis* L. *Can. J. Res.* 41, 1529–30.

Rao, A. V. N. (1967) Embryological studies in *Cleome monophylla* Linn. *Proc. Indian Acad. Sci.* B 65, 249–256.

Rao, D. (1968) A contribution to the embryology of *Erythroxylaceae*. *Proc. nat. Inst. Sci. India* **38**, 53–65.

Rao, V. S. and K. Gupte (1957) The pistil of *Ochna sqarrosa* L. *Curr. Sci.* **26**, 216.

Rau, M. A. and V. K. Sharma (1962) Embryology of *Lagotes glauca* Gaertn. (*Selaginaceae*) *Proc. 49th Indian Sci. Congr.* Part 3, p. 276.

Reeves, R. G. (1936) Comparative anatomy of the seeds of cotton and other Malvaceous plants 1. Malveae and Ureneae. 2, Hibisceae. *Am. J. Bot.* **23**, 291–296, 394–405.

Reeves, R. G. and C. C. Valle (1932) Anatomy and microchemistry of the cotton seed. *Bot. Gaz.* **93**, 259–277.

Renard, A. Le (1913) Rapports anatomiques du genre *Arfeuillea*. *Annls Sci. nat. (Bot.)* ser. 9, **17**, 353–389.

Robyns, A. (1970) Revision of the genus *Cullenia* Wight. *Bull. Jard. bot. nat. Belg.* **40**, 240–254.

Rodenburg, W. F. (1971) A revision of the genus *Trimenia* (Trimeniaceae). *Blumea* **19**, 3–15.

Röder, I. (1958) Anatomische und fluoreszenzoptische Untersuchungen an Samen von Papaveraceen. *Öst. bot. Z* **104**, 370–381.

Rohweder, O. (1971) Centrospermen-Studien 4, 5. *Bot. Jb.* **90**, 201–271, 447–468.

Rosen, W. (1932) Zur Embryologie der Campanulaceen und Lobeliaceen. *Acta Horti gothoburg.* **7**, 31–42.

Roy, S. K. (1962) A contribution to the embryology of *Myrtus communis* L. *Proc. nat. Inst. Sci. India* **32**, 305–311.

Roy, S. K. and R. Sahai (1962) The embryo-sac and embryo of *Syzygium caryophyllifolium* DC. *J. Indian bot. Soc.* **41**, 45–51.

Rutherford, R. J. (1970) The anatomy and cytology of *Pilostyles thurberi* Gray (Rafflesiaceae). *Aliso* **7**, 263–288.

Saber, A. H., S. I. Balbaz and A. T. Awad (1962) A botanical study of *Cassia obovata* Coll. growing in Egypt: 2, The fruit. *J. Bot. un. Arab Repub.* **5**, 85–106.

Sabet, Y. S. (1931) Development of the embryo-sac in *Calotropis procera* with special reference to endosperm formation. *Ann. bot. Lond.* **45**, 503–518.

Sachar, R. C. (1955) The embryology of *Argemone mexicana* L. *Phytomorph.* **5**, 200–218.

(1956) The embryology of *Isomeris* – a re-investigation. *Phytomorph.* **6**, 346–363.

Sachar, R. C. and H. Y. Mohan Ram (1958) The embryology of *Eschscholzia californica* Cham. *Phytomorph.* **8**, 114–124.

Saksena, H. B. (1954) Floral morphology and embryology of *Fumaria parviflora* Lamk. *Phytomorph.* **4**, 409–417.

Salisbury, E. J. (1967) On the reproduction and biology of *Elatine hexandra* (Lapierre) DC (Elatinaceae). *Kew Bull.* **21**, 139–149.

(1969) A note on fertile seed production by *Hypericum calycinum*. *Watsonia* **7**, 24.

(1972) *Ludwigia palustris* (L.) in England with special reference to its dispersal and germination. *Watsonia* **9**, 33–37.

Sampson, F. B. (1963) The floral morphology of *Pseudowintera*. *Phytomorph.* **13**, 402–423.

Sastri, R. L. N. (1952) Studies in Lauraceae 1. Floral anatomy of *Cinnamomum iners* Reinw. and *Cassytha filiformis* L. *J. Indian bot. Soc.* **31**, 240–246.

(1954) Embryological studies in Menispermaceae: 1, *Tiliacora racemosa* Coleb. *Proc. nat. Inst. Sci. India* **20**, 494–502.

(1958a) Studies in Lauraceae: 2, Embryology of *Cinnamomum* and *Litsea*. *J. Indian bot. Soc.* **37**, 266–278.

(1958b) Floral morphology and embryology of some Dilleniaceae. *Bot. Notiser* **111**, 495–511.

(1959a) Vascularization of the carpel of *Myristica fragrans*. *Bot. Gaz.* **121**, 92–94.

(1959b) Vascularization of the carpel in some Ranales. *New Phytol.* **58**, 306–309.

(1962) Studies in Lauraceae: 3, Embryology of *Cassytha*. *Bot. Gaz.* **123**, 197–206.

(1963) ibid. 4. *Ann. Bot. Lond.* n.s. **27**, 425–433.

(1964) Embryological studies in the Menispermaceae: 2, Embryo and seed development. *Bull. Torrey bot. Club* **91**, 79–85.

(1965) Studies in Lauraceae: 5, Comparative morphology of the flower. *Ann. Bot. Lond.* n.s. **29**, 39–44.

(1969) Floral morphology, embryology and relationships of Berberidaceae. *Aust. J. Bot.* **17**, 69–79.

Sato, Y. (1972) Development of the embryo sac of *Daphniphyllum macropodum* var. *humile* (Maxim.) Rosenth. *Sci. Rep. Tôhoku Univ.* ser. 4 (Biol.) **36**, 129–133.

Sawada, M. (1971) Floral vascularization of *Paeonia japonica* with some consideration on systematic position of the Paeoniaceae. *Bot. Mag. Tokyo* **84**, 51–60.

Saxena, N. P. (1969) Studies in the family Saxifragaceae: 4, A contribution to the embryology of *Bergenia ciliata* (Royle) Raizada. *Proc. Indian Acad. Sci.* **70**, 104–110.

Saxena, T. and D. Singh (1969) Comparative embryology and seed structure of *Solanum nigrum* complex. In Johri, Kapil and Rashid, *Symposium on morphology, anatomy and embryology of land plants*, pp. 77–78. University of Delhi.

Schmid, R. (1964) Die Systematische Stellung der Dioncophyllaceen. *Bot. Jb.* **83**, 1–56.

(1972) A resolution of the *Eugenia-Syzygium* controversy (Myrtaceae). *Am. J. Bot.* **59**, 423–436.

Schnarf, K. (1924) Bemerkungen zur Stellung der Gattung *Saurauia* im System. *Sber. Akad. Wiss. Wien Math.-nat. Kl.* **133** (1), 17–28.

(1937) *Anatomie der Gymnospermen-Samen.* Linsbauer *Handb. Pfl. Anat.* vol. 10, part 1.

Schölch, H. F. (1963) Die systematische Stellung der Didieraceen im Lichte neuer Untersuchungen über ihrer Blüten bereich. *Ber. dt. bot. Ges.* **76**, 49–55.

Schrock, G. F. and Barbara F. Palser (1967) Floral development, anatomy and embryology of *Collinsia heterophylla* with some notes on ten other species of *Collinsia* and on *Tonella tenella.* *Bot. Gaz.* **128**, 83–104.

Schroeder, C. A. (1952) Floral development, sporogenesis and embryology in the avocado, *Persea americana Bot. Gaz.* **113**, 270–278.

Sehgal, C. B. (1965) The embryology of *Cuminum cyminum* L. and *Trachyspermum ammi* (L.) Sprague (=*Carum copticum* Clarke). *Proc. nat. Inst. Sci. India* B **35**, 175–201.

(1966) Morphological and embryological studies on *Erigeron bonariensis* L. *Beitr. Biol. Pfl.* **42**, 161–183.

Sell, Y. (1969) La dissémination des Acanthacées. *Rev. gen. Bot.* **76**, 417–453.

Sethi, S. B. (1965) Structure and development of the seed in *Camellia sinensis* (L.) O.K. *Proc. nat. Inst. Sci. India* B **31**, 24–33.

Sharma, D. R. and S. R. Upadhyay (1962) Studies in the embryology of *Canscora diffusa* L. *Proc. 49th Indian Sci. Congr.* Part. 3, p. 275.

Sharma, V. K. (1968a) Floral morphology, anatomy and embryology of *Coriaria nepalensis* Wall. with a discussion on the inter-relationships of the family Coriariaceae. *Phytomorph.* **18**, 143–153.

(1968b) Morphology, floral anatomy and embryology of *Parnassia nubicola* Wall. *Phytomorph.* **18**, 193–204.

Shaw, C. H. (1904) Note on the sexual generation and development of the seed-coats in certain of the Papaveraceae. *Bull. Torrey bot. Club* **31**, 429–433.

Siddiqui, S. A. and S. B. Siddiqui (1968a) Studies in Rubiaceae: 1, A contribution to the embryology of *Oldenlandia dichotoma* Hook. f. *Beitr. Biol. Pfl.* **44**, 343–351.

(1968b) ibid. 2, A contribution to the embryology of *Borreria stricta* Linn. *Beitr. Biol. Pfl.* **44**, 353–360.

Sinclair, J. (1955) A revision of the Malayan Annonaceae. *Gdns' Bull. Singapore* **14**, 149–516.

(1958) A revision of the Malayan Myristicaceae. *Gdns' Bull. Singapore* **16**, 205–466.

Singh, B. (1936) The life-history of *Ranunculus sceleratus* Linn. *Proc. Indian Acad. Sci.* **4**, 75–91.

(1952) Studies on the structure and development of seeds of the Cucurbitaceae: 1. Seeds of *Echinocystis wrightii* Cogn. *Phytomorph.* **2**, 201–209.

(1953) Studies on the structure and development of seeds of Cucurbitaceae. *Phytomorph.* **3**, 224–239.

(1964) Development and structure of Angiosperm seed, 1. *Bull. natn. bot. Gdns India* **89**, 1–115.

(1968) The structure and development of *Abelmoschus moschatus* Medic. seed. *Phytomorph.* **17**, 282–290.

Singh, D. (1960) Studies on endosperm and development of seeds of *Carica papaya* L. *Hort. Adv.* **4**, 89–96.

(1962a) The structure and development of ovule and seed of *Passiflora foetida* Linn. *Proc. 49th Indian Sci. Congr.* Part 3, p. 262.

(1962b) Structure and development of ovule and seed of *Viola tricolor* and *Ionidium suffruticosum* Ging. *Proc. 49th Indian Sci. Congr.* Part 3, p. 264.

(1963) Structure and development of ovule and seed of *Viola tricolor* L. and *Ionidium suffruticosum* Ging. *J. Indian bot. Soc.* **42**, 448–462.

(1965) Ovule and seed of *Dicoelospermum* C. B. Clarke, together with a note on its systematic position. *J. Indian bot. Soc.* **44**, 183–190.

(1967) Structure and development of seed coat in Cucurbitaceae. 1. Seeds of *Biswarea* Cogn., *Edgaria* Clarke, and *Herpetospermum* Hook. f. *Proc. Indian Acad. Sci.* B **65**, 267–274.

Singh, D. and A. S. R. Dathan (1969) Structure and development of seed-coat in Cucurbitaceae: 5. Seeds of *Melothria* Linn. *Proc. 56th Indian Sci. Congr.* Part 3 (4), p. 393.

(1971a) Morphology and embryology of *Marah macrocarpa* Greene. *Proc. Indian Acad. Sci.* B **73**, 241–249.

(1971b) Structure and development of seed-coat in Cucurbitaceae: 7, Seeds of *Cucumis* spp. *Proc. 58th Indian Sci. Congr.* Part 3 (4), pp. 443–44.

Singh, D. and S. Gupta (1967) The seeds of the Violaceae and Resedacaceae – a comparison. *J. Indian bot. Soc.* **46**, 248–256.

Singh, D. and D. S. Negi (1962) A contribution to the morphology and embryology of *Dicentra scandens* Walp. *Proc. 49th Indian Sci. Congr.* Part 3, p. 263.

Singh, R. P. (1954) Structure and development of seeds in Euphorbiaceae: *Ricinus communis* L. *Phytomorph.* **4**, 118–123.

(1961) ibid.: *Antidesma menasu* Miquel. *Proc. 48th Indian Sci. Congr.* Part 3, p. 275.

(1962a) Forms of ovules in Euphorbiaceae. In *Plant Embryology, a Symposium*, pp. 124–128. CSIR, Delhi.

(1962b) Structure and development of seeds in Euphorbiaceae *Phyllanthus niruri* L. *Proc. 49th Indian Sci. Congr.* Part 3, p. 279.

(1962c) ibid.: *Putranjiva roxburghii* Wall. *Proc. 49th Indian Sci. Congr.* Part 3, p. 279.

(1962d) ibid.: *Bischofia javanica* Blume. *Proc. 49th Indian Sci. Congr.* Part 3, p. 280.

(1965) Structure and development of seeds in *Codiaeum variegatum* Blume. *J. Indian bot. Soc.* **44**, 205–210.

(1968) Structure and development of seeds in Euphorbiaceae *Melanthesia rhamnoides* Wt. *Beitr. Biol. Pfl.* **45**, 127–133.

(1969) Structure and development of seeds in *Euphorbia heliosopia* L. *Bot. Mag. Tokyo* **82**, 287–293.

(1970a) Structure and development of seeds in Euphorbiaceae. *Beitr. Biol. Pfl.* **47**, 79–90.

(1970b) Structure and development of seeds in *Putranjiva roxburghii* Wall. *J. Indian bot. Soc.* **49**, 99–105.

Singh, R. P. and S. Chopra (1970) Structure and development of seeds in *Croton bonplandianum*. *Phytomorph.* **20**, 83–87.

Singh, S. P. (1959) Structure and development of seeds in *Euphorbia geniculata* Orteg. *J. Indian bot. Soc.* **38**, 103–108.

(1960) Morphological studies in some members of the family Pedaliaceae: 1, *Sesamum indicum* DC. *Phytomorph.* **10**, 65–81.

Sleumer, H. (1955) Proteaceae. *Flora Malesiana* ser. 1, **5**, 147–206.

Smith, A. C. (1945) Taxonomic review of *Trochodendron* and *Tetracentron*. *J. Arnold Arb.* **26**, 123–142.

(1946) A taxonomic review of *Euptelea*. *J. Arnold Arb.* **27**, 175–185.

Smith, C. M. (1929) Development of *Dionaea muscipula*: 1, Flower and seed. *Bot. Gaz.* **87**, 508–530.

Smith, D. L. (1964) The evolution of the ovule. *Biol. Rev.* **39**, 137–159.

Smith, L. S. (1957) New species of and notes on Queensland plants: 11. *Proc. R. Soc. Queensl.* **68**, 43–50.

Smith, O. (1935) Pollination and life-history studies of the tomato (*Lycopersicum esculentum* Mill.). *Mem. Cornell Univ. agric. Exp. Stn* **184**, 1–16.

Sporne, K. R. (1956) The phylogenetic classification of the Angiosperms. *Biol. Rev.* **31**, 1–29.

(1969) The ovule as an indicator of evolutionary status in Angiosperms. *New Phytol.* **68**, 555–566.

Sprecher, A. (1919) Etude sur la semence et la germination du *Garcinia mangostana* L. *Rev. gen. bot.* **31**, 513–531, 609–633.

Srinath, K. V. (1940) Morphological studies in the genera *Calceolaria* and *Herpestis*. *Proc. Linn. Soc. Lond.* sess. 152, p. 152.

Srinivasachar, D. (1940) Embryological studies of some members of Rhamnaceae. *Proc. Indian Acad. Sci.* B **11**, 107–115.

Srinivasan, V. K. (1940) Morphological and cytological studies in the Scrophulariaceae: 2, *Angelonia grandiflora* C. Morr. and related genera. *J. Indian bot. Soc.* **19**, 197–222.

Sripleng, A. and F. H. Smith (1960) Anatomy of the seed of *Convolvulus arvensis* Am. *J. Bot.* **47**, 386–392.

Stafleu, F. A. (1952) A monograph of Vochysiaceae 2. *Acta bot. neerl.* **1**, 222–242.

Stairs, G. R. (1964) Microsporogenesis and embryogenesis in *Quercus*. *Bot. Gaz.* **125**, 115–121.

van Steenis, C. G. G. J. (1957) Scyphostegiaceae. *Flora Malesiana* ser. 1, **5**, 297–299.

(1968) Notes on *Bredemeyera* (*Comosperma*). *Acta bot. neerl.* **17**, 377–384.

Steiner, M. und I. Jancke (1955) Sind die Malpighischen Zellen die Epidermis der Leguminosen testa? *Ost. bot. Z.* **102**, 542–550.

Stemmerik, J. F. (1964) Nyctaginaceae. *Flora Malesiana* ser. 1, **6**, 450–468.

Stephens, E. L. (1910) The development of the seedcoat of *Carica papaya*. *Ann. Bot. Lond.* n.s. **24**, 607–610.

Sterling C. (1964–66) Comparative morphology of the carpel in the Rosaceae. *Am. J. Bot.* **51**, 36–44, 354–360, 705–712; **52**, 47–54, 418–426, 938–946; **53**, 225–231, 521–530, 951–960.

Stidd, B. M. and J. W. Hall (1970) The natural affinity of the carboniferous seed *Callospermarion*. *Am. J. Bot.* **57**, 827–836.

Stopes, M. C. (1905) On the double nature of the Cycadalean integument. *Ann. Bot. Lond.* **19**, 561–566.

Stushnoff, C. and Barbara F. Palser (1970) Embryology of *Vaccinium* taxa including diploid, tetraploid and hexaploid species or cultivars. *Phytomorph.* **19**, 312–331.

Subramanyam, K. (1942) Gametogenesis and embryogeny in a few members of the Melastomataceae. *J. Indian bot. Soc.* **21**, 69–85.

(1948a) An embryological study of *Melastoma malabathricum* L. *J. Indian bot. Soc.* **27**, 11–19.

(1948b) A contribution to the embryology of *Wahlenbergia gracilis* Schrad. *Proc. nat. Inst. Sci. India* **14**, 359–366.

(1949) An embryological study of *Lobelia pyramidalis* Wall. with special reference to the mechanism of nutrition of the embryo in the family Lobeliaceae. *New Phytol.* **48**, 365–373.

(1950a) An embryological study of *Levenhookia dubia* Sond. in Lehm. *Proc. nat. Inst. Sci. India* **16**, 245–253.

(1950b) A contribution to our knowledge of the systematic position of the Sphenocleaceae. *Proc. Indian Acad. Sci.* B **31**, 1–6.

(1951a) A morphological study of *Stylidium graminifolium*. *Lloydia* **14**, 65–81.

(1951b) Embryology of *Oxyspora paniculata* DC. *Phytomorph.* 1, 205–212.

(1951c) Flower structure and seed development in *Isotoma fluviatilis* F.v.M. *Proc. nat. Inst. Sci. India* 17, 275–285.

(1960a) Floral morphology. *Mem. Indian bot. Soc.* 3, 173–178.

(1960b) Nutritional mechanism of the seed. *J. Madras Univ.* 30, 29–56.

(1963) Embryology of *Sedum ternatum* Michx. *J. Indian bot. Soc.* 42A, 259–275.

(1968) Some aspects of the embryology of *Sedum chrysanthum* (Boiss.) Rayomond-Hamlet with a discussion on its sytematic position. *Phytomorph.* 17, 240–247.

Subramanyam, K. and L. L. Narayana (1968) Floral anatomy and embryology of *Primula floribunda* Wall. *Phytomorph.* 18, 105–113.

Sunder Rao, Y. (1940) Male and female gametophytes of *Polemonium coeruleum* L. *Proc. nat. Inst. Sci. India* B 6, 695–704.

Suessenguth, K. (1953) *Leeaceae, Vitaceae.* In Engler's *Die Naturlichen Pflanzenfamilien*, 2nd edn, vol. 20d, p. 372.

Svedelius, N. (1911) Uber den Samenbau bei den Gattungen *Wormia* und *Dillenia*. *Svensk bot. Tidskr.* 5, 152–171.

Svensson, H. G. (1928) Zur Entwickelungsgeschichte der Blüten und Samen von *Limosella aquatica*. *Svensk bot. Tidskr.* 22, 465–476.

Swamy, B. G. L. (1948a) A contribution to the life-history of *Casuarina*. *Proc. Am. Acad. Arts Sci.* 77, 1–32.

(1948b) A contribution to the embryology of the Marcgraviaceae. *Am. J. Bot.* 35, 628–633.

(1949) Further contributions to the morphology of the Degeneriaceae. *J. Arnold Arb.* 30, 10–38.

(1952) Some aspects in the embryology of *Zygogynum bailloni*. *Proc. nat. Inst. Sci. India* B 18, 399–404.

(1953a) The morphology and relationships of the Chloranthaceae. *J. Arnold Arb.* 34, 375–408.

(1953b) Some observations on the embryology of *Decaisnea insignis* Hook. et Thom. *Proc. nat. Inst. Sci. India* 19, 307–310.

(1953c) On the floral structure of *Scyphostegia*. *Proc. nat. Inst. Sci. India* 19, 127–142.

(1967) Casuarinaceae. In Johri (ed.) *Seminar on Comparative Embryology of Angiosperms*, p. 1. University of Delhi.

Swamy, B. G. L. and I. W. Bailey (1949) The morphology and relationship of *Cercidiphyllum*. *J. Arnold Arb.* 30, 187–210.

Swamy, B. G. L. and N. Parameswaran (1960) A contribution to the embryology of *Begonia crenata*. *J. Indian bot. Soc.* 39, 140–148.

Swamy, B. G. L. and K. Periasamy (1955) Contributions to the embryology of *Acrotrema arnottianum* Wight. *Phytomorph.* 5, 301–314.

Swamy, R. L. N. (1969) Comparative morphology and phylogeny of the Ranales. *Biol. Rev.* 44, 291–319.

Symington, C. F. (1941) *Upuna*, a new genus of the Dipterocarpaceae. *Bull. bot. Gdn. Buitenz.* ser. 3, 17, 88–95.

Szemes, G. (1943) Zur Entwickelung der Elaiosoms von *Chelidonium majus*. *Ost. bot. Z.* 92, 215–219.

Takao, S. (1968) A study in the development of embryo sac in *Impatiens textori*. *Bot. Mag. Tokyo* 18, 310–317.

Takeda, H. (1936) On the coma or hairy tufts on the seed of willow. *Bot. Mag. Tokyo* 50, 283–289.

Takhtajan, A. (1969) *Flowering plants; origin and dispersal*. Translated by C. Jeffrey. Oliver & Boyd, Edinburgh.

Tamura, M. (1972) Morphology and phyletic relationship of the Glaucidiaceae. *Bot. Mag. Tokyo* 85, 29–41.

Tandon, S. R. and J. M. Herr (1971) Embryological features of taxonomic significance in the genus *Nyssa*. *Can. J. Bot.* 49, 505–514.

Taylor, P. (1964) The genus *Utricularia* L. (Lentibulariaceae) in Africa (south of the Sahara) and Madagascar. *Kew Bull.* 18, 1.

Thathachar, T. (1942) Studies in Oxalidaceae. *J. Indian bot. Soc.* 21, 21–31.

Thompson, J. Mc-L. (1924) *The Amherstieae*. Publ. Hartley bot. Labs. no. 1. Liverpool.

(1925) *The Cassieae*. Publ. Hartley bot. Labs. no. 2.

Thomson, J. R. (1960) Morphology and anatomy of the seed and fruit of *Onobrychis viciifolia* Scop. *Proc. int. Seed Test. Ass.* 25, 848–864.

Thorne, R. F. (1968) Synopsis of a putatively phylogenetic classification of the flowering plants. *Aliso* 6, 57–66.

Tiagi, B. (1951a) A contribution to the morphology and embryology of *Cuscuta hyalina* Roth and *C. planiflora* Tenone *Phytomorph.* 1, 9–21.

(1951b) Studies in the family Orobanchaceae: 3, A contribution to the embryology of *Orobanche cernua* Loeffl. and *O. aegyptiaca* Pers. *Phytomorph.* 1, 158–169.

(1952a) ibid. 1. A contribution to the embryology of *Cistanche tubulosa* Wight. *Lloydia* 15, 129–148.

(1952b) ibid. 2, A contribution to the embryology of *Aeginetia indica* Linn. *Bull. Torrey bot. Club.* 79, 63–78.

(1956a) A contribution to the embryology of *Striga orobanchoides* Benth. and *Striga euphrasioides* Benth. *Bull. Torrey bot. Club* 83, 154–170.

(1956b) Polyembryony in *Mammellaria tenuis* DC. *Bull. bot. Soc. Univ. Saugar* 8, 25–27.

(1957) Studies in floral morphology: 3, A contribution to the floral morphology of *Mammillaria tenuis* DC. *J. Univ. Saugar* B **6**, 7–31.

van Tieghem, P. (1872) Sur les diverses modes de nervation de l'ovule et de la graine. Annls Sci. nat. (Bot.) ser 5, **16**, 228.

(1900) Sur les dicotylédones du groupe des Homoxylées. *J. Bot. Paris* **14**, 259–297, 330–361.

Tiwary, N. K. (1926) On the occurrence of polyembryony in the genus *Eugenia*. *J. Indian bot. Soc.* **5**, 124–136.

Topham, Pauline B. (1970) The histology of seed development in diploid and tetraploid raspberries (*Rubus idaeus* L.). *Ann. Bot. Lond.* **34**, 123–135, 137–145.

Tsan-Iang Chuang and L. R. Heckard (1973) Seed-coat morphology in *Cordylanthus* (Scrophulariaceae) and its taxonomic significance. *Am. J. Bot.* **59**, 258–265.

Tucker, S. C. and E. M. Gifford Jr. (1966a) Organogenesis in the carpellate flower of *Drimys lanceolata*. *Am. J. Bot.* **53**, 433–442.

(1966b) Carpel development in *Drimys lanceolata*. *Am. J. Bot.* **53**, 671–678.

Uphof, J. C. Th. (1959) *Myristicaceae*. In Engler's *Die Natürlichen Pflanzenfamilien*, 2nd edn. vol. 17a, part 2, pp. 177–220.

Vassal, J. (1968) Graines albumineusées chez les Acacias. *Trav. Lab. forest. Toulouse* t.1, vol. 7, art. 4.

(1971) Contribution à l'étude morphologique des graines d'*Acacia*. *Bull. Soc. Hist. nat. Toulouse* **107**, 191–246.

Vaughan, J. G. (1970) *The structure and utilization of oil seeds*. Chapman & Hall Ltd., London.

Vaughan, J. G. and J. A. Rest (1969) Note on the testa structure of *Panda* Pierre, *Galearia* Zoll. et Mor. and *Microdesmis* Hook.f. (Pandaceae). *Kew Bull.* **23**, 215–218.

Vaughan, J. G. and J. M. Whitehouse (1971) Seed structure and the taxonomy of Cruciferae. *Bot. J. Linn. Soc.* **64**, 383–409.

Veldkamp, J. F. (1967) A revision of *Sarcotheca* Bl. and *Dapania* Korth. (Oxalidaceae). *Blumea* **15**, 519–543.

Venkata Rao, C. (1950) Contribution to the embryology of Sterculiaceae: 2, *Waltheria indica* Linn. *J. Indian bot. Soc.* **29**, 163–176.

(1951) ibid. 3, *Melochia corchorifolia* L. *J. Indian bot. Soc.* **30**, 122–131.

(1952a) The embryology of *Muntingia calabura* L. *J. Indian bot. Soc.* **31**, 87–101.

(1952b) Contributions to the embryology of the Sterculiaceae: 4, Development of the gametophytes in *Pterospermum suberifolium* Lam. *J. Indian bot. Soc.* **31**, 250–260.

(1953a) Floral anatomy and embryology of two species of *Elaeocarpus*. *J. Indian bot. Soc.* **32**, 21–33.

(1953b) Contributions to the embryology of Sterculiaceae 5. *J. Indian bot. Soc.* **32**, 208–238.

(1954) A contribution to the embryology of Bombacaceae. *Proc. Indian Acad. Sci.* **39**, 51–75.

(1955) Embryological studies in Malvaceae: 2, Fertilization and seed development. *Proc. nat. Inst. Sci. India* B **21**, 53–67.

(1960) Studies in the Proteaceae: 1, Tribe Persoonieae. *Proc. nat. Inst. Sci. India* B **26**, 300–338.

(1961) ibid. 2, Tribes Placospermeae and Conospermeae. *Proc. nat. Inst. Sci. India* B **27**, 126–151.

(1962) Morphology and embryology of *Lomatia* Br. with a discussion on its probable origin. In *Plant Embryology, a Symposium*, pp. 261–272. CSIR, Delhi.

(1963) Studies in the Proteaceae: 3, Tribe Oriteae. *Proc. nat. Inst. Sci. India* B **29**, 489–510.

(1964) ibid. 4, Tribes Banksieae, Musgravieae and Embothrieae. *Proc. nat. Inst. Sci. India* B **30**, 197–244.

(1965) ibid. 6, Tribe Franklandieae. *J. Indian bot. Soc.* **44**, 479–494.

(1967) ibid. 8, Morphology, floral anatomy and embryology of *Grevillea* R. Br. *Proc. nat. Inst. Sci. India* B **33**, 162–199.

(1969a) ibid. 9, Australian Proteaceae. *Proc. nat. Inst. Sci. India* B **35**, 205–229.

(1969b) ibid. 13, *Proc. nat. Inst. Sci. India* B **35**, 471–486.

(1970) ibid. 14, Tribe Macadamieae. *Proc. nat. Inst. Sci. India* B **36**, 345–363.

Venkata Rao, C. and S. Rama Rao (1954) Embryology of *Cryptostegia grandiflora* R. Br. and *Caralluma attenuata* Wt. *J. Indian bot. Soc.* **33**, 453–472.

Venkata Rao, C. and K. V. Sambasiva Rao (1952) A contribution to the embryology of *Triumfetta rhomboidea* Jacq. and *Corchorus acutangulus* L. *J. Indian bot. Soc.* **31**, 56–68.

Venkatasubban, K. R. (1951) Studies in the Droseraceae 2. *Proc. Indian Acad. Sci.* B **32**, 23–38.

Venkateswarlu, J. (1937a) A contribution to the embryology of Sonneratiaceae. *Proc. Indian Acad. Sci* **5**, 206–223.

(1937b) Structure and development of the embryo sac of *Pemphis acidula* Forst. *J. Indian bot. Soc.* **16**, 259–262.

(1945) Embryological studies in the Thymelaeaceae 1. *Thymelaea arvensis* Lamk. *J. Indian bot. Soc.* **24**, 45–66.

(1947) ibid. 2, *Daphne cannabina* Wall. and *Wikstroemia canescens* Meissn. *J. Indian bot. Soc.* **13**–39.

(1948) A contribution to the embryology of *Pisonia aculeata* Linn. *J. Indian bot. Soc.* **26**, 183–194.

(1952a) Embryological studies in Lecythidaceae, I, *J. Indian bot. Soc.* **31**, 103–116.

(1952b) Contributions to the embryology of Combretaceae: I, *Poivrea coccinea* DC. *Phytomorph.* **2**, 231–240.

Venkateswarlu, J. and L. Lakshminarayana (1958) A contribution to the embryology of *Hydrocera triflora* W. et A. *Phytomorph.* **7**, 194–203.

Verhoog, V. (1968) A contribution towards the developmental gynoecium morphology of *Engelhardtia spicata* Lechen. ex Blume (Juglandaceae). *Acta bot. neerl.* **17**, 137–150.

Vijayaraghavan, M. R. (1964) Morphology and embryology of a vesselless dicotyledon – *Sarcandra irvingbaileyi* Swamy, and sytematic position of Chloranthaceae. *Phytomorph.* **14**, 429–441.

(1965) Morphology and embryology of *Actinidia polygama* Franch. et Sav. and systematic position of the family Actinidiaceae. *Phytomorph.* **15**, 224–235.

Vijayaraghavan, M. R. and N. N. Bhandari (1970) Studies in the family Ranunculaceae: embryology of *Thalictrum javanicum* Bl. *Flora* **159**, 450–458.

Vijayaraghavan, M. R. and D. Kaur (1967) Morphology and embryology of *Turnera ulmifolia* L. and affinities of the family Turneraceae. *Phytomorph.* **16**, 539–553.

Vijayaraghavan, M. R. and K. N. Marwah (1969) Studies in the family Ranunculaceae: morphology and embryology of *Nigella damascena*. *Phytomorph.* **19**, 147–153.

Vijayaraghavan, M. R. and U. Padmanaban (1969) Morphology and embryology of *Centaurium ramosissimum* Druce and affinities of the family Gentianaceae. *Beitr. Biol. Pfl.* **46**, 15–37.

Vink, W. (1970) The Winteraceae of the Old World: I, *Pseudowintera* and *Drimys*. *Blumea* **18**, 225–354.

Voigt, A. (1888) Untersuchungen über Bau und Entwickelung von Samen mit ruminiertens Endosperm aus den Familien des Palmen, Myristicaceen und Anonaceen. *Annls Jard. bot. Buitenz.* **7**, 151–190.

Walia, K. and R. N. Kapil (1965) Embryology of *Frankenia* Linn. with some comments on the systematic position of the Frankeniaceae. *Bot. Notiser* **118**, 412–429.

Walker, J. W. (1971) Unique type of angiosperm pollen from the family Annonaceae. *Science N.Y.* **172**, 565–567.

Walker, R. I. (1947) Megasporogenesis and embryo development in *Tropaeolum majus* L. *Bull. Torrey bot. Club* **74**, 240–249.

Wallich, N. (1830) *Plantae Asiaticae Rariores* vol. I, p. 5, t.5.

Watanabe, K. (1933) Biology of *Mitrastemon yamamotoi*: I, Fruit and seed. *Bot. Mag. Tokyo* **47**, 398–405.

Wellendorf, M. (1964) Morphology and anatomy of *Crotalaria* seeds. *Lloydia* **27**, 251–253.

Whiffin, T. and A. Spencer Tomb (1972) The systematic significance of seed morphology in the neotropical capsular-fruited Melastomataceae. *Am. J. Bot.* **59**, 411–422.

Wiehr, E. (1930) Beiträge zur Kenntnis der Anatomie der wichtigsten Euphorbiaceen Samen unter besonderer Berücksichtigung ihrer Erkennungsmerkmale in Futtermitteln. *Landw. Versuchsstationen* **110**, 313–398.

Wiggins, I. L. (1959) Development of the ovule and megagametophyte in *Saxifraga hieracifolia*. *Am. J. Bot.* **46**, 692–697.

Wilson, P. G. (1970) A taxonomic revision of the genera *Crowea*, *Eriostemon* and *Phlebalium* (Rutaceae). *Nuytsia* **1**, 6–154.

Wilson, T. K. and L. M. Maculans (1967) The morphology of the Myristicaceae: I, Flowers. *Am. J. Bot.* **52**, 214–220.

Winkler, H. (1927) Ueber eine *Rafflesia* aus zentral Borneo. *Planta* **4**, 1–97.

Winter, D. M. (1960) The development of the seed of *Abutilon theophrasti* L. *Am. J. Bot.* **47**, 8–14, 157–162.

Winton, A. L. and K. B. Winton (1939) *The structure and composition of foods*. Wiley, New York.

de Wit, H. C. D. (1952) A revision of the genus *Archidendron* F. Muell. (Mimosaceae). *Reinwardtia* **2**, 69–96.

(1955) A revision of the genus *Cassia* (Caesalp.) as occurring in Malaysia. *Webbia* **11**, 197–292.

(1956) A revision of Malayan Bauhinieae. *Reinwardtia* **3**, 381–541.

Woodcock, E. F. (1925) Observations on the morphology of the seed in *Phytolacca*. *Pap. Mich. Acad. Sci. Arts Lett.* **4**, 413–418.

(1926) Morphology of the seed of *Claytonia virginica*. *Pap. Mich. Acad. Sci. Arts Lett.* **5**, 195–200.

(1927) Morphological studies of the seed of *Alsine media* L. *Pap. Mich. Acad. Sci. Arts Lett.* **6**, 397–402.

(1928) Observations on the morphology of the seed of *Cerastium vulgatum* L. *Pap. Mich. Acad. Sci. Arts Lett.* **8**, 233–238.

(1929a) Seed development in *Thelygonium cynocrambe* L. *Pap. Mich. Acad. Sci. Arts Lett.* **9**, 341.

(1929b) Seed studies in Nyctaginaceae. *Pap. Mich. Acad. Sci. Arts Lett.* **9**, 495–502.

(1931) Morphological studies on the seed of *Mesembryanthemum crystallinum* L. *Pap. Mich. Acad. Sci. Arts Lett.* **13**, 221–264.

(1932) Seed development in *Amaranthus caudatus* L. *Pap. Mich. Acad. Sci. Arts Lett.* **15**, 173–177.

(1933) Seed studies in *Cyclamen persicum*. *Pap. Mich. Acad. Sci. Arts Lett.* **17**, 415–420.

(1939) Morphological studies on the seed of snap-dragon (*Antirrhinum majus* L.). *Pap. Mich Acad. Sci. Arts Lett.* **25**, 139–141.

(1944) Seed development in the Morning Glory (*Ipomaea rubro-caerulea* Hook.). *Pap. Mich. Acad. Sci. Arts Lett.* **28**, 209–212.

Wunderlich, Rosalie (1968) Some remarks on the taxonomic significance of the seed coat. *Phytomorph.* **17**, 301–311.

Wyatt, R. L. (1955) An embryological study of four species of *Asarum. J. Elisha Mitchell sci. Soc.* **71**, 64–82.

van de Wyk, R. W. and J. E. Canright (1956) The anatomy and relationships of the Annonaceae. *Trop. Woods* **104**, 1–24.

Wynne, F. E. (1944) *Drosera* in Eastern North America. *Bull. Torrey bot. Club* **71**, 166–174.

Yamazaki, T. (1953) On the floral structure, seed development and affinities of *Deinostema*, a new genus of Scrophulariaceae. *Bot. Mag. Tokyo* **66**, 141–149.

(1957) Seed formation of *Ellisiophyllum pinnatum* var. *reptans. Bot. Mag. Tokyo* **70**, 162–168.

Yen, T. K. (1936) Floral development and vascular anatomy of the fruit of *Ribes aureum. Bot. Gaz.* **98**, 105–120.

Yoshida, O. (1957) Embryologische Studien über die Ordnung Piperales: 1, Embryologie von *Chloranthus japonicus* Sieb. *J. Coll. Arts Sci. Chiba Univ.* **2**, 172–178.

(1959) ibid. 2, Embryologie von *Chloranthus serratus* Roem. et Schult. *J. Coll. Arts Sci. Chiba Univ.* **2**, 295–303.

(1960a) ibid. 3, Embryologie von *Sarcandra glabra* Nakai. *J. Coll. Arts Sci. Chiba Univ.* **3**, 55–60.

(1960b) ibid. 4, Embryologie von *Piper futokazura* Sieb. et Zucc. *J. Coll. Arts Sci. Chiba Univ.* **3**, 155–162.

(1961) ibid. 5, Embryologie von *Saururus loureiri* Decne. *J. Coll. Arts Sci. Chiba Univ.* **3**, 311–316.

(1962) Embryologische Studien über *Schisandra chinensis* Baillon. *J. Coll. Arts Sci. Chiba Univ.*, **3**, 459–462.

Ziegler, A. (1925) Beiträge zur Kenntnis des Andröceums u. der Samenentwickelung einiger Melastomaceen. *Bot. Arch.* **9**, 398–467.

Index